$\mathscr{S}$TUDENT
SOLUTIONS MANUAL

Prepared with the Assistance of
Kevin M. Bodden

EIGHTH EDITION

$\mathscr{C}$ALCULUS

Varberg   Purcell   Rigdon

PRENTICE HALL, Upper Saddle River, NJ 07458

Executive Editor: George Lobell
Editorial Assistant: Gale A. Epps
Special Projects Manager: Barbara A. Murray
Production Editor: Dawn Murrin
Supplement Cover Manager: Paul Gourhan
Supplement Cover Designer: PM Workshop Inc.
Manufacturing Buyer: Alan Fischer

ISBN 0-13-085151-5

Prentice-Hall International (UK) Limited, London
Prentice-Hall of Australia Pty. Limited, Sydney
Prentice-Hall Canada, Inc., Toronto
Prentice-Hall Hispanoamericana, S.A., Mexico
Prentice-Hall of India Private Limited, New Delhi
Pearson Education Asia Pte. Ltd., Singapore
Prentice-Hall of Japan, Inc., Tokyo
Editora Prentice-Hall do Brazil, Ltda., Rio de Janeiro

Contents

1.1 Concepts Review

1. rational   **3.** real

Problem Set 1.1

1. $4 - 2(8 - 11) + 6 = 4 - 2(-3) + 6$
$$= 4 + 6 + 6 = 16$$

3. $-4[5(-3 + 12 - 4) + 2(13 - 7)]$
$$= -4[5(5) + 2(6)] = -4[25 + 12]$$
$$= -4(37) = -148$$

5. $\dfrac{5}{7} - \dfrac{1}{13} = \dfrac{65}{91} - \dfrac{7}{91} = \dfrac{58}{91}$

7. $\dfrac{1}{3}\left[\dfrac{1}{2}\left(\dfrac{1}{4} - \dfrac{1}{3}\right) + \dfrac{1}{6}\right] = \dfrac{1}{3}\left[\dfrac{1}{2}\left(\dfrac{3-4}{12}\right) + \dfrac{1}{6}\right]$
$$= \dfrac{1}{3}\left[\dfrac{1}{2}\left(-\dfrac{1}{12}\right) + \dfrac{1}{6}\right]$$
$$= \dfrac{1}{3}\left[-\dfrac{1}{24} + \dfrac{4}{24}\right]$$
$$= \dfrac{1}{3}\left(\dfrac{3}{24}\right) = \dfrac{1}{24}$$

9. $\dfrac{14}{21}\left(\dfrac{2}{5 - \frac{1}{3}}\right)^2 = \dfrac{14}{21}\left(\dfrac{2}{\frac{14}{3}}\right)^2 = \dfrac{14}{21}\left(\dfrac{6}{14}\right)^2$
$$= \dfrac{14}{21}\left(\dfrac{3}{7}\right)^2 = \dfrac{2}{3}\left(\dfrac{9}{49}\right) = \dfrac{6}{49}$$

11. $\dfrac{\frac{11}{7} - \frac{12}{21}}{\frac{11}{7} + \frac{12}{21}} = \dfrac{\frac{11}{7} - \frac{4}{7}}{\frac{11}{7} + \frac{4}{7}} = \dfrac{\frac{7}{7}}{\frac{15}{7}} = \dfrac{7}{15}$

13. $1 - \dfrac{1}{1 + \frac{1}{2}} = 1 - \dfrac{1}{\frac{3}{2}} = 1 - \dfrac{2}{3} = \dfrac{3}{3} - \dfrac{2}{3} = \dfrac{1}{3}$

15. $\left(\sqrt{5} + \sqrt{3}\right)\left(\sqrt{5} - \sqrt{3}\right) = \left(\sqrt{5}\right)^2 - \left(\sqrt{3}\right)^2$
$$= 5 - 3 = 2$$

17. $3\sqrt{2}\left(\sqrt{2} - \sqrt{8}\right) = 3\sqrt{4} - 3\sqrt{16}$
$$= 3 \cdot 2 - 3 \cdot 4$$
$$= 6 - 12 = -6$$

19. $\left(\dfrac{7}{4} + \dfrac{1}{2}\right)^{-2} = \left(\dfrac{7}{4} + \dfrac{2}{4}\right)^{-2} = \left(\dfrac{9}{4}\right)^{-2}$
$$= \dfrac{1}{\left(\frac{9}{4}\right)^2} = \dfrac{1}{\frac{81}{16}} = \dfrac{16}{81}$$

21. $(3x - 4)(x + 1) = 3x^2 + 3x - 4x - 4$
$$= 3x^2 - x - 4$$

23. $(3x - 9)(2x + 1) = 6x^2 + 3x - 18x - 9$
$$= 6x^2 - 15x - 9$$

25. $(3t^2 - t + 1)^2$
$$= (3t^2 - t + 1)(3t^2 - t + 1)$$
$$= 9t^4 - 3t^3 + 3t^2 - 3t^3 + t^2 - t + 3t^2 - t + 1$$
$$= 9t^4 - 6t^3 + 7t^2 - 2t + 1$$

27. $\dfrac{x^2 - 4}{x - 2} = \dfrac{(x - 2)(x + 2)}{x - 2} = x + 2, \ x \neq 2$

29. $\dfrac{t^2 - 4t - 21}{t + 3} = \dfrac{(t + 3)(t - 7)}{t + 3} = t - 7, \ t \neq -3$

31. $\dfrac{12}{x^2 + 2x} + \dfrac{4}{x} + \dfrac{2}{x + 2}$
$$= \dfrac{12}{x(x + 2)} + \dfrac{4(x + 2)}{x(x + 2)} + \dfrac{2x}{x(x + 2)}$$
$$= \dfrac{12 + 4x + 8 + 2x}{x(x + 2)}$$
$$= \dfrac{6x + 20}{x(x + 2)}$$
$$= \dfrac{2(3x + 10)}{x(x + 2)}$$

33. $\dfrac{t^2+t-12}{x^2-1}\cdot\dfrac{x^2-6x-7}{8t-t^2-15}$

$=\dfrac{(t+4)(t-3)(x-7)(x+1)}{-(x+1)(x-1)(t-3)(t-5)}$

$=-\dfrac{(t+4)(x-7)}{(x-1)(t-5)}$

35. a. $0\cdot 0=0$ **b.** $\dfrac{0}{0}$ is undefined.

c. $\dfrac{0}{17}=0$ **d.** $\dfrac{3}{0}$ is undefined.

e. $0^5=0$ **f.** $17^0=1$

37. a. $-3<-7$; False   **b.** $-1>-17$; True

c. $-3<-\dfrac{22}{7};-\dfrac{21}{7}<-\dfrac{22}{7}$; False

d. $-5>-\sqrt{26};-\sqrt{25}>-\sqrt{26}$; True

e. $\dfrac{6}{7}<\dfrac{34}{39};\dfrac{234}{273}<\dfrac{238}{273}$; True

f. $-\dfrac{5}{7}<-\dfrac{44}{59};-\dfrac{295}{413}<-\dfrac{308}{413}$; False

39. $a<b$; $2a<a+b$ and $a+b<2b$, so

$2a<a+b<2b$; $a<\dfrac{a+b}{2}<b$

41. a. True; If x is positive, then x^2 is positive.

b. False;   Take $x=-2$. Then

$x^2>0$ but $x<0$.

c. False;   Take $x=\frac{1}{2}$. Then $x^2 =$

d. True; $n=2$ is even and prim

e. True;   Let x be any number. Take

$y=x^2+1$. Then $y>x^2$.

f. False;   There is no number larger than

every x^2.

g. True; $1/n$ can be made arbitrarily

close to 0.

h. True 2^{-n} can be made arbitrarily

close to 0.

43. a. $243=3\cdot 3\cdot 3\cdot 3\cdot 3$

b. $127=1\cdot 127$

c. $5100=2\cdot 2550=2$ $75=2\cdot 2\cdot 3\cdot 425$

$=2\cdot 2\cdot 3\cdot 5\cdot 85=2\cdot 2\cdot 3\cdot 5\cdot 5\cdot 17$

d. $346=2\cdot 173$

45. $\sqrt{2}=\dfrac{p}{q};2=\dfrac{p^2}{q^2};2q^2=p^2$; Since the prime

factors of p^2 must occur an even number of

times, $2q^2$ would not be valid and $\dfrac{p}{q}=\sqrt{2}$

must be irrational.

47. Let a, b, p, and q be natural numbers, so $\dfrac{a}{b}$ and

$\dfrac{p}{q}$ are rational. $\dfrac{a}{b}+\dfrac{p}{q}=\dfrac{aq+bp}{bq}$ This sum is

the quotient of natural numbers, so it is also

rational.

49. a. $-\sqrt{9}=-3$; rational

b. $0.375=\dfrac{3}{8}$; rational

c. $1-\sqrt{2}$; irrational

d. $(1+\sqrt{3})^2=1+2\sqrt{3}+3=4+2\sqrt{3}$;

irrational

e. $(3\sqrt{2})(5\sqrt{2})=15\sqrt{4}=30$; rational

f. $5\sqrt{2}$; irrational

51. If m were a perfect square, its prime factors

would occur even numbers of times. If m is not

a perfect square, some factors will occur an odd

number of times and $\sqrt{m}$ will be irrational.

53. $\sqrt{2}-\sqrt{3}+\sqrt{6}\approx 2.13165$ is irrational.

55. a. Converse: If I get an A in this course,

then I do all of the homework.

Contrapositive: If I don't get an A in this

course, then I don't do all of the

homework.

b. Converse: If x is an integer, then x is a real number.

Contrapositive: If x is not an integer, then x is not a real number.

c. Converse: If $\triangle ABC$ is an isosceles triangle, then $\triangle ABC$ is an equilateral triangle.

Contrapositive: If $\triangle ABC$ is not an isosceles triangle, then $\triangle ABC$ is not an equilateral triangle.

1.2 Concepts Review

1. 0.333..... (3s repeat); 0.200... (0s repeat); 3.14159...

3. rational; irrational

Problem Set 1.2

1.

$$
\begin{array}{r}
.0833 \\
12\overline{)1.0000} \\
96 \\
\hline
40 \\
36 \\
\hline
40
\end{array}
$$

3.

$$
\begin{array}{r}
.846153 \\
13\overline{)11.000000} \\
10\ 4 \\
\hline
60 \\
52 \\
\hline
80 \\
78 \\
\hline
20 \\
13 \\
\hline
70 \\
65 \\
\hline
50 \\
39 \\
\hline
11
\end{array}
$$

5.

$$
\begin{array}{r}
3.66 \\
3\overline{)11.00} \\
9 \\
\hline
20 \\
18 \\
\hline
20 \\
18 \\
\hline
2
\end{array}
$$

7. $x = 0.123123123...$

$1000x = 123.123123...$

$\underline{x = 0.123123...}$

$999x = 123$

$x = \dfrac{123}{999} = \dfrac{41}{333}$

9. $x = 2.56565656...$

$100x = 256.565656...$

$\underline{x = 2.565656...}$

$99x = 254$

$x = \dfrac{254}{99}$

11. $x = 0.199999...$

$100x = 19.99999...$

$\underline{10x = 1.99999...}$

$90x = 18$

$x = \dfrac{18}{90} = \dfrac{1}{5}$

13. Those rational numbers that can be expressed by a terminating decimal followed by zeros.

15. Answers will vary. Possible answer: 0.000001, $\dfrac{1}{\pi^{12}} \approx 0.0000010819...$

17. Answers will vary. Possible answer: 3.14159101001...

19. Irrational

21. $(\sqrt{3}+1)^3 \approx 20.39230485$

23. $\sqrt[4]{1.123} - \sqrt[3]{1.09} \approx 0.00028307388$

25. $\dfrac{\sqrt{130} - \sqrt{5}}{3^{1.2} - 3} \approx 12.43322783$

27. $\sqrt{8.9\pi^2 + 1} - 3\pi \approx 0.000691744752$

29. Let a and b be real numbers with $a < b$. Let n be a natural number that satisfies $1/n < b - a$. Let $S = \{k : k/n > b\}$. Since a nonempty set of integers that is bounded below contains a least element, there is a $k_0 \in S$ such that $k_0/n > b$ but $(k_0 - 1)/n \leq b$. Then

$$\frac{k_0 - 1}{n} = \frac{k_0}{n} - \frac{1}{n} > b - \frac{1}{n} > a$$

Thus, $a < \frac{k_0 - 1}{n} \leq b$. If $\frac{k_0 - 1}{n} < b$, then choose $r = \frac{k_0 - 1}{n}$. Otherwise, choose $r = \frac{k_0 - 2}{n}$.

31. $r = 4000 \text{ mi} \times 5280 \dfrac{\text{ft}}{\text{mi}} = 21{,}120{,}000 \text{ ft}$

equator $= 2\pi r = 2\pi(21{,}120{,}000) \approx 132{,}700{,}874 \text{ ft}$

33. $V = \pi r^2 h = \pi\left(\dfrac{16}{2} \cdot 12\right)^2 (270 \cdot 12)$

$\approx 93{,}807{,}453.98 \text{ in.}^3$

volume of one board foot (in inches):
$1 \cdot 3 \cdot 12 \cdot 3 \cdot 12 = 144 \text{ in.}^3$

number of board feet:
$\dfrac{93{,}807{,}453.98}{144} \approx 651{,}441 \text{ board ft}$

35. a. At $x = 2\pi$: 286.866542

   b. At $x = 2.15$: 9.16925

   c. At $x = 2.71828$: 16.34874967

   d. At $x = 1.1$: 4.292

37. a. -2 **b.** -2

   c. $x = 2.4444...$;
   $10x = 24.4444...$
   $$\underline{\quad x = 2.4444...\quad}$$
   $9x = 22$
   $$x = \frac{22}{9}$$

   d. 1

   e. $n = 1: x = 0,\ n = 2:\ x = \dfrac{3}{2},\ n = 3:\ x = -\dfrac{2}{3},$

   $n = 4:\ x = \dfrac{5}{4}$

   The upper bound is $\dfrac{3}{2}$.

   f. $\sqrt{2}$

1.3 Concepts Review

1. interval; intervals   **3.** $b > 0; b < 0$

Problem Set 1.3

1. a.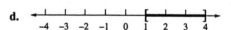

   b. (number line)

   c. (number line)

d.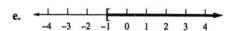

e. (number line)

f. (number line)

3. $x - 7 < 2x - 5$
   $-2 < x; (-2, \infty)$
   (number line)

5. $7x - 2 \le 9x + 3$

$-5 \le 2x$

$x \ge -\dfrac{5}{2}; \left[-\dfrac{5}{2}, \infty\right)$

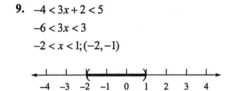

7. $10x + 1 > 8x + 5$

$2x > 4$

$x > 2; (2, \infty)$

9. $-4 < 3x + 2 < 5$

$-6 < 3x < 3$

$-2 < x < 1; (-2, -1)$

11. $-3 < 1 - 6x \le 4$

$-4 < -6x \le 3$

$\dfrac{2}{3} > x \ge -\dfrac{1}{2}; \left[-\dfrac{1}{2}, \dfrac{2}{3}\right)$

13. $2 + 3x < 5x + 1 < 16$

$2 + 3x < 5x + 1$ and $5x + 1 < 16$

$1 < 2x$ and $5x < 15$

$x > \dfrac{1}{2}$ and $x < 3; \left(\dfrac{1}{2}, 3\right)$

15. $x^2 + 2x - 12 < 0$;

$x = \dfrac{-2 \pm \sqrt{(2)^2 - 4(1)(-12)}}{2(1)} = \dfrac{-2 \pm \sqrt{52}}{2}$

$= -1 \pm \sqrt{13}$

$\left[x - \left(-1 + \sqrt{13}\right)\right]\left[x - \left(-1 - \sqrt{13}\right)\right] < 0;$

$\left(-1 - \sqrt{13}, -1 + \sqrt{13}\right)$

17. $2x^2 + 5x - 3 > 0; (2x - 1)(x + 3) > 0;$

$(-\infty, -3) \cup \left(\dfrac{1}{2}, \infty\right)$

19. $\dfrac{x + 4}{x - 3} \le 0; \ [-4, 3)$

21.

$\dfrac{2}{x} < 5$

$\dfrac{2}{x} - 5 < 0$

$\dfrac{2 - 5x}{x} < 0; \ (-\infty, 0) \cup \left(\dfrac{2}{5}, \infty\right)$

23.

$\dfrac{1}{3x - 2} \le 4$

$\dfrac{1}{3x - 2} - 4 \le 0$

$\dfrac{1 - 4(3x - 2)}{3x - 2} \le 0$

$\dfrac{9 - 12x}{3x - 2} \le 0; \left(-\infty, \dfrac{2}{3}\right) \cup \left[\dfrac{3}{4}, \infty\right)$

25. $(x + 2)(x - 1)(x - 3) > 0; (-2, 1) \cup (3, 8)$

27. $(2x - 3)(x - 1)^2(x - 3) \ge 0; \left(-\infty, \dfrac{3}{2}\right] \cup [3, \infty)$

29. $x^3 - 5x^2 - 6x < 0$

$x(x^2 - 5x - 6) < 0$

$x(x + 1)(x - 6) < 0;$

$(-\infty, -1) \cup (0, 6)$

31. a. $3x + 7 > 1$ and $2x + 1 < 3$
$3x > -6$ and $2x < 2$
$x > -2$ and $x < 1$; $(-2, 1)$

b. $3x + 7 > 1$ and $2x + 1 > -4$
$3x > -6$ and $2x > -5$
$x > -2$ and $x > -\dfrac{5}{2}$; $(-2, \infty)$

c. $3x + 7 > 1$ and $2x + 1 < -4$
$x > -2$ and $x < -\dfrac{5}{2}$; $\varnothing$

33. a. $(x+1)(x^2 + 2x - 7) \ge x^2 - 1$
$x^3 + 3x^2 - 5x - 7 \ge x^2 - 1$
$x^3 + 2x^2 - 5x - 6 \ge 0$
$(x+3)(x+1)(x-2) \ge 0$
$[-3, -1] \cup [2, \infty)$

b. $\qquad x^4 - 2x^2 \ge 8$
$\qquad x^4 - 2x^2 - 8 \ge 0$
$\qquad (x^2 - 4)(x^2 + 2) \ge 0$
$(x^2 + 2)(x + 2)(x - 2) \ge 0$
$(-\infty, 12] \cup [2, \infty)$

c. $\quad (x^2 + 1)^2 - 7(x^2 + 1) + 10 < 0$
$\quad [(x^2 + 1) - 5][(x^2 + 1) - 2] < 0$
$\qquad\qquad (x^2 - 4)(x^2 - 1) < 0$
$(x + 2)(x + 1)(x - 1)(x - 2) < 0$
$(-2, -1) \cup (1, 2)$

35. a. $1.99 < \dfrac{1}{x} < 2.01$

$1.99x < 1 < 2.01x$

$1.99x < 1$ and $1 < 2.01x$

$x < \dfrac{1}{1.99}$ and $x > \dfrac{1}{2.01}$

$\dfrac{1}{2.01} < x < \dfrac{1}{1.99}$

$\left(\dfrac{1}{2.01}, \dfrac{1}{1.99} \right)$

b. $2.99 < \dfrac{1}{x + 2} < 3.01$

$2.99(x + 2) < 1 < 3.01(x + 2)$

$2.99x + 5.98 < 1$ and

$1 < 3.01x + 6.02$

$x < \dfrac{-4.98}{2.99}$ and $x > \dfrac{-5.02}{3.01}$

$-\dfrac{5.02}{3.01} < x < -\dfrac{4.98}{2.99}$

$\left(-\dfrac{5.02}{3.01}, -\dfrac{4.98}{2.99} \right)$

c. $3 - \varepsilon < \dfrac{1}{x + 2} < 3 + \varepsilon$

$(3 - \varepsilon)(x + 2) < 1 < (3 + \varepsilon)(x + 2)$

$(3 - \varepsilon)x + (3 - \varepsilon)2 < 1 < (3 + \varepsilon)x + (3 + \varepsilon)2$

$x < \dfrac{1 - 2(3 - \varepsilon)}{3 - \varepsilon}$ and $x > \dfrac{1 - 2(3 + \varepsilon)}{3 + \varepsilon}$

$\left(\dfrac{1 - 2(3 + \varepsilon)}{3 + \varepsilon}, \dfrac{1 - 2(3 - \varepsilon)}{3 - \varepsilon} \right)$

37. $\dfrac{1}{R} \le \dfrac{1}{10} + \dfrac{1}{20} + \dfrac{1}{30}$

$\dfrac{1}{R} \le \dfrac{6 + 3 + 2}{60}$

$\dfrac{1}{R} \le \dfrac{11}{60}$

$R \ge \dfrac{60}{11}$

$\dfrac{1}{R} \ge \dfrac{1}{20} + \dfrac{1}{30} + \dfrac{1}{40}$

$\dfrac{1}{R} \ge \dfrac{6 + 4 + 3}{120}$

$\dfrac{1}{R} \ge \dfrac{13}{120}$

$R \le \dfrac{120}{13}$

$\dfrac{60}{11} \le R \le \dfrac{120}{13}$

1.4 Concepts Review

1. -1; 5   3. b, c

Problem Set 1.4

1. $|x+2| < 1;$
   $-1 < x+2 < 1$
   $-3 < x < -1$
   $(-3, -1)$

3. $|2x-1| > 2;$
   $2x - 1 < -2 \text{ or } 2x - 1 > 2$
   $2x < -1 \text{ or } 2x > 3;$
   $x < -\dfrac{1}{2} \text{ or } x > \dfrac{3}{2}, \left(-\infty, -\dfrac{1}{2}\right) \cup \left(\dfrac{3}{2}, \infty\right)$

5. $\left|\dfrac{x}{4} + 1\right| < 1$

   $-1 < \dfrac{x}{4} + 1 < 1$

   $-2 < \dfrac{x}{4} < 0;$

   $-8 < x < 0; (-8, 0)$

7. $|2x-7| > 3;$
   $2x - 7 < -3 \text{ or } 2x - 7 > 3$
   $2x < 4 \text{ or } 2x > 10$
   $x < 2 \text{ or } x > 5; (-\infty, 2) \cup (5, \infty)$

9. $|4x+2| \geq 10;$
   $4x + 2 \leq -10 \text{ or } 4x + 2 \geq 10$
   $4x \leq -12 \text{ or } 4x \geq 8$
   $x \leq -3 \text{ or } x \geq 2$
   $(-\infty, -3] \cup [2, \infty)$

11. $\left|2 + \dfrac{5}{x}\right| > 1;$

   $2 + \dfrac{5}{x} < -1 \text{ or } 2 + \dfrac{5}{x} > 1$

   $3 + \dfrac{5}{x} < 0 \text{ or } 1 + \dfrac{5}{x} > 0$

   $\dfrac{3x+5}{x} < 0 \text{ or } \dfrac{x+5}{x} > 0;$

   $(-\infty, -5) \cup \left(-\dfrac{5}{3}, 0\right) \cup (0, \infty)$

13. $x^2 - 3x - 4 \geq 0;$

   $x = \dfrac{3 \pm \sqrt{(-3)^2 - 4(1)(-4)}}{2(1)} = \dfrac{3 \pm 5}{2} = -1, 4$

   $(x+1)(x-4) = 0; (-\infty, -1] \cup [4, \infty)$

15. $3x^2 + 17x - 6 > 0;$

   $x = \dfrac{-17 \pm \sqrt{(17)^2 - 4(3)(-6)}}{2(3)} = \dfrac{-17 \pm 19}{6} = -6, \dfrac{1}{3}($

   $3x - 1)(x + 6) > 0; (-\infty, -6) \cup \left(\dfrac{1}{3}, \infty\right)$

17. $|x-3| < 0.5 \Leftrightarrow 5|x-3| < 5(0.5) \Leftrightarrow |5x - 15| < 2.5$

19. $|x-2| < \dfrac{\varepsilon}{6} \Leftrightarrow 6|x-2| < \varepsilon \Leftrightarrow |6x - 12| < \varepsilon$

21. $|3x - 15| < \varepsilon \Leftrightarrow |3(x-5)| < \varepsilon \quad\quad \Leftrightarrow 3|x - 5| < \varepsilon$

   $\Leftrightarrow |x - 5| < \dfrac{\varepsilon}{3}; \delta = \dfrac{\varepsilon}{3}$

23. $|6x + 36| < \varepsilon \Leftrightarrow |6(x+6)| < \varepsilon$

   $\Leftrightarrow 6|x + 6| < \varepsilon$

   $\Leftrightarrow |x + 6| < \dfrac{\varepsilon}{6}; \delta = \dfrac{\varepsilon}{6}$

25. $C = \pi d$

   $|C - 10| \leq 0.02$

   $|\pi d - 10| \leq 0.02$

   $\left|\pi\left(d - \dfrac{10}{\pi}\right)\right| \leq 0.02$

   $\left|d - \dfrac{10}{\pi}\right| \leq \dfrac{0.02}{\pi} \approx 0.0064$

   We must measure the diameter to an accuracy of 0.0064 in.

27. $|x-1| < 2|x-3|$

   $|x-1| < |2x-6|$

   $(x-1)^2 < (2x-6)^2$

   $x^2 - 2x + 1 < 4x^2 - 24x + 36$

   $3x^2 - 22x + 35 > 0$

   $(3x-7)(x-5) > 0;$

   $\left(-\infty, \dfrac{7}{3}\right) \cup (5, \infty)$

29. $2|2x-3| < |x+10|$

   $|4x-6| < |x+10|$

   $(4x-6)^2 < (x+10)^2$

   $16x^2 - 48x + 36 < x^2 + 20x + 100$

   $15x^2 - 68x - 64 < 0$

   $(5x+4)(3x-16) < 0;$

   $\left(-\dfrac{4}{5}, \dfrac{16}{3}\right)$

31. $|x| < |y| \Rightarrow |x||x| \le |x||y|$ and $|x||y| < |y||y|$ Order property: $x < y \Leftrightarrow xz < yz$ when z is positive.

$\qquad \Rightarrow |x|^2 < |y|^2$ Transitivity

$\qquad \Rightarrow x^2 < y^2$ $\left(|x|^2 = x^2\right)$

Conversely,

$x^2 < y^2 \Rightarrow |x|^2 < |y|^2$ $\left(x^2 = |x|^2\right)$

$\qquad \Rightarrow |x|^2 - |y|^2 < 0$ Subtract $|y|^2$ from each side.

$\qquad \Rightarrow (|x| - |y|)(|x| + |y|) < 0$ Factor the difference of two squares.

$\qquad \Rightarrow |x| - |y| < 0$ This is the only factor that can be negative.

$\qquad \Rightarrow |x| < |y|$ Add $|y|$ to each side.

33. a. $|a - b| = |a + (-b)| \le |a| + |-b| = |a| + |b|$

b. $|a - b| \ge ||a| - |b|| \ge |a| - |b|$ Use Property 4 of absolute values.

c. $|a + b + c| = |(a + b) + c| \le |a + b| + |c|$

$\qquad \le |a| + |b| + |c|$

35. $\left|\dfrac{x-2}{x^2+9}\right| = \left|\dfrac{x+(-2)}{x^2+9}\right|$

$\left|\dfrac{x-2}{x^2+9}\right| \le \left|\dfrac{x}{x^2+9}\right| + \left|\dfrac{-2}{x^2+9}\right|$

$\left|\dfrac{x-2}{x^2+9}\right| \le \dfrac{|x|}{x^2+9} + \dfrac{2}{x^2+9} = \dfrac{|x|+2}{x^2+9}$

Since $x^2 + 9 \ge 9$, $\dfrac{1}{x^2+9} \le \dfrac{1}{9}$

$\dfrac{|x|+2}{x^2+9} \le \dfrac{|x|+2}{9}$

$\left|\dfrac{x-2}{x^2+9}\right| \le \dfrac{|x|+2}{9}$

37. $\left|x^4 + \dfrac{1}{2}x^3 + \dfrac{1}{4}x^2 + \dfrac{1}{8}x + \dfrac{1}{16}\right|$

$\le |x^4| + \dfrac{1}{2}|x^3| + \dfrac{1}{4}|x^2| + \dfrac{1}{8}|x| + \dfrac{1}{16}$

$\le 1 + \dfrac{1}{2} + \dfrac{1}{4} + \dfrac{1}{8} + \dfrac{1}{16}$ since $|x| \le 1$.

So $\left|x^4 + \dfrac{1}{2}x^3 + \dfrac{1}{4}x^2 + \dfrac{1}{8}x + \dfrac{1}{16}\right| \le 1.9375 < 2.$

39. $a \ne 0 \Rightarrow$

$0 \le \left(a - \dfrac{1}{a}\right)^2 = a^2 - 2 + \dfrac{1}{a^2}$

so, $2 \le a^2 + \dfrac{1}{a^2}$ or $a^2 + \dfrac{1}{a^2} \ge 2$

41. $0 < a < b$

$a^2 < ab$ and $ab < b^2$

$a^2 < ab < b^2$

$a < \sqrt{ab} < b$

43. For a rectangle the area is ab, while for a square the area is $a^2 = \left(\dfrac{a+b}{2}\right)^2$. From Problem 42,

$\sqrt{ab} \le \dfrac{1}{2}(a+b) \Leftrightarrow ab \le \left(\dfrac{a+b}{2}\right)^2$ so the square has the largest area.

1.5 Concepts Review

1. II; IV

3. $(x+4)^2 + (y-2)^2 = 25$

Problem Set 1.5

1.

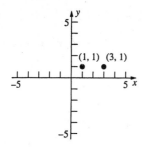

$$d = \sqrt{(3-1)^2 + (1-1)^2} = \sqrt{4} = 2$$

3.

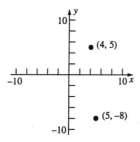

$$d = \sqrt{(4-5)^2 + (5+8)^2} = \sqrt{170} \approx 13.04$$

5.

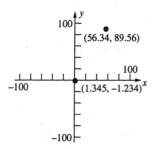

$$d = \sqrt{(1.345 - 56.34)^2 + (-1.234 - 89.56)^2}$$
$$\approx 106.151$$

7. $d_1 = \sqrt{(5+2)^2 + (3-4)^2} = \sqrt{49+1} = \sqrt{50}$

$d_2 = \sqrt{(5-10)^2 + (3-8)^2} = \sqrt{25+25} = \sqrt{50}$

$d_3 = \sqrt{(-2-10)^2 + (4-8)^2} = \sqrt{144+16} = \sqrt{160}$

$d_1 = d_2$ so the triangle is isosceles.

9. $(-1,-1), (-1,3); (7,-1), (7,3); (1,1), (5,1)$

11. $\left(\dfrac{-2+4}{2}, \dfrac{-2+3}{2}\right) = \left(1, \dfrac{1}{2}\right);$

$d = \sqrt{(1+2)^2 + \left(\dfrac{1}{2} - 3\right)^2} = \sqrt{9 + \dfrac{25}{4}} \approx 3.91$

13. $(x-1)^2 + (y-1)^2 = 1$

15. $(x-2)^2 + (y+1)^2 = r^2$

$(5-2)^2 + (3+1)^2 = r^2$

$r^2 = 9 + 16 = 25$

$(x-2)^2 + (y+1)^2 = 25$

17. center $= \left(\dfrac{1+3}{2}, \dfrac{3+7}{2}\right) = (2, 5)$

radius $= \dfrac{1}{2}\sqrt{(1-3)^2 + (3-7)^2} = \dfrac{1}{2}\sqrt{4+16}$

$\qquad = \dfrac{1}{2}\sqrt{20} = \sqrt{5}$

$(x-2)^2 + (y-5)^2 = 5$

19. Substitute $x = \dfrac{1}{4}$ into the equation and solve for y.

$\left(-\dfrac{3}{4}\right)^2 + (y-1)^2 = 1$

$(y-1)^2 = \dfrac{7}{16}$

$y - 1 = \pm\dfrac{\sqrt{7}}{4}$

$y = 1 \pm \dfrac{\sqrt{7}}{4}$

21. $x^2 + 2x + 10 + y^2 - 6y - 10 = 0$

$x^2 + 2x + y^2 - 6y = 0$

$(x^2 + 2x + 1) + (y^2 - 6y + 9) = 1 + 9$

$(x+1)^2 + (y-3)^2 = 10$

center $= (-1, 3)$; radius $= \sqrt{10}$

23. $x^2 + y^2 - 12x + 35 = 0$

$x^2 - 12x + y^2 = -35$

$(x^2 - 12x + 36) + y^2 = -35 + 36$

$(x-6)^2 + y^2 = 1$

center $= (6, 0)$; radius $= 1$

25. $4x^2 + 16x + 15 + 4y^2 + 6y = 0$

$4(x^2 + 4x + 4) + 4\left(y^2 + \dfrac{3}{2}y + \dfrac{9}{16}\right) = -15 + 16 + \dfrac{9}{4}$

$4(x+2)^2 + 4\left(y + \dfrac{3}{4}\right)^2 = \dfrac{13}{4}$

$(x+2)^2 + \left(y + \dfrac{3}{4}\right)^2 = \dfrac{13}{16}$

center $= \left(-2, -\dfrac{3}{4}\right)$; radius $= \dfrac{\sqrt{13}}{4}$

27. center: $\left(\dfrac{2+6}{2}, \dfrac{-1+3}{2}\right) = (4,1)$

midpoint $= \left(\dfrac{2+6}{2}, \dfrac{3+3}{2}\right) = (4,3)$

inscribed circle: radius $= \sqrt{(4-4)^2 + (1-3)^2}$

$\quad\quad = \sqrt{4} = 2$

$(x-4)^2 + (y-1)^2 = 4$

circumscribed circle:

radius $= \sqrt{(4-2)^2 + (1-3)^2} = \sqrt{8}$

$(x-4)^2 + (y-1)^2 = 8$

29. $AC = \sqrt{AB^2 + BC^2} = \sqrt{(214)^2 + (179)^2}$

$\quad = \sqrt{77{,}837} \approx 278.99$

Cost by truck $= 3.71(214 + 179) = \$1458.03$
Cost by plane $= 4.82(279) = \$1344.78$; cheaper by plane.

31. Put the vertex of the right angle at the origin with the other vertices at $(a, 0)$ and $(0, b)$. The midpoint of the hypotenuse is $\left(\dfrac{a}{2}, \dfrac{b}{2}\right)$. The distances from the vertices are

$\sqrt{\left(a-\dfrac{a}{2}\right)^2 + \left(0-\dfrac{b}{2}\right)^2} = \sqrt{\dfrac{a^2}{4} + \dfrac{b^2}{4}}$

$\quad\quad = \dfrac{1}{2}\sqrt{a^2 + b^2},$

$\sqrt{\left(0-\dfrac{a}{2}\right)^2 + \left(b-\dfrac{b}{2}\right)^2} = \sqrt{\dfrac{a^2}{4} + \dfrac{b^2}{4}}$

$\quad\quad = \dfrac{1}{2}\sqrt{a^2 + b^2},$ and

$\sqrt{\left(0-\dfrac{a}{2}\right)^2 + \left(0-\dfrac{b}{2}\right)^2} = \sqrt{\dfrac{a^2}{4} + \dfrac{b^2}{4}}$

$\quad\quad = \dfrac{1}{2}\sqrt{a^2 + b^2},$

which are all the same.

33. $x^2 + y^2 - 4x - 2y - 11 = 0$

$(x^2 - 4x + 4) + (y^2 - 2y + 1) = 11 + 4 + 1$

$\quad (x-2)^2 + (y-1)^2 = 16$

$x^2 + y^2 + 20x - 12y + 72 = 0$

$(x^2 + 20x + 100) + (y^2 - 12y + 36)$

$\quad\quad\quad = -72 + 100 + 36$

$\quad (x+10)^2 + (y-6)^2 = 64$

center of first circle: $(2, 1)$
center of second circle: $(-10, 6)$

$d = \sqrt{(2+10)^2 + (1-6)^2} = \sqrt{144 + 25}$

$\quad = \sqrt{169} = 13$

However, the radii only sum to $4 + 8 = 12$, so the circles must not intersect if the distance between their centers is 13.

35. Label the points C, P, Q, and R as shown in the figure below. Let $d = |OP|$, $h = |OQ|$, and $a = |PR|$. Triangles $\triangle OPR$ and $\triangle CQR$ are similar because each contains a right angle and they share angle $\angle QRC$. For an angle of $30°$, $d/h = \sqrt{3}/2$. Thus, using a property of similar triangles,

$|QC|/|RC| = \sqrt{3}/2$

$\dfrac{2}{a-2} = \dfrac{\sqrt{3}}{2}$

$a = 2 + 4/\sqrt{3}$

Thus, $h = 2a = 2(2 + 4\sqrt{3}) = 4(1 + 2\sqrt{3})$.
By the Pythagorean Theorem, we have

$d = \sqrt{h^2 - a^2} = \sqrt{3}a = 2\sqrt{3} + 4 \approx 7.464$

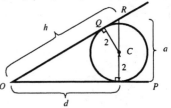

37. The centers of the circles are

$\sqrt{(10-2)^2 + (8-2)^2} = \sqrt{100} = 10$ units apart, so the belts cross at a point 5 units from each center. The belt makes a right angle with the radius at point B, as shown in the figure.

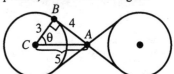

From the Pythagorean Theorem, the length of the belt from A to B is $\sqrt{5^2 - 3^2} = \sqrt{16} = 4$.

$\sin\theta = \dfrac{4}{5} \Rightarrow \theta \approx 0.93$ radians. The situation on the lower half of the wheel is identical, and the two wheels are identical, so the length of the belt around each wheel is $3(2\pi - 1.86) \approx 13.3$ units. The length of the belt is $2(13.3) + 4(4) \approx 42.6$ units.

39. Let a, b, and c be the lengths of the sides of the right triangle, with c the length of the hypotenuse. Then the Pythagorean Theorem says that $a^2 + b^2 = c^2$

Thus, $\dfrac{\pi a^2}{8} + \dfrac{\pi b^2}{8} = \dfrac{\pi c^2}{8}$ or

$$\frac{1}{2}\pi\left(\frac{a}{2}\right)^2 + \frac{1}{2}\pi\left(\frac{b}{2}\right)^2 = \frac{1}{2}\pi\left(\frac{c}{2}\right)^2$$

$\dfrac{1}{2}\pi\left(\dfrac{x}{2}\right)^2$ is the area of a semicircle with

diameter x, so the circles on the legs of the triangle have total area equal to the area of the semicircle on the hypotenuse.

From $a^2 + b^2 = c^2$,

$$\frac{\sqrt{3}}{4}a^2 + \frac{\sqrt{3}}{4}b^2 = \frac{\sqrt{3}}{4}c^2$$

$\dfrac{\sqrt{3}}{4}x^2$ is the area of an equilateral triangle with

sides of length x, so the equilateral triangles on the legs of the right triangle have total area equal to the area of the equilateral triangle on the hypotenuse of the right triangle.

41. The lengths A, B, and C are the same as the corresponding distances between the centers of the circles:

$A = \sqrt{(-2)^2 + (8)^2} = \sqrt{68} \approx 8.2$

$B = \sqrt{(6)^2 + (8)^2} = \sqrt{100} = 10$

$C = \sqrt{(8)^2 + (0)^2} = \sqrt{64} = 8$

Each circle has radius 2, so the part of the belt around the wheels is

$2(2\pi - a - \pi) + 2(2\pi - b - \pi) + 2(2\pi - c - \pi)$

$= 2[3\pi - (a + b + c)] = 2(2\pi) = 4\pi$

Since $a + b + c = \pi$, the sum of the angles of a triangle.

The length of the belt is $\approx 8.2 + 10 + 8 + 4\pi$

≈ 38.8 units.

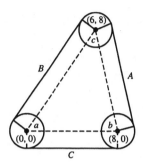

1.6 Concepts Review

1. $\dfrac{(d-b)}{(c-a)}$

3. $y = mx + b; \ x = k$

Problem Set 1.6

1. $\dfrac{2-1}{2-1} = 1$

3. $\dfrac{-6-3}{-5-2} = \dfrac{9}{7}$

5. $\dfrac{5-0}{0-3} = -\dfrac{5}{3}$

7. $\dfrac{-3.456 - 5.678}{7.654 + 1.234} \approx -1.028$

9. $\quad y - 2 = -1(x - 2)$
   $\quad\quad y - 2 = -x + 2$
   $\quad x + y - 4 = 0$

11. $\quad\quad y = 2x + 3$
    $\quad 2x - y + 3 = 0$

13. $m = \dfrac{8-3}{4-2} = \dfrac{5}{2}$;

    $y - 3 = \dfrac{5}{2}(x - 2)$

    $2y - 6 = 5x - 10$

    $5x - 2y - 4 = 0$

15. $m = \dfrac{5+3}{2-2}$: undefined; $x + 0y - 2 = 0$

17. $3y = -2x + 1; \ y = -\dfrac{2}{3}x + \dfrac{1}{3}$; slope $= -\dfrac{2}{3}$;

    y-intercept $= \dfrac{1}{3}$

19. $6 - 2y = 10x - 2$
    $\quad -2y = 10x - 8$
    $\quad\quad y = -5x + 4$;
    slope $= -5$; y-intercept $= 4$

21. a. $m = 2$;
    $\quad y + 3 = 2(x - 3)$
    $\quad\quad y = 2x - 9$

b. $m = -\dfrac{1}{2}$;

$$y + 3 = -\dfrac{1}{2}(x - 3)$$

$$y = -\dfrac{1}{2}x - \dfrac{3}{2}$$

c. $2x + 3y = 6$

$$3y = -2x + 6$$

$$y = -\dfrac{2}{3}x + 2;$$

$$m = -\dfrac{2}{3};$$

$$y + 3 = -\dfrac{2}{3}(x - 3)$$

$$y = -\dfrac{2}{3}x - 1$$

d. $m = \dfrac{3}{2}$;

$$y + 3 = \dfrac{3}{2}(x - 3)$$

$$y = \dfrac{3}{2}x - \dfrac{15}{2}$$

e. $m = \dfrac{-1 - 2}{3 + 1} = -\dfrac{3}{4}$;

$$y + 3 = -\dfrac{3}{4}(x - 3)$$

$$y = -\dfrac{3}{4}x - \dfrac{3}{4}$$

f. $x = 3$ **g.** $y = -3$

23. $m = \dfrac{3}{2}$;

$$y + 1 = \dfrac{3}{2}(x + 2)$$

$$y = \dfrac{3}{2}x + 2$$

25. $y = 3(3) - 1 = 8$; $(3, 9)$ is above the line.

27. $2x + 3y = 4$

$\quad -3x + y = 5$

$2x + 3y = 4$

$9x - 3y = -15$

$\overline{11x = -11}$

$\quad x = -1$

$-3(-1) + y = 5$

$\quad y = 2$

Point of intersection: $(-1, 2)$

$$3y = -2x + 4$$

$$y = -\dfrac{2}{3}x + \dfrac{4}{3}$$

$$m = \dfrac{3}{2}$$

$$y - 2 = \dfrac{3}{2}(x + 1)$$

$$y = \dfrac{3}{2}x + \dfrac{7}{2}$$

29. $3x - 4y = 5$

$\quad 2x + 3y = 9$

$9x - 12y = 15$

$8x + 12y = 36$

$\overline{17x = 51}$

$\quad x = 3$

$3(3) - 4y = 5$

$\quad -4y = -4$

$\quad y = 1$

Point of intersection: $(3, 1)$; $3x - 4y = 5$;

$\quad -4y = -3x + 5$

$$y = \dfrac{3}{4}x - \dfrac{5}{4}$$

$$m = -\dfrac{4}{3}$$

$$y - 1 = -\dfrac{4}{3}(x - 3)$$

$$y = -\dfrac{4}{3}x + 5$$

31. $A = 3, B = 4, C = -6$ $d = \dfrac{\left|3(-3) + 4(2) + (-6)\right|}{\sqrt{(3)^2 + (4)^2}} = \dfrac{7}{5}$

33. $A = 12, B = -5, C = 1$

$$d = \dfrac{\left|12(-2) - 5(-1) + 1\right|}{\sqrt{(12)^2 + (-5)^2}} = \dfrac{18}{13}$$

35. $2x + 4(0) = 5$

$\quad x = \dfrac{5}{2}$ $d = \dfrac{\left|2\left(\frac{5}{2}\right) + 4(0) - 7\right|}{\sqrt{(2)^2 + (4)^2}} = \dfrac{2}{\sqrt{20}} = \dfrac{\sqrt{5}}{5}$

37. $120{,}000(0.08) = 9600$; $V = 120{,}000 - 9600t$

39. $(0, 700{,}000), (10, 820{,}000)$

$$m = \dfrac{820{,}000 - 700{,}000}{10 - 0} = 12{,}000$$

$N = 12{,}000n + 700{,}000$

At $n = 25$: $N = 12{,}000(25) + 700{,}000 = 1{,}000{,}000$

41. a. When $x = 0$, $P = -2000$, which indicates that the company loses money if no items are sold.

b. Slope = 450; this is the amount of profit gained with the sale of each item.

43. If (x_0, y_0) is on both lines, then
$2x_0 - y_0 + 4 = 0$ and $x_0 + 3y_0 - 6 = 0$ so
$2x_0 - y_0 + 4 + k(x_0 + 3y_0 - 6) = 0 + 0 \cdot k = 0$,
which means that (x_0, y_0) is on the line
$2x - y + 4 + k(x + 3y - 6) = 0$ regardless of the value of k.

45. $m = \dfrac{-2-3}{1+2} = -\dfrac{5}{3}$; $m = \dfrac{3}{5}$; passes through
$$\left(\dfrac{-2+1}{2}, \dfrac{3-2}{2}\right) = \left(-\dfrac{1}{2}, \dfrac{1}{2}\right)$$
$$y - \dfrac{1}{2} = \dfrac{3}{5}\left(x + \dfrac{1}{2}\right)$$
$$y = \dfrac{3}{5}x + \dfrac{4}{5}$$

47. Let the origin be at the vertex as shown in the figure below. The center of the circle is then $(4-r, r)$, so it has equation
$(x - (4-r))^2 + (y-r)^2 = r^2$. Along the side of length 5, the y-coordinate is always $\dfrac{3}{4}$ times the x-coordinate. Thus, we need to find the value of r for which there is exactly one x-solution to
$(x - 4 + r)^2 + (y - r)^2 = r^2$. Solving for x in this equation gives
$$x = \dfrac{4}{25}\left(16 - r \pm \sqrt{24\left(-r^2 + 7r - 6\right)}\right).$$ There is exactly one solution when $-r^2 + 7r - 6 = 0$, that is, when $r = 1$ or $r = 6$. The root $r = 6$ is extraneous. Thus, the largest circle that can be inscribed in this triangle has radius $r = 1$.

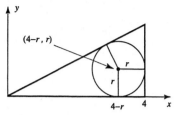

$(4-r, r)$

49. $12a + 0b = 36$
$a = 3$
$3^2 + b^2 = 36$
$b = \pm 3\sqrt{3}$
$3x - 3\sqrt{3}y = 36$
$x - \sqrt{3}y = 12$
$3x + 3\sqrt{3}y = 36$
$x + \sqrt{3}y = 12$

51. See the figure below. The midpoints of the sides are
$$P\left(\dfrac{x_1 + x_2}{2}, \dfrac{y_1 + y_2}{2}\right), \ Q\left(\dfrac{x_2 + x_3}{2}, \dfrac{y_2 + y_3}{2}\right),$$
$$R\left(\dfrac{x_3 + x_4}{2}, \dfrac{y_3 + y_4}{2}\right), \text{ and } S\left(\dfrac{x_1 + x_4}{2}, \dfrac{y_1 + y_4}{2}\right).$$
The slope of PS is
$$\dfrac{\frac{1}{2}[y_1 + y_4 - (y_1 + y_2)]}{\frac{1}{2}[x_1 + x_4 - (x_1 + x_2)]} = \dfrac{y_4 - y_2}{x_4 - x_2}. \text{ The slope of }$$
QR is
$$\dfrac{\frac{1}{2}[y_3 + y_4 - (y_2 + y_3)]}{\frac{1}{2}[x_3 + x_4 - (x_2 + x_3)]} = \dfrac{y_4 - y_2}{x_4 - x_2}. \text{ Thus } PS$$
and QR are parallel. The slopes of SR and PQ are both $\dfrac{y_3 - y_1}{x_3 - x_1}$, so $PQRS$ is a parallelogram.

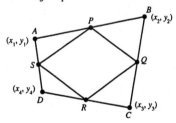

1.7 Concepts Review

1. y-axis

3. $8; -2, 1, 4$

Problem Set 1.7

1. $y = -x^2 + 1$; y-intercept $= 1$; $y = (1 + x)(1 - x)$;
x-intercepts $= -1, 1$
Symmetric with respect to the y-axis

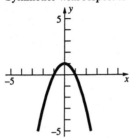

3. $x = -4y^2 - 1$; x-intercept $= -1$
Symmetric with respect to the x-axis

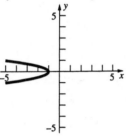

5. $x^2 + y = 0$; $y = -x^2$
x-intercept $= 0$, y-intercept $= 0$
Symmetric with respect to the y-axis

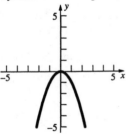

7. $7x^2 + 3y = 0$; $3y = -7x^2$; $y = -\dfrac{7}{3}x^2$

x-intercept $= 0$, y-intercept $= 0$
Symmetric with respect to the y-axis

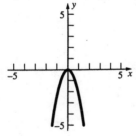

9. $x^2 + y^2 = 4$
x-intercepts $= -2, 2$; y-intercepts $= -2, 2$
Symmetric with respect to the x-axis, y-axis, and origin

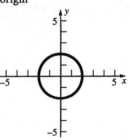

11. $y = -x^2 - 2x + 2$: y-intercept $= 2$

x-intercepts $= \dfrac{2 \pm \sqrt{4+8}}{-2} = \dfrac{2 \pm 2\sqrt{3}}{-2} = -1 \pm \sqrt{3}$

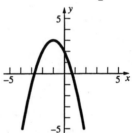

13. $x^2 - y^2 = 4$
x-intercept $= -2, 2$
Symmetric with respect to the x-axis, y-axis, and origin

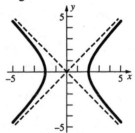

15. $4(x - 1)^2 + y^2 = 36$;
y-intercepts $= \pm\sqrt{32} = \pm 4\sqrt{2}$
x-intercepts $= -2, 4$
Symmetric with respect to the x-axis

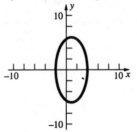

17. $x^2 + 9(y + 2)^2 = 36$; y-intercepts $= -4, 0$
x-intercept $= 0$
Symmetric with respect to the y-axis

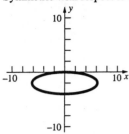

19. $x^4 + y^4 = 16$; y-intercepts $= -2, 2$
x-intercepts $= -2, 2$
Symmetric with respect to the y-axis, x-axis and
origin

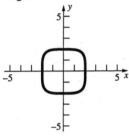

21. $y = \dfrac{1}{x^2 + 1}$; y-intercept $= 1$
Symmetric with respect to the y-axis

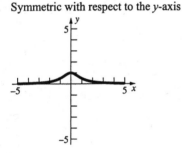

23. $2x^2 - 4x + 3y^2 + 12y = -2$

$2(x^2 - 2x + 1) + 3(y^2 + 4y + 4) = -2 + 2 + 12$

$2(x - 1)^2 + 3(y + 2)^2 = 12$

y-intercepts $= -2 \pm \dfrac{\sqrt{30}}{3}$

x-intercept $= 1$

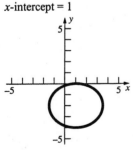

25. $y = (x - 1)(x - 2)(x - 3)$; y-intercept $= -6$
x-intercepts $= 1, 2, 3$

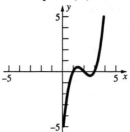

27. $y = x^2(x - 1)^2$; y-intercept $= 0$
x-intercepts $= 0, 1$

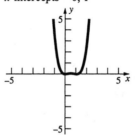

29. $|x| + |y| = 1$; y-intercepts $= -1, 1$;
x-intercepts $= -1, 1$
Symmetric with respect to the x-axis, y-axis and
origin

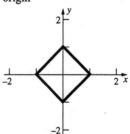

31. $-x + 1 = (x + 1)^2$

$-x + 1 = x^2 + 2x + 1$

$x^2 + 3x = 0$

$x(x + 3) = 0$

$x = 0, -3$

Intersection points: $(0, 1)$ and $(-3, 4)$

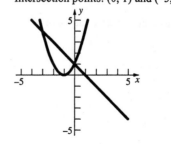

33. $-2x+3 = -2(x-4)^2$

$-2x+3 = -2x^2+16x-32$

$2x^2-18x+35 = 0$

$x = \dfrac{18 \pm \sqrt{324-280}}{4} = \dfrac{18 \pm 2\sqrt{11}}{4} = \dfrac{9 \pm \sqrt{11}}{2};$

Intersection points: $\left(\dfrac{9-\sqrt{11}}{2}, -6+\sqrt{11}\right),$

$\left(\dfrac{9+\sqrt{11}}{2}, -6-\sqrt{11}\right)$

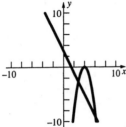

35. $x^2+x^2 = 4$

$x^2 = 2$

$x = \pm\sqrt{2}$

Intersection points: $\left(-\sqrt{2}, -\sqrt{2}\right), \left(\sqrt{2}, \sqrt{2}\right)$

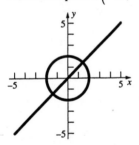

37. $y = 3x+1$

$x^2+2x+(3x+1)^2 = 15$

$x^2+2x+9x^2+6x+1 = 15$

$10x^2+8x-14 = 0$

$2(5x^2+4x-7) = 0$

$x \approx -1.65, 0.85$

Intersection points: $(-1.65, -3.95)$ and $(0.85, 3.55)$

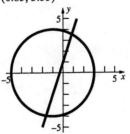

39. a. $y = x^2;$ (2)

b. ax^3+bx^2+cx+d, with $a > 0$: (1)

c. ax^3+bx^2+cx+d, with $a < 0$: (3)

d. $y = ax^3$, with $a > 0$: (4)

41. $x^2+2x+y^2-2y = 20;$ $\left(-2, 1+\sqrt{21}\right),$

$\left(-2, 1-\sqrt{21}\right), \left(2, 1+\sqrt{13}\right), \left(2, 1-\sqrt{13}\right)$

$d_1 = \sqrt{(-2-2)^2 + \left[1+\sqrt{21}-\left(1+\sqrt{13}\right)\right]^2}$

$= \sqrt{16+\left(\sqrt{21}-\sqrt{13}\right)^2}$

$= \sqrt{50-2\sqrt{273}} \approx 4.12$

$d_2 = \sqrt{(-2-2)^2 + \left[1+\sqrt{21}-\left(1-\sqrt{13}\right)\right]^2}$

$= \sqrt{16+\left(\sqrt{21}+\sqrt{13}\right)^2}$

$= \sqrt{50+2\sqrt{273}} \approx 9.11$

$d_3 = \sqrt{(-2+2)^2 + \left[1+\sqrt{21}-\left(1-\sqrt{21}\right)\right]^2}$

$= \sqrt{0+\left(\sqrt{21}+\sqrt{21}\right)^2} = \sqrt{\left(2\sqrt{21}\right)^2} = 2\sqrt{21} \approx 9.17$

$d_4 = \sqrt{(-2-2)^2 + \left[1-\sqrt{21}-(1+\sqrt{13})\right]^2}$

$= \sqrt{16+\left(-\sqrt{21}-\sqrt{13}\right)^2} = \sqrt{50+2\sqrt{273}} \approx 9.11$

$d_5 = \sqrt{(-2-2)^2 + \left[1-\sqrt{21}-\left(1-\sqrt{13}\right)\right]^2}$

$= \sqrt{16+\left(\sqrt{13}-\sqrt{21}\right)^2} = \sqrt{50-2\sqrt{273}} \approx 4.12$

$d_6 = \sqrt{(2-2)^2 + \left[1+\sqrt{13}-\left(1-\sqrt{13}\right)\right]^2}$

$= \sqrt{0+\left(\sqrt{13}+\sqrt{13}\right)^2} = \sqrt{\left(2\sqrt{13}\right)^2} = 2\sqrt{13} \approx 7.21$

Four such distances ($d_2 = d_4$ and $d_1 = d_5$)

1.8 Chapter Review

Concepts Test

1. False: p and q must be integers.

3. False: If the numbers are opposites $(-\pi$ and π) then the sum is 0, which is rational.

5. False: 0.999... is equal to 1.

7. False: $(a \cdot b) \cdot c = a^{bc}; a \cdot (b \cdot c) = a^{b^c}$

9. True: If x was not 0, then $\varepsilon = \dfrac{|x|}{2}$ would be a positive number less than $|x|$.

11. True: $a < b < 0; a < b; \dfrac{a}{b} > 1; \dfrac{1}{b} < \dfrac{1}{a}$

13. True: If (a, b) and (c, d) share a point then $c < b$ so they share the infinitely many points between b and c.

15. False: If $x = 0$, the number has no sign.

17. True: $|x| < |y| \Leftrightarrow |x|^4 < |y|^4$
$|x|^4 = x^4$ and $|y|^4 = y^4$, so $x^4 < y^4$.

19. True: If $r = 0$, then
$$\frac{1}{1+|r|} = \frac{1}{1-r} = \frac{1}{1-|r|} = 1.$$
For any r, $1+|r| \geq 1-|r|$.
Since $|r| < 1, 1-|r| > 0$ so
$$\frac{1}{1+|r|} \leq \frac{1}{1-|r|}; \quad \text{also, } -1 < r < 1.$$

If $-1 < r < 0$, then $|r| = -r$ and
$1-r = 1+|r|$, so
$$\frac{1}{1+|r|} = \frac{1}{1-r} \leq \frac{1}{1-|r|}.$$
If $0 < r < 1$, then $|r| = r$ and
$1-r = 1-|r|$, so
$$\frac{1}{1+|r|} \leq \frac{1}{1-r} = \frac{1}{1-|r|}.$$

21. True: If x and y are the same sign, then $\big||x|-|y|\big| = |x-y|$. $|x-y| \leq |x+y|$ when x and y are the same sign, so

$\big||x|-|y|\big| \leq |x+y|$. If x and y have opposite signs then either
$\big||x|-|y|\big| = |x-(-y)| = |x+y|$
$(x > 0, y < 0)$ or
$\big||x|-|y|\big| = |-x-y| = |x+y|$
$(x < 0, y > 0)$. In either case
$\big||x|-|y|\big| = |x+y|$.
If either $x = 0$ or $y = 0$, the inequality is easily seen to be true.

23. True: For every real number y, whether it is positive, zero, or negative, the cube root $x = \sqrt[3]{y}$ satisfies $x^3 = \left(\sqrt[3]{y}\right)^3 = y$

25. True:
$$x^2 + ax + y^2 + y = 0$$
$$x^2 + ax + \frac{a^2}{4} + y^2 + y + \frac{1}{4} = \frac{a^2}{4} + \frac{1}{4}$$
$$\left(x + \frac{a}{2}\right)^2 + \left(y + \frac{1}{2}\right)^2 = \frac{a^2+1}{4}$$
is a circle for all values of a.

27. True; $y - b = \dfrac{3}{4}(x - a)$
$$y = \frac{3}{4}x - \frac{3a}{4} + b;$$

If $x = a + 4$:
$$y = \frac{3}{4}(a+4) - \frac{3a}{4} + b$$
$$= \frac{3a}{4} + 3 - \frac{3a}{4} + b = b + 3$$

29. True: If $ab > 0$, a and b have the same sign, so (a, b) is in either the first or third quadrant.

31. True: If $ab = 0$, a or b is 0, so (a, b) lies on the x-axis or the y-axis. If $a = b = 0$, (a, b) is the origin.

33. True: $d = \sqrt{[(a+b)-(a-b)]^2 + (a-a)^2}$
$$= \sqrt{(2b)^2} = |2b|$$

35. True: This is the general linear equation.

37. False: The slopes of perpendicular lines are negative reciprocals.

39. False: $ax + y = c \Rightarrow y = -ax + c$
$ax - y = c \Rightarrow y = ax - c$
$(a)(-a) \neq -1$.

Sample Test Problems

1. a. $\left(n+\dfrac{1}{n}\right)^n$; $\left(1+\dfrac{1}{1}\right)^1 = 2$; $\left(2+\dfrac{1}{2}\right)^2 = \dfrac{25}{4}$;

$\left(-2+\dfrac{1}{-2}\right)^{-2} = \dfrac{4}{25}$

b. $(n^2 - n + 1)^2$; $\left[(1)^2 - (1) + 1\right]^2 = 1$;

$\left[(2)^2 - (2) + 1\right]^2 = 9$;

$\left[(-2)^2 - (-2) + 1\right]^2 = 49$

c. $4^{3/n}$; $4^{3/1} = 64$; $4^{3/2} = 8$; $4^{-3/2} = \dfrac{1}{8}$

d. $\sqrt[n]{\left|\dfrac{1}{n}\right|}$; $\sqrt[1]{\left|\dfrac{1}{1}\right|} = 1$; $\sqrt{\left|\dfrac{1}{2}\right|} = \dfrac{1}{\sqrt{2}} = \dfrac{\sqrt{2}}{2}$;

$\sqrt[-2]{\left|\dfrac{1}{-2}\right|} = \sqrt{2}$

3. Let a, b, c, and d be integers.

$\dfrac{\frac{a}{b} + \frac{c}{d}}{2} = \dfrac{a}{2b} + \dfrac{c}{2d} = \dfrac{ad + bc}{2bd}$ which is rational.

5. Answers will vary. Possible answer:

$\sqrt{\dfrac{13}{50}} \approx 0.50990...$

7. $\left(\pi - \sqrt{2.0}\right)^{2.5} - \sqrt[3]{2.0} \approx 2.66$

9. $1 - 3x > 0$

$3x < 1$

$x < \dfrac{1}{3}$

$\left(-\infty, \dfrac{1}{3}\right)$

11. $3 - 2x \le 4x + 1 \le 2x + 7$

$3 - 2x \le 4x + 1$ and $4x + 1 \le 2x + 7$

$6x \ge 2$ and $2x \ge 6$

$x \ge \dfrac{1}{3}$ and $x \le 3$; $\left[\dfrac{1}{3}, 3\right]$

13. $21t^2 - 44t + 12 \le -3$; $21t^2 - 44t + 15 \le 0$;

$t = \dfrac{44 \pm \sqrt{44^2 - 4(21)(15)}}{2(21)} = \dfrac{44 \pm 26}{42} = \dfrac{3}{7}, \dfrac{5}{3}$

$\left(t - \dfrac{3}{7}\right)\left(t - \dfrac{5}{3}\right) \le 0$; $\left[\dfrac{3}{7}, \dfrac{5}{3}\right]$

15. $(x + 4)(2x - 1)^2(x - 3) \le 0$; $[-4, 3]$

17. $\dfrac{3}{1 - x} \le 2$

$\dfrac{3}{1 - x} - 2 \le 0$

$\dfrac{3 - 2(1 - x)}{1 - x} \le 0$

$\dfrac{2x + 1}{1 - x} \le 0$;

$\left(-\infty, -\dfrac{1}{2}\right] \cup (1, \infty)$

19. For example, if $x = -2$, $\left|-(-2)\right| = 2 \ne -2$

$\left|-x\right| \ne x$ for any $x < 0$

21. $|t - 5| = |-(5 - t)| = |5 - t|$

If $|5 - t| = 5 - t$, then $5 - t \ge 0$.

$t \le 5$

23. If $|x| \le 2$, then

$0 \le \left|2x^2 + 3x + 2\right| \le \left|2x^2\right| + |3x| + 2 \le 8 + 6 + 2 = 16$

also $\left|x^2 + 2\right| \ge 2$ so $\dfrac{1}{\left|x^2 + 2\right|} \le \dfrac{1}{2}$. Thus

$\left|\dfrac{2x^2 + 3x + 2}{x^2 + 2}\right| = \left|2x^2 + 3x + 2\right|\left|\dfrac{1}{x^2 + 2}\right| \le 16\left(\dfrac{1}{2}\right)$

$= 8$

25.

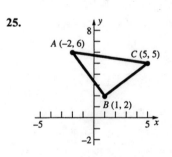

$$d(A, B) = \sqrt{(1+2)^2 + (2-6)^2} = \sqrt{9+16} = 5$$
$$d(B, C) = \sqrt{(5-1)^2 + (5-2)^2} = \sqrt{16+9} = 5$$
$$d(A, C) = \sqrt{(5+2)^2 + (5-6)^2} = \sqrt{49+1} = \sqrt{50}$$
$$= 5\sqrt{2}$$

$(AB)^2 + (BC)^2 = (AC)^2$, so $\triangle ABC$ is a right triangle.

27. center $= \left(\dfrac{2+10}{2}, \dfrac{0+4}{2}\right) = (6, 2)$

radius $= \dfrac{1}{2}\sqrt{(10-2)^2 + (4-0)^2} = \dfrac{1}{2}\sqrt{64+16}$

$= 2\sqrt{5}$

circle: $(x-6)^2 + (y-2)^2 = 20$

29.
$$x^2 - 2x + y^2 + 2y = 2$$
$$x^2 - 2x + 1 + y^2 + 2y + 1 = 2 + 1 + 1$$
$$(x-1)^2 + (y+1)^2 = 4$$
center $= (1, -1)$
$$x^2 + 6x + y^2 - 4y = -7$$
$$x^2 + 6x + 9 + y^2 - 4y + 4 = -7 + 9 + 4$$
$$(x+3)^2 + (y-2)^2 = 6$$
center $= (-3, 2)$
$$d = \sqrt{(-3-1)^2 + (2+1)^2} = \sqrt{16+9} = 5$$

31. a. $m = \dfrac{3-1}{7+2} = \dfrac{2}{9}$;

$y - 1 = \dfrac{2}{9}(x+2)$

$y = \dfrac{2}{9}x + \dfrac{13}{9}$

b. $3x - 2y = 5$

$-2y = -3x + 5$

$y = \dfrac{3}{2}x - \dfrac{5}{2}$;

$m = \dfrac{3}{2}$

$y - 1 = \dfrac{3}{2}(x+2)$

$y = \dfrac{3}{2}x + 4$

c. $3x + 4y = 9$

$4y = -3x + 9$;

$y = -\dfrac{3}{4}x + \dfrac{9}{4}$; $m = \dfrac{4}{3}$

$y - 1 = \dfrac{4}{3}(x+2)$

$y = \dfrac{4}{3}x + \dfrac{11}{3}$

d. $x = -2$

e. contains $(-2, 1)$ and $(0, 3)$; $m = \dfrac{3-1}{0+2}$;

$y = x + 3$

33. The figure is a cubic with respect to y.
The equation is **(b)** $x = y^3$.

35.

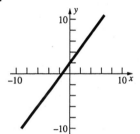

37.

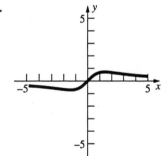

39. $y = x^2 - 2x + 4$ and $y - x = 4$;

$x + 4 = x^2 - 2x + 4$

$x^2 - 3x = 0$

$x(x-3) = 0$

points of intersection: $(0, 4)$ and $(3, 7)$

CHAPTER

2

Functions and Limits

2.1 Concepts Review

1. Domain; range

3. Asymptote

Problem Set 2.1

1. **a.** $f(1) = 1 - 1^2 = 0$

   b. $f(-2) = 1 - (-2)^2 = -3$

   c. $f(0) = 1 - 0^2 = 1$

   d. $f(k) = 1 - k^2$

   e. $f(-5) = 1 - (-5)^2 = -24$

   f. $f\left(\dfrac{1}{4}\right) = 1 - \left(\dfrac{1}{4}\right)^2 = 1 - \dfrac{1}{16} = \dfrac{15}{16}$

   g. $f(3t) = 1 - (3t)^2 = 1 - 9t^2$

   h. $f(2x) = 1 - (2x)^2 = 1 - 4x^2$

   i. $f\left(\dfrac{1}{t}\right) = 1 - \left(\dfrac{1}{t}\right)^2 = 1 - \dfrac{1}{t^2} = \dfrac{t^2 - 1}{t^2}$

3. **a.** $G(0) = \dfrac{1}{0 - 1} = -1$

   b. $G(0.999) = \dfrac{1}{0.999 - 1} = -1000$

   c. $G(1.01) = \dfrac{1}{1.01 - 1} = 100$

   d. $G(y^2) = \dfrac{1}{y^2 - 1}$

   e. $G(-x) = \dfrac{1}{-x - 1} = -\dfrac{1}{x + 1}$

   f. $G\left(\dfrac{1}{x^2}\right) = \dfrac{1}{\frac{1}{x^2} - 1} = \dfrac{x^2}{1 - x^2}$

5. **a.** $f(0.25) = \dfrac{1}{\sqrt{0.25 - 3}} = \dfrac{1}{\sqrt{-2.75}}$ is not defined

   b. $f(x) = \dfrac{1}{\sqrt{\pi - 3}} \approx 2.658$

   c. $f(3 + \sqrt{2}) = \dfrac{1}{\sqrt{3 + \sqrt{2} - 3}} = \dfrac{1}{\sqrt{\sqrt{2}}}$
   $= 2^{-0.25} \approx 0.841$

7. **a.** $x^2 + y^2 = 1$
   $y^2 = 1 - x^2$
   $y = \pm\sqrt{1 - x^2}$; not a function

   b. $xy + y + x = 1$
   $y(x + 1) = 1 - x$
   $y = \dfrac{1 - x}{x + 1}; f(x) = \dfrac{1 - x}{x + 1}$

   c. $x = \sqrt{2y + 1}$
   $x^2 = 2y + 1$
   $y = \dfrac{x^2 - 1}{2}; f(x) = \dfrac{x^2 - 1}{2}$

   d. $x = \dfrac{y}{y + 1}$
   $xy + x = y$
   $x = y - xy$
   $x = y(1 - x)$
   $y = \dfrac{x}{1 - x}; f(x) = \dfrac{x}{1 - x}$

9. $\dfrac{f(a + h) - f(a)}{h} = \dfrac{[2(a + h)^2 - 1] - (2a^2 - 1)}{h}$

   $= \dfrac{4ah + 2h^2}{h} = 4a + 2h$

11. $\dfrac{g(x+h)-g(x)}{h} = \dfrac{\frac{3}{x+h-2}-\frac{3}{x-2}}{h}$

$= \dfrac{\dfrac{3x-6-3x-3h+6}{x^2-4x+hx-2h+4}}{h} = \dfrac{-3h}{h(x^2-4x+hx-2h+4)}$

$= -\dfrac{3}{x^2-4x+hx-2h+4}$

13. a. $F(z) = \sqrt{2z+3}$

$2z+3 \geq 0;\ z \geq -\dfrac{3}{2}$

Domain: $\left\{ z \in \mathbb{R} : z \geq -\dfrac{3}{2} \right\}$

b. $g(v) = \dfrac{1}{4v-1}$

$4v-1 = 0;\ v = \dfrac{1}{4}$

Domain: $\left\{ v \in \mathbb{R} : v \neq \dfrac{1}{4} \right\}$

c. $\psi(x) = \sqrt{x^2-9}$

$x^2-9 \geq 0;\ x^2 \geq 9;\ |x| \geq 3$

Domain: $\{ x \in \mathbb{R} : |x| \geq 3 \}$

d. $H(y) = -\sqrt{625-y^4}$

$625-y^4 \geq 0;\ 625 \geq y^4;\ |y| \leq 5$

Domain: $\{ y \in \mathbb{R} : |y| \leq 5 \}$

15. $f(x) = -4;\ f(-x) = -4$; even function

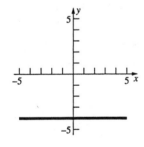

17. $F(x) = 2x+1;\ F(-x) = -2x+1$; neither

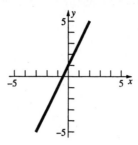

19. $g(x) = 3x^2+2x-1;\ g(-x) = 3x^2-2x-1$; neither

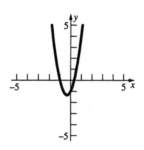

21. $g(x) = \dfrac{x}{x^2-1};\ g(-x) = \dfrac{-x}{x^2-1}$; odd

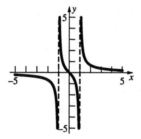

23. $f(w) = \sqrt{w-1};\ f(-w) = \sqrt{-w-1}$; neither

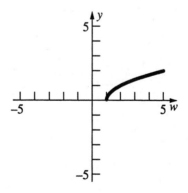

25. $f(x)=|2x|; f(-x)=|-2x|=|2x|;$ even function

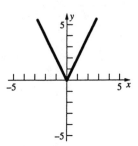

27. $g(x)=\left[\!\left[\dfrac{x}{2}\right]\!\right]; g(-x)=\left[\!\left[-\dfrac{x}{2}\right]\!\right];$ neither

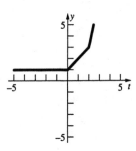

29. $g(t)=\begin{cases}1 & \text{if } t\le 0\\ t+1 & \text{if } 0<t<2\\ t^2-1 & \text{if } t\ge 2\end{cases}$ neither

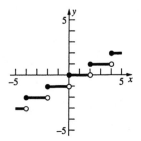

31. $T(x)=5000+805x$
Domain: $\{x\in \text{integers}: 0\le x\le 100\}$

$u(x)=\dfrac{T(x)}{x}=\dfrac{5000}{x}+805$

Domain: $\{x\in \text{integers}: 0<x\le 100\}$

33. $E(x)=x-x^2$

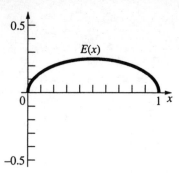

$\dfrac{1}{2}$ exceeds its square by the maximum amount.

35. a. $E(x)=24+0.40x$

b. $120=24+0.40x$
$0.40x=96; x=240$ mi

37. The area of the two semicircular ends is $\dfrac{\pi d^2}{4}$.

The length of each parallel side is $\dfrac{1-\pi d}{2}$.

$A(d)=\dfrac{\pi d^2}{4}+d\left(\dfrac{1-\pi d}{2}\right)=\dfrac{\pi d^2}{4}+\dfrac{d-\pi d^2}{2}$

$=\dfrac{2d-\pi d^2}{4}$

Since the track is one mile long, $\pi d<1$, so

$d<\dfrac{1}{\pi}$. Domain: $\left\{d\in\mathbb{R}:0<d<\dfrac{1}{\pi}\right\}$

39. a. $B(0)=0$

b. $B\left(\dfrac{1}{2}\right)=\dfrac{1}{2}B(1)=\dfrac{1}{2}\cdot\dfrac{1}{6}=\dfrac{1}{12}$

c.

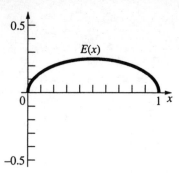

41. For any x, $x + 0 = x$, so
$f(x) = f(x + 0) = f(x) + f(0)$, hence $f(0) = 0$.
Let m be the value of $f(1)$. For p in N,
$p = p \cdot 1 = 1 + 1 + \ldots + 1$, so
$f(p) = f(1 + 1 + \ldots + 1) = f(1) + f(1) + \ldots + f(1)$
$= pf(1) = pm$.

$1 = p\left(\dfrac{1}{p}\right) = \dfrac{1}{p} + \dfrac{1}{p} + \ldots + \dfrac{1}{p}$, so

$m = f(1) = f\left(\dfrac{1}{p} + \dfrac{1}{p} + \ldots + \dfrac{1}{p}\right)$

$= f\left(\dfrac{1}{p}\right) + f\left(\dfrac{1}{p}\right) + \ldots + f\left(\dfrac{1}{p}\right) = pf\left(\dfrac{1}{p}\right)$, hence

$f\left(\dfrac{1}{p}\right) = \dfrac{m}{p}$. Any rational number can be written

as $\dfrac{p}{q}$ with p, q in N.

$\dfrac{p}{q} = p\left(\dfrac{1}{q}\right) = \dfrac{1}{q} + \dfrac{1}{q} + \ldots + \dfrac{1}{q}$,

so $f\left(\dfrac{p}{q}\right) = f\left(\dfrac{1}{q} + \dfrac{1}{q} + \ldots + \dfrac{1}{q}\right)$

$= f\left(\dfrac{1}{q}\right) + f\left(\dfrac{1}{q}\right) + \ldots + f\left(\dfrac{1}{q}\right)$

$= pf\left(\dfrac{1}{q}\right) = p\left(\dfrac{m}{q}\right) = m\left(\dfrac{p}{q}\right)$

43. **a.** $f(1.38) \approx 0.2994$
$f(4.12) \approx 3.6852$

b.

x	$f(x)$
-4	-4.05
-3	-3.1538
-2	-2.375
-1	-1.8
0	-1.25
1	-0.2

2.2 Concepts Review

1. $(x^2 + 1)^3$ **3.** 2; left

Problem Set 2.2

1. a. $(f + g)(2) = (2 + 3) + 2^2 = 9$

b. $(f \cdot g)(0) = (0 + 3)(0^2) = 0$

2	1.125
3	2.3846
4	3.55

45.

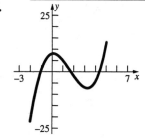

a. Range: $\{y \in R: -22 \le y \le 13\}$

b. $f(x) = 0$ when $x \approx -1.1, 1.7, 4.3$
$f(x) \ge 0$ on $[-1.1, 1.7] \cup [4.3, 5]$

47.

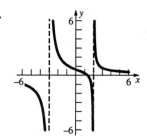

a. x-intercept: $3x - 4 = 0$; $x = \dfrac{4}{3}$

y-intercept: $\dfrac{3 \cdot 0 - 4}{0^2 + 0 - 6} = \dfrac{2}{3}$

b. $\mathbb{R}$

c. $x^2 + x - 6 = 0$; $(x + 3)(x - 2) = 0$
Vertical asymptotes at $x = -3$, $x = 2$

d. Horizontal asymptote at $y = 0$

c. $(g/f)(3) = \dfrac{3^2}{3 + 3} = \dfrac{9}{6} = \dfrac{3}{2}$

d. $(f \circ g)(1) = f(1^2) = 1 + 3 = 4$

e. $(g \circ f)(1) = g(1 + 3) = 4^2 = 16$

f. $(g \circ f)(-8) = g(-8 + 3) = (-5)^2 = 25$

3. a. $(\Phi + \Psi)(t) = t^3 + 1 + \dfrac{1}{t}$

b. $(\Phi \circ \Psi)(r) = \Phi\left(\dfrac{1}{r}\right) = \left(\dfrac{1}{r}\right)^3 + 1 = \dfrac{1}{r^3} + 1$

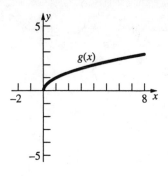

c. $(\Psi \circ \Phi)(r) = \Psi(r^3 + 1) = \dfrac{1}{r^3 + 1}$

d. $\Phi^3(z) = (z^3 + 1)^3$

e. $(\Phi - \Psi)(5t) = [(5t)^3 + 1] - \dfrac{1}{5t}$

$\qquad = 125t^3 + 1 - \dfrac{1}{5t}$

f. $((\Phi - \Psi) \circ \Psi)(t) = (\Phi - \Psi)\left(\dfrac{1}{t}\right)$

$\qquad = \left(\dfrac{1}{t}\right)^3 + 1 - \dfrac{1}{\frac{1}{t}} = \dfrac{1}{t^3} + 1 - t$

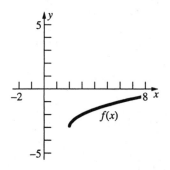

5. $(f \circ g)(x) = f\left(|1 + x|\right) = \sqrt{|1 + x|^2 - 4}$

$\qquad = \sqrt{x^2 + 2x - 3}$

$\qquad (g \circ f)(x) = g\left(\sqrt{x^2 - 4}\right) = \left|1 + \sqrt{x^2 - 4}\right|$

$\qquad = 1 + \sqrt{x^2 - 4}$

7. $g(3.141) \approx 1.188$

9. $\left[g^2(\pi) - g(\pi)\right]^{1/3} = \left[(11 - 7\pi)^2 - |11 - 7\pi|\right]^{1/3}$

$\qquad \approx 4.789$

11. a. $g(x) = \sqrt{x},\ f(x) = x + 7$

b. $g(x) = x^{15},\ f(x) = x^2 + x$

13. $p = f \circ g \circ h$ if $f(x) = \log x,\ g(x) = \sqrt{x},$
$h(x) = x^2 + 1$
$p = f \circ g \circ h$ if $f(x) = \log \sqrt{x},\ g(x) = x + 1,$
$h(x) = x^2$

15. Translate the graph of $g(x) = \sqrt{x}$ to the right 2 units and down 3 units.

17. Translate the graph of $y = x^2$ to the right 2 units and down 4 units.

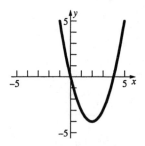

19. $(f + g)(x) = \dfrac{x - 3}{2} + \sqrt{x}$

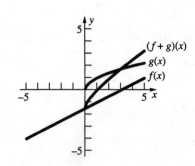

21. $F(t) = \dfrac{|t| - t}{t}$

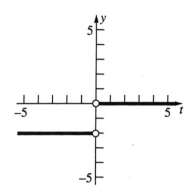

23. a. Even; $(f + g)(-x) = f(-x) + g(-x) = f(x) + g(x)$
$= (f + g)(x)$ if f and g are both even functions.

b. Odd; $(f + g)(-x) = f(-x) + g(-x) = -f(x) - g(x)$
$= -(f + g)(x)$ if f and g are both odd functions.

c. Even;
$(f \cdot g)(-x) = [f(-x)][g(-x)] = [f(x)][g(x)]$
$= (f \cdot g)(x)$ if f and g are both even functions.

d. Even; $(f \cdot g)(-x) = [f(-x)][g(-x)]$
$= [-f(x)][-g(x)] = [f(x)][g(x)] = (f \cdot g)(x)$

if f and g are both odd functions.

e. Odd; $(f \cdot g)(-x) = [f(-x)][g(-x)]$
$= [f(x)][-g(x)] = -[f(x)][g(x)] = -(f \cdot g)(x)$
if f is an even function and g is an odd function.

25. Not every polynomial of even degree is an even function. For example $f(x) = x^2 + x$ is neither even nor odd. Not every polynomial of odd degree is an odd function. For example $g(x) = x^3 + x^2$ is neither even nor odd.

27. a. $P = \sqrt{29 - 3(2 + \sqrt{t}) + (2 + \sqrt{t})^2}$
$= \sqrt{t + \sqrt{t} + 27}$

b. When $t = 15$, $P = \sqrt{15 + \sqrt{15} + 27} \approx 7$

29. $D(t) = \begin{cases} 400t & \text{if } 0 \le t \le 1 \\ \sqrt{(400t)^2 + [300(t-1)]^2} & \text{if } t > 1 \end{cases}$

$D(t) = \begin{cases} 400t & \text{if } 0 \le t \le 1 \\ \sqrt{250,000t^2 - 180,000t + 90,000} & \text{if } t > 1 \end{cases}$

31. $f(f(x)) = f\left(\dfrac{ax+b}{cx-a}\right) = \dfrac{a\left(\frac{ax+b}{cx-a}\right) + b}{c\left(\frac{ax+b}{cx-a}\right) - a}$

$= \dfrac{a^2 x + ab + bcx - ab}{acx + bc - acx + a^2} = \dfrac{x(a^2 + bc)}{a^2 + bc} = x$

If $a^2 + bc = 0$, $f(f(x))$ is undefined, while if $x = \frac{a}{c}$, $f(x)$ is undefined.

33. a. $f\left(\dfrac{1}{x}\right) = \dfrac{\frac{1}{x}}{\frac{1}{x} - 1} = \dfrac{1}{1 - x}$

b. $f(f(x)) = f\left(\dfrac{x}{x-1}\right) = \dfrac{\frac{x}{x-1}}{\frac{x}{x-1} - 1}$

$= \dfrac{x}{x - x + 1} = x$

c. $f\left(\dfrac{1}{f(x)}\right) = f\left(\dfrac{x-1}{x}\right) = \dfrac{\frac{x-1}{x}}{\frac{x-1}{x} - 1} = \dfrac{x-1}{x - 1 - x}$

$= 1 - x$

35. $f_1(f_1(x)) = x;$

$f_1(f_2(x)) = \dfrac{1}{x};$

$f_1(f_3(x)) = 1 - x;$

$f_1(f_4(x)) = \dfrac{1}{1-x};$

$f_1(f_5(x)) = \dfrac{x-1}{x};$

$f_1(f_6(x)) = \dfrac{x}{x-1};$

$f_2(f_1(x)) = \dfrac{1}{x};$

$f_2(f_2(x)) = \dfrac{1}{\frac{1}{x}} = x;$

$f_2(f_3(x)) = \dfrac{1}{1-x};$

$f_2(f_4(x)) = \dfrac{1}{\frac{1}{1-x}} = 1 - x;$

$f_2(f_5(x)) = \dfrac{1}{\frac{x-1}{x}} = \dfrac{x}{x-1};$

$$f_2(f_6(x)) = \frac{1}{\frac{x}{x-1}} = \frac{x-1}{x};$$

$$f_3(f_1(x)) = 1 - x;$$

$$f_3(f_2(x)) = 1 - \frac{1}{x} = \frac{x-1}{x};$$

$$f_3(f_3(x)) = 1 - (1-x) = x;$$

$$f_3(f_4(x)) = 1 - \frac{1}{1-x} = \frac{x}{x-1};$$

$$f_3(f_5(x)) = 1 - \frac{x-1}{x} = \frac{1}{x};$$

$$f_3(f_6(x)) = 1 - \frac{x}{x-1} = \frac{1}{1-x};$$

$$f_4(f_1(x)) = \frac{1}{1-x};$$

$$f_4(f_2(x)) = \frac{1}{1-\frac{1}{x}} = \frac{x}{x-1};$$

$$f_4(f_3(x)) = \frac{1}{1-(1-x)} = \frac{1}{x};$$

$$f_4(f_4(x)) = \frac{1}{1-\frac{1}{1-x}} = \frac{1-x}{1-x-1} = \frac{x-1}{x};$$

$$f_4(f_5(x)) = \frac{1}{1-\frac{x-1}{x}} = \frac{x}{x-(x-1)} = x;$$

$$f_4(f_6(x)) = \frac{1}{1-\frac{x}{x-1}} = \frac{x-1}{x-1-x} = 1-x;$$

$$f_5(f_1(x)) = \frac{x-1}{x};$$

$$f_5(f_2(x)) = \frac{\frac{1}{x}-1}{\frac{1}{x}} = 1-x;$$

$$f_5(f_3(x)) = \frac{1-x-1}{1-x} = \frac{x}{x-1};$$

$$f_5(f_4(x)) = \frac{\frac{1}{1-x}-1}{\frac{1}{1-x}} = \frac{1-(1-x)}{1} = x;$$

$$f_5(f_5(x)) = \frac{\frac{x-1}{x}-1}{\frac{x-1}{x}} = \frac{x-1-x}{x-1} = \frac{1}{1-x};$$

$$f_5(f_6(x)) = \frac{\frac{x}{x-1}-1}{\frac{x}{x-1}} = \frac{x-(x-1)}{x} = \frac{1}{x};$$

$$f_6(f_1(x)) = \frac{x}{x-1};$$

$$f_6(f_2(x)) = \frac{\frac{1}{x}}{\frac{1}{x}-1} = \frac{1}{1-x};$$

$$f_6(f_3(x)) = \frac{1-x}{1-x-1} = \frac{x-1}{x};$$

$$f_6(f_4(x)) = \frac{\frac{1}{1-x}}{\frac{1}{1-x}-1} = \frac{1}{1-(1-x)} = \frac{1}{x};$$

$$f_6(f_5(x)) = \frac{\frac{x-1}{x}}{\frac{x-1}{x}-1} = \frac{x-1}{x-1-x} = 1-x;$$

$$f_6(f_6(x)) = \frac{\frac{x}{x-1}}{\frac{x}{x-1}-1} = \frac{x}{x-(x-1)} = x$$

$\circ$	f_1	f_2	f_3	f_4	f_5	f_6
f_1	f_1	f_2	f_3	f_4	f_5	f_6
f_2	f_2	f_1	f_4	f_3	f_6	f_5
f_3	f_3	f_5	f_1	f_6	f_2	f_4
f_4	f_4	f_6	f_2	f_5	f_1	f_3
f_5	f_5	f_3	f_6	f_1	f_4	f_2
f_6	f_6	f_4	f_5	f_2	f_3	f_1

a. $f_3 \circ f_3 \circ f_3 \circ f_3 \circ f_3$
$= ((((f_3 \circ f_3) \circ f_3) \circ f_3) \circ f_3)$
$= (((f_1 \circ f_3) \circ f_3) \circ f_3)$
$= ((f_3 \circ f_3) \circ f_3)$
$= f_1 \circ f_3 = f_3$

b. $f_1 \circ f_2 \circ f_3 \circ f_4 \circ f_5 \circ f_6$
$= (((((f_1 \circ f_2) \circ f_3) \circ f_4) \circ f_5) \circ f_6)$
$= ((((f_2 \circ f_3) \circ f_4) \circ f_5) \circ f_6)$
$= (f_4 \circ f_4) \circ (f_5 \circ f_6)$
$= f_5 \circ f_2 = f_3$

c. If $F \circ f_6 = f_1$, then $F = f_6$.

d. If $G \circ f_3 \circ f_6 = f_1$, then $G \circ f_4 = f_1$ so $G = f_5$.

e. If $f_2 \circ f_5 \circ H = f_5$, then $f_6 \circ H = f_5$ so $H = f_3$.

37.

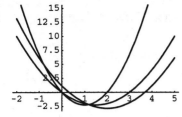

39.

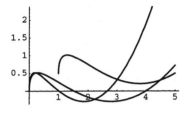

41. a.

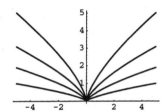

b.

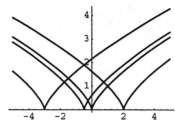

c.

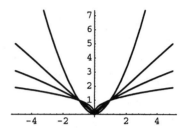

2.3 Concepts Review

1. $(-\infty, \infty)$; $[-1, 1]$ **3.** odd; even

Problem Set 2.3

1. a. $30\left(\dfrac{\pi}{180}\right) = \dfrac{\pi}{6}$

  b. $45\left(\dfrac{\pi}{180}\right) = \dfrac{\pi}{4}$

  c. $-60\left(\dfrac{\pi}{180}\right) = -\dfrac{\pi}{3}$

  d. $240\left(\dfrac{\pi}{180}\right) = \dfrac{4\pi}{3}$

  e. $-370\left(\dfrac{\pi}{180}\right) = -\dfrac{37\pi}{18}$

  f. $10\left(\dfrac{\pi}{180}\right) = \dfrac{\pi}{18}$

  g. $22\dfrac{1}{2}\left(\dfrac{\pi}{180}\right) = \dfrac{\pi}{8}$

  h. $600\left(\dfrac{\pi}{180}\right) = \dfrac{10\pi}{3}$

  i. $-120\left(\dfrac{\pi}{180}\right) = -\dfrac{2\pi}{3}$

3. a. $33.3\left(\dfrac{\pi}{180}\right) \approx 0.5812$

  b. $46\left(\dfrac{\pi}{180}\right) \approx 0.8029$

  c. $-66.6\left(\dfrac{\pi}{180}\right) \approx -1.1624$

  d. $240.11\left(\dfrac{\pi}{180}\right) \approx 4.1907$

  e. $-369\left(\dfrac{\pi}{180}\right) \approx -6.4403$

f. $11\left(\dfrac{\pi}{180}\right) \approx 0.1920$

g. $22.5\left(\dfrac{\pi}{180}\right) \approx 0.3927$

h. $359\left(\dfrac{\pi}{180}\right) \approx 6.2657$

i. $-121.35\left(\dfrac{\pi}{180}\right) \approx -2.1180$

5. a. $\dfrac{56.4\tan 34.2°}{\sin 34.1°} \approx 68.37$

b. $\dfrac{5.34\tan 21.3°}{\sin 3.1°+\cot 23.5°} \approx 0.8845$

c. $\tan(0.452) \approx 0.4855$

d. $\sin(-0.361) \approx -0.3532$

e. $\cos(-0.361) \approx 0.9355$

f. $\tan(-0.361) \approx -0.3775$

7. a. $\dfrac{56.3\tan 34.2°}{\sin 56.1°} \approx 46.097$

b. $\left(\dfrac{\sin 35°}{\sin 26°+\cos 26°}\right)^3 \approx 0.0789$

9. a. $\tan\left(\dfrac{\pi}{6}\right) = \dfrac{\sin\left(\dfrac{\pi}{6}\right)}{\cos\left(\dfrac{\pi}{6}\right)} = \dfrac{\sqrt{3}}{3}$

b. $\sec(\pi) = \dfrac{1}{\cos(\pi)} = -1$

c. $\sec\left(\dfrac{3\pi}{4}\right) = \dfrac{1}{\cos\left(\frac{3\pi}{4}\right)} = -\sqrt{2}$

d. $\csc\left(\dfrac{\pi}{2}\right) = \dfrac{1}{\sin\left(\frac{\pi}{2}\right)} = 1$

e. $\cot\left(\dfrac{\pi}{4}\right) = \dfrac{\cos\left(\frac{\pi}{4}\right)}{\sin\left(\frac{\pi}{4}\right)} = 1$

f. $\tan\left(-\dfrac{\pi}{4}\right) = \dfrac{\sin\left(-\frac{\pi}{4}\right)}{\cos\left(-\frac{\pi}{4}\right)} = -1$

11. a. $(1+\sin z)(1-\sin z) = 1-\sin^2 z$

$$= \cos^2 z = \dfrac{1}{\sec^2 z}$$

b. $(\sec t - 1)(\sec t + 1) = \sec^2 t - 1 = \tan^2 t$

c. $\sec t - \sin t \tan t = \dfrac{1}{\cos t} - \dfrac{\sin^2 t}{\cos t}$

$$= \dfrac{1-\sin^2 t}{\cos t} = \dfrac{\cos^2 t}{\cos t} = \cos t$$

d. $\dfrac{\sec^2 t - 1}{\sec^2 t} = \dfrac{\tan^2 t}{\sec^2 t} = \dfrac{\frac{\sin^2 t}{\cos^2 t}}{\frac{1}{\cos^2 t}} = \sin^2 t$

13. a. $\dfrac{\sin u}{\csc u} + \dfrac{\cos u}{\sec u} = \sin^2 u + \cos^2 u = 1$

b. $(1-\cos^2 x)(1+\cot^2 x) = (\sin^2 x)(\csc^2 x)$

$$= \sin^2 x\left(\dfrac{1}{\sin^2 x}\right) = 1$$

c. $\sin t(\csc t - \sin t) = \sin t\left(\dfrac{1}{\sin t} - \sin t\right)$

$$= 1 - \sin^2 t = \cos^2 t$$

d. $\dfrac{1-\csc^2 t}{\csc^2 t} = -\dfrac{\cot^2 t}{\csc^2 t} = -\dfrac{\frac{\cos^2 t}{\sin^2 t}}{\frac{1}{\sin^2 t}}$

$$= -\cos^2 t = -\dfrac{1}{\sec^2 t}$$

15. a. $y = \csc t$

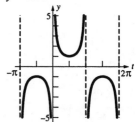

b. $y = 2\cos t$

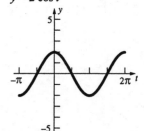

c. $y = \cos 3t$

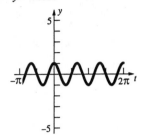

d. $y = \cos\left(t + \dfrac{\pi}{3}\right)$

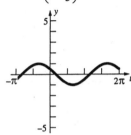

17. $y = 2\sin 2x$
   Period $= \pi$, amplitude $= 2$

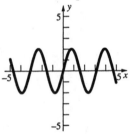

19. $y = 2 + \dfrac{1}{6}\cot(2x)$

Period $= \dfrac{\pi}{2}$, shift: 2 units up

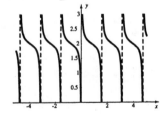

21. $y = 21 + 7\sin(2x + 3)$
   Period $= \pi$, amplitude $= 7$, shift: 21 units up,
   $\dfrac{3}{2}$ units left

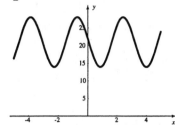

23. $y = \tan\left(2x - \dfrac{\pi}{3}\right)$

   Period $= \dfrac{\pi}{2}$, shift: $\dfrac{\pi}{6}$ units right

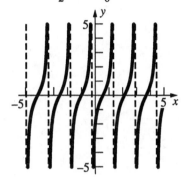

25. a. $-t\sin(-t) = t\sin t$; even

   b. $\sin^2(-t) = \sin^2 t$; even

   c. $\csc(-t) = \dfrac{1}{\sin(-t)} = -\csc t$; odd

   d. $|\sin(-t)| = |-\sin t| = |\sin t|$; even

   e. $\sin(\cos(-t)) = \sin(\cos t)$; even

   f. $-x + \sin(-x) = -x - \sin x = -(x + \sin x)$; odd

27. $\cos^2\dfrac{\pi}{3} = \dfrac{1 + \cos 2\left(\frac{\pi}{3}\right)}{2} = \dfrac{1 + \cos\frac{2\pi}{3}}{2} = \dfrac{1 - \frac{1}{2}}{2} = \dfrac{1}{4}$

29. $\sin^3\dfrac{\pi}{6} = \left(\sin^2\dfrac{\pi}{6}\right)\left(\sin\dfrac{\pi}{6}\right) = \dfrac{1}{4}\cdot\dfrac{1}{2} = \dfrac{1}{8}$

31. $\sin^2\dfrac{\pi}{8} = \dfrac{1 - \cos 2\left(\frac{\pi}{8}\right)}{2} = \dfrac{1 - \cos\frac{\pi}{4}}{2} = \dfrac{1 - \frac{\sqrt{2}}{2}}{2}$

$= \dfrac{2 - \sqrt{2}}{4}$

33. $\tan(t+\pi) = \dfrac{\tan t + \tan \pi}{1 - \tan t \tan \pi} = \dfrac{\tan t + 0}{1 - (\tan t)(0)}$

$= \tan t$

35. $s = rt = (2.5 \text{ ft})(2\pi \text{ rad}) = 5\pi$ ft, so the tire goes 5π feet per revolution, or $\dfrac{1}{5\pi}$ revolutions per foot.

$\left(\dfrac{1}{5\pi}\dfrac{\text{rev}}{\text{ft}}\right)\left(60\dfrac{\text{mi}}{\text{hr}}\right)\left(\dfrac{1}{60}\dfrac{\text{hr}}{\text{min}}\right)\left(5280\dfrac{\text{ft}}{\text{mi}}\right)$

≈ 336 rev/min

37. $r_1 t_1 = r_2 t_2; 6(2\pi)t_1 = 8(2\pi)(21)$
$t_1 = 28$ rev/sec

39. a. $\tan \alpha = \sqrt{3}$

$\alpha = \dfrac{\pi}{3}$

b. $\sqrt{3}x + 3y = 6$

$3y = -\sqrt{3}x + 6$

$y = -\dfrac{\sqrt{3}}{3}x + 2; m = -\dfrac{\sqrt{3}}{3}$

$\tan \alpha = -\dfrac{\sqrt{3}}{3}$

$\alpha = -\dfrac{\pi}{6} = \dfrac{5\pi}{6}$

41. a. $\tan \theta = \dfrac{3-2}{1+3(2)} = \dfrac{1}{7}$

$\theta \approx 0.1419$

b. $\tan \theta = \dfrac{-1-\frac{1}{2}}{1+\left(\frac{1}{2}\right)(-1)} = -3$

$\theta \approx 1.8925$

c. $2x - 6y = 12 \qquad 2x + y = 0$
$-6y = -2x + 12 \qquad y = -2x$

$y = \dfrac{1}{3}x - 2$

$m_1 = \dfrac{1}{3}, \ m_2 = -2$

$\tan \theta = \dfrac{-2-\frac{1}{3}}{1+\left(\frac{1}{3}\right)(-2)} = -7; \theta \approx 1.7127$

43. $A = \dfrac{1}{2}(2)(5)^2 = 25 \text{cm}^2$

45. The base of the triangle is the side opposite the angle t. Then the base has length $2r\sin\dfrac{t}{2}$ (similar to Problem 45). The radius of the semicircle is $r\sin\dfrac{t}{2}$ and the height of the triangle is $r\cos\dfrac{t}{2}$.

$A = \dfrac{1}{2}\left(2r\sin\dfrac{t}{2}\right)\left(r\cos\dfrac{t}{2}\right) + \dfrac{\pi}{2}\left(r\sin\dfrac{t}{2}\right)^2$

$= r^2 \sin\dfrac{t}{2}\cos\dfrac{t}{2} + \dfrac{\pi r^2}{2}\sin^2\dfrac{t}{2}$

47.
$S_1(n) = 1 + 2 + 3 + \ldots + n$
$+ \quad S_1(n) = n + (n-1) + (n-2) + \ldots + 1$
$\overline{2 \cdot S_1(n) = (n+1) + (n+1) + (n+1) + \ldots + (n+1)}$

Thus, $2 \cdot S_1(n) = n(n+1)$, so $S_1(n) = \dfrac{n(n+1)}{2}$

49. $(x+1)^4 - x^4 = 4x^3 + 6x^2 + 4x + 1$
Add up $(x+1)^4 - x^4$ for $x = 0, 1, 2, \ldots, n$ to obtain
$1^4 - 0^4 + 2^4 - 1^4 + 3^4 - 2^4 + \ldots$
$+ (n+1)^4 - n^4 = (n+1)^4$
Now add up $4x^3 + 6x^2 + 4x + 1$ for $x = 0, 1, 2, \ldots, n$ to obtain
$4(0^3 + 1^3 + 2^3 + \ldots + n^3)$
$+ 6(0^2 + 1^2 + 2^2 + \ldots + n^2)$
$+ 4(0 + 1 + 2 + \ldots + n) + 1 + 1 + 1 + \ldots + 1$
$= 4S_3(n) + 6S_2(n) + 4S_1(n) + n + 1$
Thus, since $(x+1)^4 - x^4 = 4x^3 + 6x^2 + 4x + 1$,
$(n+1)^4 = 4S_3(n) + 6S_2(n) + 4S_1(n) + n + 1$
Solve this equation for $S_3(n)$:

$S_3(n) = \dfrac{1}{4}\left[(n+1)^4 - 6S_2(n) - 4S_1(n) - (n+1)\right]$

$= \dfrac{1}{4}\left[(n+1)^4 - 6\cdot\dfrac{1}{6}n(n+1)(2n+1)\right.$

$\left. -4\cdot\dfrac{1}{2}n(n+1) - (n+1)\right]$

$= \dfrac{1}{4}(n+1)[(n+1)^3 - n(2n+1) - 2n - 1]$

$= \dfrac{1}{4}(n+1)(n^3 + 3n^2 + 3n + 1 - 2n^2 - n - 2n - 1)$

$= \dfrac{1}{4}(n+1)(n^3 + n^2) = \dfrac{1}{4}n^2(n+1)^2$

51. The temperature function is

$T(t) = 80 + 25\sin\left(\dfrac{2\pi}{12}\left(t - \dfrac{7}{2}\right)\right).$

The normal high temperature for November 15th is then $T(10.5) = 67.5 \, °\text{F}$.

53. As t increases, the point on the rim of the wheel will move around the circle of radius 2.

a. $x(2) \approx 1.902$
$y(2) \approx 0.618$
$x(6) \approx -1.176$
$y(6) \approx -1.618$
$x(10) = 0$
$y(10) = 2$
$x(0) = 0$
$y(0) = 2$

b. $x(t) = -2\sin\left(\dfrac{\pi}{5}t\right), y(t) = 2\cos\left(\dfrac{\pi}{5}t\right)$

c. The point is at $(2, 0)$ when $\dfrac{\pi}{5}t = \dfrac{\pi}{2}$; that is,

when $t = \dfrac{5}{2}$

55. a. $C\sin(\omega t + \phi) = (C\sin \omega t)\cos\phi + (C\cos\omega t)\sin\phi$. Thus $A = C\cos\omega t$ and $B = C\sin\omega t$.

b. $A^2 + B^2 = (C\cos\omega t)^2 + (C\sin\omega t)^2$
$= C^2(\cos^2\omega t) + C^2(\sin^2\omega t) = C^2$

c. $A_1\sin(\omega t + \phi_1) + A_2\sin(\omega t + \phi_2) + A_3(\sin\omega t + \phi_3)$
$= A_1(\sin\omega t\cos\phi_1 + \cos\omega t\sin\phi_1)$
$+ A_2(\sin\omega t\cos\phi_2 + \cos\omega t\sin\phi_2)$
$+ A_3(\sin\omega t\cos\phi_3 + \cos\omega t\sin\phi_3)$
$= (A_1\cos\phi_1 + A_2\cos\phi_2 + A_3\cos\phi_3)\sin\omega t$
$+ (A_1\sin\phi_1 + A_2\sin\phi_2 + A_3\sin\phi_3)\cos\omega t$

d. Written response. Answers will vary.

57. a.

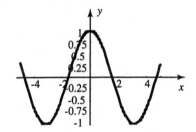

b.

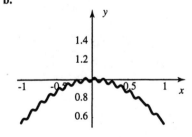

c.

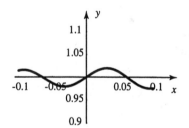

The plot in **(a)** shows the long term behavior of the function, but not the short term behavior, whereas the plot in **(c)** shows the short term behavior, but not the long term behavior. The plot in **(b)** shows a little of each.

2.4 Concepts Review

1. L; c

3. L; right

Problem Set 2.4

1. $\lim_{x\to 3}(x-5)=-2$

3. $\lim_{x\to -2}(x^2+2x-1)=(-2)^2+2(-2)-1=-1$

5. $\lim_{t\to -1}\dfrac{1-2t}{\sqrt{3t+21}}=\dfrac{1-2(-1)}{\sqrt{3(-1)+21}}=\dfrac{3}{\sqrt{18}}=\dfrac{1}{\sqrt{2}}=\dfrac{\sqrt{2}}{2}$

7. $\lim_{x\to 2}\dfrac{x^2-4}{x-2}=\lim_{x\to 2}\dfrac{(x-2)(x+2)}{x-2}=\lim_{x\to 2}(x+2)$
$=2+2=4$

9. $\lim_{x\to -1}\dfrac{x^3-4x^2+x+6}{x+1}$
$=\lim_{x\to -1}\dfrac{(x+1)(x^2-5x+6)}{x+1}$
$=\lim_{x\to -1}(x^2-5x+6)=(-1)^2-5(-1)+6=12$

11. $\lim_{x\to -t}\dfrac{x^2-t^2}{x+t}=\lim_{x\to -t}\dfrac{(x+t)(x-t)}{x+t}=\lim_{x\to -t}(x-t)$
$=-t-t=-2t$

13. $\lim_{t\to 2}\dfrac{\sqrt{(t+4)(t-2)^4}}{(3t-6)^2}=\lim_{t\to 2}\dfrac{(t-2)^2\sqrt{t+4}}{9(t-2)^2}$
$=\lim_{t\to 2}\dfrac{\sqrt{t+4}}{9}=\dfrac{\sqrt{2+4}}{9}=\dfrac{\sqrt{6}}{9}$

15. $\lim_{x\to 3}\dfrac{x^4-18x^2+81}{(x-3)^2}=\lim_{x\to 3}\dfrac{(x^2-9)^2}{(x-3)^2}$
$=\lim_{x\to 3}\dfrac{(x-3)^2(x+3)^2}{(x-3)^2}=\lim_{x\to 3}(x+3)^2=(3+3)^2$
$=36$

17.
$\lim_{h\to 0}\dfrac{(2+h)^2-4}{h}=\lim_{h\to 0}\dfrac{4+4h+h^2-4}{h}$
$=\lim_{h\to 0}\dfrac{h^2+4h}{h}=\lim_{h\to 0}(h+4)=4$

19.

x	$\frac{\sin x}{2x}$
1.	0.420735
0.1	0.499167
0.01	0.499992
0.001	0.5
−1.	0.420735
−0.1	0.499167
−0.01	0.499992
−0.001	0.5

$\lim_{x\to 0}\dfrac{\sin x}{2x}=0.5$

21.

x	$(x-\sin x)^2/x^2$
1.	0.0251314
0.1	2.775×10^{-6}
0.01	2.77775×10^{-10}
0.001	2.77778×10^{-14}
−1.	0.0251314
−0.1	2.775×10^{-6}
−0.01	2.77775×10^{-10}
−0.001	2.77778×10^{-14}

$\lim_{x\to 0}\dfrac{(x-\sin x)^2}{x^2}=0$

23.

t	$(t^2-1)/(\sin(t-1))$
2.	3.56519
1.1	2.1035
1.01	2.01003
1.001	2.001
0	1.1884
0.9	1.90317
0.99	1.99003
0.999	1.999

$\lim_{t\to 1}\dfrac{t^2-1}{\sin(t-1)}=2$

25.

x	$(1+\sin(x-3\pi/2))/(x-\pi)$
$1.+\pi$	0.4597
$0.1+\pi$	0.0500
$0.01+\pi$	0.0050
$0.001+\pi$	0.0005
$-1.+\pi$	-0.4597
$-0.1+\pi$	-0.0500
$-0.01+\pi$	-0.0050
$-0.001+\pi$	-0.0005

$$\lim_{x\to\pi}\frac{1+\sin\left(x-\frac{3\pi}{2}\right)}{x-\pi}=0$$

27.

x	$(x-\pi/4)^2/(\tan x-1)^2$
$1.+\frac{\pi}{4}$	0.0320244
$0.1+\frac{\pi}{4}$	0.201002
$0.01+\frac{\pi}{4}$	0.245009
$0.001+\frac{\pi}{4}$	0.2495
$-1.+\frac{\pi}{4}$	0.674117
$-0.1+\frac{\pi}{4}$	0.300668
$-0.01+\frac{\pi}{4}$	0.255008
$-0.001+\frac{\pi}{4}$	0.2505

$$\lim_{x\to\frac{\pi}{4}}\frac{\left(x-\frac{\pi}{4}\right)^2}{(\tan x-1)^2}=0.25$$

29. a. $\lim_{x\to-3} f(x)=2$

  b. $f(-3)=1$

  c. $f(-1)$ does not exist.

  d. $\lim_{x\to-1} f(x)=\dfrac{5}{2}$

  e. $f(1)=2$

  f. $\lim_{x\to1} f(x)$ does not exist.

  g. $\lim_{x\to1^-} f(x)=2$

  h. $\lim_{x\to1^+} f(x)=1$

31.

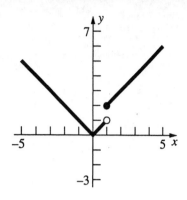

  a. $\lim_{x\to0} f(x)=0$

  b. $\lim_{x\to1} f(x)$ does not exist.

  c. $f(1)=2$

  d. $\lim_{x\to1^+} f(x)=2$

33. $f(x)=x-[\![x]\!]$

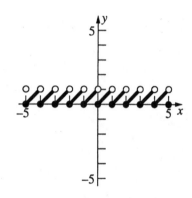

  a. $f(0)=0$

  b. $\lim_{x\to0} f(x)$ does not exist.

  c. $\lim_{x\to0^-} f(x)=1$

  d. $\lim_{x\to\frac{1}{2}} f(x)=\dfrac{1}{2}$

35. $\lim_{x\to1}\dfrac{x^2-1}{|x-1|}$ does not exist.

$$\lim_{x\to1^-}\frac{x^2-1}{|x-1|}=-2 \quad\text{and}\quad \lim_{x\to1^+}\frac{x^2-1}{|x-1|}=2$$

37. a. $\lim_{x\to1} f(x)$ does not exist.

b. $\displaystyle\lim_{x\to0} f(x) = 0$

39. $\displaystyle\lim_{x\to a} f(x)$ exists for $a = -1, 0, 1.$

41. a. $\displaystyle\lim_{x\to1} \frac{|x-1|}{x-1}$ does not exist.

$\displaystyle\lim_{x\to1^-} \frac{|x-1|}{x-1} = -1$ and $\displaystyle\lim_{x\to1^+} \frac{|x-1|}{x-1} = 1$

b. $\displaystyle\lim_{x\to1^-} \frac{|x-1|}{x-1} = -1$

c. $\displaystyle\lim_{x\to1^-} \frac{x^2 - |x-1| - 1}{|x-1|} = -3$

d. $\displaystyle\lim_{x\to1^-} \left[\frac{1}{x-1} - \frac{1}{|x-1|}\right]$ does not exist.

43. $\displaystyle\lim_{x\to0} \sqrt{x}$ does not exist since $\sqrt{x}$ is not defined for $x < 0.$

45. $\displaystyle\lim_{x\to0} \sqrt{|x|} = 0$

47. $\displaystyle\lim_{x\to0} \frac{\sin 2x}{4x} = \frac{1}{2}$

49. $\displaystyle\lim_{x\to0} \cos\left(\frac{1}{x}\right)$ does not exist.

51. $\displaystyle\lim_{x\to1} \frac{x^3 - 1}{\sqrt{2x+2} - 2} = 6$

53. $\displaystyle\lim_{x\to2^-} \frac{x^2 - x - 2}{|x-2|} = -3$

55. $\displaystyle\lim_{x\to0} \sqrt{x};$ The computer gives a value of 0, but $\displaystyle\lim_{x\to0^-} \sqrt{x}$ does not exist.

2.5 Concepts Review

1. $L - \varepsilon;\ L + \varepsilon$ **3.** $\dfrac{\varepsilon}{3}$

Problem Set 2.5

1. $0 < |t - a| < \delta \Rightarrow |f(t) - M| < \varepsilon$

3. $0 < |z - d| < \delta \Rightarrow |h(z) - P| < \varepsilon$

5. $0 < c - x < \delta \Rightarrow |f(x) - L| < \varepsilon$

7. $0 < |x - 0| < \delta \Rightarrow |(2x-1) - (-1)| < \varepsilon$

$|2x - 1 + 1| < \varepsilon \Leftrightarrow |2x| < \varepsilon$

$\Leftrightarrow 2|x| < \varepsilon$

$\Leftrightarrow |x| < \dfrac{\varepsilon}{2}$

$\delta = \dfrac{\varepsilon}{2}; 0 < |x - 0| < \delta$

$|(2x-1) - (-1)| = |2x| = 2|x| < 2\delta = \varepsilon$

9. $0 < |x - 5| < \delta \Rightarrow \left|\dfrac{x^2 - 25}{x - 5} - 10\right| < \varepsilon$

$\left|\dfrac{x^2 - 25}{x - 5} - 10\right| < \varepsilon \Leftrightarrow \left|\dfrac{(x-5)(x+5)}{x-5} - 10\right| < \varepsilon$

$\Leftrightarrow |x + 5 - 10| < \varepsilon$

$\Leftrightarrow |x - 5| < \varepsilon$

$\delta = \varepsilon; 0 < |x - 5| < \delta$

$\left|\dfrac{x^2 - 25}{x - 5} - 10\right| = \left|\dfrac{(x-5)(x+5)}{x-5} - 10\right| = |x + 5 - 10|$

$= |x - 5| < \delta = \varepsilon$

11. $0 < |x - 5| < \delta \Rightarrow \left|\dfrac{2x^2 - 11x + 5}{x - 5} - 9\right| < \varepsilon$

$\left|\dfrac{2x^2 - 11x + 5}{x - 5} - 9\right| < \varepsilon \Leftrightarrow \left|\dfrac{(2x-1)(x-5)}{x-5} - 9\right| < \varepsilon$

$\Leftrightarrow |2x - 1 - 9| < \varepsilon$

$\Leftrightarrow |2(x - 5)| < \varepsilon$

$\Leftrightarrow |x - 5| < \dfrac{\varepsilon}{2}$

$\delta = \dfrac{\varepsilon}{2}; 0 < |x - 5| < \delta$

$$\left|\frac{2x^2 - 11x + 5}{x - 5} - 9\right| = \left|\frac{(2x-1)(x-5)}{x-5} - 9\right|$$

$$= |2x - 1 - 9| = |2(x - 5)| = 2|x - 5| < 2\delta = \varepsilon$$

13. $0 < |x - 4| < \delta \Rightarrow \left|\dfrac{\sqrt{2x-1}}{\sqrt{x-3}} - \sqrt{7}\right| < \varepsilon$

$$\left|\frac{\sqrt{2x-1}}{\sqrt{x-3}} - \sqrt{7}\right| < \varepsilon \Leftrightarrow \left|\frac{\sqrt{2x-1} - \sqrt{7(x-3)}}{\sqrt{x-3}}\right| < \varepsilon$$

$$\Leftrightarrow \left|\frac{(\sqrt{2x-1} - \sqrt{7(x-3)})(\sqrt{2x-1} + \sqrt{7(x-3)})}{\sqrt{x-3}(\sqrt{2x-1} + \sqrt{7(x-3)})}\right| < \varepsilon$$

$$\Leftrightarrow \left|\frac{2x - 1 - (7x - 21)}{\sqrt{x-3}(\sqrt{2x-1} + \sqrt{7(x-3)})}\right| < \varepsilon$$

$$\Leftrightarrow \left|\frac{-5(x-4)}{\sqrt{x-3}(\sqrt{2x-1} + \sqrt{7(x-3)})}\right| < \varepsilon$$

$$\Leftrightarrow |x - 4|\left|\frac{5}{\sqrt{x-3}(\sqrt{2x-1} + \sqrt{7(x-3)})}\right| < \varepsilon$$

To bound $\dfrac{5}{\sqrt{x-3}(\sqrt{2x-1} + \sqrt{7(x-3)})}$, agree

that $\delta \le \dfrac{1}{2}$. If $\delta \le \dfrac{1}{2}$, then $\dfrac{7}{2} < x < \dfrac{9}{2}$, so

$$0.65 < \frac{5}{\sqrt{x-3}(\sqrt{2x-1} + \sqrt{7(x-3)})} < 1.65 \text{ and}$$

hence $|x - 4|\left|\dfrac{5}{\sqrt{x-3}(\sqrt{2x-1} + \sqrt{7(x-3)})}\right| < \varepsilon$

$$\Leftrightarrow |x - 4| < \frac{\varepsilon}{1.65}$$

For whatever ε is chosen, let δ be the smaller of
$\dfrac{1}{2}$ and $\dfrac{\varepsilon}{1.65}$.

$$\delta = \min\left\{\frac{1}{2}, \frac{\varepsilon}{1.65}\right\}, \quad 0 < |x - 4| < \delta$$

$$\left|\frac{\sqrt{2x-1}}{\sqrt{x-3}} - \sqrt{7}\right| = |x - 4|\left|\frac{5}{\sqrt{x-3}(\sqrt{2x-1} + \sqrt{7(x-3)})}\right|$$

$$< |x - 4|(1.65) < 1.65\delta \le \varepsilon$$

since $\delta = \dfrac{1}{2}$ only when $\dfrac{1}{2} \le \dfrac{\varepsilon}{1.65}$ so $1.65\delta \le \varepsilon$.

15. $0 < |x - 1| < \delta \Rightarrow \left|\dfrac{10x^3 - 26x^2 + 22x - 6}{(x-1)^2} - 4\right| < \varepsilon$

$$\left|\frac{10x^3 - 26x^2 + 22x - 6}{(x-1)^2} - 4\right| < \varepsilon$$

$$\Leftrightarrow \left|\frac{(10x - 6)(x-1)^2}{(x-1)^2} - 4\right| < \varepsilon$$

$$\Leftrightarrow |10x - 6 - 4| < \varepsilon$$

$$\Leftrightarrow |10(x - 1)| < \varepsilon$$

$$\Leftrightarrow 10|x - 1| < \varepsilon$$

$$\Leftrightarrow |x - 1| < \frac{\varepsilon}{10}$$

$$\delta = \frac{\varepsilon}{10}; \, 0 < |x - 1| < \delta$$

$$\left|\frac{10x^3 - 26x^2 + 22x - 6}{(x-1)^2} - 4\right|$$

$$= \left|\frac{(10x - 6)(x-1)^2}{(x-1)^2} - 4\right|$$

$$= |10x - 6 - 4| = |10(x - 1)|$$

$$= 10|x - 1| < 10\delta = \varepsilon$$

17. $0 < |x + 1| < \delta \Rightarrow |(x^2 - 2x - 1) - 2| < \varepsilon$

$$|x^2 - 2x - 1 - 2| = |x^2 - 2x - 3| = |x + 1||x - 3|$$

To bound $|x - 3|$, agree that $\delta \le 1$.

$|x + 1| < \delta$ implies

$|x - 3| = |x + 1 - 4|$

$\le |x + 1| + |-4|$

$< 1 + 4 = 5$

$$\delta \le \frac{\varepsilon}{5}; \, \delta = \min\left\{1, \frac{\varepsilon}{5}\right\}; \, 0 < |x + 1| < \delta$$

$$|(x^2 - 2x - 1) - 2| = |x^2 - 2x - 3|$$

$$= |x + 1||x - 3| < 5 \cdot \frac{\varepsilon}{5} = \varepsilon$$

19. Choose $\varepsilon > 0$. Then since $\lim\limits_{x \to c} f(x) = L$, there is
some $\delta_1 > 0$ such that
$0 < |x - c| < \delta_1 \Rightarrow |f(x) - L| < \varepsilon$.
Since $\lim\limits_{x \to c} f(x) = M$, there is some $\delta_2 > 0$ such
that $0 < |x - c| < \delta_2 \Rightarrow |f(x) - L| < \varepsilon$.
Let $\delta = \min\{\delta_1, \delta_2\}$ and choose x_0 such that
$0 < |x_0 - c| < \delta$.
Thus, $|f(x_0) - L| < \varepsilon \Rightarrow -\varepsilon < f(x_0) - L < \varepsilon$
$\Rightarrow -f(x_0) - \varepsilon < -L < -f(x_0) + \varepsilon$
$\Rightarrow f(x_0) - \varepsilon < L < f(x_0) + \varepsilon$.
Similarly,
$f(x_0) - \varepsilon < M < f(x_0) + \varepsilon$.
Thus,

$-2\varepsilon < L - M < 2\varepsilon$. As $\varepsilon \Rightarrow 0$, $L - M \rightarrow 0$, so $L = M$.

21. For all $x \neq 0$, $0 \leq \sin^2\left(\dfrac{1}{x}\right) \leq 1$ so

$x^4 \sin^2\left(\dfrac{1}{x}\right) \leq x^4$ for all $x \neq 0$. By Problem 18,

$\lim_{x\to 0} x^4 = 0$, so, by Problem 20,

$\lim_{x\to 0} x^4 \sin^2\left(\dfrac{1}{x}\right) = 0$.

23. $\lim_{x\to 0^+} |x| : 0 < x < \delta \Rightarrow \left||x| - 0\right| < \varepsilon$

For $x \geq 0$, $|x| = x$.

$\delta = \varepsilon; 0 < x < \delta \Rightarrow \left||x| - 0\right| = |x| = x < \delta = \varepsilon$

Thus, $\lim_{x\to 0^+} |x| = 0$.

$\lim_{x\to 0^-} |x| : 0 < 0 - x < \delta \Rightarrow \left||x| - 0\right| < \varepsilon$

For $x < 0$, $|x| = -x$; note also that $\left||x|\right| = |x|$ since $|x| \geq 0$.

$\delta = \varepsilon; 0 < -x < \delta \Rightarrow \left||x|\right| = |x| = -x < \delta = \varepsilon$

Thus, $\lim_{x\to 0^-} |x| = 0$,

since $\lim_{x\to 0^+} |x| = \lim_{x\to 0^-} |x| = 0$, $\lim_{x\to 0} |x| = 0$.

25. Choose $\varepsilon > 0$. Since $\lim_{x\to a} f(x) = L$, there is a $\delta > 0$ such that for $0 < |x - a| < \delta$, $|f(x) - L| < \varepsilon$.

That is, for

$a - \delta < x < a$ or $a < x < a + \delta$,

$L - \varepsilon < f(x) < L + \varepsilon$.

Let $f(a) = A$,

$M = \max\left\{|L - \varepsilon|, |L + \varepsilon|, |A|\right\}$, $c = a - \delta$,

$d = a + \delta$. Then for x in (c, d), $|f(x)| \leq M$, since either $x = a$, in which case

$|f(x)| = |f(a)| = |A| \leq M$ or $0 < |x - a| < \delta$ so

$L - \varepsilon < f(x) < L + \varepsilon$ and $|f(x)| < M$.

27. (b) and (c) are equivalent to the definition of limit.

29. a. $g(x) = \dfrac{x^3 - x^2 - 2x - 4}{x^4 - 4x^3 + x^2 + x + 6}$

b. No, because $\dfrac{x + 6}{x^4 - 4x^3 + x^2 + x + 6} + 1$ has an asymptote at $x \approx 3.49$.

c. If $\delta \leq \dfrac{1}{4}$, then $2.75 < x < 3$ or $3 < x < 3.25$ and by graphing

$y = |g(x)| = \left|\dfrac{x^3 - x^2 - 2x - 4}{x^4 - 4x^3 + x^2 + x + 6}\right|$

on the interval $[2.75, 3.25]$, we see that

$0 < \left|\dfrac{x^3 - x^2 - 2x - 4}{x^4 - 4x^3 + x^2 + x + 6}\right| < 3$

so m must be at least three.

2.6 Concepts Review

1. 48

3. $-8; -4 + 5c$

Problem Set 2.6

1. $\lim_{x\to 1}(2x + 1)$ 4

$= \lim_{x\to 1} 2x + \lim_{x\to 1} 1$ 3

$= 2 \lim_{x\to 1} x + \lim_{x\to 1} 1$ 2,1

$= 2(1) + 1 = 3$

3. $\lim_{x\to 0}[(2x + 1)(x - 3)]$ 6

$= \lim_{x\to 0}(2x + 1) \cdot \lim_{x\to 0}(x - 3)$ 4, 5

$= \left(\lim_{x\to 0} 2x + \lim_{x\to 0} 1\right) \cdot \left(\lim_{x\to 0} x - \lim_{x\to 0} 3\right)$ 3

$= \left(2 \lim_{x\to 0} x + \lim_{x\to 0} 1\right) \cdot \left(\lim_{x\to 0} x - \lim_{x\to 0} 3\right)$ 2, 1

$= [2(0) + 1](0 - 3) = -3$

5. $\lim_{x\to 2} \dfrac{2x + 1}{5 - 3x}$ 7

$= \dfrac{\lim_{x\to 2}(2x + 1)}{\lim_{x\to 2}(5 - 3x)}$ 4, 5

$= \dfrac{\lim_{x\to 2} 2x + \lim_{x\to 2} 1}{\lim_{x\to 2} 5 - \lim_{x\to 2} 3x}$ 3, 1

$= \dfrac{2 \lim_{x\to 2} x + 1}{5 - 3 \lim_{x\to 2} x}$ 2

$= \dfrac{2(2) + 1}{5 - 3(2)} = -5$

7. $\lim_{x \to 3} \sqrt{3x-5}$ 9

$= \sqrt{\lim_{x \to 3}(3x-5)}$ 5, 3

$= \sqrt{3 \lim_{x \to 3} x - \lim_{x \to 3} 5}$ 2, 1

$= \sqrt{3(3)-5} = 2$

9. $\lim_{t \to -2}(2t^3+15)^{13}$ 8

$= \left[\lim_{t \to -2}(2t^3+15)\right]^{13}$ 4, 3

$= \left[2 \lim_{t \to -2} t^3 + \lim_{t \to -2} 15\right]^{13}$ 8

$= \left[2\left(\lim_{t \to -2} t\right)^3 + \lim_{t \to -2} 15\right]^{13}$ 2, 1

$= [2(-2)^3+15]^{13} = -1$

11. $\lim_{y \to 2}\left(\dfrac{4y^3+8y}{y+4}\right)^{1/3}$ 9

$= \left(\lim_{y \to 2}\dfrac{4y^3+8y}{y+4}\right)^{1/3}$ 7

$= \left[\dfrac{\lim_{y \to 2}(4y^3+8y)}{\lim_{y \to 2}(y+4)}\right]^{1/3}$ 4, 3

$= \left(\dfrac{4\lim_{y \to 2} y^3 + 8\lim_{y \to 2} y}{\lim_{y \to 2} y + \lim_{y \to 2} 4}\right)^{1/3}$ 8, 1

$= \left[\dfrac{4\left(\lim_{y \to 2} y\right)^3 + 8\lim_{y \to 2} y}{\lim_{y \to 2} y + 4}\right]^{1/3}$ 2

$= \left[\dfrac{4(2)^3+8(2)}{2+4}\right]^{1/3} = 2$

13. $\lim_{x \to -1}\dfrac{(x-1)(x-2)(x-3)}{(x-1)(x-2)(x+7)} = \lim_{x \to -1}\dfrac{x-3}{x+7}$

$= \dfrac{-1-3}{-1+7} = -\dfrac{2}{3}$

15. $\lim_{x \to 1}\dfrac{x^2+x-2}{x^2-1} = \lim_{x \to 1}\dfrac{(x+2)(x-1)}{(x+1)(x-1)}$

$= \lim_{x \to 1}\dfrac{x+2}{x+1} = \dfrac{1+2}{1+1} = \dfrac{3}{2}$

17. $\lim_{u \to -2}\dfrac{u^2-ux+2u-2x}{u^2-u-6} = \lim_{u \to -2}\dfrac{(u+2)(u-x)}{(u+2)(u-3)}$

$= \lim_{u \to -2}\dfrac{u-x}{u-3} = \dfrac{x+2}{5}$

19. $\lim_{x \to \pi}\dfrac{2x^2-6x\pi+4\pi^2}{x^2-\pi^2} = \lim_{x \to \pi}\dfrac{2(x-\pi)(x-2\pi)}{(x-\pi)(x+\pi)}$

$= \lim_{x \to \pi}\dfrac{2(x-2\pi)}{x+\pi} = \dfrac{2(\pi-2\pi)}{\pi+\pi} = -1$

21. $\lim_{x \to a}\sqrt{f^2(x)+g^2(x)}$

$= \sqrt{\lim_{x \to a} f^2(x) + \lim_{x \to a} g^2(x)}$

$= \sqrt{\left(\lim_{x \to a} f(x)\right)^2 + \left(\lim_{x \to a} g(x)\right)^2}$

$= \sqrt{(3)^2+(-1)^2} = \sqrt{10}$

23. $\lim_{x \to a}\sqrt[3]{g(x)}[f(x)+3] = \lim_{x \to a}\sqrt[3]{g(x)} \cdot \lim_{x \to a}[f(x)+3]$

$= \sqrt[3]{\lim_{x \to a} g(x)} \cdot \left[\lim_{x \to a} f(x) + \lim_{x \to a} 3\right] = \sqrt[3]{-1} \cdot (3+3)$

$= -6$

25. $\lim_{t \to a}\left[|f(t)|+|3g(t)|\right] = \lim_{t \to a}|f(t)| + 3\lim_{t \to a}|g(t)|$

$= \left|\lim_{t \to a} f(t)\right| + 3\left|\lim_{t \to a} g(t)\right|$

$= |3| + 3|-1| = 6$

27. $\lim_{x \to 2}\dfrac{3x^2-12}{x-2} = \lim_{x \to 2}\dfrac{3(x-2)(x+2)}{x-2}$

$= 3\lim_{x \to 2}(x+2) = 3(2+2) = 12$

29. $\lim_{x \to 2}\dfrac{\frac{1}{x}-\frac{1}{2}}{x-2} = \lim_{x \to 2}\dfrac{\frac{2-x}{2x}}{x-2} = \lim_{x \to 2}\dfrac{-\frac{x-2}{2x}}{x-2}$

$= \lim_{x \to 2}-\dfrac{1}{2x} = \dfrac{-1}{2\lim_{x \to 2} x} = \dfrac{-1}{2(2)} = -\dfrac{1}{4}$

31. Suppose $\lim_{x \to c} f(x) = L$ and $\lim_{x \to c} g(x) = M$.

$$|f(x)g(x) - LM| \le |g(x)||f(x) - L| + |L||g(x) - M|$$

as shown in the text. Choose $\varepsilon_1 = 1$. Since $\lim_{x \to c} g(x) = M$, there is some $\delta_1 > 0$ such that if

$0 < |x - c| < \delta_1$, $|g(x) - M| < \varepsilon_1 = 1$ or $M - 1 <$

$g(x) < M + 1$, $|M - 1| \le |M| + 1$ and

$|M + 1| \le |M| + 1$ so for

$0 < |x - c| < \delta_1, |g(x)| < |M| + 1$. Choose $\varepsilon > 0$.

Since $\lim_{x \to c} f(x) = L$ and $\lim_{x \to c} g(x) = M$, there

exist δ_2 and δ_3 such that $0 < |x - c| < \delta_2 \Rightarrow$

$$|f(x) - L| < \frac{\varepsilon}{|L| + |M| + 1} \text{ and } 0 < |x - c| < \delta_3 \Rightarrow$$

$$|g(x) - M| < \frac{\varepsilon}{|L| + |M| + 1}. \text{ Let}$$

$\delta = \min\{\delta_1, \delta_2, \delta_3\}$, then $0 < |x - c| < \delta \Rightarrow$

$$|f(x)g(x) - LM| \le |g(x)||f(x) - L| + |L||g(x) - M|$$

$$< (|M| + 1)\frac{\varepsilon}{|L| + |M| + 1} + |L|\frac{\varepsilon}{|L| + |M| + 1} = \varepsilon$$

Hence,

$$\lim_{x \to c} f(x)g(x) = LM = \left(\lim_{x \to c} f(x)\right)\left(\lim_{x \to c} g(x)\right)$$

33. $\lim_{x \to c} f(x) = L \Leftrightarrow \lim_{x \to c} f(x) = \lim_{x \to c} L$

$\Leftrightarrow \lim_{x \to c} f(x) - \lim_{x \to c} L = 0$

$\Leftrightarrow \lim_{x \to c} [f(x) - L] = 0$

35. $\lim_{x \to c} |x| = \sqrt{\left(\lim_{x \to c} |x|\right)^2} = \sqrt{\lim_{x \to c} |x|^2} = \sqrt{\lim_{x \to c} x^2}$

$= \sqrt{\left(\lim_{x \to c} x\right)^2} = \sqrt{c^2} = |c|$

37. $\lim_{x \to -3^+} \frac{\sqrt{3 + x}}{x} = \frac{\sqrt{3 - 3}}{-3} = 0$

39. $\lim_{x \to 3^+} \frac{x - 3}{\sqrt{x^2 - 9}} = \lim_{x \to 3^+} \frac{(x - 3)\sqrt{x^2 - 9}}{x^2 - 9}$

$= \lim_{x \to 3^+} \frac{(x - 3)\sqrt{x^2 - 9}}{(x - 3)(x + 3)} = \lim_{x \to 3^+} \frac{\sqrt{x^2 - 9}}{x + 3}$

$= \frac{\sqrt{3^2 - 9}}{3 + 3} = 0$

41. $\lim_{x \to 2^+} \frac{(x^2 + 1)[\![x]\!]}{(3x - 1)^2} = \frac{(2^2 + 1)[\![2]\!]}{(3 \cdot 2 - 1)^2} = \frac{5 \cdot 2}{5^2} = \frac{2}{5}$

43. $\lim_{x \to 0^-} \frac{x}{|x|} = -1$

45. $f(x)g(x) = 1; g(x) = \frac{1}{f(x)}$

$\lim_{x \to a} g(x) = 0 \Leftrightarrow \lim_{x \to a} \frac{1}{f(x)} = 0$

$\Leftrightarrow \frac{1}{\lim_{x \to a} f(x)} = 0$

No value satisfies this equation, so $\lim_{x \to a} f(x)$ must not exist.

47. a. $NO = \sqrt{(0 - 0)^2 + (1 - 0)^2} = 1$

$OP = \sqrt{(x - 0)^2 + (y - 0)^2} = \sqrt{x^2 + y^2}$

$= \sqrt{x^2 + x}$

$NP = \sqrt{(x - 0)^2 + (y - 1)^2} = \sqrt{x^2 + y^2 - 2y + 1}$

$= \sqrt{x^2 + x - 2\sqrt{x} + 1}$

$MO = \sqrt{(1 - 0)^2 + (0 - 0)^2} = 1$

$MP = \sqrt{(x - 1)^2 + (y - 0)^2} = \sqrt{y^2 + x^2 - 2x + 1}$

$= \sqrt{x^2 - x + 1}$

$\lim_{x \to 0^+} \frac{\text{perimeter of } \triangle NOP}{\text{perimeter of } \triangle MOP}$

$= \lim_{x \to 0^+} \frac{1 + \sqrt{x^2 + x} + \sqrt{x^2 + x - 2\sqrt{x} + 1}}{1 + \sqrt{x^2 + x} + \sqrt{x^2 - x + 1}}$

$= \frac{1 + \sqrt{1}}{1 + \sqrt{1}} = 1$

b. Area of $\triangle NOP = \frac{1}{2}(1)(x) = \frac{x}{2}$

Area of $\triangle MOP = \frac{1}{2}(1)(y) = \frac{\sqrt{x}}{2}$

$\lim_{x \to 0^+} \frac{\text{area of } \triangle NOP}{\text{area of } \triangle MOP} = \lim_{x \to 0^+} \frac{\frac{x}{2}}{\frac{\sqrt{x}}{2}} = \lim_{x \to 0^+} \frac{x}{\sqrt{x}}$

$= \lim_{x \to 0^+} \sqrt{x} = 0$

2.7 Concepts Review

1. 0

3. the denominator is 0 when $t = 0$.

Problem Set 2.7

1. $\lim\limits_{x \to 0} \dfrac{\cos x}{x+1} = \dfrac{1}{1} = 1$

3.
$$\lim_{t \to 0} \frac{\cos^2 t}{1+\sin t} = \lim_{t \to 0} \frac{1-\sin^2 t}{1+\sin t}$$
$$= \lim_{t \to 0} \frac{(1-\sin t)(1+\sin t)}{1+\sin t}$$
$$= \lim_{t \to 0}(1-\sin t) = 1$$

5. $\lim\limits_{x \to 0} \dfrac{\sin x}{2x} = \dfrac{1}{2} \lim\limits_{x \to 0} \dfrac{\sin x}{x} = \dfrac{1}{2} \cdot 1 = \dfrac{1}{2}$

7.
$$\lim_{\theta \to 0} \frac{\sin 3\theta}{\tan \theta} = \lim_{\theta \to 0} \frac{\sin 3\theta}{\frac{\sin \theta}{\cos \theta}} = \lim_{\theta \to 0} \frac{\cos \theta \sin 3\theta}{\sin \theta}$$
$$= \lim_{\theta \to 0}\left[\cos \theta \cdot 3 \cdot \frac{\sin 3\theta}{3\theta} \cdot \frac{1}{\frac{\sin \theta}{\theta}}\right]$$
$$= 3 \lim_{\theta \to 0}\left[\cos \theta \cdot \frac{\sin 3\theta}{3\theta} \cdot \frac{1}{\frac{\sin \theta}{\theta}}\right]$$
$$= 3 \cdot 1 \cdot 1 \cdot 1 = 3$$

9.
$$\lim_{\theta \to 0} \frac{\cot \pi\theta \sin \theta}{2 \sec \theta} = \lim_{\theta \to 0} \frac{\frac{\cos \pi\theta}{\sin \pi\theta}\sin \theta}{\frac{2}{\cos \theta}}$$
$$= \lim_{\theta \to 0} \frac{\cos \pi\theta \sin \theta \cos \theta}{2 \sin \pi\theta}$$
$$= \lim_{\theta \to 0}\left[\frac{\cos \pi\theta \cos \theta}{2} \cdot \frac{\sin \theta}{\theta} \cdot \frac{1}{\pi} \cdot \frac{\pi\theta}{\sin \pi\theta}\right]$$
$$= \frac{1}{2\pi} \lim_{\theta \to 0}\left[\cos \pi\theta \cos \theta \cdot \frac{\sin \theta}{\theta} \cdot \frac{\pi\theta}{\sin \pi\theta}\right]$$
$$= \frac{1}{2\pi} \cdot 1 \cdot 1 \cdot 1 = \frac{1}{2\pi}$$

11.
$$\lim_{t \to 0} \frac{\tan^2 3t}{2t} = \lim_{t \to 0} \frac{\sin^2 3t}{(2t)(\cos^2 3t)}$$
$$= \lim_{t \to 0} \frac{3(\sin 3t)}{2\cos^2 3t} \cdot \frac{\sin 3t}{3t} = 0 \cdot 1 = 0$$

13.
$$\lim_{t \to 0} \frac{\sin(3t)+4t}{t \sec t} = \lim_{t \to 0}\left(\frac{\sin 3t}{t \sec t} + \frac{4t}{t \sec t}\right)$$
$$= \lim_{t \to 0} \frac{\sin 3t}{t \sec t} + \lim_{t \to 0} \frac{4t}{t \sec t}$$
$$= \lim_{t \to 0} 3\cos t \cdot \frac{\sin 3t}{3t} + \lim_{t \to 0} 4\cos t$$
$$= 3 \cdot 1 + 4 = 7$$

15. $\lim\limits_{t \to c} \tan t = \lim\limits_{t \to c} \dfrac{\sin t}{\cos t} = \dfrac{\lim\limits_{t \to c}\sin t}{\lim\limits_{t \to c}\cos t} = \dfrac{\sin c}{\cos c} = \tan c$

$\lim\limits_{t \to c} \cot t = \lim\limits_{t \to c} \dfrac{\cos t}{\sin t} = \dfrac{\lim\limits_{t \to c}\cos t}{\lim\limits_{t \to c}\sin t} = \dfrac{\cos c}{\sin c} = \cot c$

17. $\overline{BP} = \sin t, \overline{OB} = \cos t$

area$(\triangle OBP) \le$ area (sector OAP)

$\le$ area $(\triangle OBP) +$ area$(ABPQ)$

$$\frac{1}{2}\overline{OB} \cdot \overline{BP} \le \frac{1}{2}t(1)^2 \le \frac{1}{2}\overline{OB} \cdot \overline{BP} + (1-\overline{OB})\overline{BP}$$

$$\frac{1}{2}\sin t \cos t \le \frac{1}{2}t \le \frac{1}{2}\sin t \cos t + (1-\cos t)\sin t$$

$$\cos t \le \frac{t}{\sin t} \le 2 - \cos t$$

$$\frac{1}{2-\cos t} \le \frac{\sin t}{t} \le \frac{1}{\cos t}$$

$$\lim_{t \to 0} \frac{1}{2-\cos t} \le \lim_{t \to 0} \frac{\sin t}{t} \le \lim_{t \to 0} \frac{1}{\cos t}$$

$$1 \le \lim_{t \to 0} \frac{\sin t}{t} \le 1$$

Thus, $\lim\limits_{t \to 0} \dfrac{\sin t}{t} = 1$.

2.8 Concepts Review

1. x increases without bound; $f(x)$ gets close to L as x increases without bound

3. $y = 6$; horizontal

Problem Set 2.8

1. $\lim\limits_{x \to \infty} \dfrac{x}{x-5} = \lim\limits_{x \to \infty} \dfrac{1}{1-\frac{5}{x}} = 1$

3. $\lim\limits_{t \to -\infty} \dfrac{t^2}{7-t^2} = \lim\limits_{t \to -\infty} \dfrac{1}{\frac{7}{t^2}-1} = -1$

5. $\displaystyle\lim_{x\to\infty}\frac{x^2}{(x-5)(3-x)}=\lim_{x\to\infty}\frac{x^2}{-x^2+8x-15}$

$\quad=\displaystyle\lim_{x\to\infty}\frac{1}{-1+\dfrac{8}{x}-\dfrac{15}{x^2}}=-1$

7. $\displaystyle\lim_{x\to\infty}\frac{x^3}{2x^3-100x^2}=\lim_{x\to\infty}\frac{1}{2-\dfrac{100}{x}}=\frac{1}{2}$

9. $\displaystyle\lim_{x\to\infty}\frac{3x^3-x^2}{\pi x^3-5x^2}=\lim_{x\to\infty}\frac{3-\dfrac{1}{x}}{\pi-\dfrac{5}{x}}=\frac{3}{\pi}$

11. $\displaystyle\lim_{x\to\infty}\frac{3\sqrt{x^3}+3x}{\sqrt{2x^3}}=\lim_{x\to\infty}\frac{3x^{3/2}+3x}{\sqrt{2}x^{3/2}}$

$\quad=\displaystyle\lim_{x\to\infty}\frac{3+\dfrac{3}{\sqrt{x}}}{\sqrt{2}}=\frac{3}{\sqrt{2}}$

13. $\displaystyle\lim_{x\to\infty}\sqrt[3]{\frac{1+8x^2}{x^2+4}}=\sqrt[3]{\lim_{x\to\infty}\frac{1+8x^2}{x^2+4}}$

$\quad=\displaystyle\sqrt[3]{\lim_{x\to\infty}\frac{\dfrac{1}{x^2}+8}{1+\dfrac{4}{x^2}}}=\sqrt[3]{8}=2$

15. For $x>0$, $x=\sqrt{x^2}$.

$\displaystyle\lim_{x\to\infty}\frac{2x+1}{\sqrt{x^2+3}}=\lim_{x\to\infty}\frac{2+\dfrac{1}{x}}{\dfrac{\sqrt{x^2+3}}{\sqrt{x^2}}}=\lim_{x\to\infty}\frac{2+\dfrac{1}{x}}{\sqrt{1+\dfrac{3}{x^2}}}$

$\quad=\dfrac{2}{\sqrt{1}}=2$

17. $\displaystyle\lim_{x\to\infty}\left(\sqrt{2x^2+3}-\sqrt{2x^2-5}\right)$

$\quad=\displaystyle\lim_{x\to\infty}\frac{\left(\sqrt{2x^2+3}-\sqrt{2x^2-5}\right)\left(\sqrt{2x^2+3}+\sqrt{2x^2-5}\right)}{\sqrt{2x^2+3}+\sqrt{2x^2-5}}$

$\quad=\displaystyle\lim_{x\to\infty}\frac{2x^2+3-(2x^2-5)}{\sqrt{2x^2+3}+\sqrt{2x^2-5}}$

$\quad=\displaystyle\lim_{x\to\infty}\frac{8}{\sqrt{2x^2+3}+\sqrt{2x^2-5}}$

$\quad=\displaystyle\lim_{x\to\infty}\frac{\dfrac{8}{x}}{\dfrac{\sqrt{2x^2+3}+\sqrt{2x^2-5}}{\sqrt{x^2}}}=\lim_{x\to\infty}\frac{\dfrac{8}{x}}{\sqrt{2+\dfrac{3}{x^2}}+\sqrt{2-\dfrac{5}{x^2}}}=0$

19. $\displaystyle\lim_{y\to-\infty}\frac{9y^3+1}{y^2-2y+2}=\lim_{y\to-\infty}\frac{9y+\dfrac{1}{y^2}}{1-\dfrac{2}{y}+\dfrac{2}{y^2}}=-\infty$

21. As $x\to4^+$, $x\to4$ while $x-4\to0^+$.

$\displaystyle\lim_{x\to4^+}\frac{x}{x-4}=\infty$

23. As $t\to3^-$, $t^2\to9$ while $9-t^2\to0^+$.

$\displaystyle\lim_{t\to3^-}\frac{t^2}{9-t^2}=\infty$

25. As $x\to5^-$, $x^2\to25$, $x-5\to0^-$, and $3-x\to-2$.

$\displaystyle\lim_{x\to5^-}\frac{x^2}{(x-5)(3-x)}=\infty$

27. As $x\to3^-$, $x^3\to27$, while $x-3\to0^-$.

$\displaystyle\lim_{x\to3^-}\frac{x^3}{x-3}=-\infty$

29. $\displaystyle\lim_{x\to3^-}\frac{x^2-x-6}{x-3}=\lim_{x\to3^-}\frac{(x+2)(x-3)}{x-3}$

$\quad=\displaystyle\lim_{x\to3^-}(x+2)=5$

31. For $0\le x<1$, $[\![x]\!]=0$, so for $0<x<1$, $\dfrac{[\![x]\!]}{x}=0$

thus $\displaystyle\lim_{x\to0^+}\frac{[\![x]\!]}{x}=0$

33. For $x<0$, $|x|=-x$, thus

$\displaystyle\lim_{x\to0^-}\frac{|x|}{x}=\lim_{x\to0^-}\frac{-x}{x}=-1$

35. As $x\to0^-$, $1+\cos x\to2$ while $\sin x\to0^-$.

$\displaystyle\lim_{x\to0^-}\frac{1+\cos x}{\sin x}=-\infty$

37. $\displaystyle\lim_{x\to\infty}\frac{3}{x+1}=0$, $\lim_{x\to-\infty}\frac{3}{x+1}=0$;

Horizontal asymptote $y=0$.

$\displaystyle\lim_{x\to-1^+}\frac{3}{x+1}=\infty$, $\lim_{x\to-1^-}\frac{3}{x+1}=-\infty$;

Vertical asymptote $x=-1$

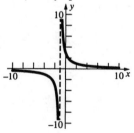

39. $\lim\limits_{x\to\infty} \dfrac{2x}{x-3} = \lim\limits_{x\to\infty} \dfrac{2}{1-\frac{3}{x}} = 2,$

$\lim\limits_{x\to-\infty} \dfrac{2x}{x-3} = \lim\limits_{x\to-\infty} \dfrac{2}{1-\frac{3}{x}} = 2,$

Horizontal asymptote $y = 2$

$\lim\limits_{x\to3^+} \dfrac{2x}{x-3} = \infty, \ \lim\limits_{x\to3^-} \dfrac{2x}{x-3} = -\infty;$

Vertical asymptote $x = 3$

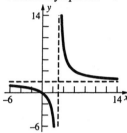

41. $\lim\limits_{x\to\infty} \dfrac{14}{2x^2+7} = 0, \ \lim\limits_{x\to-\infty} \dfrac{14}{2x^2+7} = 0;$

Horizontal asymptote $y = 0$

Since $2x^2 + 7 > 0$ for all x, $g(x)$ has no vertical asymptotes.

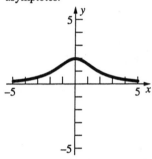

43. $f(x) = 2x + 3 - \dfrac{1}{x^3-1}$, thus

$\lim\limits_{x\to\infty}[f(x)-(2x+3)] = \lim\limits_{x\to\infty}\left[-\dfrac{1}{x^3-1}\right] = 0$

The oblique asymptote is $y = 2x + 3$.

45. a. We say that $\lim\limits_{x\to c^+} f(x) = -\infty$ if for each negative number M there corresponds a $\delta > 0$ such that $0 < x - c < \delta \Rightarrow f(x) < M$.

b. We say that $\lim\limits_{x\to c^-} f(x) = \infty$ if for each positive number M there corresponds a $\delta > 0$ such that $0 < c - x < \delta \Rightarrow f(x) > M$.

47. Let $\varepsilon > 0$ be given. Since $\lim\limits_{x\to\infty} f(x) = A$, there is a corresponding number M_1 such that

$x > M_1 \Rightarrow |f(x) - A| < \dfrac{\varepsilon}{2}$. Similarly, there is a

number M_2 such that $x > M_2 \Rightarrow |g(x) - B| < \dfrac{\varepsilon}{2}$.

Let $M = \max\{M_1, M_2\}$, then

$x > M \Rightarrow |f(x) + g(x) - (A+B)|$

$= |f(x) - A + g(x) - B| \le |f(x) - A| + |g(x) - B|$

$< \dfrac{\varepsilon}{2} + \dfrac{\varepsilon}{2} = \varepsilon$

Thus, $\lim\limits_{x\to\infty}[f(x) + g(x)] = A + B$

49. a. $\lim\limits_{x\to\infty} \sin x$ does not exist as $\sin x$ oscillates between -1 and 1 as x increases.

b. Let $u = \dfrac{1}{x}$, then as $x \to \infty, u \to 0^+$.

$\lim\limits_{x\to\infty} \sin\dfrac{1}{x} = \lim\limits_{u\to0^+} \sin u = 0$

c. Let $u = \dfrac{1}{x}$, then as $x \to \infty, u \to 0^+$.

$\lim\limits_{x\to\infty} x\sin\dfrac{1}{x} = \lim\limits_{u\to0^+} \dfrac{1}{u}\sin u = \lim\limits_{u\to0^+} \dfrac{\sin u}{u} = 1$

d. Let $u = \dfrac{1}{x}$, then

$\lim\limits_{x\to\infty} x^{3/2}\sin\dfrac{1}{x} = \lim\limits_{u\to0^+} \left(\dfrac{1}{u}\right)^{3/2}\sin u$

$= \lim\limits_{u\to0^+}\left[\left(\dfrac{1}{\sqrt{u}}\right)\left(\dfrac{\sin u}{u}\right)\right] = \infty$

e. As $x \to \infty$, $\sin x$ oscillates between -1 and 1, while $x^{-1/2} = \dfrac{1}{\sqrt{x}} \to 0$.

$\lim\limits_{x\to\infty} x^{-1/2}\sin x = 0$

f. Let $u = \dfrac{1}{x}$, then

$\lim\limits_{x\to\infty} \sin\left(\dfrac{\pi}{6} + \dfrac{1}{x}\right) = \lim\limits_{u\to0^+} \sin\left(\dfrac{\pi}{6} + u\right)$

$= \sin\dfrac{\pi}{6} = \dfrac{1}{2}$

g. As $x \to \infty, x + \dfrac{1}{x} \to \infty$, so $\lim\limits_{x\to\infty} \sin\left(x + \dfrac{1}{x}\right)$ does not exist. (See part a.)

h. $\sin\left(x+\dfrac{1}{x}\right)=\sin x\cos\dfrac{1}{x}+\cos x\sin\dfrac{1}{x}$

$\displaystyle\lim_{x\to\infty}\left[\sin\left(x+\dfrac{1}{x}\right)-\sin x\right]$

$=\displaystyle\lim_{x\to\infty}\left[\sin x\left(\cos\dfrac{1}{x}-1\right)+\cos x\sin\dfrac{1}{x}\right]$

As $x\to\infty,\ \cos\dfrac{1}{x}\to 1$ so $\cos\dfrac{1}{x}-1\to 0.$

From part **b.**, $\displaystyle\lim_{x\to\infty}\sin\dfrac{1}{x}=0.$

As $x\to\infty$ both $\sin x$ and $\cos x$ oscillate between -1 and 1.

$\displaystyle\lim_{x\to\infty}\left[\sin\left(x+\dfrac{1}{x}\right)-\sin x\right]=0.$

51. $\displaystyle\lim_{x\to\infty}\dfrac{3x^2+x+1}{2x^2-1}=\dfrac{3}{2}$

53. $\displaystyle\lim_{x\to-\infty}\left(\sqrt{2x^2+3x}-\sqrt{2x^2-5}\right)=-\dfrac{3}{2\sqrt{2}}$

55. $\displaystyle\lim_{x\to\infty}\left(1+\dfrac{1}{x}\right)^{10}=1$

57. $\displaystyle\lim_{x\to\infty}\left(1+\dfrac{1}{x}\right)^{x^2}=\infty$

59. $\displaystyle\lim_{x\to3^-}\dfrac{\sin|x-3|}{x-3}=-1$

61. $\displaystyle\lim_{x\to3^-}\dfrac{\cos(x-3)}{x-3}=-\infty$

63. $\displaystyle\lim_{x\to0^+}(1+\sqrt{x})^{\frac{1}{\sqrt{x}}}=e\approx2.718$

65. $\displaystyle\lim_{x\to0^+}(1+\sqrt{x})^x=1$

2.9 Concepts Review

1. $\displaystyle\lim_{x\to c}f(x)$

3. $\displaystyle\lim_{x\to a^+}f(x)=f(a);\ \lim_{x\to b^-}f(x)=f(b)$

Problem Set 2.9

1. $\displaystyle\lim_{x\to3}[(x-3)(x-4)]=0=f(3);$ continuous

3. $\displaystyle\lim_{x\to3}\dfrac{3}{x-3}$ and $h(3)$ do not exist, so $h(x)$ is not continuous at 3.

5. $\displaystyle\lim_{t\to3}\dfrac{|t-3|}{t-3}$ and $h(3)$ do not exist, so $h(t)$ is not continuous at 3.

7. $\displaystyle\lim_{t\to3}|t|=3=f(3);$ continuous

9. $h(3)$ does not exist, so $h(t)$ is not continuous at 3.

11. $\displaystyle\lim_{t\to3}\dfrac{t^3-27}{t-3}=\lim_{t\to3}\dfrac{(t-3)(t^2+3t+9)}{t-3}$

$=\displaystyle\lim_{t\to3}(t^2+3t+9)=(3)^2+3(3)+9=27=r(3)$

continuous

13. $\displaystyle\lim_{t\to3^+}f(t)=\lim_{t\to3^+}(3-t)=0$

$\displaystyle\lim_{t\to3^-}f(t)=\lim_{t\to3^-}(t-3)=0$

$\displaystyle\lim_{t\to3}f(t)=f(3);$ continuous

15. $\displaystyle\lim_{t\to3}f(x)=-2=f(3);$ continuous

17. h is continuous on the intervals

$(-\infty,-5)\cup\left[\dfrac{-5}{2},4\right]\cup(4,6)\cup[6,8]\cup(8,\infty)$

19. $\displaystyle\lim_{x\to3}\dfrac{2x^2-18}{3-x}=\lim_{x\to3}\dfrac{2(x+3)(x-3)}{3-x}$

$=\displaystyle\lim_{x\to3}[-2(x+3)]=-2(3+3)=-12$

Define $f(3)=-12.$

21. $\lim\limits_{t\to 1}\dfrac{\sqrt{t}-1}{t-1}=\lim\limits_{t\to 1}\dfrac{(\sqrt{t}-1)(\sqrt{t}+1)}{(t-1)(\sqrt{t}+1)}$

$=\lim\limits_{t\to 1}\dfrac{t-1}{(t-1)(\sqrt{t}+1)}=\lim\limits_{t\to 1}\dfrac{1}{\sqrt{t}+1}=\dfrac{1}{2}$

Define $H(1)=\dfrac{1}{2}$.

23. $\lim\limits_{x\to -1}\sin\left(\dfrac{x^2-1}{x+1}\right)=\lim\limits_{x\to -1}\sin\left(\dfrac{(x-1)(x+1)}{x+1}\right)$

$=\lim\limits_{x\to -1}\sin(x-1)=\sin(-1-1)=\sin(-2)=-\sin 2$

Define $F(-1)=-\sin 2$.

25. $f(x)=\dfrac{33-x^2}{(\pi-x)(x-3)}$

Discontinuous at $x=3,\pi$

27. Discontinuous at all $\theta=n\pi+\dfrac{\pi}{2}$ where n is any integer.

29. Discontinuous at $u=-1$

31. $G(x)=\dfrac{1}{\sqrt{(2-x)(2+x)}}$

Discontinuous on $(-\infty,-2]\cup[2,\infty)$

33. $\lim\limits_{x\to 0}g(x)=0=g(0)$

$\lim\limits_{x\to 1^+}g(x)=1,\ \lim\limits_{x\to 1^-}g(x)=-1$

$\lim\limits_{x\to 1}g(x)$ does not exist, so $g(x)$ is discontinuous

at

$x=1$.

35. Discontinuous at $t=n+\dfrac{1}{2}$ where n is a *y* integer

37.

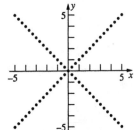

Discontinuous at all points except $x=0$, because $\lim\limits_{x\to c}f(x)\ne f(c)$ for $c\ne 0$. $\lim\limits_{x\to c}f(x)$ exists only

at $c=0$ and $\lim\limits_{x\to 0}f(x)=0=f(0)$.

39. Because the function is continuous on $[0,2\pi]$ and

$(\cos 0)0^3+6\sin^5 0-3=-3<0,$

$(\cos 2\pi)(2\pi)^3+6\sin^5(2\pi)-3=8\pi^3-3>0,$ there is at least one number c between 0 and 2π such that $(\cos t)t^3+6\sin^5 t-3=0.$

41. Suppose that f is continuous at c, so $\lim\limits_{x\to c}f(x)=f(c)$. Let $x=t+c$, so $t=x-c$, then

as $x\to c,\ t\to 0$ and the statement $\lim\limits_{x\to c}f(x)=f(c)$ becomes $\lim\limits_{t\to 0}f(t+c)=f(c).$

Suppose that $\lim\limits_{t\to 0}f(t+c)=f(c)$ and let $x=t+c$, so $t=x-c$. Since c is fixed, $t\to 0$ means that $x\to c$ and the statement $\lim\limits_{t\to 0}f(t+c)=f(c)$

becomes $\lim\limits_{x\to c}f(x)=f(c)$, so f is continuous at c.

43. Let $g(x)=x-f(x)$. Then,

$g(0)=0-f(0)=-f(0)\le 0$ and $g(1)=1-f(1)\ge 0$ since $0\le f(x)\le 1$ on $[0,1]$. If $g(0)=0$, then $f(0)=0$ and $c=0$ is a fixed point of f. If $g(1)=0$, then $f(1)=1$ and $c=1$ is a fixed point of f. If neither $g(0)=0$ nor $g(1)=0$, then $g(0)<0$ and $g(1)>0$ so there is some c in $[0,1]$ such that $g(c)=0$. If $g(c)=0$ then $c-f(c)=0$ or $f(c)=c$ and c is a fixed point of f.

45. For x in $[0,1]$, let $f(x)$ indicate where the string originally at x ends up. Thus $f(0)=a, f(1)=b$. $f(x)$ is continuous since the string is unbroken. Since $0\le a,\ b\le 1$, $f(x)$ satisfies the conditions of Problem 43, so there is some c in $[0,1]$ with $f(c)=c$, i.e., the point of string originally at c ends up at c.

47. Let $f(x)$ be the difference in times on the hiker's watch where x is a point on the path, and suppose $x=0$ at the bottom and $x=1$ at the top of the mountain.

So $f(x)=$ (time on watch on the way up) – (time on watch on the way down).

$f(0)=4-11=-7, f(1)=12-5=7$. Since time is continuous, $f(x)$ is continuous, hence there is some c between 0 and 1 where $f(c)=0$. This c is the point where the hiker's watch showed the same time on both days.

49. a. $f(x)=f(x+0)=f(x)+f(0)$, so $f(0)=0$. We want to prove that $\lim\limits_{x\to c}f(x)=f(c)$, or, equivalently, $\lim\limits_{x\to c}[f(x)-f(c)]=0$. But $f(x)-f(c)=f(x-c)$, so $\lim\limits_{x\to c}[f(x)-f(c)]=\lim\limits_{x\to c}f(x-c)$. Let $h=x-c$ then as $x\to c,\ h\to 0$ and

$\lim_{x\to c} f(x-c) = \lim_{h\to 0} f(h) = f(0) = 0$. Hence,

$\lim_{x\to c} f(x) = f(c)$ and f is continuous at c.

Thus, f is continuous everywhere, since c was arbitrary.

b. By Problem 41 of Section 2.1, $f(t) = mt$ for all t in Q. Since $g(t) = mt$ is a polynomial function, it is continuous for all real numbers. $f(t) = g(t)$ for all t in Q, thus $f(t) = g(t)$ for all t in R, i.e. $f(t) = mt$.

51. $y_1(x) = m_1 x + b_1$, $y_2(x) = m_2 x + b_2$

$y_1(y_2(x)) = y_1(m_2 x + b_2) = m_1(m_2 x + b_2) + b_1$
$= m_1 m_2 x + m_1 b_2 + b_1 = (m_1 m_2)x + (m_1 b_2 + b_1)$
This is a linear function.

53. $y_1(x) = m_1 x + b_1$, $y_2(x) = m_2 x + b_2$

$\dfrac{y_1(x)}{y_2(x)} = \dfrac{m_1 x + b_1}{m_2 x + b_2}$

This is a linear function only when $m_2 = 0$ and $b_2 \neq 0$.

55. Suppose $f(x) = \begin{cases} 1 & \text{if } x \geq 0 \\ -1 & \text{if } x < 0 \end{cases}$. $f(x)$ is

discontinuous at $x = 0$, but $g(x) = |f(x)| = 1$ is continuous everywhere.

57. a.

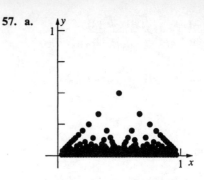

b. If r is any rational number, then any deleted interval about r contains an irrational number. Thus, if $f(r) = \dfrac{1}{q}$, any deleted interval about r contains at least one point c such that $|f(r) - f(c)| = \left| \dfrac{1}{q} - 0 \right| = \dfrac{1}{q}$. Hence,

$\lim_{x\to r} f(x)$ does not exist.

If c is any irrational number in $(0, 1)$, then as $x = \dfrac{p}{q} \to c$ (where $\dfrac{p}{q}$ is the reduced form of the rational number) $q \to \infty$, so

$f(x) \to 0$ as $x \to c$. Thus,

$\lim_{x\to c} f(x) = 0 = f(c)$ for any irrational number c.

2.10 Chapter Review

Concepts Test

1. True: $xy + x^2 = 3y$

$xy - 3y = -x^2$

$y = -\dfrac{x^2}{x-3}$

3. True: $\theta \sin\theta + t - \cos\theta = 0$

$t = \cos\theta - \theta\sin\theta$

5. False: The equation determines T as a function of θ.

7. True: $f(x) = \sqrt{-(x^2 + 4x + 3)}$

$= \sqrt{-(x+3)(x+1)}$

$-(x^2 + 4x + 3) \geq 0$ on $-3 \leq x \leq -1$.

9. True: The domain is $(-\infty, \infty)$ and the range is $[-6, \infty)$.

11. False: The range $(-\infty, \infty)$.

13. True: If $f(x)$ and $g(x)$ are odd functions, $f(-x) + g(-x) = -f(x) - g(x)$ $= -[f(x) + g(x)]$, so $f(x) + g(x)$ is odd.

15. True: If $f(x)$ is even and $g(x)$ is odd, $f(-x)g(-x) = f(x)[-g(x)]$ $= -f(x)g(x)$, so $f(x)g(x)$ is odd.

17. False: If $f(x)$ and $g(x)$ are odd functions, $f(g(-x)) = f(-g(x)) = -f(g(x))$, so $f(g(x))$ is odd.

19. True: $f(-t) = \dfrac{(\sin(-t))^2 + \cos(-t)}{\tan(-t)\csc(-t)}$

$= \dfrac{(-\sin t)^2 + \cos t}{-\tan t(-\csc t)} = \dfrac{(\sin t)^2 + \cos t}{\tan t \csc t}$

21. False: $f(x) = c$ has domain $(-\infty, \infty)$, yet the range has only one value, c.

23. True:
$$(f \circ g)(x) = (x^3)^2 = x^6$$
$$(g \circ f)(x) = (x^2)^3 = x^6$$

25. False:
The domain of $\dfrac{f}{g}$ excludes any values where $g = 0$.

27. True:
$$\cot x = \frac{\cos x}{\sin x}$$
$$\cot(-x) = \frac{\cos(-x)}{\sin(-x)} = \frac{\cos x}{-\sin x} = -\cot x$$

29. False:
The cosine function is periodic, so $\cos s = \cos t$ does not necessarily imply $s = t$; e.g., $\cos 0 = \cos 2\pi = 1$, but $0 \neq 2\pi$.

31. False:
If $f(c)$ is not defined, $\lim\limits_{x \to c} f(x)$ might exist; e.g., $f(x) = \dfrac{x^2 - 4}{x + 2}$.
$f(-2)$ does not exist,
but $\lim\limits_{x \to -2} \dfrac{x^2 - 4}{x + 2} = -4$.

33. True:
Substitution Theorem

35. False:
The tangent function is not defined for all values of c.

37. False:
Since $-1 \leq \sin x \leq 1$ for all x and $\lim\limits_{x \to \infty} \dfrac{1}{x} = 0$, we get $\lim\limits_{x \to \infty} \dfrac{\sin x}{x} = 0$.

39. True:
As $x \to 1^+$ both the numerator and denominator are positive. Since the numerator approaches a constant and the denominator approaches zero, the limit goes to $+\infty$.

41. True:
$\lim\limits_{x \to c} f(x) = f\left(\lim\limits_{x \to c} x\right) = f(c)$, so f is continuous at $x = c$.

43. True:
Choose $\varepsilon = 0.001 f(2)$ then since $\lim\limits_{x \to 2} f(x) = f(2)$, there is some δ such that $0 < |x - 2| < \delta \Rightarrow$ $|f(x) - f(2)| < 0.001 f(2)$, or

$$-0.001 f(2) < f(x) - f(2)$$
$$< 0.001 f(2)$$
Thus, $0.999 f(2) < f(x) < 1.001 f(2)$ and $f(x) < 1.001 f(2)$ for $0 < |x - 2| < \delta$. Since $f(2) < 1.001 f(2)$, as $f(2) > 0$, $f(x) < 1.001 f(2)$ on $(2 - \delta, \ 2 + \delta)$.

45. True:
Squeeze Theorem

47. False:
That $f(x) \neq g(x)$ for all x does not imply that $\lim\limits_{x \to c} f(x) \neq \lim\limits_{x \to c} g(x)$. For example, if $f(x) = \dfrac{x^2 + x - 6}{x - 2}$ and $g(x) = \dfrac{5}{2} x$, then $f(x) \neq g(x)$ for all x, but $\lim\limits_{x \to 2} f(x) = \lim\limits_{x \to 2} g(x) = 5$.

49. True:
$$\lim_{x \to a} |f(x)| = \lim_{x \to a} \sqrt{f^2(x)}$$
$$= \sqrt{\left[\lim_{x \to a} f(x)\right]^2} = \sqrt{(b)^2} = |b|$$

Sample Test Problems

1. a. $f(1) = \dfrac{1}{1+1} - \dfrac{1}{1} = -\dfrac{1}{2}$

b. $f\left(-\dfrac{1}{2}\right) = \dfrac{1}{-\frac{1}{2}+1} - \dfrac{1}{-\frac{1}{2}} = 4$

c. $f(-1)$ does not exist.

d. $f(t-1) = \dfrac{1}{t-1+1} - \dfrac{1}{t-1} = \dfrac{1}{t} - \dfrac{1}{t-1}$

e. $f\left(\dfrac{1}{t}\right) = \dfrac{1}{\frac{1}{t}+1} - \dfrac{1}{\frac{1}{t}} = \dfrac{t}{1+t} - t$

3. a. $\{x \in \mathbb{R} : x \neq -1, 1\}$

b. $\{x \in \mathbb{R} : |x| \leq 2\}$

c. $\left\{x \in \mathbb{R} : x \neq -\dfrac{3}{2}\right\}$

5. a. $f(x) = x^2 - 1$

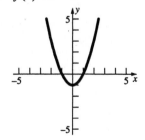

b. $g(x) = \dfrac{x}{x^2+1}$

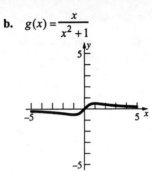

c. $h(x) = \begin{cases} x^2 & \text{if } 0 \le x \le 2 \\ 6-x & \text{if } x > 2 \end{cases}$

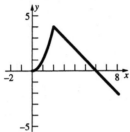

7. $V(x) = x(32-2x)(24-2x)$
Domain $[0, 12]$

9. a. $y = \dfrac{1}{4}x^2$

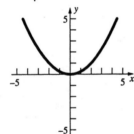

b. $y = \dfrac{1}{4}(x+2)^2$

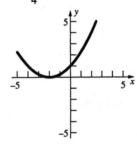

c. $y = -1 + \dfrac{1}{4}(x+2)^2$

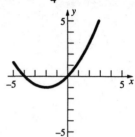

11. $f(x) = \sqrt{x},\, g(x) = 1+x,\, h(x) = x^2,$

$k(x) = \sin x,\ F(x) = \sqrt{1+\sin^2 x} = f \circ g \circ h \circ k$

13. a. $\sin(-t) = -\sin t = -0.8$

b. $\sin^2 t + \cos^2 t = 1$

$\cos^2 t = 1 - (0.8)^2 = 0.36$

$\cos t = -0.6$

c. $\sin 2t = 2\sin t \cos t = 2(0.8)(-0.6) = -0.96$

d. $\tan t = \dfrac{\sin t}{\cos t} = \dfrac{0.8}{-0.6} = -\dfrac{4}{3} \approx -1.333$

e. $\cos\left(\dfrac{\pi}{2} - t\right) = \sin t = 0.8$

f. $\sin(\pi + t) = -\sin t = -0.8$

15. $s = rt$

$= 9\left(20\,\dfrac{\text{rev}}{\text{min}}\right)\left(2\pi\,\dfrac{\text{rad}}{\text{rev}}\right)\left(\dfrac{1\ \text{min}}{60\ \text{sec}}\right)(1\ \text{sec}) = 6\pi$

≈ 18.85 in.

17. $\lim\limits_{u \to 1} \dfrac{u^2 - 1}{u-1} = \lim\limits_{u \to 1} \dfrac{(u-1)(u+1)}{u-1} = \lim\limits_{u \to 1}(u+1)$

$= 1 + 1 = 2$

19. $\lim\limits_{x \to 2} \dfrac{1 - \frac{2}{x}}{x^2 - 4} = \lim\limits_{x \to 2} \dfrac{\frac{x-2}{x}}{(x-2)(x+2)} = \lim\limits_{x \to 2} \dfrac{1}{x(x+2)}$

$= \dfrac{1}{2(2+2)} = \dfrac{1}{8}$

21. $\lim\limits_{x \to 0} \dfrac{\tan x}{\sin 2x} = \lim\limits_{x \to 0} \dfrac{\frac{\sin x}{\cos x}}{2\sin x \cos x} = \lim\limits_{x \to 0} \dfrac{1}{2\cos^2 x}$

$= \dfrac{1}{2\cos^2 0} = \dfrac{1}{2}.$

23. $\lim\limits_{x \to 4} \dfrac{x-4}{\sqrt{x}-2} = \lim\limits_{x \to 4} \dfrac{(\sqrt{x}-2)(\sqrt{x}+2)}{\sqrt{x}-2}$

$= \lim\limits_{x \to 4}(\sqrt{x}+2) = \sqrt{4}+2 = 4$

25. $\displaystyle\lim_{x\to 0^-}\frac{|x|}{x}=\lim_{x\to 0^-}\frac{-x}{x}=\lim_{x\to 0^-}(-1)=-1$

27. $\displaystyle\lim_{t\to 2^-}\left(\llbracket t\rrbracket - t\right)=\lim_{t\to 2^-}\llbracket t\rrbracket-\lim_{t\to 2^-}t=1-2=-1$

29. a. f is discontinuous at $x=1$ because $f(1)=0$, but $\displaystyle\lim_{x\to 1}f(x)$ does not exist. f is discontinuous at $x=-1$ because $f(-1)$ does not exist.

    b. Define $f(-1)=-1$

31. a. $\displaystyle\lim_{x\to 3}[2f(x)-4g(x)]$

      $=2\displaystyle\lim_{x\to 3}f(x)-4\lim_{x\to 3}g(x)$

      $=2(3)-4(-2)=14$

    b. $\displaystyle\lim_{x\to 3}g(x)\frac{x^2-9}{x-3}=\lim_{x\to 3}g(x)(x+3)$

      $=\displaystyle\lim_{x\to 3}g(x)\cdot\lim_{x\to 3}(x+3)=-2\cdot(3+3)=-12$

    c. $g(3)=-2$

    d. $\displaystyle\lim_{x\to 3}g(f(x))=g\left(\lim_{x\to 3}f(x)\right)=g(3)=-2$

    e. $\displaystyle\lim_{x\to 3}\sqrt{f^2(x)-8g(x)}$

      $=\sqrt{\left[\displaystyle\lim_{x\to 3}f(x)\right]^2-8\lim_{x\to 3}g(x)}$

      $=\sqrt{(3)^2-8(-2)}=5$

    f. $\displaystyle\lim_{x\to 3}\frac{|g(x)-g(3)|}{f(x)}=\frac{|-2-g(3)|}{3}=\frac{|-2-(-2)|}{3}$

      $=0$

33. $a(0)+b=-1$ and $a(1)+b=1$

    $b=-1;\quad a+b=1$

               $a-1=1$

                  $a=2$

3.1 Concepts Review

1. tangent line

3. $\dfrac{f(c+h)-f(c)}{h}$

Problem Set 3.1

1. Slope $=\dfrac{5-3}{2-\frac{3}{2}}=4$

3.

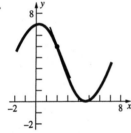

Slope ≈ -2

5.

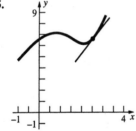

Slope $\approx \dfrac{5}{2}$

7. $y=x^2+1$

a., b.

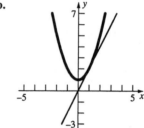

c. $m_{\tan}=2$

d. $m_{\sec}=\dfrac{(1.01)^2+1.0-2}{1.01-1}=\dfrac{0.0201}{.01}=2.01$

e. $m_{\tan}=\lim\limits_{h\to 0}\dfrac{f(1+h)-f(1)}{h}$

$=\lim\limits_{h\to 0}\dfrac{[(1+h)^2+1]-(1^2+1)}{h}$

$=\lim\limits_{h\to 0}\dfrac{2+2h+h^2-2}{h}=\lim\limits_{h\to 0}\dfrac{h(2+h)}{h}$

$=\lim\limits_{h\to 0}(2+h)=2$

9. $f(x)=x^2-1$

$m_{\tan}=\lim\limits_{h\to 0}\dfrac{f(c+h)-f(c)}{h}$

$=\lim\limits_{h\to 0}\dfrac{[(c+h)^2-1]-(c^2-1)}{h}$

$=\lim\limits_{h\to 0}\dfrac{c^2+2ch+h^2-1-c^2+1}{h}$

$=\lim\limits_{h\to 0}\dfrac{h(2c+h)}{h}=2c$

At $x=-2,\ m_{\tan}=-4$

$x=-1,\ m_{\tan}=-2$

$x=1,\ m_{\tan}=2$

$x=2,\ m_{\tan}=4$

11.

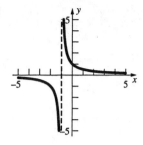

$$f(x) = \frac{1}{x+1}$$

$$m_{\tan} = \lim_{h \to 0} \frac{f(1+h) - f(1)}{h}$$

$$= \lim_{h \to 0} \frac{\frac{1}{2+h} - \frac{1}{2}}{h} = \lim_{h \to 0} \frac{-\frac{h}{2(2+h)}}{h} = \lim_{h \to 0} -\frac{1}{2(2+h)}$$

$$= -\frac{1}{4}$$

$$y - \frac{1}{2} = -\frac{1}{4}(x-1)$$

13. a. $16(1^2) - 16(0^2) = 16$ ft

b. $16(2^2) - 16(1^2) = 48$ ft

c. $V_{ave} = \frac{144 - 64}{3 - 2} = 80$ ft/sec

d. $V_{ave} = \frac{16(3.01)^2 - 16(3)^2}{3.01 - 3} = \frac{0.9616}{0.01}$

$= 96.16$ ft/s

e. $f(t) = 16t^2; v = 32c$

$v = 32(3) = 96$ ft/s

15. a. $v = \lim_{h \to 0} \frac{f(\alpha + h) - f(\alpha)}{h}$

$$= \lim_{h \to 0} \frac{\sqrt{2(\alpha + h) + 1} - \sqrt{2\alpha + 1}}{h}$$

$$= \lim_{h \to 0} \frac{\sqrt{2\alpha + 2h + 1} - \sqrt{2\alpha + 1}}{h}$$

$$= \lim_{h \to 0} \frac{(\sqrt{2\alpha + 2h + 1} - \sqrt{2\alpha + 1})(\sqrt{2\alpha + 2h + 1} + \sqrt{2\alpha + 1})}{h(\sqrt{2\alpha + 2h + 1} + \sqrt{2\alpha + 1})}$$

$$= \lim_{h \to 0} \frac{2h}{h(\sqrt{2\alpha + 2h + 1} + \sqrt{2\alpha + 1})}$$

$$= \frac{2}{\sqrt{2\alpha + 1} + \sqrt{2\alpha + 1}} = \frac{1}{\sqrt{2\alpha + 1}}$$ ft/s

b. $\frac{1}{\sqrt{2\alpha + 1}} = \frac{1}{2}$

$\sqrt{2\alpha + 1} = 2$

$2\alpha + 1 = 4; \; \alpha = \frac{3}{2}$

The object reaches a velocity of $\frac{1}{2}$ ft/s when $t = \frac{3}{2}$.

17. a. $\left[\frac{1}{2}(2.01)^2 + 1\right] - \left[\frac{1}{2}(2)^2 + 1\right] = 0.02005$ g

b. $r_{ave} = \frac{0.02005}{2.01 - 2} = 2.005$ g/hr

c. $f(t) = \frac{1}{2}t^2 + 1$

$$r = \lim_{h \to 0} \frac{\left[\frac{1}{2}(2+h)^2 + 1\right] - \left[\frac{1}{2}2^2 + 1\right]}{h}$$

$$= \lim_{h \to 0} \frac{2 + 2h + \frac{1}{2}h^2 + 1 - 2 - 1}{h}$$

$$= \lim_{h \to 0} \frac{h\left(2 + \frac{1}{2}h\right)}{h} = 2$$

At $t = 2, r = 2$

21. $a = \lim_{h \to 0} \frac{2(1+h)^2 - 2(1)^2}{h}$

$$= \lim_{h \to 0} \frac{2 + 4h + 2h^2 - 2}{h}$$

$$= \lim_{h \to 0} \frac{h(4 + 2h)}{h} = 4$$

23. $r_{ave} = \frac{100 - 800}{24 - 0} = -\frac{175}{6} \approx -29.167$

29,167 gal/hr

At 8 o'clock, $r \approx \frac{700 - 400}{6 - 10} \approx -75$

75,000 gal/hr

25. a. A tangent line at $t = 91$ has slope approximately $(63 - 48)/(91 - 61) = 0.5$. The normal high temperature increases at the rate of 0.5 degree F per day.

b. A tangent line at $t = 191$ has approximate slope $(90 - 88)/30 \approx 0.067$. The normal high temperature increases at the rate of 0.067 degree per day.

c. There is a time in January, about January 15, when the rate of change is zero. There is also a time in July, about July 15, when the rate of change is zero.

3.2 Concepts Review

1. $\dfrac{f(c+h) - f(c)}{h}$; $\dfrac{f(t) - f(c)}{t - c}$

3. continuous; $f(x) = |x|$

19. a. $d_{ave} = \dfrac{5^3 - 3^3}{5 - 3} = \dfrac{98}{2} = 49$ g/cm

b. $f(x) = x^3$

$$d = \lim_{h \to 0} \frac{(3+h)^3 - 3^3}{h}$$

$$= \lim_{h \to 0} \frac{27 + 27h + 9h^2 + h^3 - 27}{h}$$

$$= \lim_{h \to 0} \frac{h(27 + 9h + h^2)}{h} = 27 \text{ g/cm}$$

d. The greatest rate of increase occurs around day 61, that is, some time in March. The greatest rate of decrease occurs between day 301 and 331, that is, sometime in November.

27. In both (a) and (b), the tangent line is always positive. In (a) the tangent line becomes steeper and steeper as t increases; thus, the velocity is increasing. In (b) the tangent line becomes flatter and flatter as t increases; thus, the velocity is decreasing.

29. $A = \pi r^2$, $r = 2t$

$A = 4\pi t^2$

$$\text{rate} = \lim_{h \to 0} \frac{4\pi(3+h)^2 - 4\pi(3)^2}{h}$$

$$= \lim_{h \to 0} \frac{h(24\pi + 4\pi h)}{h} = 24\pi \text{ km}^2/\text{day}$$

31. $y = f(x) = x^3 - 2x^2 + 1$

a. $m_{\tan} = 7$ **b.** $m_{\tan} = 0$

c. $m_{\tan} = -1$ **d.** $m_{\tan} = 17.92$

33. $s = f(t) = t + t\cos^2 t$

At $t = 3, v \approx 2.818$

Problem Set 3.2

1. $f'(1) = \lim_{h \to 0} \dfrac{f(1+h) - f(1)}{h}$

$$= \lim_{h \to 0} \frac{(1+h)^2 - 1^2}{h} = \lim_{h \to 0} \frac{2h + h^2}{h}$$

$$= \lim_{h \to 0} (2 + h) = 2$$

3. $f'(3) = \lim\limits_{h \to 0} \dfrac{f(3+h) - f(3)}{h}$

$\quad = \lim\limits_{h \to 0} \dfrac{[(3+h)^2 - (3+h)] - (3^2 - 3)}{h}$

$\quad = \lim\limits_{h \to 0} \dfrac{5h + h^2}{h} = \lim\limits_{h \to 0}(5+h) = 5$

5. $s'(x) = \lim\limits_{h \to 0} \dfrac{s(x+h) - s(x)}{h}$

$\quad = \lim\limits_{h \to 0} \dfrac{[2(x+h) + 1] - (2x+1)}{h}$

$\quad = \lim\limits_{h \to 0} \dfrac{2h}{h} = 2$

7. $r'(x) = \lim\limits_{h \to 0} \dfrac{r(x+h) - r(x)}{h}$

$\quad = \lim\limits_{h \to 0} \dfrac{[3(x+h)^2 + 4] - (3x^2 + 4)}{h}$

$\quad = \lim\limits_{h \to 0} \dfrac{6xh + 3h^2}{h} = \lim\limits_{h \to 0}(6x + 3h) = 6x$

9. $f'(x) = \lim\limits_{h \to 0} \dfrac{f(x+h) - f(x)}{h}$

$\quad = \lim\limits_{h \to 0} \dfrac{[a(x+h)^2 + b(x+h) + c] - (ax^2 + bx + c)}{h}$

$\quad = \lim\limits_{h \to 0} \dfrac{2axh + ah^2 + bh}{h} = \lim\limits_{h \to 0}(2ax + ah + b)$

$\quad = 2ax + b$

11. $f'(x) = \lim\limits_{h \to 0} \dfrac{f(x+h) - f(x)}{h}$

$\quad = \lim\limits_{h \to 0} \dfrac{[(x+h)^3 + 2(x+h)^2 + 1] - (x^3 + 2x^2 + 1)}{h}$

$\quad = \lim\limits_{h \to 0} \dfrac{3hx^2 + 3h^2x + h^3 + 4hx + 2h^2}{h}$

$\quad = \lim\limits_{h \to 0}(3x^2 + 3hx + h^2 + 4x + 2h) = 3x^2 + 4x$

13. $h'(x) = \lim\limits_{h \to 0} \dfrac{h(x+h) - h(x)}{h}$

$\quad = \lim\limits_{h \to 0}\left[\left(\dfrac{2}{x+h} - \dfrac{2}{x}\right) \cdot \dfrac{1}{h}\right]$

$\quad = \lim\limits_{h \to 0}\left[\dfrac{-2h}{x(x+h)} \cdot \dfrac{1}{h}\right] = \lim\limits_{h \to 0} \dfrac{-2}{x(x+h)}$

$\quad = -\dfrac{2}{x^2}$

15. $F'(x) = \lim\limits_{h \to 0} \dfrac{F(x+h) - F(x)}{h}$

$\quad = \lim\limits_{h \to 0}\left[\left(\dfrac{6}{(x+h)^2 + 1} - \dfrac{6}{x^2 + 1}\right) \cdot \dfrac{1}{h}\right]$

$\quad = \lim\limits_{h \to 0}\left[\dfrac{6(x^2 + 1) - 6(x^2 + 2hx + h^2 + 1)}{(x^2 + 1)(x^2 + 2hx + h^2 + 1)} \cdot \dfrac{1}{h}\right]$

$\quad = \lim\limits_{h \to 0}\left[\dfrac{-12hx - 6h^2}{(x^2 + 1)(x^2 + 2hx + h^2 + 1)} \cdot \dfrac{1}{h}\right]$

$\quad = \lim\limits_{h \to 0} \dfrac{-12x - 6h}{(x^2 + 1)(x^2 + 2hx + h^2 + 1)} = -\dfrac{12x}{(x^2 + 1)^2}$

17. $G'(x) = \lim\limits_{h \to 0} \dfrac{G(x+h) - G(x)}{h}$

$\quad = \lim\limits_{h \to 0}\left[\left(\dfrac{2(x+h) - 1}{x+h-4} - \dfrac{2x-1}{x-4}\right) \cdot \dfrac{1}{h}\right]$

$\quad = \lim\limits_{h \to 0}\left[\dfrac{2x^2 + 2hx - 9x - 8h + 4 - (2x^2 + 2hx - 9x - h + 4)}{(x+h-4)(x-4)} \cdot \dfrac{1}{h}\right] = \lim\limits_{h \to 0}\left[\dfrac{-7h}{(x+h-4)(x-4)} \cdot \dfrac{1}{h}\right]$

$\quad = \lim\limits_{h \to 0} \dfrac{-7}{(x+h-4)(x-4)} = -\dfrac{7}{(x-4)^2}$

19. $g'(x) = \lim\limits_{h \to 0} \dfrac{g(x+h) - g(x)}{h}$

$\quad = \lim\limits_{h \to 0} \dfrac{\sqrt{3(x+h)} - \sqrt{3x}}{h}$

$$= \lim_{h \to 0} \frac{(\sqrt{3x+3h} - \sqrt{3x})(\sqrt{3x+3h} + \sqrt{3x})}{h(\sqrt{3x+3h} + \sqrt{3x})}$$

$$= \lim_{h \to 0} \frac{3h}{h(\sqrt{3x+3h} + \sqrt{3x})} = \lim_{h \to 0} \frac{3}{\sqrt{3x+3h} + \sqrt{3x}} = \frac{3}{2\sqrt{3x}}$$

21. $H'(x) = \lim\limits_{h \to 0} \dfrac{H(x+h) - H(x)}{h}$

$$= \lim_{h \to 0} \left[\left(\frac{3}{\sqrt{x+h-2}} - \frac{3}{\sqrt{x-2}} \right) \cdot \frac{1}{h} \right]$$

$$= \lim_{h \to 0} \left[\frac{3\sqrt{x-2} - 3\sqrt{x+h-2}}{\sqrt{(x+h-2)(x-2)}} \cdot \frac{1}{h} \right]$$

$$= \lim_{h \to 0} \frac{3(\sqrt{x-2} - \sqrt{x+h-2})(\sqrt{x-2} + \sqrt{x+h-2})}{h\sqrt{(x+h-2)(x-2)}(\sqrt{x-2} + \sqrt{x+h-2})}$$

$$= \lim_{h \to 0} \frac{-3h}{h[(x-2)\sqrt{x+h-2} + (x+h-2)\sqrt{x-2}]}$$

$$= \lim_{h \to 0} \frac{-3}{(x-2)\sqrt{x+h-2} + (x+h-2)\sqrt{x-2}}$$

$$= -\frac{3}{2(x-2)\sqrt{x-2}} = -\frac{3}{2(x-2)^{3/2}}$$

23. $f'(x) = \lim\limits_{t \to x} \dfrac{f(t) - f(x)}{t - x}$

$$= \lim_{t \to x} \frac{(t^2 - 3t) - (x^2 - 3x)}{t - x}$$

$$= \lim_{t \to x} \frac{t^2 - x^2 - (3t - 3x)}{t - x}$$

$$= \lim_{t \to x} \frac{(t-x)(t+x) - 3(t-x)}{t - x}$$

$$= \lim_{t \to x} \frac{(t-x)(t+x-3)}{t - x} = \lim_{t \to x}(t + x - 3)$$

$$= 2x - 3$$

25. $f'(x) = \lim\limits_{t \to x} \dfrac{f(t) - f(x)}{t - x}$

$$= \lim_{t \to x} \left[\left(\frac{t}{t-5} - \frac{x}{x-5} \right) \left(\frac{1}{t-x} \right) \right]$$

$$= \lim_{t \to x} \frac{tx - 5t - tx + 5x}{(t-5)(x-5)(t-x)}$$

$$= \lim_{t \to x} \frac{-5(t-x)}{(t-5)(x-5)(t-x)} = \lim_{t \to x} \frac{-5}{(t-5)(x-5)}$$

$$= -\frac{5}{(x-5)^2}$$

27. $f(x) = 2x^3$ at $x = 5$

29. $f(x) = x^2$ at $x = 2$

31. $f(x) = x^2$ at x

33. $f(t) = \dfrac{2}{t}$ at t

35. $f(x) = \cos x$ at x

37. $f'(0) \approx -\dfrac{1}{2}; \; f'(2) \approx 1$

$$f'(5) \approx \frac{2}{3}; \; f'(7) \approx -3$$

39.

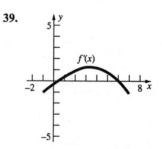

41. a. $f(2) \approx \dfrac{5}{2}; \; f'(2) \approx \dfrac{3}{2}$

$$f(0.5) \approx 1.8; \; f'(0.5) \approx -0.6$$

b. $\dfrac{2.9-1.9}{2.5-0.5} = 0.5$

c. $x = 5$

d. $x = 3, 5$

e. $x = 1, 3, 5$

f. $x = 0$

g. $x \approx -0.7, \dfrac{3}{2}$ and $5 < x < 7$

43. The derivative is 0 at approximately $t = 15$ and $t = 201$. The greatest rate of increase occurs at about $t = 61$ and it is about 0.5 degree F per day. The greatest rate of decrease occurs at about $t = 320$ and it is about 0.5 degree F per day. The derivative is positive on $(15,201)$ and negative on $(0,15)$ and $(201,365)$.

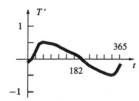

45. The short-dash function has a tangent line with zero slope at about $x = 2.1$, where the solid function is zero. The solid function has a tangent line with zero slope at about $x = 0.4, 1.2$ and 3.5.

The long-dash function is zero at these points. The graph shows that the solid function is positive (negative) when the slope of the tangent line of the short-dash function is positive (negative). Also, the long-dash function is positive (negative) when the slope of the tangent line of the solid function is positive (negative). Thus, the short-dash function is f, the solid function is $f' = g$, and the dash function is g'.

47. If f is differentiable everywhere, then it is continuous everywhere, so
$$\lim_{x \to 2^-} f(x) = \lim_{x \to 2^-}(mx+b) = 2m+b = f(2) = 4$$
and $b = 4 - 2m$.
For f to be differentiable everywhere,
$$f'(2) = \lim_{x \to 2}\frac{f(x)-f(2)}{x-2} \text{ must exist.}$$
$$\lim_{x \to 2^+}\frac{f(x)-f(2)}{x-2} = \lim_{x \to 2^+}\frac{x^2-4}{x-2} = \lim_{x \to 2^+}(x+2) = 4$$
$$\lim_{x \to 2^-}\frac{f(x)-f(2)}{x-2} = \lim_{x \to 2^-}\frac{mx+b-4}{x-2}$$

$$= \lim_{x \to 2^-}\frac{mx+4-2m-4}{x-2} = \lim_{x \to 2^-}\frac{m(x-2)}{x-2} = m$$
Thus $m = 4$ and $b = 4 - 2(4) = -4$

49. $f'(x_0) = \lim\limits_{t \to x_0}\dfrac{f(t)-f(x_0)}{t-x_0}$, so

$$f'(-x_0) = \lim_{t \to -x_0}\frac{f(t)-f(-x_0)}{t-(-x_0)}$$

$$= \lim_{t \to -x_0}\frac{f(t)-f(-x_0)}{t+x_0}$$

a. If f is an odd function,
$$f'(-x_0) = \lim_{t \to -x_0}\frac{f(t)-[-f(-x_0)]}{t+x_0}$$
$$= \lim_{t \to -x_0}\frac{f(t)+f(-x_0)}{t+x_0}.$$
Let $u = -t$. As $t \to -x_0$, $u \to x_0$ and so
$$f'(-x_0) = \lim_{u \to x_0}\frac{f(-u)+f(x_0)}{-u+x_0}$$
$$= \lim_{u \to x_0}\frac{-f(u)+f(x_0)}{-(u-x_0)} = \lim_{u \to x_0}\frac{-[f(u)-f(x_0)]}{-(u-x_0)}$$
$$= \lim_{u \to x_0}\frac{f(u)-f(x_0)}{u-x_0} = f'(x_0) = m.$$

b. If f is an even function,
$$f'(-x_0) = \lim_{t \to -x_0}\frac{f(t)-f(x_0)}{t+x_0}. \text{ Let } u = -t, \text{ as}$$
above, then $f'(-x_0) = \lim\limits_{u \to x_0}\dfrac{f(-u)-f(x_0)}{-u+x_0}$
$$= \lim_{u \to x_0}\frac{f(u)-f(x_0)}{-(u-x_0)} = -\lim_{u \to x_0}\frac{f(u)-f(x_0)}{u-x_0}$$
$$= -f'(x_0) = -m.$$

51.

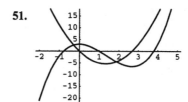

a. $0 < x < \dfrac{8}{3}$ **b.** $0 < x < \dfrac{8}{3}$

c. A function $f(x)$ decreases as x increases when $f'(x) < 0$.

3.3 Concepts Review

1. the derivative of the second; second; $f(x)g'(x)+g(x)f'(x)$

3. $nx^{n-1}h$; nx^{n-1}

Problem Set 3.3

1. $D_x(2x^2)=2D_x(x^2)=2\cdot 2x=4x$

3. $D_x(\pi x)=\pi D_x(x)=\pi\cdot 1=\pi$

5. $D_x(2x^{-2})=2D_x(x^{-2})=2(-2x^{-3})=-4x^{-3}$

7. $D_x\left(\dfrac{\pi}{x}\right)=\pi D_x(x^{-1})=\pi(-1x^{-2})=-\pi x^{-2}$

$=-\dfrac{\pi}{x^2}$

9. $D_x\left(\dfrac{100}{x^5}\right)=100D_x(x^{-5})=100(-5x^{-6})$

$=-500x^{-6}=-\dfrac{500}{x^6}$

11. $D_x(x^2+2x)=D_x(x^2)+2D_x(x)=2x+2$

13. $D_x(x^4+x^3+x^2+x+1)$

$=D_x(x^4)+D_x(x^3)+D_x(x^2)+D_x(x)+D_x(1)$

$=4x^3+3x^2+2x+1$

15. $D_x(\pi x^7-2x^5-5x^{-2})$

$=\pi D_x(x^7)-2D_x(x^5)-5D_x(x^{-2})$

$=\pi(7x^6)-2(5x^4)-5(-2x^{-3})$

$=7\pi x^6-10x^4+10x^{-3}$

17. $D_x\left(\dfrac{3}{x^3}+x^{-4}\right)=3D_x(x^{-3})+D_x(x^{-4})$

$=3(-3x^{-4})+(-4x^{-5})=-\dfrac{9}{x^4}-4x^{-5}$

19. $D_x\left(\dfrac{2}{x}-\dfrac{1}{x^2}\right)=2D_x(x^{-1})-D_x(x^{-2})$

$=2(-1x^{-2})-(-2x^{-3})=-\dfrac{2}{x^2}+\dfrac{2}{x^3}$

21. $D_x\left(\dfrac{1}{2x}+2x\right)=\dfrac{1}{2}D_x(x^{-1})+2D_x(x)$

$=\dfrac{1}{2}(-1x^{-2})+2(1)=-\dfrac{1}{2x^2}+2$

23. $D_x[x(x^2+1)]=xD_x(x^2+1)+(x^2+1)D_x(x)$

$=x(2x)+(x^2+1)(1)=3x^2+1$

25. $D_x[(2x+1)^2]$

$=(2x+1)D_x(2x+1)+(2x+1)D_x(2x+1)$

$=(2x+1)(2)+(2x+1)(2)=8x+4$

27. $D_x[(x^2+2)(x^3+1)]$

$=(x^2+2)D_x(x^3+1)+(x^3+1)D_x(x^2+2)$

$=(x^2+2)(3x^2)+(x^3+1)(2x)$

$=3x^4+6x^2+2x^4+2x$

$=5x^4+6x^2+2x$

29. $D_x[(x^2+17)(x^3-3x+1)]$

$=(x^2+17)D_x(x^3-3x+1)+(x^3-3x+1)D_x(x^2+17)$

$=(x^2+17)(3x^2-3)+(x^3-3x+1)(2x)$

$=3x^4+48x^2-51+2x^4-6x^2+2x$

$=5x^4+42x^2+2x-51$

31. $D_x[(5x^2-7)(3x^2-2x+1)]=(5x^2-7)D_x(3x^2-2x+1)+(3x^2-2x+1)D_x(5x^2-7)$

$=(5x^2-7)(6x-2)+(3x^2-2x+1)(10x)$

$=60x^3-30x^2-32x+14$

33. $D_x\left(\dfrac{1}{3x^2+1}\right) = \dfrac{(3x^2+1)D_x(1)-(1)D_x(3x^2+1)}{(3x^2+1)^2}$

$= \dfrac{(3x^2+1)(0)-(6x)}{(3x^2+1)^2} = -\dfrac{6x}{(3x^2+1)^2}$

35. $D_x\left(\dfrac{1}{4x^2-3x+9}\right) = \dfrac{(4x^2-3x+9)D_x(1)-(1)D_x(4x^2-3x+9)}{(4x^2-3x+9)^2}$

$= \dfrac{(4x^2-3x+9)(0)-(8x-3)}{(4x^2-3x+9)^2} = -\dfrac{8x-3}{(4x^2-3x+9)^2}$

$= \dfrac{-8x+3}{(4x^2-3x+9)^2}$

37. $D_x\left(\dfrac{x-1}{x+1}\right) = \dfrac{(x+1)D_x(x-1)-(x-1)D_x(x+1)}{(x+1)^2}$

$= \dfrac{(x+1)(1)-(x-1)(1)}{(x+1)^2} = \dfrac{2}{(x+1)^2}$

39. $D_x\left(\dfrac{2x^2-1}{3x+5}\right) = \dfrac{(3x+5)D_x(2x^2-1)-(2x^2-1)D_x(3x+5)}{(3x+5)^2}$

$= \dfrac{(3x+5)(4x)-(2x^2-1)(3)}{(3x+5)^2}$

$= \dfrac{6x^2+20x+3}{(3x+5)^2}$

41. $D_x\left(\dfrac{2x^2-3x+1}{2x+1}\right) = \dfrac{(2x+1)D_x(2x^2-3x+1)-(2x^2-3x+1)D_x(2x+1)}{(2x+1)^2}$

$= \dfrac{(2x+1)(4x-3)-(2x^2-3x+1)(2)}{(2x+1)^2}$

$= \dfrac{4x^2+4x-5}{(2x+1)^2}$

43. $D_x\left(\dfrac{x^2-x+1}{x^2+1}\right) = \dfrac{(x^2+1)D_x(x^2-x+1)-(x^2-x+1)D_x(x^2+1)}{(x^2+1)^2}$

$= \dfrac{(x^2+1)(2x-1)-(x^2-x+1)(2x)}{(x^2+1)^2}$

$= \dfrac{x^2-1}{(x^2+1)^2}$

45. a. $(f \cdot g)'(0) = f(0)g'(0)+g(0)f'(0)$
$= 4(5)+(-3)(-1) = 23$

b. $(f+g)'(0) = f'(0)+g'(0) = -1+5 = 4$

c. $(f/g)'(0) = \dfrac{g(0)f'(0) - f(0)g'(0)}{g^2(0)}$

$\qquad = \dfrac{-3(-1) - 4(5)}{(-3)^2} = -\dfrac{17}{9}$

47. $D_x[f(x)]^2 = D_x[f(x)f(x)]$
$\qquad = f(x)D_x[f(x)] + f(x)D_x[f(x)]$
$\qquad = 2 \cdot f(x) \cdot D_x f(x)$

49. $D_x(x^2 - 2x + 2) = 2x - 2$
At $x = 1$: $m_{\tan} = 2(1) - 2 = 0$
Tangent line: $y = 1$

51. $D_x(x^3 - x^2) = 3x^2 - 2x$
The tangent line is horizontal when $m_{\tan} = 0$:
$m_{\tan} = 3x^2 - 2x = 0$
$x(3x - 2) = 0$
$x = 0$ and $x = \dfrac{2}{3}$
$(0, 0)$ and $\left(\dfrac{2}{3}, -\dfrac{4}{27}\right)$

53. $\quad y = 100/x^5 = 100x^{-5}$
$\qquad y' = -500x^{-6}$

Set y' equal to -1, the negative reciprocal of the slope of the line $y = x$. Solving for x gives
$x = \pm 500^{1/6} \approx \pm 2.817$
$y = \pm 100(500)^{-5/6} \approx \pm 0.563$

The points are $(2.817, 0.563)$ and $(-2.817, -0.563)$.

55. a. $\quad D_t(-16t^2 + 40t + 100) = -32t + 40$
$\qquad v = -32(2) + 40 = -24$ ft/s

b. $\quad v = -32t + 40 = 0$
$\qquad t = \dfrac{5}{4}$ s

57. $m_{\tan} = D_x(4x - x^2) = 4 - 2x$
The line through $(2,5)$ and (x_0, y_0) has slope
$\dfrac{y_0 - 5}{x_0 - 2}$.

$4 - 2x_0 = \dfrac{4x_0 - x_0^2 - 5}{x_0 - 2}$

$-2x_0^2 + 8x_0 - 8 = -x_0^2 + 4x_0 - 5$

$x_0^2 - 4x_0 + 3 = 0$
$(x_0 - 3)(x_0 - 1) = 0$
$x_0 = 1, \; x_0 = 3$
At $x_0 = 1$: $y_0 = 4(1) - (1)^2 = 3$
$m_{\tan} = 4 - 2(1) = 2$
Tangent line: $y - 3 = 2(x - 1)$; $y = 2x + 1$
At $x_0 = 3$: $y_0 = 4(3) - (3)^2 = 3$
$m_{\tan} = 4 - 2(3) = -2$
Tangent line: $y - 3 = -2(x - 3)$; $y = -2x + 9$

59. $D_x(7 - x^2) = -2x$
The line through $(4, 0)$ and (x_0, y_0) has slope $\dfrac{y_0 - 0}{x_0 - 4}$. If the fly is at (x_0, y_0) when the spider sees it, then $m_{\tan} = -2x_0 = \dfrac{7 - x_0^2 - 0}{x_0 - 4}$.

$-2x_0^2 + 8x_0 = 7 - x_0^2$
$x_0^2 - 8x_0 + 7 = 0$
$(x_0 - 7)(x_0 - 1) = 0$
At $x_0 = 1$: $y_0 = 6$
$d = \sqrt{(4 - 1)^2 + (0 - 6)^2} = \sqrt{9 + 36} = \sqrt{45} = 3\sqrt{5}$
≈ 6.7
They are 6.7 units apart when they see each other.

61. The watermelon has volume $\dfrac{4}{3}\pi r^3$; the volume of the rind is
$V = \dfrac{4}{3}\pi r^3 - \dfrac{4}{3}\pi\left(r - \dfrac{r}{10}\right)^3 = \dfrac{271}{750}\pi r^3$.

At the end of the fifth week $r = 10$, so
$D_r V = \dfrac{271}{250}\pi r^2 = \dfrac{271}{250}\pi(10)^2 = \dfrac{542\pi}{5} \approx 340$ cm^3
per cm of radius growth. Since the radius is growing 2 cm per week, the volume of the rind is growing at the rate of $\dfrac{542\pi}{5}(2) \approx 681$ cm^3 per week.

3.4 Concepts Review

1. $\dfrac{\sin(x+h)-\sin(x)}{h}$

3. $\cos x;\ -\sin x$

Problem Set 3.4

1. $D_x(2\sin x + 3\cos x) = 2\,D_x(\sin x) + 3\,D_x(\cos x)$
   $= 2\cos x - 3\sin x$

3. $D_x(\sin^2 x + \cos^2 x) = D_x(1) = 0$

5. $D_x(\sec x) = D_x\left(\dfrac{1}{\cos x}\right)$

   $= \dfrac{\cos x\, D_x(1) - (1)D_x(\cos x)}{\cos^2 x}$

   $= \dfrac{\sin x}{\cos^2 x} = \dfrac{1}{\cos x}\cdot\dfrac{\sin x}{\cos x} = \sec x \tan x$

7. $D_x(\tan x) = D_x\left(\dfrac{\sin x}{\cos x}\right)$

   $= \dfrac{\cos x\, D_x(\sin x) - \sin x\, D_x(\cos x)}{\cos^2 x}$

   $= \dfrac{\cos^2 x + \sin^2 x}{\cos^2 x} = \dfrac{1}{\cos^2 x} = \sec^2 x$

9. $D_x\left(\dfrac{\sin x + \cos x}{\cos x}\right)$

   $= \dfrac{\cos x\, D_x(\sin x + \cos x) - (\sin x + \cos x)D_x(\cos x)}{\cos^2 x}$

   $= \dfrac{\cos x(\cos x - \sin x) - (-\sin^2 x - \sin x\cos x)}{\cos^2 x}$

   $= \dfrac{\cos^2 x + \sin^2 x}{\cos^2 x} = \dfrac{1}{\cos^2 x} = \sec^2 x$

11. $D_x(x^2\cos x) = x^2 D_x(\cos x) + \cos x\, D_x(x^2)$
    $= -x^2\sin x + 2x\cos x$

13. $y = \tan^2 x = (\tan x)(\tan x)$
    $y' = (\tan x)(\sec^2 x) + (\tan x)(\sec^2 x)$
    $= 2\tan x\sec^2 x$

15. $D_x(\cos x) = -\sin x$
    At $x = 1$: $m_{\tan} = -\sin 1 \approx -0.8415$
    $y = \cos 1 \approx 0.5403$
    Tangent line: $y - 0.5403 = -0.8415(x - 1)$

17. $D_t(30\cos 2t) = 30D_t(\cos^2 t - \sin^2 t)$

    $= 30[\cos t\, D_t(\cos t) + \cos t\, D_t(\cos t) - D_t(\sin^2 t)]$

    $= 30[-2\sin t\cos t - \sin t D_t(\sin t) - \sin t D_t(\sin t)]$

    $= 30[-2\sin t\cos t - 2\sin t\cos t]$

    $= -60\sin 2t$

    At $t = \dfrac{\pi}{4}$; $-60\sin\left(2\cdot\dfrac{\pi}{4}\right) = -60$ ft/s

    The seat is moving to the left at the rate of 60 ft/s.

19. $y = \tan x$
    $y' = \sec^2 x$
    When $y = 0$, $y = \tan 0 = 0$ and $y' = \sec^2 0 = 1$.
    The tangent line at $x = 0$ is $y = x$.

21. $y = 9\sin x\cos x$
    $y' = 9\left[\sin x(-\sin x) + \cos x(\cos x)\right]$
    $= 9\left[\sin^2 x - \cos^2 x\right]$
    $= 9\left[-\cos 2x\right]$
    The tangent line is horizontal when $y' = 0$ or, in this case, where $\cos 2x = 0$. This occurs when
    $x = \dfrac{\pi}{4} + k\dfrac{\pi}{2}$ where k is an integer.

23. The curves intersect when $\sqrt{2}\sin x = \sqrt{2}\cos x$,
    $\sin x = \cos x$ at $x = \dfrac{\pi}{4}$ for $0 < x < \dfrac{\pi}{2}$.

    $D_x(\sqrt{2}\sin x) = \sqrt{2}\cos x$; $\sqrt{2}\cos\dfrac{\pi}{4} = 1$

    $D_x(\sqrt{2}\cos x) = -\sqrt{2}\sin x$; $-\sqrt{2}\sin\dfrac{\pi}{4} = -1$

    $1(-1) = -1$ so the curves intersect at right angles.

25. $D_x(\sin x^2) = \lim_{h\to 0} \dfrac{\sin(x+h)^2 - \sin x^2}{h}$

$= \lim_{h\to 0} \dfrac{\sin(x^2 + 2xh + h^2) - \sin x^2}{h}$

$= \lim_{h\to 0} \dfrac{\sin x^2 \cos(2xh + h^2) + \cos x^2 \sin(2xh + h^2) - \sin x^2}{h} = \lim_{h\to 0} \dfrac{\sin x^2 [\cos(2xh + h^2) - 1] + \cos x^2 \sin(2xh + h^2)}{h}$

$= \lim_{h\to 0}(2x + h)\left[\sin x^2 \dfrac{\cos(2xh + h^2) - 1}{2xh + h^2} + \cos x^2 \dfrac{\sin(2xh + h^2)}{2xh + h^2} \right] = 2x(\sin x^2 \cdot 0 + \cos x^2 \cdot 1) = 2x\cos x^2$

27. $\sin x_0 = \sin 2x_0$

$\sin x_0 = 2\sin x_0 \cos x_0$

$\cos x_0 = \dfrac{1}{2}$ [if $\sin x_0 \neq 0$]

$x_0 = \dfrac{\pi}{3}$

$D_x(\sin x) = \cos x$, $D_x(\sin 2x) = 2\cos 2x$, so at x_0, the tangent lines to $y = \sin x$ and $y = \sin 2x$ have

slopes of $m_1 = \dfrac{1}{2}$ and $m_2 = 2\left(-\dfrac{1}{2}\right) = -1$,

respectively. From Problem 40 of Section 2.3,

$\tan\theta = \dfrac{m_2 - m_1}{1 + m_1 m_2}$ where θ is the angle between

the tangent lines. $\tan\theta = \dfrac{-1 - \frac{1}{2}}{1 + \left(\frac{1}{2}\right)(-1)} = \dfrac{-\frac{3}{2}}{\frac{1}{2}} = -3$,

so $\theta \approx -1.25$. The curves intersect at an angle of 1.25 radians.

29. $f(x) = x\sin x$

a.

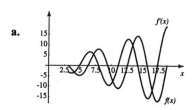

b. $f(x) = 0$ has 6 solutions on $[\pi, 6\pi]$
$f'(x) = 0$ has 5 solutions on $[\pi, 6\pi]$

c. $f(x) = x\sin x$ is a counterexample.

d. The maximum value of $|f(x) - f'(x)|$ on $[\pi, 6\pi]$ is about 24.93.

3.5 Concepts Review

1. $D_t u$; $f'(g(t))g'(t)$

3. $(f(x))^2$; $(f(x))^2$

Problem Set 3.5

1. $y = u^{15}$ and $u = 1 + x$
$D_x y = D_u y \cdot D_x u$
$= (15u^{14})(1)$
$= 15(1 + x)^{14}$

3. $y = u^5$ and $u = 3 - 2x$
$D_x y = D_u y \cdot D_x u$
$= (5u^4)(-2) = -10(3 - 2x)^4$

5. $y = u^{11}$ and $u = x^3 - 2x^2 + 3x + 1$
$D_x y = D_u y \cdot D_x u$
$= (11u^{10})(3x^2 - 4x + 3)$
$= 11(3x^2 - 4x + 3)(x^3 - 2x^2 + 3x + 1)^{10}$

7. $y = u^{111}$ and $u = x^3 - 2x^2 + 3x + 1$
$D_x y = D_u y \cdot D_x u$
$= (111u^{110})(3x^2 - 4x + 3)$
$= 111(3x^2 - 4x + 3)(x^3 - 2x^2 + 3x + 1)^{110}$

9. $y = u^{-5}$ and $u = x + 3$
$D_x y = D_u y \cdot D_x u$
$= (-5u^{-6})(1) = -5(x + 3)^{-6} = -\dfrac{5}{(x + 3)^6}$

11. $y = \sin u$ and $u = x^2 + x$

$D_x y = D_u y \cdot D_x u$

$= (\cos u)(2x + 1)$

$= (2x + 1)\cos(x^2 + x)$

13. $y = u^3$ and $u = \cos x$

$D_x y = D_u y \cdot D_x u$

$= (3u^2)(-\sin x)$

$= -3\sin x \cos^2 x$

15. $y = u^3$ and $u = \dfrac{x+1}{x-1}$

$D_x y = D_u y \cdot D_x u$

$= (3u^2)\dfrac{(x-1)D_x(x+1) - (x+1)D_x(x-1)}{(x-1)^2}$

$= 3\left(\dfrac{x+1}{x-1}\right)^2 \left(\dfrac{-2}{(x-1)^2}\right) = -\dfrac{6(x+1)^2}{(x-1)^4}$

17. $y = \cos u$ and $u = \dfrac{3x^2}{x+2}$

$D_x y = D_u y \cdot D_x u = (-\sin u)\dfrac{(x+2)D_x(3x^2) - (3x^2)D_x(x+2)}{(x+2)^2}$

$= -\sin\left(\dfrac{3x^2}{x+2}\right)\dfrac{(x+2)(6x) - (3x^2)(1)}{(x+2)^2} = -\dfrac{3x^2 + 12x}{(x+2)^2}\sin\left(\dfrac{3x^2}{x+2}\right)$

19. $D_x[(3x-2)^2(3-x^2)^2] = (3x-2)^2 D_x(3-x^2)^2 + (3-x^2)^2 D_x(3x-2)^2$

$= (3x-2)^2(2)(3-x^2)(-2x) + (3-x^2)^2(2)(3x-2)(3)$

$= 2(3x-2)(3-x^2)[(3x-2)(-2x) + (3-x^2)(3)] = 2(3x-2)(3-x^2)(9 + 4x - 9x^2)$

21. $D_x\left[\dfrac{(x+1)^2}{3x-4}\right] = \dfrac{(3x-4)D_x(x+1)^2 - (x+1)^2 D_x(3x-4)}{(3x-4)^2} = \dfrac{(3x-4)(2)(x+1)(1) - (x+1)^2(3)}{(3x-4)^2} = \dfrac{3x^2 - 8x - 11}{(3x-4)^2}$

$= \dfrac{(x+1)(3x-11)}{(3x-4)^2}$

23. $D_t\left(\dfrac{3t-2}{t+5}\right)^3 = 3\left(\dfrac{3t-2}{t+5}\right)^2 \dfrac{(t+5)D_t(3t-2) - (3t-2)D_t(t+5)}{(t+5)^2}$

$= 3\left(\dfrac{3t-2}{t+5}\right)^2 \dfrac{(t+5)(3) - (3t-2)(1)}{(t+5)^2} = \dfrac{51(3t-2)^2}{(t+5)^4}$

25. $D_t\left(\dfrac{(3t-2)^3}{t+5}\right) = \dfrac{(t+5)D_t(3t-2)^3 - (3t-2)^3 D_t(t+5)}{(t+5)^2} = \dfrac{(t+5)(3)(3t-2)^2(3) - (3t-2)^3(1)}{(t+5)^2}$

$= \dfrac{(6t+47)(3t-2)^2}{(t+5)^2}$

27. $D_x\left(\dfrac{\sin x}{\cos 2x}\right)^3 = 3\left(\dfrac{\sin x}{\cos 2x}\right)^2 \dfrac{(\cos 2x)D_x(\sin x) - (\sin x)D_x(\cos 2x)}{\cos^2 2x}$

$= 3\left(\dfrac{\sin x}{\cos 2x}\right)^2 \dfrac{\cos x \cos 2x + 2\sin x \sin 2x}{\cos^2 2x} = \dfrac{3\sin^2 x \cos x \cos 2x + 6\sin^3 x \sin 2x}{\cos^4 2x}$

$= \dfrac{3(\sin^2 x)(\cos x \cos 2x + 2\sin x \sin 2x)}{\cos^4 2x}$

29. $f'(x) = 3\left(\dfrac{x^2+1}{x+2}\right)^2 \dfrac{(x+2)D_x(x^2+1)-(x^2+1)D_x(x+2)}{(x+2)^2}$

$= 3\left(\dfrac{x^2+1}{x+2}\right)^2 \dfrac{2x^2+4x-x^2-1}{(x+2)^2} = \dfrac{3(x^2+1)^2(x^2+4x-1)}{(x+2)^4}$

$f'(3) = 9.6$

31. $F'(t) = [\cos(t^2+3t+1)](2t+3) = (2t+3)\cos(t^2+3t+1)$

$F'(1) = 5\cos 5 \approx 1.4183$

33. $D_x[\sin^4(x^2+3x)] = 4\sin^3(x^2+3x)D_x\sin(x^2+3x) = 4\sin^3(x^2+3x)\cos(x^2+3x)D_x(x^2+3x)$

$= 4\sin^3(x^2+3x)\cos(x^2+3x)(2x+3) = 4(2x+3)\sin^3(x^2+3x)\cos(x^2+3x)$

35. $D_t[\sin^3(\cos t)] = 3\sin^2(\cos t)D_t\sin(\cos t) = 3\sin^2(\cos t)\cos(\cos t)D_t(\cos t)$

$= 3\sin^2(\cos t)\cos(\cos t)(-\sin t) = -3\sin t\sin^2(\cos t)\cos(\cos t)$

37. $D_\theta[\cos^4(\sin\theta^2)] = 4\cos^3(\sin\theta^2)D_\theta\cos(\sin\theta^2) = 4\cos^3(\sin\theta^2)[-\sin(\sin\theta^2)]D_\theta(\sin\theta^2)$

$= -4\cos^3(\sin\theta^2)\sin(\sin\theta^2)(\cos\theta^2)D_\theta(\theta^2) = -8\theta\cos^3(\sin\theta^2)\sin(\sin\theta^2)(\cos\theta^2)$

39. $D_x\{\sin[\cos(\sin 2x)]\} = \cos[\cos(\sin 2x)]D_x\cos(\sin 2x) = \cos[\cos(\sin 2x)][-\sin(\sin 2x)]D_x(\sin 2x)$

$= -\cos[\cos(\sin 2x)]\sin(\sin 2x)(\cos 2x)D_x(2x) = -2\cos[\cos(\sin 2x)]\sin(\sin 2x)(\cos 2x)$

41. $D_xy = (x^2+1)^3 D_x(x^4+1)^2 + (x^4+1)^2 D_x(x^2+1)^3 = (x^2+1)^3(2)(x^4+1)(4x^3) + (x^4+1)^2(3)(x^2+1)(2x)$

$= 8x^3(x^2+1)^3(x^4+1) + 6x(x^2+1)(x^4+1)^2$

At $x = 1$: $m_{\tan} = 224$

Tangent line: $y - 32 = 224(x-1)$

43. a. $(10\cos 8\pi t, 10\sin 8\pi t)$

b. $D_t(10\sin 8\pi t) = 10\cos(8\pi t)D_t(8\pi t)$

$= 80\pi\cos(8\pi t)$

At $t = 1$: rate $= 80\pi \approx 251$ cm/s

P is rising at the rate of 251 cm/s.

c. $D_t\left(\sin 2\pi t + \sqrt{25-\cos^2 2\pi t}\right)$

$= 2\pi\cos 2\pi t$

$+\dfrac{1}{2\sqrt{25-\cos^2 2\pi t}}\cdot 4\pi\cos 2\pi t\sin 2\pi t$

$= 2\pi\cos 2\pi t\left(1+\dfrac{\sin 2\pi t}{\sqrt{25-\cos^2 2\pi t}}\right)$

45. 60 revolutions per minute is 120π radians per minute or 2π radians per second.

a. $(\cos 2\pi t, \sin 2\pi t)$

b. $(0-\cos 2\pi t)^2 + (y-\sin 2\pi t)^2 = 5^2$, so

$y = \sin 2\pi t + \sqrt{25-\cos^2 2\pi t}$

47. $V(t) = \dfrac{4}{3}\pi[r(t)]^3$

$[r(t)]^3 = \dfrac{3V(t)}{4\pi}$

$r(t) = \left(\dfrac{3V(t)}{4\pi}\right)^{1/3}$

$r'(t) = \dfrac{1}{3}\left(\dfrac{3V(t)}{4\pi}\right)^{-2/3}\dfrac{3}{4\pi}V'(t)$

Let t_0 be the time when $V = 60$.

$r'(t_0) = \left(\dfrac{3\cdot 60}{4\pi}\right)^{-2/3}\dfrac{1}{4\pi}\cdot 3$

≈ 0.0405 cm/sec

49. a. $D_x\left|x^2-1\right| = \dfrac{\left|x^2-1\right|}{x^2-1}D_x(x^2-1)$

$= \dfrac{\left|x^2-1\right|}{x^2-1}(2x) = \dfrac{2x\left|x^2-1\right|}{x^2-1}$

b. $D_x\left|\sin x\right| = \dfrac{\left|\sin x\right|}{\sin x}D_x(\sin x)$

$= \dfrac{\left|\sin x\right|}{\sin x}\cos x = \cot x\left|\sin x\right|$

51. $[f(f(f(f(0)))))]' = f'(f(f(f(0)))) \cdot$

$f'(f(f(0))) \cdot f'(f(0)) \cdot f'(0)$

$= 2 \cdot 2 \cdot 2 \cdot 2 = 16$

53. $f(x) = \sin(\sin(\sin(\sin x)))$

a.

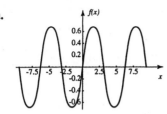

Odd function

$f(-x) = \sin(\sin(\sin(\sin(-x))))$

$= \sin(\sin(\sin(-\sin x))) = \sin(\sin(-\sin(\sin x)))$

$= \sin(-\sin(\sin(\sin x))) = -\sin(\sin(\sin(\sin x)))$

b.

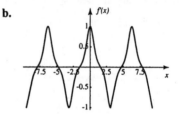

Even; the derivative of an odd function is even.

c. Largest value of $f(x) \approx 0.678$

d. Largest value of $\left|f'(x)\right| = 1$

3.6 Concepts Review

1. Increment; $\dfrac{\Delta y}{\Delta x}$; $\dfrac{dy}{dx}$

3. $\dfrac{dy}{du}\dfrac{du}{dx}$

Problem Set 3.6

1. $\Delta y = [3(1.5)+2]-[3(1)+2] = 1.5$

3. $\Delta y = \dfrac{3}{2.31+1}-\dfrac{3}{2.34+1} \approx 0.0081$

5. $\dfrac{\Delta y}{\Delta x} = \dfrac{(x+\Delta x)^2-x^2}{\Delta x} = \dfrac{2x\Delta x+(\Delta x)^2}{\Delta x} = 2x+\Delta x$

$\dfrac{dy}{dx} = \lim_{\Delta x \to 0}(2x+\Delta x) = 2x$

7. $\dfrac{\Delta y}{\Delta x} = \dfrac{\frac{1}{x+\Delta x+1}-\frac{1}{x+1}}{\Delta x}$

$= \left(\dfrac{x+1-(x+\Delta x+1)}{(x+\Delta x+1)(x+1)}\right)\left(\dfrac{1}{\Delta x}\right)$

$= \dfrac{-\Delta x}{(x+\Delta x+1)(x+1)\Delta x}$

$= -\dfrac{1}{(x+\Delta x+1)(x+1)}$

$\dfrac{dy}{dx} = \lim_{\Delta x \to 0}\left[-\dfrac{1}{(x+\Delta x+1)(x+1)}\right] = -\dfrac{1}{(x+1)^2}$

9. $\dfrac{dy}{dx} = \dfrac{dy}{du}\dfrac{du}{dx} = (2u)(\cos x) = 2\sin x\cos x = \sin 2x$

11. $y = \tan u$ and $u = x^2$

$\dfrac{dy}{dx} = \dfrac{dy}{du}\dfrac{du}{dx} = (\sec^2 u)(2x) = 2x\sec^2(x^2)$

13. $y = u^4$ and $u = \dfrac{x^2+1}{\cos x}$

$\dfrac{dy}{dx} = \dfrac{dy}{du}\dfrac{du}{dx} = (4u^3)\dfrac{(\cos x)(2x)-(x^2+1)(-\sin x)}{\cos^2 x}$

$= 4\left(\dfrac{x^2+1}{\cos x}\right)^3\left(\dfrac{2x\cos x+\sin x+x^2\sin x}{\cos^2 x}\right)$

$= \dfrac{4(x^2+1)^3(2x\cos x+\sin x+x^2\sin x)}{\cos^5 x}$

15. $\dfrac{dy}{dx} = \cos(x^2)\dfrac{d}{dx}(\sin^2 x) + \sin^2 x\dfrac{d}{dx}[\cos(x^2)]$

$\qquad = \cos(x^2)2\sin x\cos x + \sin^2 x[-\sin(x^2)](2x)$

$\qquad = \sin 2x\cos(x^2) - 2x\sin^2 x\sin(x^2)$

17. $y = u^4$, $u = \sin v$, and $v = x^2 + 3$

$\qquad \dfrac{dy}{dx} = \dfrac{dy}{du}\dfrac{du}{dv}\dfrac{dv}{dx} = (4u^3)(\cos v)(2x)$

$\qquad = 4\sin^3(x^2 + 3)\cos(x^2 + 3)(2x)$

$\qquad = 8x\sin^3(x^2 + 3)\cos(x^2 + 3)$

19. $y = u^2$, $u = \cos v$, and $v = \dfrac{x^2 + 2}{x^2 - 2}$

$\qquad \dfrac{dy}{dx} = \dfrac{dy}{du}\dfrac{du}{dv}\dfrac{dv}{dx}$

$\qquad = (2u)(-\sin v)\dfrac{(x^2 - 2)(2x) - (x^2 + 2)(2x)}{(x^2 - 2)^2}$

$\qquad = -2\cos\left(\dfrac{x^2 + 2}{x^2 - 2}\right)\sin\left(\dfrac{x^2 + 2}{x^2 - 2}\right)\left[\dfrac{-8x}{(x^2 - 2)^2}\right]$

$\qquad = \dfrac{16x}{(x^2 - 2)^2}\cos\left(\dfrac{x^2 + 2}{x^2 - 2}\right)\sin\left(\dfrac{x^2 + 2}{x^2 - 2}\right)$

21. $\dfrac{d}{dt}(\sin^3 t + \cos^3 t) = 3\sin^2 t\cos t + 3\cos^2 t(-\sin t)$

$\qquad = 3(\sin^2 t\cos t - \sin t\cos^2 t)$

23. $D_r[\pi(r + 3)^2 - 3\pi r(r + 2)^2]$

$\qquad = 2\pi(r + 3)(1) - [3\pi r(2)(r + 2) + (r + 2)^2(3\pi)]$

$\qquad = 2\pi r + 6\pi - 6\pi r^2 - 12\pi r - 3\pi r^2 - 12\pi r - 12\pi$

$\qquad = -9\pi r^2 - 22\pi r - 6\pi$

25. $f'(x) = 4\left(x + \dfrac{1}{x}\right)^3[1 + (-1)x^{-2}]$

$\qquad = 4\left(x + \dfrac{1}{x}\right)^3\left(1 - \dfrac{1}{x^2}\right)$

$\qquad f'(2) = 4\left(\dfrac{5}{2}\right)^3\left(\dfrac{3}{4}\right) = \dfrac{1500}{32} = 46.875$

27. a. $(f + g)'(3) = f'(3) + g'(3) = -1 + (-4) = -5$

b. $(f \cdot g)'(3) = f(3)g'(3) + g(3)f'(3)$

$\qquad = (2)(-4) + (3)(-1) = -11$

c. $(f/g)'(3) = \dfrac{g(3)f'(3) - f(3)g'(3)}{g^2(3)}$

$\qquad = \dfrac{(3)(-1) - (2)(-4)}{(3)^2} = \dfrac{5}{9}$

d. $(f \circ g)'(3) = f'(g(3)) \cdot g'(3)$

$\qquad = f'(3) \cdot g'(3) = (-1)(-4) = 4$

29. a. $(f + g)'(4) = f'(4) + g'(4)$

$\qquad \approx \dfrac{1}{2} + \dfrac{3}{2} \approx 2$

b. $(f \circ g)'(6) = f'(g(6))g'(6)$

$\qquad = f'(2)g'(6) \approx (1)(-1) = -1$

31. a. $V = s^3$ where s is the length of an edge of the cube. $\dfrac{ds}{dt} = 16$ cm/min.

$\qquad \dfrac{dV}{dt} = \dfrac{dV}{ds}\dfrac{ds}{dt} = (3s^2)(16) = 48s^2 \text{ cm}^3/\text{min}$.

$\qquad$ When $s = 20$,

$\qquad \dfrac{dV}{dt} = 48(20)^2 = 19,200 \text{ cm}^3/\text{min}$.

b. $A = 6s^2 \text{ cm}^2$.

$\qquad \dfrac{dA}{dt} = \dfrac{dA}{ds}\dfrac{ds}{dt} = (12s)(16) = 192s \text{ cm}^2/\text{min}$.

$\qquad$ When $s = 15$,

$\qquad \dfrac{dA}{dt} = 192(15) = 2880 \text{ cm}^2/\text{min}$.

33. $\dfrac{dy}{dx} = (x^2)(2)\sin(x^2)\cos(x^2)(2x) + 2x\sin^2(x^2)$

$\qquad = 4x^3\sin(x^2)\cos(x^2) + 2x\sin^2(x^2)$

$\qquad$ At $x = \sqrt{\dfrac{\pi}{2}}: m_{\text{tan}} = \sqrt{2\pi} \approx 2.507$

$\qquad y = \dfrac{\pi}{2}$

$\qquad$ Tangent line: $\quad y - \dfrac{\pi}{2} = \sqrt{2\pi}\left(x - \sqrt{\dfrac{\pi}{2}}\right)$

$\qquad$ The line intersects the x-axis when

$\qquad -\dfrac{\pi}{2} = \sqrt{2\pi}x - \pi$

$\qquad \dfrac{\pi}{2} = \sqrt{2\pi}x$

$\qquad x = \dfrac{\sqrt{2\pi}}{4}$

35. $g'(x) = f'(f(f(f(x)))) \cdot f'(f(f(x))) \cdot f'(f(x)) \cdot f'(x)$

$g'(x_1) = f'(f(f(f(x_1)))) \cdot f'(f(f(x_1))) \cdot f'(f(x_1)) \cdot f'(x_1)$

$= f'(f(f(x_2))) \cdot f'(f(x_2)) \cdot f'(x_2) \cdot f'(x_1)$

$= f'(f(x_1)) \cdot f'(x_1) \cdot f'(x_2) \cdot f'(x_1)$

$= f'(x_2)f'(x_1)f'(x_2)f'(x_1) = [f'(x_1)f'(x_2)]^2$

$g'(x_2) = f'(f(f(f(x_2)))) \cdot f'(f(f(x_2))) \cdot f'(f(x_2)) \cdot f'(x_2)$

$= f'(f(f(x_1))) \cdot f'(f(x_1)) \cdot f'(x_1) \cdot f'(x_2)$

$= f'(f(x_2)) \cdot f'(x_2) \cdot f'(x_1) \cdot f'(x_2)$

$= f'(x_1)f'(x_2)f'(x_1)f'(x_2) = [f'(x_1)f'(x_2)]^2$

37. The minute hand makes 1 revolution every hour, so at t minutes after noon it makes an angle of $\dfrac{\pi t}{30}$ radians with the vertical. Similarly, at t minutes after noon the hour hand makes an angle of $\dfrac{\pi t}{360}$ with the vertical. Thus, by the Law of Cosines, the distance between the tips of the hands is

$$s = \sqrt{6^2 + 8^2 - 2 \cdot 6 \cdot 8 \cos\left(\frac{\pi t}{30} - \frac{\pi t}{360}\right)}$$

$$= \sqrt{100 - 96\cos\frac{11\pi t}{360}}$$

$$\frac{ds}{dt} = \frac{1}{2\sqrt{100 - 96\cos\frac{11\pi t}{360}}} \cdot \frac{44\pi}{15}\sin\frac{11\pi t}{360}$$

$$= \frac{22\pi\sin\frac{11\pi t}{360}}{15\sqrt{100 - 96\cos\frac{11\pi t}{360}}}$$

At 12:20,

$$\frac{ds}{dt} = \frac{22\pi\sin\frac{11\pi}{18}}{15\sqrt{100 - 96\cos\frac{11\pi}{18}}} \approx 0.38 \text{ in./min}$$

3.7 Concepts Review

1. $f'''(x), D_x^3 y, \dfrac{d^3 y}{dx^3}$

3. $0; < 0$

Problem Set 3.7

1. $\dfrac{dy}{dx} = 3x^2 + 6x + 6$

$\dfrac{d^2 y}{dx^2} = 6x + 6$

$\dfrac{d^3 y}{dx^3} = 6$

3. $\dfrac{dy}{dx} = 3(3x + 5)^2(3) = 9(3x + 5)^2$

$\dfrac{d^2 y}{dx^2} = 18(3x + 5)(3) = 162x + 270$

$\dfrac{d^3 y}{dx^3} = 162$

5. $\dfrac{dy}{dx} = 7\cos(7x)$

$\dfrac{d^2 y}{dx^2} = -7^2 \sin(7x)$

$\dfrac{d^3 y}{dx^3} = -7^3 \cos(7x) = -343\cos(7x)$

7. $\dfrac{dy}{dx} = \dfrac{(x-1)(0) - (1)(1)}{(x-1)^2} = -\dfrac{1}{(x-1)^2}$

$\dfrac{d^2 y}{dx^2} = -\dfrac{(x-1)^2(0) - 2(x-1)}{(x-1)^4} = \dfrac{2}{(x-1)^3}$

$\dfrac{d^3 y}{dx^3} = \dfrac{(x-1)^3(0) - 2[3(x-1)^2]}{(x-1)^6}$

$= -\dfrac{6}{(x-1)^4}$

9. $f'(x) = 2x; f''(x) = 2; f''(2) = 2$

11. $f'(t) = -\dfrac{2}{t^2}$

$f''(t) = \dfrac{4}{t^3}$

$f''(2) = \dfrac{4}{8} = \dfrac{1}{2}$

13. $f'(\theta) = -2(\cos\theta\pi)^{-3}(-\sin\theta\pi)\pi = 2\pi(\cos\theta\pi)^{-3}(\sin\theta\pi)$

$f''(\theta) = 2\pi[(\cos\theta\pi)^{-3}(\pi)(\cos\theta\pi) + (\sin\theta\pi)(-3)(\cos\theta\pi)^{-4}(-\sin\theta\pi)(\pi)] = 2\pi^2[(\cos\theta\pi)^{-2} + 3\sin^2\theta\pi(\cos\theta\pi)^{-4}]$

$f''(2) = 2\pi^2[1 + 3(0)(1)] = 2\pi^2$

15. $f'(s) = s(3)(1-s^2)^2(-2s) + (1-s^2)^3 = -6s^2(1-s^2)^2 + (1-s^2)^3 = -7s^6 + 15s^4 - 9s^2 + 1$

$f''(s) = -42s^5 + 60s^3 - 18s$

$f''(2) = -900$

17. $D_x(x^n) = nx^{n-1}$

$D_x^2(x^n) = n(n-1)x^{n-2}$

$D_x^3(x^n) = n(n-1)(n-2)x^{n-3}$

$D_x^4(x^n) = n(n-1)(n-2)(n-3)x^{n-4}$

$\vdots$

$D_x^{n-1}(x^n) = n(n-1)(n-2)(n-3)\ldots(2)x$

$D_x^n(x^n) = n(n-1)(n-2)(n-3)\ldots2(1)x^0$

$= n!$

19. a. $D_x^4(3x^3 + 2x - 19) = 0$

b. $D_x^{12}(100x^{11} - 79x^{10}) = 0$

c. $D_x^{11}(x^2 - 3)^5 = 0$

21. $f'(x) = 3x^2 + 6x - 45 = 3(x+5)(x-3)$

$3(x+5)(x-3) = 0$

$x = -5, x = 3$

$f''(x) = 6x + 6$

$f''(-5) = -24$

$f''(3) = 24$

23. a. $v(t) = \dfrac{ds}{dt} = 12 - 4t$

$a(t) = \dfrac{d^2s}{dt^2} = -4$

b. $12 - 4t > 0$

$4t < 12$

$t < 3; \ (-\infty, 3)$

c. $12 - 4t < 0$

$t > 3; \ (3, \infty)$

d. $a(t) = -4 < 0$ for all t

e.

25. a. $v(t) = \dfrac{ds}{dt} = 3t^2 - 18t + 24$

$a(t) = \dfrac{d^2s}{dt^2} = 6t - 18$

b. $3t^2 - 18t + 24 > 0$

$3(t-2)(t-4) > 0$

$(-\infty, 2) \cup (4, \infty)$

c. $3t^2 - 18t + 24 < 0$

$(2, 4)$

d. $6t - 18 < 0$

$6t < 18$

$t < 3; \ (-\infty, 3)$

e.

27. a. $v(t) = \dfrac{ds}{dt} = 2t - \dfrac{16}{t^2}$

$a(t) = \dfrac{d^2s}{dt^2} = 2 + \dfrac{32}{t^3}$

b. $2t - \dfrac{16}{t^2} > 0$

$\dfrac{2t^3 - 16}{t^2} > 0; \ (2, \infty)$

c. $2t - \dfrac{16}{t^2} < 0; \ (0, 2)$

d. $2 + \dfrac{32}{t^3} < 0$

$\dfrac{2t^3 + 32}{t^3} < 0$; The acceleration is not negative for any positive t.

e.

29. $v(t) = \dfrac{ds}{dt} = 2t^3 - 15t^2 + 24t$

$a(t) = \dfrac{d^2 s}{dt^2} = 6t^2 - 30t + 24$

$6t^2 - 30t + 24 = 0$
$6(t - 4)(t - 1) = 0$
$t = 4, 1$
$v(4) = -16, \ v(1) = 11$

31. $v_1(t) = \dfrac{ds_1}{dt} = 4 - 6t$

$v_2(t) = \dfrac{ds_2}{dt} = 2t - 2$

a. $4 - 6t = 2t - 2$
$8t = 6$
$t = \dfrac{3}{4}$ sec

b. $|4 - 6t| = |2t - 2|; \quad 4 - 6t = -2t + 2$

$t = \dfrac{1}{2}$ sec and $t = \dfrac{3}{4}$ sec

c. $4t - 3t^2 = t^2 - 2t$
$4t^2 - 6t = 0$
$2t(2t - 3) = 0$
$t = 0$ sec and $t = \dfrac{3}{2}$ sec

33. a. $v(t) = -32t + 48$
initial velocity = $v_0 = 48$ ft/sec

b. $-32t + 48 = 0$
$t = \dfrac{3}{2}$ sec

c. $s = -16(1.5)^2 + 48(1.5) + 256 = 292$ ft

d. $-16t^2 + 48t + 256 = 0$

$t = \dfrac{-48 \pm \sqrt{48^2 - 4(-16)(256)}}{-32} \approx -2.77, 5.77$

The object hits the ground at $t = 5.77$ sec.

e. $v(5.77) \approx -137$ ft/sec;
speed = $|-137| = 137$ ft/sec.

35. $v(t) = v_0 - 32t$

$v_0 - 32t = 0$

$t = \dfrac{v_0}{32}$

$v_0 \left(\dfrac{v_0}{32}\right) - 16 \left(\dfrac{v_0}{32}\right)^2 = 5280$

$\dfrac{v_0^2}{32} - \dfrac{v_0^2}{64} = 5280$

$\dfrac{v_0^2}{64} = 5280$

$v_0 = \sqrt{337{,}920} \approx 581$ ft/sec

37. $v(t) = 3t^2 - 6t - 24$

$\dfrac{d}{dt} \left| 3t^2 - 6t - 24 \right| = \dfrac{\left| 3t^2 - 6t - 24 \right|}{3t^2 - 6t - 24}(6t - 6)$

$= \dfrac{\left| (t - 4)(t + 2) \right|}{(t - 4)(t + 2)}(6t - 6)$

$\dfrac{\left| (t - 4)(t + 2) \right| (6t - 6)}{(t - 4)(t + 2)} < 0$

$t < -2, \ 1 < t < 4; \ (-\infty, -2) \cup (1, 4)$

39. Let s be the distance traveled. Then $\dfrac{ds}{dt}$ is the speed of the car.

a. $\dfrac{ds}{dt} = ks, \ k$ a constant

b. $\dfrac{d^2 s}{dt^2} > 0$

c. $\dfrac{d^3 s}{dt^3} < 0, \dfrac{d^2 s}{dt^2} > 0$

d. $\dfrac{d^2 s}{dt^2} = 10$ mph/min

e. $\dfrac{ds}{dt}$ and $\dfrac{d^2 s}{dt^2}$ are approaching zero.

f. $\dfrac{ds}{dt}$ is constant.

41. a. $\dfrac{dC}{dt} > 0, \dfrac{d^2 C}{dt^2} > 0$, where C is the car's cost.

b. $f(t)$ is oil consumption at time t.

$$\frac{df}{dt} < 0, \frac{d^2f}{dt^2} > 0$$

c. $\frac{dP}{dt} > 0, \frac{d^2P}{dt^2} < 0$, where P is world population.

d. $\frac{d\theta}{dt} > 0, \frac{d^2\theta}{dt^2} > 0$, where θ is the angle that the tower makes with the vertical.

e. $P = f(t)$ is profit at time t.

$$\frac{dP}{dt} > 0, \frac{d^2P}{dt^2} < 0$$

f. R is revenue at time t.

$$R < 0, \frac{dR}{dt} > 0$$

43. $D_x(uv) = uv' + u'v$

$D_x^2(uv) = uv'' + u'v' + u'v' + u''v$
$\quad = uv'' + 2u'v' + u''v$

$D_x^3(uv) = uv''' + u'v'' + 2(u'v'' + u''v') + u''v' + u'''v$
$\quad = uv''' + 3u'v'' + 3u''v' + u'''v$

$D_x^n(uv) = \sum_{k=0}^{n} \binom{n}{k} D_x^{n-k}(u) D_x^k(v)$

where $\binom{n}{k}$ is the binomial coefficient

$$\frac{n!}{(n-k)!k!}.$$

45. a.

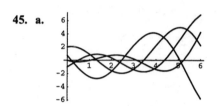

b. $f'''(2.13) \approx -1.2826$

3.8 Concepts Review

1. $\dfrac{9}{x^3 - 3}$

3. $x(2y)\dfrac{dy}{dx} + y^2 + 3y^2\dfrac{dy}{dx} - \dfrac{dy}{dx} = 3x^2$

Problem Set 3.8

1. $2y D_x y - 2x = 0$

$$D_x y = \frac{2x}{2y} = \frac{x}{y}$$

3. $x D_x y + y = 0$

$$D_x y = -\frac{y}{x}$$

5. $x(2y)D_x y + y^2 = 1$

$$D_x y = \frac{1 - y^2}{2xy}$$

7. $12x^2 + 7x(2y)D_x y + 7y^2 = 6y^2 D_x y$

$12x^2 + 7y^2 = 6y^2 D_x y - 14xy D_x y$

$$D_x y = \frac{12x^2 + 7y^2}{6y^2 - 14xy}$$

9. $\dfrac{1}{2\sqrt{5xy}} \cdot (5x D_x y + 5y) + 2 D_x y$

$= 2y D_x y + x(3y^2)D_x y + y^3$

$\dfrac{5x}{2\sqrt{5xy}} D_x y + 2 D_x y - 2y D_x y - 3xy^2 D_x y$

$= y^3 - \dfrac{5y}{2\sqrt{5xy}}$

$$D_x y = \frac{y^3 - \dfrac{5y}{2\sqrt{5xy}}}{\dfrac{5x}{2\sqrt{5xy}} + 2 - 2y - 3xy^2}$$

11. $x D_x y + y + \cos(xy)(x D_x y + y) = 0$

$x D_x y + x\cos(xy)D_x y = -y - y\cos(xy)$

$$D_x y = \frac{-y - y\cos(xy)}{x + x\cos(xy)} = -\frac{y}{x}$$

13. $x^3y' + 3x^2y + y^3 + 3xy^2y' = 0$

$y'(x^3 + 3xy^2) = -3x^2y - y^3$

$y' = \dfrac{-3x^2y - y^3}{x^3 + 3xy^2}$

At $(1, 3)$, $y' = -\dfrac{36}{28} = -\dfrac{9}{7}$

Tangent line: $y - 3 = -\dfrac{9}{7}(x - 1)$

15. $\cos(xy)(xy' + y) = y'$

$y'[x\cos(xy) - 1] = -y\cos(xy)$

$y' = \dfrac{-y\cos(xy)}{x\cos(xy) - 1} = \dfrac{y\cos(xy)}{1 - x\cos(xy)}$

At $\left(\dfrac{\pi}{2}, 1\right)$, $y' = 0$

Tangent line: $y - 1 = 0\left(x - \dfrac{\pi}{2}\right)$

$y = 1$

17. $\dfrac{2}{3}x^{-1/3} - \dfrac{2}{3}y^{-1/3}y' - 2y' = 0$

$\dfrac{2}{3}x^{-1/3} = y'\left(\dfrac{2}{3}y^{-1/3} + 2\right)$

$y' = \dfrac{\frac{2}{3}x^{-1/3}}{\frac{2}{3}y^{-1/3} + 2}$

At $(1, -1)$, $y' = \dfrac{\frac{2}{3}}{\frac{4}{3}} = \dfrac{1}{2}$

Tangent line: $y + 1 = \dfrac{1}{2}(x - 1)$

19. $\dfrac{dy}{dx} = 5x^{2/3} + \dfrac{1}{2\sqrt{x}}$

21. $\dfrac{dy}{dx} = \dfrac{1}{3}x^{-2/3} - \dfrac{1}{3}x^{-4/3} = \dfrac{1}{3\sqrt[3]{x^2}} - \dfrac{1}{3\sqrt[3]{x^4}}$

23. $\dfrac{dy}{dx} = \dfrac{1}{4}(3x^2 - 4x)^{-3/4}(6x - 4)$

$= \dfrac{6x - 4}{4\sqrt[4]{(3x^2 - 4x)^3}} = \dfrac{3x - 2}{2\sqrt[4]{(3x^2 - 4x)^3}}$

25. $\dfrac{dy}{dx} = \dfrac{d}{dx}[(x^3 + 2x)^{-2/3}]$

$= -\dfrac{2}{3}(x^3 + 2x)^{-5/3}(3x^2 + 2) = -\dfrac{6x^2 + 4}{3\sqrt[3]{(x^3 + 2x)^5}}$

27. $\dfrac{dy}{dx} = \dfrac{1}{2\sqrt{x^2 + \sin x}}(2x + \cos x)$

$= \dfrac{2x + \cos x}{2\sqrt{x^2 + \sin x}}$

29. $\dfrac{dy}{dx} = \dfrac{d}{dx}[(x^2\sin x)^{-1/3}]$

$= -\dfrac{1}{3}(x^2\sin x)^{-4/3}(x^2\cos x + 2x\sin x)$

$= -\dfrac{x^2\cos x + 2x\sin x}{3\sqrt[3]{(x^2\sin x)^4}}$

31. $\dfrac{dy}{dx} = \dfrac{[1 + \cos(x^2 + 2x)]^{-3/4}[-\sin(x^2 + 2x)(2x + 2)]}{4}$

$= -\dfrac{(x + 1)\sin(x^2 + 2x)}{2\sqrt[4]{[1 + \cos(x^2 + 2x)]^3}}$

33. $s^2 + 2st\dfrac{ds}{dt} + 3t^2 = 0$

$\dfrac{ds}{dt} = \dfrac{-s^2 - 3t^2}{2st} = -\dfrac{s^2 + 3t^2}{2st}$

$s^2\dfrac{dt}{ds} + 2st + 3t^2\dfrac{dt}{ds} = 0$

$\dfrac{dt}{ds}(s^2 + 3t^2) = -2st$

$\dfrac{dt}{ds} = -\dfrac{2st}{s^2 + 3t^2}$

35.

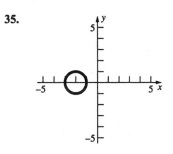

$2x + 4 + 2y\dfrac{dy}{dx} = 0$

$\dfrac{dy}{dx} = -\dfrac{2x + 4}{2y} = -\dfrac{x + 2}{y}$

The tangent line at (x_0, y_0) has equation

$y - y_0 = -\dfrac{x_0 + 2}{y_0}(x - x_0)$ which simplifies to

$2x_0 - yy_0 - 2x - xx_0 + y_0^2 + x_0^2 = 0$. Since

(x_0, y_0) is on the circle, $x_0^2 + y_0^2 = -3 - 4x_0$,

so the equation of the tangent line is

$-yy_0 - 2x_0 - 2x - xx_0 = 3$.

If $(0, 0)$ is on the tangent line, then $x_0 = -\dfrac{3}{2}$.

Solve for y_0 in the equation of the circle to get $y_0 = \pm\dfrac{\sqrt{3}}{2}$. Put these values into the equation of the tangent line to get that the tangent lines are $\sqrt{3}y + x = 0$ and $\sqrt{3}y - x = 0$.

37. a. $\quad xy' + y + 3y^2 y' = 0$

$$y'(x + 3y^2) = -y$$

$$y' = -\frac{y}{x + 3y^2}$$

b. $\quad xy'' + \left(\dfrac{-y}{x + 3y^2}\right) + \left(\dfrac{-y}{x + 3y^2}\right) + 3y^2 y''$

$$+ 6y\left(\frac{-y}{x + 3y^2}\right)^2 = 0$$

$$xy'' + 3y^2 y'' - \frac{2y}{x + 3y^2} + \frac{6y^3}{(x + 3y^2)^2} = 0$$

$$y''(x + 3y^2) = \frac{2y}{x + 3y^2} - \frac{6y^3}{(x + 3y^2)^2}$$

$$y''(x + 3y^2) = \frac{2xy}{(x + 3y^2)^2}$$

$$y'' = \frac{2xy}{(x + 3y^2)^3}$$

39. $\quad 2(x^2 y' + 2xy) - 12y^2 y' = 0$

$$2x^2 y' - 12y^2 y' = -4xy$$

$$y' = \frac{2xy}{6y^2 - x^2}$$

$$2(x^2 y'' + 2xy' + 2xy' + 2y) - 12[y^2 y'' + 2y(y')^2] = 0$$

$$2x^2 y'' - 12y^2 y'' = -8xy' - 4y + 24y(y')^2$$

$$y''(2x^2 - 12y^2) = -\frac{16x^2 y}{6y^2 - x^2} - 4y + \frac{96x^2 y^3}{(6y^2 - x^2)^2}$$

$$y''(2x^2 - 12y^2) = \frac{12x^4 y + 48x^2 y^3 - 144y^5}{(6y^2 - x^2)^2}$$

$$y''(6y^2 - x^2) = \frac{72y^5 - 6x^4 y - 24x^2 y^3}{(6y^2 - x^2)^2}$$

$$y'' = \frac{72y^5 - 6x^4 y - 24x^2 y^3}{(6y^2 - x^2)^3}$$

At $(2, 1)$, $y'' = \dfrac{-120}{8} = -15$

41. $\quad 3x^2 + 3y^2 y' = 3(xy' + y)$

$$y'(3y^2 - 3x) = 3y - 3x^2$$

$$y' = \frac{y - x^2}{y^2 - x}$$

At $\left(\dfrac{3}{2}, \dfrac{3}{2}\right)$, $y' = -1$

Slope of the normal line is 1.

Normal line: $y - \dfrac{3}{2} = 1\left(x - \dfrac{3}{2}\right)$; $y = x$

This line includes the point $(0, 0)$.

43. Implicitly differentiate the first equation.

$$4x + 2yy' = 0$$

$$y' = -\frac{2x}{y}$$

Implicitly differentiate the second equation.

$$2yy' = 4$$

$$y' = \frac{2}{y}$$

Solve for the points of intersection.

$$2x^2 + 4x = 6$$

$$2(x^2 + 2x - 3) = 0$$

$$(x + 3)(x - 1) = 0$$

$$x = -3, x = 1$$

$x = -3$ is extraneous, and $y = -2, 2$ when $x = 1$.
The graphs intersect at $(1, -2)$ and $(1, 2)$.
At $(1, -2)$: $m_1 = 1, m_2 = -1$
At $(1, 2)$: $m_1 = -1, m_2 = 1$

45. $\quad x^2 - x(2x) + 2(2x)^2 = 28$

$$7x^2 = 28$$

$$x^2 = 4$$

$$x = -2, 2$$

Intersection point in first quadrant: $(2, 4)$

$$y_1' = 2$$

$$2x - xy_2' - y + 4yy_2' = 0$$

$$y_2'(4y - x) = y - 2x$$

$$y_2' = \frac{y - 2x}{4y - x}$$

At $(2, 4)$: $m_1 = 2, m_2 = 0$

$$\tan\theta = \frac{0 - 2}{1 + (0)(2)} = -2; \theta = \pi + \tan^{-1}(-2) \approx 2.034$$

47. $x^2 - xy + y^2 = 16$, when $y = 0$,

$x^2 = 16$

$x = -4, 4$

The ellipse intersects the x-axis at $(-4, 0)$ and $(4, 0)$.

$2x - xy' - y + 2yy' = 0$

$y'(2y - x) = y - 2x$

$y' = \dfrac{y - 2x}{2y - x}$

At $(-4, 0)$, $y' = 2$

At $(4, 0)$, $y' = 2$

Tangent lines: $y = 2(x + 4)$ and $y = 2(x - 4)$

49. $2x + 2y\dfrac{dy}{dx} = 0; \dfrac{dy}{dx} = -\dfrac{x}{y}$

The tangent line at (x_0, y_0) has slope $-\dfrac{x_0}{y_0}$,

hence the equation of the tangent line is

$y - y_0 = -\dfrac{x_0}{y_0}(x - x_0)$ which simplifies to

$yy_0 + xx_0 - (x_0{}^2 + y_0{}^2) = 0$ or $yy_0 + xx_0 = 1$

since (x_0, y_0) is on $x^2 + y^2 = 1$. If $(1.25, 0)$ is

on the tangent line through (x_0, y_0), $x_0 = 0.8$.

Put this into $x^2 + y^2 = 1$ to get $y_0 = 0.6$, since

$y_0 > 0$. The line is $6y + 8x = 10$. When $x = -2$,

$y = \dfrac{13}{3}$, so the light bulb must be $\dfrac{13}{3}$ units high.

3.9 Concepts Review

1. $\dfrac{du}{dt}; t = 2$

3. negative

Problem Set 3.9

1. $V = x^3; \dfrac{dx}{dt} = 3$

$\dfrac{dV}{dt} = 3x^2 \dfrac{dx}{dt}$

When $x = 12$, $\dfrac{dV}{dt} = 3(12)^2(3) = 1296$ in.^{3}/s.

3. $y^2 = x^2 + 1^2; \dfrac{dx}{dt} = 400$

$2y\dfrac{dy}{dt} = 2x\dfrac{dx}{dt}$

$\dfrac{dy}{dt} = \dfrac{x}{y}\dfrac{dx}{dt}$ mi/hr

When $x = 5, y = \sqrt{26}, \dfrac{dy}{dt} = \dfrac{5}{\sqrt{26}}(400)$

≈ 392 mi/h.

5. $s^2 = (x + 300)^2 + y^2; \dfrac{dx}{dt} = 300, \dfrac{dy}{dt} = 400,$

$2s\dfrac{ds}{dt} = 2(x + 300)\dfrac{dx}{dt} + 2y\dfrac{dy}{dt}$

$s\dfrac{ds}{dt} = (x + 300)\dfrac{dx}{dt} + y\dfrac{dy}{dt}$

When $x = 300, y = 400, s = 200\sqrt{13}$, so

$200\sqrt{13}\dfrac{ds}{dt} = (300 + 300)(300) + 400(400)$

$\dfrac{ds}{dt} \approx 471$ mi/h

7. $20^2 = x^2 + y^2; \dfrac{dx}{dt} = 1$

$0 = 2x\dfrac{dx}{dt} + 2y\dfrac{dy}{dt}$

When $x = 5$, $y = \sqrt{375} = 5\sqrt{15}$, so

$\dfrac{dy}{dt} = -\dfrac{x}{y}\dfrac{dx}{dt} = -\dfrac{5}{5\sqrt{15}}(1) \approx -0.258$ ft/s

The top of the ladder is moving down at 0.258 ft/s.

9. $V = \dfrac{1}{3}\pi r^2 h; h = \dfrac{d}{4} = \dfrac{r}{2}, r = 2h$

$V = \dfrac{1}{3}\pi(2h)^2 h = \dfrac{4}{3}\pi h^3; \dfrac{dV}{dt} = 16$

$\dfrac{dV}{dt} = 4\pi h^2 \dfrac{dh}{dt}$

When $h = 4$, $16 = 4\pi(4)^2 \dfrac{dh}{dt}$

$\dfrac{dh}{dt} = \dfrac{1}{4\pi} \approx 0.0796$ ft/s

11. $V = \dfrac{hx}{2}(20); \dfrac{40}{5} = \dfrac{x}{h}, x = 8h$

$V = 10h(8h) = 80h^2; \dfrac{dV}{dt} = 40$

$$\frac{dV}{dt} = 160h\frac{dh}{dt}$$

When $h = 3$, $40 = 160(3)\frac{dh}{dt}$

$$\frac{dh}{dt} = \frac{1}{12} \text{ ft/s}$$

13. $A = \pi r^2; \frac{dr}{dt} = 0.02$

$$\frac{dA}{dt} = 2\pi r\frac{dr}{dt}$$

When $r = 8.1$, $\frac{dA}{dt} = 2\pi(0.02)(8.1) = 0.324\pi$

$\approx 1.018 \text{ in.}^2/\text{s}$

15. Let x be the distance from the beam to the point opposite the lighthouse and θ be the angle between the beam and the line from the lighthouse to the point opposite.

$$\tan\theta = \frac{x}{1}; \frac{d\theta}{dt} = 2(2\pi) = 4\pi \text{ rad/min},$$

$$\sec^2\theta\frac{d\theta}{dt} = \frac{dx}{dt}$$

At $x = \frac{1}{2}, \theta = \tan^{-1}\frac{1}{2}$ and $\sec^2\theta = \frac{5}{4}$.

$$\frac{dx}{dt} = \frac{5}{4}(4\pi) \approx 15.71 \text{ km/min}$$

17. **a.** Let x be the distance along the ground from the light pole to Chris, and let s be the distance from Chris to the tip of his shadow. By similar triangles, $\frac{6}{s} = \frac{30}{x+s}$, so $s = \frac{x}{4}$

and $\frac{ds}{dt} = \frac{1}{4}\frac{dx}{dt}$. $\frac{dx}{dt} = 2$ ft/s, hence

$\frac{ds}{dt} = \frac{1}{2}$ ft/s no matter how far from the light pole Chris is.

b. Let $l = x + s$, then

$$\frac{dl}{dt} = \frac{dx}{dt} + \frac{ds}{dt} = 2 + \frac{1}{2} = \frac{5}{2} \text{ ft/s}.$$

c. The angular rate at which Chris must lift his head to follow his shadow is the same as the rate at which the angle that the light makes with the ground is decreasing. Let θ be the angle that the light makes with the ground at the tip of Chris' shadow.

$$\tan\theta = \frac{6}{s} \text{ so } \sec^2\theta\frac{d\theta}{dt} = -\frac{6}{s^2}\frac{ds}{dt} \text{ and}$$

$$\frac{d\theta}{dt} = -\frac{6\cos^2\theta}{s^2}\frac{ds}{dt}. \frac{ds}{dt} = \frac{1}{2} \text{ ft/s}$$

When $s = 6$, $\theta = \frac{\pi}{4}$, so

$$\frac{d\theta}{dt} = -\frac{6\left(\frac{1}{\sqrt{2}}\right)^2}{6^2}\left(\frac{1}{2}\right) = -\frac{1}{24}.$$

Chris must lift his head at the rate of $\frac{1}{24}$ rad/s.

19. Let p be the point on the bridge directly above the railroad tracks. If a is the distance between p and the automobile, then $\frac{da}{dt} = 66$ ft/s. If l is the distance between the train and the point directly below p, then $\frac{dl}{dt} = 88$ ft/s. The distance from the train to p is $\sqrt{100^2 + l^2}$, while the distance from p to the automobile is a. The distance between the train and automobile is

$$D = \sqrt{a^2 + \left(\sqrt{100^2 + l^2}\right)^2} = \sqrt{a^2 + l^2 + 100^2}.$$

$$\frac{dD}{dt} = \frac{1}{2\sqrt{a^2 + l^2 + 100^2}} \cdot \left(2a\frac{da}{dt} + 2l\frac{dl}{dt}\right)$$

$$= \frac{a\frac{da}{dt} + l\frac{dl}{dt}}{\sqrt{a^2 + l^2 + 100^2}}. \text{ After 10 seconds, } a = 660$$

and $l = 880$, so

$$\frac{dD}{dt} = \frac{660(66) + 880(88)}{\sqrt{660^2 + 880^2 + 100^2}} \approx 110 \text{ ft/s}.$$

21. $V = \pi h^2\left[r - \frac{h}{3}\right]; \frac{dV}{dt} = -2, r = 8$

$$V = \pi r h^2 - \frac{\pi h^3}{3} = 8\pi h^2 - \frac{\pi h^3}{3}$$

$$\frac{dV}{dt} = 16\pi h\frac{dh}{dt} - \pi h^2\frac{dh}{dt}$$

When $h = 3$, $-2 = \frac{dh}{dt}[16\pi(3) - \pi(3)^2]$

$$\frac{dh}{dt} = \frac{-2}{39\pi} \approx -0.016 \text{ ft/hr}$$

23. $PV = k$

$$P\frac{dV}{dt} + V\frac{dP}{dt} = 0$$

At $t = 6.5$, $P \approx 67$, $\frac{dP}{dt} \approx -30$, $V = 300$

$$\frac{dV}{dt} = -\frac{V}{P}\frac{dP}{dt} = -\frac{300}{67}(-30) \approx 134 \text{ in.}^3/\text{min}$$

25. Assuming that the tank is now in the shape of an upper hemisphere with radius r, we again let t be

the number of hours past midnight and h be the height of the water at time t. The volume, V, of water in the tank at that time is given by

$$V = \frac{2}{3}\pi r^3 - \frac{\pi}{3}(r-h)^2(2r+h)$$

and so $V = \frac{16000}{3}\pi - \frac{\pi}{3}(20-h)^2(40+h)$

from which

$$\frac{dV}{dt} = -\frac{\pi}{3}(20-h)^2\frac{dh}{dt} + \frac{2\pi}{3}(20-h)(40+h)\frac{dh}{dt}$$

At $t = 7$, $\frac{dV}{dt} \approx -525\pi \approx -1649$

Thus Webster City residents were using water at the rate of $2400 + 1649 = 4049$ cubic feet per hour at 7:00 A.M.

27. a. Let x be the distance from the bottom of the wall to the end of the ladder on the ground, so $\frac{dx}{dt} = 2$ ft/s. Let y be the height of the opposite end of the ladder. By similar triangles, $\frac{y}{12} = \frac{18}{\sqrt{144+x^2}}$, so

$$y = \frac{216}{\sqrt{144+x^2}}.$$

$$\frac{dy}{dt} = -\frac{216}{2(144+x^2)^{3/2}}2x\frac{dx}{dt}$$

$$= -\frac{216x}{(144+x^2)^{3/2}}\frac{dx}{dt}$$

When the ladder makes an angle of $60°$ with the ground, $x = 4\sqrt{3}$ and

$$\frac{dy}{dt} = -\frac{216(4\sqrt{3})}{(144+48)^{3/2}}\cdot 2 = -1.125 \text{ ft/s.}$$

b.

$$\frac{d^2y}{dt^2} = \frac{d}{dt}\left(-\frac{216x}{(144+x^2)^{3/2}}\frac{dx}{dt}\right)$$

$$= \frac{d}{dt}\left(-\frac{216x}{(144+x^2)^{3/2}}\right)\frac{dx}{dt} - \frac{216x}{(144+x^2)^{3/2}}\cdot\frac{d^2x}{dt^2}$$

Since $\frac{dx}{dt} = 2, \frac{d^2x}{dt^2} = 0$, thus

$$\frac{d^2y}{dt^2} = \frac{dx}{dt}\cdot\frac{-216(144+x^2)^{3/2}\frac{dx}{dt}}{(144+x^2)^3}$$

$$+ \frac{dx}{dt}\cdot\frac{216x\left(\frac{3}{2}\right)\sqrt{144+x^2}(2x)\frac{dx}{dt}}{(144+x^2)^3}$$

$$= \frac{-216(144+x^2)+648x^2}{(144+x^2)^{5/2}}\left(\frac{dx}{dt}\right)^2$$

$$= \frac{432x^2 - 31{,}104}{(144+x^2)^{5/2}}\left(\frac{dx}{dt}\right)^2$$

When the ladder makes an angle of $60°$ with the ground,

$$\frac{d^2y}{dt^2} = \frac{432\cdot 48 - 31{,}104}{(144+48)^{5/2}}(2)^2 \approx -0.08 \text{ ft/s}^2$$

29. $\frac{dV}{dt} = k(4\pi r^2)$

a. $V = \frac{4}{3}\pi r^3$

$$\frac{dV}{dt} = 4\pi r^2\frac{dr}{dt}$$

$$k(4\pi r^2) = 4\pi r^2\frac{dr}{dt}$$

$$\frac{dr}{dt} = k$$

b. If the original volume was V_0, the volume after 1 hour is $\frac{8}{27}V_0$. The original radius was $r_0 = \sqrt[3]{\frac{3}{4\pi}V_0}$ while the radius after 1 hour is $r_1 = \sqrt[3]{\frac{8}{27}V_0 \cdot \frac{3}{4\pi}} = \frac{2}{3}r_0$. Since $\frac{dr}{dt}$ is constant, $\frac{dr}{dt} = -\frac{1}{3}r_0$ unit/hr. The snowball will take 3 hours to melt completely.

31. Let l be the distance along the ground from the brother to the tip of the shadow. The shadow is controlled by both siblings when $\frac{3}{l} = \frac{5}{l+4}$ or $l = 6$. Again using similar triangles, this occurs when $\frac{y}{20} = \frac{6}{3}$, so $y = 40$. Thus, the girl controls the tip of the shadow when $y \geq 40$ and the boy controls it when $y < 40$.

Let x be the distance along the ground from the light pole to the girl. $\dfrac{dx}{dt} = -4$

When $y \geq 40$, $\dfrac{20}{y} = \dfrac{5}{y-x}$ or $y = \dfrac{4}{3}x$.

When $y < 40$, $\dfrac{20}{y} = \dfrac{3}{y-(x+4)}$ or $y = \dfrac{20}{17}(x+4)$.

$x = 30$ when $y = 40$. Thus,

$$y = \begin{cases} \dfrac{4}{3}x & \text{if } x \geq 30 \\ \dfrac{20}{17}(x+4) & \text{if } x < 30 \end{cases}$$

and

$$\dfrac{dy}{dt} = \begin{cases} \dfrac{4}{3}\dfrac{dx}{dt} & \text{if } x \geq 30 \\ \dfrac{20}{17}\dfrac{dx}{dt} & \text{if } x < 30 \end{cases}$$

Hence, the tip of the shadow is moving at the rate of $\dfrac{4}{3}(4) = \dfrac{16}{3}$ ft/s when the girl is at least 30 feet from the light pole, and it is moving $\dfrac{20}{17}(4) = \dfrac{80}{17}$ ft/s when the girl is less than 30 ft from the light pole.

3.10 Concepts Review

1. $f'(x)dx$
3. Δx is small.

Problem Set 3.10

1. $dy = (2x+1)dx$

3. $dy = -4(2x+3)^{-5}(2)dx = -8(2x+3)^{-5}dx$

5. $dy = 3(\sin x + \cos x)^2(\cos x - \sin x)dx$

7. $dy = -\dfrac{3}{2}(7x^2+3x-1)^{-5/2}(14x+3)dx$

$= -\dfrac{3}{2}(14x+3)(7x^2+3x-1)^{-5/2}dx$

9. $ds = \dfrac{3}{2}(t^2 - \cot t + 2)^{1/2}(2t + \csc^2 t)dt$

$= \dfrac{3}{2}(2t + \csc^2 t)\sqrt{t^2 - \cot t + 2}\,dt$

11.

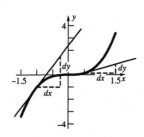

13.

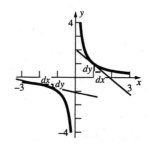

15. **a.** $\Delta y = \dfrac{1}{1.5} - \dfrac{1}{1} = -\dfrac{1}{3}$

b. $\Delta y = \dfrac{1}{-1.25} + \dfrac{1}{2} = -0.3$

17. **a.** $\Delta y = [(3)^4 + 2(3)] - [(2)^4 + 2(2)] = 67$
$dy = (4x^3 + 2)dx = [4(2)^3 + 2](1) = 34$

b. $\Delta y = [(2.005)^4 + 2(2.005)] - [(2)^4 + 2(2)]$
≈ 0.1706
$dy = (4x^3 + 2)dx = [4(2)^3 + 2](0.005) = 0.17$

19. $y = \sqrt{x}; dy = \dfrac{1}{2\sqrt{x}}dx; x = 36, dx = -0.1$

$dy = \dfrac{1}{2\sqrt{36}}(-0.1) \approx -0.0083$

$\sqrt{35.9} \approx \sqrt{36} + dy = 6 - 0.0083 = 5.9917$

21. $V = \dfrac{4}{3}\pi r^3; r = 5, dr = 0.125$

$dV = 4\pi r^2 dr = 4\pi(5)^2(0.125) \approx 39.27 \text{ cm}^3$

23. $V = \frac{4}{3}\pi r^3; r = 6\,\text{ft} = 72\,\text{in.}, dr = -0.3$

$dV = 4\pi r^2 dr = 4\pi(72)^2(-0.3) \approx -19{,}543$

$V \approx \frac{4}{3}\pi(72)^3 - 19{,}543$

$\approx 1{,}543{,}915\,\text{in}^3 \approx 893\,\text{ft}^3$

25. $C = 2\pi r\,; r = 4000\,\text{mi} = 21{,}120{,}000\,\text{ft}, dr = 2$

$dC = 2\pi\,dr = 2\pi(2) = 4\pi \approx 12.6\,ft$

27. $V = \frac{4}{3}\pi r^3 = \frac{4}{3}\pi(10)^3 \approx 4189$

$dV = 4\pi r^2 dr = 4\pi(10)^2(0.05) \approx 62.8$ The

volume is $4189 \pm 62.8\,\text{cm}^3$.

The absolute error is ≈ 62.8 while the relative

error is $62.8/4189 \approx 0.015$ or 1.5% .

29. $s = \sqrt{a^2 + b^2 - 2ab\cos\theta}$

$= \sqrt{151^2 + 151^2 - 2(151)(151)\cos 0.53} \approx 79.097$

$s = \sqrt{45{,}602 - 45{,}602\cos\theta}$

$ds = \dfrac{1}{2\sqrt{45{,}602 - 45{,}602\cos\theta}} \cdot 45{,}602\sin\theta\,d\theta$

$= \dfrac{22{,}801\sin\theta}{\sqrt{45{,}602 - 45{,}602\cos\theta}}d\theta$

$= \dfrac{22{,}801\sin 0.53}{\sqrt{45{,}602 - 45{,}602\cos 0.53}}(0.005) \approx 0.729$

$s \approx 79.097 \pm 0.729\,\text{cm}$

The absolute error is ≈ 0.729 while the relative

error is $0.729/79.097 \approx 0.0092$ or 0.92% .

31. $y = 3x^2 - 2x + 11; x = 2, dx = 0.001$

$dy = (6x - 2)dx = [6(2) - 2](0.001) = 0.01$

$\dfrac{d^2 y}{dx^2} = 6,$ so with $\Delta x = 0.001$,

$\left|\Delta y - dy\right| \le \dfrac{1}{2}(6)(0.001)^2 = 0.000003$

33. $V = \pi r^2 h + \dfrac{4}{3}\pi r^3$

$V = 100\pi r^2 + \dfrac{4}{3}\pi r^3; r = 10, dr = 0.1$

$dV = (200\pi r + 4\pi r^2)dr$

$= (2000\pi + 400\pi)(0.1) = 240\pi \approx 754\ \text{cm}^3$

35. The percent increase in mass is $\dfrac{dm}{m}$.

$dm = -\dfrac{m_0}{2}\left(1 - \dfrac{v^2}{c^2}\right)^{-3/2}\left(-\dfrac{2v}{c^2}\right)dv$

$= \dfrac{m_0 v}{c^2}\left(1 - \dfrac{v^2}{c^2}\right)^{-3/2}dv$

$\dfrac{dm}{m} = \dfrac{v}{c^2}\left(1 - \dfrac{v^2}{c^2}\right)^{-1}dv = \dfrac{v}{c^2}\left(\dfrac{c^2}{c^2 - v^2}\right)dv$

$= \dfrac{v}{c^2 - v^2}dv$

$v = 0.9c, dv = 0.02c$

$\dfrac{dm}{m} = \dfrac{0.9c}{c^2 - 0.81c^2}(0.02c) = \dfrac{0.018}{0.19} \approx 0.095$

The percent increase in mass is about 9.5.

37. a. $h(x) = \sin x;\ h'(x) = \cos x;\ a = 0$

The linear approximation is then

$L(x) = 0 + 1(x - 0) = x$

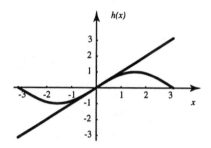

b. $F(x) = 3x + 4;\ F'(x) = 3;\ a = 3$

The linear approximation is then

$L(x) = 13 + 3(x - 3) = 13 + 3x - 9$

$= 3x + 4$

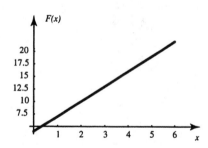

3.11 Chapter Review

Concepts Test

1. False: If $f(x) = x^3$, $f'(x) = 3x^2$ and the tangent line $y = 0$ at $x = 0$ crosses the curve at the point of tangency.

3. True: $m_{\tan} = 4x^3$, which is unique for each value of x.

5. True: If the velocity is negative and increasing, the speed is decreasing.

7. True: If the tangent line is horizontal, the slope must be 0.

9. True: $D_x f(g(x)) = f'(g(x))g'(x)$; since $g(x) = x$, $g'(x) = 1$, so $D_x f(g(x)) = f'(g(x))$.

11. True: Theorem 3.2.A

13. False: $(f \cdot g)'(x) = f(x)g'(x) + g(x)f'(x)$

15. True: If $f(x) = x^3 g(x)$, then
$$D_x f(x) = x^3 g'(x) + 3x^2 g(x)$$
$$= x^2[xg'(x) + 3g(x)].$$

17. False: $D_x y = f(x)g'(x) + g(x)f'(x)$
$$D_x^2 y = f(x)g''(x) + g'(x)f'(x)$$
$$+ g(x)f''(x) + f'(x)g'(x)$$
$$= f(x)g''(x) + 2f'(x)g'(x) + f''(x)g(x)$$

19. True: $f(x) = ax^n$; $f'(x) = anx^{n-1}$

21. True: $h'(x) = f(x)g'(x) + g(x)f'(x)$
$$h'(c) = f(c)g'(c) + g(c)f'(c)$$
$$= f(c)(0) + g(c)(0) = 0$$

23. True: $D^2(kf) = kD^2 f$ and
$$D^2(f + g) = D^2 f + D^2 g$$

25. True: $(f \circ g)'(2) = f'(g(2)) \cdot g'(2)$
$$= f'(2) \cdot g'(2) = 2 \cdot 2 = 4$$

27. False: The rate of volume change depends on the radius of the sphere.

29. True: $D_x(\sin x) = \cos x$;
$$D_x^2(\sin x) = -\sin x;$$
$$D_x^3(\sin x) = -\cos x;$$
$$D_x^4(\sin x) = \sin x;$$
$$D_x^5(\sin x) = \cos x$$

31. True: $\lim_{x \to 0} \dfrac{\tan x}{3x} = \dfrac{1}{3} \lim_{x \to 0} \dfrac{\sin x}{x \cos x}$
$$= \dfrac{1}{3} \cdot 1 = \dfrac{1}{3}$$

33. True: $V = \dfrac{4}{3}\pi r^3$

$$\dfrac{dV}{dt} = 4\pi r^2 \dfrac{dr}{dt}$$

If $\dfrac{dV}{dt} = 3$, then $\dfrac{dr}{dt} = \dfrac{3}{4\pi r^2}$ so

$$\dfrac{dr}{dt} > 0.$$

$$\dfrac{d^2 r}{dt^2} = -\dfrac{3}{2\pi r^3}\dfrac{dr}{dt} \text{ so } \dfrac{d^2 r}{dt^2} < 0$$

35. True: $V = \dfrac{4}{3}\pi r^3$, $S = 4\pi r^2$

$$dV = 4\pi r^2 dr = S \cdot dr$$
If $\Delta r = dr$, then $dV = S \cdot \Delta r$

37. False: The slope of the linear approximation is equal to
$$f'(a) = f'(0) = -\sin(0) = 0.$$

Sample Test Problems

1. a. $f'(x) = \lim_{h \to 0} \dfrac{3(x+h)^3 - 3x^3}{h} = \lim_{h \to 0} \dfrac{9x^2 h + 9xh^2 + 3h^3}{h} = \lim_{h \to 0}(9x^2 + 9xh + 3h^2) = 9x^2$

b. $f'(x) = \lim_{h \to 0} \dfrac{[2(x+h)^5 + 3(x+h)] - (2x^5 + 3x)}{h} = \lim_{h \to 0} \dfrac{10x^4 h + 20x^3 h^2 + 20x^2 h^3 + 10xh^4 + 2h^5 + 3h}{h}$

$= \lim_{h \to 0}(10x^4 + 20x^3 h + 20x^2 h^2 + 10xh^3 + 2h^4 + 3) = 10x^4 + 3$

c. $f'(x) = \lim\limits_{h \to 0} \dfrac{\frac{1}{3(x+h)} - \frac{1}{3x}}{h} = \lim\limits_{h \to 0}\left[-\dfrac{h}{3(x+h)x}\right]\dfrac{1}{h} = \lim\limits_{h \to 0}-\left(\dfrac{1}{3x(x+h)}\right) = -\dfrac{1}{3x^2}$

d. $f'(x) = \lim\limits_{h \to 0}\left[\left(\dfrac{1}{3(x+h)^2 + 2} - \dfrac{1}{3x^2 + 2}\right)\dfrac{1}{h}\right] = \lim\limits_{h \to 0}\left[\dfrac{3x^2 + 2 - 3(x+h)^2 - 2}{(3(x+h)^2 + 2)(3x^2 + 2)} \cdot \dfrac{1}{h}\right]$

$= \lim\limits_{h \to 0}\left[\dfrac{-6xh - 3h^2}{(3(x+h)^2 + 2)(3x^2 + 2)} \cdot \dfrac{1}{h}\right] = \lim\limits_{h \to 0}\dfrac{-6x - 3h}{(3(x+h)^2 + 2)(3x^2 + 2)} = -\dfrac{6x}{(3x^2 + 2)^2}$

e. $f'(x) = \lim\limits_{h \to 0}\dfrac{\sqrt{3(x+h)} - \sqrt{3x}}{h} = \lim\limits_{h \to 0}\dfrac{(\sqrt{3x + 3h} - \sqrt{3x})(\sqrt{3x + 3h} + \sqrt{3x})}{h(\sqrt{3x + 3h} + \sqrt{3x})}$

$= \lim\limits_{h \to 0}\dfrac{3h}{h(\sqrt{3x + 3h} + \sqrt{3x})} = \lim\limits_{h \to 0}\dfrac{3}{\sqrt{3x + 3h} + \sqrt{3x}} = \dfrac{3}{2\sqrt{3x}}$

f. $f'(x) = \lim\limits_{h \to 0}\dfrac{\sin[3(x+h)] - \sin 3x}{h} = \lim\limits_{h \to 0}\dfrac{\sin(3x + 3h) - \sin 3x}{h}$

$= \lim\limits_{h \to 0}\dfrac{\sin 3x \cos 3h + \sin 3h \cos 3x - \sin 3x}{h} = \lim\limits_{h \to 0}\dfrac{\sin 3x(\cos 3h - 1)}{h} + \lim\limits_{h \to 0}\dfrac{\sin 3h \cos 3x}{h}$

$= 3\sin 3x \lim\limits_{h \to 0}\dfrac{\cos 3h - 1}{3h} + \cos 3x \lim\limits_{h \to 0}\dfrac{\sin 3h}{h} = (3\sin 3x)(0) + (\cos 3x)3\lim\limits_{h \to 0}\dfrac{\sin 3h}{3h} = (\cos 3x)(3)(1) = 3\cos 3x$

g. $f'(x) = \lim\limits_{h \to 0}\dfrac{\sqrt{(x+h)^2 + 5} - \sqrt{x^2 + 5}}{h} = \lim\limits_{h \to 0}\dfrac{\left(\sqrt{(x+h)^2 + 5} - \sqrt{x^2 + 5}\right)\left(\sqrt{(x+h)^2 + 5} + \sqrt{x^2 + 5}\right)}{h\left(\sqrt{(x+h)^2 + 5} + \sqrt{x^2 + 5}\right)}$

$= \lim\limits_{h \to 0}\dfrac{2xh + h^2}{h\left(\sqrt{(x+h)^2 + 5} + \sqrt{x^2 + 5}\right)} = \lim\limits_{h \to 0}\dfrac{2x + h}{\sqrt{(x+h)^2 + 5} + \sqrt{x^2 + 5}} = \dfrac{2x}{2\sqrt{x^2 + 5}} = \dfrac{x}{\sqrt{x^2 + 5}}$

h. $f'(x) = \lim\limits_{h \to 0}\dfrac{\cos[\pi(x+h)] - \cos \pi x}{h} = \lim\limits_{h \to 0}\dfrac{\cos(\pi x + \pi h) - \cos \pi x}{h} = \lim\limits_{h \to 0}\dfrac{\cos \pi x \cos \pi h - \sin \pi x \sin \pi h - \cos \pi x}{h}$

$= \lim\limits_{h \to 0}\left(-\pi \cos \pi x \dfrac{1 - \cos \pi h}{\pi h}\right) - \lim\limits_{h \to 0}\left(\pi \sin \pi x \dfrac{\sin \pi h}{\pi h}\right) = (-\pi \cos \pi x)(0) - (\pi \sin \pi x) = -\pi \sin \pi x$

3. a. $f(x) = 3x$ at $x = 1$

b. $f(x) = 4x^3$ at $x = 2$

c. $f(x) = \sqrt{x^3}$ at $x = 1$

d. $f(x) = \sin x$ at $x = \pi$

e. $f(x) = \dfrac{4}{x}$ at x

f. $f(x) = -\sin 3x$ at x

g. $f(x) = \tan x$ at $x = \dfrac{\pi}{4}$

h. $f(x) = \dfrac{1}{\sqrt{x}}$ at $x = 5$

5. $D_x(3x^5) = 15x^4$

7. $D_z(z^3 + 4z^2 + 2z) = 3z^2 + 8z + 2$

9. $D_t\left(\dfrac{4t - 5}{6t^2 + 2t}\right) = \dfrac{(6t^2 + 2t)(4) - (4t - 5)(12t + 2)}{(6t^2 + 2t)^2}$

$= \dfrac{-24t^2 + 60t + 10}{(6t^2 + 2t)^2}$

11. $\dfrac{d}{dx}\left(\dfrac{4x^2-2}{x^3+x}\right) = \dfrac{(x^3+x)(8x)-(4x^2-2)(3x^2+1)}{(x^3+x)^2}$

$= \dfrac{-4x^4+10x^2+2}{(x^3+x)^2}$

13. $\dfrac{d}{dx}\left(\dfrac{1}{\sqrt{x^2+4}}\right) = \dfrac{d}{dx}(x^2+4)^{-1/2}$

$= -\dfrac{1}{2}(x^2+4)^{-3/2}(2x)$

$= -\dfrac{x}{\sqrt{(x^2+4)^3}}$

15. $D_\theta(\sin\theta+\cos^3\theta) = \cos\theta+3\cos^2\theta(-\sin\theta)$

$= \cos\theta-3\sin\theta\cos^2\theta$

$D_\theta^2(\sin\theta+\cos^3\theta)$

$= -\sin\theta-3[\sin\theta(2)(\cos\theta)(-\sin\theta)+\cos^3\theta]$

$= -\sin\theta+6\sin^2\theta\cos\theta-3\cos^3\theta$

17. $D_\theta[\sin(\theta^2)] = \cos(\theta^2)(2\theta) = 2\theta\cos(\theta^2)$

19. $\dfrac{d}{d\theta}[\sin^2(\sin(\pi\theta))] = 2\sin(\sin(\pi\theta))\cos(\sin(\pi\theta))(\cos(\pi\theta))(\pi) = 2\pi\sin(\sin(\pi\theta))\cos(\sin(\pi\theta))\cos(\pi\theta)$

21. $D_\theta\tan3\theta = (\sec^2 3\theta)(3) = 3\sec^2 3\theta$

23. $f'(x) = (x^2-1)^2(9x^2-4)+(3x^3-4x)(2)(x^2-1)(2x) = (x^2-1)^2(9x^2-4)+4x(x^2-1)(3x^3-4x)$

$f'(2) = 672$

25. $\dfrac{d}{dx}\left(\dfrac{\cot x}{\sec x^2}\right) = \dfrac{(\sec x^2)(-\csc^2 x)-(\cot x)(\sec x^2)(\tan x^2)(2x)}{\sec^2 x^2} = \dfrac{-\csc^2 x-2x\cot x\tan x^2}{\sec x^2}$

27. $f'(x) = (x-1)^3 2(\sin\pi x-x)(\pi\cos\pi x-1)+(\sin\pi x-x)^2 3(x-1)^2$

$= 2(x-1)^3(\sin\pi x-x)(\pi\cos\pi x-1)+3(\sin\pi x-x)^2(x-1)^2$

$f'(2) = 16-4\pi \approx 3.43$

29. $g'(r) = 3(\cos^2 5r)(-\sin 5r)(5) = -15\cos^2 5r\sin 5r$

$g''(r) = -15[(\cos^2 5r)(\cos 5r)(5)+(\sin 5r)2(\cos 5r)(-\sin 5r)(5)] = -15[5\cos^3 5r-10(\sin^2 5r)(\cos 5r)]$

$g'''(r) = -15[5(3)(\cos^2 5r)(-\sin 5r)(5)-(10\sin^2 5r)(-\sin 5r)(5)-(\cos 5r)(20\sin 5r)(\cos 5r)(5)]$

$= -15[-175(\cos^2 5r)(\sin 5r)+50\sin^3 5r]$

$g'''(1) \approx 458.8$

31. $G'(x) = F'(r(x)+s(x))(r'(x)+s'(x))+s'(x)$

$G''(x) = F'(r(x)+s(x))(r''(x)+s''(x))+(r'(x)+s'(x))F''(r(x)+s(x))(r'(x)+s'(x))+s''(x)$

$= F'(r(x)+s(x))(r''(x)+s''(x))+(r'(x)+s'(x))^2 F''(r(x)+s(x))+s''(x)$

33. $F'(z) = r'(s(z))s'(z) = [3\cos(3s(z))](9z^2)$

$= 27z^2\cos(9z^3)$

35. One possibility:

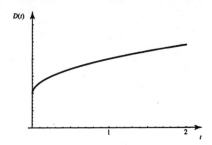

37. One possibility:

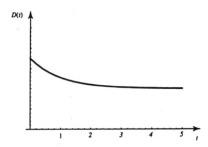

39. $V = \dfrac{4}{3}\pi r^3$

$\dfrac{dV}{dr} = 4\pi r^2$

When $r = 5$, $\dfrac{dV}{dr} = 4\pi(5)^2 = 100\pi \approx 314$ m³ per

meter of increase in the radius.

41. $V = \dfrac{1}{2}bh(12)$; $\dfrac{6}{4} = \dfrac{b}{h}$; $b = \dfrac{3h}{2}$

$V = 6\left(\dfrac{3h}{2}\right)h = 9h^2$; $\dfrac{dV}{dt} = 9$

$\dfrac{dV}{dt} = 18h\dfrac{dh}{dt}$

When $h = 3$, $9 = 18(3)\dfrac{dh}{dt}$

$\dfrac{dh}{dt} = \dfrac{1}{6} \approx 0.167$ ft/min

43. $s = t^3 - 6t^2 + 9t$

$v(t) = \dfrac{ds}{dt} = 3t^2 - 12t + 9$

$a(t) = \dfrac{d^2s}{dt^2} = 6t - 12$

a. $3t^2 - 12t + 9 < 0$
$3(t - 3)(t - 1) < 0$
$1 < t < 3$; $(1,3)$

b. $3t^2 - 12t + 9 = 0$
$3(t - 3)(t - 1) = 0$
$t = 1, 3$
$a(1) = -6$, $a(3) = 6$

c. $6t - 12 > 0$
$t > 2$; $(2,\infty)$

45. a. $2(x - 1) + 2y\dfrac{dy}{dx} = 0$

$\dfrac{dy}{dx} = \dfrac{-(x - 1)}{y} = \dfrac{1 - x}{y}$

b. $x(2y)\dfrac{dy}{dx} + y^2 + y(2x) + x^2\dfrac{dy}{dx} = 0$

$\dfrac{dy}{dx}(2xy + x^2) = -(y^2 + 2xy)$

$\dfrac{dy}{dx} = -\dfrac{y^2 + 2xy}{x^2 + 2xy}$

c. $3x^2 + 3y^2\dfrac{dy}{dx} = x^3(3y^2)\dfrac{dy}{dx} + 3x^2y^3$

$\dfrac{dy}{dx}(3y^2 - 3x^3y^2) = 3x^2y^3 - 3x^2$

$\dfrac{dy}{dx} = \dfrac{3x^2y^3 - 3x^2}{3y^2 - 3x^3y^2} = \dfrac{x^2y^3 - x^2}{y^2 - x^3y^2}$

d. $x\cos(xy)\left[x\dfrac{dy}{dx} + y\right] + \sin(xy) = 2x$

$x^2\cos(xy)\dfrac{dy}{dx} = 2x - \sin(xy) - xy\cos(xy)$

$\dfrac{dy}{dx} = \dfrac{2x - \sin(xy) - xy\cos(xy)}{x^2\cos(xy)}$

e. $x\sec^2(xy)\left(x\dfrac{dy}{dx} + y\right) + \tan(xy) = 0$

$x^2\sec^2(xy)\dfrac{dy}{dx} = -[\tan(xy) + xy\sec^2(xy)]$

$\dfrac{dy}{dx} = -\dfrac{\tan(xy) + xy\sec^2(xy)}{x^2\sec^2(xy)}$

47. $dy = [\pi\cos(\pi x) + 2x]dx$; $x = 2$, $dx = 0.01$
$dy = [\pi\cos(2\pi) + 2(2)](0.01) = (4 + \pi)(0.01)$
≈ 0.0714

49. a. $\dfrac{d}{dx}[f^2(x)+g^3(x)]$

$= 2f(x)f'(x)+3g^2(x)g'(x)$

$2f(2)f'(2)+3g^2(2)g'(2)$

$= 2(3)(4)+3(2)^2(5) = 84$

b. $\dfrac{d}{dx}[f(x)g(x)] = f(x)g'(x)+g(x)f'(x)$

$f(2)g'(2)+g(2)f'(2) = (3)(5)+(2)(4) = 23$

c. $\dfrac{d}{dx}[f(g(x))] = f'(g(x))g'(x)$

$f'(g(2))g'(2) = f'(2)g'(2) = (4)(5) = 20$

d. $D_x[f^2(x)] = 2f(x)f'(x)$

$D_x^2[f^2(x)] = 2[f(x)f''(x)+f'(x)f'(x)]$

$2f(2)f''(2)+2[f'(2)]^2$

$= 2(3)(-1)+2(4)^2 = 26$

3.12 Additional Problem Set

1. $g(t)$ is not differentiable at $t = \alpha$ if

$\displaystyle\lim_{h\to 0}\dfrac{g(\alpha+h)-g(\alpha)}{h}$ does not exist.

3. $f(f(x)) = f((1-x^n)^{1/n})$

$= \left\{1-[(1-x^n)^{1/n}]^n\right\}^{1/n} = \{1-(1-x^n)\}^{1/n}$

$= \{x^n\}^{1/n} = x$

$f'(x) = \dfrac{1}{n}(1-x^n)^{\frac{1-n}{n}}(-nx^{n-1})$

$= -x^{n-1}(1-x^n)^{\frac{1-n}{n}}$

$f'(f(x))f'(x) = f'\left((1-x^n)^{1/n}\right)f'(x)$

$= -[(1-x^n)^{1/n}]^{n-1}\{1-[(1-x^n)^{1/n}]^n\}^{\frac{1-n}{n}} \cdot$

$\qquad\qquad (-1)x^{n-1}(1-x^n)^{\frac{1-n}{n}}$

$= (1-x^n)^{\frac{n-1}{n}+\frac{1-n}{n}}\{1-(1-x^n)\}^{\frac{1-n}{n}}x^{n-1}$

$= \{x^n\}^{\frac{1-n}{n}}x^{n-1} = x^{1-n+n-1} = 1$

5. a. Using the chain rule,

$D_r[(r+1)^n] = n(r+1)^{n-1}$

51. $\sin 15° = \dfrac{y}{x}, \dfrac{dx}{dt} = 400$

$y = x\sin 15°$

$\dfrac{dy}{dt} = \sin 15°\dfrac{dx}{dt}$

$\dfrac{dy}{dt} = 400\sin 15° \approx 104$ mi/hr

53. a. $D_\theta|\sin\theta| = \dfrac{|\sin\theta|}{\sin\theta}\cos\theta = \cot\theta|\sin\theta|$

b. $D_\theta|\cos\theta| = \dfrac{|\cos\theta|}{\cos\theta}(-\sin\theta) = -\tan\theta|\cos\theta|$

b. $(r+1)^n = C_n^n r^n + C_n^{n-1}r^{n-1}+\ldots+C_n^1 r + C_n^0$

so

$D_r[(r+1)^n] = D_r[C_n^n r^n + C_n^{n-1}r^{n-1}+\ldots$

$\qquad\qquad +C_n^1 r + C_n^0]$

$= nC_n^n r^{n-1}+(n-1)C_n^{n-1}r^{n-2}+\ldots+C_n^1$

$(r+1)^{n-1} = C_{n-1}^{n-1}r^{n-1}+C_{n-1}^{n-2}r^{n-2}+\ldots$

$\qquad\qquad +C_{n-1}^1 r + C_{n-1}^0$

so

$D_r[(r+1)^n] = n(r+1)^{n-1}$

$= nC_{n-1}^{n-1}r^{n-1}+nC_{n-1}^{n-2}r^{n-2}+nC_{n-1}^1 r+nC_{n-1}^0$

Equating like powers of r,

$nC_n^n = nC_{n-1}^{n-1}, (n-1)C_n^{n-1} = nC_{n-1}^{n-2},$

$(n-2)C_n^{n-2} = nC_{n-1}^{n-3},\ldots, C_n^1 = nC_{n-1}^0$

or

$mC_n^m = nC_{n-1}^{m-1}$

c. $C_1^1 = 1$, thus $2\cdot C_2^2 = 2C_1^1, C_2^2 = C_1^1 = 1$

Similarly, $C_3^3 = C_4^4 = C_n^n = 1$

For $m < n$, then using part b repeatedly,

$$C_n^m = \frac{n}{m}C_{n-1}^{m-1} = \frac{n}{m}\cdot\frac{n-1}{m-1}C_{n-2}^{m-2}$$

$$= \frac{n\cdot(n-1)}{m\cdot(m-1)}\cdot\frac{n-2}{m-2}C_{n-3}^{m-3}$$

$$= ... = \frac{n\cdot(n-1)...(n-m+2)}{m\cdot(m-1)...2}C_{n-m+1}^1$$

$$= \frac{n\cdot(n-1)...(n-m+1)}{m\cdot(m-1)...2\cdot1}C_{n-m}^0$$

$$= \frac{n\cdot(n-1)...(n-m+1)}{1\cdot2\cdot...\cdot m}$$

7. No. The linear approximation is $L(x) = x$ for both $y = \sin x$ and $y = \tan x$.
For $\sin x$, the error is negative when $x > 0$ because the approximation lies above the curve; the error is positive when $x < 0$ because the approximation lies below the curve.
For $\tan x$, the error is positive when $x > 0$ because the approximation lies below the curve; the error is negative when $x < 0$ because the approximation lies above the curve.

9. a. $f(3) = \sqrt{4} - 2.1 = -0.1$,
$f(8) = \sqrt{9} - 2.1 = 0.9$ so there must be at least one zero of $f(x)$ between $x = 3$ and $x = 8$.

b. $f'(x) = \dfrac{1}{2\sqrt{1+x}}; f'(8) = \dfrac{1}{6}$

$L(x) = 0.9 + \dfrac{1}{6}(x-8)$ or $L(x) = \dfrac{1}{6}x - \dfrac{13}{30}$

$L(x) = 0$ when $x = \dfrac{13}{5} = 2.6$ which is less than three, so must be discarded.

c. $f'(3) = \dfrac{1}{4}$

$L(x) = -0.1 + \dfrac{1}{4}(x-3)$ or

$L(x) = 0.25x - 0.85$

$L(x) = 0$ when $x = 3.4$

d. $f(x) = 0$ at $x = 3.41$

e.

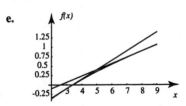

11. a. $\dfrac{5.04}{93.5} \approx 0.054$ ¢/in.^{2}

b. $dP = P'(A)dA$
$P'(93.5) = 10.5$ while
$dA = 8.7\cdot11.2 - 93.5 = 3.94$ in.^{2}
$dP = 10.5\cdot3.94 = 41.37$ ¢/pg

13. Measure r in feet for consistency.

a. $\dfrac{GM_1}{R^2} = 32$ ft/s^2

$\Delta g = g(3956 \text{ mi}) - g(R)$

$= \dfrac{GM_1(20,887,680)}{R^3} - \dfrac{GM_1}{R^2}$

$= \dfrac{GM_1}{R^2}\left(\dfrac{20,887,680}{20,908,800} - 1\right)$

$= 32\left(-\dfrac{21,120}{20,908,800}\right)$

$= -\dfrac{528}{16,355} \approx -0.0323$ ft/s^2

$\dfrac{\Delta g}{g} = \dfrac{-\frac{528}{16,355}}{32} = -\dfrac{33}{32,710} \approx -0.0010$

$\dfrac{\Delta g}{g} \times 100 = -0.10$

$dg \dfrac{GM_1}{R^3} dr = \dfrac{32}{R}(-21,120) = -\dfrac{528}{16,355} = \Delta g$

The exact and approximate values of the absolute, relative, and percent change in g are the same since $g(r)$ is linear with respect to r for $r < R$.

b. $\Delta g = g(R+30,000) - g(R)$

$= \dfrac{GM_1}{(R+30,000)^2} - \dfrac{GM_1}{R^2}$

$= \dfrac{GM_1}{R^2}\left[\dfrac{R^2}{(R+30,000)^2} - 1\right]$

$= 32\left[\dfrac{R^2 - (R+30,000)^2}{(R+30,000)^2}\right]$

$= 32\left(\dfrac{-60,000R - 900,000,000}{(R+30,000)^2}\right) \approx -0.0916$

$\dfrac{\Delta g}{g} = \dfrac{-60,000R - 900,000,000}{(R+30,000)^2} \approx -0.0029$

$\dfrac{\Delta g}{g} \times 100 \approx -0.29$

$dg = -\dfrac{2GM_1}{R^3} dr = -\dfrac{2}{R}\left(\dfrac{GM_1}{R^2}\right)(30,000)$

$$= \frac{(-60,000)(32)}{20,908,800} \approx -0.0918$$

$$\frac{dg}{g} = \frac{-60,000}{20,908,800} \approx -0.0029$$

$$\frac{dg}{g} \times 100 \approx -0.29$$

c. $\Delta g = g(R+2000) - g(R)$

$$= \frac{GM_1}{(R+2000)^2} - \frac{GM_1}{R^2}$$

$$= \frac{GM_1}{R^2} \left[\frac{R^2 - (R+2000)^2}{(R+2000)^2} \right]$$

$$= 32 \left[\frac{-4000R - 4,000,000}{(R+2000)^2} \right] \approx -0.00612$$

$$\frac{\Delta g}{g} = \frac{-4000R - 4,000,000}{(R+2000)^2} \approx -1.913 \times 10^{-4}$$

$$\frac{\Delta g}{g} \times 100 \approx -1.913 \times 10^{-2} = -0.019$$

$$dg = -\frac{2GM_1}{R^3} dr = -\frac{2}{R}(32)(2000)$$

$$= \frac{(-4000)(32)}{20,908,800} \approx -0.00612$$

$$\frac{dg}{g} = \frac{-4000}{20,908,800} \approx -1.913 \times 10^{-4}$$

$$\frac{dg}{g} \times 100 \approx -1.913 \times 10^{-2} = -0.019$$

CHAPTER 4

Applications of the Derivative

4.1 Concepts Review

1. continuous; closed

3. endpoints; stationary points; singular points

Problem Set 4.1

1. $f'(x) = 2x + 4; 2x + 4 = 0$ when $x = -2$.
   Critical points: $-4, -2, 0$
   $f(-4) = 4, f(-2) = 0, f(0) = 4$
   Maximum value = 4, minimum value = 0

3. $\Psi'(x) = 2x + 3; \ 2x + 3 = 0$ when $x = -\dfrac{3}{2}$.

   Critical points: $-2, -\dfrac{3}{2}, 1$

   $\Psi(-2) = -2, \Psi\left(-\dfrac{3}{2}\right) = -\dfrac{9}{4}, \Psi(1) = 4$

   Maximum value = 4, minimum value = $-\dfrac{9}{4}$

5. $f'(x) = 3x^2 - 3; 3x^2 - 3 = 0$ when $x = -1, 1$.
   Critical points: $-1, 1$
   $f(-1) = 3, f(1) = -1$
   No maximum value, minimum value = -1

   (See graph.)

   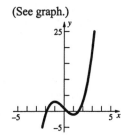

7. $h'(r) = -\dfrac{1}{r^2}; h'(r)$ is never 0; $h'(r)$ is not defined

   when $r = 0$, but $r = 0$ is not in the domain on
   $[-1, 3]$ since $h(0)$ is not defined.
   Critical points: $-1, 3$
   Note that $\lim\limits_{x \to 0^-} h(r) = -\infty$ and $\lim\limits_{x \to 0^+} h(x) = \infty$.
   No maximum value, no minimum value.

9. $g'(x) = -\dfrac{2x}{(1+x^2)^2}; \ -\dfrac{2x}{(1+x^2)^2} = 0$ when $x = 0$.
   Critical point: 0
   $g(0) = 1$
   As $x \to \infty, g(x) \to 0^+$; as $x \to -\infty, g(x) \to 0^+$.
   Maximum value = 1, no minimum value
   (See graph.)

   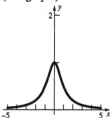

11. $r'(\theta) = \cos\theta; \ \cos\theta = 0$ when $\theta = \dfrac{\pi}{2} + k\pi$

    Critical points: $-\dfrac{\pi}{4}, \dfrac{\pi}{6}$

    $r\left(-\dfrac{\pi}{4}\right) = -\dfrac{1}{\sqrt{2}}, r\left(\dfrac{\pi}{6}\right) = \dfrac{\sqrt{3}}{2}$

    Maximum value $= \dfrac{\sqrt{3}}{2}$, minimum value $= -\dfrac{1}{\sqrt{2}}$

13. $a'(x) = \dfrac{x-1}{|x-1|}; a'(x)$ does not exist when $x = 1$.

    Critical points: 0, 1, 3
    $a(0) = 1, a(1) = 0, a(3) = 2$
    Maximum value = 2, minimum value = 0

15. $g'(x) = \dfrac{1}{3x^{2/3}}; f'(x)$ does not exist when $x = 0$.

    Critical points: $-1, 0, 27$
    $g(-1) = -1, g(0) = 0, g(27) = 3$
    Maximum value = 3, minimum value = -1

17. $f(x) = x(10 - x) = 10x - x^2; f'(x) = 10 - 2x;$
    $10 - 2x = 0; x = 5$
    Critical points: 0, 5, 10
    $f(0) = 0, f(5) = 25, f(10) = 0$; the numbers are $x = 5$
    and $10 - x = 5$.

19. Let x represent the width, then $100 - x$ represents the length.

$f(x) = x(100 - x) = 100x - x^2$; $f'(x) = 100 - 2x$;

$100 - 2x = 0$; $x = 50$

Critical points: 0, 50, 100

$f(0) = f(100) = 0$; $f(50) = 50(50) = 2500$

The dimensions should be 50 ft by 50 ft.

21. Let x be the width of the square to be cut out and V the volume of the resulting open box.

$V = x(24 - 2x)^2 = 4x^3 - 96x^2 + 576x$

$\dfrac{dV}{dx} = 12x^2 - 192x + 576 = 12(x - 12)(x - 4)$;

$12(x - 12)(x - 4) = 0$; $x = 12$ or $x = 4$.

Critical points: 0, 4, 12

At $x = 0$ or 12, $V = 0$; at $x = 4$, $V = 1024$.

The volume of the largest box is 1024 in.^{3}

23. Let A be the area of the pen.

$A = x(80 - 2x) = 80x - 2x^2$; $\dfrac{dA}{dx} = 80 - 4x$;

$80 - 4x = 0$; $x = 20$

Critical points: 0, 20, 40.

At $x = 0$ or 40, $A = 0$; at $x = 20$, $A = 800$.

The dimensions are 20 ft by $80 - 2(20) = 40$ ft, with the length along the barn being 40 ft.

25. Let A be the area of the pen. The perimeter is $100 + 180 = 280$ ft.

$y + y - 100 + 2x = 180$; $y = 140 - x$

$A = x(140 - x) = 140x - x^2$; $\dfrac{dA}{dx} = 140 - 2x$;

$140 - 2x = 0$; $x = 70$

Since $0 \le x \le 40$, the critical points are 0 and 40.

When $x = 0$, $A = 0$. When $x = 40$, $A = 4000$. The dimensions are 40 ft by 100 ft.

27. x is limited by $0 \le x \le \sqrt{12}$.

$A = 2x(12 - x^2) = 24x - 2x^3$; $\dfrac{dA}{dx} = 24 - 6x^2$;

$24 - 6x^2 = 0$; $x = -2, 2$

Critical points: 0, 2, $\sqrt{12}$.

When $x = 0$ or $\sqrt{12}$, $A = 0$.

When $x = 2$, $y = 12 - (2)^2 = 8$.

The dimensions are $2x = 2(2) = 4$ by 8.

29. The carrying capacity of the gutter is maximized when the area of the vertical end of the gutter is maximized. The height of the gutter is $3 \sin \theta$. The area is

$A = 3(3 \sin \theta) + 2\left(\dfrac{1}{2}\right)(3 \cos \theta)(3 \sin \theta)$

$= 9 \sin \theta + 9 \cos \theta \sin \theta$.

$\dfrac{dA}{d\theta} = 9 \cos \theta + 9(-\sin \theta) \sin \theta + 9 \cos \theta \cos \theta$

$= 9(\cos \theta - \sin^2 \theta + \cos^2 \theta)$

$= 9(2 \cos^2 \theta + \cos \theta - 1)$

$2 \cos^2 \theta + \cos \theta - 1 = 0$; $\cos \theta = -1, \dfrac{1}{2}$; $\theta = \pi, \dfrac{\pi}{3}$

Since $0 \le \theta \le \dfrac{\pi}{2}$, the critical points are

$0, \dfrac{\pi}{3}$, and $\dfrac{\pi}{2}$.

When $\theta = 0$, $A = 0$.

When $\theta = \dfrac{\pi}{3}$, $A = \dfrac{27\sqrt{3}}{4} \approx 11.7$.

When $\theta = \dfrac{\pi}{2}$, $A = 9$.

The carrying capacity is maximized when $\theta = \dfrac{\pi}{3}$.

31. Let c be the cost of driving the truck in cents. It takes $\dfrac{400}{x}$ hours to drive 400 miles.

$c = \dfrac{400}{x}(1200) + 400\left(25 + \dfrac{x}{4}\right)$

$= 480{,}000x^{-1} + 10{,}000 + 100x$

$\dfrac{dc}{dx} = -480{,}000x^{-2} + 100$;

$-480{,}000x^{-2} + 100 = 0$; $x \approx 69$

Since $40 \le x \le 55$, the critical points are 40 and 55.

At $x = 40$, $c = 26{,}000$.

At $x = 55$, $c \approx 24{,}227$.

The most economical allowed speed is at 55 mph.

33. Let D be the square of the distance.

$D = (x - 0)^2 + (y - 4)^2 = x^2 + \left(\dfrac{x^2}{4} - 4\right)^2$

$= \dfrac{x^4}{16} - x^2 + 16$

$\dfrac{dD}{dx} = \dfrac{x^3}{4} - 2x$; $\dfrac{x^3}{4} - 2x = 0$; $x(x^2 - 8) = 0$

$x = 0, x = \pm 2\sqrt{2}$

Critical points: $0, 2\sqrt{2}, 2\sqrt{3}$

At $x = 0$, $y = 0$, and $D = 16$. At $x = 2\sqrt{2}$, $y = 2$, and $D = 12$. At $x = 2\sqrt{3}$, $y = 3$, and $D = 13$.

$P\left(2\sqrt{2}, 2\right), Q(0, 0)$

35. Let V be the volume. $y = 4 - x$ and $z = 5 - 2x$. x is limited by $\le x \le 2.5$.

$$V = x(4 - x)(5 - 2x) = 20x - 13x^2 + 2x^3$$

$$\frac{dV}{dx} = 20 - 26x + 6x^2; \, 2(3x^2 - 13x + 10) = 0;$$

$$2(3x - 10)(x - 1) = 0;$$

$$x = 1, \frac{10}{3}$$

Critical points: 0, 1, 2.5
At $x = 0$ or 2.5, $V = 0$. At $x = 1$, $V = 9$.
Maximum volume when $x = 1$, $y = 4 - 1 = 3$, and $z = 5 - 2(1) = 3$.

37. a. $f'(x) = x \cos x$; on $[-1, 5]$, $x \cos x = 0$ when

$$x = 0, \, x = \frac{\pi}{2}, x = \frac{3\pi}{2}$$

Critical points: $-1, 0, \dfrac{\pi}{2}, \dfrac{3\pi}{2}, 5$

$$f(-1) \approx 3.38, f(0) = 3, \, f\left(\frac{\pi}{2}\right) \approx 3.57,$$

$$f\left(\frac{3\pi}{2}\right) \approx -2.71, \, f(5) \approx -2.51$$

Maximum value ≈ 3.57,
minimum value ≈ -2.71

b. $g'(x) = \dfrac{(\cos x + x \sin x + 2)(x \cos x)}{|\cos x + x \sin x + 2|}$;

$g'(x) = 0$ when $x = 0$, $x = \dfrac{\pi}{2}$, $x = \dfrac{3\pi}{2}$

$g'(x)$ does not exist when $f(x) = 0$;
on $[-1, 5]$, $f(x) = 0$ when $x \approx 3.45$

Critical points: $-1, 0, \dfrac{\pi}{2}, 3.45, \dfrac{3\pi}{2}, 5$

$$g(-1) \approx 3.38, g(0) = 3,$$

$$g\left(\frac{\pi}{2}\right) \approx 3.57,$$

$$g(3.45) = 0, \, g\left(\frac{3\pi}{2}\right) \approx 2.71,$$

$$g(5) \approx 2.51$$

Maximum value ≈ 3.57;
minimum value $= 0$

4.2 Concepts Review

1. Increasing; concave up

3. An inflection point

Problem Set 4.2

1. $f'(x) = 3$; $3 > 0$ for all x. $f(x)$ is increasing for all x.

3. $h'(t) = 2t + 2$; $2t + 2 > 0$ when $t > -1$. $h(t)$ is increasing on $[-1, \infty)$ and decreasing on $(-\infty, -1]$.

5. $G'(x) = 6x^2 - 18x + 12 = 6(x - 2)(x - 1)$
Split the x-axis into the intervals $(-\infty, 1)$, $(1, 2)$, $(2, \infty)$.

Test points: $x = 0, \dfrac{3}{2}, 3; G'(0) = 12, \, G'\left(\dfrac{3}{2}\right) = -\dfrac{3}{2},$

$G'(3) = 12$

$G(x)$ is increasing on $(-\infty, 1] \cup [2, \infty)$ and decreasing on $[1, 2]$.

7. $h'(z) = z^3 - 2z^2 = z^2(z - 2)$
Split the x-axis into the intervals $(-\infty, 0)$, $(0, 2)$, $(2, \infty)$.
Test points: $z = -1, 1, 3$; $h'(-1) = -3$, $h'(1) = -1$, $h'(3) = 9$
$h(z)$ is increasing on $[2, \infty)$ and decreasing on $(-\infty, 2]$.

9. $H'(t) = \cos t$; $H'(t) > 0$ when $0 \le t < \dfrac{\pi}{2}$ and

$$\frac{3\pi}{2} < t \le 2\pi.$$

$H(t)$ is increasing on $\left[0, \dfrac{\pi}{2}\right] \cup \left[\dfrac{3\pi}{2}, 2\pi\right]$ and

decreasing on $\left[\dfrac{\pi}{2}, \dfrac{3\pi}{2}\right]$.

11. $f''(x) = 2$; $2 > 0$ for all x. $f(x)$ is concave up for all x; no inflection points.

13. $T''(t) = 18t$; $18t > 0$ when $t > 0$. $T(t)$ is concave up on $(0, \infty)$ and concave down on $(-\infty, 0)$; $(0, 0)$ is the only inflection point.

15. $q''(x) = 12x^2 - 36x - 48$; $q''(x) > 0$ when $x < -1$ and $x > 4$. $q(x)$ is concave up on $(-\infty, -1) \cup (4, \infty)$ and concave down on $(-1, 4)$; inflection points are $(-1, -19)$ and $(4, -499)$.

17. $F''(x) = 2\sin^2 x - 2\cos^2 x + 4 = 6 - 4\cos^2 x;$

$6 - 4\cos^2 x > 0$ for all x since $0 \le \cos^2 x \le 1$.
$F(x)$ is concave up for all x; no inflection points.

19. $f'(x) = 3x^2 - 12; 3x^2 - 12 > 0$ when
$x < -2$ or $x > 2$.
$f(x)$ is increasing on $(-\infty, -2] \cup [2, \infty)$ and
decreasing on $[-2, 2]$.
$f''(x) = 6x;\ 6x > 0$ when $x > 0$. $f(x)$ is concave up
on $(0, \infty)$ and concave down on $(-\infty, 0)$.

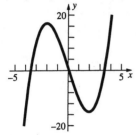

21. $g'(x) = 12x^3 - 12x^2 = 12x^2(x-1);\ g'(x) > 0$
when $x > 1$. $g(x)$ is increasing on $[1, \infty)$ and
decreasing on $(-\infty, 1]$.

$g''(x) = 36x^2 - 24x = 12x(3x - 2);\ g''(x) > 0$

when $x < 0$ or $x > \dfrac{2}{3}$. $g(x)$ is concave up on

$(-\infty, 0) \cup \left(\dfrac{2}{3}, \infty\right)$ and concave down on $\left(0, \dfrac{2}{3}\right)$.

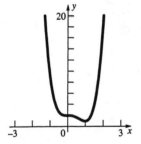

23. $G'(x) = 15x^4 - 15x^2 = 15x^2(x^2 - 1);\ G'(x) > 0$
when $x < -1$ or $x > 1$. $G(x)$ is increasing on
$(-\infty, -1] \cup [1, \infty)$ and decreasing on $[-1, 1]$.
$G''(x) = 60x^3 - 30x = 30x(2x^2 - 1);$

Split the x-axis into the intervals $\left(-\infty, -\dfrac{1}{\sqrt{2}}\right),$

$\left(-\dfrac{1}{\sqrt{2}}, 0\right), \left(0, \dfrac{1}{\sqrt{2}}\right), \left(\dfrac{1}{\sqrt{2}}, \infty\right).$

Test points: $x = -1, -\dfrac{1}{2}, \dfrac{1}{2}, 1;\ G''(-1) = -30,$

$G''\left(-\dfrac{1}{2}\right) = \dfrac{15}{2}, G''\left(\dfrac{1}{2}\right) = -\dfrac{15}{2}, G''(1) = 30.$

$G(x)$ is concave up on $\left(-\dfrac{1}{\sqrt{2}}, 0\right) \cup \left(\dfrac{1}{\sqrt{2}}, \infty\right)$ and

concave down on $\left(-\infty, -\dfrac{1}{\sqrt{2}}\right) \cup \left(0, \dfrac{1}{\sqrt{2}}\right).$

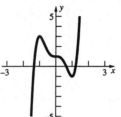

25. $f'(x) = \dfrac{\cos x}{2\sqrt{\sin x}};\ f'(x) > 0$ when $0 < x < \dfrac{\pi}{2}$. $f(x)$

is increasing on $\left[0, \dfrac{\pi}{2}\right]$ and decreasing on

$\left[\dfrac{\pi}{2}, \pi\right].$

$f''(x) = \dfrac{-\cos^2 x - 2\sin^2 x}{4\sin^{3/2} x};\ f''(x) < 0$ for all x in

$(0, \infty)$. $f(x)$ is concave down on $(0, \pi)$.

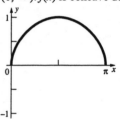

27. $f'(x) = \dfrac{2 - 5x}{3x^{1/3}};\ 2 - 5x > 0$ when $x < \dfrac{2}{5},\ f'(x)$

does not exist at $x = 0$.
Split the x-axis into the intervals $(-\infty, 0),$

$\left(0, \dfrac{2}{5}\right), \left(\dfrac{2}{5}, \infty\right).$

Test points: $-1, \dfrac{1}{5}, 1; f'(-1) = -\dfrac{7}{3},$

$f'\left(\dfrac{1}{5}\right) = \dfrac{\sqrt[3]{5}}{3}, f'(1) = -1.$

$f(x)$ is increasing on $\left[0, \dfrac{2}{5}\right]$ and decreasing on

$(-\infty, 0] \cup \left[\dfrac{2}{5}, \infty\right).$

$f''(x) = \dfrac{-2(5x+1)}{9x^{4/3}};\ -2(5x + 1) > 0$ when

$x < -\dfrac{1}{5}, f''(x)$ does not exist at $x = 0$.

At the top right:

$G(x)$ is concave up on $\left(-\dfrac{1}{\sqrt{2}}, 0\right) \cup \left(\dfrac{1}{\sqrt{2}}, \infty\right)$ and

concave down on $\left(-\infty, -\dfrac{1}{\sqrt{2}}\right) \cup \left(0, \dfrac{1}{\sqrt{2}}\right).$

Test points: $-1, -\dfrac{1}{10}, 1; f''(-1) = \dfrac{8}{9}$,

$f''\left(-\dfrac{1}{10}\right) = -\dfrac{10^{4/3}}{9}, f(1) = -\dfrac{4}{3}.$

$f(x)$ is concave up on $\left(-\infty, -\dfrac{1}{5}\right)$ and concave

down on $\left(-\dfrac{1}{5}, 0\right) \cup (0, \infty).$

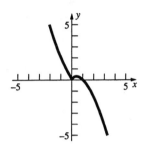

29.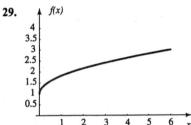

31.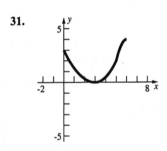

33.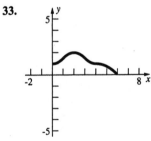

35. $f(x) = ax^2 + bx + c; f'(x) = 2ax + b;$
$f''(x) = 2a$

An inflection point would occur where $f''(x) = 0$, or $2a = 0$. This would only occur when $a = 0$, but if $a = 0$, the equation is not quadratic. Thus, quadratic functions have no points of inflection.

37. Suppose that there are points x_1 and x_2 in I where $f'(x_1) > 0$ and $f'(x_2) < 0$. Since f' is continuous on I, the Intermediate Value Theorem says that there is some number c between x_1 and x_2 such that $f'(c) = 0$, which is a contradiction. Thus, either $f'(x) > 0$ for all x in I and f is increasing throughout I or $f'(x) < 0$ for all x in I and f is decreasing throughout I.

39. a. Let $f(x) = x^2$ and let $I = [0, a], a > y$.
$f'(x) = 2x > 0$ on I. Therefore, $f(x)$ is increasing on I, so $f(x) < f(y)$ for $x < y$.

   b. Let $f(x) = \sqrt{x}$ and let $I = [0, a], a > y$.
$f'(x) = \dfrac{1}{2\sqrt{x}} > 0$ on I. Therefore, $f(x)$ is increasing on I, so $f(x) < f(y)$ for $x < y$.

   c. Let $f(x) = \dfrac{1}{x}$ and let $I = [0, a], a > y$.
$f'(x) = -\dfrac{1}{x^2} < 0$ on I. Therefore $f(x)$ is decreasing on I, so $f(x) > f(y)$ for $x < y$.

41. $f''(x) = \dfrac{3b - ax}{4x^{5/2}}$. If $(4, 13)$ is an inflection point then $13 = 2a + \dfrac{b}{2}$ and $\dfrac{3b - 4a}{4 \cdot 32} = 0$. Solving these equations simultaneously, $a = \dfrac{39}{8}$ and $b = \dfrac{13}{2}$.

43. a. $[f(x) + g(x)]' = f'(x) + g'(x).$
Since $f'(x) > 0$ and $g'(x) > 0$ for all x, $f'(x) + g'(x) > 0$ for all x. No additional conditions are needed.

   b. $[f(x) \cdot g(x)]' = f(x)g'(x) + f'(x)g(x).$
$f(x)g'(x) + f'(x)g(x) > 0$ if
$f(x) > -\dfrac{f'(x)}{g'(x)} g(x)$ for all x.

c. $[f(g(x))]' = f'(g(x))g'(x)$.

Since $f'(x) > 0$ and $g'(x) > 0$ for all x, $f'(g(x))g'(x) > 0$ for all x. No additional conditions are needed.

45. a.

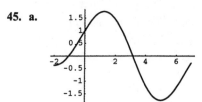

b. $f'(x) < 0 : (1.3, 5.0)$

c. $f''(x) < 0 : (-0.25, 3.1) \cup (6.5, 7]$

d. $f'(x) = \cos x - \dfrac{1}{2}\sin\dfrac{x}{2}$

e. $f''(x) = -\sin x - \dfrac{1}{4}\cos x$

47. $f'(x) > 0$ on $(-0.598, 0.680)$

f is increasing on $[-0.598, 0.680]$.

49. $\dfrac{dV}{dt} = 2 \text{ in}^3 / \text{sec}$

The cup is a portion of a cone with the bottom cut off. If we let x represent the height of the missing cone, we can use similar triangles to show that

$$\frac{x}{3} = \frac{x+5}{3.5}$$

$$3.5x = 3x + 15$$

$$.5x = 15$$

$$x = 30 .$$

4.3 Concepts Review

1. maximum

3. maximum

Similar triangles can be used again to show that, at any given time, the radius of the cone at water level is

$$r = \frac{h+30}{20}$$

Therefore, the volume of water can be expressed as

$$V = \frac{\pi(h+30)^3}{1200} - \frac{45\pi}{2} .$$

We also know that $V = 2t$ from above. Setting the two volume equations equal to each other and solving for h gives

$$h = \sqrt[3]{\frac{2400}{\pi}t + 27000} - 30 .$$

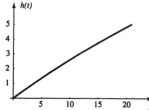

51. $V = 3t,\ 0 \le t \le 8$. The height is always increasing, so $h'(t) > 0$. The rate of change of the height decreases from time $t = 0$ until time t_1 when the water reaches the middle of the rounded bottom part. The rate of change then increases until time $t_2 = 2t_1$ when the water reaches the neck of the vase. Thus $h''(t) < 0$ for $0 < t < t_1$ and $h''(t) > 0$ for $t_1 < t < t_2$. For $t_2 < t < t_3$, the rate of change in the height increases until the water reaches the middle of the neck. Then the rate of change decreases until $t = 8$ and the vase is full. Thus, $h''(t) > 0$ for $t_2 < t < t_3$ and $h''(t) < 0$ for $t_3 < t < 8$.

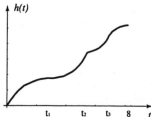

Problem Set 4.3

1. $f'(x) = 3x^2 - 12x = 3x(x-4)$

Critical points: 0, 4

$f'(x) > 0$ on $(-\infty, 0)$, $f'(x) < 0$ on $(0, 4)$,

$f'(x) > 0$ on $(4, \infty)$

$f''(x) = 6x - 12; \ f''(0) = -12, \ f''(4) = 12.$

Local minimum at $x = 4$; local maximum at $x = 0$

3. $f'(\theta) = 2\cos 2\theta; \ 2\cos 2\theta \neq 0$ on $\left(0, \dfrac{\pi}{4}\right)$

No critical points; no local maxima or minima on $\left(0, \dfrac{\pi}{4}\right)$.

5. $\Psi'(\theta) = 2\sin\theta\cos\theta$

$-\dfrac{\pi}{2} < \theta < \dfrac{\pi}{2}$

Critical point: 0

$\Psi'(\theta) < 0$ on $\left(-\dfrac{\pi}{2}, 0\right), \ \Psi'(\theta) > 0$ on $\left(0, \dfrac{\pi}{2}\right),$

$\Psi''(\theta) = 2\cos^2\theta - 2\sin^2\theta; \ \Psi''(0) = 2$

Local minima at $\theta = 0$

7. $f'(x) = 3x^2 - 3 = 3(x^2 - 1)$

Critical points: $-1, 1$

$f''(x) = 6x; \ f''(-1) = -6, \ f''(1) = 6$

Local minimum value $f(1) = -2$;

local maximum value $f(-1) = 2$

9. $H'(x) = 4x^3 - 6x^2 = 2x^2(2x - 3)$

Critical points: $0, \dfrac{3}{2}$

$H''(x) = 12x^2 - 12x = 12x(x - 1); \ H''(0) = 0,$

$H''\left(\dfrac{3}{2}\right) = 9$

$H'(x) < 0$ on $(-\infty, 0), H'(x) < 0$ on $\left(0, \dfrac{3}{2}\right)$

Local minimum value $H\left(\dfrac{3}{2}\right) = -\dfrac{27}{16}$; no local

maximum values ($x = 0$ is neither a local minimum nor maximum)

11. $g'(t) = -\dfrac{2}{3(t-2)^{1/3}}; g'(t)$ does not exist at $t = 2$.

Critical point: 2

$g'(1) = \dfrac{2}{3}, g'(3) = -\dfrac{2}{3}$

No local minimum values; local maximum value $g(2) = \pi$.

13. $f'(t) = 1 + \dfrac{1}{t^2}$

No critical points

No local minimum or maximum values

15. $\Lambda'(\theta) = -\dfrac{1}{1+\sin\theta}; \Lambda'(\theta)$ does not exist at

$\theta = \dfrac{3\pi}{2},$ but $\Lambda(\theta)$ does not exist at that point either.

No critical points

No local minimum or maximum values

17. $F'(x) = \dfrac{3}{\sqrt{x}} - 4; \dfrac{3}{\sqrt{x}} - 4 = 0$ when $x = \dfrac{9}{16}$

Critical points: $0, \dfrac{9}{16}, 4$

$F(0) = 0, \ F\left(\dfrac{9}{16}\right) = \dfrac{9}{4}, \ F(4) = -4$

Minimum value $F(4) = -4$; maximum value $F\left(\dfrac{9}{16}\right) = \dfrac{9}{4}$

19. $f'(x) = 64(-1)(\sin x)^{-2}\cos x$

$+ 27(-1)(\cos x)^{-2}(-\sin x)$

$= -\dfrac{64\cos x}{\sin^2 x} + \dfrac{27\sin x}{\cos^2 x}$

$= \dfrac{(3\sin x - 4\cos x)(9\sin^2 x + 12\cos x\sin x + 16\cos^2 x)}{\sin^2 x\cos^2 x}$

On $\left(0, \dfrac{\pi}{2}\right), f'(x) = 0$ only where $3\sin x = 4\cos x$;

$\tan x = \dfrac{4}{3}$;

$x = \tan^{-1}\dfrac{4}{3} \approx 0.9273$

Critical point: 0.9273

For $0 < x < 0.9273, \ f'(x) < 0,$ while for

$0.9273 < x < \dfrac{\pi}{2}, f'(x) > 0$

Minimum value $f\left(\tan^{-1}\dfrac{4}{3}\right) = \dfrac{64}{\frac{4}{5}} + \dfrac{27}{\frac{3}{5}} = 125$;

no maximum value

21. $g'(x) = 2x + \dfrac{(8-x)^2(32x) - (16x^2)2(8-x)(-1)}{(8-x)^4}$

$= 2x + \dfrac{256x}{(8-x)^3} = \dfrac{2x[(8-x)^3 + 128]}{(8-x)^3}$

For $x > 8, \ g'(x) = 0$ when $(8-x)^3 + 128 = 0$;

$(8-x)^3 = -128; 8 - x = -\sqrt[3]{128}$;

$x = 8 + 4\sqrt[3]{2} \approx 13.04$

$g'(x) < 0$ on $(8, 8 + 4\sqrt[3]{2}),$

$g'(x) > 0$ on $(8 + 4\sqrt[3]{2}, \infty)$

$g(13.04) \approx 277$ is the minimum value

23. $f'(x) = 0$ when $x = 0$ and $x = 1$. On the interval $(-\infty, 0)$ we get $f'(x) < 0$. On $(0, \infty)$, we get $f'(x) > 0$. Thus there is a local min at $x = 0$ but no local max.

25. $f'(x) = 0$ at $x = 1, 2, 3, 4$; $f'(x)$ is negative on $(3, 4)$ and positive on $(-\infty, 1) \cup (1, 2) \cup (2, 3) \cup (4, \infty)$ Thus, the function has a local minimum at $x = 4$ and a local maximum at $x = 3$.

27. Critical points: $-2, -1, 2, 3$
Split the x-axis into the intervals $(-\infty, -2)$, $(-2, -1), (-1, 2), (2, 3), (3, \infty)$.
Use test values: $-3, -\dfrac{3}{2}, 0, \dfrac{5}{2}$, and 4 to find that
$f'(x) < 0$ on $(-2, -1) \cup (-1, 2) \cup (2, 3)$ and
$f'(x) > 0$ on $(-\infty, -2) \cup (3, \infty)$.
Local maximum at $x = -2$; local minimum at $x = 3$.

29. Written response (graph)

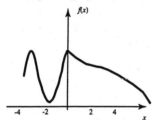

a. Increasing: $(-\infty, -3] \cup [-1, 0)$
Decreasing: $[-3, -1] \cup (0, \infty)$

b. Concave up: $(-2, 0) \cup (0, 2)$
Concave down: $(-\infty, -2) \cup (2, \infty)$

c. Local maximum at $x = -3$; local minimum at $x = -1$
$f'(0), f''(0)$ do not exist, but $f(x)$ is continuous. $x = 0$ is a critical point.
$f'(0) > 0$ on $[-1, 0]$, and $f'(0) < 0$ on $(0, \infty)$.
Thus, at $x = 0$ there is a local maximum.

4.4 Concepts Review

1. $0 < x < \infty$

3. $3x + \dfrac{300}{x}$

d. Inflection points at $x = -2, 2$

31. $f'(x) = 5x^4 - 15x^2$
$f''(x) = -30x + 20x^3$
Solving $f'(x) = 0$ gives critical numbers of $x = 0, \pm\sqrt{3}$. Using the second derivative we get that $f''(0) = 0$, $f''(\sqrt{3}) > 0$, and $f''(-\sqrt{3}) < 0$. The second derivative fails at $x = 0$ (there is actually a point of inflection there) but indicates that we have a local maximum $f(-\sqrt{3}) \approx 14.392$ and a local minimum $f(\sqrt{3}) \approx -6.392$ (also the global minimum). The global maximum, $f(2.5) \approx 23.531$, on $[-2, 2.5]$ occurs at the right endpoint.

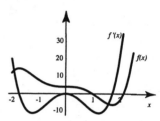

33. $f'(x) = (x^3 - x)\sec^2 x + (3x^2 - 1)\tan x$
$f''(x) = 6x \tan x$
$\qquad + 2(\sec^2 x)(3x^2 - 1 + x(x^2 - 1)\tan x)$

Graphing the first derivative and zooming in gives critical numbers of $x = 0$ and $x \approx -0.745, 0.745$. Using the second derivative gives

$f''(-0.745) > 0; f''(0) < 0; f''(0.745) > 0$. So have a local maximum of $f(0) = -2$ and local minimums of $f(-0.745) = f(0.745) \approx -0.306$ (also the global minimum). There is no global maximum since the graph blows up at the endpoints.

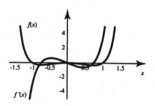

Problem Set 4.4

1. Let x be one number, y be the other, and Q be the sum of the squares.
$xy = -16$

$$y = -\frac{16}{x}$$

The possible values for x are in $(-\infty, 0)$ or $(0, \infty)$.

$$Q = x^2 + y^2 = x^2 + \frac{256}{x^2}$$

$$\frac{dQ}{dx} = 2x - \frac{512}{x^3}$$

$$2x - \frac{512}{x^3} = 0$$

$$x^4 = 256$$

$$x = \pm 4$$

The critical points are –4, 4.

$\frac{dQ}{dx} < 0$ on $(-\infty, -4)$ and $(0, 4)$. $\frac{dQ}{dx} > 0$ on $(-4, 0)$ and $(4, \infty)$.

When $x = -4$, $y = 4$ and when $x = 4$, $y = -4$. The two numbers are –4 and 4.

3. Let x be the number.

$$Q = \sqrt[4]{x} - 2x$$

x will be in the interval $(0, \infty)$.

$$\frac{dQ}{dx} = \frac{1}{4}x^{-3/4} - 2$$

$$\frac{1}{4}x^{-3/4} - 2 = 0$$

$$x^{-3/4} = 8$$

$$x = \frac{1}{16}$$

$\frac{dQ}{dx} > 0$ on $\left(0, \frac{1}{16}\right)$ and $\frac{dQ}{dx} < 0$ on $\left(\frac{1}{16}, \infty\right)$

Q attains its maximum value at $x = \frac{1}{16}$.

5. Let Q be the square of the distance between (x, y) and $(0, 5)$.

$$Q = (x - 0)^2 + (y - 5)^2 = x^2 + (x^2 - 5)^2$$

$$= x^4 - 9x^2 + 25$$

$$\frac{dQ}{dx} = 4x^3 - 18x$$

$$4x^3 - 18x = 0$$

$$2x(2x^2 - 9) = 0$$

$$x = 0, \pm \frac{3}{\sqrt{2}}$$

$\frac{dQ}{dx} < 0$ on $\left(-\infty, -\frac{3}{\sqrt{2}}\right)$ and $\left(0, \frac{3}{\sqrt{2}}\right)$.

$\frac{dQ}{dx} > 0$ on $\left(-\frac{3}{\sqrt{2}}, 0\right)$ and $\left(\frac{3}{\sqrt{2}}, \infty\right)$.

When $x = -\frac{3}{\sqrt{2}}$, $y = \frac{9}{2}$ and when $x = \frac{3}{\sqrt{2}}$,

$$y = \frac{9}{2}.$$

The points are $\left(-\frac{3}{\sqrt{2}}, \frac{9}{2}\right)$ and $\left(\frac{3}{\sqrt{2}}, \frac{9}{2}\right)$.

7. $xy = 900$; $y = \frac{900}{x}$

The possible values for x are in $(0, \infty)$.

$$Q = 4x + 3y = 4x + 3\left(\frac{900}{x}\right) = 4x + \frac{2700}{x}$$

$$\frac{dQ}{dx} = 4 - \frac{2700}{x^2}$$

$$4 - \frac{2700}{x^2} = 0$$

$$x^2 = 675$$

$$x = \pm 15\sqrt{3}$$

$x = 15\sqrt{3}$ is the only critical point in $(0, \infty)$.

$\frac{dQ}{dx} < 0$ on $(0, 15\sqrt{3})$ and

$\frac{dQ}{dx} > 0$ on $(15\sqrt{3}, \infty)$.

When $x = 15\sqrt{3}$, $y = \frac{900}{15\sqrt{3}} = 20\sqrt{3}$.

Q has a minimum when $x = 15\sqrt{3} \approx 25.98$ ft and $y = 20\sqrt{3} \approx 34.64$ ft.

9. $xy = 300$; $y = \frac{300}{x}$

The possible values for x are in $(0, \infty)$.

$$Q = 3(6x + 2y) + 2(2y) = 18x + 10y = 18x + \frac{3000}{x}$$

$$\frac{dQ}{dx} = 18 - \frac{3000}{x^2}$$

$$18 - \frac{3000}{x^2} = 0$$

$$x^2 = \frac{500}{3}$$

$$x = \pm \frac{10\sqrt{5}}{\sqrt{3}}$$

$x = \frac{10\sqrt{5}}{\sqrt{3}}$ is the only critical point in $(0, \infty)$.

$\frac{dQ}{dx} < 0$ on $\left(0, \frac{10\sqrt{5}}{\sqrt{3}}\right)$ and

$\frac{dQ}{dx} > 0$ on $\left(\frac{10\sqrt{5}}{\sqrt{3}}, \infty\right)$.

When $x = \dfrac{10\sqrt{5}}{\sqrt{3}}, y = \dfrac{300}{\dfrac{10\sqrt{5}}{\sqrt{3}}} = 6\sqrt{15}$

Q has a minimum when $x = \dfrac{10\sqrt{5}}{\sqrt{3}} \approx 12.91$ ft and

11. We are interested in the global extrema for the distance of the object from the observer. We obtain the same extrema by considering the squared distance

$$D(x) = (x - 2.6656)^2 + (42 + x - .08x^2)^2$$

The first derivative is given by

$$D'(x) = \dfrac{16}{625}x^3 - \dfrac{12}{25}x^2 - \dfrac{236}{25}x - \dfrac{49168}{625}.$$

This function can be plotted on a graphing calculator and then analyzed with the TRACE and ZOOM features. Doing so yields critical numbers of $x \approx 6.8267, 28.0$. The second derivative is

$$D''(x) = \dfrac{48}{625}x^2 - \dfrac{24}{25}x - \dfrac{236}{25}$$

which gives $D''(6.8267) < 0$ and $D''(28) > 0$. So we have a local maximum at $x \approx 6.827$ and a local minimum at $x \approx 28.0$. Thus, the object is at the point $(28, 7.28)$ when it is closest to the observer and at the point $(6.8267, 45.098)$ when it is farthest from the observer.

13. Let x be the distance from P to where the woman lands the boat. She must row a distance of $\sqrt{x^2 + 4}$ miles and walk $10 - x$ miles. This will take her $T(x) = \dfrac{\sqrt{x^2 + 4}}{3} + \dfrac{10 - x}{4}$ hours;

$0 \le x \le 10$. $T'(x) = \dfrac{x}{3\sqrt{x^2 + 4}} - \dfrac{1}{4}; T'(x) = 0$

when $x = \dfrac{6}{\sqrt{7}}$.

$T(0) = \dfrac{19}{6}$ hr $= 3$ hr, 10 min;

$T\left(\dfrac{6}{\sqrt{7}}\right) = \dfrac{15 + \sqrt{7}}{6} \approx 2.94$ hr,

$T(10) = \dfrac{\sqrt{104}}{3} \approx 3.40$ hr

She should land the boat $\dfrac{6}{\sqrt{7}} \approx 2.27$ mi down the shore from P.

15. $T(x) = \dfrac{\sqrt{x^2 + 4}}{20} + \dfrac{10 - x}{4}, 0 \le x \le 10$.

$T'(x) = \dfrac{x}{20\sqrt{x^2 + 4}} - \dfrac{1}{4}; T'(x) = 0$ has no solution.

$T(0) = \dfrac{2}{20} + \dfrac{10}{4} = \dfrac{13}{5}$ hr $= 2$ hr, 36 min

$T(10) = \dfrac{\sqrt{104}}{20} \approx 0.5$ hr

She should take the boat all the way to town.

17. Let the coordinates of the first ship at 7:00 a.m. be $(0, 0)$. Thus, the coordinates of the second ship at 7:00 a.m. are $(-60, 0)$. Let t be the time in hours since 7:00 a.m. The coordinates of the first and second ships at t are $(-20t, 0)$ and $\left(-60 + 15\sqrt{2}t, -15\sqrt{2}t\right)$ respectively. Let D be the square of the distances at t.

$$D = \left(-20t + 60 - 15\sqrt{2}t\right)^2 + \left(0 + 15\sqrt{2}t\right)^2$$

$$= \left(1300 + 600\sqrt{2}\right)t^2 - \left(2400 + 1800\sqrt{2}\right)t + 3600$$

$$\dfrac{dD}{dt} = 2\left(1300 + 600\sqrt{2}\right)t - \left(2400 + 1800\sqrt{2}\right)$$

$2\left(1300 + 600\sqrt{2}\right)t - \left(2400 + 1800\sqrt{2}\right) = 0$ when

$$t = \dfrac{12 + 9\sqrt{2}}{13 + 6\sqrt{2}} \approx 1.15 \text{ hrs or 1 hr, 9 min}$$

D is the minimum at $t = \dfrac{12 + 9\sqrt{2}}{13 + 6\sqrt{2}}$ since $\dfrac{d^2 D}{dt^2} > 0$

for all t.
The ships are closest at 8:09 A.M.

19. Let x be the radius of the base of the cylinder and h the height.

$$V = \pi x^2 h; r^2 = x^2 + \left(\dfrac{h}{2}\right)^2; x^2 = r^2 - \dfrac{h^2}{4}$$

$$V = \pi\left(r^2 - \dfrac{h^2}{4}\right)h = \pi h r^2 - \dfrac{\pi h^3}{4}$$

$$\dfrac{dV}{dh} = \pi r^2 - \dfrac{3\pi h^2}{4}; V' = 0 \text{ when } h = \pm\dfrac{2\sqrt{3}r}{3}$$

Since $\dfrac{d^2 V}{dh^2} = -\dfrac{3\pi h}{2}$, the volume is maximized

when $h = \dfrac{2\sqrt{3}r}{3}$.

$$V = \pi\left(\dfrac{2\sqrt{3}}{3}r\right)r^2 - \dfrac{\pi\left(\dfrac{2\sqrt{3}}{3}r\right)^3}{4}$$

$$= \dfrac{2\pi\sqrt{3}}{3}r^3 - \dfrac{2\pi\sqrt{3}}{9}r^3 = \dfrac{4\pi\sqrt{3}}{9}r^3$$

21. Let x be the radius of the cylinder, r the radius of the sphere, and h the height of the cylinder.

$$A = 2\pi x h; \quad r^2 = x^2 + \frac{h^2}{4}; \quad x = \sqrt{r^2 - \frac{h^2}{4}}$$

$$A = 2\pi\sqrt{r^2 - \frac{h^2}{4}}\,h = 2\pi\sqrt{h^2 r^2 - \frac{h^4}{4}}$$

$$\frac{dA}{dh} = \frac{\pi\left(2r^2 h - h^3\right)}{\sqrt{h^2 r^2 - \frac{h^4}{4}}}; \quad A' = 0 \text{ when } h = 0, \pm\sqrt{2}r$$

$\frac{dA}{dh} > 0$ on $(0, \sqrt{2}r)$ and $\frac{dA}{dh} < 0$ on $(\sqrt{2}r, 2r)$,

so A is a maximum when $h = \sqrt{2}r$.

The dimensions are $h = \sqrt{2}r, x = \dfrac{r}{\sqrt{2}}$.

23. Let x be the length of a side of the square, so $\dfrac{100 - 4x}{3}$ is the side of the triangle, $0 \le x \le 25$

$$A = x^2 + \frac{1}{2}\left(\frac{100 - 4x}{3}\right)\frac{\sqrt{3}}{2}\left(\frac{100 - 4x}{3}\right)$$

$$= x^2 + \frac{\sqrt{3}}{4}\left(\frac{10{,}000 - 800x + 16x^2}{9}\right)$$

$$\frac{dA}{dx} = 2x - \frac{200\sqrt{3}}{9} + \frac{8\sqrt{3}}{9}x; \quad A' = 0 \text{ when } x \approx 10.87$$

Critical points: $x = 0, 10.87, 25$
At $x = 0$, $A \approx 481$; at $x = 10.87$, $A \approx 272$; at $x = 25$, $A = 625$.

a. For minimum area, the cut should be approximately $4(10.87) = 43.48$ cm from one end and the shorter length should be bent to form the square.

b. For maximum area, the wire should not be cut, it should be bent to form a square.

25. Let r be the radius of the cylinder and h the height of the cylinder.

$$V = \pi r^2 h + \frac{2}{3}\pi r^3; \quad h = \frac{V - \frac{2}{3}\pi r^3}{\pi r^2} = \frac{V}{\pi r^2} - \frac{2}{3}r$$

Let k be the cost per square foot of the cylindrical wall. The cost is

$$C = k(2\pi r h) + 2k(2\pi r^2)$$

$$= k\left(2\pi r\left(\frac{V}{\pi r^2} - \frac{2}{3}r\right) + 4\pi r^2\right) = k\left(\frac{2V}{r} + \frac{8\pi r^2}{3}\right)$$

$$\frac{dC}{dr} = k\left(-\frac{2V}{r^2} + \frac{16\pi r}{3}\right); \quad k\left(-\frac{2V}{r^2} + \frac{16\pi r}{3}\right) = 0$$

when $r^3 = \dfrac{3V}{8\pi}, r = \dfrac{1}{2}\left(\dfrac{3V}{\pi}\right)^{1/3}$

$$h = \frac{4V}{\pi\left(\frac{3V}{\pi}\right)^{2/3}} - \frac{1}{3}\left(\frac{3V}{\pi}\right)^{1/3} = \left(\frac{3V}{\pi}\right)^{1/3}$$

For a given volume V, the height of the cylinder is $\left(\dfrac{3V}{\pi}\right)^{1/3}$ and the radius is $\dfrac{1}{2}\left(\dfrac{3V}{\pi}\right)^{1/3}$.

27. $A = \dfrac{r^2\theta}{2}; \quad \theta = \dfrac{2A}{r^2}$

The perimeter is

$$Q = 2r + r\theta = 2r + \frac{2Ar}{r^2} = 2r + \frac{2A}{r}$$

$$\frac{dQ}{dr} = 2 - \frac{2A}{r^2}; Q' = 0 \text{ when } r = \sqrt{A}$$

$$\theta = \frac{2A}{\left(\sqrt{A}\right)^2} = 2$$

$$\frac{d^2Q}{dr^2} = \frac{4A}{r^3} > 0, \text{ so this minimizes the perimeter.}$$

29. Let d_1 be the distance that the light travels in medium 1 and let d_2 be the distance the light travels in medium 2.

$$d_1 = \sqrt{a^2 + x^2}, d_2 = \sqrt{b^2 + (d - x)^2}$$

The time that it takes the light to travel from A to B is

$$t = \frac{d_1}{c_1} + \frac{d_2}{c_2} = \frac{\sqrt{a^2 + x^2}}{c_1} + \frac{\sqrt{b^2 + (d - x)^2}}{c_2}$$

$$\frac{dt}{dx} = \frac{x}{c_1\sqrt{a^2 + x^2}} - \frac{d - x}{c_2\sqrt{b^2 + (d - x)^2}}$$

$$= \frac{\sin\theta_1}{c_1} - \frac{\sin\theta_2}{c_2}$$

$$\frac{dt}{dx} = 0 \text{ implies } \frac{\sin\theta_1}{c_1} = \frac{\sin\theta_2}{c_2}.$$

31. Consider the following sketch.

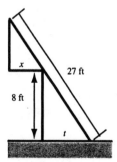

By similar triangles, $\dfrac{x}{27-\sqrt{t^2+64}}=\dfrac{t}{\sqrt{t^2+64}}$.

$$x=\frac{27t}{\sqrt{t^2+64}}-t$$

$$\frac{dx}{dt}=\frac{27\sqrt{t^2+64}-\frac{27t^2}{\sqrt{t^2+64}}}{t^2+64}-1=\frac{1728}{(t^2+64)^{3/2}}-1$$

$$\frac{1728}{(t^2+64)^{3/2}}-1=0 \text{ when } t=4\sqrt{5}$$

$$\frac{d^2x}{dt^2}=\frac{-5184t}{(t^2+64)^{5/2}};\ \left.\frac{d^2x}{dt^2}\right|_{t=4\sqrt{5}}<0$$

Therefore

$$x=\frac{27\left(4\sqrt{5}\right)}{\sqrt{\left(4\sqrt{5}\right)^2+64}}-4\sqrt{5}=5\sqrt{5}\approx 11.18\ \text{ft is the}$$

maximum horizontal overhang.

33. Consider the figure below.

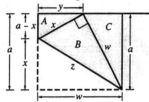

a. $y=\sqrt{x^2-(a-x)^2}=\sqrt{2ax-a^2}$

Area of $A=A=\dfrac{1}{2}(a-x)y$

$=\dfrac{1}{2}(a-x)\sqrt{2ax-a^2}$

$$\frac{dA}{dx}=-\frac{1}{2}\sqrt{2ax-a^2}+\frac{\frac{1}{2}(a-x)\left(\frac{1}{2}\right)(2a)}{\sqrt{2ax-a^2}}$$

$$=\frac{a^2-\frac{3}{2}ax}{\sqrt{2ax-a^2}}$$

$$\frac{a^2-\frac{3}{2}ax}{\sqrt{2ax-a^2}}=0 \text{ when } x=\frac{2a}{3}.$$

$\dfrac{dA}{dx}>0$ on $\left(\dfrac{a}{2},\dfrac{2a}{3}\right)$ and $\dfrac{dA}{dx}<0$ on $\left(\dfrac{2a}{3},a\right)$,

so $x=\dfrac{2a}{3}$ maximizes the area of triangle A.

b. Triangle A is similar to triangle C, so

$$w=\frac{ax}{y}=\frac{ax}{\sqrt{2ax-a^2}}$$

Area of $B=B=\dfrac{1}{2}xw=\dfrac{ax^2}{2\sqrt{2ax-a^2}}$

$$\frac{dB}{dx}=\frac{a}{2}\left(\frac{2x\sqrt{2ax-a^2}-x^2\frac{a}{\sqrt{2ax-a^2}}}{2ax-a^2}\right)$$

$$=\frac{a}{2}\left(\frac{2x(2ax-a^2)-ax^2}{(2ax-a^2)^{3/2}}\right)=\frac{a}{2}\left(\frac{3ax^2-2xa^2}{(2ax-a^2)^{3/2}}\right)$$

$$\frac{a^2}{2}\left(\frac{3x^2-2xa}{(2ax-a^2)^{3/2}}\right)=0 \text{ when } x=0,\frac{2a}{3}$$

Since $x=0$ is not possible, $x=\dfrac{2a}{3}$.

$\dfrac{dB}{dx}<0$ on $\left(\dfrac{a}{2},\dfrac{2a}{3}\right)$ and $\dfrac{dB}{dx}>0$ on $\left(\dfrac{2a}{3},a\right)$,

so $x=\dfrac{2a}{3}$ minimizes the area of triangle B.

c. $z=\sqrt{x^2+w^2}=\sqrt{x^2+\dfrac{a^2x^2}{2ax-a^2}}$

$$=\sqrt{\frac{2ax^3}{2ax-a^2}}$$

$$\frac{dz}{dx}=\frac{1}{2}\sqrt{\frac{2ax-a^2}{2ax^3}}\left(\frac{6ax^2(2ax-a^2)-2ax^3(2a)}{(2ax-a^2)^2}\right)$$

$$=\frac{4a^2x^3-3a^3x^2}{\sqrt{2ax^3(2ax-a^2)^3}}$$

$\dfrac{dz}{dx}=0$ when $x=0,\dfrac{3a}{4}$

$x=\dfrac{3a}{4}$

$\dfrac{dz}{dx}<0$ on $\left(\dfrac{a}{2},\dfrac{3a}{4}\right)$ and $\dfrac{dz}{dx}>0$ on $\left(\dfrac{3a}{4},a\right)$,

so $x=\dfrac{3a}{4}$ minimizes length z.

35. a. $L'(\theta) = 15(9 + 25 - 30\cos\theta)^{-1/2}\sin\theta = 15(34 - 30\cos\theta)^{-1/2}\sin\theta$

$L''(\theta) = -\dfrac{15}{2}(34 - 30\cos\theta)^{-3/2}(30\sin\theta)\sin\theta + 15(34 - 30\cos\theta)^{-1/2}\cos\theta$

$= -225(34 - 30\cos\theta)^{-3/2}\sin^2\theta + 15(34 - 30\cos\theta)^{-1/2}\cos\theta$

$= 15(34 - 30\cos\theta)^{-3/2}[-15\sin^2\theta + (34 - 30\cos\theta)\cos\theta]$

$= 15(34 - 30\cos\theta)^{-3/2}[-15\sin^2\theta + 34\cos\theta - 30\cos^2\theta]$

$= 15(34 - 30\cos\theta)^{-3/2}[-15 + 34\cos\theta - 15\cos^2\theta]$

$= -15(34 - 30\cos\theta)^{-3/2}[15\cos^2\theta - 34\cos\theta + 15]$

$L'' = 0$ when $\cos\theta = \dfrac{34 \pm \sqrt{(34)^2 - 4(15)(15)}}{2(15)} = \dfrac{5}{3}, \dfrac{3}{5}$

$\theta = \cos^{-1}\left(\dfrac{3}{5}\right)$

$L'\left(\cos^{-1}\left(\dfrac{3}{5}\right)\right) = 15\left(9 + 25 - 30\left(\dfrac{3}{5}\right)\right)^{-1/2}\left(\dfrac{4}{5}\right) = 3$

$L\left(\cos^{-1}\left(\dfrac{3}{5}\right)\right) = \left(9 + 25 - 30\left(\dfrac{3}{5}\right)\right)^{1/2} = 4$

$\phi = 90°$ since the resulting triangle is a 3-4-5 right triangle.

b. $L'(\theta) = 65(25 + 169 - 130\cos\theta)^{-1/2}\sin\theta = 65(194 - 130\cos\theta)^{-1/2}\sin\theta$

$L''(\theta) = -\dfrac{65}{2}(194 - 130\cos\theta)^{-3/2}(130\sin\theta)\sin\theta + 65(194 - 130\cos\theta)^{-1/2}\cos\theta$

$= -4225(194 - 130\cos\theta)^{-3/2}\sin^2\theta + 65(194 - 130\cos\theta)^{-1/2}\cos\theta$

$= 65(194 - 130\cos\theta)^{-3/2}[-65\sin^2\theta + (194 - 130\cos\theta)\cos\theta]$

$= 65(194 - 130\cos\theta)^{-3/2}[-65\sin^2\theta + 194\cos\theta - 130\cos^2\theta]$

$= 65(194 - 130\cos\theta)^{-3/2}[-65\cos^2\theta + 194\cos\theta - 65]$

$= -65(194 - 130\cos\theta)^{-3/2}[65\cos^2\theta - 194\cos\theta + 65]$

$L'' = 0$ when $\cos\theta = \dfrac{194 \pm \sqrt{(194)^2 - 4(65)(65)}}{2(65)} = \dfrac{13}{5}, \dfrac{5}{13} \Rightarrow \theta = \cos^{-1}\left(\dfrac{5}{13}\right)$

$L'\left(\cos^{-1}\left(\dfrac{5}{13}\right)\right) = 65\left(25 + 169 - 130\left(\dfrac{5}{13}\right)\right)^{1/2}\left(\dfrac{12}{13}\right) = 5$

$L\left(\cos^{-1}\left(\dfrac{5}{13}\right)\right) = \left(25 + 169 - 130\left(\dfrac{5}{13}\right)\right)^{1/2} = 12$

$\phi = 90°$ since the resulting triangle is a 5-12-13 right triangle.

c. When the tips are separating most rapidly, $\phi = 90°$, $L = \sqrt{m^2 - h^2}$, $L' = h$

d. $L'(\theta) = hm(h^2 + m^2 - 2hm\cos\theta)^{-1/2}\sin\theta$

$L''(\theta) = -h^2m^2(h^2 + m^2 - 2hm\cos\theta)^{-3/2}\sin^2\theta + hm(h^2 + m^2 - 2hm\cos\theta)^{-1/2}\cos\theta$

$= hm(h^2 + m^2 - 2hm\cos\theta)^{-3/2}[-hm\sin^2\theta + (h^2 + m^2)\cos\theta - 2hm\cos^2\theta]$

$= hm(h^2 + m^2 - 2hm\cos\theta)^{-3/2}[-hm\cos^2\theta + (h^2 + m^2)\cos\theta - hm]$

$= -hm(h^2 + m^2 - 2hm\cos\theta)^{-3/2}[hm\cos^2\theta - (h^2 + m^2)\cos\theta + hm]$

$L'' = 0$ when $hm\cos^2\theta - (h^2 + m^2)\cos\theta + hm = 0$

$(h\cos\theta - m)(m\cos\theta - h) = 0$

$\cos\theta = \dfrac{m}{h}, \dfrac{h}{m}$

Since $h < m$, $\cos\theta = \dfrac{h}{m}$ so $\theta = \cos^{-1}\left(\dfrac{h}{m}\right)$.

$L'\left(\cos^{-1}\left(\dfrac{h}{m}\right)\right) = hm\left(h^2 + m^2 - 2hm\left(\dfrac{h}{m}\right)\right)^{-1/2}\dfrac{\sqrt{m^2 - h^2}}{m} = hm(m^2 - h^2)^{-1/2}\dfrac{\sqrt{m^2 - h^2}}{m} = h$

$L\left(\cos^{-1}\left(\dfrac{h}{m}\right)\right) = \left(h^2 + m^2 - 2hm\left(\dfrac{h}{m}\right)\right)^{1/2} = \sqrt{m^2 - h^2}$

Since $h^2 + L^2 = m^2$, $\phi = 90°$.

37. Here we are interested in minimizing the distance between the earth and the asteroid. Using the coordinates P and Q for the two bodies, we can use the distance formula to obtain a suitable equation. However, for simplicity, we will minimize the squared distance to find the critical points. The squared distance between the objects is given by

$$D(t) = (93\cos(2\pi t) - 60\cos[2\pi(1.51t - 1)])^2$$
$$+ (93\sin(2\pi t) - 120\sin[2\pi(1.51t - 1)])^2$$

The first derivative is

$$D'(t) \approx -34359[\cos(2\pi t)][\sin(9.48761t)]$$
$$+ [\cos(9.48761t)][(204932\sin(9.48761t)$$
$$- 141643\sin(2\pi t))]$$

Plotting the function and its derivative reveal a periodic relationship due to the orbiting of the objects. Careful examination of the graphs reveals that there is indeed a minimum squared distance (and hence a minimum distance) that occurs only once. The critical value for this occurrence is $t \approx 13.8279$. This value gives a distance between the objects of

≈ 0.047851 million miles $= 47,851$ miles

39. **a.** $\dfrac{dS}{db} = \dfrac{d}{db}\sum_{i=1}^{n}[y_i - (5 + bx_i)]^2$

$= \sum_{i=1}^{n}\dfrac{d}{db}[y_i - (5 + bx_i)]^2$

$= \sum_{i=1}^{n}2(y_i - 5 - bx_i)(-x_i)$

$= 2\left[\sum_{i=1}^{n}\left(-x_i y_i + 5x_i + bx_i^2\right)\right]$

$= -2\sum_{i=1}^{n}x_i y_i + 10\sum_{i=1}^{n}x_i + 2b\sum_{i=1}^{n}x_i^2$

Setting $\dfrac{dS}{db} = 0$ gives

$0 = -2\sum_{i=1}^{n}x_i y_i + 10\sum_{i=1}^{n}x_i + 2b\sum_{i=1}^{n}x_i^2$

$0 = -\sum_{i=1}^{n}x_i y_i + 5\sum_{i=1}^{n}x_i + b\sum_{i=1}^{n}x_i^2$

$b\sum_{i=1}^{n}x_i^2 = \sum_{i=1}^{n}x_i y_i - 5\sum_{i=1}^{n}x_i$

$b = \dfrac{\displaystyle\sum_{i=1}^{n}x_i y_i - 5\sum_{i=1}^{n}x_i}{\displaystyle\sum_{i=1}^{n}x_i^2}$

You should check that this is indeed the value of b that minimizes the sum. Taking the second derivative yields

$\dfrac{d^2 S}{db^2} = 2\sum_{i=1}^{n}x_i^2$

which is always positive (unless all the x values are zero). Therefore, the value for b above does minimize the sum as required.

b. Using the formula from **a.**, we get that

$b = \dfrac{(2037) - 5(52)}{590} \approx 3.0119$

c. The Least Squares Regression line is

$y = 5 + 3.0119x$

Using this line, the predicted total number of labor hours to produce a lot of 15 brass bookcases is

$y = 5 + 3.0119(15)$

≈ 50.179 hours

4.5 Concepts Review

1. continuous

3. fixed; variable

Problem Set 4.5

1. Profit is $P(y) = ny - 10n$

$$= \frac{100y}{y-10} + 20y(100-y) - \frac{1000}{y-10} - 200(100-y)$$

$$= 20(-y^2 + 110y - 995)$$

$$\frac{dP}{dy} = 20(-2y + 110) = 40(55 - y)$$

$P'(y) = 0$ when $y = 55$

$P''(y) = -40$ so the maximum profit is when the items are sold at \$55 each.

3. $n = 100 + 10\dfrac{250 - p(n)}{5}$ so $p(n) = 300 - \dfrac{n}{2}$

$$R(n) = np(n) = 300n - \frac{n^2}{2}$$

5.

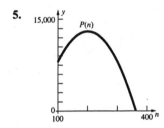

Estimate $n \approx 200$

$P'(n) = 200 - n$; $200 - n = 0$ when $n = 200$.

$P''(n) = -1$, so profit is maximum at $n = 200$.

7. $\dfrac{C(n)}{n} = \dfrac{1000}{n} + \dfrac{n}{1200}$

When $n = 800$, $\dfrac{C(n)}{n} \approx 1.9167$ or \$1.92 per unit.

$$\frac{dC}{dn} = \frac{n}{600}$$

$C'(800) \approx 1.333$ or \$1.33

9. a. $R(x) = xp(x) = 20x + 4x^2 - \dfrac{x^3}{3}$

$$\frac{dR}{dx} = 20 + 8x - x^2$$

b. Increasing when $\dfrac{dR}{dx} > 0$

$20 + 8x - x^2 > 0$ on $[0, 10)$

Total revenue is increasing if $0 \le x \le 10$.

c. $\dfrac{d^2R}{dx^2} = 8 - 2x$; $\dfrac{d^2R}{dx^2} = 0$ when $x = 4$

$\dfrac{d^3R}{dx^3} = -2$; $\dfrac{dR}{dx}$ is maximum at $x = 4$.

11. $R(x) = \dfrac{800x}{x+3} - 3x$

$$\frac{dR}{dx} = \frac{(x+3)(800) - 800x}{(x+3)^2} - 3 = \frac{2400}{(x+3)^2} - 3;$$

$\dfrac{dR}{dx} = 0$ when $x = 20\sqrt{2} - 3 \approx 25$

$x_1 = 25$; $R(25) \approx 639.29$

At x_1, $\dfrac{dR}{dx} = 0$.

13. $x = 4000 + \dfrac{6 - p(x)}{0.15}(250)$;

$$p(x) = 6 - (0.15)\frac{(x - 4000)}{250} = 8.4 - 0.0006x$$

$$R(x) = 8.4x - 0.0006x^2$$

$\dfrac{dR}{dx} = 8.4 - 0.0012x$; $\dfrac{dR}{dx} = 0$ when $x = 7000$

Revenue is maximum when

$p(x) = 8.4 - 0.0006(7000) = \$4.20$ per yard.

15. a. $C(x) = \begin{cases} 6000 + 1.40x & \text{if } 0 \le x \le 4500 \\ 6000 + 1.60x & \text{if } 4500 < x \end{cases}$

b. $x = 4000 + \dfrac{7 - p(x)}{0.10}(100)$;

$$p(x) = 7 - (0.10)\frac{x - 4000}{100}$$

$$p(x) = 11 - 0.001x$$

c. $R(x) = 11x - 0.001x^2$

For $0 \le x \le 4500$,

$P(x) = (11x - 0.001x^2) - (6000 + 1.40x)$

$= -6000 + 9.6x - 0.001x^2$

$\dfrac{dP}{dx} = 9.6 - 0.002x$; $\dfrac{dP}{dx} = 0$ when $x = 4800$;

this is not in the interval $[0, 4500]$.

The critical numbers are 0 and 4500.

$P(0) = -6000$ and $P(4500) = 16,950$

For $4500 < x$,

$P(x) = (11x - 0.001x^2) - (6000 + 1.60x)$

$$= -6000 + 9.4x - 0.001x^2$$

$$\frac{dP}{dx} = 9.4 - 0.002x; \frac{dP}{dx} = 0 \text{ when } x = 4700$$

$P(4700) = 16{,}090$

Therefore, the number of units for maximum profit is $x = 4500$.

17. $C(x) = \begin{cases} 200 + 4x - 0.01x^2 & \text{if } 0 \le x \le 300 \\ 800 + 3x - 0.01x^2 & \text{if } 300 < x \le 450 \end{cases}$

$P(x) = \begin{cases} -200 + 6x + 0.009x^2 & \text{if } 0 \le x \le 300 \\ -800 + 7x + 0.009x^2 & \text{if } 300 < x \le 450 \end{cases}$

There are no stationary points on the interval [0, 300]. On [300, 450]:

$$\frac{dP}{dx} = 7 + 0.018x; \frac{dP}{dx} = 0 \text{ when } x \approx -389$$

The critical numbers are 0, 300, 450.
$P(0) = -200$, $P(300) = 2410$, $P(450) = 4172.5$
Monthly profit is maximized at $x = 450$,

$P(450) = 4172.50$

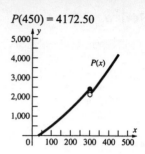

19. The number of lots ordered per year is $\dfrac{N}{x}$. Let C represent the inventory cost.

$$C = \frac{N}{x}(F + Bx) + \frac{x}{2}A = \frac{FN}{x} + BN + \frac{Ax}{2}$$

$$\frac{dC}{dx} = -\frac{FN}{x^2} + \frac{A}{2}; -\frac{FN}{x^2} + \frac{A}{2} = 0 \text{ when}$$

$$x = \sqrt{\frac{2FN}{A}}$$

4.6 Concepts Review

1. $f(x); -f(x)$

3. $x = -1, x = 2, x = 3; y = 1$

Problem Set 4.6

1. Domain: $(-\infty, \infty)$; range: $(-\infty, \infty)$
Neither an even nor an odd function.
y-intercept: 5; x-intercept: ≈ -2.3
$f'(x) = 3x^2 - 3; 3x^2 - 3 = 0$ when $x = -1, 1$
Critical points: $-1, 1$
$f'(x) > 0$ when $x < -1$ or $x > 1$
$f(x)$ is increasing on $(-\infty, -1] \cup [1, \infty)$ and decreasing on $[-1, 1]$.
Local minimum $f(1) = 3$;
local maximum $f(-1) = 7$
$f''(x) = 6x; f''(x) > 0$ when $x > 0$.
$f(x)$ is concave up on $(0, \infty)$ and concave down on $(-\infty, 0)$; inflection point $(0, 5)$.

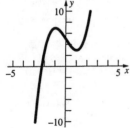

3. Domain: $(-\infty, \infty)$; range: $(-\infty, \infty)$
Neither an even nor an odd function.
y-intercept: 3; x-intercepts: $\approx -2.0, 0.2, 3.2$
$f'(x) = 6x^2 - 6x - 12 = 6(x - 2)(x + 1)$;
$f'(x) = 0$ when $x = -1, 2$
Critical points: $-1, 2$
$f'(x) > 0$ when $x < -1$ or $x > 2$
$f(x)$ is increasing on $(-\infty, -1] \cup [2, \infty)$ and decreasing on $[-1, 2]$.
Local minimum $f(2) = -17$;
local maximum $f(-1) = 10$
$f''(x) = 12x - 6 = 6(2x - 1)$;

$$f''(x) > 0 \text{ when } x > \frac{1}{2}.$$

$f(x)$ is concave up on $\left(\dfrac{1}{2}, \infty\right)$ and concave down

on $\left(-\infty, \dfrac{1}{2}\right)$; inflection point: $\left(\dfrac{1}{2}, -\dfrac{7}{2}\right)$

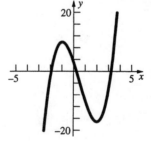

5. Domain: $(-\infty, \infty)$; range: $[0, \infty)$

Neither an even nor an odd function.

y-intercept: 1; x-intercept: 1

$G'(x) = 4(x-1)^3$; $G'(x) = 0$ when $x = 1$

Critical point: 1

$G'(x) > 0$ for $x > 1$

$G(x)$ is increasing on $[1, \infty)$ and decreasing on $(-\infty, 1]$.

Global minimum $f(1) = 0$; no local maxima

$G''(x) = 12(x-1)^2$; $G''(x) > 0$ for all $x \neq 1$

$G(x)$ is concave up on $(-\infty, 1) \cup (1, \infty)$; no inflection points

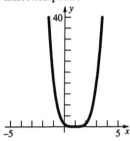

7. Domain: $(-\infty, \infty)$; range: $(-\infty, \infty)$

Neither an even nor an odd function.

y-intercept: 10; x-intercept: $1 - 11^{1/3} \approx -1.2$

$f'(x) = 3x^2 - 6x + 3 = 3(x-1)^2$; $f'(x) = 0$ when $x = 1$.

Critical point: 1

$f'(x) > 0$ for all $x \neq 1$.

$f(x)$ is increasing on $(-\infty, \infty)$ and decreasing nowhere.

No local maxima or minima

$f''(x) = 6x - 6 = 6(x-1)$; $f''(x) > 0$ when $x > 1$.

$f(x)$ is concave up on $(1, \infty)$ and concave down on $(-\infty, 1)$; inflection point $(1, 11)$

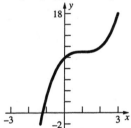

9. Domain: $(-\infty, -1) \cup (-1, \infty)$;

range: $(-\infty, 1) \cup (1, \infty)$

Neither an even nor an odd function

y-intercept: 0; x-intercept: 0

$g'(x) = \dfrac{1}{(x+1)^2}$; $g'(x)$ is never 0.

No critical points

$g'(x) > 0$ for all $x \neq -1$.

$g(x)$ is increasing on $(-\infty, -1) \cup (-1, \infty)$.

No local minima or maxima

$g''(x) = -\dfrac{2}{(x+1)^3}$; $g''(x) > 0$ when $x < -1$.

$g(x)$ is concave up on $(-\infty, -1)$ and concave down on $(-1, \infty)$; no inflection points (-1 is not in the domain of g).

$\displaystyle \lim_{x \to \infty} \frac{x}{x+1} = \lim_{x \to \infty} \frac{1}{1+\frac{1}{x}} = 1$;

$\displaystyle \lim_{x \to -\infty} \frac{x}{x+1} = \lim_{x \to -\infty} \frac{1}{1+\frac{1}{x}} = 1$;

horizontal asymptote: $y = 1$

As $x \to -1^-$, $x+1 \to 0^-$ so $\displaystyle \lim_{x \to -1^-} \frac{x}{x+1} = \infty$;

as $x \to -1^+$, $x+1 \to 0^+$ so $\displaystyle \lim_{x \to -1^+} \frac{x}{x+1} = -\infty$;

vertical asymptote: $x = -1$

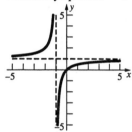

11. Domain: $(-\infty, \infty)$; range: $\left[-\dfrac{1}{4}, \dfrac{1}{4}\right]$

$f(-x) = \dfrac{-x}{(-x)^2 + 4} = -\dfrac{x}{x^2 + 4} = -f(x)$; odd

function; symmetric with respect to the origin.

y-intercept: 0; x-intercept: 0

$f'(x) = \dfrac{4 - x^2}{(x^2 + 4)^2}$; $f'(x) = 0$ when $x = -2, 2$

Critical points: $-2, 2$

$f'(x) > 0$ for $-2 < x < 2$

$f(x)$ is increasing on $[-2, 2]$ and decreasing on $(-\infty, -2] \cup [2, \infty)$.

Global minimum $f(-2) = -\dfrac{1}{4}$; global maximum $f(2) = \dfrac{1}{4}$

$f''(x) = \dfrac{2x(x^2 - 12)}{(x^2 + 4)^3}$; $f''(x) > 0$ when

$-2\sqrt{3} < x < 0$ or $x > 2\sqrt{3}$

$f(x)$ is concave up on $(-2\sqrt{3}, 0) \cup (2\sqrt{3}, \infty)$ and concave down on $(-\infty, -2\sqrt{3}) \cup (0, 2\sqrt{3})$;

inflection points $\left(-2\sqrt{3}, -\dfrac{\sqrt{3}}{8}\right)$, $(0, 0)$,

$$\left(2\sqrt{3}, \frac{\sqrt{3}}{8}\right)$$

$$\lim_{x\to\infty} \frac{x}{x^2+4} = \lim_{x\to\infty} \frac{\frac{1}{x}}{1+\frac{4}{x^2}} = 0;$$

$$\lim_{x\to-\infty} \frac{x}{x^2+4} = \lim_{x\to-\infty} \frac{\frac{1}{x}}{1+\frac{4}{x^2}} = 0;$$

$y = 0$ is a horizontal asymptote.
No vertical asymptotes

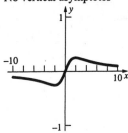

13. Domain: $(-\infty, 1) \cup (1, \infty)$;
range $(-\infty, 1) \cup (1, \infty)$
Neither an even nor an odd function
y-intercept: 0; x-intercept: 0

$$h(x) = -\frac{1}{(x-1)^2}; h'(x) \text{ is never } 0.$$

No critical points
$h'(x) < 0$ for all $x \neq 1$.
$h(x)$ is increasing nowhere and
decreasing on $(-\infty, 1) \cup (1, \infty)$.
No local maxima or minima

$$h''(x) = \frac{2}{(x-1)^3}; h''(x) > 0 \text{ when } x > 1$$

$h'(x)$ is concave up on $(1, \infty)$ and concave
down on $(-\infty, 1)$; no inflection points (1 is not in
the domain of $h(x)$)

$$\lim_{x\to\infty} \frac{x}{x-1} = \lim_{x\to\infty} \frac{1}{1-\frac{1}{x}} = 1;$$

$$\lim_{x\to-\infty} \frac{x}{x-1} = \lim_{x\to-\infty} \frac{1}{1-\frac{1}{x}} = 1;$$

$y = 1$ is a horizontal asymptote.

As $x \to 1^-, x-1 \to 0^-$ so $\lim_{x\to1^-} \frac{x}{x-1} = -\infty$;

as $x \to 1^+, x-1 \to 0^+$ so $\lim_{x\to1^+} \frac{x}{x-1} = \infty$;

$x = 1$ is a vertical asymptote.

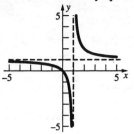

15. Domain: $(-\infty, -1) \cup (-1, 2) \cup (2, \infty)$;
range: $(-\infty, \infty)$
Neither an even nor an odd function
y-intercept: $-\frac{3}{2}$; x-intercepts: 1, 3

$$f'(x) = \frac{3x^2 - 10x + 11}{(x+1)^2(x-2)^2}; f'(x) \text{ is never } 0.$$

No critical points
$f'(x) > 0$ for all $x \neq -1, 2$
$f(x)$ is increasing on
$(-\infty, -1) \cup (-1, 2) \cup (2, \infty)$.
No local minima or maxima

$$f''(x) = \frac{-6x^3 + 30x^2 - 66x + 42}{(x+1)^3(x-2)^3}; f''(x) > 0 \text{ when}$$

$x < -1$ or $1 < x < 2$
$f(x)$ is concave up on $(-\infty, -1) \cup (1, 2)$ and
concave down on $(-1, 1) \cup (2, \infty)$;
inflection point $f(1) = 0$

$$\lim_{x\to\infty} \frac{(x-1)(x-3)}{(x+1)(x-2)} = \lim_{x\to\infty} \frac{x^2-4x+3}{x^2-x-2}$$

$$= \lim_{x\to\infty} \frac{1-\frac{4}{x}+\frac{3}{x^2}}{1-\frac{1}{x}-\frac{2}{x^2}} = 1;$$

$$\lim_{x\to-\infty} \frac{(x-1)(x-3)}{(x+1)(x-2)} = \lim_{x\to-\infty} \frac{1-\frac{4}{x}+\frac{3}{x^2}}{1-\frac{1}{x}-\frac{2}{x^2}} = 1;$$

$y = 1$ is a horizontal asymptote.
As $x \to -1^-, x-1 \to -2, x-3 \to -4$,
$x-2 \to -3$, and $x+1 \to 0^-$ so $\lim_{x\to-1^-} f(x) = \infty$;

as $x \to -1^+, x-1 \to -2, x-3 \to -4$,
$x-2 \to -3$, and $x+1 \to 0^+$, so
$\lim_{x\to-1^+} f(x) = -\infty$

As $x \to 2^-, x-1 \to 1, x-3 \to -1, x+1 \to 3$, and
$x-2 \to 0^-$, so $\lim_{x\to2^-} f(x) = \infty$; as

$x \to 2^+, x-1 \to 1, x-3 \to -1, x+1 \to 3$, and
$x-2 \to 0^+$, so $\lim_{x\to2^+} f(x) = -\infty$

$x = -1$ and $x = 2$ are vertical asymptotes.

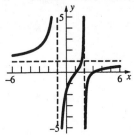

17. Domain: $(-\infty, 1) \cup (1, \infty)$

Range: $(-\infty, \infty)$

Neither even nor odd function.

y-intercept: $y = 6$; x-intercept: $x = -3, 2$

$g'(x) = \dfrac{x^2 - 2x + 5}{(x-1)^2}$; $g'(x)$ is never zero. No critical points.

$g'(x) > 0$ over the entire domain so the function is always increasing. No local extrema.

$f''(x) = \dfrac{-8}{(x-1)^3}$; $f''(x) > 0$ when

$x < 1$ (concave up) and $f''(x) < 0$ when

$x > 1$ (concave down); no inflection points. No horizontal asymptote; $x = 1$ is a vertical asymptote; the line $y = x + 2$ is an oblique (or slant) asymptote.

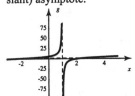

19. Domain: $(-\infty, \infty)$; range: $(-\infty, \infty)$

$R(-z) = -z|-z| = -z|z| = -R(z)$; odd function; symmetric with respect to the origin.

y-intercept: 0; z-intercept: 0

$R'(z) = |z| + \dfrac{z^2}{|z|} = 2|z|$ since $z^2 = |z|^2$ for all z;

$R'(z) = 0$ when $z = 0$

Critical point: 0

$R'(z) > 0$ when $z \neq 0$

$R(z)$ is increasing on $(-\infty, \infty)$ and decreasing nowhere.

No local minima or maxima

$R''(z) = \dfrac{2z}{|z|}$; $R''(z) > 0$ when $z > 0$.

$R(z)$ is concave up on $(0, \infty)$ and concave down

on $(-\infty, 0)$; inflection point $(0, 0)$.

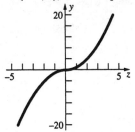

21. Domain: $(-\infty, \infty)$; range: $[0, \infty)$

Neither an even nor an odd function.

Note that for $x \leq 0$, $|x| = -x$ so $|x| + x = 0$, while

for $x > 0$, $|x| = x$ so $\dfrac{|x| + x}{2} = x$.

$g(x) = \begin{cases} 0 & \text{if } x \leq 0 \\ 3x^2 + 2x & \text{if } x > 0 \end{cases}$

y-intercept: 0; x-intercepts: $(-\infty, 0]$

$g'(x) = \begin{cases} 0 & \text{if } x \leq 0 \\ 6x + 2 & \text{if } x > 0 \end{cases}$

No critical points for $x > 0$.

$g(x)$ is increasing on $[0, \infty)$ and decreasing nowhere.

$g''(x) = \begin{cases} 0 & \text{if } x \leq 0 \\ 6 & \text{if } x > 0 \end{cases}$

$g(x)$ is concave up on $(0, \infty)$; no inflection points

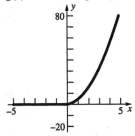

23. Domain: $(-\infty, \infty)$; range: $[0, 1]$

$f(-x) = |\sin(-x)| = |-\sin x| = |\sin x| = f(x)$; even function; symmetric with respect to the y-axis.

y-intercept: 0; x-intercepts: $k\pi$ where k is any integer.

$f'(x) = \dfrac{\sin x}{|\sin x|} \cos x$; $f'(x) = 0$ when $x = \dfrac{\pi}{2} + k\pi$

and $f'(x)$ does not exist when $x = k\pi$, where k is any integer.

Critical points: $\dfrac{k\pi}{2}$, where k is any integer;

$f'(x) > 0$ when $\sin x$ and $\cos x$ are either both positive or both negative.

$f(x)$ is increasing on $\left[k\pi, k\pi + \dfrac{\pi}{2} \right]$ and decreasing on $\left[k\pi + \dfrac{\pi}{2}, (k+1)\pi \right]$ where k is any integer.

Global minima $f(k\pi) = 0$; global maxima

$f\left(k\pi + \dfrac{\pi}{2} \right) = 1$, where k is any integer.

$$f''(x) = \dfrac{\cos^2 x}{|\sin x|} - \dfrac{\sin^2 x}{|\sin x|}$$

$$+ \sin x \cos x \left(-\dfrac{1}{|\sin x|^2} \right) \left(\dfrac{\sin x}{|\sin x|} \right) (\cos x)$$

$$= \dfrac{\cos^2 x}{|\sin x|} - \dfrac{\sin^2 x}{|\sin x|} - \dfrac{\cos^2 x}{|\sin x|}$$

$$= -\dfrac{\sin^2 x}{|\sin x|} = -|\sin x|$$

$f''(x) < 0$ when $x \neq k\pi$, k any integer

$f(x)$ is never concave up and concave down on $(k\pi, (k+1)\pi)$ where k is any integer.

No inflection points

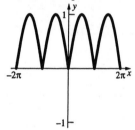

25. Domain: $(-\infty, \infty)$

Range: $[0,1]$

Even function since

$$h(-t) = \cos^2(-t) = \cos^2 t = h(t)$$

so the function is symmetric with respect to the y-axis.

y-intercept: $y = 1$; t-intercepts: $x = \dfrac{\pi}{2} + k\pi$

where k is any integer.

$h'(t) = -2\cos t \sin t$; $h'(t) = 0$ at $t = \dfrac{k\pi}{2}$.

Critical points: $t = \dfrac{k\pi}{2}$

$h'(t) > 0$ when $k\pi + \dfrac{\pi}{2} < t < (k+1)\pi$. The

function is increasing on the intervals

$\left[k\pi + (\pi/2), (k+1)\pi \right]$ and decreasing on the

intervals $\left[k\pi, k\pi + (\pi/2) \right]$.

Global maxima $h(k\pi) = 1$

Global minima $h\left(\dfrac{\pi}{2} + k\pi \right) = 0$

$h''(t) = 2\sin^2 t - 2\cos^2 t = -2(\cos 2t)$

$h''(t) < 0$ on $\left(k\pi - \dfrac{\pi}{4}, k\pi + \dfrac{\pi}{4} \right)$ so h is concave

down, and $h''(t) > 0$ on $\left(k\pi + \dfrac{\pi}{4}, k\pi + \dfrac{3\pi}{4} \right)$ so h

is concave up.

Inflection points: $\left(\dfrac{k\pi}{2} + \dfrac{\pi}{4}, \dfrac{1}{2} \right)$

No vertical asymptotes; no horizontal asymptotes.

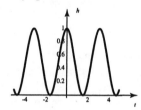

27. Domain: $\approx (-\infty, 0.44) \cup (0.44, \infty)$;

range: $(-\infty, \infty)$

Neither an even nor an odd function

y-intercept: 0; x-intercepts: 0, ≈ 0.24

$$f'(x) = \dfrac{74.6092x^3 - 58.2013x^2 + 7.82109x}{(7.126x - 3.141)^2};$$

$f'(x) = 0$ when $x = 0$, ≈ 0.17, ≈ 0.61

Critical points: 0, ≈ 0.17, ≈ 0.61

$f'(x) > 0$ when $0 < x < 0.17$ or $0.61 < x$

$f(x)$ is increasing on $\approx [0, 0.17] \cup [0.61, \infty)$

and decreasing on

$(-\infty, 0] \cup [0.17, 0.44) \cup (0.44, 0.61]$

Local minima $f(0) = 0$, $f(0.61) \approx 0.60$; local

maximum $f(0.17) \approx 0.01$

$$f''(x) = \frac{531.665x^3 - 703.043x^2 + 309.887x - 24.566}{(7.126x - 3.141)^3};$$

$f''(x) > 0$ when $x < 0.10$ or $x > 0.44$

$f(x)$ is concave up on $(-\infty, 0.10) \cup (0.44, \infty)$

and concave down on $(0.10, 0.44)$;

inflection point $\approx (0.10, 0.003)$

$$\lim_{x \to \infty} \frac{5.235x^3 - 1.245x^2}{7.126x - 3.141} = \lim_{x \to \infty} \frac{5.235x^2 - 1.245x}{7.126 - \frac{3.141}{x}} = \infty$$

so $f(x)$ does not have a horizontal asymptote.

As $x \to 0.44^-$, $5.235x^3 - 1.245x^2 \to 0.20$ while

$7.126x - 3.141 \to 0^-$, so $\displaystyle\lim_{x \to 0.44^-} f(x) = -\infty$;

as $x \to 0.44^+$, $5.235x^3 - 1.245x^2 \to 0.20$ while

$7.126x - 3.141 \to 0^+$, so $\displaystyle\lim_{x \to 0.44^+} f(x) = \infty$;

$x \approx 0.44$ is a vertical asymptote of $f(x)$.

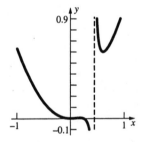

29.

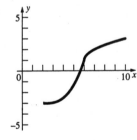

31.

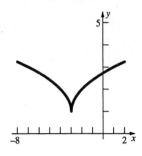

33.

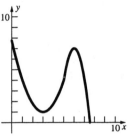

35.

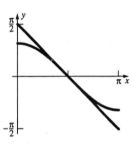

37.

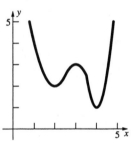

39.

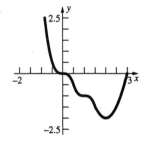

41. Let $f(x) = ax^3 + bx^2 + cx + d$, then

$f'(x) = 3ax^2 + 2bx + c$ and $f''(x) = 6ax + 2b$. As long as $a \neq 0$, $f''(x)$ will be positive on one side of $x = \dfrac{b}{3a}$ and negative on the other side.

$x = \dfrac{b}{3a}$ is the only inflection point.

43. Since the c term is squared, the only difference occurs when $c = 0$. When $c = 0$,

$y = x^2 \sqrt{x^2} = |x|^3$ which has domain $(-\infty, \infty)$ and range $[0, \infty)$. When $c \neq 0$, $y = x^2 \sqrt{x^2 - c^2}$ has domain $(-\infty, -|c|] \cup [|c|, \infty)$ and range $[0, \infty)$.

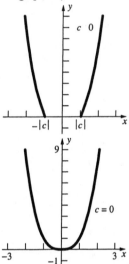

45. Let $f(x) = \dfrac{1}{(cx^2 - 4)^2 + cx^2}$, then

$f'(x) = \dfrac{2cx(7 - 2cx^2)}{[(cx^2 - 4)^2 + cx^2]^2};$

If $c > 0$, $f'(x) = 0$ when $x = 0, \pm\sqrt{\dfrac{7}{2c}}$.

If $c < 0$, $f'(x) = 0$ when $x = 0$.

Note that $f(x) = \dfrac{1}{16}$ (a horizontal line) if $c = 0$.

If $c > 0$, $f'(x) > 0$ when $x < -\sqrt{\dfrac{7}{2c}}$ and

$0 < x < \sqrt{\dfrac{7}{2c}}$, so $f(x)$ is increasing on

$\left(-\infty, -\sqrt{\dfrac{7}{2c}}\right] \cup \left[0, \sqrt{\dfrac{7}{2c}}\right]$ and decreasing on

$\left[-\sqrt{\dfrac{7}{2c}}, 0\right] \cup \left[\sqrt{\dfrac{7}{2c}}, \infty\right)$. Thus, $f(x)$ has local

maxima $f\left(-\sqrt{\dfrac{7}{2c}}\right) = \dfrac{4}{15}$, $f\left(\sqrt{\dfrac{7}{2c}}\right) = \dfrac{4}{15}$ and

local minimum $f(0) = \dfrac{1}{16}$. If $c < 0$, $f'(x) > 0$ when $x < 0$, so $f(x)$ is increasing on $(-\infty, 0]$ and decreasing on $[0, \infty)$. Thus, $f(x)$ has a local

maximum $f(0) = \dfrac{1}{16}$. Note that $f(x) > 0$ and has horizontal asymptote $y = 0$.

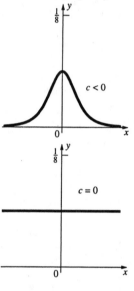

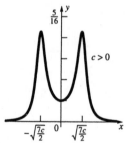

47.

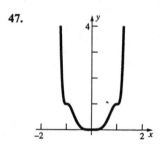

Justification:

$f(1) = g(1) = 1$

$f(-x) = g((-x)^4) = g(x^4) = f(x)$

f is an even function; symmetric with respect to the y-axis.

$f'(x) = g'(x^4)4x^3$

$f'(x) > 0$ for x on $(0,1) \cup (1,\infty)$

$f'(x) < 0$ for x on $(-\infty,-1) \cup (-1,0)$

$f'(x) = 0$ for $x = -1,0,1$ since f' is continuous.

$f''(x) = g''(x^4)16x^6 + g'(x)12x^2$

$f''(x) = 0$ for $x = -1,0,1$

$f''(x) > 0$ for x on $(0,x_0) \cup (1,\infty)$

$f''(x) < 0$ for x on $(x_0,1)$

Where x_0 is a root of $f''(x) = 0$ (assume that there is only one root on $(0,1)$).

49. a. Not possible; $F'(x) > 0$ means that $F(x)$ is increasing. $F''(x) > 0$ means that the rate at which $F(x)$ is increasing never slows down. Thus the values of F must eventually become positive.

b. Not possible; If $F(x)$ is concave down for all x, then $F(x)$ cannot always be positive.

c.

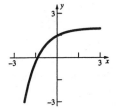

51. a.

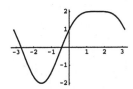

$f'(x) = 2\cos x - 2\cos x \sin x$
$= 2\cos x(1 - \sin x);$

$f'(x) = 0$ when $x = -\dfrac{\pi}{2}, \dfrac{\pi}{2}$

$f''(x) = -2\sin x - 2\cos^2 x + 2\sin^2 x$

$= 4\sin^2 x - 2\sin x - 2; \ f''(x) = 0$ when

$\sin x = -\dfrac{1}{2}$ or $\sin x = 1$ which occur when

$x = -\dfrac{\pi}{6}, -\dfrac{5\pi}{6}, \dfrac{\pi}{2}$

Global minimum $f\left(-\dfrac{\pi}{2}\right) = -2;$ global

maximum $f\left(\dfrac{\pi}{2}\right) = 2;$ inflection points

$f\left(-\dfrac{\pi}{6}\right) = -\dfrac{1}{4}, f\left(-\dfrac{5\pi}{6}\right) = -\dfrac{1}{4}$

b.

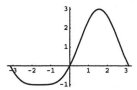

$f'(x) = 2\cos x + 2\sin x \cos x$
$= 2\cos x(1 + \sin x); \ f'(x) = 0$ when

$x = -\dfrac{\pi}{2}, \dfrac{\pi}{2}$

$f''(x) = -2\sin x + 2\cos^2 x - 2\sin^2 x$

$= -4\sin^2 x - 2\sin x + 2; \ f''(x) = 0$ when

$\sin x = -1$ or $\sin x = \dfrac{1}{2}$ which occur when

$x = -\dfrac{\pi}{2}, \dfrac{\pi}{6}, \dfrac{5\pi}{6}$

Global minimum $f\left(-\dfrac{\pi}{2}\right) = -1;$ global

maximum $f\left(\dfrac{\pi}{2}\right) = 3;$ inflection points

$f\left(\dfrac{\pi}{6}\right) = \dfrac{5}{4}, f\left(\dfrac{5\pi}{6}\right) = \dfrac{5}{4}.$

c.

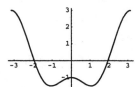

$f'(x) = -2\sin 2x + 2\sin x$
$= -4\sin x \cos x + 2\sin x = 2\sin x(1 - 2\cos x);$

$f'(x) = 0$ when $x = -\pi, -\dfrac{\pi}{3}, 0, \dfrac{\pi}{3}, \pi$

$f''(x) = -4\cos 2x + 2\cos x; \ f''(x) = 0$ when

$x \approx -2.2, -0.6, 0.6, 2.2$

Global minimum $f\left(-\dfrac{\pi}{3}\right) = f\left(\dfrac{\pi}{3}\right) = -1.5;$

Global maximum $f(-\pi) = f(\pi) = 3;$
inflection points $f(-2.2) \approx 0.87,$
$f(-0.6) \approx -1.29, f(0.6) \approx -1.29,$
$f(2.2) \approx 0.87$

d.

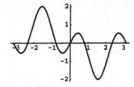

$f'(x) = 3\cos 3x - \cos x; \ f'(x) = 0$ when
$3\cos 3x = \cos x$ which occurs when

$x = -\dfrac{\pi}{2}, \dfrac{\pi}{2}$ and when

$x \approx -2.7, -0.4, 0.4, 2.7$
$f''(x) = -9\sin 3x + \sin x$ which occurs when
$x = -\pi, 0, \pi$ and when
$x \approx -2.13, -1.02, 1.02, 2.13$

Global minimum $f\left(\dfrac{\pi}{2}\right) = -2;$

global maximum $f\left(-\dfrac{\pi}{2}\right) = 2;$

inflection points $f(-2.13) \approx 0.7,$
$f(-1.02) \approx 0.8; f(0) = 0, f(1.02) \approx -0.8,$
$f(2.13) \approx -0.7$

e.

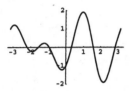

$f'(x) = 2\cos 2x + 3\sin 3x$

Using the graphs, $f(x)$ has a global minimum
at $f(2.17) \approx -1.9$ and a global maximum at
$f(0.97) \approx 1.9$
$f''(x) = -4\sin 2x + 9\cos 3x; \ f''(x) = 0$ when

$x = -\dfrac{\pi}{2}, \dfrac{\pi}{2}$ and when

$x \approx -2.47, -0.67, 0.41, 2.73.$
Inflection points $f(-2.47) \approx 0.54,$

$f\left(-\dfrac{\pi}{2}\right) = 0, f(-0.67) \approx -0.55,$

$f(0.41) \approx 0.40, \ f\left(\dfrac{\pi}{2}\right) = 0, \ f(2.73) \approx -0.40$

53.

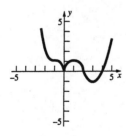

55. a.

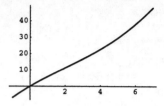

$f'(x) = \dfrac{2x^2 - 9x + 40}{\sqrt{x^2 - 6x + 40}}; \ f'(x)$ is never 0,

and always positive, so $f(x)$ is increasing for
all x. Thus, on $[-1, 7]$, the global minimum is
$f(-1) \approx -6.9$ and the global maximum if
$f(7) \approx 48.0.$

$f''(x) = \dfrac{2x^3 - 18x^2 + 147x - 240}{(x^2 - 6x + 40)^{3/2}}; \ f''(x) = 0$

when $x \approx 2.02;$ inflection point
$f(2.02) \approx 11.4$

b.

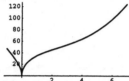

Global minimum $f(0) = 0;$ global maximum
$f(7) \approx 124.4;$ inflection point at $x \approx 2.34,$
$f(2.34) \approx 48.09$

c.

No global minimum or maximum; no
inflection points

d.

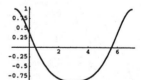

Global minimum $f(3) \approx -0.9;$
global maximum $f(-1) \approx 1.0$ or $f(7) \approx 1.0;$
Inflection points at $x \approx 0.05$ and $x \approx 5.9,$
$f(0.05) \approx 0.3, f(5.9) \approx 0.3.$

4.7 Concepts Review

1. continuous; (a, b); $f(b) - f(a) = f'(c)(b - a)$

3. $F(x) = G(x) + C$

Problem Set 4.7

1. $f'(x) = \dfrac{x}{|x|}$

$\dfrac{f(2) - f(1)}{2 - 1} = \dfrac{2 - 1}{1} = 1$

$\dfrac{c}{|c|} = 1$ for all $c > 0$, hence for all c in $(1, 2)$

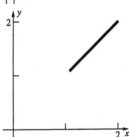

3. $f'(x) = 2x + 1$

$\dfrac{f(2) - f(-2)}{2 - (-2)} = \dfrac{6 - 2}{4} = 1$

$2c + 1 = 1$ when $c = 0$

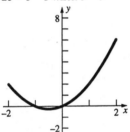

5. $H'(s) = 2s + 3$

$\dfrac{H(1) - H(-3)}{1 - (-3)} = \dfrac{3 - (-1)}{1 - (-3)} = 1$

$2c + 3 = 1$ when $c = -1$

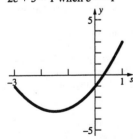

7. $f'(z) = \dfrac{1}{3}(3z^2 + 1) = z^2 + \dfrac{1}{3}$

$\dfrac{f(2) - f(-1)}{2 - (-1)} = \dfrac{2 - (-2)}{3} = \dfrac{4}{3}$

$c^2 + \dfrac{1}{3} = \dfrac{4}{3}$ when $c = -1, 1$, but -1 is not in

$(-1, 2)$ so $c = 1$ is the only solution.

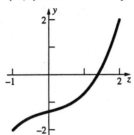

9. $h'(x) = -\dfrac{3}{(x - 3)^2}$

$\dfrac{h(2) - h(0)}{2 - 0} = \dfrac{-2 - 0}{2} = -1$

$-\dfrac{3}{(c - 3)^2} = -1$ when $c = 3 \pm \sqrt{3}$,

$c = 3 - \sqrt{3} \approx 1.27$ ($3 + \sqrt{3}$ is not in $(0, 2)$.)

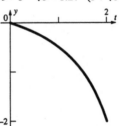

11. $h'(t) = \dfrac{2}{3t^{1/3}}$

$\dfrac{h(2) - h(0)}{2 - 0} = \dfrac{2^{2/3} - 0}{2} = 2^{-1/3}$

$\dfrac{2}{3c^{1/3}} = 2^{-1/3}$ when $c = \dfrac{16}{27} \approx 0.59$

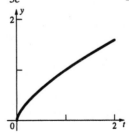

13. $g'(x) = \frac{5}{3}x^{2/3}$

$\dfrac{g(1) - g(0)}{1 - 0} = \dfrac{1 - 0}{1} = 1$

$\frac{5}{3}c^{2/3} = 1$ when $c = \pm\left(\frac{3}{5}\right)^{3/2}$,

$c = \left(\frac{3}{5}\right)^{3/2} \approx 0.46, \left(-\left(\frac{3}{5}\right)^{3/2} \text{ is not in } (0, 1).\right)$

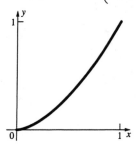

15. $S'(\theta) = \cos\theta$

$\dfrac{S(\pi) - S(-\pi)}{\pi - (-\pi)} = \dfrac{0 - 0}{2\pi} = 0$

$\cos c = 0$ when $c = \pm\dfrac{\pi}{2}$.

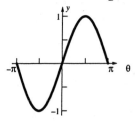

17. The Mean Value Theorem does not apply

because $T(\theta)$ is not continuous at $\theta = \dfrac{\pi}{2}$.

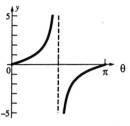

19. $f'(x) = 1 - \dfrac{1}{x^2}$

$\dfrac{f(2) - f(1)}{2 - 1} = \dfrac{\frac{5}{2} - 2}{1} = \dfrac{1}{2}$

$1 - \dfrac{1}{c^2} = \dfrac{1}{2}$ when $c = \pm\sqrt{2}$, $c = \sqrt{2} \approx 1.41$

($c = -\sqrt{2}$ is not in $(1, 2)$.)

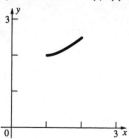

21. The Mean Value Theorem does not apply because f is not differentiable at $x = 0$.

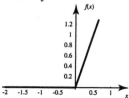

23. $\dfrac{f(8) - f(0)}{8 - 0} = -\dfrac{1}{4}$

There are three values for c such that

$f'(c) = -\dfrac{1}{4}$.

They are approximately 1.5, 3.75, and 7.

25. By the Monotonicity Theorem, f is increasing on the intervals (a, x_0) and (x_0, b).

To show that $f(x_0) > f(x)$ for x in (a, x_0), consider f on the interval $(a, x_0]$.

f satisfies the conditions of the Mean Value Theorem on the interval $[x, x_0]$ for x in (a, x_0).

So for some c in (x, x_0),

$f(x_0) - f(x) = f'(c)(x_0 - x)$.

Because

$f'(c) > 0$ and $x_0 - x > 0$, $f(x_0) - f(x) > 0$,

so $f(x_0) > f(x)$.

Similar reasoning shows that

$f(x) > f(x_0)$ for x in (x_0, b).

Therefore, f is increasing on (a, b).

27. $s(t)$ is defined in any interval not containing $t = 0$.

$s'(c) = -\dfrac{1}{c^2} < 0$ for all $c \neq 0$. For any a, b with

$a < b$ and both either positive or negative, the Mean Value Theorem says

$s(b) - s(a) = s'(c)(b - a)$ for some c in (a, b).

Since $a < b$, $b - a > 0$ while $s'(c) < 0$, hence

$s(b) - s(a) < 0$, or $s(b) < s(a)$.

Thus, $s(t)$ is decreasing on any interval not containing $t = 0$.

29. $F'(x) = 0$ and $G(x) = 0$; $G'(x) = 0$.
By Theorem B,
$F(x) = G(x) + C$, so $F(x) = 0 + C = C$.

31. Let $G(x) = Dx$; $F'(x) = D$ and $G'(x) = D$.
By Theorem B, $F(x) = G(x) + C$; $F(x) = Dx + C$.

33. Since $f(a)$ and $f(b)$ have opposite signs, 0 is between $f(a)$ and $f(b)$. $f(x)$ is continuous on $[a, b]$, since it has a derivative. Thus, by the Intermediate Value Theorem, there is at least one point c,
$a < c < b$ with $f(c) = 0$.
Suppose there are two points, c and c', $c < c'$ in (a, b) with $f(c) = f(c') = 0$. Then by Rolle's Theorem, there is at least one number d in (c, c') with $f'(d) = 0$. This contradicts the given information that $f'(x) \neq 0$ for all x in $[a, b]$, thus there cannot be more than one x in $[a, b]$ where $f(x) = 0$.

35. Suppose there is more than one zero between successive distinct zeros of f'. That is, there are a and b such that $f(a) = f(b) = 0$ with a and b between successive distinct zeros of f'. Then by Rolle's Theorem, there is a c between a and b such that $f'(c) = 0$. This contradicts the supposition that a and b lie between successive distinct zeros.

37. $f(x)$ is a polynomial function so it is continuous on $[0, 4]$ and $f''(x)$ exists for all x on $(0, 4)$.
$f(1) = f(2) = f(3) = 0$, so by Problem 36, there are at least two values of x in $[0, 4]$ where $f'(x) = 0$ and at least one value of x in $[0, 4]$ where $f''(x) = 0$.

39. $f'(x) = 2\cos 2x$; $|f'(x)| \leq 2$
$$\frac{|f(x_2) - f(x_1)|}{|x_2 - x_1|} = |f'(x)|; \frac{|f(x_2) - f(x_1)|}{|x_2 - x_1|} \leq 2$$
$$|f(x_2) - f(x_1)| \leq 2|x_2 - x_1|;$$
$$|\sin 2x_2 - \sin 2x_1| \leq 2|x_2 - x_1|$$

41. Suppose $f'(x) \geq 0$. Let a and b lie in the interior of I such that $b > a$. By the Mean Value Theorem, there is a point c between a and b such that
$$f'(c) = \frac{f(b) - f(a)}{b - a}; \frac{f(b) - f(a)}{b - a} \geq 0.$$
Since $a < b$, $f(b) \geq f(a)$, so f is nondecreasing.
Suppose $f'(x) \leq 0$. Let a and b lie in the interior of I such that $b > a$. By the Mean Value Theorem, there is a point c between a and b such

that $f'(c) = \frac{f(b) - f(a)}{b - a}$; $\frac{f(b) - f(a)}{b - a} \leq 0$.
Since
$a < b$, $f(a) \geq f(b)$, so f is nonincreasing.

43. Let $f(x) = h(x) - g(x)$.
$f'(x) = h'(x) - g'(x)$; $f'(x) \geq 0$ for all x in (a, b) since $g'(x) \leq h'(x)$ for all x in (a, b), so f is nondecreasing on (a, b) by Problem 41. Thus
$x_1 < x_2 \Rightarrow f(x_1) \leq f(x_2)$;
$h(x_1) - g(x_1) \leq h(x_2) - g(x_2)$;
$g(x_2) - g(x_1) \leq h(x_2) - h(x_1)$ for all x_1 and x_2 in (a, b).

45. Let $f(x) = \sin x$. $f'(x) = \cos x$, so
$|f'(x)| = |\cos x| \leq 1$ for all x.
By the Mean Value Theorem,
$$\frac{f(x) - f(y)}{x - y} = f'(c) \text{ for some } c \text{ in } (x, y).$$
Thus, $\dfrac{|f(x) - f(y)|}{|x - y|} = |f'(c)| \leq 1$;
$$|\sin x - \sin y| \leq |x - y|.$$

47. Let s be the difference in speeds between horse A and horse B as function of time t.
Then s' is the difference in accelerations.
Let t_2 be the time in Problem 46 at which the horses had the same speeds and let t_1 be the finish time of the race.
$s(t_2) = s(t_1) = 0$
By the Mean Value Theorem,
$$\frac{s(t_1) - s(t_2)}{t_1 - t_2} = s'(c) \text{ for some } c \text{ in } (t_2, t_1).$$
Therefore $s'(c) = 0$ for some c in (t_2, t_1).

49. If $|f(y) - f(x)| \leq M(y - x)^2$, then
$\dfrac{|f(y) - f(x)|}{|y - x|} \leq M|y - x|$. By the Mean Value Theorem, there is some c between x and y such that $\dfrac{|f(y) - f(x)|}{|y - x|} = |f'(c)| \leq M|y - x|$. Since the given inequality is true for all x and y, we can choose x and y so close together that $M|y - x|$ is arbitrarily close to 0. Thus, in order for the last inequality to be true, $f'(c)$ must be 0, hence $f(x)$ is a constant function.

51. Let $f(t)$ be the distance traveled at time t.

$$\frac{f(2)-f(0)}{2-0}=\frac{112-0}{2}=56$$

By the Mean Value Theorem, there is a time c such that $f'(c)=56$.

At some time during the trip, Johnny must have gone 56 miles per hour.

53. Since the car is stationary at $t=0$, and since v is continuous, there exists a δ such that $v(t)<\dfrac{1}{2}$ for all t in the interval $[0,\delta]$. $v(t)$ is therefore less than $\dfrac{1}{2}$ and $s(\delta)<\delta\cdot\dfrac{1}{2}=\dfrac{\delta}{2}$. By the Mean Value Theorem, there exists a c in the interval $(\delta,20)$ such that

$$v(c)=s'(c)=\frac{\left(20-\dfrac{\delta}{2}\right)}{(20-\delta)}$$

$$>\frac{20-\delta}{20-\delta}$$

$$=1\text{ mile per minute}$$

$$=60\text{ miles per hour}$$

4.8 Chapter Review

Concepts Test

1. True: Max-Min Existence Theorem

3. True: For example, let $f(x)=\sin x$.

5. True: $f'(x)=18x^5+16x^3+4x$;
   $f''(x)=90x^4+48x^2+4$, which is greater than zero for all x.

7. True: When $f'(x)>0$, $f(x)$ is increasing.

9. True: $f(x)=ax^2+bx+c$;
   $f'(x)=2ax+b$; $f''(x)=2a$

11. False: $\tan^2 x$ has a minimum value of 0. This occurs whenever $x=k\pi$ where k is an integer.

13. True: $\displaystyle\lim_{x\to\frac{\pi}{2}}(2x^3+x+\tan x)=\infty$ while
    $\displaystyle\lim_{x\to-\frac{\pi}{2}}(2x^3+x+\tan x)=-\infty$.

15. True: $\displaystyle\lim_{x\to\infty}\frac{x^2+1}{1-x^2}=\lim_{x\to\infty}\frac{1+\frac{1}{x^2}}{\frac{1}{x^2}-1}$
    $=\dfrac{1}{-1}=-1$ and

$$\lim_{x\to-\infty}\frac{x^2+1}{1-x^2}=\lim_{x\to-\infty}\frac{1+\frac{1}{x^2}}{\frac{1}{x^2}-1}$$
$$=\frac{1}{-1}=-1.$$

17. True: The function is differentiable on $(0, 2)$.

19. False: There are two points: $x=-\dfrac{\sqrt{3}}{3},\dfrac{\sqrt{3}}{3}$.

21. False: For example if $f(x)=x^4$,
    $f'(0)=f''(0)=0$ but f has a minimum at $x=0$.

23. True: $A=xy=K$; $y=\dfrac{K}{x}$
    $P=2x+\dfrac{2K}{x}$; $\dfrac{dP}{dx}=2-\dfrac{2K}{x^2}$; $\dfrac{dP}{dx}=0$
    when $x=\sqrt{K}, y=\sqrt{K}$

25. True: If $f(x_1)<f(x_2)$ and $g(x_1)<g(x_2)$ for $x_1<x_2$,
    $f(x_1)+g(x_1)<f(x_2)+g(x_2)$, so $f+g$ is increasing.

27. True: Since $f''(x)>0$, $f'(x)$ is increasing for $x\geq 0$. Therefore, $f'(x)>0$ for x in $[0,\infty)$, so $f(x)$ is increasing.

29. True: If the function is nondecreasing, $f'(x)$ must be greater than or equal to zero, and if $f'(x)\geq 0$, f is

nondecreasing. This can be seen using the Mean Value Theorem.

31. False: For example, let $f(x) = e^x$.

$\lim\limits_{x \to -\infty} e^x = 0$, so $y = 0$ is a horizontal asymptote.

33. True: $f'(x) = 3ax^2 + 2bx + c$; $f'(x) = 0$

when $x = \dfrac{-b \pm \sqrt{b^2 - 3ac}}{3a}$ by the

Quadratic Formula. $f''(x) = 6ax + 2b$

so

$$f''\left(\dfrac{-b \pm \sqrt{b^2 - 3ac}}{3a}\right) = \pm 2\sqrt{b^2 - 3ac}.$$

Thus, if $b^2 - 3ac > 0$, one critical point is a local maximum and the other is a local minimum.

(If $b^2 - 3ac = 0$ the only critical point is an inflection point while if $b^2 - 3ac < 0$ there are no critical points.)
On an open interval, no local maxima can come from endpoints, so there can be at most one local maximum in an open interval.

Sample Test Problems

1. $f'(x) = 2x - 2$; $2x - 2 = 0$ when $x = 1$.
Critical points: 0, 1, 4
$f(0) = 0, f(1) = -1, f(4) = 8$
Global minimum $f(1) = -1$;
global maximum $f(4) = 8$

3. $f'(z) = -\dfrac{2}{z^3}$; $-\dfrac{2}{z^3}$ is never 0.

Critical points: $-2, -\dfrac{1}{2}$

$f(-2) = \dfrac{1}{4}, f\left(-\dfrac{1}{2}\right) = 4$

Global minimum $f(-2) = \dfrac{1}{4}$;

global maximum $f\left(-\dfrac{1}{2}\right) = 4$.

5. $f'(x) = \dfrac{x}{|x|}$; $f'(x)$ does not exist at $x = 0$.

Critical points: $-\dfrac{1}{2}, 0, 1$

$f\left(-\dfrac{1}{2}\right) = \dfrac{1}{2}, f(0) = 0, f(1) = 1$

Global minimum $f(0) = 0$;
global maximum $f(1) = 1$

7. $f'(x) = 12x^3 - 12x^2 = 12x^2(x - 1)$; $f'(x) = 0$
when $x = 0, 1$
Critical points: $-2, 0, 1, 3$
$f(-2) = 80, f(0) = 0, f(1) = -1, f(3) = 135$
Global minimum $f(1) = -1$;
global maximum $f(3) = 135$

9. $f'(x) = 10x^4 - 20x^3 = 10x^3(x - 2)$;
$f'(x) = 0$ when $x = 0, 2$
Critical points: $-1, 0, 2, 3$
$f(-1) = 0, f(0) = 7, f(2) = -9, f(3) = 88$
Global minimum $f(2) = -9$;
global maximum $f(3) = 88$

11. $f'(\theta) = \cos\theta$; $f'(\theta) = 0$ when $\theta = \dfrac{\pi}{2}$ in

$\left[\dfrac{\pi}{4}, \dfrac{4\pi}{3}\right]$

Critical points: $\dfrac{\pi}{4}, \dfrac{\pi}{2}, \dfrac{4\pi}{3}$

$f\left(\dfrac{\pi}{4}\right) = \dfrac{1}{\sqrt{2}} \approx 0.71, f\left(\dfrac{\pi}{2}\right) = 1,$

$f\left(\dfrac{4\pi}{3}\right) = -\dfrac{\sqrt{3}}{2} \approx -0.87$

Global minimum $f\left(\dfrac{4\pi}{3}\right) \approx -0.87$;

global maximum $f\left(\dfrac{\pi}{2}\right) = 1$

13. $f'(x) = 3 - 2x$; $f'(x) > 0$ when $x < \dfrac{3}{2}$.
$f''(x) = -2$; $f''(x)$ is always negative.
$f(x)$ is increasing on $\left(-\infty, \dfrac{3}{2}\right]$ and concave down
on $(-\infty, \infty)$.

15. $f'(x) = 3x^2 - 3 = 3(x^2 - 1)$; $f'(x) > 0$ when
$x < -1$ or $x > 1$.
$f''(x) = 6x$; $f''(x) < 0$ when $x < 0$.
$f(x)$ is increasing on $(-\infty, -1] \cup [1, \infty)$ and
concave down on $(-\infty, 0)$.

17. $f'(x) = 4x^3 - 20x^4 = 4x^3(1-5x); f'(x) > 0$

when $0 < x < \dfrac{1}{5}$.

$f''(x) = 12x^2 - 80x^3 = 4x^2(3-20x); f''(x) < 0$

when $x > \dfrac{3}{20}$.

$f(x)$ is increasing on $\left[0, \dfrac{1}{5}\right]$ and concave down on

$\left(\dfrac{3}{20}, \infty\right)$.

19. $f'(x) = 3x^2 - 4x^3 = x^2(3-4x); f'(x) > 0$ when

$x < \dfrac{3}{4}$.

$f''(x) = 6x - 12x^2 = 6x(1-2x); f''(x) < 0$ when

$x < 0$ or $x > \dfrac{1}{2}$.

$f(x)$ is increasing on $\left(-\infty, \dfrac{3}{4}\right]$ and concave down

on $(-\infty, 0) \cup \left(\dfrac{1}{2}, \infty\right)$.

21. $f'(x) = 2x(x-4) + x^2 = 3x^2 - 8x = x(3x-8);$

$f'(x) > 0$ when $x < 0$ or $x > \dfrac{8}{3}$

$f(x)$ is increasing on $(-\infty, 0] \cup \left[\dfrac{8}{3}, \infty\right)$ and

decreasing on $\left[0, \dfrac{8}{3}\right]$

Local minimum $f\left(\dfrac{8}{3}\right) = -\dfrac{256}{27} \approx -9.48;$

local maximum $f(0) = 0$

$f''(x) = 6x - 8; f''(x) > 0$ when $x > \dfrac{4}{3}$.

$f(x)$ is concave up on $\left(\dfrac{4}{3}, \infty\right)$ and concave down

on $\left(-\infty, \dfrac{4}{3}\right)$; inflection point $\left(\dfrac{4}{3}, -\dfrac{128}{27}\right)$

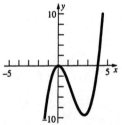

23. $f'(x) = 4x^3 - 2; f'(x) = 0$ when $x = \dfrac{1}{\sqrt[3]{2}}$.

$f''(x) = 12x^2; f''(x) = 0$ when $x = 0$.

$f''\left(\dfrac{1}{\sqrt[3]{2}}\right) = \dfrac{12}{2^{2/3}} > 0$, so

$f\left(\dfrac{1}{\sqrt[3]{2}}\right) = \dfrac{1}{2^{4/3}} - \dfrac{2}{2^{1/3}} = -\dfrac{3}{2^{4/3}}$ is a global

minimum.

$f''(x) > 0$ for all $x \neq 0$; no inflection points

No horizontal or vertical asymptotes

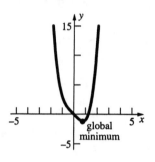

25. $f'(x) = \dfrac{3x-6}{2\sqrt{x-3}}; f'(x) = 0$ when $x = 2$, but $x = 2$

is not in the domain of $f(x)$. $f'(x)$ does not exist

when $x = 3$.

$f''(x) = \dfrac{3(x-4)}{4(x-3)^{3/2}}; f''(x) = 0$ when $x = 4$.

Global minimum $f(3) = 0$; no local maxima

Inflection point $(4, 4)$

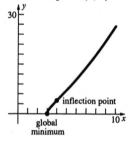

27. $f'(x) = 12x^3 - 12x^2 = 12x^2(x-1); f'(x) = 0$
when $x = 0, 1$.

$f''(x) = 36x^2 - 24x = 12x(3x-2); f''(x) = 0$

when $x = 0, \dfrac{2}{3}$.

$f''(1) = 12$, so $f(1) = -1$ is a minimum.

Global minimum $f(1) = -1$; no local maxima

Inflection points $(0, 0)$, $\left(\dfrac{2}{3}, -\dfrac{16}{27}\right)$

No horizontal or vertical asymptotes.

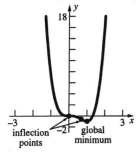

29. $f'(x) = 3 + \dfrac{1}{x^2}; f'(x) > 0$ for all $x \neq 0$.

$f''(x) = -\dfrac{2}{x^3}; f''(x) > 0$ when $x < 0$ and

$f''(x) < 0$ when $x > 0$

No local minima or maxima

No inflection points

$f(x) = 3x - \dfrac{1}{x}$, so

$\displaystyle\lim_{x \to \infty} [f(x) - 3x] = \lim_{x \to \infty}\left(-\dfrac{1}{x}\right) = 0$ and $y = 3x$ is an

oblique asymptote.

Vertical asymptote $x = 0$

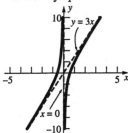

31. $f'(x) = -\sin x - \cos x; f'(x) = 0$ when

$x = -\dfrac{\pi}{4}, \dfrac{3\pi}{4}$.

$f''(x) = -\cos x + \sin x; f''(x) = 0$ when

$x = -\dfrac{3\pi}{4}, \dfrac{\pi}{4}$.

$f''\left(-\dfrac{\pi}{4}\right) = -\sqrt{2}, f''\left(\dfrac{3\pi}{4}\right) = \sqrt{2}$

Global minimum $f\left(\dfrac{3\pi}{4}\right) = -\sqrt{2}$;

global maximum $f\left(-\dfrac{\pi}{4}\right) = \sqrt{2}$

Inflection points $\left(-\dfrac{3\pi}{4}, 0\right), \left(\dfrac{\pi}{4}, 0\right)$

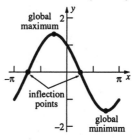

33. $f'(x) = x\sec^2 x + \tan x; f'(x) = 0$ when $x = 0$

$f''(x) = 2\sec^2 x(1 + x\tan x); f''(x)$ is never 0 on

$\left(-\dfrac{\pi}{2}, \dfrac{\pi}{2}\right)$.

$f''(0) = 2$

Global minimum $f(0) = 0$

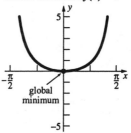

35. $f'(x) = \cos x - 2\cos x\sin x = \cos x(1 - 2\sin x)$;

$f'(x) = 0$ when $x = -\dfrac{\pi}{2}, \dfrac{\pi}{6}, \dfrac{\pi}{2}, \dfrac{5\pi}{6}$

$f''(x) = -\sin x + 2\sin^2 x - 2\cos^2 x; f''(x) = 0$

when $x \approx -2.51, -0.63, 1.00, 2.14$

$f''\left(-\dfrac{\pi}{2}\right) = 3, f''\left(\dfrac{\pi}{6}\right) = -\dfrac{3}{2}, f''\left(\dfrac{\pi}{2}\right) = 1$,

$f''\left(\dfrac{5\pi}{6}\right) = -\dfrac{3}{2}$

Global minimum $f\left(-\dfrac{\pi}{2}\right) = -2$,

local minimum $f\left(\dfrac{\pi}{2}\right) = 0$;

global maxima $f\left(\dfrac{\pi}{6}\right) = \dfrac{1}{4}, f\left(\dfrac{5\pi}{6}\right) = \dfrac{1}{4}$

Inflection points $(-2.51, -0.94)$,

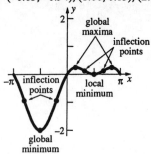

37.

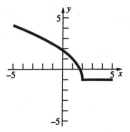

39.

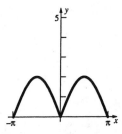

41. Let p be the length of the plank and let x be the distance from the fence to where the plank touches the ground.
See the figure below.

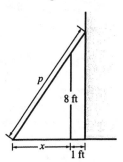

By properties of similar triangles,

$$\frac{p}{x+1} = \frac{\sqrt{x^2 + 64}}{x}$$

$$p = \left(1 + \frac{1}{x}\right)\sqrt{x^2 + 64}$$

Minimize p:

$$\frac{dp}{dx} = -\frac{1}{x^2}\sqrt{x^2 + 64} + \left(1 + \frac{1}{x}\right)\frac{x}{\sqrt{x^2 + 64}}$$

$$= \frac{1}{x^2\sqrt{x^2 + 64}}\left(-(x^2 + 64) + \left(1 + \frac{1}{x}\right)x^3\right)$$

$$= \frac{x^3 - 64}{x^2\sqrt{x^2 + 64}}$$

$$\frac{x^3 - 64}{x^2\sqrt{x^2 + 64}} = 0; x = 4$$

$$\frac{dp}{dx} < 0 \text{ if } x < 4, \frac{dp}{dx} > 0 \text{ if } x > 4$$

When $x = 4$, $p = \left(1 + \frac{1}{4}\right)\sqrt{16 + 64} \approx 11.18$ ft.

43. $\dfrac{1}{2}\pi r^2 h = 128\pi$

$$h = \frac{256}{r^2}$$

Let S be the surface area of the trough.

$$S = \pi r^2 + \pi rh = \pi r^2 + \frac{256\pi}{r}$$

$$\frac{dS}{dr} = 2\pi r - \frac{256\pi}{r^2}$$

$$2\pi r - \frac{256\pi}{r^2} = 0; r^3 = 128, r = 4\sqrt[3]{2}$$

Since $\dfrac{d^2 S}{dr^2} > 0$ when $r = 4\sqrt[3]{2}$, $r = 4\sqrt[3]{2}$
minimizes S.

$$h = \frac{256}{\left(4\sqrt[3]{2}\right)^2} = 8\sqrt[3]{2}$$

45. a. $f'(x) = x^2$

$$\frac{f(3) - f(-3)}{3 - (-3)} = \frac{9 + 9}{6} = 3$$

$$c^2 = 3; c = -\sqrt{3}, \sqrt{3}$$

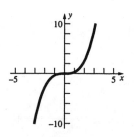

b. The Mean Value Theorem does not apply because $F'(0)$ does not exist.

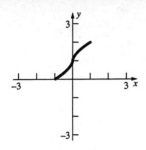

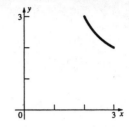

47.

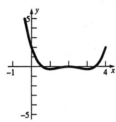

c. $g'(x) = \dfrac{(x-1)-(x+1)}{(x-1)^2} = \dfrac{-2}{(x-1)^2}$

$\dfrac{g(3)-g(2)}{3-2} = \dfrac{2-3}{1} = -1$

$\dfrac{-2}{(c-1)^2} = -1; c = 1 \pm \sqrt{2}$

Only $c = 1 + \sqrt{2}$ is in the interval $(2, 3)$.

4.9 Additional Problem Set

1. a. $ab \le \left(\dfrac{a+b}{2}\right)^2 = \dfrac{a^2 + 2ab + b^2}{4}$

$= \dfrac{a^2}{4} + \dfrac{1}{2}ab + \dfrac{b^2}{4}$

This is true if

$0 \le \dfrac{a^2}{4} - \dfrac{1}{2}ab + \dfrac{b^2}{4} = \left(\dfrac{a}{2} - \dfrac{b}{2}\right)^2 = \left(\dfrac{a-b}{2}\right)^2$

Since a square can never be negative, this is always true.

b. $F(b) = \dfrac{a^2 + 2ab + b^2}{4b}$

As $b \to 0^+, a^2 + 2ab + b^2 \to a^2$ while

$4b \to 0^+$, thus $\lim\limits_{b \to 0^+} F(b) = \infty$ which is not

close to a.

$\lim\limits_{b \to \infty} \dfrac{a^2 + 2ab + b^2}{4b} = \lim\limits_{b \to \infty} \dfrac{\dfrac{a^2}{b} + 2a + b}{4} = \infty$,

so when b is very large,
$F(b)$ is not close to a.

$F'(b) = \dfrac{2(a+b)(4b) - 4(a+b)^2}{16b^2}$

$= \dfrac{4b^2 - 4a^2}{16b^2} = \dfrac{b^2 - a^2}{4b^2}$;

$F'(b) = 0$ when $b^2 = a^2$ or $b = a$ since a and

b are both positive.

$F(a) = \dfrac{(a+a)^2}{4a} = \dfrac{4a^2}{4a} = a$

Thus $a \le \dfrac{(a+b)^2}{4b}$ for all $b > 0$ or

$ab \le \dfrac{(a+b)^2}{4}$ which leads to $\sqrt{ab} \le \dfrac{a+b}{2}$.

c. Let $F(b) = \dfrac{1}{b}\left(\dfrac{a+b+c}{3}\right)^3 = \dfrac{(a+b+c)^3}{27b}$

$F'(b) = \dfrac{3(a+b+c)^2(27b) - 27(a+b+c)^3}{27^2 b^2}$

$= \dfrac{(a+b+c)^2[3b - (a+b+c)]}{27b^2}$

$= \dfrac{(a+b+c)^2(2b - a - c)}{27b^2}$;

$F'(b) = 0$ when $b = \dfrac{a+c}{2}$.

$F\left(\dfrac{a+c}{2}\right) = \dfrac{2}{a+c} \cdot \left(\dfrac{a+c}{3} + \dfrac{a+c}{6}\right)^3$

$= \dfrac{2}{a+c}\left(\dfrac{3(a+c)}{6}\right)^3 = \dfrac{2}{a+c}\left(\dfrac{a+c}{2}\right)^3 = \left(\dfrac{a+c}{2}\right)^2$

Thus $\left(\dfrac{a+c}{2}\right)^2 \le \dfrac{1}{b}\left(\dfrac{a+b+c}{3}\right)^3$ for all $b > 0$.

From Problem 1b, $ac \le \left(\dfrac{a+c}{2}\right)^2$, thus

$ac \le \dfrac{1}{b}\left(\dfrac{a+b+c}{3}\right)^3$ or $abc \le \left(\dfrac{a+b+c}{3}\right)^3$

which gives the desired result

$(abc)^{1/3} \le \dfrac{a+b+c}{3}$.

3. (Proof by induction.)

Suppose that the result holds for all m with $1 \le m \le N$. Now let $n = N+1$, and

$$F(x_1) = \frac{1}{x_1}\left(\frac{x_1 + x_2 + \ldots + x_n}{n}\right)^n = \frac{(x_1 + x_2 + \ldots + x_n)^n}{n^n x_1}$$

$$F'(x_1) = \frac{n(x_1 + x_2 + \ldots + x_n)^{n-1}(n^n x_1) - n^n(x_1 + x_2 + \ldots + x_n)^n}{n^n x_1} = \frac{(x_1 + x_2 + \ldots + x_n)^{n-1}[nx_1 - (x_1 + x_2 + \ldots + x_n)]}{x_1}$$

$$= \frac{(x_1 + x_2 + \ldots + x_n)^{n-1}[(n-1)x_1 - x_2 - \ldots - x_n]}{x_1}$$

$$F'(x_1) = 0 \text{ when } x_1 = \frac{x_2 + x_3 + \ldots + x_n}{n-1}$$

$$F\left(\frac{x_2 + x_3 + \ldots + x_n}{n-1}\right) = \frac{n-1}{x_2 + x_3 + \ldots + x_n}\left(\frac{x_2 + x_3 + \ldots + x_n}{n(n-1)} + \frac{x_2 + x_3 + \ldots + x_n}{n}\right)^n$$

$$= \frac{n-1}{x_2 + x_3 + \ldots + x_n}\left(\frac{x_2 + x_3 + \ldots + x_n}{n-1}\right)^n = \left(\frac{x_2 + x_3 + \ldots x_n}{n-1}\right)^{n-1}$$

Thus, $\left(\dfrac{x_2 + x_3 + \ldots + x_n}{n-1}\right)^{n-1} \le \dfrac{1}{x_1}\left(\dfrac{x_1 + x_2 + \ldots + x_n}{n}\right)^n$ for all $x_1 > 0$.

But, by assmption, $(x_2 x_3 \ldots x_n)^{\frac{1}{n-1}} \le \dfrac{x_2 + x_3 + \ldots + x_n}{n-1}$ with equality holding when $x_2 = x_3 = \ldots = x_n$.

Thus, $x_2 x_3 \ldots x_n \le \left(\dfrac{x_2 + x_3 + \ldots + x_3}{n-1}\right)^{n-1} \le \dfrac{1}{x_1}\left(\dfrac{x_1 + x_2 + \ldots + x_n}{n}\right)^n$, and combining the first and last terms

$(x_1 x_2 \ldots x_n)^{\frac{1}{n}} \le \dfrac{x_1 + x_2 + \ldots + x_n}{n}$ with equality holding when $x_2 = x_3 = \ldots = x_n$ and

$$x_1 = \frac{x_2 + x_3 + \ldots + x_n}{n-1} = \frac{(n-1)x_2}{n-1} = x_2 \text{ or } x_1 = x_2 = \ldots = x_n.$$

The case where $n = 1$ is clear, $n = 2$ and $n = 3$ were proved in Problems 1b and 1c, so the relation holds for any set of positive x_j.

5. $P_{ns}(p) = \dfrac{n!}{s!(n-s)!}p^s(1-p)^{n-s}$

$P_{ns}'(p) = \dfrac{n!}{s!(n-s)!}[sp^{s-1}(1-p)^{n-s}$

$\qquad\qquad\qquad - (n-s)p^s(1-p)^{n-s-1}]$

$= \dfrac{n!}{s!(n-s)!}p^{s-1}(1-p)^{n-s-1}[s(1-p) - (n-s)p]$

$= \dfrac{n!}{s!(n-s)!}p^{s-1}(1-p)^{n-s-1}(s - np)$

$P_{ns}'(p) = 0$ when $p = 0, \dfrac{s}{n}, 1$

$P_{ns}(0) = P_{ns}(1) = 0$

$P_{ns}\left(\dfrac{s}{n}\right) = \dfrac{n!}{s!(n-s)!}\left(\dfrac{s}{n}\right)^s\left(1 - \dfrac{s}{n}\right)^{n-s} > 0,$

thus P_{ns} is a maximum when $p = \dfrac{s}{n}$.

7. If the end of the cylinder has radius r and h is the height of the cylinder, the surface area is

$$A = 2\pi r^2 + 2\pi rh \text{ so } h = \frac{A}{2\pi r} - r.$$

The volume is

$$V = \pi r^2 h = \pi r^2\left(\frac{A}{2\pi r} - r\right) = \frac{Ar}{2} - \pi r^3.$$

$$V'(r) = \frac{A}{2} - 3\pi r^2; V'(r) = 0 \text{ when } r = \sqrt{\frac{A}{6\pi}},$$

$V''(r) = -6\pi r$, so the volume is maximum when

$$r = \sqrt{\frac{A}{6\pi}}.$$

$$h = \frac{A}{2\pi r} - r = 2\sqrt{\frac{A}{6\pi}} = 2r$$

CHAPTER

5

The Integral

5.1 Concepts Review

1. $rx^{r-1}; \dfrac{x^{r+1}}{r+1} + C, r \neq -1$

3. $u = x^4 + 3x^2 + 1, du = (4x^3 + 6x)dx$

$\displaystyle\int (x^4 + 3x^2 + 1)^8 (4x^3 + 6x)dx = \int u^8 du$

$= \dfrac{u^9}{9} + C = \dfrac{(x^4 + 3x^2 + 1)^9}{9} + C$

Problem Set 5.1

1. $\displaystyle\int 5 dx = 5x + C$

3. $\displaystyle\int (x^2 + \pi)dx = \int x^2 dx + \pi \int 1 dx = \dfrac{x^3}{3} + \pi x + C$

5. $\displaystyle\int x^{5/4} dx = \dfrac{x^{9/4}}{\frac{9}{4}} + C = \dfrac{4}{9} x^{9/4} + C$

7. $\displaystyle\int \dfrac{1}{\sqrt[3]{x^2}} dx = \int x^{-2/3} dx = 3x^{1/3} + C = 3\sqrt[3]{x} + C$

9. $\displaystyle\int (x^2 - x)dx = \int x^2 dx - \int x dx = \dfrac{x^3}{3} - \dfrac{x^2}{2} + C$

11. $\displaystyle\int (4x^5 - x^3)dx = 4\int x^5 dx - \int x^3 dx$

$= 4\left(\dfrac{x^6}{6} + C_1\right) - \left(\dfrac{x^4}{4} + C_2\right)$

$= \dfrac{2x^6}{3} - \dfrac{x^4}{4} + C$

13. $\displaystyle\int (27x^7 + 3x^5 - 45x^3 + \sqrt{2}\,x)dx$

$= 27\int x^7 dx + 3\int x^5 dx - 45\int x^3 dx + \sqrt{2}\int x dx$

$= \dfrac{27x^8}{8} + \dfrac{x^6}{2} - \dfrac{45x^4}{4} + \dfrac{\sqrt{2}\,x^2}{2} + C$

15. $\displaystyle\int \left(\dfrac{3}{x^2} - \dfrac{2}{x^3}\right)dx = \int (3x^{-2} - 2x^{-3})dx$

$= 3\int x^{-2} dx - 2\int x^{-3} dx$

$= \dfrac{3x^{-1}}{-1} - \dfrac{2x^{-2}}{-2} + C$

$= -\dfrac{3}{x} + \dfrac{1}{x^2} + C$

17. $\displaystyle\int \dfrac{4x^6 + 3x^4}{x^3} dx = \int (4x^3 + 3x)dx$

$= 4\int x^3 dx + 3\int x dx$

$= x^4 + \dfrac{3x^2}{2} + C$

19. $\displaystyle\int (x^2 + x)\,dx = \int x^2 dx + \int x dx = \dfrac{x^3}{3} + \dfrac{x^2}{2} + C$

21. Let $u = x + 1$; then $du = dx$.

$\displaystyle\int (x+1)^2 dx = \int u^2 du = \dfrac{u^3}{3} + C = \dfrac{(x+1)^3}{3} + C$

23. $\displaystyle\int \dfrac{(z^2 + 1)^2}{\sqrt{z}} dz = \int \dfrac{z^4 + 2z^2 + 1}{\sqrt{z}} dz$

$= \int z^{7/2} dz + 2\int z^{3/2} dz + \int z^{-1/2} dz$

$= \dfrac{2}{9} z^{9/2} + \dfrac{4}{5} z^{5/2} + 2z^{1/2} + C$

25. $\displaystyle\int (\sin\theta - \cos\theta)d\theta = \int \sin\theta\, d\theta - \int \cos\theta\, d\theta$

$= -\cos\theta - \sin\theta + C$

27. Let $g(x) = \sqrt{2}\,x + 1$; then $g'(x) = \sqrt{2}$.

$\displaystyle\int \left(\sqrt{2}\,x+1\right)^3 \sqrt{2}\, dx = \int [g(x)]^3\, g'(x)dx$

$= \dfrac{[g(x)]^4}{4} + C = \dfrac{\left(\sqrt{2}\,x+1\right)^4}{4} + C$

29. Let $u = 5x^3 + 3x - 8$; then $du = (15x^2 + 3)\,dx$.

$$\int (5x^2 + 1)(5x^3 + 3x - 8)^6\,dx$$

$$= \int \frac{1}{3}(15x^2 + 3)(5x^3 + 3x - 8)^6\,dx$$

$$= \frac{1}{3}\int u^6\,du = \frac{1}{3}\left(\frac{u^7}{7} + C_1\right)$$

$$= \frac{(5x^3 + 3x - 8)^7}{21} + C$$

31. Let $u = 2t^2 - 11$; then $du = 4t\,dt$.

$$\int 3t\sqrt[3]{2t^2 - 11}\,dt = \int \frac{3}{4}(4t)(2t^2 - 11)^{1/3}\,dt$$

$$= \frac{3}{4}\int u^{1/3}\,du = \frac{3}{4}\left(\frac{3}{4}u^{4/3} + C_1\right)$$

$$= \frac{9}{16}\sqrt[3]{(2t^2 - 11)^4} + C$$

33. $f'(x) = \int (3x + 1)\,dx = \frac{3}{2}x^2 + x + C_1$

$$f(x) = \int \left(\frac{3}{2}x^2 + x + C_1\right)dx$$

$$= \frac{1}{2}x^3 + \frac{1}{2}x^2 + C_1 x + C_2$$

35. $f'(x) = \int x^{1/2}\,dx = \frac{2}{3}x^{3/2} + C_1$

$$f(x) = \int \left(\frac{2}{3}x^{3/2} + C_1\right)dx$$

$$= \frac{4}{15}x^{5/2} + C_1 x + C_2$$

37. $f''(x) = x + x^{-3}$

$$f'(x) = \int (x + x^{-3})\,dx = \frac{x^2}{2} - \frac{x^{-2}}{2} + C_1$$

$$f(x) = \int \left(\frac{1}{2}x^2 - \frac{1}{2}x^{-2} + C_1\right)dx$$

$$= \frac{1}{6}x^3 + \frac{1}{2}x^{-1} + C_1 x + C_2$$

$$= \frac{1}{6}x^3 + \frac{1}{2x} + C_1 x + C_2$$

39. The Product Rule for derivatives says

$$\frac{d}{dx}[f(x)g(x) + C] = f(x)g'(x) + f'(x)g(x).$$

Thus,

$$\int [f(x)g'(x) + f'(x)g(x)]\,dx = f(x)g(x) + C.$$

41. Let $f(x) = x^2, g(x) = \sqrt{x - 1}$.

$$f'(x) = 2x, g'(x) = \frac{1}{2\sqrt{x - 1}}$$

$$\int \left[\frac{x^2}{2\sqrt{x - 1}} + 2x\sqrt{x - 1}\right]dx$$

$$= \int [f(x)g'(x) + f'(x)g(x)]\,dx = f(x)g(x) + C$$

$$= x^2\sqrt{x - 1} + C$$

43. $\int f''(x)\,dx = \int \frac{d}{dx}f'(x)\,dx = f'(x) + C$

$$f'(x) = \sqrt{x^3 + 1} + \frac{3x^3}{2\sqrt{x^3 + 1}} = \frac{5x^3 + 2}{2\sqrt{x^3 + 1}} \text{ so}$$

$$\int f''(x)\,dx = \frac{5x^3 + 2}{2\sqrt{x^3 + 1}} + C.$$

45. The Product Rule for derivatives says that

$$\frac{d}{dx}[f^m(x)g^n(x) + C]$$

$$= f^m(x)[g^n(x)]' + [f^m(x)]'g^n(x)$$

$$= f^m(x)[ng^{n-1}(x)g'(x)] + [mf^{m-1}(x)f'(x)]g^n(x)$$

$$= f^{m-1}(x)g^{n-1}(x)[nf(x)g'(x) + mg(x)f'(x)].$$

Thus,

$$\int f^{m-1}(x)g^{n-1}(x)[nf(x)g'(x) + mg(x)f'(x)]\,dx$$

$$= f^m(x)g^n(x) + C.$$

47. If $x \geq 0$, then $|x| = x$ and $\int |x|\,dx = \frac{1}{2}x^2 + C$.

If $x < 0$, then $|x| = -x$ and $\int |x|\,dx = -\frac{1}{2}x^2 + C$.

$$\int |x|\,dx = \begin{cases} \dfrac{1}{2}x^2 + C & \text{if } x \geq 0 \\[2mm] -\dfrac{1}{2}x^2 + C & \text{if } x < 0 \end{cases}$$

49. Different software may produce different, but equivalent answers. These answers were produced by Mathematica.

a. $\int 6\sin(3(x - 2))\,dx = -2\cos(3(x - 2)) + C$

b. $\int \sin^3\left(\frac{x}{6}\right)dx = \frac{1}{2}\cos\left(\frac{x}{2}\right) - \frac{9}{2}\cos\left(\frac{x}{6}\right) + C$

c. $\int (x^2\cos 2x + x\sin 2x)\,dx = \frac{x^2\sin 2x}{2} + C$

5.2 Concepts Review

1. differential equation

3. separate variables

Problem Set 5.2

1. $\dfrac{dy}{dx} = \dfrac{-2x}{2\sqrt{1-x^2}} = \dfrac{-x}{\sqrt{1-x^2}}$

$\dfrac{dy}{dx} + \dfrac{x}{y} = \dfrac{-x}{\sqrt{1-x^2}} + \dfrac{x}{\sqrt{1-x^2}} = 0$

3. $\dfrac{dy}{dx} = C_1 \cos x - C_2 \sin x;$

$\dfrac{d^2 y}{dx^2} = -C_1 \sin x - C_2 \cos x$

$\dfrac{d^2 y}{dx^2} + y$

$= (-C_1 \sin x - C_2 \cos x) + (C_1 \sin x + C_2 \cos x) = 0$

5. $\dfrac{dy}{dx} = x^2 + 1$

$dy = (x^2 + 1)\,dx$

$\int dy = \int (x^2 + 1)\,dx$

$y + C_1 = \dfrac{x^3}{3} + x + C_2$

$y = \dfrac{x^3}{3} + x + C$

At $x = 1, y = 1$:

$1 = \dfrac{1}{3} + 1 + C; C = -\dfrac{1}{3}$

$y = \dfrac{x^3}{3} + x - \dfrac{1}{3}$

7. $\dfrac{dy}{dx} = \dfrac{x}{y}$

$\int y\,dy = \int x\,dx$

$\dfrac{y^2}{2} + C_1 = \dfrac{x^2}{2} + C_2$

$y^2 = x^2 + C$

$y = \pm\sqrt{x^2 + C}$

At $x = 1, y = 1$:

$1 = \pm\sqrt{1 + C}; C = 0$ and the square root is positive.

$y = \sqrt{x^2}$ or $y = |x|$

9. $\dfrac{dz}{dt} = t^2 z^2$

$\int z^{-2}\,dz = \int t^2\,dt$

$-z^{-1} + C_1 = \dfrac{t^3}{3} + C_2$

$\dfrac{1}{z} = -\dfrac{t^3}{3} + C_3 = \dfrac{C - t^3}{3}$

$z = \dfrac{3}{C - t^3}$

At $t = 1, z = \dfrac{1}{3}$:

$\dfrac{1}{3} = \dfrac{3}{C-1}; C - 1 = 9; C = 10$

$z = \dfrac{3}{10 - t^3}$

11. $\dfrac{ds}{dt} = 16t^2 + 4t - 1$

$\int ds = \int (16t^2 + 4t - 1)\,dt$

$s + C_1 = \dfrac{16}{3}t^3 + 2t^2 - t + C_2$

$s = \dfrac{16}{3}t^3 + 2t^2 - t + C$

At $t = 0, s = 100$:

$C = 100$

$s = \dfrac{16}{3}t^3 + 2t^2 - t + 100$

13. $\dfrac{dy}{dx} = (2x + 1)^4$

$y = \int (2x+1)^4\,dx = \dfrac{1}{2}\int (2x+1)^4 2\,dx$

$= \dfrac{1}{2}\dfrac{(2x+1)^5}{5} + C = \dfrac{(2x+1)^5}{10} + C$

At $x = 0, y = 6$:

$6 = \dfrac{1}{10} + C; C = \dfrac{59}{10}$

$y = \dfrac{(2x+1)^5}{10} + \dfrac{59}{10} = \dfrac{(2x+1)^5 + 59}{10}$

15. $\dfrac{dy}{dx} = 3x$

$y = \int 3x\,dx = \dfrac{3}{2}x^2 + C$

At $(1, 2)$:

$2 = \dfrac{3}{2} + C$

$$C = \frac{1}{2}$$

$$y = \frac{3}{2}x^2 + \frac{1}{2} = \frac{3x^2 + 1}{2}$$

17. $v = \int t \, dt = \frac{t^2}{2} + v_0$

$$v = \frac{t^2}{2} + 3$$

$$s = \int \left(\frac{t^2}{2} + 3 \right) dt = \frac{t^3}{6} + 3t + s_0$$

$$s = \frac{t^3}{6} + 3t + 0 = \frac{t^3}{6} + 3t$$

At $t = 2$:

$v = 5$ cm/s

$$s = \frac{22}{3} \text{ cm}$$

19. $v = \int (2t+1)^{1/3} \, dt = \frac{1}{2} \int (2t+1)^{1/3} \, 2dt$

$$= \frac{3}{8}(2t+1)^{4/3} + C_1$$

$$v_0 = 0 : 0 = \frac{3}{8} + C_1; C_1 = -\frac{3}{8}$$

$$v = \frac{3}{8}(2t+1)^{4/3} - \frac{3}{8}$$

$$s = \frac{3}{8}\int (2t+1)^{4/3} \, dt - \frac{3}{8}\int 1 \, dt$$

$$= \frac{3}{16}\int (2t+1)^{4/3} \, 2dt - \frac{3}{8}\int 1 \, dt$$

$$= \frac{9}{112}(2t+1)^{7/3} - \frac{3}{8}t + C_2$$

$$s_0 = 10 : 10 = \frac{9}{112} + C_2; C_2 = \frac{1111}{112}$$

$$s = \frac{9}{112}(2t+1)^{7/3} - \frac{3}{8}t + \frac{1111}{112}$$

At $t = 2$: $v = \frac{3}{8}(5)^{4/3} - \frac{3}{8} \approx 2.83$

$$s = \frac{9}{112}(5)^{7/3} - \frac{6}{8} + \frac{1111}{112} \approx 12.6$$

21. $v = -32t + 96,$

$s = -16t^2 + 96t + s_0 = -16t^2 + 96t$

$v = 0$ at $t = 3$

At $t = 3$, $s = -16(3^2) + 96(3) = 144$ ft

23. $\frac{dv}{dt} = -5.28$

$$\int dv = -\int 5.28 dt$$

$$v = \frac{ds}{dt} = -5.28t + v_0 = -5.28t + 56$$

$$\int ds = \int (-5.28t + 56) dt$$

$$s = -2.64t^2 + 56t + s_0 = -2.64t^2 + 56t + 1000$$

When $t = 4.5$, $v = 32.24$ ft/s and $s = 1198.54$ ft

25. $\frac{dV}{dt} = -kS$

Since $V = \frac{4}{3}\pi r^3$ and $S = 4\pi r^2$,

$$4\pi r^2 \frac{dr}{dt} = -k4\pi r^2 \text{ so } \frac{dr}{dt} = -k .$$

$$\int dr = -\int k \, dt$$

$r = -kt + C$

$2 = -k(0) + C$ and $0.5 = -k(10) + C$, so

$C = 2$ and $k = \frac{3}{20}$. Then, $r = -\frac{3}{20}t + 2$.

27. $v_{esc} = \sqrt{2gR}$

For the Moon, $v_{esc} \approx \sqrt{2(0.165)(32)(1080 \cdot 5280)}$

≈ 7760 ft/s ≈ 1.470 mi/s.

For Venus, $v_{esc} \approx \sqrt{2(0.85)(32)(3800 \cdot 5280)}$

$\approx 33{,}038$ ft/s ≈ 6.257 mi/s.

For Jupiter, $v_{esc} \approx 194{,}369$ ft/s ≈ 36.812 mi/s.

For the Sun, $v_{esc} \approx 2{,}021{,}752$ ft/s

≈ 382.908 mi/s.

29. $a = \frac{dv}{dt} = \frac{\Delta v}{\Delta t} = \frac{60 - 45}{10} = 1.5$ mi/h/s $= 2.2$ ft/s^2

31. For the first 10 s, $a = \frac{dv}{dt} = 6t, v = 3t^2$, and

$s = t^3$. So $v(10) = 300$ and $s(10) = 1000$. After

10 s, $a = \frac{dv}{dt} = -10$, $v = -10(t - 10) + 300$, and

$s = -5(t - 10)^2 + 300(t - 10) + 1000$. $v = 0$ at

$t = 40$, at which time $s = 5500$ m.

33. a.

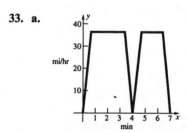

b. Since the trip that involves 1 min more travel time at speed v_m is 0.6 mi longer,

$v_m = 0.6$ mi/min $= 36$ mi/h.

c. From part b, $v_m = 0.6$ mi/min. Note that the average speed during acceleration and deceleration is $\frac{v_m}{2} = 0.3$ mi/min. Let t be the time spent between stop C and stop D at the constant speed v_m, so $0.6t + 0.3(4 - t) = 2$ miles. Therefore,

$t = 2\frac{2}{3}$ min and the time spent accelerating

is $\dfrac{4 - 2\frac{2}{3}}{2} = \dfrac{2}{3}$ min.

$a = \dfrac{0.6 - 0}{\frac{2}{3}} = 0.9$ mi/min^2.

35. a. $\dfrac{dV}{dt} = C_1\sqrt{h}$ where h is the depth of the water. Here, $V = \pi r^2 h = 100h$, so $h = \dfrac{V}{100}$.

Hence $\dfrac{dV}{dt} = C_1\dfrac{\sqrt{V}}{10}$, $V(0) = 1600$,

$V(40) = 0$.

b. $\int 10V^{-1/2}\,dV = \int C_1\,dt;\ 20\sqrt{V} = C_1 t + C_2;$

$V(0) = 1600:\ C_2 = 20 \cdot 40 = 800;$

$V(40) = 0:\ C_1 = -\dfrac{800}{40} = -20$

$V(t) = \dfrac{1}{400}(-20t + 800)^2$

c. $V(10) = \dfrac{1}{400}(-200 + 800)^2 = 900$ cm^3

37. Initially, $v = -32t$ and $s = -16t^2 + 16$. $s = 0$ when $t = 1$. Later, the ball falls 9 ft in a time given by

$0 = -16t^2 + 9$, or $\dfrac{3}{4}$ s, and on impact has a

velocity of $-32\left(\dfrac{3}{4}\right) = -24$ ft/s. By symmetry,

24 ft/s must be the velocity right after the first bounce. So

a. $v(t) = \begin{cases} -32t & \text{for } 0 \le t \le 1 \\ -32(t-1) + 24 & \text{for } 1 < t \le 2.5 \end{cases}$

b. $9 = -16t^2 + 16 \Rightarrow t \approx 0.66$ sec; s also equals 9 at the apex of the first rebound at $t = 1.75$ sec.

5.3 Concepts Review

1. $2 \cdot \dfrac{5(6)}{2} = 30;\ 2(5) = 10$

3. $1 - \dfrac{1}{10} = 0.9$

Problem Set 5.3

1. $\displaystyle\sum_{k=1}^{6}(k-1) = \sum_{k=1}^{6}k - \sum_{k=1}^{6}1 = \dfrac{6(7)}{2} - 6(1) = 15$

3. $\displaystyle\sum_{k=1}^{7}\dfrac{1}{k+1} = \dfrac{1}{1+1} + \dfrac{1}{2+1} + \dfrac{1}{3+1}$

$+ \dfrac{1}{4+1} + \dfrac{1}{5+1} + \dfrac{1}{6+1} + \dfrac{1}{7+1}$

$= \dfrac{1}{2} + \dfrac{1}{3} + \dfrac{1}{4} + \dfrac{1}{5} + \dfrac{1}{6} + \dfrac{1}{7} + \dfrac{1}{8} = \dfrac{1443}{840} = \dfrac{481}{280}$

5. $\displaystyle\sum_{m=1}^{8}(-1)^m 2^{m-2} = (-1)^1 2^{-1} + (-1)^2 2^0 + (-1)^3 2^1$

$+ (-1)^4 2^2 + (-1)^5 2^3 + (-1)^6 2^4$

$+ (-1)^7 2^5 + (-1)^8 2^6$

$= -\dfrac{1}{2} + 1 - 2 + 4 - 8 + 16 - 32 + 64 = \dfrac{85}{2}$

7. $\displaystyle\sum_{n=1}^{6} n\cos(n\pi) = \cos\pi + 2\cos 2\pi + 3\cos 3\pi$

$+ 4\cos 4\pi + 5\cos 5\pi + 6\cos 6\pi$

$= -1 + 2 - 3 + 4 - 5 + 6 = 3$

9. $1 + 2 + 3 + \cdots + 41 = \displaystyle\sum_{i=1}^{41} i$

11. $1 + \dfrac{1}{2} + \dfrac{1}{3} + \cdots + \dfrac{1}{100} = \displaystyle\sum_{i=1}^{100}\dfrac{1}{i}$

13. $a_1 + a_3 + a_5 + a_7 + \cdots + a_{99} = \sum_{i=1}^{50} a_{2i-1}$

15. $f(c_1) + f(c_2) + \cdots + f(c_n) = \sum_{i=1}^{n} f(c_i)$

17. $\sum_{i=1}^{10} (a_i + b_i) = \sum_{i=1}^{10} a_i + \sum_{i=1}^{10} b_i = 40 + 50 = 90$

19. $\sum_{p=0}^{9} (a_{p+1} - b_{p+1}) = \sum_{p=1}^{10} a_p - \sum_{p=1}^{10} b_p = 40 - 50$
$= -10$

21. $\sum_{k=1}^{40} \left(\frac{1}{k} - \frac{1}{k+1} \right)$
$= \left(1 - \frac{1}{2} \right) + \left(\frac{1}{2} - \frac{1}{3} \right) + \left(\frac{1}{3} - \frac{1}{4} \right) + \cdots + \left(\frac{1}{40} - \frac{1}{41} \right)$
$= 1 - \frac{1}{41} = \frac{40}{41}$

23. $\sum_{k=3}^{20} \left(\frac{1}{(k+1)^2} - \frac{1}{k^2} \right)$
$= \left(\frac{1}{4^2} - \frac{1}{3^2} \right) + \left(\frac{1}{5^2} - \frac{1}{4^2} \right) + \cdots\cdots + \left(\frac{1}{21^2} - \frac{1}{20^2} \right)$
$= -\frac{1}{3^2} + \frac{1}{21^2} = -\frac{49}{441} + \frac{1}{441} = -\frac{48}{441} = -\frac{16}{147}$

25. $\sum_{i=1}^{100} (3i - 2) = 3 \sum_{i=1}^{100} i - \sum_{i=1}^{100} 2 = 3(5050) - 2(100)$
$= 14{,}950$

27. $\sum_{k=1}^{10} (k^3 - k^2) = \sum_{k=1}^{10} k^3 - \sum_{k=1}^{10} k^2 = 3025 - 385$
$= 2640$

29. $\sum_{i=1}^{n} (2i^2 - 3i + 1) = 2\sum_{i=1}^{n} i^2 - 3\sum_{i=1}^{n} i + \sum_{i=1}^{n} 1$
$= \frac{2n(n+1)(2n+1)}{6} - \frac{3n(n+1)}{2} + n$
$= \frac{2n^3 + 3n^2 + n}{3} - \frac{3n^2 + 3n}{2} + n = \frac{4n^3 - 3n^2 - n}{6}$

31. $\sum_{i=3}^{19} i(i-2) = \sum_{k=1}^{17} (k+2)k$

33. $\sum_{k=0}^{10} \frac{k}{k+1} = \sum_{i=1}^{11} \frac{i-1}{i}$

35. $\sum_{i=1}^{10} \frac{3i}{5} \cdot \frac{1}{5} = \frac{3}{25} \sum_{i=1}^{10} i = \frac{33}{5}$

37. a. $\sum_{k=0}^{10} \left(\frac{1}{2} \right)^k = \frac{1 - \left(\frac{1}{2} \right)^{11}}{\frac{1}{2}} = 2 - \left(\frac{1}{2} \right)^{10}$, so
$\sum_{k=1}^{10} \left(\frac{1}{2} \right)^k = 1 - \left(\frac{1}{2} \right)^{10}$.

b. $\sum_{k=0}^{10} 2^k = \frac{1 - 2^{11}}{-1} = 2^{11} - 1$, so
$\sum_{k=1}^{10} 2^k = 2^{11} - 2$.

39.
$S = a + (a+d) + (a+2d) + \cdots\cdots + [a + (n-2)d] + [a + (n-1)d] + (a + nd)$
$\underline{+\,S = (a+nd) + [a+(n-1)d] + [a+(n-2)d] + \cdots + (a+2d) + (a+d) + a}$
$2S = (2a+nd) + (2a+nd) + (2a+nd) + \cdots + (2a+nd) + (2a+nd) + (2a+nd)$
$2S = (n+1)(2a+nd)$
$S = \frac{(n+1)(2a+nd)}{2}$

41. $\bar{x} = \frac{1}{7}(2+5+7+8+9+10+14) = \frac{55}{7} \approx 7.86$

$s^2 = \frac{1}{7}\left[\left(2 - \frac{55}{7}\right)^2 + \left(5 - \frac{55}{7}\right)^2 + \left(7 - \frac{55}{7}\right)^2 + \left(8 - \frac{55}{7}\right)^2 \right.$
$\left. + \left(9 - \frac{55}{7}\right)^2 + \left(10 - \frac{55}{7}\right)^2 + \left(14 - \frac{55}{7}\right)^2 \right] \approx 12.4$

43. a. $\sum_{i=1}^{n} (x_i - \bar{x}) = \sum_{i=1}^{n} x_i - \sum_{i=1}^{n} \bar{x} = n\bar{x} - n\bar{x} = 0$

b. $s^2 = \frac{1}{n} \sum_{i=1}^{n} (x_i - \bar{x})^2 = \frac{1}{n} \sum_{i=1}^{n} (x_i^2 - 2\bar{x} x_i + \bar{x}^2)$
$= \frac{1}{n} \sum_{i=1}^{n} x_i^2 - \frac{2\bar{x}}{n} \sum_{i=1}^{n} x_i + \frac{1}{n} \sum_{i=1}^{n} \bar{x}^2$

$$= \frac{1}{n}\sum_{i=1}^{n} x_i^2 - \frac{2\bar{x}}{n}(n\bar{x}) + \frac{1}{n}(n\bar{x}^2)$$

$$= \left(\frac{1}{n}\sum_{i=1}^{n} x_i^2\right) - 2\bar{x}^2 + \bar{x}^2 = \left(\frac{1}{n}\sum_{i=1}^{n} x_i^2\right) - \bar{x}^2$$

45. Let $S(c) = \sum_{i=1}^{n}(x_i - c)^2$. Then

$$S'(c) = \frac{d}{dc}\sum_{i=1}^{n}(x_i - c)^2$$

$$= \sum_{i=1}^{n}\frac{d}{dc}(x_i - c)^2$$

$$= \sum_{i=1}^{n} 2(x_i - c)(-1)$$

$$= -2\sum_{i=1}^{n} x_i + 2nc$$

$S''(c) = 2n$

Set $S'(c) = 0$ and solve for c:

$$-2\sum_{i=1}^{n} x_i + 2nc = 0$$

$$c = \frac{1}{n}\sum_{i=1}^{n} x_i = \bar{x}$$

Since $S''(\bar{x}) = 2n > 0$ we know that $\bar{x}$ minimizes $S(c)$.

47.

$$(i+1)^4 - i^4 = 4i^3 + 6i^2 + 4i + 1$$

$$\sum_{i=1}^{n}\left[(i+1)^4 - i^4\right] = \sum_{i=1}^{n}\left(4i^3 + 6i^2 + 4i + 1\right)$$

$$(n+1)^4 - 1^4 = 4\sum_{i=1}^{n} i^3 + 6\sum_{i=1}^{n} i^2 + 4\sum_{i=1}^{n} i + \sum_{i=1}^{n} 1$$

$$n^4 + 4n^3 + 6n^2 + 4n = 4\sum_{i=1}^{n} i^3 + 6\frac{n(n+1)(2n+1)}{6} + 4\frac{n(n+1)}{2} + n$$

Solving for $\sum_{i=1}^{n} i^3$ gives

$$4\sum_{i=1}^{n} i^3 = n^4 + 4n^3 + 6n^2 + 4n - \left(2n^3 + 3n^2 + n\right) - \left(2n^2 + 2n\right) - n$$

$$4\sum_{i=1}^{n} i^3 = n^4 + 2n^3 + n^2$$

$$\sum_{i=1}^{n} i^3 = \frac{n^4 + 2n^3 + n^2}{4} = \left[\frac{n(n+1)}{2}\right]^2$$

49.

$$\sum_{i=1}^{n}(a_i t + b_i)^2 = \sum_{i=1}^{n}(a_i^2 t^2 + 2a_i b_i t + b_i^2)$$

$$= t^2 \sum_{i=1}^{n} a_i^2 + 2t \sum_{i=1}^{n} a_i b_i + \sum_{i=1}^{n} b_i^2 = At^2 + 2Bt + C$$

Because $\sum_{i=1}^{n}(a_i t + b_i)^2 \geq 0$, $At^2 + 2Bt + C \geq 0$.

51. $\displaystyle\sum_{i=1}^{n}(a_i - a_{i-1})b_{i-1} = \sum_{i=1}^{n}a_i b_{i-1} - \sum_{i=1}^{n}a_{i-1}b_{i-1}$

$\displaystyle = \sum_{i=1}^{n}a_i b_{i-1} - \sum_{i=0}^{n-1}a_i b_i$

$\displaystyle = \sum_{i=1}^{n}a_i b_{i-1} - \sum_{i=1}^{n}a_i b_i + a_n b_n - a_0 b_0$

$\displaystyle = a_n b_n - a_0 b_0 + \sum_{i=1}^{n}a_i (b_{i-1} - b_i)$

$\displaystyle = a_n b_n - a_0 b_0 - \sum_{i=1}^{n}a_i (b_i - b_{i-1})$

53. Suppose we have a $(n+1)\times n$ grid. Shade in $n + 1 - k$ boxes in the kth column. There are n columns, and the shaded area is $1+2+\cdots+n$. The shaded area is also half the area of the grid or $\dfrac{n(n+1)}{2}$. Thus, $1+2+\cdots+n = \dfrac{n(n+1)}{2}$.

Suppose we have a square grid with sides of length $1+2+\cdots+n = \dfrac{n(n+1)}{2}$. From the diagram the area is $1^3 + 2^3 + \cdots + n^3$ or $\left[\dfrac{n(n+1)}{2}\right]^2$. Thus,

$1^3 + 2^3 + \cdots + n^3 = \left[\dfrac{n(n+1)}{2}\right]^2$.

55. The bottom layer contains $10 \cdot 16 = 160$ oranges, the next layer contains $9 \cdot 15 = 135$ oranges, the third layer contains $8 \cdot 14 = 112$ oranges, and so on, up to the top layer, which contains $1 \cdot 7 = 7$ oranges. The stack contains

$1 \cdot 7 + 2 \cdot 8 + \cdots + 9 \cdot 15 + 10 \cdot 16$

$\displaystyle = \sum_{i=1}^{10}i(6+i)$

$\displaystyle = \sum_{i=1}^{10}i(16-10+i) = 715$ oranges.

If the bottom layer is 50 oranges by 60 oranges, the stack contains

$\displaystyle\sum_{i=1}^{50}i(60-50+i) = \sum_{i=1}^{50}i(10+i) = 55{,}675.$

For a general stack whose base is m rows of n oranges with $m \le n$, the stack contains

$\displaystyle\sum_{i=1}^{m}i(n-m+i) = (n-m)\sum_{i=1}^{m}i + \sum_{i=1}^{m}i^2$

$\displaystyle = (n-m)\frac{m(m+1)}{2} + \frac{m(m+1)(2m+1)}{6}$

$\displaystyle = \frac{m(m+1)(3n-m+1)}{6}$

5.4 Concepts Review

1. 16

3. $\dfrac{1}{2}(4)(4) = 8;\ 8 + \dfrac{1}{2}(2)(2) = 10$

5. $A = \dfrac{1}{2}\left[\left(\dfrac{1}{2}\cdot 0^2 + 1\right) + \left(\dfrac{1}{2}\cdot\left(\dfrac{1}{2}\right)^2 + 1\right) + \left(\dfrac{1}{2}\cdot 1^2 + 1\right) + \left(\dfrac{1}{2}\cdot\left(\dfrac{3}{2}\right)^2 + 1\right)\right] = \dfrac{1}{2}\left(1 + \dfrac{9}{8} + \dfrac{3}{2} + \dfrac{17}{8}\right) = \dfrac{23}{8}$

7.

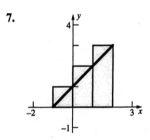

$A = 1(1 + 2 + 3) = 6$

Problem Set 5.4

1. $A = \dfrac{1}{2}\left[1 + \dfrac{3}{2} + 2 + \dfrac{5}{2}\right] = \dfrac{7}{2}$

3. $A = \dfrac{1}{2}\left[\dfrac{3}{2} + 2 + \dfrac{5}{2} + 3\right] = \dfrac{9}{2}$

9.

$$A = \frac{1}{6}\left[\left(\left(\frac{13}{6}\right)^2 - 1\right) + \left(\left(\frac{7}{3}\right)^2 - 1\right) + \left(\left(\frac{5}{2}\right)^2 - 1\right) + \left(\left(\frac{8}{3}\right)^2 - 1\right) + \left(\left(\frac{17}{6}\right)^2 - 1\right) + (3^2 - 1)\right]$$

$$= \frac{1}{6}\left(\frac{133}{36} + \frac{40}{9} + \frac{21}{4} + \frac{55}{9} + \frac{253}{36} + 8\right) = \frac{1243}{216}$$

11. $\Delta x = \frac{1}{n}, \; x_i = \frac{i}{n}$

$$f(x_i)\Delta x = \left(\frac{i}{n} + 2\right)\left(\frac{1}{n}\right) = \frac{i}{n^2} + \frac{2}{n}$$

$$A(S_n) = \left[\left(\frac{1}{n^2} + \frac{2}{n}\right) + \left(\frac{2}{n^2} + \frac{2}{n}\right) + \cdots + \left(\frac{n}{n^2} + \frac{2}{n}\right)\right] = \frac{1}{n^2}(1 + 2 + 3 + \cdots + n) + 2 = \frac{n(n+1)}{2n^2} + 2 = \frac{1}{2n} + \frac{5}{2}$$

$$\lim_{n\to\infty} A(S_n) = \lim_{n\to\infty}\left(\frac{1}{2n} + \frac{5}{2}\right) = \frac{5}{2}$$

13. $\Delta x = \frac{2}{n}, \; x_i = -1 + \frac{2i}{n}$

$$f(x_i)\Delta x = \left[2\left(-1 + \frac{2i}{n}\right) + 2\right]\left(\frac{2}{n}\right) = \frac{8i}{n^2}$$

$$A(S_n) = \left[\left(\frac{8}{n^2}\right) + \left(\frac{16}{n^2}\right) + \cdots + \left(\frac{8n}{n^2}\right)\right]$$

$$= \frac{8}{n^2}(1 + 2 + 3 + \cdots + n) = \frac{8}{n^2}\left[\frac{n(n+1)}{2}\right]$$

$$= 4\left[\frac{n^2 + n}{n^2}\right] = 4 + \frac{4}{n}$$

$$\lim_{n\to\infty} A(S_n) = \lim_{n\to\infty}\left(4 + \frac{4}{n}\right) = 4$$

15. $\Delta x = \frac{1}{n}, \; x_i = \frac{i}{n}$

$$f(x_i)\Delta x = \left(\frac{i}{n}\right)^3\left(\frac{1}{n}\right) = \frac{i^3}{n^4}$$

$$A(S_n) = \left[\frac{1}{n^4}(1^3) + \frac{1}{n^4}(2^3) + \cdots + \frac{1}{n^4}(n^3)\right]$$

$$= \frac{1}{n^4}(1^3 + 2^3 + \cdots + n^3) = \frac{1}{n^4}\left[\frac{n(n+1)}{2}\right]^2$$

$$= \frac{1}{n^4}\left[\frac{n^4 + 2n^3 + n^2}{4}\right] = \frac{1}{4}\left[1 + \frac{2}{n} + \frac{1}{n^2}\right]$$

$$\lim_{n\to\infty} A(S_n) = \lim_{n\to\infty}\frac{1}{4}\left[1 + \frac{2}{n} + \frac{1}{n^2}\right] = \frac{1}{4}$$

17. $f(t_i)\Delta t = \left[\frac{i}{n} + 2\right]\frac{1}{n} = \frac{i}{n^2} + \frac{2}{n}$

$$A(S_n) = \sum_{i=1}^{n}\left(\frac{i}{n^2} + \frac{2}{n}\right) = \frac{1}{n^2}\sum_{i=1}^{n} i + \sum_{i=1}^{n}\frac{2}{n}$$

$$= \frac{1}{n^2}\left[\frac{n(n+1)}{2}\right] + 2$$

$$= \left[\frac{n^2 + n}{2n^2}\right] + 2$$

$$= \left(\frac{1}{2} + \frac{1}{2n}\right) + 2$$

$$\lim_{n\to\infty} A(S_n) = \frac{1}{2} + 2 = \frac{5}{2}$$

The object traveled $2\frac{1}{2}$ ft.

19. a. $f(x_i)\Delta x = \left(\dfrac{ib}{n}\right)^2 \left(\dfrac{b}{n}\right) = \dfrac{b^3 i^2}{n^3}$

$A_0^b = \dfrac{b^3}{n^3}\sum_{i=1}^{n} i^2 = \dfrac{b^3}{n^3}\left[\dfrac{n(n+1)(2n+1)}{6}\right]$

$= \dfrac{b^3}{6}\left[2 + \dfrac{3}{n} + \dfrac{1}{n^2}\right]$

$\lim_{n\to\infty} A_0^b = \dfrac{2b^3}{6} = \dfrac{b^3}{3}$

b. Since $a \geq 0$, $A_0^b = A_0^a + A_a^b$, or

$A_a^b = A_0^b - A_0^a = \dfrac{b^3}{3} - \dfrac{a^3}{3}$.

21. a. $A_0^5 = \dfrac{5^3}{3} = \dfrac{125}{3}$

b. $A_1^4 = \dfrac{4^3}{3} - \dfrac{1^3}{3} = \dfrac{63}{3} = 21$

c. $A_2^5 = \dfrac{5^3}{3} - \dfrac{2^3}{3} = \dfrac{117}{3} = 39$

23. a. $A_0^2(x^3) = \dfrac{2^{3+1}}{3+1} = 4$

b. $A_1^2(x^3) = \dfrac{2^{3+1}}{3+1} - \dfrac{1^{3+1}}{3+1} = 4 - \dfrac{1}{4} = \dfrac{15}{4}$

c. $A_1^2(x^5) = \dfrac{2^{5+1}}{5+1} - \dfrac{1^{5+1}}{5+1} = \dfrac{32}{3} - \dfrac{1}{6} = \dfrac{63}{6}$

$= \dfrac{21}{2} = 10.5$

d. $A_0^2(x^9) = \dfrac{2^{9+1}}{9+1} = \dfrac{1024}{10} = 102.4$

5.5 Concepts Review

1. Riemann sum　　　**3.** $A_{\text{up}} - A_{\text{down}}$

Problem Set 5.5

1. $R_P = f(2)(2.5-1) + f(3)(3.5-2.5) + f(4.5)(5-3.5) = 4(1.5) + 3(1) + (-2.25)(1.5) = 5.625$

3. $R_P = \sum_{i=1}^{5} f(\overline{x}_i)\Delta x_i = f(3)(3.75-3) + f(4)(4.25-3.75) + f(4.75)(5.5-4.25) + f(6)(6-5.5) + f(6.5)(7-6)$

$= 2(0.75) + 3(0.5) + 3.75(1.25) + 5(0.5) + 5.5(1) = 15.6875$

5. $R_P = \sum_{i=1}^{8} f(\overline{x}_i)\Delta x_i = [f(-1.75) + f(-1.25) + f(-0.75) + f(-0.25) + f(0.25) + f(0.75) + f(1.25) + f(1.75)](0.5)$

$= [-0.21875 - 0.46875 - 0.46875 - 0.21875 + 0.28125 + 1.03125 + 2.03125 + 3.28125](0.5) = 2.625$

7. $\displaystyle\int_1^3 x^3\, dx$

9. $\displaystyle\int_{-1}^{1} \dfrac{x^2}{1+x}\, dx$

11. $\Delta x = \dfrac{2}{n}, \ \overline{x}_i = \dfrac{2i}{n}$

$f(\overline{x}_i) = \overline{x}_i + 1 = \dfrac{2i}{n} + 1$

$\sum_{i=1}^{n} f(\overline{x}_i)\Delta x = \sum_{i=1}^{n}\left[1 + i\left(\dfrac{2}{n}\right)\right]\dfrac{2}{n}$

$$= \frac{2}{n}\sum_{i=1}^{n}1 + \frac{4}{n^2}\sum_{i=1}^{n}i = \frac{2}{n}(n) + \frac{4}{n^2}\left[\frac{n(n+1)}{2}\right]$$

$$= 2 + 2\left(1 + \frac{1}{n}\right)$$

$$\int_0^2 (x+1)dx = \lim_{n\to\infty}\left[2 + 2\left(1 + \frac{1}{n}\right)\right] = 4$$

13. $\Delta x = \frac{3}{n}, \bar{x}_i = -2 + \frac{3i}{n}$

$$f(\bar{x}_i) = 2\left(-2 + \frac{3i}{n}\right) + \pi = \pi - 4 + \frac{6i}{n}$$

$$\sum_{i=1}^{n}f(\bar{x}_i)\Delta x = \sum_{i=1}^{n}\left[\pi - 4 + \frac{6i}{n}\right]\frac{3}{n}$$

$$= \frac{3}{n}\sum_{i=1}^{n}(\pi - 4) + \frac{18}{n^2}\sum_{i=1}^{n}i = 3(\pi - 4) + \frac{18}{n^2}\left[\frac{n(n+1)}{2}\right]$$

$$= 3\pi - 12 + 9\left(1 + \frac{1}{n}\right)$$

$$\int_{-2}^1 (2x+\pi)dx = \lim_{n\to\infty}\left[3\pi - 12 + 9\left(1 + \frac{1}{n}\right)\right]$$

$$= 3\pi - 3$$

15. $\Delta x = \frac{5}{n}, \bar{x}_i = \frac{5i}{n}$

$$f(\bar{x}_i) = 1 + \frac{5i}{n}$$

$$\sum_{i=1}^{n}f(\bar{x}_i)\Delta x = \sum_{i=1}^{n}\left[1 + i\left(\frac{5}{n}\right)\right]\frac{5}{n}$$

$$= \frac{5}{n}\sum_{i=1}^{n}1 + \frac{25}{n^2}\sum_{i=1}^{n}i = 5 + \frac{25}{n^2}\left[\frac{n(n+1)}{2}\right]$$

$$= 5 + \frac{25}{2}\left(1 + \frac{1}{n}\right)$$

$$\int_0^5 (x+1)dx = \lim_{n\to\infty}\left[5 + \frac{25}{2}\left(1 + \frac{1}{n}\right)\right] = \frac{35}{2}$$

17.

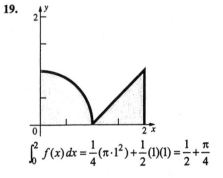

$$\int_0^5 f(x)dx$$

$$= \frac{1}{2}(1)(2) + 1(2) + 3(2) + \frac{1}{2}(3)(3) = \frac{27}{2}$$

19.

$$\int_0^2 f(x)dx = \frac{1}{4}(\pi \cdot 1^2) + \frac{1}{2}(1)(1) = \frac{1}{2} + \frac{\pi}{4}$$

21. The area under the curve is equal to the area of a
semi-circle: $\int_{-A}^{A}\sqrt{A^2 - x^2}\,dx = \frac{1}{2}\pi A^2$.

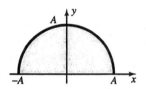

23. a. $\int_{-3}^3 [\![x]\!]dx = (-3 - 2 - 1 + 0 + 1 + 2)(1) = -3$

b. $\int_{-3}^3 [\![x]\!]^2\,dx = [(-3)^2 + (-2)^2$

$$+ (-1)^2 + 0 + 1 + 4](1) = 19$$

c. $\int_{-3}^3 (x - [\![x]\!])dx = 6\left[\frac{1}{2}(1)(1)\right] = 3$

d. $\int_{-3}^3 (x - [\![x]\!])^2 = \int_{-3}^3 (x^2 - 2x[\![x]\!] + [\![x]\!]^2)\,dx$

$$= \int_{-3}^3 x^2 dx - 2\int_{-3}^3 x[\![x]\!]dx + \int_{-3}^3 [\![x]\!]^2\,dx$$

$$= 2\frac{3^3}{3} - 2\left[-3\int_{-3}^{-2}x\,dx - 2\int_{-2}^{-1}x\,dx - \int_{-1}^0 x\,dx\right.$$

$$\left. + 0\int_0^1 x\,dx + \int_1^2 x\,dx + 2\int_2^3 x\,dx\right] + 19$$

$$= 18 - 2\left[(-3)\left(1(-2) - \frac{1}{2}(1)(1)\right)\right.$$
$$-2\left(1(-1) - \frac{1}{2}(1)(1)\right) - \left(-\frac{1}{2}(1)(1)\right) + 0$$
$$\left. + \left(1 \cdot 1 + \frac{1}{2}(1)(1)\right) + 2\left(1 \cdot 2 + \frac{1}{2}(1)(1)\right)\right] + 19$$
$$= 37 - 2\left[(-3)\left(-\frac{5}{2}\right) - 2\left(-\frac{3}{2}\right) + \frac{1}{2} + \frac{3}{2} + 2\left(\frac{5}{2}\right)\right]$$
$$= 37 - 35 = 2$$

e. $\int_{-3}^{3} |x|\, dx = \frac{1}{2}(3)(3) + \frac{1}{2}(3)(3) = 9$

f. $\int_{-3}^{3} x|x|\, dx = \frac{(-3)^3}{3} + \frac{(3)^3}{3} = 0$

g. $\int_{-1}^{2} |x| [\![x]\!]\, dx = -\int_{-1}^{0} |x|\, dx + 0 \int_{0}^{1} |x|\, dx + \int_{1}^{2} |x|\, dx$
$$= -\frac{1}{2}(1)(1) + 1(1) + \frac{1}{2}(1)(1) = 1$$

h. $\int_{-1}^{2} x^2 [\![x]\!]\, dx = -\int_{-1}^{0} x^2\, dx + 0 \int_{0}^{1} x^2\, dx$
$$+ \int_{1}^{2} x^2\, dx$$
$$= -\frac{1^3}{3} + \left(\frac{2^3}{3} - \frac{1^3}{3}\right) = 2$$

25. $R_P = \frac{1}{2}\sum_{i=1}^{n}(x_i + x_{i-1})(x_i - x_{i-1})$
$$= \frac{1}{2}\sum_{i=1}^{n}\left(x_i^2 - x_{i-1}^2\right)$$
$$= \frac{1}{2}\left[(x_1^2 - x_0^2) + (x_2^2 - x_1^2) + (x_3^2 - x_2^2)\right.$$
$$\left. + \cdots + (x_n^2 - x_{n-1}^2)\right]$$

$$= \frac{1}{2}(x_n^2 - x_0^2)$$
$$= \frac{1}{2}(b^2 - a^2)$$
$$\lim_{n \to \infty} \frac{1}{2}(b^2 - a^2) = \frac{1}{2}(b^2 - a^2)$$

27. Left: $\int_{0}^{2} (x^3 + 1)\, dx \approx 5.24$

Right: $\int_{0}^{2} (x^3 + 1)\, dx \approx 6.84$

Midpoint: $\int_{0}^{2} (x^3 + 1)\, dx \approx 5.98$

29. Left: $\int_{0}^{1} \cos x\, dx \approx 0.8638$

Right: $\int_{0}^{1} \cos x\, dx \approx 0.8178$

Midpoint: $\int_{0}^{1} \cos x\, dx \approx 0.8418$

31. $\int_{-2}^{4} (-1 + |x|)\, dx = 4$

33. $\int_{-1}^{2} (x^4 - 3x^2 + 1)\, dx = 0.6$

35. $\int_{0}^{1} \left(\frac{1}{x}\right) dx$ is undefined; $\frac{1}{x}$ is not bounded near $x = 0$.

5.6 Concepts Review

1. $4(4 - 2) = 8$; $16(4 - 2) = 32$

3. $\int_{1}^{4} f(x)\, dx$; $\int_{2}^{5} \sqrt{x}\, dx$

Problem Set 5.6

1. $A(x) = \frac{1}{2}(x - 1)(-1 + x)$, $x > 1$

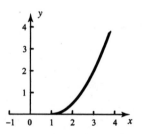

3.
$$A(x) = \begin{cases} 2x & 0 \le x \le 1 \\ 2+(x-1) & 1 < x \le 2 \\ 3+2(x-2) & 2 < x \le 3 \\ 5+(x-3) & 3 < x \le 4 \\ \text{etc.} \end{cases}$$

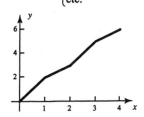

5. $\displaystyle \int_1^2 2f(x)\,dx = 2\int_1^2 f(x)\,dx = 2(3) = 6$

7. $\displaystyle \int_0^2 [2f(x)+g(x)]\,dx = 2\int_0^2 f(x)\,dx + \int_0^2 g(x)\,dx$

$\displaystyle = 2\left[\int_0^1 f(x)\,dx + \int_1^2 f(x)\,dx\right] + \int_0^2 g(x)\,dx$

$= 2(2+3) + 4 = 14$

9. $\displaystyle \int_2^1 [2f(s)+5g(s)]\,ds = -2\int_1^2 f(s)\,ds - 5\int_1^2 g(s)\,ds$

$\displaystyle = -2(3) - 5\left[\int_0^2 g(s)\,ds - \int_0^1 g(s)\,ds\right]$

$= -6 - 5[4+1] = -31$

11. $\displaystyle \int_0^2 [3f(t)+2g(t)]\,dt$

$\displaystyle = 3\left[\int_0^1 f(t)\,dt + \int_1^2 f(t)\,dt\right] + 2\int_0^2 g(t)\,dt$

$= 3(2+3) + 2(4) = 23$

13. $\displaystyle G'(x) = D_x\left[\int_1^x 2t\,dt\right] = 2x$

15. $\displaystyle G'(x) = D_x\left[\int_0^x \left(2t^2 + \sqrt{t}\right)dt\right] = 2x^2 + \sqrt{x}$

17. $\displaystyle G'(x) = D_x\left[\int_x^{\pi/2} (s-2)\cot(2s)\,ds\right]$

$\displaystyle = D_x\left[-\int_{\pi/2}^x (s-2)\cot(2s)\,ds\right] = -(x-2)\cot(2x)$

19. $\displaystyle G'(x) = D_x\left[\int_1^{x^2} \sin t\,dt\right] = 2x\sin(x^2)$

21.
$$G(x) = \int_{-x^2}^x \frac{t^2}{1+t^2}\,dt = \int_{-x^2}^0 \frac{t^2}{1+t^2}\,dt + \int_0^x \frac{t^2}{1+t^2}\,dt$$

$$= -\int_0^{-x^2} \frac{t^2}{1+t^2}\,dt + \int_0^x \frac{t^2}{1+t^2}\,dt$$

$$G'(x) = -\frac{\left(-x^2\right)^2}{1+\left(-x^2\right)^2}(-2x) + \frac{x^2}{1+x^2}$$

$$= \frac{2x^5}{1+x^4} + \frac{x^2}{1+x^2}$$

23. $\displaystyle f'(x) = \frac{x}{\sqrt{a^2 + x^2}}$

$\displaystyle f''(x) = \frac{1}{\sqrt{a^2+x^2}} - \frac{x^2}{\sqrt{(a^2+x^2)^3}}$

$\displaystyle = \frac{a^2}{\sqrt{(a^2+x^2)^3}}$, which is always positive.

25.

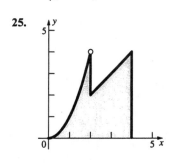

$\displaystyle \int_0^4 f(x)\,dx = \int_0^2 x^2\,dx + \int_2^4 x\,dx$

$\displaystyle = \left[\frac{x^3}{3}\right]_0^2 + \left[\frac{x^2}{2}\right]_2^4 = \left(\frac{8}{3} - 0\right) + (8-2) = \frac{26}{3}$

27.

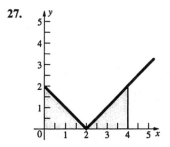

$\displaystyle \int_0^4 f(x)\,dx = \int_0^2 (2-x)\,dx + \int_2^4 (x-2)\,dx$

$\displaystyle = 2\int_0^2 dx - \int_0^2 x\,dx + \int_2^4 x\,dx - 2\int_2^4 dx$

$$= 2[x]_0^2 - \left[\frac{x^2}{2}\right]_0^2 + \left[\frac{x^2}{2}\right]_2^4 - 2[x]_2^4$$

$$= 2(2-0) - (2-0) + (8-2) - 2(4-2) = 4$$

29. **a.**
$$F(0) = \int_0^0 \left(t^4 + 1\right) dt = 0$$

b.
$$y = F(x)$$

$$\frac{dy}{dx} = F'(x) = x^4 + 1$$

$$dy = \left(x^4 + 1\right) dx$$

$$y = \frac{1}{5}x^5 + x + C$$

c. Now apply the initial condition $y(0) = 0$:
$$0 = \frac{1}{5}0^5 + 0 + C$$

$$C = 0$$

Thus $y = F(x) = \frac{1}{5}x^5 + x$

d.
$$\int_0^1 \left(x^4 + 1\right) dx = F(1) = \frac{1}{5}1^5 + 1 = \frac{6}{5}.$$

31. $1 \le \sqrt{1+x^4} \le 1 + x^4$, so

$$\int_0^1 dx \le \int_0^1 \sqrt{1+x^4}\, dx \le \int_0^1 (1+x^4)\, dx$$

$$\int_0^1 dx \le \int_0^1 \sqrt{1+x^4}\, dx \le \int_0^1 dx + \int_0^1 x^4 dx$$

$$[x]_0^1 \le \int_0^1 \sqrt{1+x^4}\, dx \le [x]_0^1 + \left[\frac{x^5}{5}\right]_0^1$$

$$(1-0) \le \int_0^1 \sqrt{1+x^4}\, dx \le (1-0) + \left(\frac{1}{5} - 0\right)$$

$$1 \le \int_0^1 \sqrt{1+x^4}\, dx \le \frac{6}{5}$$

33. $5 \le f(x) \le 69$ so

$$4 \cdot 5 \le \int_0^4 \left(5+x^3\right) dx \le 4 \cdot 69$$

$$20 \le \int_0^4 \left(5+x^3\right) dx \le 276$$

35. On [1,5],

$$3 + \frac{2}{5} \le 3 + \frac{2}{x} \le 3 + \frac{2}{1}$$

$$4\left(\frac{17}{5}\right) \le \int_1^5 \left(3 + \frac{2}{x}\right) dx \le 4 \cdot 5$$

$$\frac{68}{5} \le \int_1^5 \left(3 + \frac{2}{x}\right) dx \le 20$$

37. On $[4\pi, 8\pi]$

$$5 \le 5 + \frac{1}{20}\sin^2 x \le 5 + \frac{1}{20}x dx$$

$$(4\pi)(5) \le \int_{4\pi}^{8\pi} \left(5 + \frac{1}{20}\sin^2 x\right) dx \le (4\pi)\left(5 + \frac{1}{20}\right)$$

$$20\pi \le \int_{4\pi}^{8\pi} \left(5 + \frac{1}{20}\sin^2 x\right) dx \le \frac{101}{5}\pi$$

39. Let $F(x) = \int_0^x \frac{1+t}{2+t} dt$. Then

$$\lim_{x \to 0} \frac{1}{x} \int_0^x \frac{1+t}{2+t} dt = \lim_{x \to 0} \frac{F(x) - F(0)}{x - 0}$$

$$= F'(0) = \frac{1+0}{2+0} = \frac{1}{2}$$

41. $\int_1^x f(t)\, dt = 2x - 2$

Differentiate both sides with respect to x:

$$\frac{d}{dx} \int_1^x f(t)\, dt = \frac{d}{dx}(2x - 2)$$

$$f(x) = 2$$

If such a function exists, it must satisfy $f(x) = 2$, but both sides of the first equality may differ by a constant yet still have equal derivatives. When $x = 1$ the left side is $\int_1^1 f(t)\, dt = 0$ and the right side is $2 \cdot 1 - 2 = 0$. Thus the function $f(x) = 2$ satisfies $\int_1^x f(t)\, dt = 2x - 2$.

43. $\int_0^{x^2} f(t)\, dt = \frac{1}{3}x^3$

Differentiate both sides with respect to x:

$$\frac{d}{dx}\int_0^{x^2} f(t)\,dt = \frac{d}{dx}\left(\tfrac{1}{3}x^3\right)$$

$$f\!\left(x^2\right)(2x) = x^2$$

$$f\!\left(x^2\right) = \frac{x}{2}$$

$$f(x) = \frac{\sqrt{x}}{2}$$

If such a function exists, it must satisfy

$f(x) = \dfrac{\sqrt{x}}{2}$, but both sides of the first equality

may differ by a constant yet still have equal derivatives. When $x = 0$ the left side is

$\displaystyle\int_0^{0^2} f(t)\,dt = 0$ and the right side is $\dfrac{\sqrt{0}}{2} = 0$.

Thus the function $f(x) = \dfrac{\sqrt{x}}{2}$ satisfies

$$\int_0^{x^2} f(t)\,dt = \tfrac{1}{3}x^3.$$

45. True. $f(x)$ will have a minimum, m on $[a, b]$,

$m \ge 0$, so $\displaystyle\int_a^b f(x)\,dx \ge \int_a^b m\,dx = m(b-a) \ge 0$.

5.7 Concepts Review

1. antiderivative; $F(b) - F(a)$

3. $\dfrac{(3)^3}{3} - \dfrac{(1)^3}{3} = \dfrac{26}{3}$

Problem Set 5.7

1. $\displaystyle\int_0^2 x^3\,dx = \left[\frac{x^4}{4}\right]_0^2 = 4 - 0 = 4$

3. $\displaystyle\int_{-1}^2 (3x^2 - 2x + 3)\,dx = \left[x^3 - x^2 + 3x\right]_{-1}^2$
$= (8 - 4 + 6) - (-1 - 1 - 3) = 15$

5. $\displaystyle\int_1^4 \frac{1}{w^2}\,dw = \left[-\frac{1}{w}\right]_1^4 = \left(-\frac{1}{4}\right) - (-1) = \frac{3}{4}$

7. $\displaystyle\int_0^4 \sqrt{t}\,dt = \left[\frac{2}{3}t^{3/2}\right]_0^4 = \left(\frac{2}{3}\cdot 8\right) - 0 = \frac{16}{3}$

9. $\displaystyle\int_{-4}^{-2} \left(y^2 + \frac{1}{y^3}\right)dy = \left[\frac{y^3}{3} - \frac{1}{2y^2}\right]_{-4}^{-2}$
$= \left(-\frac{8}{3} - \frac{1}{8}\right) - \left(-\frac{64}{3} - \frac{1}{32}\right) = \frac{1783}{96}$

47. False. $a = -1$, $b = 1$, $f(x) = x$ is a counterexample.

49. True. $\displaystyle\int_a^b f(x)\,dx - \int_a^b g(x)\,dx$
$= \displaystyle\int_a^b [f(x) - g(x)]\,dx$

51. $-|f(x)| \le f(x) \le |f(x)|$, so

$$\int_a^b -|f(x)|\,dx \le \int_a^b f(x)\,dx \Rightarrow$$

$$\int_a^b |f(x)|\,dx \ge -\int_a^b f(x)\,dx$$

and combining this with

$$\int_a^b |f(x)|\,dx \ge \int_a^b f(x)\,dx,$$

we can conclude that

$$\left|\int_a^b f(x)\,dx\right| \ge \int_a^b |f(x)|\,dx$$

11. $\displaystyle\int_0^{\pi/2} \cos x\,dx = \left[\sin x\right]_0^{\pi/2} = 1 - 0 = 1$

13. $\displaystyle\int_0^1 (2x^4 - 3x^2 + 5)\,dx = \left[\frac{2}{5}x^5 - x^3 + 5x\right]_0^1$
$= \left(\frac{2}{5} - 1 + 5\right) - 0 = \frac{22}{5}$

15. $u = x^2 + 1$, $du = 2x\,dx$

$$\int (x^2 + 1)^{10}(2x)\,dx = \int u^{10}\,du = \frac{u^{11}}{11} + C$$

$$= \frac{1}{11}(x^2 + 1)^{11} + C$$

$$\int_0^1 (x^2 + 1)^{10}(2x)\,dx = \left[\frac{1}{11}(x^2 + 1)^{11}\right]_0^1$$

$$= \left[\frac{1}{11}(2)^{11}\right] - \left[\frac{1}{11}(1)^{11}\right] = \frac{2047}{11}$$

17. $u = t + 2$, $du = dt$

$$\int \frac{1}{(t+2)^2}\,dt = \int u^{-2}\,du = -\frac{1}{u} + C = -\frac{1}{t+2} + C$$

$$\int_{-1}^3 \frac{1}{(t+2)^2}\,dt = \left[-\frac{1}{t+2}\right]_{-1}^3 = \left[-\frac{1}{5}\right] - [-1] = \frac{4}{5}$$

19. $u = 3x + 1$, $du = 3\,dx$

$$\int \sqrt{3x+1}\,dx = \frac{1}{3}\int \sqrt{3x+1}\,3dx = \frac{1}{3}\int \sqrt{u}\,du$$

$$= \frac{2}{9}u^{3/2} + C = \frac{2}{9}(3x+1)^{3/2} + C$$

$$\int_5^8 \sqrt{3x+1}\, dx = \left[\frac{2}{9}(3x+1)^{3/2}\right]_5^8$$

$$= \left[\frac{2}{9}(125)\right] - \left[\frac{2}{9}(64)\right] = \frac{122}{9}$$

21. $u = 7 + 2t^2,\ du = 4t\, dt$

$$\int \sqrt{7+2t^2}\,(8t)dt = 2\int \sqrt{7+2t^2}\,(4t)dt$$

$$= 2\int \sqrt{2}\, du = \frac{4}{3}u^{3/2} + C = \frac{4}{3}(7+2t^2)^{3/2} + C$$

$$\int_{-3}^3 \sqrt{7+2t^2}\,(8t)dt = \left[\frac{4}{3}(7+2t^2)^{3/2}\right]_{-3}^3$$

$$= \left[\frac{4}{3}(125)\right] - \left[\frac{4}{3}(125)\right] = 0$$

23. $u = \cos x,\ du = -\sin x\, dx$

$$\int \cos^2 x \sin x\, dx = -\int \cos^2 x(-\sin x)\, dx$$

$$= -\int u^2\, du = -\frac{u^3}{3} + C = -\frac{1}{3}\cos^3 x + C$$

$$\int_0^{\pi/2} \cos^2 x \sin x\, dx = \left[-\frac{1}{3}\cos^3 x\right]_0^{\pi/2}$$

$$= 0 - \left(-\frac{1}{3}\right) = \frac{1}{3}$$

25. $\int_0^{\pi/2} (2x + \sin x)\, dx$

$$= 2\int_0^{\pi/2} x\, dx + \int_0^{\pi/2} \sin x\, dx$$

$$= 2\left[\frac{x^2}{2}\right]_0^{\pi/2} + \left[-\cos x\right]_0^{\pi/2}$$

$$= 2\left(\frac{\pi^2}{8} - 0\right) + (0+1) = \frac{\pi^2}{4} + 1$$

27. $\int_0^4 \left[\sqrt{x} + \sqrt{2x+1}\right] dx = \int_0^4 \sqrt{x}\, dx + \int_0^4 \sqrt{2x+1}\, dx$

$$= \left[\frac{2}{3}x^{3/2}\right]_0^4 + \frac{1}{2}\left[\frac{2}{3}(2x+1)^{3/2}\right]_0^4$$

$$= \left(\frac{16}{3} - 0\right) + \frac{1}{2}\left(18 - \frac{2}{3}\right) = 14$$

29. $\int_0^1 (x^2 + 2x)^2\, dx = \int_0^1 (x^4 + 4x^3 + 4x^2)\, dx$

$$= \int_0^1 x^4\, dx + 4\int_0^1 x^3\, dx + 4\int_0^1 x^2\, dx$$

$$= \left[\frac{x^5}{5}\right]_0^1 + 4\left[\frac{x^4}{4}\right]_0^1 + 4\left[\frac{x^3}{3}\right]_0^1$$

$$= \left(\frac{1}{5} - 0\right) + 4\left(\frac{1}{4} - 0\right) + 4\left(\frac{1}{3} - 0\right) = \frac{38}{15}$$

31. $\int_1^x t^3\, dt = \left[\frac{t^4}{4}\right]_1^x = \frac{x^4}{4} - \frac{1}{4}$

33. $\frac{1}{3-1}\int_1^3 4x^3\, dx = \frac{1}{2}\left[x^4\right]_1^3 = 40$

35.

$$\frac{1}{1+2}\int_{-2}^1 (2 + |x|)\, dx$$

$$= \frac{1}{3}\left[\int_{-2}^0 (2 - x)\, dx + \int_0^1 (2 + x)\, dx\right]$$

$$= \frac{1}{3}\left\{\left[2x - \frac{1}{2}x^2\right]_{-2}^0 + \left[2x + \frac{1}{2}x^2\right]_0^1\right\}$$

$$= \frac{1}{3}\left(-2(-2) + \frac{1}{2}(-2)^2 + 2 + \frac{1}{2}\right) = \frac{17}{6}$$

37. $\frac{1}{\pi}\int_0^\pi \cos x\, dx = \frac{1}{\pi}\left[\sin x\right]_0^\pi$

$$= \frac{1}{\pi}\left[\sin \pi - \sin 0\right] = 0$$

39. Using $c = \pi$ yields $2\pi(5)^4 = 1250\pi \approx 3927$

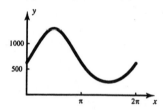

41. Using $c = 0.5$ yields

$$2\frac{2}{1+0.5^2} = 3.2$$

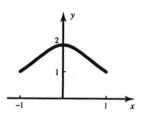

43.

$$f(c)(3-0) = \int_0^3 \sqrt{x+1}\, dx$$

$$3\sqrt{c+1} = \frac{2}{3}\left[(x+1)^{3/2}\right]_0^3$$

$$\sqrt{c+1} = \frac{2}{9}(8-1)$$

$$c = \left(\frac{14}{9}\right)^2 - 1 = \frac{115}{81}$$

45.

$$2f(c) = \int_0^2 |x|\, dx$$

$$2|c| = \int_0^2 x\, dx$$

$$2|c| = \left[\frac{x^2}{2}\right]_0^2$$

$$2|c| = 2$$

$$c = \pm 1$$

Only $c = 1$ is in the interval $[0,2]$.

47.

$$\int_0^3 x^2\, dx = \left[\frac{x^3}{3}\right]_0^3 = 9 - 0 = 9$$

49.

$$\int_0^\pi \sin x\, dx = [-\cos x]_0^\pi = 1 + 1 = 2$$

51. The right-endpoint Riemann sum is

$$\sum_{i=1}^n \left(0 + \frac{1-0}{n}i\right)^2 \left(\frac{1}{n}\right) = \frac{1}{n^3}\sum_{i=1}^n i^2, \text{ which for}$$

$n = 10$ equals $\dfrac{77}{200} = 0.385$.

$$\int_0^1 x^2\, dx = \left[\frac{1}{3}x^3\right]_0^1 = \frac{1}{3} = 0.\overline{333}$$

53.

$$\frac{d}{dx}\left(\frac{1}{2}x|x|\right) = \frac{1}{2}x\left(\frac{|x|}{x}\right) + \frac{|x|}{2} = |x|$$

$$\int_a^b |x|\, dx = \left[\frac{1}{2}x|x|\right]_a^b = \frac{1}{2}(b|b| - a|a|)$$

If b is not an integer, let $n = [\![b]\!]$. Then

$$\int_0^b [\![x]\!]\, dx = 0 + 1 + 2 + \cdots + (n-1) + n(b-n)$$

$$= \frac{(n-1)n}{2} + n(b-n)$$

$$= \frac{([\![b]\!]-1)[\![b]\!]}{2} + [\![b]\!](b - [\![b]\!]).$$

55. a. $\displaystyle \int_a^b x^n\, dx = B_n;\ \int_{a^n}^{b^n} \sqrt[n]{y}\, dy = A_n$

Using Figure 3 of the text,

$$(a)(a^n) + A_n + B_n = (b)(b^n) \text{ or}$$

$$B_n + A_n = b^{n+1} - a^{n+1}. \text{ Thus}$$

$$\int_a^b x^n\, dx + \int_{a^n}^{b^n} \sqrt[n]{y}\, dy = b^{n+1} - a^{n+1}$$

b. $\displaystyle \int_a^b x^n\, dx + \int_{a^n}^{b^n} \sqrt[n]{y}\, dy$

$$= \left[\frac{x^{n+1}}{n+1}\right]_a^b + \left[\frac{n}{n+1}y^{(n+1)/n}\right]_{a^n}^{b^n}$$

$$= \left(\frac{b^{n+1}}{n+1} - \frac{a^{n+1}}{n+1}\right) + \left(\frac{n}{n+1}b^{n+1} - \frac{n}{n+1}a^{n+1}\right)$$

$$= \frac{(n+1)b^{n+1} - (n+1)a^{n+1}}{n+1} = b^{n+1} - a^{n+1}$$

c. $\displaystyle B_n = \int_a^b x^n\, dx = \frac{1}{n+1}\left[x^{n+1}\right]_a^b$

$$= \frac{1}{n+1}(b^{n+1} - a^{n+1})$$

$$A_n = \int_{a^n}^{b^n} \sqrt[n]{y}\, dy = \left[\frac{n}{n+1}y^{(n+1)/n}\right]_{a^n}^{b^n}$$

$$= \frac{n}{n+1}(b^{n+1} - a^{n+1})$$

$$nB_n = \frac{n}{n+1}(b^{n+1} - a^{n+1}) = A_n$$

57. a. Let c be in $[a,b]$. Then $G'(c) = f(c)$ by the First Fundamental Theorem of Calculus. Since G is differentiable at c, G is continuous there. Now suppose $c = a$. Then

$$\lim_{x \to c} G(x) = \lim_{x \to a} \int_a^x f(t)\, dt. \text{ Since } f \text{ is}$$

continuous on $[a,b]$, there exist (by the Min-Max Existence Theorem) m and M such that $f(m) \le f(x) \le f(M)$ for all x in $[a,b]$. Then

$$\int_a^x f(m)\, dt \le \int_a^x f(t)\, dt \le \int_a^x f(M)\, dt$$

$$(x-a)f(m) \le G(x) \le (x-a)f(M)$$

By the Squeeze Theorem

$$\lim_{x \to a^+}(x-a)f(m) \le \lim_{x \to a^+} G(x)$$

$$\le \lim_{x \to a^+}(x-a)f(M)$$

Thus

$$\lim_{x \to a^+} G(x) = 0 = \int_a^a f(t)\, dt = G(a)$$

Therefore G is right-continuous at $x = a$.

Now, suppose $c = b$. Then

$$\lim_{x \to b^-} G(x) = \lim_{x \to b^-} \int_x^b f(t)\,dt$$

As before,

$$(b-x)f(m) \le G(x) \le (b-x)f(M)$$

so we can apply the Squeeze Theorem again to obtain

$$\lim_{x \to b^-} (b-x)f(m) \le \lim_{x \to b^-} G(x)$$

$$\le \lim_{x \to b^-} (b-x)f(M)$$

Thus

$$\lim_{x \to b^-} G(x) = 0 = \int_b^b f(t)\,dt = G(b)$$

Therefore, G is left-continuous at $x = b$.

b. Let F be any antiderivative of f. Note that G is also an antiderivative of f. Thus, $F(x) = G(x) + C$. We know from part (a) that $G(x)$ is continuous on $[a,b]$. Thus $F(x)$, being equal to $G(x)$ plus a constant, is also continuous on $[a,b]$.

59. a. Between 0 and 3, $f(x) > 0$. Thus,

$$\int_0^3 f(x)\,dx > 0.$$

b. Since f is an antiderivative of f',

$$\int_0^3 f'(x)\,dx = f(3) - f(0)$$

$$= 0 - 2 = -2 < 0$$

c. $\int_0^3 f''(x)\,dx = f'(3) - f'(0)$

$$= -1 - 0 = -1 < 0$$

d. $\int_0^3 f'''(x)\,dx = f''(3) - f''(0)$

$$= 0 - (\text{negative number}) > 0$$

Since f is concave down at 0, $f''(0) < 0$.

61. Let $y = G(x) = \int_a^x f(t)\,dt$. Then

$$\frac{dy}{dx} = G'(x) = f(x)$$

$$dy = f(x)\,dx$$

Let F be any antiderivative of f. Then $y = F(x) + C$. When $x = a$, we must have $y = 0$. Thus, $C = -F(a)$ and $y = F(x) - F(a)$.

Now choose $x = b$ to obtain

$$\int_a^b f(t)\,dt = y = F(b) - F(a)$$

63. A rectangle with height 25 and width 7 has approximately the same area as that under the curve. Thus

$$\frac{1}{7}\int_0^7 H(t)\,dt \approx 25$$

5.8 Concepts Review

1. $\dfrac{1}{3}\displaystyle\int_1^2 u^4\,du$

3. $\displaystyle\int_0^2 (2 + x^2)\,dx\,;\ 0$

Problem Set 5.8

1. $u = 3x + 2,\ du = 3\,dx$

$$\int \sqrt{u} \cdot \frac{1}{3}\,du = \frac{2}{9}u^{3/2} + C = \frac{2}{9}(3x+2)^{3/2} + C$$

3. $u = 6x - 7,\ du = 6\,dx$

$$\int u^{1/8} \cdot \frac{1}{6}\,du = \frac{4}{27}u^{9/8} + C = \frac{4}{27}(6x-7)^{9/8} + C$$

5. $u = 3x + 2,\ du = 3\,dx$

$$\int \cos(u) \cdot \frac{1}{3}\,du = \frac{1}{3}\sin u + C = \frac{1}{3}\sin(3x+2) + C$$

7. $u = 6x - 7,\ du = 6\,dx$

$$\int \sin u \cdot \frac{1}{6}\,du = -\frac{1}{6}\cos u + C$$

$$= -\frac{1}{6}\cos(6x-7) + C$$

9. $u = x^2 + 4,\ du = 2x\,dx$

$$\int \sqrt{u} \cdot \frac{1}{2}\,du = \frac{1}{3}u^{3/2} + C = \frac{1}{3}(x^2+4)^{3/2} + C$$

11. $u = x^2 + 3,\ du = 2x\,dx$

$$\int u^{-12/7} \cdot \frac{1}{2}\,du = -\frac{7}{10}u^{-5/7} + C$$

$$= -\frac{7}{10}(x^2+3)^{-5/7} + C$$

13. $u = x^2 + 4,\ du = 2x\,dx$

$$\int \sin(u) \cdot \frac{1}{2}\,du = -\frac{1}{2}\cos u + C$$

$$= -\frac{1}{2}\cos(x^2+4) + C$$

15. $y = 6x^3 - 7, du = 18x^2\, dx$

$$\int \sin u \cdot \frac{1}{18}\, du = -\frac{1}{18}\cos u + C$$

$$= -\frac{1}{18}\cos(6x^3 - 7) + C$$

17. $u = \sqrt{x^2 + 4}, du = \frac{x}{\sqrt{x^2 + 4}}\, dx$

$$\int \sin u\, du = -\cos u + C = -\cos\sqrt{x^2 + 4} + C$$

19. $u = (x^3 + 5)^9,$

$$du = 9(x^3 + 5)^8 (3x^2)dx = 27x^2(x^3 + 5)^8\, dx$$

$$\int \cos u \cdot \frac{1}{27}\, du = \frac{1}{27}\sin u + C$$

$$= \frac{1}{27}\sin\left[(x^3 + 5)^9\right] + C$$

21. $u = \sin(x^2 + 4), du = 2x\cos(x^2 + 4)\, dx$

$$\int \sqrt{u} \cdot \frac{1}{2}\, du = \frac{1}{3}u^{3/2} + C$$

$$= \frac{1}{3}\left[\sin(x^2 + 4)\right]^{3/2} + C$$

23. $u = \cos(x^3 + 5), du = -3x^2\sin(x^3 + 5)\, dx$

$$\int u^9 \cdot \left(-\frac{1}{3}\right)du = -\frac{1}{30}u^{10} + C$$

$$= -\frac{1}{30}\cos^{10}(x^3 + 5) + C$$

25. $\int (x+1)\sec^{1/2}(x^2 + 2x)\tan(x^2 + 2x)dx$

$$= \int (x+1)\frac{\sin(x^2 + 2x)}{\cos^{3/2}(x^2 + 2x)}\, dx$$

$$u = \cos(x^2 + 2x), du = -(2x + 2)\sin(x^2 + 2x)\, dx$$

$$\int u^{-3/2} \cdot -\frac{1}{2}\, du = u^{-1/2} + C$$

$$= \frac{1}{\cos^{1/2}(x^2 + 2x)} + C = \sec^{1/2}(x^2 + 2x) + C$$

27. $u = 3x + 1, du = 3dx$

$$\frac{1}{3}\int_1^4 u^3\, du = \left[\frac{1}{12} \cdot u^4\right]_1^4 = \left(\frac{64}{3} - \frac{1}{12}\right) = \frac{85}{4}$$

29. $u = t^2 + 9, du = 2t\, dt$

$$\frac{1}{2}\int_9^{13}\frac{1}{u^2}\, du = -\frac{1}{2}\left[\frac{1}{u}\right]_9^{13} = -\frac{1}{2}\left(\frac{1}{13} - \frac{1}{9}\right) = \frac{2}{117}$$

31. $u = x^2 + 4x + 1, du = (2x + 4)\, dx$

$$\frac{1}{2}\int_1^6 u^{-2}\, du = -\frac{1}{2}\left[\frac{1}{u}\right]_1^6 = -\frac{1}{2}\left(\frac{1}{6} - 1\right) = \frac{5}{12}$$

33. $u = \sin\theta, du = \cos\theta\, d\theta$

$$\int_0^{1/2} u^3\, du = \left[\frac{u^4}{4}\right]_0^{1/2} = \frac{1}{64} - 0 = \frac{1}{64}$$

35. $u = 3x - 3, du = 3dx$

$$\frac{1}{3}\int_{-3}^0 \cos u\, du = \frac{1}{3}[\sin u]_{-3}^0 = \frac{1}{3}(0 - \sin(-3))$$

$$= \frac{\sin 3}{3}$$

37. $u = \pi x^2, du = 2\pi x\, dx$

$$\frac{1}{2\pi}\int_0^\pi \sin u\, du = -\frac{1}{2\pi}[\cos u]_0^\pi = -\frac{1}{2\pi}(-1 - 1)$$

$$= \frac{1}{\pi}$$

39. $u = 2x, du = 2dx$

$$\frac{1}{2}\int_0^{\pi/2}\cos u\, du + \frac{1}{2}\int_0^{\pi/2}\sin u\, du$$

$$= \frac{1}{2}[\sin u]_0^{\pi/2} - \frac{1}{2}[\cos u]_0^{\pi/2}$$

$$= \frac{1}{2}(1 - 0) - \frac{1}{2}(0 - 1) = 1$$

41. $u = \cos x, du = -\sin x\, dx$

$$-\int_1^0 \sin u\, du = [\cos u]_1^0 = 1 - \cos 1$$

43. $u = \cos(x^2), du = -2x\sin(x^2)dx$

$$-\frac{1}{2}\int_1^{\cos 1} u^3\, du = -\frac{1}{2}\left[\frac{u^4}{4}\right]_1^{\cos 1} = -\frac{\cos^4 1}{8} + \frac{1}{8}$$

$$= \frac{1 - \cos^4 1}{8}$$

45. $u = \sqrt{t} + 1, du = \frac{1}{2\sqrt{t}}\, dt$

$$2\int_2^3 u^{-3}\, du = \left[-u^{-2}\right]_2^3 = \frac{5}{36}$$

47.

$$\int_{-\pi}^\pi (\sin x + \cos x)\, dx = \int_{-\pi}^\pi \sin x\, dx + 2\int_0^\pi \cos x\, dx$$

$$= 0 + 2[\sin x]_0^\pi = 0$$

49. $\int_{-\pi/2}^{\pi/2} \dfrac{\sin x}{1+\cos x}\,dx = 0$, since

$$\frac{\sin(-x)}{1+\cos(-x)} = -\frac{\sin x}{1+\cos x}$$

51. $\int_{-\pi}^{\pi} (\sin x + \cos x)^2\,dx$

$= \int_{-\pi}^{\pi} (\sin^2 x + 2\sin x \cos x + \cos^2 x)\,dx$

$= \int_{-\pi}^{\pi} (1 + 2\sin x \cos x)\,dx = \int_{-\pi}^{\pi} dx + \int_{-\pi}^{\pi} \sin 2x\,dx$

$= 2\int_{0}^{\pi} dx + 0 = 2[x]_0^{\pi} = 2\pi$

53. $\int_{-1}^{1} (1 + x + x^2 + x^3)\,dx$

$= \int_{-1}^{1} dx + \int_{-1}^{1} x\,dx + \int_{-1}^{1} x^2\,dx + \int_{-1}^{1} x^3\,dx$

$= 2[x]_0^1 + 0 + 2\left[\dfrac{x^3}{3}\right]_0^1 + 0 = \dfrac{8}{3}$

55. $\int_{-1}^{1} \left(\left|x^3\right| + x^3\right)dx = 2\int_{0}^{1} \left|x^3\right|dx + \int_{-1}^{1} x^3\,dx$

$= 2\left[\dfrac{x^4}{4}\right]_0^1 + 0 = \dfrac{1}{2}$

57. $\int_{-b}^{-a} f(x)\,dx = \int_{a}^{b} f(x)\,dx$ when f is even.

$\int_{-b}^{-a} f(x)\,dx = -\int_{a}^{b} f(x)\,dx$ when f is odd.

59. $\int_{0}^{4\pi} |\cos x|\,dx = \int_{0}^{\pi} |\cos x|\,dx + \int_{\pi}^{2\pi} |\cos x|\,dx$

$\quad + \int_{2\pi}^{3\pi} |\cos x|\,dx + \int_{3\pi}^{4\pi} |\cos x|\,dx$

$= \int_{0}^{\pi} |\cos x|\,dx + \int_{0}^{\pi} |\cos x|\,dx + \int_{0}^{\pi} |\cos x|\,dx$

$\quad + \int_{0}^{\pi} |\cos x|\,dx$

$= 4\int_{0}^{\pi} |\cos x|\,dx = 8\int_{0}^{\pi/2} |\cos x|\,dx$

$= 8[\sin x]_0^{\pi/2} = 8$

61. $\int_{1}^{1+\pi} |\sin x|\,dx = \int_{0}^{\pi} |\sin x|\,dx = \int_{0}^{\pi} \sin x\,dx$

$= [-\cos x]_0^{\pi} = 2$

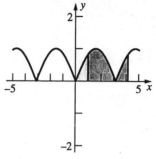

63. $\dfrac{1}{12}\int_{6}^{18} T(t)\,dt = \left[70t - \dfrac{96}{\pi}\cos\left[\dfrac{\pi}{12}(t-9)\right]\right]_6^{18}$

$= 70(18-6) - \dfrac{96}{\pi}\left(\cos\dfrac{3\pi}{4} - \cos\dfrac{\pi}{4}\right)$

$= 840 + \dfrac{96\sqrt{2}}{\pi} \approx 883.2$

65. $y' = -k^2 \sin kx$

At $x = \dfrac{\pi}{2k}$, $y' = -k^2$. The equation of the tangent

line at $x = \dfrac{\pi}{2k}$ is $y = -k^2\left(x - \dfrac{\pi}{2k}\right)$.

When $x = 0$, $y = \dfrac{k\pi}{2}$.

$A = \dfrac{1}{2}\left(\dfrac{\pi}{k}\right)\left(\dfrac{k\pi}{2}\right) - \int_{-\pi/2k}^{\pi/2k} k\cos kx\,dx$

$= \dfrac{\pi^2}{4} - [\sin kx]_{-\pi/2k}^{\pi/2k} = \dfrac{\pi^2}{4} - 2$

67. a. Even

b. 2π

c.

Interval	Value of Integral
$\left[0, \dfrac{\pi}{2}\right]$	0.46
$\left[-\dfrac{\pi}{2}, \dfrac{\pi}{2}\right]$	0.92
$\left[0, \dfrac{3\pi}{2}\right]$	−0.46
$\left[-\dfrac{3\pi}{2}, \dfrac{3\pi}{2}\right]$	−0.92
$[0, 2\pi]$	0

$\left[\dfrac{\pi}{6}, \dfrac{13\pi}{6}\right]$	0
$\left[\dfrac{\pi}{6}, \dfrac{4\pi}{3}\right]$	–0.44
$\left[\dfrac{13\pi}{6}, \dfrac{10\pi}{3}\right]$	–0.44

69. Using a right-Riemann sum,

$$\text{Distance} = \int_0^{24} v(t)\, dt$$

$$\approx \sum_{i=1}^{8} v(t_i)\, \Delta t$$

$$= (31 + 54 + 53 + 52 + 35 + 31 + 28)\frac{3}{60}$$

5.9 Chapter Review

Concepts Test

1. True: Theorem 5.1.C

3. True: $(-\sin x)^2 = \sin^2 x = 1 - \cos^2 x$

5. True: If $F(x) = \int f(x)\, dx$, $f(x)$ is a derivative of $F(x)$.

7. False: The two sides will in general differ by a constant term.

9. True: $a_1 + a_0 + a_2 + a_1 + a_3 + a_2$
$$+ \cdots + a_{n-1} + a_{n-2} + a_n + a_{n-1}$$
$$= a_0 + 2a_1 + 2a_2 + \cdots + 2a_{n-1} + a_n$$

11. True: $\displaystyle\sum_{i=1}^{10}(a_i + 1)^2 = \sum_{i=1}^{10} a_i^2 + 2\sum_{i=1}^{10} a_i + \sum_{i=1}^{100} 1$
$$= 100 + 2(20) + 10 = 150$$

13. True: The area of a vertical line segment is 0.

15. True: $[f(x)]^2 \geq 0$, and if $[f(x)]^2$ is greater than 0 on $[a, b]$, the integral will be also.

17. True: $\sin x + \cos x$ has period 2π, so
$$\int_x^{x+2\pi} (\sin x + \cos x)\, dx$$
is independent of x.

19. True: $\sin^{13} x$ is an odd function.

21. False: The statement is not true if $c > d$.

23. True: Both sides equal 4.

$$= \frac{852}{60} = 14.2 \text{ miles}$$

71. If f is odd, then $f(-x) = -f(x)$ and we can write

$$\int_{-a}^{0} f(x)\, dx = \int_{-a}^{0} \left[-f(-x)\right] dx$$

$$= \int_{a}^{0} f(u)\, du$$

$$= -\int_{0}^{a} f(u)\, du$$

$$= -\int_{0}^{a} f(x)\, dx$$

On the second line, we have made the substitution $u = -x$.

25. True: The derivatives of even functions are odd.

27. False: $f(x) = x^2$ is a counterexample.

29. False: $f(x) = x^2$, $v(x) = 2x + 1$ is a counterexample.

31. False: $f(x) = \sqrt{x}$ is a counterexample.

33. True: $F(b) - F(a) = \displaystyle\int_a^b F'(x)\, dx$
$$= \int_a^b G'(x)\, dx = G(b) - G(a)$$

35. False: $z(t) = t^2$ is a counterexample.

37. True: Odd-exponent terms cancel themselves out over the interval, since they are odd.

39. False: $a = 0$, $b = 1$, $f(x) = -1$, $g(x) = 0$ is a counterexample.

41. True: See Problem 28 of Section 5.7.

43. True: Definition of Definite Integral

Sample Test Problems

1. $\left[\dfrac{1}{4}x^4 - x^3 + 2x^{3/2}\right]_0^1 = \dfrac{5}{4}$

3. $\dfrac{1}{3}y^3 + 9\cos y - \dfrac{26}{y} + C$

5. $\frac{1}{4} \cdot \frac{3}{4}(2z^2-3)^{4/3}+C$

$= \frac{3}{16}(2z^2-3)^{4/3}+C$

7. $u = \tan(3x^2+6x), du = (6x+6)\sec^2(3x^2+6x)$

$\frac{1}{6}\int u^2\,du = \frac{1}{18}u^3+C$

$\frac{1}{18}\left[\tan^3(3x^2+6x)\right]_0^\pi = \frac{1}{18}\tan^3(3\pi^2)$

9. $\frac{1}{5}\left[\frac{3}{5}(t^5+5)^{5/3}\right]_1^2 \approx 46.9$

11. $u = 2y^3+3y^2+6y, du = (6y^2+6y+6)\,dy$

$\frac{1}{6}\int u^{-1/5}\,du = \frac{5}{24}(2y^3+3y^2+6y)^{4/5}+C$

13. $\int dy = \int \frac{1}{\sqrt{x+1}}\,dx$

$y = 2\sqrt{x+1}+C$

$y = 2\sqrt{x+1}+14$

15. $\int dy = \int\sqrt{2t-1}\,dt$

$y = \frac{1}{3}(2t-1)^{3/2}+C$

$y = \frac{1}{3}(2t-1)^{3/2}-1$

17. $\int 2y\,dy = \int(6x-x^3)\,dx$

$y^2 = 3x^2-\frac{1}{4}x^4+C$

$y^2 = 3x^2-\frac{1}{4}x^4+9$

$y = \sqrt{3x^2-\frac{1}{4}x^4+9}$

19. On $xy=2$, $\frac{dy}{dx} = -\frac{2}{x^2}$, so the curve has slope

$\frac{dy}{dx} = \frac{x^2}{2}$.

$\int dy = \int \frac{x^2}{2}\,dx$

$y = \frac{1}{6}x^3+C$

$-\frac{1}{3} = -\frac{4}{3}+C$

$y = \frac{1}{6}x^3+1$

21. $s = -16t^2+48t+448$; $s=0$ at $t=7$;
when $t=7$, $v = -32(7)+48 = -176$ ft/s

23. $\sum_{i=1}^{4}\left[\left(\frac{i}{2}\right)^2-1\right]\left(\frac{1}{2}\right) = \frac{7}{4}$

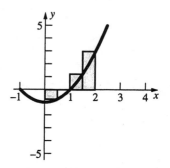

25. $\int_0^3(2-\sqrt{x+1})^2\,dx = \int_0^3\left(x+5-4\sqrt{x+1}\right)dx$

$= \left[\frac{1}{2}x^2+5x-\frac{8}{3}(x+1)^{3/2}\right]_0^3 = \frac{5}{6}$

27. $\int_2^4\left(5-\frac{1}{x^2}\right)dx = \left[5x+\frac{1}{x}\right]_2^4 = \frac{39}{4}$

29. $\sum_{i=1}^{10}(6i^2-8i) = 6\sum_{i=1}^{10}i^2-8\sum_{i=1}^{10}i$

$= 6\left[\frac{10(11)(21)}{6}\right]-8\left[\frac{10(11)}{2}\right] = 1870$

31. a. $\sum_{n=2}^{78}\frac{1}{n}$

b. $\sum_{n=1}^{50}nx^{2n}$

33. a. $\int_1^2 f(x)\,dx = \int_1^0 f(x)\,dx + \int_0^2 f(x)\,dx$

$= -4+2 = -2$

b. $\int_1^0 f(x)\,dx = -\int_0^1 f(x)\,dx = -4$

c. $\int_0^2 3f(u)\,du = 3\int_0^2 f(u)\,du = 3(2) = 6$

d. $\int_0^2 [2g(x)-3f(x)]dx$

$= 2\int_0^2 g(x)-3\int_0^2 f(x)\,dx$

$= 2(-3)-3(2)=-12$

e. $\int_0^{-2} f(-x)\,dx = \int_0^2 f(x)\,dx = 2$

35. a. $\int_{-2}^2 f(x)\,dx = 2\int_0^2 f(x)\,dx = 2(-4)=-8$

b. $\int_{-2}^2 |f(x)|\,dx = 2\int_0^2 |f(x)|\,dx = 2|-4|=8$

c. $\int_{-2}^2 g(x)\,dx = 0$ **d.**

$\int_{-2}^2 [f(x)+f(-x)]dx$

$= 2\int_0^2 f(x)\,dx+2\int_0^2 f(x)\,dx$

$= 4(-4)=-16$

e. $\int_0^2 [2g(x)+3f(x)]dx$

$= 2\int_0^2 g(x)\,dx+3\int_0^2 f(x)\,dx$

$= 2(5)+3(-4)=-2$

f. $\int_{-2}^0 g(x)\,dx = \int_2^0 g(-x)\cdot -dx = \int_2^0 g(x)\,dx$

$= -\int_0^2 g(x)\,dx = -5$

37. $\int_{-4}^{-1} 3x^2\,dx = 3c^2(-1+4)$

$\left[x^3\right]_{-4}^{-1} = 9c^2$

$c^2 = 7$

$c = -\sqrt{7} \approx -2.65$

39. a. $G'(x) = \sin^2 x$

b. $G'(x) = f(x+1)-f(x)$

c. $G'(x) = -\dfrac{1}{x^2}\int_0^x f(z)\,dz+\dfrac{1}{x}f(x)$

d. $G'(x) = \int_0^x f(t)\,dt$

e. $G(x) = \int_0^{g(x)} \dfrac{dg(u)}{du}\,du = [g(u)]_0^{g(x)}$

$= g(g(x))-g(0)$

$G'(x) = g'(g(x))g'(x)$

f. $G(x) = \int_0^{-x} f(-t)\,dt = \int_0^x f(u)(-du)$

$= -\int_0^x f(u)\,du$

$G'(x) = -f(x)$

41. $f(x) = \int_{2x}^{5x} \dfrac{1}{t}\,dt = \int_0^{5x} \dfrac{1}{t}\,dt - \int_0^{2x} \dfrac{1}{t}\,dt$

$f'(x) = \dfrac{1}{5x}\cdot 5 - \dfrac{1}{2x}\cdot 2 = 0$

5.10 Additional Problem Set

1. Recall that by implicitly differentiating the equation of the circle $x^2+y^2=r^2$, we get

$\dfrac{dy}{dx} = -\dfrac{x}{y}$. Thus, the solution must be a graph of

a circle. Since the point (0, 1) must lie on the graph, the answer is Figure 1.

3. a. $30[88+80+66+51+37+26+14+5]$
  $= 11{,}010$ feet

b. $30[80+66+51+37+26+14+5+0]$
  $= 8370$ feet

c. $30\left[84+73+\dfrac{117}{2}+44+\dfrac{63}{2}+20+\dfrac{19}{2}+\dfrac{5}{2}\right]$
  $= 9690$ feet

d. Part **a.** uses the largest value of each pair and **b.** uses the smallest. Part **c.** averages each pair, so the total value is the average of **a.** and **b.**

5. a. $\bar{u}+\bar{v} = \dfrac{1}{b-a}\int_a^b u\,dx+\dfrac{1}{b-a}\int_a^b v\,dx$

$= \dfrac{1}{b-a}\int_a^b (u+v)\,dx = \overline{u+v}$

b. $k\bar{u} = \dfrac{k}{b-a}\int_a^b u\,dx = \dfrac{1}{b-a}\int_a^b ku\,dx = \overline{ku}$

c. $\bar{u} = \dfrac{1}{b-a}\int_a^b u\,dx \le \dfrac{1}{b-a}\int_a^b v\,dx = \bar{v}$

7. a. Local minima at 0, ≈ 3.8, ≈ 5.8, ≈ 7.9, ≈ 9.9, ≈ 10;
  local maxima at ≈ 3.1, ≈ 5, ≈ 7.1, ≈ 9

b. Absolute minimum at 0, absolute maximum at ≈ 9

c. $\approx (0.7, 1.5), (2.5, 3.5), (4.5, 5.5), (6.5, 7.5),$ $(8.5, 9.5)$

d.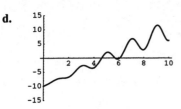

9. a. $f(x) > 0$ so $\displaystyle\int_{-1}^{3} f(x)\,dx > 0$

b. $\displaystyle\int_{-1}^{3} f'(x)\,dx = f(3) - f(-1) = 1 - 2 = -1 < 0$

c. $\displaystyle\int_{-1}^{3} f''(x)\,dx = f'(3) - f'(1) = -1 - 1 = -2 < 0$

d. $\displaystyle\int_{-1}^{3} f'''(x)\,dx = f''(3) - f''(1)$

$= \text{positive} - \text{negative} > 0$

Applications of the Integral

6.1 Concepts Review

1. $\int_a^b f(x)dx; -\int_a^b f(x)dx$

3. $g(x) - f(x); f(x) = g(x)$

Problem Set 6.1

1. Slice vertically.

$$\Delta A \approx (x^2 + 1)\Delta x$$

$$A = \int_{-1}^2 (x^2 + 1)dx = \left[\frac{1}{3}x^3 + x\right]_{-1}^2$$

$$= \left(\frac{8}{3} + 2\right) - \left(-\frac{1}{3} - 1\right) = 6$$

3. Slice vertically.

$$\Delta A \approx \left[(x^2 + 2) - (-x)\right]\Delta x = (x^2 + x + 2)\Delta x$$

$$A = \int_{-2}^2 (x^2 + x + 2)dx = \left[\frac{1}{3}x^3 + \frac{1}{2}x^2 + 2x\right]_{-2}^2$$

$$= \left(\frac{8}{3} + 2 + 4\right) - \left(-\frac{8}{3} + 2 - 4\right) = \frac{40}{3}$$

5. To find the intersection points, solve $2 - x^2 = x$.

$$x^2 + x - 2 = 0$$
$$(x + 2)(x - 1) = 0$$
$$x = -2, 1$$

Slice vertically.

$$\Delta A \approx \left[(2 - x^2) - x\right]\Delta x = (-x^2 - x - 2)\Delta x$$

$$A = \int_{-2}^1 (-x^2 - x + 2)dx = \left[-\frac{1}{3}x^3 - \frac{1}{2}x^2 + 2x\right]_{-2}^1$$

$$= \left(-\frac{1}{3} - \frac{1}{2} + 2\right) - \left(\frac{8}{3} - 2 - 4\right) = \frac{9}{2}$$

7. Solve $x^3 - x^2 - 6x = 0$.

$$x(x^2 - x - 6) = 0$$
$$x(x + 2)(x - 3) = 0$$
$$x = -2, 0, 3$$

Slice vertically.

$$\Delta A_1 \approx (x^3 - x^2 - 6x)\Delta x$$

$$\Delta A_2 \approx -(x^3 - x^2 - 6x)\Delta x = (-x^3 + x^2 + 6x)\Delta x$$

$$A = A_1 + A_2$$

$$= \int_{-2}^0 (x^3 - x^2 - 6x)dx + \int_0^3 (-x^3 + x^2 + 6x)dx$$

$$= \left[\frac{1}{4}x^4 - \frac{1}{3}x^3 - 3x^2\right]_{-2}^0$$

$$+ \left[-\frac{1}{4}x^4 + \frac{1}{3}x^3 + 3x^2\right]_0^3$$

$$= \left[0 - \left(4 + \frac{8}{3} - 12\right)\right] + \left[-\frac{81}{4} + 9 + 27 - 0\right]$$

$$= \frac{16}{3} + \frac{63}{4} = \frac{253}{12}$$

9. To find the intersection points, solve
$y + 1 = 3 - y^2$.

$$y^2 + y - 2 = 0$$
$$(y + 2)(y - 1) = 0$$
$$y = -2, 1$$

Slice horizontally.

$$\Delta A \approx \left[(3 - y^2) - (y + 1)\right]\Delta y = (-y^2 - y + 2)\Delta y$$

$$A = \int_{-2}^1 (-y^2 - y + 2)dy = \left[-\frac{1}{3}y^3 - \frac{1}{2}y^2 + 2y\right]_{-2}^1$$

$$= \left(-\frac{1}{3} - \frac{1}{2} + 2\right) - \left(\frac{8}{3} - 2 - 4\right) = \frac{9}{2}$$

11.

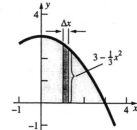

$$\Delta A \approx \left(3 - \frac{1}{3}x^2\right)\Delta x$$

$$A = \int_0^3 \left(3 - \frac{1}{3}x^2\right)dx = \left[3x - \frac{1}{9}x^3\right]_0^3 = 9 - 3 = 6$$

Estimate the area to be $(3)(2) = 6$.

13.

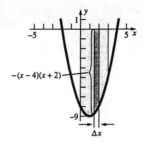

$$\Delta A \approx -(x-4)(x+2)\Delta x = (-x^2+2x+8)\Delta x$$

$$A = \int_0^3 (-x^2+2x+8)dx = \left[-\frac{1}{3}x^3+x^2+8x\right]_0^3$$

$$= -9+9+24 = 24$$

Estimate the area to be $(3)(8) = 24$.

15.

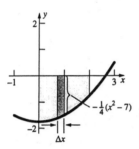

$$\Delta A \approx -\frac{1}{4}(x^2-7)\Delta x$$

$$A = \int_0^2 -\frac{1}{4}(x^2-7)dx = -\frac{1}{4}\left[\frac{1}{3}x^3-7x\right]_0^2$$

$$= -\frac{1}{4}\left(\frac{8}{3}-14\right) = \frac{17}{6} \approx 2.83$$

Estimate the area to be $(2)\left(1\frac{1}{2}\right) = 3$.

17.

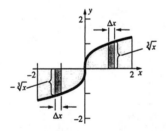

$$\Delta A_1 \approx -\sqrt[3]{x}\,\Delta x$$

$$\Delta A_2 \approx \sqrt[3]{x}\,\Delta x$$

$$A = A_1 + A_2 = \int_{-2}^0 -\sqrt[3]{x}\,dx + \int_0^2 \sqrt[3]{x}\,dx$$

$$= \left[-\frac{3}{4}x^{4/3}\right]_{-2}^0 + \left[\frac{3}{4}x^{4/3}\right]_0^2 = \left(\frac{3\sqrt[3]{2}}{2}\right) + \left(\frac{3\sqrt[3]{2}}{2}\right)$$

$$= 3\sqrt[3]{2} \approx 3.78$$

Estimate the area to be $(2)(1) + (2)(1) = 4$.

19.

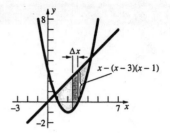

$$\Delta A \approx \left[x-(x-3)(x-1)\right]\Delta x$$

$$= \left[x-(x^2-4x+3)\right]\Delta x = (-x^2+5x-3)\Delta x$$

To find the intersection points, solve
$x = (x-3)(x-1)$.

$$x^2-5x+3 = 0$$

$$x = \frac{5\pm\sqrt{25-12}}{2}$$

$$x = \frac{5\pm\sqrt{13}}{2}$$

$$A = \int_{\frac{5-\sqrt{13}}{2}}^{\frac{5+\sqrt{13}}{2}} (-x^2+5x-3)dx$$

$$= \left[-\frac{1}{3}x^3+\frac{5}{2}x^2-3x\right]_{\frac{5-\sqrt{13}}{2}}^{\frac{5+\sqrt{13}}{2}} = \frac{13\sqrt{13}}{6} \approx 7.81$$

Estimate the area to be $\frac{1}{2}(4)(4) = 8$.

21.

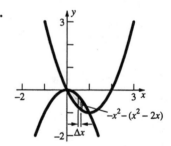

$$\Delta A \approx \left[-x^2-(x^2-2x)\right]\Delta x = (-2x^2+2x)\Delta x$$

To find the intersection points, solve
$-x^2 = x^2 - 2x$.

$$2x^2-2x = 0$$

$$2x(x-1) = 0$$

$$x = 0, x = 1$$

$$A = \int_0^1 (-2x^2+2x)dx = \left[-\frac{2}{3}x^3+x^2\right]_0^1$$

$$= -\frac{2}{3} + 1 = \frac{1}{3} \approx 0.33$$

Estimate the area to be $\left(\frac{1}{2}\right)\left(\frac{1}{2}\right) = \frac{1}{4}$.

23.

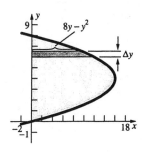

$$\Delta A \approx (8y - y^2)\Delta y$$

To find the intersection points, solve

$$8y - y^2 = 0.$$
$$y(8 - y) = 0$$
$$y = 0, 8$$

$$A = \int_0^8 (8y - y^2)\,dy = \left[4y^2 - \frac{1}{3}y^3\right]_0^8$$

$$= 256 - \frac{512}{3} = \frac{256}{3} \approx 85.33$$

Estimate the area to be $(16)(5) = 80$.

25.

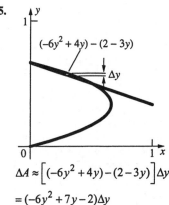

$$\Delta A \approx \left[(-6y^2 + 4y) - (2 - 3y)\right]\Delta y$$

$$= (-6y^2 + 7y - 2)\Delta y$$

To find the intersection points, solve

$$-6y^2 + 4y = 2 - 3y.$$
$$6y^2 - 7y + 2 = 0$$
$$(2y - 1)(3y - 2) = 0$$
$$y = \frac{1}{2}, \frac{2}{3}$$

$$A = \int_{1/2}^{2/3} (-6y^2 + 7y - 2)\,dy = \left[-2y^3 + \frac{7}{2}y^2 - 2y\right]_{1/2}^{2/3}$$

$$= \left(-\frac{16}{27} + \frac{14}{9} - \frac{4}{3}\right) - \left(-\frac{1}{4} + \frac{7}{8} - 1\right) = \frac{1}{216} \approx 0.0046$$

Estimate the area to be

$$\frac{1}{2}\left(\frac{1}{2}\right)\left(\frac{1}{5}\right) - \frac{1}{2}\left(\frac{1}{2}\right)\left(\frac{1}{6}\right) = \frac{1}{120}.$$

27.

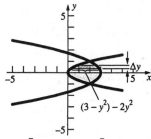

$$\Delta A \approx \left[(3 - y^2) - 2y^2\right]\Delta y = (-3y^2 + 3)\Delta y$$

To find the intersection points, solve

$$2y^2 = 3 - y^2.$$

$$3y^2 - 3 = 0$$
$$3(y + 1)(y - 1) = 0$$
$$y = -1, 1$$

$$A = \int_{-1}^{1} (-3y^2 + 3)\,dy = \left[-y^3 + 3y\right]_{-1}^{1}$$
$$= (-1 + 3) - (1 - 3) = 4$$

Estimate the value to be $(2)(2) = 4$.

29.

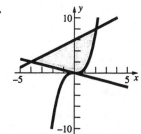

Let R_1 be the region bounded by $2y + x = 0$, $y = x + 6$, and $x = 0$.

$$A(R_1) = \int_{-4}^{0}\left[(x + 6) - \left(-\frac{1}{2}x\right)\right]dx$$

$$= \int_{-4}^{0}\left(\frac{3}{2}x + 6\right)dx$$

Let R_2 be the region bounded by $y = x + 6$, $y = x^3$, and $x = 0$.

$$A(R_2) = \int_{0}^{2}\left[(x + 6) - x^3\right]dx = \int_{0}^{2}(-x^3 + x + 6)dx$$

$$A(R) = A(R_1) + A(R_2)$$

$$= \int_{-4}^{0}\left(\frac{3}{2}x + 6\right)dx + \int_{0}^{2}(-x^3 + x + 6)dx$$

$$= \left[\frac{3}{4}x^2 + 6x\right]_{-4}^{0} + \left[-\frac{1}{4}x^4 + \frac{1}{2}x^2 + 6x\right]_{0}^{2}$$

$$= 12 + 10 = 22$$

31. $\int_{-1}^{9}(3t^2-24t+36)dt=\left[t^3-12t^2+36t\right]_{-1}^{9}=(729-972+324)-(-1-12-36)=130$

The displacement is 130 ft. Solve $3t^2-24t+36=0$.

$3(t-2)(t-6)=0$

$t=2,6$

$$|V(t)|=\begin{cases}3t^2-24t+36 & t\le 2, t\ge 6\\ -3t^2+24t-36 & 2<t<6\end{cases}$$

$\int_{-1}^{9}\left|3t^2-24t+36\right|dt=\int_{-1}^{2}(3t^2-24t+36)\,dt+\int_{2}^{6}(-3t^2+24t-36)\,dt+\int_{6}^{9}(3t^2-24t+36)\,dt$

$=\left[t^3-12t^2+36t\right]_{-1}^{2}+\left[-t^3+12t^2-36t\right]_{2}^{6}+\left[t^3-12t^2+36t\right]_{6}^{9}=81+32+81=194$

The total distance traveled is 194 feet.

33. $s(t)=\int v(t)dt=\int(2t-4)dt=t^2-4t+C$

Since $s(0)=0$, $C=0$ and $s(t)=t^2-4t$. $s=12$ when $t=6$, so it takes the object 6 seconds to get $s=12$.

$$|2t-4|=\begin{cases}4-2t & 0\le t<2\\ 2t-4 & 2\le t\end{cases}$$

$\int_{0}^{2}|2t-4|\,dt=\left[-t^2+4t\right]_{0}^{2}=4$, so the object travels a distance of 4 cm in the first two seconds.

$\int_{2}^{x}|2t-4|\,dt=\left[t^2-4t\right]_{2}^{x}=x^2-4x+4$

$x^2-4x+4=8$ when $x=2+2\sqrt{2}$, so the object takes $2+2\sqrt{2}\approx 4.83$ seconds to travel a total distance of 12 centimeters.

35. Equation of line through (–2, 4) and (3, 9):
$y=x+6$
Equation of line through (2, 4) and (–3, 9):
$y=-x+6$

$A(A)=\int_{-3}^{0}[9-(-x+6)]dx+\int_{0}^{3}[9-(x+6)]dx$

$=\int_{-3}^{0}(3+x)dx+\int_{0}^{3}(3-x)dx$

$=\left[3x+\frac{1}{2}x^2\right]_{-3}^{0}+\left[3x-\frac{1}{2}x^2\right]_{0}^{3}=\frac{9}{2}+\frac{9}{2}=9$

$A(B)$
$=\int_{-3}^{-2}[(-x+6)-x^2]dx$
$+\int_{-2}^{0}[(-x+6)-(x+6)]dx$

$=\int_{-3}^{-2}(-x^2-x+6)dx+\int_{-2}^{0}(-2x)dx$

$=\left[-\frac{1}{3}x^3-\frac{1}{2}x^2+6x\right]_{-3}^{-2}+\left[-x^2\right]_{-2}^{0}=\frac{37}{6}$

$A(C)=A(B)=\dfrac{37}{6}$ (by symmetry)

$A(D)=\int_{-2}^{0}[(x+6)-x^2]dx+\int_{0}^{2}[(-x+6)-x^2]dx$

$=\left[-\frac{1}{3}x^3+\frac{1}{2}x^2+6x\right]_{-2}^{0}+\left[-\frac{1}{3}x^3-\frac{1}{2}x^2+6x\right]_{0}^{2}$

$=\dfrac{44}{3}$

$A(A)+A(B)+A(C)+A(D)=36$

$A(A+B+C+D)=\int_{-3}^{3}(9-x^2)dx=\left[9x-\frac{1}{3}x^3\right]_{-3}^{3}$

$=36$

37. The height of the triangular region is given by
for $0\le x\le 1$. We need only show that the height of the second region is the same in order to apply Cavalieri''s Principle. The height of the second region is

$h_2=(x^2-2x+1)-(x^2-3x+1)$

$\quad=x^2-2x+1-x^2+3x-1$

$\quad=x$ for $0\le x\le 1$.

Since $h_1=h_2$ over the same closed interval, we can conclude that their areas are equal.

6.2 Concepts Review

1. $\pi r^2 h$

3. $\pi x^4 \Delta x$

Problem Set 6.2

1. Slice vertically.

$\Delta V\approx \pi(x^2+1)^2\,\Delta x=\pi(x^4+2x^2+1)\Delta x$

$V=\pi\int_{0}^{2}(x^4+2x^2+1)dx$

$$= \pi\left[\frac{1}{5}x^5 + \frac{2}{3}x^3 + x\right]_0^2 = \pi\left(\frac{32}{5} + \frac{16}{3} + 2\right) = \frac{206\pi}{15}$$

$$\approx 43.14$$

3. a. Slice vertically.

$$\Delta V \approx \pi(4-x^2)^2\Delta x = \pi(16 - 8x^2 + x^4)\Delta x$$

$$V = \pi\int_0^2 (16 - 8x^2 + x^4)dx$$

$$= \frac{256\pi}{15} \approx 53.62$$

b. Slice horizontally.

$$x = \sqrt{4-y}$$

Note that when $x = 0$, $y = 4$.

$$\Delta V \approx \pi\left(\sqrt{4-y}\right)^2\Delta y = \pi(4-y)\Delta y$$

$$V = \pi\int_0^4 (4-y)dy = \pi\left[4y - \frac{1}{2}y^2\right]_0^4$$

$$= \pi(16-8) = 8\pi \approx 25.13$$

5.

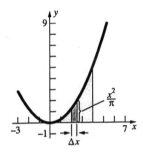

$$\Delta V \approx \pi\left(\frac{x^2}{\pi}\right)^2\Delta x = \frac{x^4}{\pi}\Delta x$$

$$V = \int_0^4 \frac{x^4}{\pi}dx = \frac{1}{\pi}\left[\frac{1}{5}x^5\right]_0^4 = \frac{1024}{5\pi} \approx 65.19$$

7.

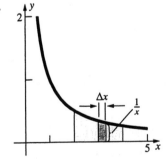

$$\Delta V \approx \pi\left(\frac{1}{x}\right)^2\Delta x = \pi\left(\frac{1}{x^2}\right)\Delta x$$

$$V = \pi\int_2^4 \frac{1}{x^2}dx = \pi\left[-\frac{1}{x}\right]_2^4 = \pi\left(-\frac{1}{4} + \frac{1}{2}\right) = \frac{\pi}{4}$$

$$\approx 0.79$$

9.

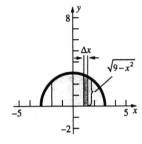

$$\Delta V \approx \pi\left(\sqrt{9-x^2}\right)^2\Delta x = \pi(9-x^2)\Delta x$$

$$V = \pi\int_{-2}^3 (9-x^2)dx = \pi\left[9x - \frac{1}{3}x^3\right]_{-2}^3$$

$$= \pi\left[(27-9) - \left(-18 + \frac{8}{3}\right)\right] = \frac{100\pi}{3} \approx 104.72$$

11.

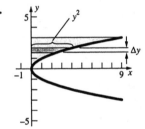

$$\Delta V \approx \pi(y^2)^2\Delta y = \pi y^4\Delta y$$

$$V = \pi\int_0^3 y^4 dy = \pi\left[\frac{1}{5}y^5\right]_0^3 = \frac{243\pi}{5} \approx 152.68$$

13.

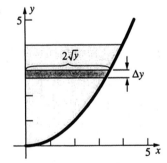

$$\Delta V \approx \pi\left(2\sqrt{y}\right)^2\Delta y = 4\pi y\Delta y$$

$$V = 4\pi\int_0^4 y\,dy = 4\pi\left[\frac{1}{2}y^2\right]_0^4 = 32\pi \approx 100.53$$

15.

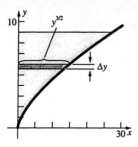

$$\Delta V \approx \pi(y^{3/2})^2 \Delta y = \pi y^3 \Delta y$$

$$V = \pi \int_0^9 y^3 dy = \pi \left[\frac{1}{4}y^4\right]_0^9 = \frac{6561\pi}{4} \approx 5153.00$$

17. The equation of the upper half of the ellipse is

$$y = b\sqrt{1-\frac{x^2}{a^2}} \text{ or } y = \frac{b}{a}\sqrt{a^2-x^2}.$$

$$V = \pi \int_{-a}^a \frac{b^2}{a^2}(a^2-x^2)dx$$

$$= \frac{b^2\pi}{a^2}\left[a^2 x - \frac{x^3}{3}\right]_{-a}^a$$

$$= \frac{b^2\pi}{a^2}\left[\left(a^3 - \frac{a^3}{3}\right) - \left(-a^3 + \frac{a^3}{3}\right)\right] = \frac{4}{3}ab^2\pi$$

19. Sketch the region.

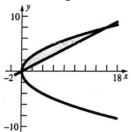

To find the intersection points, solve $\frac{x}{2} = 2\sqrt{x}$.

$$\frac{x^2}{4} = 4x$$

$$x^2 - 16x = 0$$

$$x(x-16) = 0$$

$$x = 0, 16$$

$$\Delta V \approx \pi\left[\left(2\sqrt{x}\right)^2 - \left(\frac{x}{2}\right)^2\right]\Delta x = \pi\left(4x - \frac{x^2}{4}\right)\Delta x$$

$$V = \pi \int_0^{16}\left(4x - \frac{x^2}{4}\right)dx = \pi\left[2x^2 - \frac{x^3}{12}\right]_0^{16}$$

$$= \pi\left(512 - \frac{1024}{3}\right) = \frac{512\pi}{3} \approx 536.17$$

21. Sketch the region.

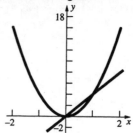

To find the intersection points, solve $\frac{y}{4} = \frac{\sqrt{y}}{2}$.

$$\frac{y^2}{16} = \frac{y}{4}$$

$$y^2 - 4y = 0$$

$$y(y-4) = 0$$

$$y = 0, 4$$

$$\Delta V \approx \pi\left[\left(\frac{\sqrt{y}}{2}\right)^2 - \left(\frac{y}{4}\right)^2\right]\Delta y = \pi\left(\frac{y}{4} - \frac{y^2}{16}\right)\Delta y$$

$$V = \pi \int_0^4\left(\frac{y}{4} - \frac{y^2}{16}\right)dy = \pi\left[\frac{y^2}{8} - \frac{y^3}{48}\right]_0^4$$

$$= \frac{2\pi}{3} \approx 2.0944$$

23.

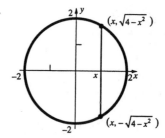

The square at x has sides of length $2\sqrt{4-x^2}$, as shown.

$$V = \int_{-2}^2 \left(2\sqrt{4-x^2}\right)^2 dx = \int_{-2}^2 4(4-x^2)dx$$

$$= 4\left[4x - \frac{x^3}{3}\right]_{-2}^2 = 4\left[\left(8 - \frac{8}{3}\right) - \left(-8 + \frac{8}{3}\right)\right] = \frac{128}{3}$$

$$\approx 42.67$$

25. The square at x has sides of length $\sqrt{\cos x}$.

$$V = \int_{-\pi/2}^{\pi/2} \cos x\, dx = [\sin x]_{-\pi/2}^{\pi/2} = 2$$

27. The square at x has sides of length $\sqrt{1-x^2}$.

$$V = \int_0^1 (1-x^2)dx = \left[x - \frac{x^3}{3}\right]_0^1 = \frac{2}{3} \approx 0.67$$

29. Using the result from Problem 28, the volume of one octant of the common region in the "+" is

$$\int_0^r (r^2 - y^2)dy = r^2 y - \frac{1}{3}y^3 \Big|_0^r$$

$$= r^3 - \frac{1}{3}r^3 = \frac{2}{3}r^3$$

Thus, the volume inside the "+" for two cylinders of radius r and length L is

V = vol. of cylinders - vol. of common region

$$= 2(\pi r^2 L) - 8\left(\frac{2}{3}r^3\right)$$

$$= 2\pi r^2 L - \frac{16}{3}r^3$$

31. From Problem 30, the general form for the volume of a "T" formed by two cylinders with the same radius is

V = vol. of cylinders - vol. of common region

$$= (\pi r^2)(L_1 + L_2) - 4\left(\frac{2}{3}r^3\right)$$

$$= \pi r^2 (L_1 + L_2) - \frac{8}{3}r^3$$

33. Sketch the region.

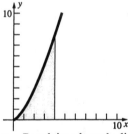

a. Revolving about the line $x = 4$, the radius of the disk at y is $4 - \sqrt[3]{y^2} = 4 - y^{2/3}$.

$$V = \pi \int_0^8 (4 - y^{2/3})^2\, dy$$

$$= \pi \int_0^8 (16 - 8y^{2/3} + y^{4/3})dy$$

$$= \pi \left[16y - \frac{24}{5}y^{5/3} + \frac{3}{7}y^{7/3}\right]_0^8$$

$$= \pi \left(128 - \frac{768}{5} + \frac{384}{7}\right)$$

$$= \frac{1024\pi}{35} \approx 91.91$$

b. Revolving about the line $y = 8$, the inner radius of the disk at x is $8 - \sqrt{x^3} = 8 - x^{3/2}$.

$$V = \pi \int_0^4 \left[8^2 - (8 - x^{3/2})^2\right]dx$$

$$= \pi \int_0^4 (16x^{3/2} - x^3)dx$$

$$= \pi\left[\frac{32}{5}x^{5/2} - \frac{1}{4}x^4\right]_0^4 = \pi\left(\frac{1024}{5} - 64\right)$$

$$= \frac{704\pi}{5} \approx 442.34$$

35. The area of a quarter circle with radius 2 is

$$\int_0^2 \sqrt{4 - y^2}\, dy = \pi.$$

$$\int_0^2 \left[2\sqrt{4 - y^2} + 4 - y^2\right]dy$$

$$= 2\int_0^2 \sqrt{4 - y^2}\, dy + \int_0^2 (4 - y^2)dy$$

$$= 2\pi + \left[4y - \frac{1}{3}y^3\right]_0^2 = 2\pi + \left(8 - \frac{8}{3}\right)$$

$$= 2\pi + \frac{16}{3} \approx 11.62$$

37. Let the x-axis lie on the base perpendicular to the diameter through the center of the base. The slice at x is a rectangle with base of length $2\sqrt{r^2 - x^2}$ and height $x\tan\theta$.

$$V = \int_0^r 2x\tan\theta\sqrt{r^2 - x^2}\, dx$$

$$= \left[-\frac{2}{3}\tan\theta (r^2 - x^2)^{3/2}\right]_0^r$$

$$= \frac{2}{3}r^3 \tan\theta$$

39. Let A lie on the xy-plane. Suppose $\Delta A = f(x)\Delta x$ where $f(x)$ is the length at x, so $A = \int f(x)dx$. Slice the general cone at height z parallel to A. The slice of the resulting region is A_z and ΔA_z is a region related to $f(x)$ and Δx by similar triangles:

$$\Delta A_z = \left(1 - \frac{z}{h}\right)f(x)\cdot\left(1 - \frac{z}{h}\right)\Delta x$$

$$= \left(1 - \frac{z}{h}\right)^2 f(x)\Delta x$$

Therefore, $A_z = \left(1 - \frac{z}{h}\right)^2 \int f(x)dx = \left(1 - \frac{z}{h}\right)^2 A.$

$$\Delta V \approx A_z \Delta z = A\left(1 - \frac{z}{h}\right)^2 \Delta z \quad V = A\int_0^h \left(1 - \frac{z}{h}\right)^2 dz$$

$$= A\left[-\frac{h}{3}\left(1 - \frac{z}{h}\right)^3\right]_0^h = \frac{1}{3}Ah.$$

a. $A = \pi r^2$

$$V = \frac{1}{3}Ah = \frac{1}{3}\pi r^2 h$$

b. A face of a regular tetrahedron is an equilateral triangle. If the side of an equilateral triangle has length r, then the area

is $A = \dfrac{1}{2}r \cdot \dfrac{\sqrt{3}}{2}r = \dfrac{\sqrt{3}}{4}r^2$.

The center of an equilateral triangle is

$\dfrac{2}{3} \cdot \dfrac{\sqrt{3}}{2}r = \dfrac{1}{\sqrt{3}}r$ from a vertex. Then the

height of a regular tetrahedron is

$$h = \sqrt{r^2 - \left(\dfrac{1}{\sqrt{3}}r\right)^2} = \sqrt{\dfrac{2}{3}r^2} = \dfrac{\sqrt{2}}{\sqrt{3}}r.$$

$$V = \dfrac{1}{3}Ah = \dfrac{\sqrt{2}}{12}r^3$$

41. First we examine the cross-sectional areas of each shape.
Hemisphere: cross-sectional shape is a circle.

The radius of the circle at height y is $\sqrt{r^2 - y^2}$.
Therefore, the cross-sectional area for the hemisphere is

$$A_h = \pi(\sqrt{r^2 - y^2})^2 = \pi(r^2 - y^2)$$

Cylinder w/o cone: cross-sectional shape is a washer. The outer radius is a constant, r. The inner radius at height y is equal to y. Therefore, the cross-sectional area is

$$A_2 = \pi r^2 - \pi y^2 = \pi(r^2 - y^2).$$

Since both cross-sectional areas are the same, we can apply Cavaleri's Principle. The volume of the hemisphere of radius r is
$V = $ vol. of cylinder - vol. of cone

$$= \pi r^2 h - \dfrac{1}{3}\pi r^2 h$$

$$= \dfrac{2}{3}\pi r^2 h$$

With the height of the cylinder and cone equal to r, the volume of the hemisphere is

$$V = \dfrac{2}{3}\pi r^2 (r) = \dfrac{2}{3}\pi r^3 .$$

6.3 Concepts Review

1. $2\pi x\, f(x)\Delta x$

3. $2\pi \displaystyle\int_0^2 (1+x)x\,dx$

Problem Set 6.3

1. a,b.

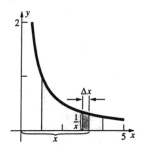

c. $\Delta V \approx 2\pi x\left(\dfrac{1}{x}\right)\Delta x = 2\pi \Delta x$

d,e. $V = 2\pi \displaystyle\int_1^4 dx = 2\pi [x]_1^4 = 6\pi \approx 18.85$

3. a,b.

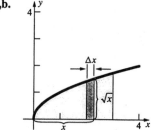

c. $\Delta V \approx 2\pi x\sqrt{x}\,\Delta x = 2\pi x^{3/2}\Delta x$

d,e. $V = 2\pi \displaystyle\int_0^3 x^{3/2}dx = 2\pi\left[\dfrac{2}{5}x^{5/2}\right]_0^3$

$= \dfrac{36\sqrt{3}}{5}\pi \approx 39.18$

5. a,b.

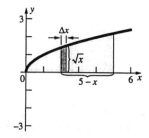

c. $\Delta V \approx 2\pi(5-x)\sqrt{x}\,\Delta x$

$= 2\pi(5x^{1/2} - x^{3/2})\Delta x$

d,e. $V = 2\pi \int_0^5 (5x^{1/2} - x^{3/2})dx$

$$= 2\pi \left[\frac{10}{3}x^{3/2} - \frac{2}{5}x^{5/2}\right]_0^5$$

$$= 2\pi \left(\frac{50\sqrt{5}}{3} - 10\sqrt{5}\right) = \frac{40\sqrt{5}}{3}\pi \approx 93.66$$

7. a,b.

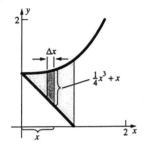

c. $\Delta V \approx 2\pi x \left[\left(\frac{1}{4}x^3 + 1\right) - (1-x)\right]\Delta x$

$$= 2\pi \left(\frac{1}{4}x^4 + x^2\right)\Delta x$$

d,e. $V = 2\pi \int_0^1 \left(\frac{1}{4}x^4 + x^2\right)dx$

$$= 2\pi \left[\frac{1}{20}x^5 + \frac{1}{3}x^3\right]_0^1 = 2\pi \left(\frac{1}{20} + \frac{1}{3}\right)$$

$$= \frac{23\pi}{30} \approx 2.41$$

9. a,b.

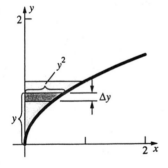

c. $\Delta V \approx 2\pi y(y^2)\Delta y = 2\pi y^3 \Delta y$

d,e. $V = 2\pi \int_0^1 y^3 dy = 2\pi \left[\frac{1}{4}y^4\right]_0^1 = \frac{\pi}{2} \approx 1.57$

11. a,b.

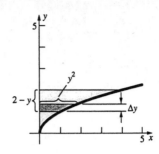

c. $\Delta V \approx 2\pi(2-y)y^2 \Delta y = 2\pi(2y^2 - y^3)\Delta y$

d,e. $V = 2\pi \int_0^2 (2y^2 - y^3)dy = 2\pi \left[\frac{2}{3}y^3 - \frac{1}{4}y^4\right]_0^2$

$$= 2\pi \left(\frac{16}{3} - 4\right) = \frac{8\pi}{3} \approx 8.38$$

13. a. $\pi \int_a^b \left[f(x)^2 - g(x)^2\right]dx$

b. $2\pi \int_a^b x[f(x) - g(x)]dx$

c. $2\pi \int_a^b (x-a)[f(x) - g(x)]dx$

d. $2\pi \int_a^b (b-x)[f(x) - g(x)]dx$

15.

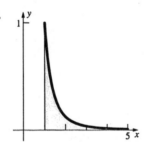

a. $A = \int_1^3 \frac{1}{x^3}dx$

b. $V = 2\pi \int_1^3 x\left(\frac{1}{x^3}\right)dx = 2\pi \int_1^3 \frac{1}{x^2}dx$

c. $V = \pi \int_1^3 \left[\left(\frac{1}{x^3} + 1\right)^2 - (-1)^2\right]dx$

$$= \pi \int_1^3 \left(\frac{1}{x^6} + \frac{2}{x^3}\right)dx$$

d. $V = 2\pi \int_1^3 (4-x)\left(\dfrac{1}{x^3}\right)dx$

$\quad = 2\pi \int_1^3 \left(\dfrac{4}{x^3} - \dfrac{1}{x^2}\right)dx$

17. To find the intersection point, solve $\sqrt{y} = \dfrac{y^3}{32}$.

$y = \dfrac{y^6}{1024}$

$y^6 - 1024y = 0$

$y(y^5 - 1024) = 0$

$y = 0, 4$

$V = 2\pi \int_0^4 y\left(\sqrt{y} - \dfrac{y^3}{32}\right)dy$

$\quad = 2\pi \int_0^4 \left(y^{3/2} - \dfrac{y^4}{32}\right)dy$

$\quad = 2\pi \left[\dfrac{2}{5}y^{5/2} - \dfrac{y^5}{160}\right]_0^4 = 2\pi\left(\dfrac{64}{5} - \dfrac{32}{5}\right) = \dfrac{64\pi}{5}$

$\quad \approx 40.21$

19. Let R be the region bounded by $y = \sqrt{b^2 - x^2}$,

$y = -\sqrt{b^2 - x^2}$, and $x = a$. When R is revolved about the y-axis, it produces the desired solid.

$V = 2\pi \int_a^b x\left(\sqrt{b^2 - x^2} + \sqrt{b^2 - x^2}\right)dx$

$\quad = 4\pi \int_a^b x\sqrt{b^2 - x^2}\,dx = 4\pi\left[-\dfrac{1}{3}(b^2 - x^2)^{3/2}\right]_a^b$

$\quad = 4\pi\left[\dfrac{1}{3}(b^2 - a^2)^{3/2}\right] = \dfrac{4\pi}{3}(b^2 - a^2)^{3/2}$

21. To find the intersection point, solve
$\sin(x^2) = \cos(x^2)$.

$\tan(x^2) = 1$

$x^2 = \dfrac{\pi}{4}$

$x = \dfrac{\sqrt{\pi}}{2}$

$V = 2\pi \int_0^{\sqrt{\pi}/2} x\left[\cos(x^2) - \sin(x^2)\right]dx$

$\quad = 2\pi \int_0^{\sqrt{\pi}/2} \left[x\cos(x^2) - x\sin(x^2)\right]dx$

$\quad = 2\pi\left[\dfrac{1}{2}\sin(x^2) + \dfrac{1}{2}\cos(x^2)\right]_0^{\sqrt{\pi}/2}$

$\quad = 2\pi\left[\left(\dfrac{1}{2\sqrt{2}} + \dfrac{1}{2\sqrt{2}}\right) - \dfrac{1}{2}\right] = \pi\left(\sqrt{2} - 1\right) \approx 1.30$

23. a. The curves intersect when $x = 0$ and $x = 1$.

$V = \pi \int_0^1 [x^2 - (x^2)^2]\,dx = \pi \int_0^1 (x^2 - x^4)\,dx$

$\quad = \pi\left[\dfrac{1}{3}x^3 - \dfrac{1}{5}x^5\right]_0^1 = \pi\left(\dfrac{1}{3} - \dfrac{1}{5}\right) = \dfrac{2\pi}{15} \approx 0.42$

b. $V = 2\pi \int_0^1 x(x - x^2)\,dx = 2\pi \int_0^1 (x^2 - x^3)\,dx$

$\quad = 2\pi\left[\dfrac{1}{3}x^3 - \dfrac{1}{4}x^4\right]_0^1 = 2\pi\left(\dfrac{1}{3} - \dfrac{1}{4}\right) = \dfrac{\pi}{6}$

$\quad \approx 0.52$

c. Slice perpendicular to the line $y = x$. At (a, a), the perpendicular line has equation $y = -(x - a) + a = -x + 2a$. Substitute

$y = -x + 2a$ into $y = x^2$ and solve for $x \geq 0$.

$x^2 + x - 2a = 0$

$x = \dfrac{-1 \pm \sqrt{1 + 8a}}{2}$

$x = \dfrac{-1 + \sqrt{1 + 8a}}{2}$

Substitute into $y = -x + 2a$, so

$y = \dfrac{1 + 4a - \sqrt{1 + 8a}}{2}$. Find an expression for

r^2, the square of the distance from (a, a) to
$\left(\dfrac{-1 + \sqrt{1 + 8a}}{2}, \dfrac{1 + 4a - \sqrt{1 + 8a}}{2}\right)$.

$r^2 = \left[a - \dfrac{-1 + \sqrt{1 + 8a}}{2}\right]^2$

$\quad + \left[a - \dfrac{1 + 4a - \sqrt{1 + 8a}}{2}\right]^2$

$\quad = \left[\dfrac{2a + 1 - \sqrt{1 + 8a}}{2}\right]^2$

$\quad + \left[-\dfrac{2a + 1 - \sqrt{1 + 8a}}{2}\right]^2$

$\quad = 2\left[\dfrac{2a + 1 - \sqrt{1 + 8a}}{2}\right]^2$

$\quad = 2a^2 + 6a + 1 - 2a\sqrt{1 + 8a} - \sqrt{1 + 8a}$

$\Delta V \approx \pi r^2 \Delta a$

$V = \pi \int_0^1 (2a^2 + 6a + 1$

$\qquad - 2a\sqrt{1 + 8a} - \sqrt{1 + 8a}\,)\,da$

$\quad = \pi\left[\dfrac{2}{3}a^3 + 3a^2 + a - \dfrac{1}{12}(1 + 8a)^{3/2}\right]_0^1$

$$-\pi \int_0^1 2a\sqrt{1+8a}\, da$$

$$= \pi \left[\left(\frac{2}{3}+3+1-\frac{9}{4} \right) - \left(-\frac{1}{12} \right) \right]$$

$$-\pi \int_0^1 2a\sqrt{1+8a}\, da$$

$$= \frac{5\pi}{2} - \pi \int_0^1 2a\sqrt{1+8a}\, da$$

To integrate $\int_0^1 2a\sqrt{1+8a}\, da$, use the substitution $u = 1 + 8a$.

$$\int_0^1 2a\sqrt{1+8a}\, da = \int_1^9 \frac{1}{4}(u-1)\sqrt{u}\,\frac{1}{8}\, du$$

$$= \frac{1}{32} \int_1^9 (u^{3/2} - u^{1/2})\, du$$

$$= \frac{1}{32} \left[\frac{2}{5}u^{5/2} - \frac{2}{3}u^{3/2} \right]_1^9$$

$$= \frac{1}{32}\left[\left(\frac{486}{5}-18 \right) - \left(\frac{2}{5}-\frac{2}{3} \right) \right] = \frac{149}{60}$$

$$V = \frac{5\pi}{2} - \frac{149\pi}{60} = \frac{\pi}{60} \approx 0.052$$

25. $\Delta V \approx \dfrac{x^2}{r^2} S\Delta x$

$$V = \frac{S}{r^2} \int_0^r x^2\, dx = \frac{S}{r^2}\left[\frac{1}{3}x^3 \right]_0^r = \frac{1}{3}rS$$

6.4 Concepts Review

1. Circle
$$x^2 + y^2 = 16\cos^2 t + 16\sin^2 t = 16$$

3. $\int_a^b \sqrt{[f'(t)]^2 + [g'(t)]^2}\, dt$

Problem Set 6.4

1. $f(x) = x^2; a = -1; b = 3$

   a. $n = 2; x = -1,1,3; y = 1,1,9$

   $$\sum_{i=1}^n \Delta w_i = \sqrt{(1+1)^2 + (1-1)^2}$$

   $$+\sqrt{(3-1)^2 + (9-1)^2}$$

   $$= \sqrt{4} + \sqrt{68} = 2 + 2\sqrt{17} \approx 10.2462$$

   b. $n = 4; x = -1,0,1,2,3; y = 1,0,1,4,9$

   $$\sum_{i=1}^n \Delta w_i = \sqrt{(0+1)^2 + (0-1)^2}$$

   $$+\sqrt{(1-0)^2 + (1-0)^2}$$

   $$+\sqrt{(2-1)^2 + (4-1)^2}$$

   $$+\sqrt{(3-2)^2 + (9-4)^2}$$

   $$= \sqrt{2} + \sqrt{2} + \sqrt{10} + \sqrt{26}$$

   $$\approx 11.0897$$

3. $f(x) = \sin x; a = 0; b = 2\pi$

   a. $n = 2; x = 0, \pi, 2\pi; y = 0,0,0$

   $$\sum_{i=1}^n \Delta w = \sqrt{(\pi - 0)^2 + (0-0)^2}$$

   $$+\sqrt{(2\pi - \pi)^2 + (0-0)^2}$$

   $$= 2\pi \approx 6.2832$$

   b. $n = 4; x = 0, \pi/2, \pi, 3\pi/2, 2\pi;$
   $y = 0,1,0,-1,0$

   $$\sum_{i=1}^n \Delta w = \sqrt{(\pi/2 - 0)^2 + (1-0)^2}$$

   $$+\sqrt{(\pi - \pi/2)^2 + (0-1)^2}$$

   $$+\sqrt{(3\pi/2 - \pi)^2 + (-1-0)^2}$$

   $$+\sqrt{(2\pi - 3\pi/2)^2 + (0+1)^2}$$

   $$\approx 7.4484$$

   c. $n = 8; x = 0, \dfrac{\pi}{4}, \dfrac{\pi}{2}, \dfrac{3\pi}{4}, \pi, \dfrac{5\pi}{4}, \dfrac{3\pi}{2}, \dfrac{7\pi}{4}, 2\pi;$

   $$y = 0, \frac{\sqrt{2}}{2}, 1, \frac{\sqrt{2}}{2}, 0, \frac{-\sqrt{2}}{2}, -1, \frac{-\sqrt{2}}{2}, 0$$

$$\sum_{i=1}^{n} \Delta w = \sqrt{(\pi/4-0)^2 + (\sqrt{2}/2-0)^2}$$

$$+\sqrt{(\pi/2-\pi/4)^2 + (1-\sqrt{2}/2)^2}$$

$$+\sqrt{(3\pi/4-\pi/2)^2 + (\sqrt{2}/2-1)^2}$$

$$+\sqrt{(\pi-3\pi/4)^2 + (0-\sqrt{2}/2)^2}$$

$$+\sqrt{(5\pi/4-\pi)^2 + (-\sqrt{2}/2-0)^2}$$

$$+\sqrt{(3\pi/2-5\pi/4)^2 + (-1+\sqrt{2}/2)^2}$$

$$+\sqrt{(7\pi/4-3\pi/2)^2 + (-\sqrt{2}/2+1)^2}$$

$$+\sqrt{(2\pi-7\pi/4)^2 + (0+\sqrt{2}/2)^2}$$

$$\approx 7.5802$$

5. $f(x) = 2x+3, f'(x) = 2$

$$L = \int_1^3 \sqrt{1+(2)^2}\, dx = \sqrt{5}\int_1^3 dx = 2\sqrt{5}$$

At $x=1, y=2(1)+3=5$.
At $x=3, y=2(3)+3=9$.

$$d = \sqrt{(3-1)^2 + (9-5)^2} = \sqrt{20} = 2\sqrt{5}$$

7. $f(x) = 4x^{3/2}, f'(x) = 6x^{1/2}$

$$L = \int_{1/3}^5 \sqrt{1+(6x^{1/2})^2}\, dx = \int_{1/3}^5 \sqrt{1+36x}\, dx$$

$$= \left[\frac{1}{36}\cdot\frac{2}{3}(1+36x)^{3/2}\right]_{1/3}^5$$

$$= \frac{1}{54}\left(181\sqrt{181} - 13\sqrt{13}\right) \approx 44.23$$

9. $f(x) = (4-x^{2/3})^{3/2}$,

$$f'(x) = \frac{3}{2}(4-x^{2/3})^{1/2}\left(-\frac{2}{3}x^{-1/3}\right)$$

$$= -x^{-1/3}(4-x^{2/3})^{1/2}$$

$$L = \int_1^8 \sqrt{1+\left[-x^{-1/3}(4-x^{2/3})^{1/2}\right]^2}\, dx$$

$$= \int_1^8 \sqrt{4x^{-2/3}}\, dx = \int_1^8 2x^{-1/3}dx$$

$$= 2\left[\frac{3}{2}x^{2/3}\right]_1^8 = 3(4-1) = 9$$

11. $g(y) = \dfrac{y^4}{16} + \dfrac{1}{2y^2}, g'(y) = \dfrac{y^3}{4} - \dfrac{1}{y^3}$

$$L = \int_{-3}^{-2}\sqrt{1+\left(\frac{y^3}{4}-\frac{1}{y^3}\right)^2}\, dy$$

$$= \int_{-3}^{-2}\sqrt{\frac{y^6}{16} + \frac{1}{2} + \frac{1}{y^6}}\, dy = \int_{-3}^{-2}\sqrt{\left(\frac{y^3}{4} + \frac{1}{y^3}\right)^2}\, dy$$

$$= \int_{-3}^{-2} -\left(\frac{y^3}{4} + \frac{1}{y^3}\right)dy = -\left[\frac{y^4}{16} - \frac{1}{2y^2}\right]_{-3}^{-2}$$

$$= -\left[\left(1-\frac{1}{8}\right) - \left(\frac{81}{16} - \frac{1}{18}\right)\right] = \frac{595}{144} \approx 4.13$$

13.

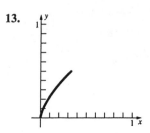

$$\frac{dx}{dt} = t^2, \frac{dy}{dt} = t$$

$$L = \int_0^1 \sqrt{(t^2)^2 + (t)^2}\, dt = \int_0^1 \sqrt{t^4 + t^2}\, dt$$

$$= \int_0^1 t\sqrt{t^2+1}\, dt = \left[\frac{1}{3}(t^2+1)^{3/2}\right]_0^1$$

$$= \frac{1}{3}\left(2\sqrt{2} - 1\right) \approx 0.61$$

15.

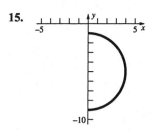

$$\frac{dx}{dt} = 4\cos t, \frac{dy}{dt} = -4\sin t$$

$$L = \int_0^\pi \sqrt{(4\cos t)^2 + (-4\sin t)^2}\, dt$$

$$= \int_0^\pi \sqrt{16\cos^2 t + 16\sin^2 t}\, dt$$

$$= \int_0^\pi 4\, dt = 4\pi \approx 12.57$$

17.

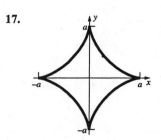

$\dfrac{dx}{dt} = 3a\cos t\sin^2 t,\ \dfrac{dy}{dt} = -3a\sin t\cos^2 t$

The first quadrant length is L

$= \displaystyle\int_0^{\pi/2}\sqrt{(3a\cos t\sin^2 t)^2 + (-3a\sin t\cos^2 t)^2}\ dt$

$= \displaystyle\int_0^{\pi/2}\sqrt{9a^2\cos^2 t\sin^4 t + 9a^2\sin^2 t\cos^4 t}\ dt$

$= \displaystyle\int_0^{\pi/2}\sqrt{9a^2\cos^2 t\sin^2 t(\sin^2 t + \cos^2 t)}\ dt$

$= \displaystyle\int_0^{\pi/2}3a\cos t\sin t\,dt = 3a\left[-\dfrac{1}{2}\cos^2 t\right]_0^{\pi/2} = \dfrac{3a}{2}$

(The integral can also be evaluated as

$3a\left[\dfrac{1}{2}\sin^2 t\right]_0^{\pi/2}$ with the same result.)

The total length is $6a$.

19. From Problem 18,

$x = a(\theta - \sin\theta),\ y = a(1-\cos\theta)$

$\dfrac{dx}{d\theta} = a(1-\cos\theta),\ \dfrac{dy}{d\theta} = a\sin\theta$ so

$\left(\dfrac{dx}{d\theta}\right)^2 + \left(\dfrac{dy}{d\theta}\right)^2 = [a(1-\cos\theta)]^2 + [a\sin\theta]^2$

$= a^2 - 2a^2\cos\theta + a^2\cos^2\theta + a^2\sin^2\theta$

$= 2a^2 - 2a^2\cos\theta = 2a^2(1-\cos\theta)$

$= 4a^2\dfrac{1-\cos\theta}{2} = 4a^2\sin^2\left(\dfrac{\theta}{2}\right).$

The length of one arch of the cycloid is

$\displaystyle\int_0^{2\pi}\sqrt{4a^2\sin^2\left(\dfrac{\theta}{2}\right)}\,d\theta = \int_0^{2\pi}2a\sin\left(\dfrac{\theta}{2}\right)d\theta$

$= 2a\left[-2\cos\dfrac{\theta}{2}\right]_0^{2\pi} = 2a(2+2) = 8a$

21. a. $\dfrac{dy}{dx} = \sqrt{x^3 - 1}$

$L = \displaystyle\int_1^2\sqrt{1 + x^3 - 1}\,dx = \int_1^2 x^{3/2}\,dx$

$= \left[\dfrac{2}{5}x^{5/2}\right]_1^2 = \dfrac{2}{5}\left(4\sqrt{2} - 1\right) \approx 1.86$

b. $f'(t) = 1-\cos t,\ g'(t) = \sin t$

$L = \displaystyle\int_0^{4\pi}\sqrt{2-2\cos t}\,dt = \int_0^{4\pi}2\left|\sin\left(\dfrac{t}{2}\right)\right|dt$

$\sin\left(\dfrac{t}{2}\right)$ is positive for $0 < t < 2\pi$, and

by symmetry, we can double the integral from 0 to 2π.

$L = 4\displaystyle\int_0^{2\pi}\sin\left(\dfrac{t}{2}\right)dt = \left[-8\cos\dfrac{t}{2}\right]_0^{2\pi}$

$= 8 + 8 = 16$

23. $f(x) = 6x,\ f'(x) = 6$

$A = 2\pi\displaystyle\int_0^1 6x\sqrt{1+36}\,dx = 12\sqrt{37}\pi\int_0^1 x\,dx$

$= 12\sqrt{37}\pi\left[\dfrac{1}{2}x^2\right]_0^1 = 6\sqrt{37}\pi \approx 114.66$

25. $f(x) = \dfrac{x^3}{3},\ f'(x) = x^2$

$A = 2\pi\displaystyle\int_1^{\sqrt{7}}\dfrac{x^3}{3}\sqrt{1+x^4}\,dx$

$= 2\pi\left[\dfrac{1}{18}(1+x^4)^{3/2}\right]_1^{\sqrt{7}} = \dfrac{\pi}{9}\left(250\sqrt{2} + 2\sqrt{2}\right)$

$= 28\sqrt{2}\pi \approx 124.40$

27. $\dfrac{dx}{dt} = 1,\ \dfrac{dy}{dt} = 3t^2$

$A = 2\pi\displaystyle\int_0^1 t^3\sqrt{1+9t^4}\,dt$

$= 2\pi\left[\dfrac{1}{54}(1+9t^4)^{3/2}\right]_0^1 = \dfrac{\pi}{27}\left(10\sqrt{10} - 1\right)$

≈ 3.56

29. $y = f(x) = \sqrt{r^2 - x^2}$

$f'(x) = -x(r^2 - x^2)^{-1/2}$

$A = 2\pi\displaystyle\int_{-r}^r\sqrt{r^2 - x^2}\sqrt{1 + \left[-x(r^2 - x^2)^{-1/2}\right]^2}\,dx$

$= 2\pi\displaystyle\int_{-r}^r\sqrt{r^2 - x^2}\sqrt{1 + x^2(r^2 - x^2)^{-1}}\,dx$

$= 2\pi\displaystyle\int_{-r}^r\sqrt{\left(r^2 - x^2\right)\left(1 + x^2(r^2 - x^2)^{-1}\right)}\,dx$

$= 2\pi\displaystyle\int_{-r}^r\sqrt{r^2 - x^2 + x^2}\,dx$

$= 2\pi\displaystyle\int_{-r}^r\sqrt{r^2}\,dx = 2\pi\int_{-r}^r r\,dx$

$= 2\pi rx\,|_{-r}^r = 4\pi r^2$

31. a. The base circumference is equal to the arc length of the sector, so $2\pi r = \theta l$. Therefore,

$\theta = \dfrac{2\pi r}{l}.$

b. The area of the sector is equal to the lateral surface area. Therefore, the lateral surface area is $\dfrac{1}{2}l^2\theta = \dfrac{1}{2}l^2\left(\dfrac{2\pi r}{l}\right) = \pi rl$.

c. Assume $r_2 > r_1$. Let l_1 and l_2 be the slant heights for r_1 and r_2, respectively. Then $A = \pi r_2 l_2 - \pi r_1 l_1 = \pi r_2(l_1 + l) - \pi r_1 l_1$.

From part a, $\theta = \dfrac{2\pi r_2}{l_2} = \dfrac{2\pi r_2}{l_1 + l} = \dfrac{2\pi r_1}{l_1}$.

Solve for $l_1 : l_1 r_2 = l_1 r_1 + l r_1$

$l_1(r_2 - r_1) = l r_1$

$l_1 = \dfrac{l r_1}{r_2 - r_1}$

$A = \pi r_2 \left(\dfrac{l r_1}{r_2 - r_1} + l \right) - \pi r_1 \left(\dfrac{l r_1}{r_2 - r_1} \right)$

$= \pi(l r_1 + l r_2) = 2\pi \left[\dfrac{r_1 + r_2}{2} \right] l$

33. a. $\dfrac{dx}{dt} = a(1 - \cos t), \dfrac{dy}{dt} = a \sin t$

$A = 2\pi \displaystyle\int_0^{2\pi} a(1 - \cos t) \cdot$

$\sqrt{a^2(1 - \cos t)^2 + a^2 \sin^2 t} \; dt$

$= 2\pi a \displaystyle\int_0^{2\pi} (1 - \cos t)\sqrt{2a^2 - 2a^2 \cos t} \; dt$

$= 2\sqrt{2}\pi a^2 \displaystyle\int_0^{2\pi} (1 - \cos t)^{3/2} \, dt$

b. $1 - \cos t = 2 \sin^2 \left(\dfrac{t}{2} \right)$, so

$A = 2\sqrt{2}\pi a^2 \displaystyle\int_0^{2\pi} 2^{3/2} \sin^3 \left(\dfrac{t}{2} \right) dt$

$= 8\pi a^2 \displaystyle\int_0^{2\pi} \sin \left(\dfrac{t}{2} \right) \sin^2 \left(\dfrac{t}{2} \right) dt$

$= 8\pi a^2 \displaystyle\int_0^{2\pi} \sin \left(\dfrac{t}{2} \right) \left[1 - \cos^2 \left(\dfrac{t}{2} \right) \right] dt$

$= 8\pi a^2 \left[-2\cos \left(\dfrac{t}{2} \right) + \dfrac{2}{3} \cos^3 \left(\dfrac{t}{2} \right) \right]_0^{2\pi}$

$= 8\pi a^2 \left[\left(2 - \dfrac{2}{3} \right) - \left(-2 + \dfrac{2}{3} \right) \right] = \dfrac{64}{3}\pi a^2$

35. a.

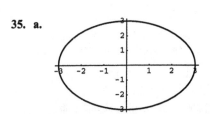

b.

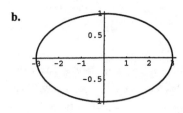

c.

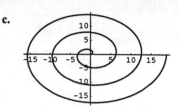

d.

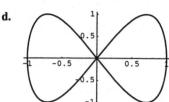

e.

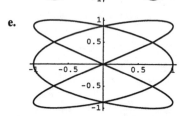

f.

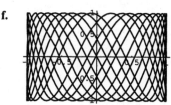

37.

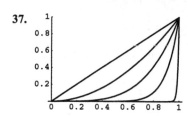

$y = x, y' = 1,$

$L = \displaystyle\int_0^1 \sqrt{2}\,dx = \left[\sqrt{2}\,x \right]_0^1 = \sqrt{2} \approx 1.41421$

$y = x^2, y' = 2x$, $L = \displaystyle\int_0^1 \sqrt{1 + 4x^2}\,dx \approx 1.47894$

$y = x^4, y' = 4x^3$, $L = \displaystyle\int_0^1 \sqrt{1 + 16x^6}\,dx \approx 1.60023$

$y = x^{10}, y' = 10x^9,$

$L = \displaystyle\int_0^1 \sqrt{1 + 100^{18}}\,dx \approx 1.75441$

$y = x^{100}, y' = 100x^{99},$

$L = \displaystyle\int_0^1 \sqrt{1 + 10{,}000x^{198}}\,dx \approx 1.95167$

When $n = 10,000$ the length will be close to 2.

6.5 Concepts Review

1. $F \cdot (b - a)$; $\displaystyle\int_a^b F(x)\,dx$

3. $30 \cdot 10 = 300$

Problem Set 6.5

1. $F\left(\dfrac{1}{2}\right) = 6; k \cdot \dfrac{1}{2} = 6, k = 12$

 $F(x) = 12x$

 $W = \displaystyle\int_0^{1/2} 12x\,dx = \left[6x^2\right]_0^{1/2} = \dfrac{3}{2} = 1.5$ ft-lb

3. $F(0.01) = 0.6; k = 60$
 $F(x) = 60x$

 $W = \displaystyle\int_0^{0.02} 60x\,dx = \left[30x^2\right]_0^{0.02} = 0.012$ Joules

5. $W = \displaystyle\int_0^d kx\,dx = \left[\dfrac{1}{2}kx^2\right]_0^d$

 $= \dfrac{1}{2}k(d^2 - 0) = \dfrac{1}{2}kd^2$

7. $W = \displaystyle\int_0^2 9s\,ds = 9\left[\dfrac{1}{2}s^2\right]_0^2 = 18$ ft-lb

9. A slab of thickness Δy at height y has width

 $4 - \dfrac{4}{5}y$ and length 10. The slab will be lifted a

 distance $10 - y$.

 $\Delta W \approx \delta \cdot 10 \cdot \left(4 - \dfrac{4}{5}y\right)\Delta y(10 - y)$

 $= 8\delta(y^2 - 15y + 50)\Delta y$

 $W = \displaystyle\int_0^5 8\delta(y^2 - 15y + 50)\,dy$

 $= 8(62.4)\left[\dfrac{1}{3}y^3 - \dfrac{15}{2}y^2 + 50y\right]_0^5$

 $= 8(62.4)\left(\dfrac{125}{3} - \dfrac{375}{2} + 250\right)$

 $= 52{,}000$ ft-lb

11. A slab of thickness Δy at height y has width

 $\dfrac{3}{4}y + 3$ and length 10. The slab will be lifted a

 distance $9 - y$. $\Delta W \approx \delta \cdot 10 \cdot \left(\dfrac{3}{4}y + 3\right)\Delta y(9 - y)$

 $= \dfrac{15}{2}\delta(36 + 5y - y^2)\Delta y$

 $W = \displaystyle\int_0^4 \dfrac{15}{2}\delta(36 + 5y - y^2)\,dy$

 $= \dfrac{15}{2}(62.4)\left[36y + \dfrac{5}{2}y^2 - \dfrac{1}{3}y^3\right]_0^4$

 $= \dfrac{15}{2}(62.4)\left(144 + 40 - \dfrac{64}{3}\right)$

 $= 76{,}128$ ft-lb

13. The volume of a disk with thickness Δy is
 $16\pi\Delta y$. If it is at height y, it will be lifted a
 distance $10 - y$.
 $\Delta W \approx \delta 16\pi\Delta y(10 - y) = 16\pi\delta(10 - y)\Delta y$

 $W = \displaystyle\int_0^{10} 16\pi\delta(10 - y)\,dy = 16\pi(50)\left[10y - \dfrac{1}{2}y^2\right]_0^{10}$

 $= 16\pi(50)(100 - 50) \approx 125{,}664$ ft-lb

15. The total force on the face of the piston is $A \cdot f(x)$
 if the piston is x inches from the cylinder head.
 The work done by moving the piston from

 x_1 to x_2 is $W = \displaystyle\int_{x_1}^{x_2} A \cdot f(x)\,dx = A\displaystyle\int_{x_1}^{x_2} f(x)\,dx$.

 This is the work done by the gas in moving the
 piston. The work done by the piston to compress

 the gas is the opposite of this or $A\displaystyle\int_{x_2}^{x_1} f(x)\,dx$.

17. $c = 40(16)^{1.4}$

 $A = 2; p(v) = cv^{-1.4}$

 $f(x) = c(2x)^{-1.4}$

 $x_1 = \dfrac{16}{2} = 8, x_2 = \dfrac{2}{2} = 1$

 $W = 2\displaystyle\int_1^8 c(2x)^{-1.4}\,dx = 2c\left[-1.25(2x)^{-0.4}\right]_1^8$

 $= 80(16)^{1.4}(-1.25)(16^{-0.4} - 2^{-0.4})$

 ≈ 2075.83 in.-lb

19. The total work is equal to the work W_1 to haul
 the load by itself and the work W_2 to haul the
 rope by itself.
 $W_1 = 200 \cdot 500 = 100{,}000$ ft-lb
 Let $y = 0$ be the bottom of the shaft. When the
 rope is at y, $\Delta W_2 \approx 2\Delta y(500 - y)$.

 $W_2 = \displaystyle\int_0^{500} 2(500 - y)\,dy = 2\left[500y - \dfrac{1}{2}y^2\right]_0^{500}$

 $= 2(250{,}000 - 125{,}000) = 250{,}000$ ft-lb
 $W = W_1 + W_2 = 100{,}000 + 250{,}000$
 $= 350{,}000$ ft-lb

21. $f(x) = \dfrac{k}{x^2}; f(4000) = 5000$

 $\dfrac{k}{4000^2} = 5000, k = 80{,}000{,}000{,}000$

$$W = \int_{4000}^{4200} \frac{80,000,000,000}{x^2}\,dx$$

$$= 80,000,000,000\left[-\frac{1}{x}\right]_{4000}^{4200}$$

$$= \frac{20,000,000}{21} \approx 952,381 \text{ mi-lb}$$

23. The relationship between the height of the bucket and time is $y = 2t$, so $t = \frac{1}{2}y$. When the bucket is a height y, the sand has been leaking out of the bucket for $\frac{1}{2}y$ seconds. The weight of the bucket and sand is $100 + 500 - 3\left(\frac{1}{2}y\right) = 600 - \frac{3}{2}y$.

$$\Delta W \approx \left(600 - \frac{3}{2}y\right)\Delta y$$

$$W = \int_0^{80}\left(600 - \frac{3}{2}y\right)dy = \left[600y - \frac{3}{4}y^2\right]_0^{80}$$

$$= 48,000 - 4800 = 43,200 \text{ ft-lb}$$

25. Let W_1 be the work to lift V to the surface and W_2 be the work to lift V from the surface to 15 feet above the surface. The volume displaced by the buoy y feet above its original position is

$$\frac{1}{3}\pi\left(a - \frac{a}{h}y\right)^2(h - y) = \frac{1}{3}\pi a^2 h\left(1 - \frac{y}{h}\right)^3.$$

The weight displaced is $\frac{\delta}{3}\pi a^2 h\left(1 - \frac{y}{h}\right)^3$.

Note by Archimede's Principle $m = \frac{\delta}{3}\pi a^2 h$ or $a^2 h = \frac{3m}{\delta\pi}$, so the displaced weight is

$$m\left(1 - \frac{y}{h}\right)^3.$$

6.6 Concepts Review

1. right; $\dfrac{4\cdot 1 + 6\cdot 3}{4 + 6} = 2.2$

3. $1; 3$

Problem Set 6.6

1. $\bar{x} = \dfrac{2\cdot 5 + (-2)\cdot 7 + 1\cdot 9}{5 + 7 + 9} = \dfrac{5}{21}$

$$\Delta W_1 \approx \left(m - m\left(1 - \frac{y}{h}\right)^3\right)\Delta y = m\left(1 - \left(1 - \frac{y}{h}\right)^3\right)\Delta y$$

$$W_1 = m\int_0^h\left(1 - \left(1 - \frac{y}{h}\right)^3\right)dy$$

$$= m\left[y + \frac{h}{4}\left(1 - \frac{y}{h}\right)^4\right]_0^h = \frac{3mh}{4}$$

$$W_2 = m\cdot 15 = 15m$$

$$W = W_1 + W_2 = \frac{3mh}{4} + 15m$$

27. First calculate the work W_1 needed to lift the contents of the bottom tank to 10 feet.

$$\Delta W_1 \approx \delta 40\Delta y(10 - y)$$

$$W_1 = \int_0^4 \delta 40(10 - y)\,dy$$

$$= (62.4)(40)\left[-\frac{1}{2}(10 - y)^2\right]_0^4$$

$$= (62.4)(40)(-18 + 50)$$

$$= 79,872 \text{ ft-lb}$$

Next calculate the work W_2 needed to fill the top tank. Let y be the distance from the bottom of the top tank.

$$\Delta W_2 \approx \delta(36\pi)\Delta y\, y$$

Solve for the height of the top tank:

$$36\pi h = 160;\ h = \frac{160}{36\pi} = \frac{40}{9\pi}$$

$$W_2 = \int_0^{40/9\pi} \delta 36\pi y\, dy$$

$$= (62.4)(36\pi)\left[\frac{1}{2}y^2\right]_0^{40/9\pi}$$

$$= (62.4)(36\pi)\left(\frac{800}{81\pi^2}\right)$$

$$\approx 7062 \text{ ft-lbs}$$

$$W = W_1 + W_2 \approx 86,934 \text{ ft-lbs}$$

3. $\bar{x} = \dfrac{\int_0^7 x\sqrt{x}\,dx}{\int_0^7 \sqrt{x}\,dx} = \dfrac{\left[\frac{2}{5}x^{5/2}\right]_0^7}{\left[\frac{2}{3}x^{3/2}\right]_0^7} = \dfrac{\frac{2}{5}\left(49\sqrt{7}\right)}{\frac{2}{3}\left(7\sqrt{7}\right)} = \dfrac{21}{5}$

5. $M_y = 1\cdot 2 + 7\cdot 3 + (-2)\cdot 4 + (-1)\cdot 6 + 4\cdot 2 = 17$

$M_x = 1\cdot 2 + 1\cdot 3 + (-5)\cdot 4 + 0\cdot 6 + 6\cdot 2 = -3$

$m = 2 + 3 + 4 + 6 + 2 = 17$

$\bar{x} = \dfrac{M_y}{m} = 1,\ \bar{y} = \dfrac{M_x}{m} = -\dfrac{3}{17}$

7. Consider two regions R_1 and R_2 such that R_1 is bounded by $f(x)$ and the x-axis, and R_2 is

bounded by $g(x)$ and the x-axis. Let R_3 be the region formed by $R_1 - R_2$. Make a regular partition of the homogeneous region R_3 such that each sub-region is of width , Δx and let x be the distance from the y-axis to the center of mass of a sub-region. The heights of R_1 and R_2 at x are approximately $f(x)$ and $g(x)$ respectively. The mass of R_3 is approximately

$$\Delta m = \Delta m_1 - \Delta m_2$$
$$\approx \delta f(x)\Delta x - \delta g(x)\Delta x$$
$$= \delta[f(x) - g(x)]\Delta x$$

where δ is the density. The moments for R_3 are approximately

$$M_x = M_x(R_1) - M_x(R_2)$$
$$\approx \frac{\delta}{2}[f(x)]^2 \Delta x - \frac{\delta}{2}[g(x)]^2 \Delta x$$
$$= \frac{\delta}{2}\Big[(f(x))^2 - (g(x))^2\Big]\Delta x$$
$$M_y = M_y(R_1) - M_y(R_2)$$
$$\approx x\delta f(x)\Delta x - x\delta g(x)\Delta x$$
$$= x\delta[f(x) - g(x)]\Delta x$$

Taking the limit of the regular partition as $\Delta x \to 0$ yields the resulting integrals in Figure 10.

9.

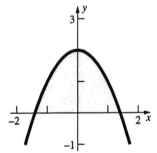

$\bar{x} = 0$ (by symmetry)

$$\bar{y} = \frac{\frac{1}{2}\int_{-\sqrt{2}}^{\sqrt{2}} (2-x^2)^2 \, dx}{\int_{-\sqrt{2}}^{\sqrt{2}} (2-x^2)\, dx}$$

$$= \frac{\frac{1}{2}\int_{-\sqrt{2}}^{\sqrt{2}} (4-4x^2+x^4)\, dx}{\left[2x - \frac{1}{3}x^3\right]_{-\sqrt{2}}^{\sqrt{2}}}$$

$$= \frac{\frac{1}{2}\left[4x - \frac{4}{3}x^3 + \frac{1}{5}x^5\right]_{-\sqrt{2}}^{\sqrt{2}}}{\frac{8\sqrt{2}}{3}} = \frac{\frac{32\sqrt{2}}{15}}{\frac{8\sqrt{2}}{3}} = \frac{4}{5}$$

11.

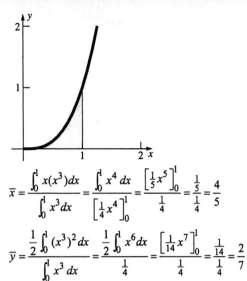

$$\bar{x} = \frac{\int_0^1 x(x^3)\, dx}{\int_0^1 x^3\, dx} = \frac{\int_0^1 x^4\, dx}{\left[\frac{1}{4}x^4\right]_0^1} = \frac{\left[\frac{1}{5}x^5\right]_0^1}{\frac{1}{4}} = \frac{\frac{1}{5}}{\frac{1}{4}} = \frac{4}{5}$$

$$\bar{y} = \frac{\frac{1}{2}\int_0^1 (x^3)^2\, dx}{\int_0^1 x^3\, dx} = \frac{\frac{1}{2}\int_0^1 x^6\, dx}{\frac{1}{4}} = \frac{\left[\frac{1}{14}x^7\right]_0^1}{\frac{1}{4}} = \frac{\frac{1}{14}}{\frac{1}{4}} = \frac{2}{7}$$

13.

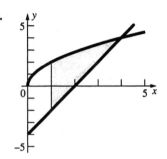

To find the intersection point, solve
$$2x - 4 = 2\sqrt{x}.$$
$$x - 2 = \sqrt{x}$$
$$x^2 - 4x + 4 = x$$
$$x^2 - 5x + 4 = 0$$
$$(x-4)(x-1) = 0$$
$$x = 4 \; (x = 1 \text{ is extraneous.})$$

$$\bar{x} = \frac{\int_1^4 x\left[2\sqrt{x} - (2x-4)\right]dx}{\int_1^4 \left[2\sqrt{x} - (2x-4)\right]dx}$$

$$= \frac{2\int_1^4 (x^{3/2} - x^2 + 2x)dx}{2\int_1^4 (x^{1/2} - x + 2)dx}$$

$$= \frac{2\left[\frac{2}{5}x^{5/2} - \frac{1}{3}x^3 + x^2\right]_1^4}{2\left[\frac{2}{3}x^{3/2} - \frac{1}{2}x^2 + 2x\right]_1^4} = \frac{\frac{64}{5}}{\frac{19}{3}} = \frac{192}{95}$$

$$\bar{y} = \frac{\frac{1}{2}\int_1^4 \left[\left(2\sqrt{x}\right)^2 - (2x-4)^2\right]dx}{\int_1^4 \left[2\sqrt{x} - (2x-4)\right]dx}$$

$$= \frac{2\int_1^4 (-x^2 + 5x - 4)dx}{\frac{19}{3}}$$

$$= \frac{2\left[-\frac{1}{3}x^3 + \frac{5}{2}x^2 - 4x\right]_1^4}{\frac{19}{3}} = \frac{9}{\frac{19}{3}} = \frac{27}{19}$$

15.

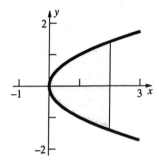

To find the intersection points, solve $y^2 = 2$.

$$y = \pm\sqrt{2}$$

$$\bar{x} = \frac{\frac{1}{2}\int_{-\sqrt{2}}^{\sqrt{2}}\left[2^2 - (y^2)^2\right]dy}{\int_{-\sqrt{2}}^{\sqrt{2}}(2 - y^2)dy} = \frac{\frac{1}{2}\int_{-\sqrt{2}}^{\sqrt{2}}(4 - y^4)dy}{\left[2y - \frac{1}{3}y^3\right]_{-\sqrt{2}}^{\sqrt{2}}}$$

$$= \frac{\frac{1}{2}\left[4y - \frac{1}{5}y^5\right]_{-\sqrt{2}}^{\sqrt{2}}}{\frac{8\sqrt{2}}{3}} = \frac{\frac{16\sqrt{2}}{5}}{\frac{8\sqrt{2}}{3}} = \frac{6}{5}$$

$\bar{y} = 0$ (by symmetry)

17. We let δ be the density of the regions and A_i be the area of region i.

Region R_1:

$$m(R_1) = \delta A_1 = \delta(1/2)(1)(1) = \frac{1}{2}\delta$$

$$\bar{x}_1 = \frac{\int_0^1 x(x)dx}{\int_0^1 xdx} = \frac{\frac{1}{3}x^3\Big|_0^1}{\frac{1}{2}x^2\Big|_0^1} = \frac{\frac{1}{3}}{\frac{1}{2}} = \frac{2}{3}$$

Since R_1 is symmetric about the line $y = 1 - x$, the centroid must lie on this line. Therefore,

$$\bar{y}_1 = 1 - \bar{x}_1 = 1 - \frac{2}{3} = \frac{1}{3}; \text{ and we have}$$

$$M_y(R_1) = \bar{x}_2 \cdot m(R_1) = \frac{1}{3}\delta$$

$$M_x(R_1) = \bar{y}_2 \cdot m(R_1) = \frac{1}{6}\delta$$

Region R_2:
$$m(R_2) = \delta A_2 = \delta(2)(1) = 2\delta$$
By symmetry we get

$$\bar{x}_2 = 2 \quad \text{and} \quad \bar{y}_2 = \frac{1}{2}.$$

Thus,
$$M_y(R_2) = \bar{x}_2 \cdot m(R_2) = 4\delta$$
$$M_x(R_2) = \bar{y}_2 \cdot m(R_2) = \delta$$

19. $m(R_1) = \delta\int_a^b (g(x) - f(x))dx$

$$m(R_2) = \delta\int_b^c (g(x) - f(x))dx$$

$$M_x(R_1) = \frac{\delta}{2}\int_a^b ((g(x))^2 - (f(x))^2)dx$$

$$M_x(R_2) = \frac{\delta}{2}\int_b^c ((g(x))^2 - (f(x))^2)dx$$

$$M_y(R_1) = \delta\int_a^b x(g(x) - f(x))dx$$

$$M_y(R_2) = \delta\int_b^c x(g(x) - f(x))dx$$

Now,

$$m(R_3) = \delta\int_a^c (g(x) - f(x))dx$$

$$= \delta\int_a^b (g(x) - f(x))dx + \delta\int_b^c (g(x) - f(x))dx$$

$$= m(R_1) + m(R_2)$$

$$M_x(R_3) = \frac{\delta}{2}\int_a^c ((g(x))^2 - (f(x))^2)dx$$

$$= \frac{\delta}{2}\int_a^b ((g(x))^2 - (f(x))^2)dx$$

$$+ \frac{\delta}{2}\int_b^c ((g(x))^2 - (f(x))^2)dx$$

$$= M_x(R_1) + M_x(R_2)$$

$$M_y(R_3) = \delta\int_a^c x(g(x) - f(x))dx$$

$$= \delta\int_a^b x(g(x) - f(x))dx$$

$$+ \delta\int_b^c x(g(x) - f(x))dx$$

$$= M_y(R_1) + M_y(R_2)$$

21. Let region 1 be the region bounded by $x = -2$, $x = 2$, $y = 0$, and $y = 1$, so $m_1 = 4 \cdot 1 = 4$.

By symmetry, $\bar{x}_1 = 0$ and $\bar{y}_1 = \frac{1}{2}$. Therefore

$M_{1y} = \bar{x}_1 m_1 = 0$ and $M_{1x} = \bar{y}_1 m_1 = 2$.

Let region 2 be the region bounded by $x = -2$, $x = 1$, $y = -1$, and $y = 0$, so $m_2 = 3 \cdot 1 = 3$.

By symmetry, $\bar{x}_2 = -\frac{1}{2}$ and $\bar{y}_2 = -\frac{1}{2}$. Therefore

$$M_{2y} = \bar{x}_2 m_2 = -\frac{3}{2} \text{ and } M_{2x} = \bar{y}_2 m_2 = -\frac{3}{2}.$$

$$\bar{x} = \frac{M_{1y} + M_{2y}}{m_1 + m_2} = \frac{-\frac{3}{2}}{7} = -\frac{3}{14}$$

$$\bar{y} = \frac{M_{1x} + M_{2x}}{m_1 + m_2} = \frac{\frac{1}{2}}{7} = \frac{1}{14}$$

23. Let region 1 be the region bounded by $x = -2$, $x = 2$, $y = 2$, and $y = 4$, so $m_1 = 4 \cdot 2 = 8$. By symmetry, $\bar{x}_1 = 0$ and $\bar{y}_1 = 3$. Therefore, $M_{1y} = \bar{x}_1 m_1 = 0$ and $M_{1x} = \bar{y}_1 m_1 = 24$. Let region 2 be the region bounded by $x = -1$, $x = 2$, $y = 0$, and $y = 2$, so $m_2 = 3 \cdot 2 = 6$. By symmetry, $\bar{x}_2 = \frac{1}{2}$ and $\bar{y}_2 = 1$. Therefore, $M_{2y} = \bar{x}_2 m_2 = 3$ and $M_{2x} = \bar{y}_2 m_2 = 6$. Let region 3 be the region bounded by $x = 2$, $x = 4$, $y = 0$, and $y = 1$, so $m_3 = 2 \cdot 1 = 2$. By symmetry, $\bar{x}_3 = 3$ and $\bar{y}_2 = \frac{1}{2}$. Therefore, $M_{3y} = \bar{x}_3 m_3 = 6$ and $M_{3x} = \bar{y}_3 m_3 = 1$.

$$\bar{x} = \frac{M_{1y} + M_{2y} + M_{3y}}{m_1 + m_2 + m_3} = \frac{9}{16}$$

$$\bar{y} = \frac{M_{1x} + M_{2x} + M_{3x}}{m_1 + m_2 + m_3} = \frac{31}{16}$$

25. $A = \int_0^1 x^3 \, dx = \left[\frac{1}{4}x^4\right]_0^1 = \frac{1}{4}$

From Problem 11, $\bar{x} = \frac{4}{5}$.

$$V = A(2\pi\bar{x}) = \frac{1}{4}\left(2\pi \cdot \frac{4}{5}\right) = \frac{2\pi}{5}$$

Using cylindrical shells:

$$V = 2\pi \int_0^1 x \cdot x^3 \, dx = 2\pi \int_0^1 x^4 \, dx = 2\pi\left[\frac{1}{5}x^5\right]_0^1 = \frac{2\pi}{5}$$

27. The volume of a sphere of radius a is $\frac{4}{3}\pi a^3$. If the semicircle $y = \sqrt{a^2 - x^2}$ is revolved about the x-axis the result is a sphere of radius a. The centroid of the region travels a distance of $2\pi\bar{y}$.

The area of the region is $\frac{1}{2}\pi a^2$. Pappus's Theorem says that

$$(2\pi\bar{y})\left(\frac{1}{2}\pi a^2\right) = \pi^2 a^2 \bar{y} = \frac{4}{3}\pi a^3.$$

$$\bar{y} = \frac{4a}{3\pi}, \quad \bar{x} = 0 \text{ (by symmetry)}$$

29. a. $\Delta V \approx 2\pi(K - y)w(y)\Delta y$

$$V = 2\pi \int_c^d (K - y)w(y)\,dy$$

b. $\Delta m \approx w(y)\Delta y$, so $m = \int_c^d w(y)\,dy = A$.

$\Delta M_x \approx yw(y)\Delta y$, so $M_x = \int_c^d yw(y)\,dy$.

$$\bar{y} = \frac{\int_c^d yw(y)\,dy}{A}$$

The distance traveled by the centroid is $2\pi(K - \bar{y})$.

$$2\pi(K - \bar{y})A = 2\pi(KA - M_x)$$
$$= 2\pi\left(\int_c^d Kw(y)\,dy - \int_c^d yw(y)\,dy\right)$$
$$= 2\pi \int_c^d (K - y)w(y)\,dy$$

Therefore, $V = 2\pi(K - \bar{y})A$.

31. a. The area of a regular polygon P of $2n$ sides is $2r^2 n \sin\frac{\pi}{2n}\cos\frac{\pi}{2n}$. (To find this consider the isosceles triangles with one vertex at the center of the polygon and the other vertices on adjacent corners of the polygon. Each such triangle has base of length $2r\sin\frac{\pi}{2n}$ and height $r\cos\frac{\pi}{2n}$.) Since P is a regular polygon the centroid is at its center. The distance from the centroid to any side is $r\cos\frac{\pi}{2n}$, so the centroid travels a distance of $2\pi r\cos\frac{\pi}{2n}$.

Thus, by Pappus's Theorem, the volume of the resulting solid is

$$\left(2\pi r\cos\frac{\pi}{2n}\right)\left(2r^2 n\sin\frac{\pi}{2n}\cos\frac{\pi}{2n}\right)$$
$$= 4\pi r^3 n\sin\frac{\pi}{2n}\cos^2\frac{\pi}{2n}.$$

b. $\lim_{n \to \infty} 4\pi r^3 n\sin\frac{\pi}{2n}\cos^2\frac{\pi}{2n}$

$$\lim_{n \to \infty} \frac{\sin\frac{\pi}{2n}}{\frac{\pi}{2n}} 2\pi^2 r^3 \cos^2\frac{\pi}{2n} = 2\pi^2 r^3$$

As $n \to \infty$, the regular polygon approaches a circle. Using Pappus's Theorem on the circle of area πr^2 whose centroid (= center) travels a distance of $2\pi r$, the volume of the solid is $(\pi r^2)(2\pi r) = 2\pi^2 r^3$ which agrees with the results from the polygon.

33. Consider the region $S - R$.

$$\bar{y}_{S-R} = \dfrac{\dfrac{1}{2}\displaystyle\int_0^1 \left[g^2(x) - f^2(x) \right] dx}{S - R} \geq \bar{y}_R$$

$$= \dfrac{\dfrac{1}{2}\displaystyle\int_0^1 f^2(x)\,dx}{R}$$

$$\frac{1}{2} R \int_0^1 \left[g^2(x) - f^2(x) \right] dx \geq \frac{1}{2}(S-R) \int_0^1 f^2(x)\,dx$$

$$\frac{1}{2} R \int_0^1 \left[g^2(x) - f^2(x) \right] dx + \frac{1}{2} R \int_0^1 f^2(x)\,dx$$

$$\geq \frac{1}{2}(S-R) \int_0^1 f^2(x)\,dx + \frac{1}{2} R \int_0^1 f^2(x)\,dx$$

$$\frac{1}{2} R \int_0^1 g^2(x)\,dx \geq \frac{1}{2} S \int_0^1 f^2(x)\,dx$$

$$\dfrac{\dfrac{1}{2}\displaystyle\int_0^1 g^2(x)\,dx}{S} \geq \dfrac{\dfrac{1}{2}\displaystyle\int_0^1 f^2(x)\,dx}{R}$$

$$\bar{y}_S \geq \bar{y}_R$$

35. First we place the lamina so that the origin is centered inside the hole. We then recompute the centroid of Problem 34 (in this position) as

$$\bar{x} \approx \dfrac{\displaystyle\sum_{i=1}^8 x_i h_i}{\displaystyle\sum_{i=1}^8 h_i} = \dfrac{(-25)(6.5) + (-15)(8) + \cdots + (5)(10) + (10)(8)}{6.5 + 8 + \cdots + 10 + 8}$$

$$= \dfrac{-480}{72.5} \approx -6.62$$

$$\bar{y} \approx \dfrac{\dfrac{1}{2}\displaystyle\sum_{i=1}^8 ((h_i - 4)^2 - (-4)^2)}{\displaystyle\sum_{i=1}^8 h_i}$$

$$= \dfrac{(1/2)((2.5^2 - (-4)^2) + \cdots + (4^2 - (-4)^2))}{6.5 + 8 + \cdots + 10 + 8}$$

$$= \dfrac{45.875}{72.5} \approx 0.633$$

6.7 Chapter Review

Concepts Test

1. False: $\displaystyle\int_0^\pi \cos x\,dx = 0$ because half of the area lies above the x-axis and half below the x-axis.

3. False: The statement would be true if either $f(x) \geq g(x)$ or $g(x) \geq f(x)$ for $a \leq x \leq b$. Consider Problem 1 with $f(x) = \cos x$ and $g(x) = 0$.

A quick computation will show that these values agree with those in Problem 34 (using a different reference point).

Now consider the whole lamina as R_3, the circular hole as R_2, and the remaining lamina as R_1. We can find the centroid of R_1 by noting that

$$M_x(R_1) = M_x(R_3) - M_x(R_2)$$

and similarly for $M_y(R_1)$.

From symmetry, we know that the centroid of a circle is at the center. Therefore, both $M_x(R_2)$ and $M_y(R_2)$ must be zero in our case.

This leads to the following equations

$$\bar{x} = \dfrac{M_y(R_3) - M_y(R_2)}{m(R_3) - m(R_2)}$$

$$= \dfrac{\delta\Delta x(-480)}{\delta\Delta x(72.5) - \delta\pi(2.5)^2}$$

$$= \dfrac{-2400}{342.87} \approx -7$$

$$\bar{y} = \dfrac{M_x(R_3) - M_x(R_2)}{m(R_3) - m(R_2)}$$

$$= \dfrac{\delta\Delta x(45.875)}{\delta\Delta x(72.5) - \delta\pi(2.5)^2}$$

$$= \dfrac{229.375}{342.87} \approx 0.669$$

5. True: Since the cross sections in all planes parallel to the bases have the same area, the integrals used to compute the volumes will be equal.

7. False: Using the method of shells,
$$V = 2\pi \int_0^1 x(-x^2 + x)\,dx .$$ To use the method of washers we need to solve $y = -x^2 + x$ for x in terms of y.

9. False: Consider the curve given by
$$x = \dfrac{\cos t}{t}, \quad y = \dfrac{\sin t}{t}, 2 \leq t < \infty .$$

11. False: If the cone-shaped tank is placed with the point downward, then the amount of water that needs to be pumped from near the bottom of the tank is much less than the amount that needs to be pumped from near the bottom of the cylindrical tank.

13. True: This is the definition of the center of mass.

15. True: By symmetry, the centroid is on the line $x = \dfrac{\pi}{2}$, so the centroid travels a distance of $2\pi\left(\dfrac{\pi}{2}\right) = \pi^2$.

17. True: Since the density is proportional to the square of the distance from the midpoint, equal masses are on either side of the midpoint.

Sample Test Problems

1. $A = \int_0^1 (x - x^2)\,dx = \left[\dfrac{1}{2}x^2 - \dfrac{1}{3}x^3\right]_0^1 = \dfrac{1}{6}$

3. $V = 2\pi \int_0^1 x(x - x^2)\,dx = 2\pi \int_0^1 (x^2 - x^3)\,dx$

$= 2\pi\left[\dfrac{1}{3}x^3 - \dfrac{1}{4}x^4\right]_0^1 = \dfrac{\pi}{6}$

5. $V = 2\pi \int_0^1 (3 - x)(x - x^2)\,dx$

$= 2\pi \int_0^1 (x^3 - 4x^2 + 3x)\,dx$

$= 2\pi\left[\dfrac{1}{4}x^4 - \dfrac{4}{3}x^3 + \dfrac{3}{2}x^2\right]_0^1 = \dfrac{5\pi}{6}$

7. From Problem 1, $A = \dfrac{1}{6}$.

From Problem 6, $\bar{x} = \dfrac{1}{2}$ and $\bar{y} = \dfrac{1}{10}$.

$V(S_1) = 2\pi\left(\dfrac{1}{10}\right)\left(\dfrac{1}{6}\right) = \dfrac{\pi}{30}$

$V(S_2) = 2\pi\left(\dfrac{1}{2}\right)\left(\dfrac{1}{6}\right) = \dfrac{\pi}{6}$

$V(S_3) = 2\pi\left(\dfrac{1}{10} + 2\right)\left(\dfrac{1}{6}\right) = \dfrac{7\pi}{10}$

$V(S_4) = 2\pi\left(3 - \dfrac{1}{2}\right)\left(\dfrac{1}{6}\right) = \dfrac{5\pi}{6}$

9. $W = \int_0^6 (62.4)(5^2)\pi(10 - y)\,dy$

$= 1560\pi \int_0^6 (10 - y)\,dy$

$= 1560\pi\left[10y - \dfrac{1}{2}y^2\right]_0^6 = 65{,}520\pi \approx 205{,}837$ ft-lb

11. a. To find the intersection points, solve
$4x = x^2$.

$x^2 - 4x = 0$
$x(x - 4) = 0$
$x = 0,\ 4$

$A = \int_0^4 (4x - x^2)\,dx = \left[2x^2 - \dfrac{1}{3}x^3\right]_0^4$

$= \left(32 - \dfrac{64}{3}\right) = \dfrac{32}{3}$

b. To find the intersection points, solve
$\dfrac{y}{4} = \sqrt{y}$.

$\dfrac{y^2}{16} = y$

$y^2 - 16y = 0$
$y(y - 16) = 0$
$y = 0,\ 16$

$A = \int_0^{16} \left(\sqrt{y} - \dfrac{y}{4}\right)dy = \left[\dfrac{2}{3}y^{3/2} - \dfrac{1}{8}y^2\right]_0^{16}$

$= \left(\dfrac{128}{3} - 32\right) = \dfrac{32}{3}$

13. $V = \pi \int_0^4 \left[(4x)^2 - (x^2)^2\right]dx$

$= \pi \int_0^4 (16x^2 - x^4)\,dx$

$= \pi\left[\dfrac{16}{3}x^3 - \dfrac{1}{5}x^5\right]_0^4 = \dfrac{2048\pi}{15}$

Using Pappus's Theorem:

From Problem 11, $A = \dfrac{32}{3}$.

From Problem 12, $\bar{y} = \dfrac{32}{5}$.

$V = 2\pi\bar{y}\cdot A = 2\pi\left(\dfrac{32}{5}\right)\left(\dfrac{32}{3}\right) = \dfrac{2048\pi}{15}$

15. $\dfrac{dy}{dx} = x^2 - \dfrac{1}{4x^2}$

$L = \int_1^3 \sqrt{1 + \left(x^2 - \dfrac{1}{4x^2}\right)^2}\,dx$

$$= \int_1^3 \sqrt{x^4 + \frac{1}{2} + \frac{1}{16x^4}}\, dx = \int_1^3 \left(x^2 + \frac{1}{4x^2}\right) dx$$

$$= \left[\frac{1}{3}x^3 - \frac{1}{4x}\right]_1^3 = \left(9 - \frac{1}{12}\right) - \left(\frac{1}{3} - \frac{1}{4}\right) = \frac{53}{6}$$

17. $V = \int_{-3}^3 \left(\sqrt{9 - x^2}\right)^2 dx = \int_{-3}^3 (9 - x^2)\, dx$

$$= \left[9x - \frac{1}{3}x^3\right]_{-3}^3 = (27 - 9) - (-27 + 9) = 36$$

19. $V = \pi \int_a^b \left[f^2(x) - g^2(x)\right] dx$

6.8 Additional Problem Set

1. a. feet/second
$F(T)$ is the change in velocity from 6 seconds to T seconds.

b. foot-pounds
$F(s)$ is the change in work from 3 feet to s feet.

c. inches
$F(r)$ is the center of mass of an object r inches long whose mass at x is $f(x)$.

3. $F(t) = \int_0^t f(u)\, du$

a. $\displaystyle\lim_{t\to\infty} F(t) = \int_0^3 A\, du = \left[Au\right]_0^3 = 3A = 1$

$A = \frac{1}{3}$

b. $\displaystyle\lim_{t\to\infty} F(t) = \int_0^7 A(u^2 - 7u)\, du$

$$= A\left[\frac{1}{3}u^3 - \frac{7}{2}u^2\right]_0^7 = \frac{343}{6}A = 1$$

$A = \frac{6}{343}$

c. $\displaystyle\lim_{t\to\infty} F(t) = \int_0^3 Au^2\, du + \int_3^9 A(9 - u)\, du$

$$= A\left[\frac{1}{3}u^3\right]_0^3 + A\left[9u - \frac{1}{2}u^2\right]_3^9$$

$$= 9A + 18A = 27A = 1$$

$A = \frac{1}{27}$

21. $M_y = \delta \int_a^b x\left[f(x) - g(x)\right] dx$

$$M_x = \frac{\delta}{2} \int_a^b \left[f^2(x) - g^2(x)\right] dx$$

23. $A_1 = 2\pi \int_a^b f(x)\sqrt{1 + \left[f'(x)\right]^2}\, dx$

$$A_2 = 2\pi \int_a^b g(x)\sqrt{1 + \left[g'(x)\right]^2}\, dx$$

$$A_3 = \pi\left[f^2(a) - g^2(a)\right]$$

$$A_4 = \pi\left[f^2(b) - g^2(b)\right]$$

Total surface area $= A_1 + A_2 + A_3 + A_4$.

5. a. We can consider the expression

$$\int_0^B t f(t)\, dt$$

as a moment if we consider that the total mass of the region is one unit. The total area under the curve must equal one as well, and so we can consider the region to have a density of $\delta = 1$. A moment is defined to be the product of a region's mass times it's directed distance from the rotation point. *f(t)* represents the height of small region and *dt* represents a very small change in t. *f(t)dt* can be considered as the area of a very small sliver of the region. Since the density is equal to 1, the quantity *f(t)dt* also represents the mass of the sliver. Multiplying by *t*, the direct distance from the y-axis, we obtain the moment for that sliver about the y-axis. The integral represents an infinite sum over all the slivers that make up the region, and represents the moment of the region about the y-axis.

b. $E = \int_0^4 \frac{15}{512} t\left[t^2(4 - t)^2\right] dt$

$$= \frac{15}{512} \int_0^4 (16t^3 - 8t^4 + t^5)\, dt$$

$$= \frac{15}{512}\left[\frac{16}{4}t^4 - \frac{8}{5}t^5 + \frac{1}{6}t^6\right]_0^4$$

$$= \frac{15}{512}\left[1024 - \frac{8192}{5} + \frac{4096}{6}\right] = 2$$

7. $F = (62.4)(10)(0.0004) = 0.2496$ lb
$p = (62.4)(10) = 624$ lb/ft^2

9. a. At height h, a strip Δh has
$F \approx 62.4(10 - h)\Delta h \cdot W(h)$.

Thus $F = 62.4 \int_0^{10} (10 - h)W(h)\, dh$.

b. $W(h) = \dfrac{2}{\sqrt{3}}h$ follows from the geometry of an equilateral triangle. (Similar triangles, 30-60-90 triangle, etc.)

c. $F = 62.4\displaystyle\int_0^{10}(10-h)(2h/\sqrt{3})dh$

$= 62.4\displaystyle\int_0^{10}(\dfrac{20}{\sqrt{3}}h - \dfrac{2}{\sqrt{3}}h^2)dh$

$= 62.4\left[\dfrac{10}{\sqrt{3}}h^2 - \dfrac{2}{3\sqrt{3}}h^3\right]_0^{10} \approx 12009$

11. Let h measure the distance from the top of the plate, where $0 \le h \le 3\sqrt{3}$. The depth at that distance is $h + 3$. The width of the plate at h is given by $W(h) = -\dfrac{2}{\sqrt{3}}(h - 3\sqrt{3})$. At distance h from the top a strip Δh has

$F \approx 62.4(h+3)\Delta h\left[-\dfrac{2}{\sqrt{3}}(h - 3\sqrt{3})\right].$

$F = -\dfrac{124.8}{\sqrt{3}}\displaystyle\int_0^{3\sqrt{3}}(h+3)(h-3\sqrt{3})dh$

$= -\dfrac{124.8}{\sqrt{3}}\displaystyle\int_0^{3\sqrt{3}}[h^2 + 3(1-\sqrt{3})h - 9\sqrt{3}]dh$

$= -\dfrac{124.8}{\sqrt{3}}\left[\dfrac{1}{3}h^3 + \dfrac{3(1-\sqrt{3})}{2}h^2 - 9\sqrt{3}h\right]_0^{3\sqrt{3}}$

$= -\dfrac{124.8}{\sqrt{3}}\left[-\dfrac{27(3+\sqrt{3})}{2}\right]$

$= 1684.8(1+\sqrt{3}) \approx 4602.96$ lb

13. Let y measure the distance from the bottom of the window, where $0 \le y \le 2$. Let h be the height of the tank. The depth at y is $h - y$. The width of the window at y is y. At y, the strip Δy has

$F \approx 62.4(h-y)\Delta y \cdot y.$

$F = 62.4\displaystyle\int_0^2 y(h-y)dy$

$= 62.4\displaystyle\int_0^2(hy - y^2)dy$

$= 62.4\left[\dfrac{h}{2}y^2 - \dfrac{1}{3}y^3\right]_0^2$

$= 62.4\left(2h - \dfrac{8}{3}\right)$

$62.4\left(2h - \dfrac{8}{3}\right) = 400$, so $h \approx 4.54$ ft.

15. a. The additional force on the car is the component of the force due to gravity along the incline.

$F = 100 \cdot 32\sin(\tan^{-1}0.05) \approx 159.8$ lb.

$P \approx 159.8 \cdot 66 = 10{,}546.8$ ft-lb/s

$\dfrac{10{,}546.8 \text{ ft-lb/s}}{550 \text{ (ft-lb/s)/hp}} \approx 19$ hp additional is required.

b. Since acceleration is not assumed to be uniform, we will calculate the minimum horsepower required, so we will assume that the additional power is constant while accelerating.

$P = 100a(t)v(t)$. Since $a(t) = \dfrac{dv(t)}{dt}$,

$P\,dt = 100v(t)\,dv(t).$

Integrate both sides to get $Pt = 50v(t)^2 + C$.

60 mi/h = 88 ft/s

At $t = 0$, $v = 66$ and at $t = 4$, $v = 88$.

Therefore, $C = -217{,}800$ and

$P = \dfrac{50(88)^2 - 217{,}800}{4} = 42{,}350$ ft-lb/s.

$\dfrac{42{,}350 \text{ ft-lb/s}}{550 \text{ (ft-lb/s)/hp}} = 77$ hp additional is required. Note that since the car is already on the incline, we do not need to consider the power needed to overcome the force due to gravity.

If we assume uniform acceleration, then

$a = \dfrac{88-66}{4} = 5.5$ ft/s^2, so the additional force is $F = 100 \cdot 5.5 = 550$ lb. The most power is needed when $v = 88$ ft/s, so $P = 550 \cdot 88 = 48{,}400$ ft-lb/s.

$\dfrac{48{,}400 \text{ ft-lb/s}}{550 \text{ (ft-ls/2)/hp}} = 88$ hp additional is required.

c. The force due to gravity along the incline is $F = 100 \cdot 32\sin(\tan^{-1}0.05) \approx 159.8$ lb. The distance traveled is (66 ft/s)(180 s) = 11,880 ft. Thus the work is $W \approx 159.8 \cdot 11{,}800 = 1{,}898{,}424$ ft-lb

Transcendental Functions

7.1 Concepts Review

1. $\int_1^x \frac{1}{t}\,dt;\ (0,\infty);\ (-\infty,\infty)$

3. $\frac{1}{x};\ \ln|x| + C$

Problem Set 7.1

1. **a.** $\ln 6 = \ln(2 \cdot 3) = \ln 2 + \ln 3$
   $= 0.693 + 1.099 = 1.792$

   b. $\ln 1.5 = \ln\left(\frac{3}{2}\right) = \ln 3 - \ln 2 = 1.099 - 0.693$
   $= 0.406$

   c. $\ln 81 = \ln 3^4 = 4\ln 3 = 4(1.099) = 4.396$

   d. $\ln\sqrt{2} = \ln 2^{1/2} = \frac{1}{2}\ln 2 = \frac{1}{2}(0.693) = 0.3465$

   e. $\ln\left(\frac{1}{36}\right) = -\ln 36 = -\ln(2^2 \cdot 3^2)$
   $= -2\ln 2 - 2\ln 3 = -2(0.693) - 2(1.099)$
   $= -3.584$

   f. $\ln 48 = \ln(2^4 \cdot 3) = 4\ln 2 + \ln 3$
   $= 4(0.693) + 1.099 = 3.871$

3. $D_x \ln(x^2 + 3x + \pi)$
   $= \frac{1}{x^2 + 3x + \pi} \cdot D_x(x^2 + 3x + \pi)$
   $= \frac{2x + 3}{x^2 + 3x + \pi}$

5. $D_x \ln(x-4)^3 = D_x 3\ln(x-4)$
   $= 3 \cdot \frac{1}{x-4} D_x(x-4) = \frac{3}{x-4}$

7. $\frac{dy}{dx} = 3 \cdot \frac{1}{x} = \frac{3}{x}$

9. $z = x^2 \ln x^2 + (\ln x)^3\ = x^2 \cdot 2\ln x + (\ln x)^3$
   $\frac{dz}{dx} = x^2 \cdot \frac{2}{x} + 2x \cdot 2\ln x + 3(\ln x)^2 \cdot \frac{1}{x}$
   $= 2x + 4x\ln x + \frac{3}{x}(\ln x)^2$

11. $g'(x) = \frac{1}{x + \sqrt{x^2+1}}\left[1 + \frac{1}{2}(x^2+1)^{-1/2} \cdot 2x\right]$
    $= \frac{1}{\sqrt{x^2+1}}$

13. $f(x) = \ln\sqrt[3]{x} = \frac{1}{3}\ln x$
    $f'(x) = \frac{1}{3} \cdot \frac{1}{x} = \frac{1}{3x}$
    $f'(81) = \frac{1}{3 \cdot 81} = \frac{1}{243}$

15. Let $u = 2x + 1$ so $du = 2\,dx$.
    $\int \frac{1}{2x+1}\,dx = \frac{1}{2}\int\frac{1}{u}\,du$
    $= \frac{1}{2}\ln|u| + C = \frac{1}{2}\ln|2x+1| + C$

17. Let $u = 3v^2 + 9v$ so $du = 6v + 9$.
    $\int \frac{6v+9}{3v^2+9v}\,dv = \int\frac{1}{u}\,du = \ln|u| + C$
    $= \ln\left|3v^2 + 9v\right| + C$

19. Let $u = \ln x$ so $du = \frac{1}{x}\,dx$
    $\int \frac{2\ln x}{x}\,dx = 2\int u\,du$
    $= u^2 + C = (\ln x)^2 + C$

21. Let $u = 2x^5 + \pi$ so $du = 10x^4\,dx$.
    $\int \frac{x^4}{2x^5 + \pi}\,dx = \frac{1}{10}\int\frac{1}{u}\,du$

$$= \frac{1}{10} \ln|u| + C = \frac{1}{10} \ln\left|2x^5 + \pi\right| + C$$

$$\int_0^3 \frac{x^4}{2x^5 + \pi} dx = \left[\frac{1}{10} \ln\left|2x^5 + \pi\right|\right]_0^3$$

$$= \frac{1}{10}[\ln(486 + \pi) - \ln \pi] = \ln\sqrt[10]{\frac{486 + \pi}{\pi}}$$

23. $2\ln(x+1) - \ln x = \ln(x+1)^2 - \ln x$

$$= \ln \frac{(x+1)^2}{x}$$

25. $\ln(x-2) - \ln(x+2) + 2\ln x$

$$= \ln(x-2) - \ln(x+2) + \ln x^2$$

$$= \ln \frac{x^2(x-2)}{x+2}$$

27. $\ln y = \ln(x+11) - \frac{1}{2}\ln(x^3 - 4)$

$$\frac{1}{y}\frac{dy}{dx} = \frac{1}{x+11} \cdot 1 - \frac{1}{2} \cdot \frac{1}{x^3 - 4} \cdot 3x^2$$

$$= \frac{1}{x+11} - \frac{3x^2}{2(x^3 - 4)}$$

$$\frac{dy}{dx} = y \cdot \left[\frac{1}{x+11} - \frac{3x^2}{2(x^3 - 4)}\right]$$

$$= \frac{x+11}{\sqrt{x^3 - 4}}\left[\frac{1}{x+11} - \frac{3x^2}{2(x^3 - 4)}\right]$$

$$= -\frac{x^3 + 33x^2 + 8}{2(x^3 - 4)^{3/2}}$$

29. $\ln y = \frac{1}{2}\ln(x+13) - \ln(x-4) - \frac{1}{3}\ln(2x+1)$

$$\frac{1}{y}\frac{dy}{dx} = \frac{1}{2(x+13)} - \frac{1}{x-4} - \frac{2}{3(2x+1)}$$

$$\frac{dy}{dx} = \frac{\sqrt{x+13}}{(x-4)\sqrt[3]{2x+1}}\left[\frac{1}{2(x+13)} - \frac{1}{x-4} - \frac{2}{3(2x+1)}\right] = -\frac{10x^2 + 219x - 118}{6(x-4)^2(x+13)^{1/2}(2x+1)^{4/3}}$$

31.

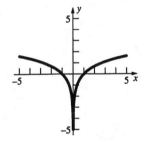

$y = \ln x$ is reflected across the y-axis.

33.

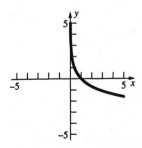

$y = \ln x$ is reflected across the x-axis since

$$\ln\left(\frac{1}{x}\right) = -\ln x.$$

35.

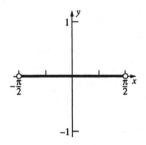

$y = \ln \cos x + \ln \sec x$

$$= \ln \cos x + \ln \frac{1}{\cos x}$$

$$= \ln \cos x - \ln \cos x = 0 \text{ on } \left(-\frac{\pi}{2}, \frac{\pi}{2}\right)$$

37. $f'(x) = 4x\ln x + 2x^2\left(\frac{1}{x}\right) - 2x = 4x\ln x$

so $f(1) = -1$ is a minimum.

39. $\ln 4 > 1$

so $\ln 4^m = m\ln 4 > m \cdot 1 = m$

Thus $x > 4^m \Rightarrow \ln x > m$

so $\lim_{x \to \infty} \ln x = \infty$

41. $\int_{1/3}^{x} \frac{1}{t} dt = 2 \int_{1}^{x} \frac{1}{t} dt$

$\int_{1/3}^{1} \frac{1}{t} dt + \int_{1}^{x} \frac{1}{t} dt = 2 \int_{1}^{x} \frac{1}{t} dt$

$\int_{1/3}^{1} \frac{1}{t} dt = \int_{1}^{x} \frac{1}{t} dt$

$-\int_{1}^{1/3} \frac{1}{t} dt = \int_{1}^{x} \frac{1}{t} dt$

$-\ln \frac{1}{3} = \ln x$

$\ln 3 = \ln x$

$x = 3$

43. $\lim_{n \to \infty} \left[\frac{1}{n+1} + \frac{1}{n+2} + \cdots + \frac{1}{2n} \right]$

$= \lim_{n \to \infty} \left[\frac{1}{1 + \frac{1}{n}} + \frac{1}{1 + \frac{2}{n}} + \cdots + \frac{1}{1 + \frac{n}{n}} \right] \cdot \frac{1}{n}$

$= \lim_{n \to \infty} \sum_{i=1}^{n} \left(\frac{1}{1 + \frac{i}{n}} \right) \cdot \frac{1}{n} = \int_{1}^{2} \frac{1}{x} dx = \ln 2 \approx 0.693$

45. a. $f(x) = \ln \left(\frac{ax-b}{ax+b} \right)^c = c \ln \left(\frac{ax-b}{ax+b} \right)$

$= \frac{a^2 - b^2}{2ab} [\ln(ax-b) - \ln(ax+b)]$

$f'(x) = \frac{a^2 - b^2}{2ab} \left[\frac{a}{ax-b} - \frac{a}{ax+b} \right]$

$= \frac{a^2 - b^2}{2ab} \left[\frac{2ab}{(ax-b)(ax+b)} \right] = \frac{a^2 - b^2}{a^2 x^2 - b^2}$

$f'(1) = \frac{a^2 - b^2}{a^2 - b^2} = 1$

b. $f'(x) = \cos^2 u \cdot \frac{du}{dx}$

$= \cos^2 [\ln(x^2 + x - 1)] \cdot \frac{2x+1}{x^2 + x - 1}$

$f'(1) = \cos^2 [\ln(1^2 + 1 - 1)] \cdot \frac{2 \cdot 1 + 1}{1^2 + 1 - 1}$

$= 3 \cos^2 (0) = 3$

47. $V = 2\pi \int_{1}^{4} x f(x) dx = \int_{1}^{4} \frac{2\pi x}{x^2 + 4} dx$

Let $u = x^2 + 4$ so $du = 2x\, dx$.

$\int \frac{2\pi x}{x^2 + 4} dx = \pi \int \frac{1}{u} du = \pi \ln |u| + C$

$= \pi \ln |x^2 + 4| + C$

$\int_{1}^{4} \frac{2\pi x}{x^2 + 4} dx = \left[\pi \ln |x^2 + 4| \right]_{1}^{4}$

$= \pi \ln 20 - \pi \ln 5 = \pi \ln 4 \approx 4.355$

49.

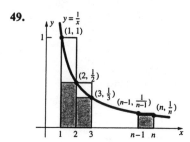

$\frac{1}{2} + \frac{1}{3} + \cdots + \frac{1}{n} = $ the lower approximate area

$1 + \frac{1}{2} + \cdots + \frac{1}{n-1} = $ the upper approximate area

$\ln n = $ the exact area under the curve

Thus,

$\frac{1}{2} + \frac{1}{3} + \cdots + \frac{1}{6} < \ln n < 1 + \frac{1}{2} + \frac{1}{3} + \cdots + \frac{1}{n-1} + \frac{1}{n}$.

51. a. $f'(x) = \frac{1}{1.5 + \sin x} \cdot \cos x = \frac{\cos x}{1.5 + \sin x}$

$f'(x) = 0$ when $\cos x = 0$.

Critical points: $0, \frac{\pi}{2}, \frac{3\pi}{2}, \frac{5\pi}{2}, 3\pi$

$f(0) \approx 0.405$,

$f\left(\frac{\pi}{2} \right) \approx 0.916, f\left(\frac{3\pi}{2} \right) \approx -0.693$,

$f\left(\frac{5\pi}{2} \right) \approx 0.916, f(3\pi) \approx 0.405$.

On $[0, 3\pi]$, the maximum value points are

$\left(\frac{\pi}{2}, 0.916 \right), \left(\frac{5\pi}{2}, 0.916 \right)$ and the minimum

value point is $\left(\frac{3\pi}{2}, -0.693 \right)$.

b. $f''(x) = -\frac{1 + 1.5 \sin x}{(1.5 + \sin x)^2}$

On $[0, 3\pi]$, $f''(x) = 0$ when $x \approx 3.871$,
5.553.
Inflection points are $(3.871, -0.182)$,
$(5.553, -0.183)$.

c. $\int_{0}^{3\pi} \ln(1.5 + \sin x) dx \approx 4.042$

53.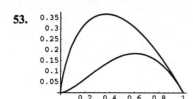

a. $\int_0^1 \left[x \ln\left(\frac{1}{x}\right) - x^2 \ln\left(\frac{1}{x}\right) \right] dx \approx 0.139$

b. Maximum of ≈ 0.260 at $x \approx 0.236$

7.2 Concepts Review

1. $f(x_1) \neq f(x_2)$

3. monotonic; increasing; decreasing

Problem Set 7.2

1. $f(x)$ is one-to-one, so it has an inverse.
Since $f(4) = 2, f^{-1}(2) = 4$.

3. $f(x)$ is not one-to-one, so it does not have an inverse.

5. $f(x)$ is one-to-one, so it has an inverse.
Since $f(-1) = 2$, $f^{-1}(2) = -1$.

7. $f'(x) = -5x^4 - 3x^2 = -(5x^4 + 3x^2) < 0$ for all $x \neq 0$. $f(x)$ is strictly decreasing at $x = 0$ because $f(x) > 0$ for $x < 0$ and $f(x) < 0$ for $x > 0$. Therefore $f(x)$ is strictly decreasing for x and so it has an inverse.

9. $f'(\theta) = -\sin\theta < 0$ for $0 < \theta < \pi$
$f(\theta)$ is decreasing at $\theta = 0$ because $f(0) = 1$ and $f(\theta) < 1$ for $0 < \theta < \pi$. $f(\theta)$ is decreasing at $\theta = \pi$ because $f(\pi) = -1$ and $f(\theta) > -1$ for $0 < \theta < \pi$. Therefore $f(\theta)$ is strictly decreasing on $0 \leq \theta \leq \pi$ and so it has an inverse.

11. $f'(z) = 2(z-1) > 0$ for $z > 1$
$f(z)$ is increasing at $z = 1$ because $f(1) = 0$ and $f(z) > 0$ for $z > 1$. Therefore, $f(z)$ is strictly increasing on $z \geq 1$ and so it has an inverse.

13. $f'(x) = \sqrt{x^4 + x^2 + 10} > 0$ for all real x. $f(x)$ is strictly increasing and so it has an inverse.

15. Step 1:
$y = x + 1$
$x = y - 1$
Step 2: $f^{-1}(y) = y - 1$
Step 3: $f^{-1}(x) = x - 1$
Check:

$f^{-1}(f(x)) = (x+1) - 1 = x$
$f(f^{-1}(x)) = (x-1) + 1 = x$

17. Step 1:
$y = \sqrt{x+1}$ (note that $y \geq 0$)
$x + 1 = y^2$
$x = y^2 - 1, y \geq 0$
Step 2: $f^{-1}(y) = y^2 - 1, y \geq 0$
Step 3: $f^{-1}(x) = x^2 - 1, x \geq 0$
Check:
$f^{-1}(f(x)) = (\sqrt{x+1})^2 - 1 = (x+1) - 1 = x$
$f(f^{-1}(x)) = \sqrt{(x^2 - 1) + 1} = \sqrt{x^2} = |x| = x$

19. Step 1:
$y = -\frac{1}{x-3}$

$x - 3 = -\frac{1}{y}$

$x = 3 - \frac{1}{y}$

Step 2: $f^{-1}(y) = 3 - \frac{1}{y}$

Step 3: $f^{-1}(x) = 3 - \frac{1}{x}$

Check:

$f^{-1}(f(x)) = 3 - \frac{1}{-\frac{1}{x-3}} = 3 + (x-3) = x$

$f(f^{-1}(x)) = -\frac{1}{\left(3 - \frac{1}{x}\right) - 3} = -\frac{1}{-\frac{1}{x}} = x$

21. Step 1:
$y = 4x^2, x \leq 0$ (note that $y \geq 0$)

$x^2 = \frac{y}{4}$

$x = -\sqrt{\frac{y}{4}} = -\frac{\sqrt{y}}{2}$, negative since $x \leq 0$

Step 2: $f^{-1}(y) = -\frac{\sqrt{y}}{2}$

Step 3: $f^{-1}(x) = -\dfrac{\sqrt{x}}{2}$

Check:

$$f^{-1}(f(x)) = -\frac{\sqrt{4x^2}}{2} = -\sqrt{x^2} = -|x| = -(-x) = x$$

$$f(f^{-1}(x)) = 4\left(-\frac{\sqrt{x}}{2}\right)^2 = 4 \cdot \frac{x}{4} = x$$

23. Step 1:

$$y = (x-1)^3$$
$$x-1 = \sqrt[3]{y}$$
$$x = 1 + \sqrt[3]{y}$$

Step 2: $f^{-1}(y) = 1 + \sqrt[3]{y}$

Step 3: $f^{-1}(x) = 1 + \sqrt[3]{x}$

Check: $f^{-1}(f(x)) = 1 + \sqrt[3]{(x-1)^3} = 1 + (x-1) = x$

$f(f^{-1}(x)) = [(1 + \sqrt[3]{x}) - 1]^3 = (\sqrt[3]{x})^3 = x$

25. Step 1:

$$y = \frac{x-1}{x+1}$$
$$xy + y = x - 1$$
$$x - xy = 1 + y$$
$$x = \frac{1+y}{1-y}$$

Step 2: $f^{-1}(y) = \dfrac{1+y}{1-y}$

Step 3: $f^{-1}(x) = \dfrac{1+x}{1-x}$

Check:

$$f^{-1}(f(x)) = \frac{1 + \frac{x-1}{x+1}}{1 - \frac{x-1}{x+1}} = \frac{x+1+x-1}{x+1-x+1} = \frac{2x}{2} = x$$

$$f(f^{-1}(x)) = \frac{\frac{1+x}{1-x} - 1}{\frac{1+x}{1-x} + 1} = \frac{1+x-1+x}{1+x+1-x} = \frac{2x}{2} = x$$

27. Step 1:

$$y = \frac{x^3 + 2}{x^3 + 1}$$
$$x^3 y + y = x^3 + 2$$
$$x^3 y - x^3 = 2 - y$$
$$x^3 = \frac{2-y}{y-1}$$
$$x = \left(\frac{2-y}{y-1}\right)^{1/3}$$

Step 2: $f^{-1}(y) = \left(\dfrac{2-y}{y-1}\right)^{1/3}$

Step 3: $f^{-1}(x) = \left(\dfrac{2-x}{x-1}\right)^{1/3}$

Check:

$$f^{-1}(f(x)) = \left(\frac{2 - \frac{x^3+2}{x^3+1}}{\frac{x^3+2}{x^3+1} - 1}\right)^{1/3} = \left(\frac{2x^3 + 2 - x^3 - 2}{x^3 + 2 - x^3 - 1}\right)^{1/3}$$

$$= \left(\frac{x^3}{1}\right)^{1/3} = x$$

$$f(f^{-1}(x)) = \frac{\left[\left(\frac{2-x}{x-1}\right)^{1/3}\right]^3 + 2}{\left[\left(\frac{2-x}{x-1}\right)^{1/3}\right]^3 + 1} = \frac{\frac{2-x}{x-1} + 2}{\frac{2-x}{x-1} + 1}$$

$$= \frac{2 - x + 2x - 2}{2 - x + x - 1} = \frac{x}{1} = x$$

29. By similar triangles $\dfrac{r}{h} = \dfrac{4}{6}$. Thus, $r = \dfrac{2h}{3}$

This gives

$$V = \frac{\pi r^2 h}{3} = \frac{\pi\left(4h^2/9\right)h}{3} = \frac{4\pi h^3}{27}$$

$$h^3 = \frac{27V}{4\pi}$$

$$h = 3\sqrt[3]{\frac{V}{4\pi}}$$

31. $f'(x) = 4x + 1$; $f'(x) > 0$ when $x > -\dfrac{1}{4}$ and

$f'(x) < 0$ when $x < -\dfrac{1}{4}$.

The function is decreasing on $\left(-\infty, -\dfrac{1}{4}\right]$ and

increasing on $\left[-\dfrac{1}{4}, \infty\right)$. Restrict the domain to

$\left(-\infty, -\dfrac{1}{4}\right]$ or restrict it to $\left[-\dfrac{1}{4}, \infty\right)$.

Then $f^{-1}(x) = \dfrac{1}{4}(-1 - \sqrt{8x + 33})$ or

$f^{-1}(x) = \dfrac{1}{4}(-1 + \sqrt{8x + 33})$.

33.

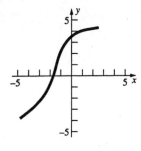

$$(f^{-1})'(3) \approx \frac{1}{10}$$

35.

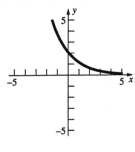

$$(f^{-1})'(3) \approx -\frac{1}{3}$$

37. $f'(x) = 15x^4 + 1$ and $y = 2$ corresponds to $x = 1$,

so $(f^{-1})'(2) = \dfrac{1}{f'(1)} = \dfrac{1}{15+1} = \dfrac{1}{16}$.

39. $f'(x) = 2\sec^2 x$ and $y = 2$ corresponds to $x = \dfrac{\pi}{4}$,

so $(f^{-1})'(2) = \dfrac{1}{f'\left(\frac{\pi}{4}\right)} = \dfrac{1}{2\sec^2\left(\frac{\pi}{4}\right)} = \dfrac{1}{2}\cos^2\left(\dfrac{\pi}{4}\right)$

$= \dfrac{1}{4}$.

41. $(g^{-1} \circ f^{-1})(h(x)) = (g^{-1} \circ f^{-1})(f(g(x)))$

$\qquad = g^{-1} \circ [f^{-1}(f(g(x)))] = g^{-1} \circ [g(x)] = x$

$$\frac{1}{p-1} = \frac{1}{\frac{q}{q-1}-1} = \frac{1}{\left[\frac{q-(q-1)}{q-1}\right]} = \frac{q-1}{1} = q-1.$$

Thus, if $y = x^{p-1}$ then $x = y^{\frac{1}{p-1}} = y^{q-1}$, so

$f^{-1}(y) = y^{q-1}$.

By Problem 44, since $f(x) = x^{p-1}$ is strictly

Similarly,

$h((g^{-1} \circ f^{-1})(x)) = f(g((g^{-1} \circ f^{-1})(x)))$

$\qquad = f(g(g^{-1}(f^{-1}(x)))) = f(f^{-1}(x)) = x$

Thus $h^{-1} = g^{-1} \circ f^{-1}$

43. f has an inverse because it is monotonic (increasing):

$$f'(x) = \sqrt{1+\cos^2 x} > 0$$

a. $(f^{-1})'(A) = \dfrac{1}{f'\left(\frac{\pi}{2}\right)} = \dfrac{1}{\sqrt{1+\cos^2\left(\frac{\pi}{2}\right)}} = 1$

b. $(f^{-1})'(B) = \dfrac{1}{f'\left(\frac{5\pi}{6}\right)} = \dfrac{1}{\sqrt{1+\cos^2\left(\frac{5\pi}{6}\right)}} = \dfrac{1}{\sqrt{\frac{7}{4}}}$

$\qquad = \dfrac{2}{\sqrt{7}}$

c. $(f^{-1})'(0) = \dfrac{1}{f'(0)} = \dfrac{1}{\sqrt{1+\cos^2(0)}} = \dfrac{1}{\sqrt{2}}$

45.

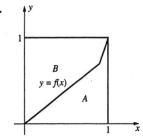

$\displaystyle\int_0^1 f^{-1}(y)\,dy = $ (Area of region B)

$= 1 - $ (Area of region A)

$= 1 - \displaystyle\int_0^1 f(x)\,dx = 1 - \dfrac{2}{5} = \dfrac{3}{5}$

47. Given $p > 1$, $q > 1$, $\dfrac{1}{p} + \dfrac{1}{q} = 1$, and $f(x) = x^{p-1}$,

solving $\dfrac{1}{p} + \dfrac{1}{q} = 1$ for p gives $p = \dfrac{q}{q-1}$, so

increasing for $p > 1$, $ab \le \displaystyle\int_0^a x^{p-1}\,dx + \int_0^b y^{q-1}\,dy$

$ab \le \left[\dfrac{x^p}{p}\right]_0^a + \left[\dfrac{y^q}{q}\right]_0^b$

$ab \le \dfrac{a^p}{p} + \dfrac{b^q}{q}$

7.3 Concepts Review

1. increasing; exp

3. x; x

Problem Set 7.3

1. **a.** 20.086

   b. 8.1662

   c. $e^{\sqrt{2}} \approx e^{1.41} = e^{1.4} \cdot e^{0.01} \approx 4.1$

   d. $e^{\cos(\ln 4)} \approx e^{0.18} \approx 1.20$

3. $e^{3\ln x} = e^{\ln x^3} = x^3$

5. $\ln e^{\cos x} = \cos x$

7. $\ln(x^3 e^{-3x}) = \ln x^3 + \ln e^{-3x} = 3\ln x - 3x$

9. $e^{\ln 3 + 2\ln x} = e^{\ln 3} \cdot e^{2\ln x} = 3 \cdot e^{\ln x^2} = 3x^2$

11. $D_x e^{x+2} = e^{x+2} D_x(x+2) = e^{x+2}$

13. $D_x e^{\sqrt{x+2}} = e^{\sqrt{x+2}} D_x \sqrt{x+2} = \dfrac{e^{\sqrt{x+2}}}{2\sqrt{x+2}}$

15. $D_x e^{2\ln x} = D_x e^{\ln x^2} = D_x x^2 = 2x$

17. $D_x(x^3 e^x) = x^3 D_x e^x + e^x D_x(x^3)$
$= x^3 e^x + e^x \cdot 3x^2 = x^2 e^x(x+3)$

19. $D_x[\sqrt{e^{x^2}} + e^{\sqrt{x^2}}] = D_x(e^{x^2})^{1/2} + D_x e^{\sqrt{x^2}}$
$= \dfrac{1}{2}(e^{x^2})^{-1/2} D_x e^{x^2} + e^{\sqrt{x^2}} D_x \sqrt{x^2}$
$= \dfrac{1}{2}(e^{x^2})^{-1/2} e^{x^2} D_x x^2 + e^{\sqrt{x^2}} \cdot \dfrac{x}{\sqrt{x^2}}$
$= \dfrac{1}{2}(e^{x^2})^{1/2} 2x + e^{\sqrt{x^2}} \cdot \dfrac{x}{|x|}$
$= x\sqrt{e^{x^2}} + \dfrac{xe^{\sqrt{x^2}}}{|x|}$

21. $D_x[e^{xy} + xy] = D_x[2]$
$e^{xy}(xD_x y + y) + (xD_x y + y) = 0$

$xe^{xy}D_x y + ye^{xy} + xD_x y + y = 0$
$xe^{xy}D_x y + xD_x y = -ye^{xy} - y$
$D_x y = \dfrac{-ye^{xy} - y}{xe^{xy} + x} = -\dfrac{y(e^{xy}+1)}{x(e^{xy}+1)} = -\dfrac{y}{x}$

23. **a.**

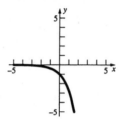

The graph of $y = e^x$ is reflected across the x-axis.

b.

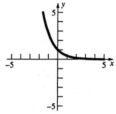

The graph of $y = e^x$ is reflected across the y-axis.

25. $f'(x) = -xe^{-x} + e^{-x} = e^{-x}(1-x)$;
$f'(x) = 0$ when $x = 1$
$f'(x) > 0$ on $(-\infty, 1)$
$f''(x) = xe^{-x} - 2e^{-x} = e^{-x}(x-2)$;
$f''(x) = 0$ when $x = 2$
$f''(x) > 0$ on $(2, \infty)$
Increasing on $(-\infty, 1]$, Decreasing on $[1, \infty)$
Maximum at $\left(1, \dfrac{1}{e}\right) \approx (1, 0.4)$
Concave up on $(2, \infty)$
Concave down on $(-\infty, 2)$
Inflection point at $\left(2, \dfrac{2}{e^2}\right) \approx (2, 0.3)$

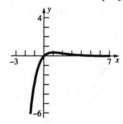

27. $f'(x) = -2(x-2)e^{-(x-2)^2}$;
$f'(x) = 0$ when $x = 2$

$f'(x) > 0$ on $(-\infty, 2)$

$f''(x) = 4(x-2)^2 e^{-(x-2)^2} - 2e^{-(x-2)^2}$

$= 2e^{-(x-2)^2}[2(x-2)^2 - 1];$

$f''(x) = 0$ when $x = 2 \pm \dfrac{1}{2}\sqrt{2}$

$f''(x) > 0$ on $\left(-\infty, 2 - \dfrac{1}{2}\sqrt{2}\right)$ and

$\left(2 + \dfrac{1}{2}\sqrt{2}, \infty\right)$

Increasing on $(-\infty, 2]$
Decreasing on $[2, \infty)$
Maximum at $(2, 1)$

Concave up on $\left(-\infty, 2 - \dfrac{1}{2}\sqrt{2}\right) \cup \left(2 + \dfrac{1}{2}\sqrt{2}, \infty\right)$

Concave down on $\left(2 - \dfrac{1}{2}\sqrt{2}, 2 + \dfrac{1}{2}\sqrt{2}\right)$

Inflection points at $\left(2 - \dfrac{1}{2}\sqrt{2}, \dfrac{1}{\sqrt{e}}\right) \approx (1.3, 0.6)$

and $\left(2 + \dfrac{1}{2}\sqrt{2}, \dfrac{1}{\sqrt{e}}\right) \approx (2.7, 0.6)$

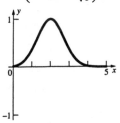

29. Let $u = 3x + 1$, so $du = 3\,dx$.

$\displaystyle \int e^{3x+1}\,dx = \frac{1}{3}\int e^{3x+1}3\,dx = \frac{1}{3}\int e^u\,du = \frac{1}{3}e^u + C$

$= \dfrac{1}{3}e^{3x+1} + C$

31. Let $u = x^2 + 6x$, so $du = (2x+6)\,dx$.

$\displaystyle \int (x+3)e^{x^2+6x}\,dx = \frac{1}{2}\int e^u\,du = \frac{1}{2}e^u + C$

$= \dfrac{1}{2}e^{x^2+6x} + C$

33. Let $u = -\dfrac{1}{x}$, so $du = \dfrac{1}{x^2}\,dx$.

$\displaystyle \int \frac{e^{-1/x}}{x^2}\,dx = \int e^u\,du = e^u + C = e^{-1/x} + C$

35. Let $u = 2x + 3$, so $du = 2\,dx$

$\displaystyle \int e^{2x+3}\,dx = \frac{1}{2}\int e^u\,du = \frac{1}{2}e^u + C = \frac{1}{2}e^{2x+3} + C$

$\displaystyle \int_0^1 e^{2x+3}\,dx = \left[\frac{1}{2}e^{2x+3}\right]_0^1 = \frac{1}{2}e^5 - \frac{1}{2}e^3$

$= \dfrac{1}{2}e^3(e^2 - 1) \approx 64.2$

37. $V = \pi \displaystyle\int_0^{\ln 3} (e^x)^2\,dx = \pi \int_0^{\ln 3} e^{2x}\,dx$

$= \pi \left[\dfrac{1}{2}e^{2x}\right]_0^{\ln 3} = \pi\left(\dfrac{1}{2}e^{2\ln 3} - \dfrac{1}{2}e^0\right) \approx 12.57$

39. The line through $(0, 1)$ and $\left(1, \dfrac{1}{e}\right)$ has slope

$\dfrac{\frac{1}{e} - 1}{1 - 0} = \dfrac{1}{e} - 1 = \dfrac{1-e}{e} \Rightarrow y - 1 = \dfrac{1-e}{e}(x-0);$

$y = \dfrac{1-e}{e}x + 1$

$\displaystyle \int_0^1 \left[\left(\frac{1-e}{e}x + 1\right) - e^{-x}\right]dx = \left[\frac{1-e}{2e}x^2 + x + e^{-x}\right]_0^1$

$= \dfrac{1-e}{2e} + 1 + \dfrac{1}{e} - 1 = \dfrac{3-e}{2e} \approx 0.052$

41. a. Exact:

$\quad 10! = 10 \cdot 9 \cdot 8 \cdot 7 \cdot 6 \cdot 5 \cdot 4 \cdot 3 \cdot 2 \cdot 1$

$\quad = 3,628,800$

Approximate:

$\quad 10! \approx \sqrt{20\pi}\left(\dfrac{10}{e}\right)^{10} \approx 3,598,696$

b. $60! \approx \sqrt{120\pi}\left(\dfrac{60}{e}\right)^{60} \approx 8.31 \times 10^{81}$

43. $x = e^t \sin t$, so $dx = (e^t \sin t + e^t \cos t)dt$

$y = e^t \cos t$, so $dy = (e^t \cos t - e^t \sin t)dt$

$ds = \sqrt{dx^2 + dy^2}$

$= e^t \sqrt{(\sin t + \cos t)^2 + (\cos t - \sin t)^2}\,dt$

$= e^t \sqrt{2\sin^2 t + 2\cos^2 t}\,dt = \sqrt{2}e^t\,dt$

The length of the curve is

$\displaystyle \int_0^\pi \sqrt{2}e^t\,dt = \sqrt{2}\left[e^t\right]_0^\pi = \sqrt{2}(e^\pi - 1) \approx 31.312$

45. a. $\displaystyle \lim_{x \to 0^+} \frac{\ln x}{1 + (\ln x)^2}$ is of the form $\dfrac{\infty}{\infty}$.

$= \displaystyle \lim_{x \to 0^+} \frac{D_x \ln x}{D_x[1 + (\ln x)^2]} = \lim_{x \to 0^+} \frac{\frac{1}{x}}{2\ln x \cdot \frac{1}{x}}$

$$= \lim_{x \to 0^+} \frac{1}{2 \ln x} = 0$$

$$\lim_{x \to \infty} \frac{\ln x}{1+(\ln x)^2} = \lim_{x \to \infty} \frac{1}{2 \ln x} = 0$$

b. $f'(x) = \dfrac{[1+(\ln x)^2] \cdot \frac{1}{x} - \ln x \cdot 2 \ln x \cdot \frac{1}{x}}{[1+(\ln x)^2]^2}$

$\quad = \dfrac{1-(\ln x)^2}{x[1+(\ln x)^2]^2}$

$f'(x) = 0$ when $\ln x = \pm 1$ so $x = e^1 = e$

or $x = e^{-1} = \dfrac{1}{e}$

$f(e) = \dfrac{\ln e}{1+(\ln e)^2} = \dfrac{1}{1+1^2} = \dfrac{1}{2}$

$f\left(\dfrac{1}{e}\right) = \dfrac{\ln \frac{1}{e}}{1+\left(\ln \frac{1}{e}\right)^2} = \dfrac{-1}{1+(-1)^2} = -\dfrac{1}{2}$

Maximum value of $\dfrac{1}{2}$ at $x = e$; minimum

value of $-\dfrac{1}{2}$ at $x = e^{-1}$.

c. $F(x) = \displaystyle\int_1^{x^2} \dfrac{\ln t}{1+(\ln t)^2} dt$

$F'(x) = \dfrac{\ln x^2}{1+(\ln x^2)^2} \cdot 2x$

$F'(\sqrt{e}) = \dfrac{\ln(\sqrt{e})^2}{1+[\ln(\sqrt{e})^2]^2} \cdot 2\sqrt{e} = \dfrac{1}{1+1^2} \cdot 2\sqrt{e}$

$\quad = \sqrt{e} \approx 1.65$

7.4 Concepts Review

1. $e^{\sqrt{3} \ln \pi}$; $e^{x \ln a}$

3. $\dfrac{\ln x}{\ln a}$

Problem Set 7.4

1. $2^x = 8 = 2^3$; $x = 3$

3. $x = 4^{3/2} = 8$

5. $\log_9 \left(\dfrac{x}{3}\right) = \dfrac{1}{2}$

47. $\displaystyle\lim_{n \to \infty} \left(e^{1/n} + e^{2/n} + \cdots + e^{n/n}\right)\left(\dfrac{1}{n}\right) = \int_0^1 e^x dx$

$\quad = \left[e^x\right]_0^1 = e - 1 \approx 1.718$

49. **a.** $\displaystyle\int_{-3}^3 \exp\left(-\dfrac{1}{x^2}\right) dx = 2 \int_0^3 \exp\left(-\dfrac{1}{x^2}\right) dx \approx 3.11$

b. $\displaystyle\int_0^{8\pi} e^{-0.1x} \sin x \, dx \approx 0.910$

51. $f(x) = e^{-x^2}$

$\quad f'(x) = -2xe^{-x^2}$

$\quad f''(x) = -2e^{-x^2} + 4x^2 e^{-x^2} = 2e^{-x^2}(2x^2 - 1)$

$\quad y = f(x)$ and $y = f''(x)$ intersect when

$\quad e^{-x^2} = 2e^{-x^2}(2x^2 - 1); 1 = 4x^2 - 2;$

$\quad 4x^2 - 3 = 0, x = \pm \dfrac{\sqrt{3}}{2}$

Both graphs are symmetric with respect to the y-axis so the area is

$2\left\{ \displaystyle\int_0^{\frac{\sqrt{3}}{2}} [e^{-x^2} - 2e^{x^2}(2x^2 - 1)] dx \right.$

$\quad \left. + \displaystyle\int_{\frac{\sqrt{3}}{2}}^3 [2e^{-x^2}(2x^2 - 1) - e^{-x^2}] dx \right\}$

≈ 4.2614

53. $\displaystyle\lim_{x \to -\infty} \ln(x^2 + e^{-x}) = \infty$ (behaves like $-x$)

$\displaystyle\lim_{x \to \infty} \ln(x^2 + e^{-x}) = \infty$ (behaves like $2 \ln x$)

$\dfrac{x}{3} = 9^{1/2} = 3$

$x = 9$

7. $\log_2(x+3) - \log_2 x = 2$

$\log_2 \dfrac{x+3}{x} = 2$

$\dfrac{x+3}{x} = 2^2 = 4$

$x + 3 = 4x$

$x = 1$

9. $\log_5 12 = \dfrac{\ln 12}{\ln 5} \approx 1.544$

11. $\log_{11}(8.12)^{1/5} = \dfrac{1}{5}\dfrac{\ln 8.12}{\ln 11} \approx 0.1747$

13. $x \ln 2 = \ln 17$

$x = \dfrac{\ln 17}{\ln 2} \approx 4.08746$

15. $(2s-3)\ln 5 = \ln 4$

$2s - 3 = \dfrac{\ln 4}{\ln 5}$

$s = \dfrac{1}{2}\left(3 + \dfrac{\ln 4}{\ln 5}\right) \approx 1.9307$

17. $D_x(6^{2x}) = 6^{2x}\ln 6 \cdot D_x(2x) = 2 \cdot 6^{2x}\ln 6$

19. $D_x \log_3 e^x = \dfrac{1}{e^x \ln 3} \cdot D_x e^x$

$= \dfrac{e^x}{e^x \ln 3} = \dfrac{1}{\ln 3} \approx 0.9102$

Alternate method:

$D_x \log_3 e^x = D_x (x \log_3 e) = \log_3 e$

$= \dfrac{\ln e}{\ln 3} = \dfrac{1}{\ln 3} \approx 0.9102$

21.

$D_z[3^z \ln(z+5)]$

$= 3^z \cdot \dfrac{1}{z+5}(1) + \ln(z+5) \cdot 3^z \ln 3$

$= 3^z\left[\dfrac{1}{z+5} + \ln(z+5)\ln 3\right]$

23. Let $u = x^2$ so $du = 2x\,dx$.

$\displaystyle\int x \cdot 2^{x^2}\,dx = \dfrac{1}{2}\int 2^u\,du = \dfrac{1}{2} \cdot \dfrac{2^u}{\ln 2} + C$

$= \dfrac{2^{x^2}}{2\ln 2} + C = \dfrac{2^{x^2-1}}{\ln 2} + C$

25. Let $u = \sqrt{x}$, so $du = \dfrac{1}{2\sqrt{x}}\,dx$.

$\displaystyle\int \dfrac{5^{\sqrt{x}}}{\sqrt{x}}\,dx = 2\int 5^u\,du = 2 \cdot \dfrac{5^u}{\ln 5} + C$

$= \dfrac{2 \cdot 5^{\sqrt{x}}}{\ln 5} + C$

$\displaystyle\int_1^4 \dfrac{5^{\sqrt{x}}}{\sqrt{x}}\,dx = 2\left[\dfrac{5^{\sqrt{x}}}{\ln 5}\right]_1^4 = 2\left(\dfrac{25}{\ln 5} - \dfrac{5}{\ln 5}\right)$

$= \dfrac{40}{\ln 5} \approx 24.85$

27. $\dfrac{d}{dx}10^{(x^2)} = 10^{(x^2)}\ln 10 \dfrac{d}{dx}x^2 = 10^{(x^2)}2x\ln 10$

$\dfrac{d}{dx}(x^2)^{10} = \dfrac{d}{dx}x^{20} = 20x^{19}$

$\dfrac{dy}{dx} = \dfrac{d}{dx}[10^{(x^2)} + (x^2)^{10}]$

$= 10^{(x^2)}2x\ln 10 + 20x^{19}$

29. $\dfrac{d}{dx}x^{\pi+1} = (\pi+1)x^{\pi}$

$\dfrac{d}{dx}(\pi+1)^x = (\pi+1)^x \ln(\pi+1)$

$\dfrac{dy}{dx} = \dfrac{d}{dx}[x^{\pi+1} + (\pi+1)^x]$

$= (\pi+1)x^{\pi} + (\pi+1)^x \ln(\pi+1)$

31. $y = (x^2+1)^{\ln x} = e^{(\ln x)\ln(x^2+1)}$

$\dfrac{dy}{dx} = e^{(\ln x)\ln(x^2+1)}\dfrac{d}{dx}[(\ln x)\ln(x^2+1)]$

$= e^{(\ln x)\ln(x^2+1)}\left[\dfrac{1}{x}\ln(x^2+1) + \ln x\dfrac{2x}{x^2+1}\right]$

$= (x^2+1)^{\ln x}\left(\dfrac{\ln(x^2+1)}{x} + \dfrac{2x\ln x}{x^2+1}\right)$

33. $f(x) = x^{\sin x} = e^{\sin x \ln x}$

$f'(x) = e^{\sin x \ln x}\dfrac{d}{dx}(\sin x \ln x)$

$= e^{\sin x \ln x}\left[(\sin x)\left(\dfrac{1}{x}\right) + (\cos x)(\ln x)\right]$

$= x^{\sin x}\left(\dfrac{\sin x}{x} + \cos x \ln x\right)$

$f'(1) = 1^{\sin 1}\left(\dfrac{\sin 1}{1} + \cos 1\ln 1\right) = \sin 1 \approx 0.8415$

35. $\log_{1/2}x = \dfrac{\ln x}{\ln\frac{1}{2}} = \dfrac{\ln x}{-\ln 2} = -\log_2 x$

37. $M = 0.67\log_{10}(0.37E) + 1.46$

$\log_{10}(0.37E) = \dfrac{M - 1.46}{0.67}$

$E = \dfrac{10^{\frac{M-1.46}{0.67}}}{0.37}$

Evaluating this expression for $M = 7$ and $M = 8$ gives $E \approx 5.017 \times 10^8$ kW-h and $E \approx 1.560 \times 10^{10}$ kW-h, respectively.

39. If r is the ratio between the frequencies of successive notes, then the frequency of $\overline{C} = r^{12}$ (the frequency of C). Since $\overline{C}$ has twice the frequency of C, $r = 2^{1/12} \approx 1.0595$
Frequency of $\overline{C} = 440(2^{1/12})^3 = 440\sqrt[4]{2} \approx 523.25$

41. If $y = A \cdot b^x$, then $\ln y = \ln A + x \ln b$, so the $\ln y$ vs. x plot will be linear.
If $y = C \cdot x^d$, then $\ln y = \ln C + d \ln x$, so the $\ln y$ vs. $\ln x$ plot will be linear.

43. $f(x) = (x^x)^x = x^{(x^2)} \neq x^{(x^x)} = g(x)$

$f(x) = x^{(x^2)} = e^{x^2 \ln x}$

$f'(x) = e^{x^2 \ln x} \dfrac{d}{dx}(x^2 \ln x) = e^{x^2 \ln x}\left(2x \ln x + x^2 \cdot \dfrac{1}{x}\right)$

$= x^{(x^2)}(2x \ln x + x)$

$g(x) = x^{(x^x)} = e^{x^x \ln x}$

Using the result from Example 5

$\left(\dfrac{d}{dx}x^x = x^x(1 + \ln x)\right)$:

$g'(x) = e^{x^x \ln x}\dfrac{d}{dx}(x^x \ln x)$

$= e^{x^x \ln x}\left[x^x(1 + \ln x)\ln x + x^x \cdot \dfrac{1}{x}\right]$

$= x^{(x^x)}x^x\left[(1 + \ln x)\ln x + \dfrac{1}{x}\right]$

$= x^{x^x + x}\left[\ln x + (\ln x)^2 + \dfrac{1}{x}\right]$

45. a. Let $g(x) = \ln f(x) = \ln\left(\dfrac{x^a}{a^x}\right) = a \ln x - x \ln a$.

$g'(x) = \left(\dfrac{a}{x}\right) - \ln a$

$g'(x) < 0$ when $x > \dfrac{a}{\ln a}$, so as $x \to \infty$ $g(x)$ is decreasing. $g''(x) = -\dfrac{a}{x^2}$, so $g(x)$ is concave down. Thus, $\lim\limits_{x \to \infty} g(x) = -\infty$, so

$\lim\limits_{x \to \infty} f(x) = \lim\limits_{x \to \infty} e^{g(x)} = 0$.

b. Again let $g(x) = \ln f(x) = a \ln x - x \ln a$. Since $y = \ln x$ is an increasing function, $f(x)$ is maximized when $g(x)$ is maximized.

$g'(x) = \left(\dfrac{a}{x}\right) - \ln a$, so $g'(x) > 0$ on $\left(0, \dfrac{a}{\ln a}\right)$

and $g'(x) < 0$ on $\left(\dfrac{a}{\ln a}, \infty\right)$.
Therefore, $g(x)$ (and hence $f(x)$) is maximized at $x_0 = \dfrac{a}{\ln a}$.

c. Note that $x^a = a^x$ is equivalent to $g(x) = 0$.
By part b., $g(x)$ is maximized at $x_0 = \dfrac{a}{\ln a}$.
If $a = e$, then

$g(x_0) = g\left(\dfrac{e}{\ln e}\right) = g(e) = e \ln e - e \ln e = 0.$

Since $g(x) < g(x_0) = 0$ for all $x \neq x_0$, the equation $g(x) = 0$ (and hence $x^a = a^x$) has just one positive solution. If $a \neq e$, then

$g(x_0) = g\left(\dfrac{a}{\ln a}\right) = a \ln\left(\dfrac{a}{\ln a}\right) - \dfrac{a}{\ln a}(\ln a)$

$= a\left[\ln\left(\dfrac{a}{\ln a}\right) - 1\right].$

Now $\dfrac{a}{\ln a} > e$ (justified below), so

$g(x_0) = a\left[\ln\dfrac{a}{\ln a} - 1\right] > a(\ln e - 1) = 0.$ Since $g'(x) > 0$ on $(0, x_0), g(x_0) > 0$, and $\lim\limits_{x \to 0} g(x) = -\infty$, $g(x) = 0$ has exactly one solution on $(0, x_0)$.
Since $g'(x) < 0$ on (x_0, ∞), $g(x_0) > 0$, and $\lim\limits_{x \to \infty} g(x) = -\infty$, $g(x) = 0$ has exactly one solution on (x_0, ∞). Therefore, the equation $g(x) = 0$ (and hence $x^a = a^x$) has exactly two positive solutions.

To show that $\dfrac{a}{\ln a} > e$ when $a \neq e$:

Consider the function $h(x) = \dfrac{x}{\ln x}$, for $x > 1$.

$h'(x) = \dfrac{\ln(x)(1) - x\left(\frac{1}{x}\right)}{(\ln x)^2} = \dfrac{\ln x - 1}{(\ln x)^2}$

Note that $h'(x) < 0$ on $(1, e)$ and $h'(x) > 0$ on (e, ∞), so $h(x)$ has its minimum at (e, e).
Therefore $\dfrac{x}{\ln x} > e$ for all $x \neq e, x > 1$.

d. For the case $a = e$, part c. shows that
$g(x) = e \ln x - x \ln e < 0$ for $x \neq e$. Therefore,
when $x \neq e$, $\ln x^e < \ln e^x$, which implies
$x^e < e^x$. In particular, $\pi^e < e^\pi$.

47. $f(x) = x^x = e^{x \ln x}$

Let $g(x) = x \ln x$.

Using L'Hôpital's Rule,

$$\lim_{x \to 0^+} g(x) = \lim_{x \to 0^+} \frac{\ln x}{\frac{1}{x}}$$

$$= \lim_{x \to 0^+} \frac{\frac{1}{x}}{-\frac{1}{x^2}} = \lim_{x \to 0^+} (-x) = 0$$

Therefore, $\lim_{x \to 0^+} x^x = e^0 = 1$.

$g'(x) = \ln x - 1$

Since $g'(x) < 0$ on $(0, e)$ and $g'(x) > 0$ on
(e, ∞), $g(x)$ has its minimum at (e, e). Therefore, $f(x)$ has
its minimum at (e, e^e).

49. $\int_0^{4\pi} x^{\sin x} dx \approx 20.2259$

7.5 Concepts Review

1. ky; $ky(L - y)$ **3.** half-life

Problem Set 7.5

1. $k = -6$, $y_0 = 4$, so $y = 4e^{-6t}$

3. $k = 0.005$, so $y = y_0 e^{0.005t}$

$y(10) = y_0 e^{0.005(10)} = y_0 e^{0.05}$

$y(10) = 2 \Rightarrow y_0 = \dfrac{2}{e^{0.05}}$

$y = \dfrac{2}{e^{0.05}} e^{0.005t} = 2e^{0.005t - 0.05} = 2e^{0.005(t-10)}$

5. $y_0 = 10,000$, $y(10) = 20,000$

$20,000 = 10,000 e^{k(10)}$

$2 = e^{10k}$

$\ln 2 = 10k$

$k = \dfrac{\ln 2}{10} \approx 0.069$

$y = 10,000 e^{0.069t}$

After 25 days, $y = 10,000 e^{0.069 \cdot 25} \approx 56,125$.

7. $3y_0 = y_0 e^{0.069t}$

$3 = e^{0.069}$

$\ln 3 = 0.069t$

$t = \dfrac{\ln 3}{0.069} \approx 15.9$ days

9. 1 year: (4.5 million) (1.032) ≈ 4.64 million

2 years: (4.5 million) $(1.032)^2 \approx 4.79$ million

10 years: (4.5 million) $(1.032)^{10} \approx 6.17$ million

100 years: (4.5 million) $(1.032)^{100} \approx 105$ million

11. $\dfrac{1}{2} = e^{k(700)}$ and $y_0 = 10$

$-\ln 2 = 700k$

$k = -\dfrac{\ln 2}{700} \approx -0.00099$

$y = 10 e^{-0.00099t}$

At $t = 300$, $y = 10 e^{-0.00099 \cdot 300} \approx 7.43$.

After 10 years there will be about 7.43 g.

13. $\dfrac{1}{2} = e^{5730k}$

$k = \dfrac{\ln\left(\frac{1}{2}\right)}{5730} \approx -1.210 \times 10^{-4}$

$0.7y_0 = y_0 e^{(-1.210 \times 10^{-4})t}$

$t = \dfrac{\ln 0.7}{-1.210 \times 10^{-4}} \approx 2950$

The fort burned down about 2950 years ago.

15. $\dfrac{dT}{dt} = k(T - 75)$

$\int \dfrac{1}{T - 75} dT = \int k\, dt$

$\ln |T - 75| = kt + C$

$|T - 75| = e^{kt + C}$

$T = 75 + Ae^{kt}$

$T(0) = 75 + A$

$T(0) = 300 \Rightarrow A = 225$

$T = 75 + 225 e^{kt}$

$T(0.5) = 200 \Rightarrow 200 = 75 + 225e^{0.5k}$;

$\dfrac{125}{225} = e^{0.5k}; k = 2\ln\dfrac{5}{9} \approx -1.176$

$T = 75 + 225e^{-1.176t}$

At $t = 3$, $T = 75 + 225e^{-1.176\cdot 3} \approx 81.6$.

After 3 hours, the temperature will be about 81.6°F.

17. a. $(\$375)(1.095)^2 \approx \$449.63$

    b. $(\$375)\left(1+\dfrac{0.095}{12}\right)^{24} \approx \$453.13$

    c. $(\$375)\left(1+\dfrac{0.095}{365}\right)^{730} \approx \$453.46$

    d. $(\$375)e^{0.095\cdot 2} \approx \$453.47$

19. a. $\left(1+\dfrac{0.12}{12}\right)^{12t} = 2$

    $1.01^{12t} = 2$

    $12t = \log_{1.01} 2$

    $t = \dfrac{1}{12}\log_{1.01} 2 = \dfrac{\ln 2}{12\ln 1.01} \approx 5.805$

    It will take about 5.805 years or
5 years, 10 months.

    b. $e^{0.12t} = 2 \Rightarrow t = \dfrac{\ln 2}{0.12} \approx 5.776$

    It will take about 5.776 years
or 5 years, 9 months, and 9 days.

21. 1626 to 2000 is 374 years.

$y = 24e^{0.06\cdot 374} \approx \$133.6$ billion

23. If t is the doubling time, then

$\left(1+\dfrac{p}{100}\right)^t = 2$

$t\ln\left(1+\dfrac{p}{100}\right) = \ln 2$

$t = \dfrac{\ln 2}{\ln\left(1+\frac{p}{100}\right)} \approx \dfrac{\ln 2}{\frac{p}{100}} = \dfrac{100\ln 2}{p} \approx \dfrac{70}{p}$

25. $y = \dfrac{16\cdot 5}{5+(16-5)e^{-16(0.00186)t}} = \dfrac{80}{5+11e^{-0.02976t}}$

27. a. $\lim\limits_{x\to 0}(1-x)^{1/x} = \lim\limits_{x\to 0}\dfrac{1}{[1+(-x)]^{1/(-x)}} = \dfrac{1}{e}$

    b. $\lim\limits_{x\to 0}(1+3x)^{1/x} = \lim\limits_{x\to 0}\left[(1+3x)^{\frac{1}{3x}}\right]^3 = e^3$

    c. $\lim\limits_{n\to\infty}\left(\dfrac{n+2}{n}\right)^n = \lim\limits_{n\to\infty}\left(1+\dfrac{2}{n}\right)^n$

    $= \lim\limits_{x\to 0^+}(1+2x)^{1/x}$

    $= \lim\limits_{x\to 0^+}\left[(1+2x)^{\frac{1}{2x}}\right]^2 = e^2$

    d. $\lim\limits_{n\to\infty}\left(\dfrac{n-1}{n}\right)^{2n} = \lim\limits_{n\to\infty}\left(1-\dfrac{1}{n}\right)^{2n}$

    $= \lim\limits_{x\to 0^+}(1-x)^{2/x}$

    $= \lim\limits_{x\to 0^+}\left[(1-x)^{\frac{1}{-x}}\right]^{-2} = \dfrac{1}{e^2}$

29. Let y = population in millions, $t = 0$ in 1985,
$a = 0.012$, $b = 0.06$, $y_0 = 10$

$\dfrac{dy}{dt} = 0.012y + 0.06$

$y = \left(10+\dfrac{0.06}{0.012}\right)e^{0.012t} - \dfrac{0.06}{0.012} = 15e^{0.012t} - 5$

From 1985 to 2010 is 25 years. At $t = 25$,

$y = 15e^{0.012\cdot 25} - 5 \approx 15.25$. The population in 2010
will be about 15.25 million.

$N(t) = L(1-e^{-0.1386t})$

$0.99L = L(1-e^{-0.1386t})$

$0.01 = e^{-0.1386t}$

$t = \dfrac{\ln 0.01}{-0.1386} \approx 33$

99% of the people will have heard about the scandal
after 33 days.

31. Maximum population:

$$13{,}500{,}000 \text{ mi}^2 \cdot \frac{640 \text{ acres}}{1 \text{ mi}^2} \cdot \frac{1 \text{ person}}{\frac{1}{2} \text{ acre}}$$

$$= 1.728 \times 10^{10} \text{ people}$$

Let $t = 0$ be in 1998.

$$(5.9 \times 10^9)e^{0.0132t} = 1.728 \times 10^{10}$$

$$t = \frac{\ln\left(\dfrac{1.728 \cdot 10^{10}}{5.9 \cdot 10^9}\right)}{0.0132} \approx 81.4 \text{ years from 1998, or}$$

sometime in the year 2079.

33. a. $k = 0.0132 - 0.0001t$

b. $y' = (0.0132 - 0.0001t)\, y$

c. $\dfrac{dy}{dt} = (0.0132 - 0.0001t)\, y$

$$\frac{dy}{y} = (0.0132 - 0.0001t)\, dt$$

$$\ln y = 0.0132t - 0.00005t^2 + C_0$$

$$y = C_1 e^{0.0132t - 0.00005t^2}$$

The initial condition $y(0) = 5.9$ implies that

$C_1 = 5.9$. Thus $y = 5.9 e^{0.0132t - 0.00005t^2}$

d.

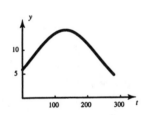

e. The maximum population will occur when

$$\frac{d}{dt}\left(0.0132t - 0.00005t^2\right) = 0$$

$$0.0132 = 0.0001t$$

$$t = 0.0132 / 0.0001 = 132$$

$t = 132$, which is year 2130.

The population will equal the 1998 value of 5.9 billion when $0.0132t - 0.00005t^2 = 0$

$t = 0$ or $t = 264$.

The model predicts that the population will return to the 1998 level in year. Hence,

$$E(x) = E_0 e^{kx} = E(0)e^{kx} = 1 \cdot e^{kx} = e^{kx}.$$

Check: $E(u + v) = e^{k(u+v)} = e^{ku + kv}$

$$= e^{ku} \cdot e^{kv} = E(u) \cdot E(v)$$

35.

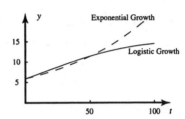

Exponential growth:
In 2010 ($t = 12$): 6.91 billion
In 2040 ($t = 42$): 10.27 billion
In 2090 ($t = 92$): 19.87 billion
Logistic growth:
In 2010 ($t = 12$): 7.29 billion
In 2040 ($t = 42$): 10.74 billion
In 2090 ($t = 92$): 14.28 billion

7.6 Concepts Review

1. $\exp\left(\int P(x)dx\right)$

3. $\dfrac{1}{x}; \dfrac{d}{dx}\left(\dfrac{y}{x}\right) = 1; x^2 + Cx$

Problem Set 7.6

1. Integrating factor is e^x.

$$D(ye^x) = 1$$

$$y = e^{-x}(x + C)$$

3. $y' + \dfrac{x}{1 - x^2} y = \dfrac{ax}{1 - x^2}$

Integrating factor:

$$\exp\int \frac{x}{1 - x^2} dx = \exp\left[\ln(1 - x^2)^{-1/2}\right]$$

$$= (1 - x^2)^{-1/2}$$

$$D[y(1 - x^2)^{-1/2}] = ax(1 - x^2)^{-3/2} \text{ (see note 2}$$

above.) Then $y(1 - x^2)^{-1/2} = a(1 - x^2)^{-1/2} + C$, so

$$y = a + C(1 - x^2)^{1/2}.$$

5. Integrating factor is $\dfrac{1}{x}$.

$$D\left[\frac{y}{x}\right] = e^x$$

$$y = xe^x + Cx$$

7. Integrating factor is x. $D[yx] = 1; \; y = 1 + Cx^{-1}$

9. $y' + f(x)y = f(x)$

Integrating factor: $e^{\int f(x)dx}$

$$D\left[ye^{\int f(x)dx}\right] = f(x)e^{\int f(x)dx}$$

Then $ye^{\int f(x)dx} = e^{\int f(x)dx} + C$, so

$y = 1 + Ce^{-\int f(x)dx}$.

11. Integrating factor is $\dfrac{1}{x}$. $D\left[\dfrac{y}{x}\right] = 3x^2$; $y = x^4 + Cx$

$y = x^4 + 2x$ goes through (1, 3).

13. Integrating factor: xe^x

$d[yxe^x] = 1$; $y = e^{-x}(1 + Cx^{-1})$; $y = e^{-x}(1 - x^{-1})$
through (1, 0).

15. Let y denote the number of pounds of chemical A after t minutes.

$$\frac{dy}{dt} = \left(2\,\frac{\text{lbs}}{\text{gal}}\right)\left(3\,\frac{\text{gal}}{\text{min}}\right) - \left(\frac{y\,\text{lbs}}{20\,\text{gal}}\right)\left(\frac{3\,\text{gal}}{\text{min}}\right)$$

$$= 6 - \frac{3y}{20}\ \text{lb/min}$$

$$y' + \frac{3}{20}y = 6$$

Integrating factor: $e^{\int (3/20)dt} = e^{3t/20}$

$D[ye^{3t/20}] = 6e^{3t/20}$

Then $ye^{3t/20} = 40e^{3t/20} + C$. $t = 0, y = 10$

$\Rightarrow C = -30$.

Therefore, $y(t) = 40 - 30e^{-3t/20}$, so

$y(20) = 40 - 30e^{-3} \approx 38.506$ lb.

17. $\dfrac{dy}{dt} = 4 - \left[\dfrac{y}{(120 - 2t)}\right](6)$ or $y' + \left[\dfrac{3}{(60 - t)}\right]y = 4$

Integrating factor is $(60 - t)^{-3}$.

$D[y(60 - t)^{-3}] = 4(60 - t)^{-3}$

$y(t) = 2(60 - t) + C(60 - t)^3$

$y(t) = 2(60 - t) - \left(\dfrac{1}{1800}\right)(60 - t)^3$ goes through

(0, 0).

19. $I' + 10^6 I = 1$

Integrating factor $= \exp(10^6 t)$

$D[I\exp(10^6 t)] = \exp(10^6 t)$

$I(t) = 10^{-6} + C\exp(-10^6 t)$

$I(t) = 10^{-6}[1 - \exp(-10^6 t)]$ goes through (0, 0).

21. $1000\,I = 120 \sin 377t$

$I(t) = 0.12 \sin 377t$

23. Let y be the number of gallons of pure alcohol in the tank at time t.

a. $y' = \dfrac{dy}{dt} = 5(0.25) - \left(\dfrac{5}{100}\right)y = 1.25 - 0.05y$

Integrating factor is $e^{0.05t}$.

$y(t) = 25 + Ce^{-0.05t}$; $y = 100, t = 0, C = 75$

$y(t) = 25 + 75e^{-0.05t}$; $y = 50, t = T$,

$T = 20(\ln 3) \approx 21.97$ min

b. Let A be the number of gallons of pure alcohol drained away.

$$(100 - A) + 0.25A = 50 \Rightarrow A = \frac{200}{3}$$

It took $\dfrac{\frac{200}{3}}{5}$ minutes for the draining and the same amount of time to refill, so

$$T = \frac{2\left(\frac{200}{3}\right)}{5} = \frac{80}{3} \approx 26.67\ \text{min}.$$

c. c would need to satisfy

$$\frac{\frac{200}{3}}{5} + \frac{\frac{200}{3}}{c} < 20(\ln 3).$$

$$c > \frac{10}{(3\ln 3 - 2)} \approx 7.7170$$

d. $y' = 4(0.25) - 0.05y = 1 - 0.05y$
Solving for y, as in part a, yields
$y = 20 + 80e^{-0.05t}$. The drain is closed when
$t = 0.8T$. We require that

$$(20 + 80e^{-0.05 \cdot 0.8T}) + 4 \cdot 0.25 \cdot 0.25 \cdot 0.8T = 50,$$

or $400e^{-0.04T} + T = 150$.

25. a. $v_\infty = -\dfrac{32}{0.05} = -640$

$v(t) = [120 - (-640)]e^{-0.05t} + (-640) = 0$ if

$t = 20\ln\left(\dfrac{19}{16}\right)$.

$y(t) = 0 + (-640)t$

$+\left(\dfrac{1}{0.05}\right)[120 - (-640)](1 - e^{-0.05t})$

$= -640t + 15,200(1 - e^{-0.05t})$

Therefore, the maximum altitude is

$y\left(20\ln\left(\dfrac{19}{16}\right)\right) = -12,800\ln\left(\dfrac{19}{16}\right) + \dfrac{45,600}{19}$

≈ 200.32

b. $-640T + 15,200(1 - e^{-0.05T}) = 0;$

$95 - 4T - 95e^{-0.05T} = 0$

27. a. $e^{-\ln x + C}\left(\dfrac{dy}{dx} - \dfrac{y}{x}\right) = x^2 e^{-\ln x + C}$

$e^{-\ln x}e^C\left(\dfrac{dy}{dx} - \dfrac{y}{x}\right) = x^2 e^C e^{-\ln x}$

$\dfrac{1}{x}e^C\dfrac{dy}{dx} - ye^C\dfrac{1}{x^2} = x^2 e^C\dfrac{1}{x}$

$\dfrac{d}{dx}\left(e^C\dfrac{1}{x}y\right) = xe^C$

b. $e^C\dfrac{y}{x} = e^C\displaystyle\int x\,dx$

$\dfrac{y}{x} = \dfrac{x^2}{2} + C_1$

$y = \dfrac{x^3}{2} + C_1 x$

7.7 Concepts Review

1. $\left[-\dfrac{\pi}{2}, \dfrac{\pi}{2}\right]$; arcsin

3. 1

Problem Set 7.7

1. $\arccos\left(\dfrac{\sqrt{2}}{2}\right) = \dfrac{\pi}{4}$ since $\cos\dfrac{\pi}{4} = \dfrac{\sqrt{2}}{2}$

3. $\sin^{-1}\left(-\dfrac{\sqrt{3}}{2}\right) = -\dfrac{\pi}{3}$ since $\sin\left(-\dfrac{\pi}{3}\right) = -\dfrac{\sqrt{3}}{2}$

5. $\arctan(\sqrt{3}) = \dfrac{\pi}{3}$ since $\tan\left(\dfrac{\pi}{3}\right) = \sqrt{3}$

7. $\arcsin\left(-\dfrac{1}{2}\right) = -\dfrac{\pi}{6}$ since $\sin\left(-\dfrac{\pi}{6}\right) = -\dfrac{1}{2}$

9. $\sin(\sin^{-1}0.4567) = 0.4567$ by definition

11. $\sin^{-1}(0.1113) \approx 0.1115$

13. $\cos(\text{arccot}\,3.212) = \cos\left(\arctan\dfrac{1}{3.212}\right)$

$\approx \cos 0.3018 \approx 0.9548$

15. $\sec^{-1}(-2.222) = \cos^{-1}\left(\dfrac{1}{-2.222}\right) \approx 2.038$

17. $\cos(\sin(\tan^{-1}2.001)) \approx 0.6259$

19. $\theta = \sin^{-1}\dfrac{x}{8}$

21. $\theta = \sin^{-1}\dfrac{5}{x}$

23. Let θ_1 be the angle opposite the side of length 3, and $\theta_2 = \theta_1 - \theta$, so $\theta = \theta_1 - \theta_2$. Then $\tan\theta_1 = \dfrac{3}{x}$

and $\tan\theta_2 = \dfrac{1}{x}$. $\theta = \tan^{-1}\dfrac{3}{x} - \tan^{-1}\dfrac{1}{x}$.

25. $\cos\left[2\sin^{-1}\left(-\dfrac{2}{3}\right)\right] = 1 - 2\sin^2\left[\sin^{-1}\left(-\dfrac{2}{3}\right)\right]$

$= 1 - 2\left(-\dfrac{2}{3}\right)^2 = \dfrac{1}{9}$

27. $\sin\left[\cos^{-1}\left(\dfrac{3}{5}\right) + \cos^{-1}\left(\dfrac{5}{13}\right)\right] = \sin\left[\cos^{-1}\left(\dfrac{3}{5}\right)\right]\cos\left[\cos^{-1}\left(\dfrac{5}{13}\right)\right] + \cos\left[\cos^{-1}\left(\dfrac{3}{5}\right)\right]\sin\left[\cos^{-1}\left(\dfrac{5}{13}\right)\right]$

$= \sqrt{1 - \left(\dfrac{3}{5}\right)^2}\cdot\dfrac{5}{13} + \dfrac{3}{5}\sqrt{1 - \left(\dfrac{5}{13}\right)^2} = \dfrac{56}{65}$

29. $\tan(\sin^{-1} x) = \dfrac{\sin(\sin^{-1} x)}{\cos(\sin^{-1} x)} = \dfrac{x}{\sqrt{1-x^2}}$

31. $\cos(2\sin^{-1} x) = 1 - 2\sin^2(\sin^{-1} x) = 1 - 2x^2$

33. a. $\displaystyle\lim_{x\to\infty} \tan^{-1} x = \dfrac{\pi}{2}$ since $\displaystyle\lim_{\theta\to\pi/2^-} \tan\theta = \infty$

   b. $\displaystyle\lim_{x\to-\infty} \tan^{-1} x = -\dfrac{\pi}{2}$ since $\displaystyle\lim_{\theta\to-\pi/2^+} \tan\theta = -\infty$

35.

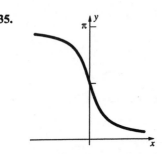

37. $\dfrac{d}{dx}\ln(\sec x + \tan x) = \dfrac{\sec x \tan x + \sec^2 x}{\sec x + \tan x}$

$= \dfrac{(\sec x)(\tan x + \sec x)}{\sec x + \tan x} = \sec x$

39. $\dfrac{d}{dx}\sin^{-1}(2x^2) = \dfrac{1}{\sqrt{1-(2x^2)^2}} \cdot 4x = \dfrac{4x}{\sqrt{1-4x^4}}$

41. $\dfrac{d}{dx}[x^3 \tan^{-1}(e^x)] = x^3 \cdot \dfrac{e^x}{1+(e^x)^2} + 3x^2 \tan^{-1}(e^x)$

$= x^2\left[\dfrac{xe^x}{1+e^{2x}} + 3\tan^{-1}(e^x)\right]$

43. $\dfrac{d}{dx}(\tan^{-1} x)^3 = 3(\tan^{-1} x)^2 \cdot \dfrac{1}{1+x^2} = \dfrac{3(\tan^{-1} x)^2}{1+x^2}$

45. $\dfrac{d}{dx}\sec^{-1}(x^3) = \dfrac{1}{|x^3|\sqrt{(x^3)^2-1}} \cdot 3x^2 = \dfrac{3}{|x|\sqrt{x^6-1}}$

47. $\dfrac{d}{dx}(1+\sin^{-1} x)^3 = 3(1+\sin^{-1} x)^2 \cdot \dfrac{1}{\sqrt{1-x^2}}$

$= \dfrac{3(1+\sin^{-1} x)^2}{\sqrt{1-x^2}}$

49. Let $u = \sin 2x$, so $du = 2\cos 2x\, dx$.

$\displaystyle\int \sin 2x \cos 2x\, dx = \dfrac{1}{2}\int \sin 2x(2\cos 2x)dx$

$= \dfrac{1}{2}\int u\, du$

$= \dfrac{u^2}{4} + C = \dfrac{1}{4}\sin^2 2x + C$

51. Let $u = e^{2x}$, so $du = 2e^{2x}dx$.

$\displaystyle\int e^{2x}\cos(e^{2x})dx = \dfrac{1}{2}\int \cos(e^{2x})(2e^{2x})dx$

$= \dfrac{1}{2}\int \cos u\, du$

$= \dfrac{1}{2}\sin u + C = \dfrac{1}{2}\sin(e^{2x}) + C$

$\displaystyle\int_0^1 e^{2x}\cos(e^{2x})\,dx = \left[\dfrac{1}{2}\sin(e^{2x})\right]_0^1$

$= \left[\dfrac{1}{2}\sin(e^2) - \dfrac{1}{2}\sin(e^0)\right]$

$= \dfrac{\sin e^2 - \sin 1}{2} \approx 0.0262$

53. $\displaystyle\int_0^{\sqrt{2}/2} \dfrac{1}{\sqrt{1-x^2}}\,dx = [\arcsin x]_0^{\sqrt{2}/2}$

$= \arcsin\dfrac{\sqrt{2}}{2} - \arcsin 0 = \dfrac{\pi}{4}$

55. $\displaystyle\int_{-1}^1 \dfrac{1}{1+x^2}\,dx = \left[\tan^{-1} x\right]_{-1}^1 = \tan^{-1} 1 - \tan^{-1}(-1)$

$= \dfrac{\pi}{4} - \left(-\dfrac{\pi}{4}\right) = \dfrac{\pi}{2}$

57. Let $u = 2x$, so $du = 2\, dx$.

$\displaystyle\int \dfrac{1}{1+4x^2}\,dx = \dfrac{1}{2}\int \dfrac{1}{1+(2x)^2}2dx$

$= \dfrac{1}{2}\int \dfrac{1}{1+u^2}\,du = \dfrac{1}{2}\arctan u + C$

$= \dfrac{1}{2}\arctan 2x + C$

59. The top of the picture is 7.6 ft above eye level, and the bottom of the picture is 2.6 ft above eye level. Let θ_1 be the angle between the viewer's line of sight to the top of the picture and the horizontal. Then call $\theta_2 = \theta_1 - \theta$, so $\theta = \theta_1 - \theta_2$.

$\tan\theta_1 = \dfrac{7.6}{b}; \tan\theta_2 = \dfrac{2.6}{b};$

$\theta = \tan^{-1}\dfrac{7.6}{b} - \tan^{-1}\dfrac{2.6}{b}$

If $b = 12.9$, $\theta \approx 0.3335$ or $19.1°$.

61. $\tan\left[2\tan^{-1}\left(\frac{1}{4}\right)\right] = \dfrac{2\tan\left[\tan^{-1}\left(\frac{1}{4}\right)\right]}{1-\tan^2\left[\tan^{-1}\left(\frac{1}{4}\right)\right]}$

$= \dfrac{2\cdot\frac{1}{4}}{1-\left(\frac{1}{4}\right)^2} = \dfrac{8}{15}$

$\tan\left[3\tan^{-1}\left(\frac{1}{4}\right)\right] = \tan\left[2\tan^{-1}\left(\frac{1}{4}\right)+\tan^{-1}\left(\frac{1}{4}\right)\right]$

$= \dfrac{\tan\left[2\tan^{-1}\left(\frac{1}{4}\right)\right]+\tan\left[\tan^{-1}\left(\frac{1}{4}\right)\right]}{1-\tan\left[2\tan^{-1}\left(\frac{1}{4}\right)\right]\tan\left[\tan^{-1}\left(\frac{1}{4}\right)\right]}$

$= \dfrac{\frac{8}{15}+\frac{1}{4}}{1-\frac{8}{15}\cdot\frac{1}{4}} = \dfrac{47}{52}$

$\tan\left[3\tan^{-1}\left(\frac{1}{4}\right)+\tan^{-1}\left(\frac{5}{99}\right)\right]$

$= \dfrac{\tan\left[3\tan^{-1}\left(\frac{1}{4}\right)\right]+\tan\left[\tan^{-1}\left(\frac{5}{99}\right)\right]}{1-\tan\left[3\tan^{-1}\left(\frac{1}{4}\right)\right]\tan\left[\tan^{-1}\left(\frac{5}{99}\right)\right]}$

$= \dfrac{\frac{47}{52}+\frac{5}{99}}{1-\frac{47}{52}\cdot\frac{5}{99}} = \dfrac{4913}{4913} = 1 = \tan\dfrac{\pi}{4}$

Thus, $3\tan^{-1}\left(\frac{1}{4}\right)+\tan^{-1}\left(\frac{5}{99}\right) = \tan^{-1}(1) = \dfrac{\pi}{4}$.

63.

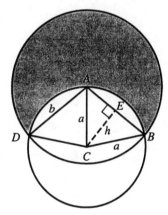

Let θ represent $\angle DAB$, then $\angle CAB$ is $\dfrac{\theta}{2}$. Since

$\triangle ABC$ is isosceles, $|AE| = \dfrac{b}{2}$, $\cos\dfrac{\theta}{2} = \dfrac{\frac{b}{2}}{a} = \dfrac{b}{2a}$ and

$\theta = 2\cos^{-1}\dfrac{b}{2a}$. Thus sector ADB has area

$\dfrac{1}{2}\left(2\cos^{-1}\dfrac{b}{2a}\right)b^2 = b^2\cos^{-1}\dfrac{b}{2a}$. Let ϕ represent

$\angle DCB$, then $\angle ACB$ is $\dfrac{\phi}{2}$ and $\angle ECA$ is $\dfrac{\phi}{4}$, so

$\sin\dfrac{\phi}{4} = \dfrac{\frac{b}{2}}{a} = \dfrac{b}{2a}$ and $\phi = 4\sin^{-1}\dfrac{b}{2a}$. Thus sector DCB

has area $\dfrac{1}{2}\left(4\sin^{-1}\dfrac{b}{2a}\right)a^2 = 2a^2\sin^{-1}\dfrac{b}{2a}$. These

sectors overlap on the triangles $\triangle DAC$ and $\triangle CAB$,
each of which has area

$\dfrac{1}{2}|AB|h = \dfrac{1}{2}b\sqrt{a^2-\left(\dfrac{b}{2}\right)^2} = \dfrac{1}{2}b\dfrac{\sqrt{4a^2-b^2}}{2}$.

The large circle has area πb^2, hence the shaded
region has area

$\pi b^2 - b^2\cos^{-1}\dfrac{b}{2a} - 2a^2\sin^{-1}\dfrac{b}{2a} + \dfrac{1}{2}b\sqrt{4a^2-b^2}$

65.

It is the same graph as $y = \arccos x$.

Conjecture: $\dfrac{\pi}{2} - \arcsin x = \arccos x$

Proof: Let $\theta = \dfrac{\pi}{2} - \arcsin x$

Then $x = \sin\left(\dfrac{\pi}{2}-\theta\right) = \cos\theta$

so $\theta = \arccos x$.

67. $\displaystyle\int\dfrac{dx}{\sqrt{a^2-x^2}} = \int\dfrac{dx}{\sqrt{a^2\left[1-\left(\frac{x}{a}\right)^2\right]}}$

$= \displaystyle\int\dfrac{1}{|a|}\cdot\dfrac{dx}{\sqrt{1-\left(\frac{x}{a}\right)^2}} = \int\dfrac{1}{a}\cdot\dfrac{dx}{\sqrt{1-\left(\frac{x}{a}\right)^2}}$ since $a > 0$

Let $u = \dfrac{x}{a}$, so $du = \dfrac{1}{a}dx$.

$\displaystyle\int\dfrac{1}{a}\cdot\dfrac{dx}{\sqrt{1-\left(\frac{x}{a}\right)^2}} = \int\dfrac{du}{\sqrt{1-u^2}} = \sin^{-1}u + C$

$= \sin^{-1}\dfrac{x}{a} + C$

69. Let $u = \dfrac{x}{a}$, so $du = \dfrac{1}{a}dx$

$$\int \frac{dx}{a^2 + x^2} = \frac{1}{a}\int \frac{1}{1+\left(\frac{x}{a}\right)^2}\frac{1}{a}dx$$

$$= \frac{1}{a}\int \frac{1}{1+u^2}du = \frac{1}{a}\tan^{-1}u + C$$

$$= \frac{1}{a}\tan^{-1}\left(\frac{x}{a}\right) + C$$

71. Note that $\dfrac{d}{dx}\sin^{-1}\left(\dfrac{x}{a}\right) = \dfrac{1}{\sqrt{a^2 - x^2}}$ (See Problem 67).

$$\frac{d}{dx}\left[\frac{x}{2}\sqrt{a^2 - x^2} + \frac{a^2}{2}\sin^{-1}\frac{x}{a} + C\right]$$

$$= \frac{1}{2}\sqrt{a^2 - x^2} + \frac{x}{2}\frac{1}{2\sqrt{a^2 - x^2}}(-2x)$$

$$+ \frac{a^2}{2}\frac{1}{\sqrt{a^2 - x^2}} + 0$$

$$= \frac{1}{2}\sqrt{a^2 - x^2} + \frac{1}{2}\frac{-x^2 + a^2}{\sqrt{a^2 - x^2}} = \sqrt{a^2 - x^2}$$

73. Let θ be the angle subtended by the viewer's eye.

$$\theta = \tan^{-1}\left(\frac{12}{b}\right) - \tan^{-1}\left(\frac{2}{b}\right)$$

$$\frac{d\theta}{db} = \frac{1}{1+\left(\frac{12}{b}\right)^2}\left(-\frac{12}{b^2}\right) - \frac{1}{1+\left(\frac{2}{b}\right)^2}\left(-\frac{2}{b^2}\right)$$

$$= \frac{2}{b^2 + 4} - \frac{12}{b^2 + 144} = \frac{10(24 - b^2)}{(b^2 + 4)(b^2 + 144)}$$

Since $\dfrac{d\theta}{db} > 0$ for b in $\left[0, 2\sqrt{6}\right)$

and $\dfrac{d\theta}{db} < 0$ for b in $\left(2\sqrt{6}, \infty\right)$, the angle is

maximized for $b = 2\sqrt{6} \approx 4.899$.
The ideal distance is about 4.9 ft from the wall.

75. Let $h(t)$ represent the height of the elevator (the number of feet above the spectator's line of sight) t seconds after the line of sight passes horizontal, and let $\theta(t)$ denote the angle of elevation.

Then $h(t) = 15t$, so $\theta(t) = \tan^{-1}\left(\dfrac{15t}{60}\right) = \tan^{-1}\left(\dfrac{t}{4}\right)$.

$$\frac{d\theta}{dt} = \frac{1}{1+\left(\frac{t}{4}\right)^2}\left(\frac{1}{4}\right) = \frac{4}{16 + t^2}$$

At $t = 6$, $\dfrac{d\theta}{dt} = \dfrac{4}{16 + 6^2} = \dfrac{1}{13}$ radians per second or about 4.41° per second.

77. Let x represent the position on the shoreline and let θ represent the angle of the beam ($x = 0$ and $\theta = 0$ when the light is pointed at P). Then

$$\theta = \tan^{-1}\left(\frac{x}{2}\right), \text{ so } \frac{d\theta}{dt} = \frac{1}{1+\left(\frac{x}{2}\right)^2}\frac{1}{2}\frac{dx}{dt} = \frac{2}{4+x^2}\frac{dx}{dt}$$

When $x = 1$,

$$\frac{dx}{dt} = 5\pi, \text{ so } \frac{d\theta}{dt} = \frac{2}{4+1^2}(5\pi) = 2\pi \text{ The beacon}$$

revolves at a rate of 2π radians per minute or 1 revolution per minute.

79. Let x represent the distance to the *center* of the earth and let θ represent the angle subtended by the earth. Then $\theta = 2\sin^{-1}\left(\dfrac{6376}{x}\right)$, so

$$\frac{d\theta}{dt} = 2\frac{1}{\sqrt{1-\left(\frac{6376}{x}\right)^2}}\left(-\frac{6376}{x^2}\right)\frac{dx}{dt}$$

$$= -\frac{12,752}{x\sqrt{x^2 - 6376^2}}\frac{dx}{dt}$$

When she is 3000 km from the surface

$x = 3000 + 6376 = 9376$ and $\dfrac{dx}{dt} = -2$. Substituting

these values, we obtain $\dfrac{d\theta}{dt} \approx 3.96\times 10^{-4}$ radians per second.

7.8 Concepts Review

1. $\dfrac{e^x - e^{-x}}{2}; \dfrac{e^x + e^{-x}}{2}$

3. the graph of $x^2 - y^2 = 1$, a hyperbola

Problem Set 7.8

1. $\cosh x + \sinh x = \dfrac{e^x + e^{-x}}{2} + \dfrac{e^x - e^{-x}}{2}$

$$= \frac{2e^x}{2} = e^x$$

3. $\cosh x - \sinh x = \dfrac{e^x + e^{-x}}{2} - \dfrac{e^x - e^{-x}}{2}$

$= \dfrac{2e^{-x}}{2} = e^{-x}$

5. $\sinh x \cosh y + \cosh x \sinh y = \dfrac{e^x - e^{-x}}{2} \cdot \dfrac{e^y + e^{-y}}{2} + \dfrac{e^x + e^{-x}}{2} \cdot \dfrac{e^y - e^{-y}}{2}$

$= \dfrac{e^{x+y} + e^{x-y} - e^{-x+y} - e^{-x-y}}{4} + \dfrac{e^{x+y} - e^{x-y} + e^{-x+y} - e^{-x-y}}{4}$

$= \dfrac{2e^{x+y} - 2e^{-(x+y)}}{4} = \dfrac{e^{x+y} - e^{-(x+y)}}{2} = \sinh(x+y)$

7. $\cosh x \cosh y + \sinh x \sinh y = \dfrac{e^x + e^{-x}}{2} \cdot \dfrac{e^y + e^{-y}}{2} + \dfrac{e^x - e^{-x}}{2} \cdot \dfrac{e^y - e^{-y}}{2}$

$= \dfrac{e^{x+y} + e^{x-y} + e^{-x+y} + e^{-x-y}}{4} + \dfrac{e^{x+y} - e^{x-y} - e^{-x+y} + e^{-x-y}}{4}$

$= \dfrac{2e^{x+y} + 2e^{-x-y}}{4} = \dfrac{e^{x+y} + e^{-(x+y)}}{2} = \cosh(x+y)$

9. $\dfrac{\tanh x + \tanh y}{1 + \tanh x \tanh y} = \dfrac{\dfrac{\sinh x}{\cosh x} + \dfrac{\sinh y}{\cosh y}}{1 + \dfrac{\sinh x}{\cosh x} \cdot \dfrac{\sinh y}{\cosh y}}$

$= \dfrac{\sinh x \cosh y + \cosh x \sinh y}{\cosh x \cosh y + \sinh x \sinh y} = \dfrac{\sinh(x+y)}{\cosh(x+y)}$

$= \tanh(x+y)$

11. $2 \sinh x \cosh x = \sinh x \cosh x + \cosh x \sinh x$
$= \sinh(x+x) = \sinh 2x$

13. $D_x \sinh^2 x = 2 \sinh x \cosh x = \sinh 2x$

15. $D_x (5 \sinh^2 x) = 10 \sinh x \cdot \cosh x = 5 \sinh 2x$

17. $D_x \cosh(3x+1) = \sinh(3x+1) \cdot 3 = 3 \sinh(3x+1)$

19. $D_x \ln(\sinh x) = \dfrac{1}{\sinh x} \cdot \cosh x = \dfrac{\cosh x}{\sinh x}$

$= \coth x$

21. $D_x (x^2 \cosh x) = x^2 \cdot \sinh x + \cosh x \cdot 2x$

$= x^2 \sinh x + 2x \cosh x$

23. $D_x (\cosh 3x \sinh x) = \cosh 3x \cdot \cosh x + \sinh x \cdot \sinh 3x \cdot 3 = \cosh 3x \cosh x + 3 \sinh 3x \sinh x$

25. $D_x (\tanh x \sinh 2x) = \tanh x \cdot \cosh 2x \cdot 2 + \sinh 2x \cdot \text{sech}^2 x = 2 \tanh x \cosh 2x + \sinh 2x \, \text{sech}^2 x$

27. $D_x \sinh^{-1}(x^2) = \dfrac{1}{\sqrt{(x^2)^2 + 1}} \cdot 2x = \dfrac{2x}{\sqrt{x^4 + 1}}$

29. $D_x \tanh^{-1}(2x-3) = \dfrac{1}{1-(2x-3)^2} \cdot 2 = \dfrac{2}{1-(4x^2-12x+9)} = \dfrac{2}{-4x^2+12x-8} = -\dfrac{1}{2(x^2-3x+2)}$

31. $D_x [x \cosh^{-1}(3x)] = x \cdot \dfrac{1}{\sqrt{(3x)^2 - 1}} \cdot 3 + \cosh^{-1}(3x) \cdot 1 = \dfrac{3x}{\sqrt{9x^2-1}} + \cosh^{-1} 3x$

33. $D_x \ln(\cosh^{-1} x) = \dfrac{1}{\cosh^{-1} x} \cdot \dfrac{1}{\sqrt{x^2 - 1}}$

$= \dfrac{1}{\sqrt{x^2 - 1}\,\cosh^{-1} x}$

35. $D_x \tanh(\cot x) = \operatorname{sech}^2(\cot x) \cdot (-\csc^2 x)$

$= -\csc^2 x\,\operatorname{sech}^2(\cot x)$

37. Area $= \displaystyle\int_0^{\ln 3} \cosh 2x\,dx = \left[\dfrac{1}{2}\sinh 2x\right]_0^{\ln 3}$

$= \dfrac{1}{2}\left(\dfrac{e^{2\ln 3} - e^{-2\ln 3}}{2} - \dfrac{e^0 - e^{-0}}{2}\right)$

$= \dfrac{1}{4}(e^{\ln 9} - e^{\ln\frac{1}{9}}) = \dfrac{1}{4}\left(9 - \dfrac{1}{9}\right) = \dfrac{20}{9}$

39. Let $u = \pi x^2 + 5$, so $du = 2\pi x\,dx$.

$\displaystyle\int x\cosh(\pi x^2 + 5)dx = \dfrac{1}{2\pi}\int \cosh u\,du$

$= \dfrac{1}{2\pi}\sinh u + C = \dfrac{1}{2\pi}\sinh(\pi x^2 + 5) + C$

41. Let $u = 2z^{1/4}$, so $du = \dfrac{1}{4}\cdot 2z^{-3/4}dz = \dfrac{1}{2\sqrt[4]{z^3}}dz$.

$\displaystyle\int \dfrac{\sinh(2z^{1/4})}{\sqrt[4]{z^3}}dz = 2\int \sinh u\,du = 2\cosh u + C$

$= 2\cosh(2z^{1/4}) + C$

43. Let $u = \sin x$, so $du = \cos x\,dx$

$\displaystyle\int \cos x\sinh(\sin x)dx = \int \sinh u\,du = \cosh u + C$

$= \cosh(\sin x) + C$

45. Let $u = \ln(\sinh x^2)$, so

$du = \dfrac{1}{\sinh x^2}\cdot \cosh x^2 \cdot 2x\,dx = 2x\coth x^2\,dx$.

53. $y = a\cosh\left(\dfrac{x}{a}\right) + C$

$\dfrac{dy}{dx} = \sinh\left(\dfrac{x}{a}\right)$

$\dfrac{d^2 y}{dx^2} = \dfrac{1}{a}\cosh\left(\dfrac{x}{a}\right)$

We need to show that $\dfrac{d^2 y}{dx^2} = \dfrac{1}{a}\sqrt{1 + \left(\dfrac{dy}{dx}\right)^2}$.

$\displaystyle\int x\coth x^2 \ln(\sinh x^2)dx = \dfrac{1}{2}\int u\,du = \dfrac{1}{2}\cdot\dfrac{u^2}{2} + C$

$= \dfrac{1}{4}[\ln(\sinh x^2)]^2 + C$

47. Note that the graphs of $y = \sinh x$ and $y = 0$ intersect at the origin.

Area $= \displaystyle\int_0^{\ln 2} \sinh x\,dx = [\cosh x]_0^{\ln 2}$

$= \dfrac{e^{\ln 2} + e^{-\ln 2}}{2} - \dfrac{e^0 + e^0}{2} = \dfrac{1}{2}\left(2 + \dfrac{1}{2}\right) - 1 = \dfrac{1}{4}$

49. Volume $= \displaystyle\int_0^1 \pi\cosh^2 x\,dx = \dfrac{\pi}{2}\int_0^1 (1 + \cosh 2x)dx$

$= \dfrac{\pi}{2}\left[x + \dfrac{\sinh 2x}{2}\right]_0^1$

$= \dfrac{\pi}{2}\left(1 + \dfrac{\sinh 2}{2} - 0\right)$

$= \dfrac{\pi}{2} + \dfrac{\pi\sinh 2}{4} \approx 4.42$

51. Note that $1 + \sinh^2 x = \cosh^2 x$ and

$\cosh^2 x = \dfrac{1 + \cosh 2x}{2}$

Surface area $= \displaystyle\int_0^1 2\pi y\sqrt{1 + \left(\dfrac{dy}{dx}\right)^2}\,dx$

$= \displaystyle\int_0^1 2\pi\cosh x\sqrt{1 + \sinh^2 x}\,dx$

$= \displaystyle\int_0^1 2\pi\cosh x\cosh x\,dx$

$= \displaystyle\int_0^1 \pi(1 + \cosh 2x)dx$

$= \left[\pi x + \dfrac{\pi}{2}\sinh 2x\right]_0^1 = \pi + \dfrac{\pi}{2}\sinh 2 \approx 8.84$

Note that $1+\sinh^2\left(\dfrac{x}{a}\right) = \cosh^2\left(\dfrac{x}{a}\right)$ and $\cosh\left(\dfrac{x}{a}\right) > 0.$ Therefore,

$$\frac{1}{a}\sqrt{1+\left(\frac{dy}{dx}\right)^2} = \frac{1}{a}\sqrt{1+\sinh^2\left(\frac{x}{a}\right)} = \frac{1}{a}\sqrt{\cosh^2\left(\frac{x}{a}\right)} = \frac{1}{a}\cosh\left(\frac{x}{a}\right) = \frac{d^2y}{dx^2}$$

55. a.

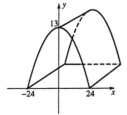

b. Area under the curve is

$$\int_{-24}^{24}\left[37-24\cosh\left(\frac{x}{24}\right)\right]dx = \left[37x-576\sinh\left(\frac{x}{24}\right)\right]_{-24}^{24} \approx 422$$

Volume is about $(422)(100) = 42{,}200$ ft^3.

c. Length of the curve is

$$\int_{-24}^{24}\sqrt{1+\left(\frac{dy}{dx}\right)^2}\,dx = \int_{-24}^{24}\sqrt{1+\sinh^2\left(\frac{x}{24}\right)}\,dx = \int_{-24}^{24}\cosh\left(\frac{x}{24}\right)dx = \left[24\sinh\left(\frac{x}{24}\right)\right]_{-24}^{24} = 48\sinh 1 \approx 56.4$$

Surface area $\approx (56.4)(100) = 5640$ ft^2

57. a. $(\sinh x + \cosh x)^r = \left(\dfrac{e^x-e^{-x}}{2}+\dfrac{e^x+e^{-x}}{2}\right)^r = \left(\dfrac{2e^x}{2}\right)^r = e^{rx}$

$\sinh rx + \cosh rx = \dfrac{e^{rx}-e^{-rx}}{2}+\dfrac{e^{rx}+e^{-rx}}{2} = \dfrac{2e^{rx}}{2} = e^{rx}$

b. $(\cosh x - \sinh x)^r = \left(\dfrac{e^x+e^{-x}}{2}-\dfrac{e^x-e^{-x}}{2}\right)^r = \left(\dfrac{2e^{-x}}{2}\right)^r = e^{-rx}$

$\cosh rx - \sinh rx = \dfrac{e^{rx}+e^{-rx}}{2}-\dfrac{e^{rx}-e^{-rx}}{2} = \dfrac{2e^{-rx}}{2} = e^{-rx}$

c. $(\cos x + i\sin x)^r = \left(\dfrac{e^{ix}+e^{-ix}}{2}+i\dfrac{e^{ix}-e^{-ix}}{2i}\right)^r = \left(\dfrac{2e^{ix}}{2}\right)^r = e^{irx}$

$\cos rx + i\sin rx = \dfrac{e^{irx}+e^{-irx}}{2}+i\dfrac{e^{irx}-e^{-irx}}{2i} = \dfrac{2e^{irx}}{2} = e^{irx}$

d. $(\cos x - i\sin x)^r = \left(\dfrac{e^{ix}+e^{-ix}}{2}-i\dfrac{e^{ix}-e^{-ix}}{2i}\right)^r = \left(\dfrac{2e^{-ix}}{2}\right)^r = e^{-irx}$

$\cos rx - i\sin rx = \dfrac{e^{irx}+e^{-irx}}{2}-i\dfrac{e^{irx}-e^{-irx}}{2i} = \dfrac{2e^{-irx}}{2} = e^{-irx}$

59. Area $= \int_0^x \cosh t \, dt = [\sinh t]_0^x = \sinh x$

Arc length $=$

$\int_0^x \sqrt{1+[D_t \cosh t]^2} \, dt = \int_0^x \sqrt{1+\sinh^2 t} \, dt$

$= \int_0^x \cosh t \, dt = [\sinh t]_0^x = \sinh x$

61.

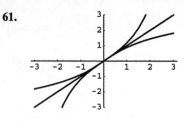

The functions $y = \sinh x$ and $y = \ln(x+\sqrt{x^2+1})$ are inverse functions.

7.9 Chapter Review

Concepts Test

1. False: $\ln 0$ is undefined.

3. True: $\int_1^{e^3} \frac{1}{t} \, dt = \left[\ln|t|\right]_1^{e^3} = \ln e^3 - \ln 1 = 3$

5. True: The range of $y = \ln x$ is the set of all real numbers.

7. False: $4\ln x = \ln(x^4)$

9. True: $f(g(x)) = 4 + e^{\ln(x-4)}$
$= 4 + (x-4) = x$
and
$g(f(x)) = \ln(4 + e^x - 4) = \ln e^x = x$

11. True: $\ln x$ is an increasing function.

13. True: $e^z > 0$ for all z.

15. True: $\lim_{x \to 0^+} (\ln \sin x - \ln x)$
$= \lim_{x \to 0^+} \ln\left(\frac{\sin x}{x}\right) = \ln 1 = 0$

17. False: $\ln \pi$ is a constant so $\frac{d}{dx} \ln \pi = 0$.

19. True: e is a number.

21. False: $D_x(x^x) = x^x(1 + \ln x)$

23. True: The integrating factor is
$e^{\int 4/x \, dx} = e^{4\ln x} = \left(e^{\ln x}\right)^4 = x^4$

25. False: $\arcsin(\sin 2\pi) = \arcsin 0 = 0$

27. False: $\cosh x$ is not increasing.

29. True: $|\sinh x| \le \frac{1}{2}e^{|x|}$ is equivalent to

$\left|e^x - e^{-x}\right| \le e^{|x|}$. When $x = 0$,

$\sinh x = 0 < \frac{1}{2}e^0 = \frac{1}{2}$. If $x > 0$,

$e^x > 1$ and $e^{-x} < 1 < e^x$, thus
$\left|e^x - e^{-x}\right| = e^x - e^{-x} < e^x = e^{|x|}$.

If $x < 0$, $e^{-x} > 1$ and $e^x < 1 < e^{-x}$, thus
$\left|e^x - e^{-x}\right| = -(e^x - e^{-x})$
$= e^{-x} - e^x < e^{-x} = e^{|x|}$.

31. False: $\cosh(\ln 3) = \dfrac{e^{\ln 3} + e^{-\ln 3}}{2}$

$= \frac{1}{2}\left(3 + \frac{1}{3}\right) = \frac{5}{3}$

33. True: $\lim_{x \to -\infty} \tan^{-1} x = -\frac{\pi}{2}$, since

$\lim_{x \to -\frac{\pi}{2}^+} \tan x = -\infty$.

35. True: $\tanh x = \dfrac{\sinh x}{\cosh x}$; $\sinh x$ is an odd function and $\cosh x$ is an even function.

37. True: $\ln 3^{100} = 100 \ln 3 > 100 \cdot 1$ since $\ln 3 > 1$.

39. True: y triples every time t increases by t_1.

41. True: $(y(t) + z(t))' = y'(t) + z'(t)$
$= ky(t) + kz(t) = k(y(t) + z(t))$

43. False: Use the substitution $u = -h$.
$\lim_{h \to 0} (1-h)^{-1/h} = \lim_{u \to 0} (1+u)^{1/u} = e$
by Theorem 7.5.A.

45. True: If $D_x(a^x) = a^x \ln a = a^x$, then $\ln a = 1$, so $a = e$.

Sample Test Problems

1. $\ln \dfrac{x^4}{2} = 4 \ln x - \ln 2$

$$\frac{d}{dx} \ln \frac{x^4}{2} = \frac{d}{dx}(4 \ln x - \ln 2) = \frac{4}{x}$$

3. $\dfrac{d}{dx} e^{x^2-4x} = e^{x^2-4x} \dfrac{d}{dx}(x^2 - 4x)$

$$= (2x - 4)e^{x^2-4x}$$

5. $\dfrac{d}{dx} \tan(\ln e^x) = \dfrac{d}{dx} \tan x = \sec^2 x$

7. $\dfrac{d}{dx} 2 \tanh \sqrt{x} = 2 \operatorname{sech}^2 \sqrt{x} \dfrac{d}{dx} \sqrt{x} = \dfrac{\operatorname{sech}^2 \sqrt{x}}{\sqrt{x}}$

9. $\dfrac{d}{dx} \sinh^{-1}(\tan x) = \dfrac{1}{\sqrt{\tan^2 x + 1}} \dfrac{d}{dx} \tan x$

$$= \frac{\sec^2 x}{\sqrt{\tan^2 x + 1}} = \frac{\sec^2 x}{\sqrt{\sec^2 x}} = |\sec x|$$

11. $\dfrac{d}{dx} \sec^{-1} e^x = \dfrac{1}{|e^x| \sqrt{(e^x)^2 - 1}} \dfrac{d}{dx} e^x$

$$= \frac{e^x}{e^x \sqrt{e^{2x} - 1}} = \frac{1}{\sqrt{e^{2x} - 1}}$$

13. $\dfrac{d}{dx} 3 \ln(e^{5x} + 1) = \dfrac{3}{e^{5x} + 1}(5e^{5x}) = \dfrac{15e^{5x}}{e^{5x} + 1}$

15. $\dfrac{d}{dx} \cos e^{\sqrt{x}} = -\sin e^{\sqrt{x}} \dfrac{d}{dx} e^{\sqrt{x}}$

$$= (-\sin e^{\sqrt{x}})e^{\sqrt{x}} \frac{d}{dx} \sqrt{x}$$

$$= -\frac{e^{\sqrt{x}} \sin e^{\sqrt{x}}}{2\sqrt{x}}$$

17. $\dfrac{d}{dx} 2 \cos^{-1} \sqrt{x} = \dfrac{-2}{\sqrt{1 - (\sqrt{x})^2}} \dfrac{d}{dx} \sqrt{x}$

$$= \frac{-2}{\sqrt{1-x}} \frac{1}{2\sqrt{x}} = -\frac{1}{\sqrt{x - x^2}}$$

19. $\dfrac{d}{dx} 2 \csc e^{\ln \sqrt{x}} = \dfrac{d}{dx} 2 \csc \sqrt{x}$

$$= -2 \csc \sqrt{x} \cot \sqrt{x} \frac{d}{dx} \sqrt{x}$$

$$= -\frac{\csc \sqrt{x} \cot \sqrt{x}}{\sqrt{x}}$$

21. $\dfrac{d}{dx} 4 \tan 5x \sec 5x$

$$= 20 \sec^2 5x \sec 5x + 20 \tan 5x \sec 5x \tan 5x$$

$$= 20 \sec 5x(\sec^2 5x + \tan^2 5x)$$

$$= 20 \sec 5x(2 \sec^2 5x - 1)$$

23. $\dfrac{d}{dx} x^{1+x} = \dfrac{d}{dx} e^{(1+x)\ln x}$

$$= e^{(1+x)\ln x} \frac{d}{dx}[(1+x)\ln x]$$

$$= x^{1+x}\left[(1)(\ln x) + (1+x)\left(\frac{1}{x}\right)\right]$$

$$= x^{1+x}\left(\ln x + 1 + \frac{1}{x}\right)$$

25. Let $u = 3x - 1$, so $du = 3\, dx$.

$$\int e^{3x-1} dx = \frac{1}{3}\int e^{3x-1} 3\, dx = \frac{1}{3}\int e^u\, du$$

$$= \frac{1}{3} e^u + C = \frac{1}{3} e^{3x-1} + C$$

Check:

$$\frac{d}{dx}\left(\frac{1}{3} e^{3x-1} + C\right) = \frac{1}{3} e^{3x-1} \frac{d}{dx}(3x - 1) = e^{3x-1}$$

27. Let $u = e^x$, so $du = e^x dx$.

$$\int e^x \sin e^x dx = \int \sin u\, du = -\cos u + C$$

$$= -\cos e^x + C$$

Check:

$$\frac{d}{dx}(-\cos e^x + C) = (\sin e^x) \frac{d}{dx} e^x = e^x \sin e^x$$

29. Let $u = e^{x+3} + 1$, so $du = e^{x+3} dx$.

$$\int \frac{e^{x+2}}{e^{x+3}+1} dx = \frac{1}{e}\int \frac{1}{e^{x+3}+1} e^{x+3} dx = \frac{1}{e}\int \frac{1}{u} du$$

$$= \frac{1}{e} \ln|u| + C = \frac{\ln(e^{x+3}+1)}{e} + C$$

Check:

$$\frac{d}{dx}\left(\frac{\ln(e^{x+3}+1)}{e}+C\right)=\frac{1}{e}\frac{1}{e^{x+3}+1}\frac{d}{dx}(e^{x+3}+1)$$

$$=\frac{e^{x+3}e^{-1}}{e^{x+3}+1}=\frac{e^{x+2}}{e^{x+3}+1}$$

31. Let $u = 2x$, so $du = 2\ dx$.

$$\int\frac{4}{\sqrt{1-4x^2}}dx=2\int\frac{1}{\sqrt{1-(2x)^2}}2dx$$

$$=2\int\frac{1}{\sqrt{1-u^2}}du$$

$$=2\sin^{-1}u+C=2\sin^{-1}2x+C$$

Check:

$$\frac{d}{dx}(2\sin^{-1}2x+C)=2\left(\frac{1}{\sqrt{1-(2x)^2}}\right)\frac{d}{dx}2x$$

$$=\frac{4}{\sqrt{1-4x^2}}$$

33. Let $u = \ln x$, so $du = \frac{1}{x}dx$.

$$\int\frac{-1}{x+x(\ln x)^2}dx=-\int\frac{1}{1+(\ln x)^2}\cdot\frac{1}{x}dx$$

$$=-\int\frac{1}{1+u^2}du=-\tan^{-1}u+C=-\tan^{-1}(\ln x)+C$$

Check:

$$\frac{d}{dx}[-\tan^{-1}(\ln x)+C]=-\frac{1}{1+(\ln x)^2}\frac{d}{dx}\ln x$$

$$=\frac{-1}{x+x(\ln x)^2}$$

35. $f'(x)=\cos x-\sin x;\ f'(x)=0$ when $\tan x=1$,

$x=\dfrac{\pi}{4}$

$f'(x)>0$ when $\cos x>\sin x$ which occurs when

$-\dfrac{\pi}{2}\le x<\dfrac{\pi}{4}$.

$f''(x)=-\sin x-\cos x;\ f''(x)=0$ when

$\tan x=-1,\ x=-\dfrac{\pi}{4}$

$f''(x)>0$ when $\cos x<-\sin x$ which occurs

when $-\dfrac{\pi}{2}\le x<-\dfrac{\pi}{4}$.

Increasing on $\left[-\dfrac{\pi}{2},\dfrac{\pi}{4}\right]$

Decreasing on $\left[\dfrac{\pi}{4},\dfrac{\pi}{2}\right]$

Concave up on $\left(-\dfrac{\pi}{2},-\dfrac{\pi}{4}\right)$

Concave down on $\left(-\dfrac{\pi}{4},\dfrac{\pi}{2}\right)$

Inflection point at $\left(-\dfrac{\pi}{4},0\right)$

Global maximum at $\left(\dfrac{\pi}{4},\sqrt{2}\right)$

Global minimum at $\left(-\dfrac{\pi}{2},-1\right)$

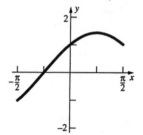

37. a. $f'(x)=5x^4+6x^2+4\ge 4>0$ for all x, so $f(x)$ is increasing.

b. $f(1)=7$, so $g(7)=f^{-1}(7)=1$.

c. $g'(7)=\dfrac{1}{f'(1)}=\dfrac{1}{15}$

39. a. $(\$100)(1.12)=\$112$

b. $(\$100)\left(1+\dfrac{0.12}{12}\right)^{12}\approx\$112.68$

c. $(\$100)\left(1+\dfrac{0.12}{365}\right)^{365}\approx\$112.75$

d. $(\$100)e^{(0.12)(1)}\approx\$112.75$

41. $y=(\cos x)^{\sin x}=e^{\sin x\,\ln(\cos x)}$

$$\frac{dy}{dx}=e^{\sin x\,\ln(\cos x)}\frac{d}{dx}[\sin x\,\ln(\cos x)]$$

$$=e^{\sin x\,\ln(\cos x)}\left[\cos x\,\ln(\cos x)+(\sin x)\left(\frac{1}{\cos x}\right)(-\sin x)\right]$$

$$=(\cos x)^{\sin x}\left[\cos x\,\ln(\cos x)-\frac{\sin^2 x}{\cos x}\right]$$

At $x=0$, $\dfrac{dy}{dx}=1^0(1\ln 1-0)=0$.

The tangent line has slope 0, so it is horizontal:
$y=1$.

43. Integrating factor is $|x|$. $D[y|x|] = 0$; $y = Cx^{-1}$

45. (Linear first-order) $y' + 2xy = 2x$

Integrating factor: $e^{\int 2x\,dx} = e^{x^2}$

$D[ye^{x^2}] = 2xe^{x^2}$; $ye^{x^2} = e^{x^2} + C$;

$y = 1 + Ce^{-x^2}$

If $x = 0$, $y = 3$, then $3 = 1 + C$, so $C = 2$.

Therefore, $y = 1 + 2e^{-x^2}$.

47. Integrating factor is e^{-2x}.

$D[ye^{-2x}] = e^{-x}$; $y = -e^{x} + Ce^{2x}$

7.10 Additional Problem Set

1. $1000e^{(0.05)(1)} = \$1051.27$

3. a. $100e^{0.05(360/365)} + 100e^{0.05(330/365)}$

$+ 100e^{0.05\cdot(300/365)} + 100e^{0.05\cdot(270/365)}$

$+ 100e^{0.05\cdot(240/365)} + 100e^{0.05\cdot(210/365)} + 100e^{0.05\cdot(180/365)} + 100e^{0.05(150/365)}$

$+ 100e^{0.05\cdot(120/365)} + 100e^{0.05\cdot(90/365)} + 100e^{0.05\cdot(60/365)} + 100e^{0.05\cdot(30/365)} \approx \$1232.61$

b. The formula in part a. is a geometric series $\sum_{k=1}^{n} ar^{k-1}$ with $a = 100e^{0.05\cdot(30/365)}$, $r = e^{0.05\cdot(30/365)}$,

and $n = 12$. The sum is $\dfrac{100e^{0.05\cdot(30/365)} - 100e^{0.05\cdot(390/365)}}{1 - e^{0.05\cdot(30/365)}}$.

5. a. If you deposit A_0 into an account earning an interest rate of $100r$ percent, compounded continuously, then the amount you have at time t is $A(t) = A_0 e^{rt}$. The money has doubled when $A(t) = 2A_0$ or $A_0 e^{rt} = 2A_0$, so T satisfies the equation $A_0 e^{rT} = 2A_0$.

$A_0 e^{r(70/100r)} = A_0 e^{0.7} \approx 2A_0$

b. Solving $A_0 e^{rT} = 2A_0$ for T to get $T = \dfrac{1}{r} \ln 2 \approx \dfrac{1}{r}(0.7) = \dfrac{70}{100r}$

c. $T \approx \dfrac{70}{100(0.07)} = 10$

Ten years

7. $f(x) = a_n x^n + a_{n-1}x^{n-1} + \cdots + a_1 x + a_0$

$\lim_{x \to \infty} \dfrac{f'(x)}{f(x)} = \lim_{x \to \infty} \dfrac{na_n x^{n-1} + (n-1)a_{n-1}x^{n-2} + \cdots + a_1}{a_n x^n + a_{n-1}x + \cdots + a_1 x + a_0} = \lim_{x \to \infty} \dfrac{\dfrac{na_n}{x} + \dfrac{(n-1)a_{n-1}}{x^2} + \cdots + \dfrac{a_1}{x^n}}{a_n + \dfrac{a_{n-1}}{x} + \cdots + \dfrac{a_1}{x^{n-1}} + \dfrac{a_0}{x^n}} = 0$

9. $\dfrac{f'(x)}{f(x)} = k < 0$ can be written as $\dfrac{1}{y}\dfrac{dy}{dx} = k$ where $y = f(x)$. $\dfrac{dy}{y} = k\,dx$ has the solution $y = Ce^{kx}$.

Thus, $f(x) = Ce^{kx}$ which represents exponential decay since $k < 0$.

11. a. In order of increasing slope, the graphs represent the curves $y = 2^x$, $y = 3^x$, and $y = 4^x$.

b. $\ln y$ is linear with respect to x, and at $x = 0$,
$y = 1$ since $C = 1$.

c. The graph passes through the points $(0.2, 4)$ and $(0.6, 8)$. Thus, $4 = Cb^{0.2}$ and $8 = Cb^{0.6}$.
Dividing the second equation by the first, gets $2 = b^{0.4}$ so $b = 2^{5/2}$.
Therefore $C = 2^{3/2}$.

13. a. $\eta = 3\operatorname{sech}^2 \dfrac{x}{2}$

$\eta' = -3\operatorname{sech}^2 \dfrac{x}{2}\tanh\dfrac{x}{2}$

$3\eta^2 = 3\cdot 9\operatorname{sech}^4 \dfrac{x}{2} = 27\operatorname{sech}^4 \dfrac{x}{2}$

$\eta^3 + 3(\eta')^2 = 27\operatorname{sech}^6 \dfrac{x}{2} + 3\left(9\operatorname{sech}^4 \dfrac{x}{2}\tanh^2 \dfrac{x}{2}\right)$

$\quad = 27\operatorname{sech}^4 \dfrac{x}{2}\left(\operatorname{sech}^2 \dfrac{x}{2} + \tanh^2 \dfrac{x}{2}\right)$

$\quad = 27\operatorname{sech}^4 \dfrac{x}{2} = 3\eta^2$

Thus, $\eta = 3\operatorname{sech}^2 \dfrac{x}{2}$ satisfies the equation.

b.

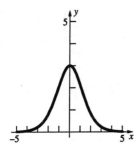

CHAPTER 8

Techniques of Integration

8.1 Concepts Review

1. $\int u^5 \, du$

3. $\int_1^2 u^3 \, du$

Problem Set 8.1

1. $\int (x-2)^5 \, dx = \frac{1}{6}(x-2)^6 + C$

3. $u = x^2 + 1, du = 2x \, dx$

   When $x = 0, u = 1$ and when $x = 1, u = 5$.

   $\int_0^2 x(x^2+1)^5 \, dx = \frac{1}{2}\int_0^2 (x^2+1)^5(2x \, dx)$

   $= \frac{1}{2}\int_1^5 u^5 \, du$

   $= \left[\frac{u^6}{12}\right]_1^5 = \frac{5^5 - 1^5}{12} = \frac{15624}{12} = 1302$

5. $\int \frac{dx}{x^2+4} = \frac{1}{2}\tan^{-1}\left(\frac{x}{2}\right) + C$

7. $u = x^2 + 4, du = 2x \, dx$

   $\int \frac{x}{x^2+4} \, dx = \frac{1}{2}\int \frac{du}{u}$

   $= \frac{1}{2}\ln|u| + C$

   $= \frac{1}{2}\ln|x^2+4| + C$

   $= \frac{1}{2}\ln(x^2+4) + C$

9. $u = 4 + z^2, du = 2z \, dz$

   $\int 6z\sqrt{4+z^2} \, dz = 3\int\sqrt{u} \, du$

   $= 2u^{3/2} + C$

   $= 2(4+z^2)^{3/2} + C$

11. $\int \frac{\tan z}{\cos^2 z} \, dz = \int \tan z \sec^2 z \, dz$

    $u = \tan z, \ du = \sec^2 z \, dz$

    $\int \tan z \sec^2 z \, dz = \int u \, du = \frac{1}{2}u^2 + C$

    $= \frac{1}{2}\tan^2 z + C$

13. $u = \sqrt{t}, du = \frac{1}{2\sqrt{t}} \, dt$

    $\int \frac{\sin\sqrt{t}}{\sqrt{t}} \, dt = 2\int \sin u \, du$

    $= -2\cos u + C$

    $= -2\cos\sqrt{t} + C$

15. $u = \sin x, du = \cos x \, dx$

    $\int_0^{\pi/4} \frac{\cos x}{1+\sin^2 x} \, dx = \int_0^{\sqrt{2}/2} \frac{du}{1+u^2}$

    $= [\tan^{-1} u]_0^{\sqrt{2}/2} = \tan^{-1}\frac{\sqrt{2}}{2}$

    ≈ 0.6155

17. $\int \frac{3x^2+2x}{x+1} \, dx = \int(3x-1)dx + \int \frac{1}{x+1} \, dx$

    $= \frac{3}{2}x^2 - x + \ln|x+1| + C$

19. $u = \ln 4x^2, du = \frac{2}{x} \, dx$

    $\int \frac{\sin(\ln 4x^2)}{x} \, dx = \frac{1}{2}\int \sin u \, du$

    $= -\frac{1}{2}\cos u + C$

    $= -\frac{1}{2}\cos(\ln 4x^2) + C$

21. $\int \frac{6e^x}{\sqrt{1-e^{2x}}} \, dx = 6\sin^{-1}(e^x) + C$

23. $u = 1 - e^{2x}, du = -2e^{2x} \, dx$

    $\int \frac{3e^{2x}}{\sqrt{1-e^{2x}}} \, dx = -\frac{3}{2}\int \frac{du}{\sqrt{u}}$

$$= -3\sqrt{u} + C$$
$$= -3\sqrt{1 - e^{2x}} + C$$

25. $\int_0^1 t3^{t^2} dt = \frac{1}{2}\int_0^1 2t3^{t^2} dt$

$$= \left[\frac{3^{t^2}}{2\ln 3}\right]_0^1 = \frac{3}{2\ln 3} - \frac{1}{2\ln 3}$$

$$= \frac{1}{\ln 3} \approx 0.9102$$

27. $\int \dfrac{\sin x - \cos x}{\sin x} dx = \int\left(1 - \dfrac{\cos x}{\sin x}\right) dx$

$u = \sin x,\ du = \cos x\, dx$

$\int \dfrac{\sin x - \cos x}{\sin x} dx = x - \int \dfrac{du}{u}$

$= x - \ln|u| + C$

$= x - \ln|\sin x| + C$

29. $u = e^x,\ du = e^x dx$

$\int e^x \sec e^x dx = \int \sec u\, du$

$= \ln|\sec u + \tan u| + C$

$= \ln|\sec e^x + \tan e^x| + C$

31. $\int \dfrac{\sec^3 x + e^{\sin x}}{\sec x} dx = \int(\sec^2 x + e^{\sin x}\cos x)\,dx$

$= \tan x + \int e^{\sin x}\cos x\, dx$

$u = \sin x,\ du = \cos x\, dx$

$\tan x + \int e^{\sin x}\cos x\, dx = \tan x + \int e^u\, du$

$= \tan x + e^u + C = \tan x + e^{\sin x} + C$

33. $u = t^3 - 2,\ du = 3t^2 dt$

$\int \dfrac{t^2 \cos(t^3 - 2)}{\sin^2(t^3 - 2)} dt = \dfrac{1}{3}\int \dfrac{\cos u}{\sin^2 u} du$

$v = \sin u,\ dv = \cos u\, du$

$\dfrac{1}{3}\int \dfrac{\cos u}{\sin^2 u} du = \dfrac{1}{3}\int v^{-2} dv = -\dfrac{1}{3}v^{-1} + C$

$= -\dfrac{1}{3\sin u} + C$

$= -\dfrac{1}{3\sin(t^3 - 2)} + C.$

35. $u = t^3 - 2,\ du = 3t^2 dt$

$\int \dfrac{t^2 \cos^2(t^3 - 2)}{\sin^2(t^3 - 2)} dt = \dfrac{1}{3}\int \dfrac{\cos^2 u}{\sin^2 u} du$

$$= \frac{1}{3}\int \cot^2 u\, du = \frac{1}{3}\int(\csc^2 u - 1)du$$

$$= \frac{1}{3}[-\cot u - u] + C_1$$

$$= \frac{1}{3}[-\cot(t^3 - 2) - (t^3 - 2)] + C_1$$

$$= -\frac{1}{3}[\cot(t^3 - 2) + t^3] + C$$

37. $u = \tan^{-1} 2t,\ du = \dfrac{2}{1 + 4t^2} dt$

$\int \dfrac{e^{\tan^{-1} 2t}}{1 + 4t^2} dt = \dfrac{1}{2}\int e^u\, du$

$= \dfrac{1}{2}e^u + C = \dfrac{1}{2}e^{\tan^{-1} 2t} + C$

39. $u = 3y^2,\ du = 6y\, dy$

$\int \dfrac{y}{\sqrt{16 - 9y^4}} dy = \dfrac{1}{6}\int \dfrac{1}{\sqrt{4^2 - u^2}} du$

$= \dfrac{1}{6}\sin^{-1}\left(\dfrac{u}{4}\right) + C$

$= \dfrac{1}{6}\sin^{-1}\left(\dfrac{3y^2}{4}\right) + C$

41. $u = x^3,\ du = 3x^2 dx$

$\int x^2 \sinh x^3 dx = \dfrac{1}{3}\int \sinh u\, du$

$= \dfrac{1}{3}\cosh u + C$

$= \dfrac{1}{3}\cosh x^3 + C$

43. $u = e^{3t},\ du = 3e^{3t} dt$

$\int \dfrac{e^{3t}}{\sqrt{4 - e^{6t}}} dt = \dfrac{1}{3}\int \dfrac{1}{\sqrt{2^2 - u^2}} du$

$= \dfrac{1}{3}\sin^{-1}\left(\dfrac{u}{2}\right) + C$

$= \dfrac{1}{3}\sin^{-1}\left(\dfrac{e^{3t}}{2}\right) + C$

45. $u = \cos x,\ du = -\sin x\, dx$

$\int_0^{\pi/2} \dfrac{\sin x}{16 + \cos^2 x} dx = -\int_1^0 \dfrac{1}{16 + u^2} du$

$= \int_0^1 \dfrac{1}{16 + u^2} du$

$= \left[\dfrac{1}{4}\tan^{-1}\left(\dfrac{u}{4}\right)\right]_0^1$

$$= \left[\frac{1}{4} \tan^{-1}\left(\frac{1}{4}\right) - \frac{1}{4}\tan^{-1} 0 \right]$$

$$= \frac{1}{4}\tan^{-1}\left(\frac{1}{4}\right) \approx 0.0612$$

47. $\displaystyle\int \frac{1}{x^2+2x+5}\,dx = \int \frac{1}{x^2+2x+1+4}\,dx$

$$= \int \frac{1}{(x+1)^2+2^2}\,d(x+1)$$

$$= \frac{1}{2}\tan^{-1}\left(\frac{x+1}{2}\right)+C$$

49. $\displaystyle\int \frac{dx}{9x^2+18x+10} = \int \frac{dx}{9x^2+18x+9+1}$

$$= \int \frac{dx}{(3x+3)^2+1^2}$$

$u = 3x+3,\ du = 3\,dx$

$$\int \frac{dx}{(3x+3)^2+1^2} = \frac{1}{3}\int \frac{du}{u^2+1^2}$$

$$= \frac{1}{3}\tan^{-1}(3x+3)+C$$

51. $\displaystyle\int \frac{x+1}{9x^2+18x+10}\,dx = \frac{1}{18}\int \frac{18x+18}{9x^2+18x+10}\,dx$

$$= \frac{1}{18}\ln\left|9x^2+18x+10\right|+C$$

53. $u = \sqrt{2}t,\ du = \sqrt{2}dt$

$$\int \frac{dt}{t\sqrt{2t^2-9}} = \int \frac{du}{u\sqrt{u^2-3^2}}$$

$$= \frac{1}{3}\sec^{-1}\left(\frac{|\sqrt{2}t|}{3}\right)+C$$

55. $\displaystyle\int x\sqrt{3x+2}\,dx$

$$= \frac{2}{15\cdot 3^2}(3\cdot 3x-2\cdot 2)(3x+2)^{3/2}+C$$

$$= \frac{2}{135}(9x-4)(3x+2)^{3/2}+C$$

Use Formula 96 with $a = 3$, $b = 2$, and $u = x$ for
$\int u\sqrt{3u+2}\,du$.

57. $u = 4x,\ du = 4\,dx$

$$\int \frac{dx}{9-16x^2} = \frac{1}{4}\int \frac{du}{3^2-u^2}\,du$$

$$= \frac{1}{4}\left[\frac{1}{2(3)}\ln\left|\frac{u+3}{u-3}\right|\right]+C$$

$$= \frac{1}{24}\ln\left|\frac{4x+3}{4x-3}\right|+C$$

Use Formula 18 with $a = 3$ for $\displaystyle\int \frac{du}{3^2-u^2}$.

59. $\displaystyle\int x^2\sqrt{9-2x^2}\,dx = \int \sqrt{2}x^2\sqrt{\frac{9-2x^2}{2}}\,dx$

$$= \sqrt{2}\int x^2\sqrt{\frac{9}{2}-x^2}\,dx = \sqrt{2}\int x^2\sqrt{\left(\frac{3}{\sqrt{2}}\right)^2-x^2}\,dx$$

$$= \sqrt{2}\left[\frac{x}{8}\left(2x^2-\frac{9}{2}\right)\sqrt{\frac{9}{2}-x^2}+\frac{\left(\frac{81}{4}\right)}{8}\sin^{-1}\left(\frac{x}{\frac{3}{\sqrt{2}}}\right)+C\right]$$

$$= \frac{x}{16}(4x^2-9)\sqrt{9-2x^2}+\frac{81\sqrt{2}}{32}\sin^{-1}\left(\frac{\sqrt{2}\,x}{3}\right)+C$$

Use Formula 57 with $a = \dfrac{3}{\sqrt{2}}$ and $u = x$ for

$$\int x^2\sqrt{\left(\frac{3}{\sqrt{2}}\right)^2-x^2}\,dx.$$

61. $u = \sqrt{3}x,\ du = \sqrt{3}\,dx$

$$\int \frac{dx}{\sqrt{5+3x^2}} = \frac{1}{\sqrt{3}}\int \frac{du}{\sqrt{\left(\sqrt{5}\right)^2+u^2}}$$

$$= \frac{1}{\sqrt{3}}\ln\left|u+\sqrt{u^2+5}\right|+C$$

$$= \frac{1}{\sqrt{3}}\ln\left|\sqrt{3}\,x+\sqrt{3x^2+5}\right|+C$$

Use Formula 45 with $a = \sqrt{5}$ for

$$\int \frac{du}{\sqrt{\left(\sqrt{5}\right)^2+u^2}}.$$

63. $u = t+1,\ du = dt$

$$\int \frac{dt}{\sqrt{t^2+2t-3}} = \int \frac{dt}{\sqrt{t^2+2t+1-4}}$$

$$= \int \frac{dt}{\sqrt{(t+1)^2-4}}$$

$$= \int \frac{du}{\sqrt{u^2-2^2}} = \ln\left|u+\sqrt{u^2-4}\right|+C$$

$$= \ln\left|t+1+\sqrt{t^2+2t-3}\right|+C$$

Use Formula 45 with $a = 2$ for $\displaystyle\int \frac{du}{\sqrt{u^2-2^2}}$.

65. $u = \sin t,\ du = \cos t\,dt$

$$\int \frac{\sin t\cos t}{\sqrt{3\sin t+5}}\,dt$$

$$= \int \frac{u}{\sqrt{3u+5}}\,du = \frac{2}{3 \cdot 3^2}(3u - 2 \cdot 5)\sqrt{3u+5} + C$$

$$= \frac{2}{27}(3\sin t - 10)\sqrt{3\sin t + 5} + C$$

Use Formula 98 with $a = 3$, and $b = 5$ for

$$\int \frac{u}{\sqrt{3u+5}}\,du.$$

67. The length is given by

$$L = \int_a^b \sqrt{1 + \left(\frac{dy}{dx}\right)^2}\,dx$$

$$= \int_0^{\pi/4} \sqrt{1 + \left[\frac{1}{\cos x}(-\sin x)\right]^2}\,dx$$

$$= \int_0^{\pi/4} \sqrt{1 + \tan^2 x}\,dx$$

$$= \int_0^{\pi/4} \sqrt{\sec^2 x}\,dx$$

$$\int_0^{\pi/4} \sec x\,dx$$

$$= \left[\ln|\sec x + \tan x|\right]_0^{\pi/4}$$

$$= \ln\left|\sqrt{2} + 1\right| - \ln|1|$$

$$= \ln\left|\sqrt{2} + 1\right| \approx 0.881$$

69. $u = x - \pi$, $du = dx$

$$\int_0^{2\pi} \frac{x|\sin x|}{1 + \cos^2 x}\,dx = \int_{-\pi}^\pi \frac{(u + \pi)|\sin(u + \pi)|}{1 + \cos^2(u + \pi)}\,du$$

$$= \int_{-\pi}^\pi \frac{(u + \pi)|\sin u|}{1 + \cos^2 u}\,du$$

$$= \int_{-\pi}^\pi \frac{u|\sin u|}{1 + \cos^2 u}\,du + \int_{-\pi}^\pi \frac{\pi|\sin u|}{1 + \cos^2 u}\,du$$

$$\int_{-\pi}^\pi \frac{u|\sin u|}{1 + \cos^2 u}\,du = 0 \text{ by symmetry.}$$

$$\int_{-\pi}^\pi \frac{\pi|\sin u|}{1 + \cos^2 u}\,du = 2\int_0^\pi \frac{\pi\sin u}{1 + \cos^2 u}\,du$$

$v = \cos u$, $dv = -\sin u\,du$

$$-2\int_1^{-1} \frac{\pi}{1 + v^2}\,dv = 2\pi \int_{-1}^1 \frac{1}{1 + v^2}\,dv$$

$$= 2\pi[\tan^{-1} v]_{-1}^1 = 2\pi\left[\frac{\pi}{4} - \left(-\frac{\pi}{4}\right)\right]$$

$$= 2\pi\left(\frac{\pi}{2}\right) = \pi^2$$

8.2 Concepts Review

1. $\int \frac{1 + \cos 2x}{2}\,dx$

3. $\int \sin^2 x(1 - \sin^2 x)\cos x\,dx$

Problem Set 8.2

1. $\int \sin^2 x\,dx = \int \frac{1 - \cos 2x}{2}\,dx$

$$= \frac{1}{2}\int dx - \frac{1}{2}\int \cos 2x\,dx$$

$$= \frac{1}{2}x - \frac{1}{4}\sin 2x + C$$

3. $\int \sin^3 x\,dx = \int \sin x(1 - \cos^2 x)\,dx$

$$= \int \sin x\,dx - \int \sin x \cos^2 x\,dx$$

$$= -\cos x + \frac{1}{3}\cos^3 x + C$$

5. $\int_0^{\pi/2} \cos^5 \theta\,d\theta = \int_0^{\pi/2}(1 - \sin^2 \theta)^2 \cos\theta\,d\theta$

$$= \int_0^{\pi/2}(1 - 2\sin^2 \theta + \sin^4 \theta)\cos\theta\,d\theta$$

$$= \left[\sin\theta - \frac{2}{3}\sin^3 \theta + \frac{1}{5}\sin^5 \theta\right]_0^{\pi/2}$$

$$= \left(1 - \frac{2}{3} + \frac{1}{5}\right) - 0 = \frac{8}{15}$$

7. $\int \sin^5 4x\cos^2 4x\,dx = \int(1 - \cos^2 4x)^2 \cos^2 4x\sin 4x\,dx = \int(1 - 2\cos^2 4x + \cos^4 4x)\cos^2 4x\sin 4x\,dx$

$$= -\frac{1}{4}\int(\cos^2 4x - 2\cos^4 4x + \cos^6 4x)(-4\sin 4x)\,dx = -\frac{1}{12}\cos^3 4x + \frac{1}{10}\cos^5 4x - \frac{1}{28}\cos^7 4x + C$$

9. $\int \cos^3 3\theta \sin^{-2} 3\theta \, d\theta = \int (1 - \sin^2 3\theta) \sin^{-2} 3\theta \cos 3\theta \, d\theta = \frac{1}{3} \int (\sin^{-2} 3\theta - 1) 3 \cos 3\theta \, d\theta$

$= -\frac{1}{3} \csc 3\theta - \frac{1}{3} \sin 3\theta + C$

11. $\int \sin^4 3t \cos^4 3t \, dt = \int \left(\frac{1 - \cos 6t}{2} \right)^2 \left(\frac{1 + \cos 6t}{2} \right)^2 dt = \frac{1}{16} \int (1 - 2\cos^2 6t + \cos^4 6t) dt$

$= \frac{1}{16} \int \left[1 - (1 + \cos 12t) + \frac{1}{4}(1 + \cos 12t)^2 \right] dt = -\frac{1}{16} \int \cos 12t \, dt + \frac{1}{64} \int (1 + 2\cos 12t + \cos^2 12t) dt$

$= -\frac{1}{192} \int 12 \cos 12t \, dt + \frac{1}{64} \int dt + \frac{1}{384} \int 12 \cos 12t \, dt + \frac{1}{128} \int (1 + \cos 24t) dt$

$= -\frac{1}{192} \sin 12t + \frac{1}{64} t + \frac{1}{384} \sin 12t + \frac{1}{128} t + \frac{1}{3072} \sin 24t + C = \frac{3}{128} t - \frac{1}{384} \sin 12t + \frac{1}{3072} \sin 24t + C$

13. $\int \sin 4y \cos 5y \, dy = \frac{1}{2} \int [\sin 9y + \sin(-y)] dy = \frac{1}{2} \int (\sin 9y - \sin y) dy$

$= \frac{1}{2} \left(-\frac{1}{9} \cos 9y + \cos y \right) + C = \frac{1}{2} \cos y - \frac{1}{18} \cos 9y + C$

15. $\int \sin^4 \left(\frac{w}{2} \right) \cos^2 \left(\frac{w}{2} \right) dw = \int \left(\frac{1 - \cos w}{2} \right)^2 \left(\frac{1 + \cos w}{2} \right) dw = \frac{1}{8} \int (1 - \cos w - \cos^2 w + \cos^3 w) dw$

$= \frac{1}{8} \int \left[1 - \cos w - \frac{1}{2}(1 + \cos 2w) + (1 - \sin^2 w) \cos w \right] dw = \frac{1}{8} \int \left[\frac{1}{2} - \frac{1}{2} \cos 2w - \sin^2 w \cos w \right] dw$

$= \frac{1}{16} w - \frac{1}{32} \sin 2w - \frac{1}{24} \sin^3 w + C$

17. $\int \tan^4 x \, dx = \int (\tan^2 x)(\tan^2 x) \, dx$

$= \int (\tan^2 x)(\sec^2 x - 1) \, dx$

$= \int (\tan^2 x \sec^2 x - \tan^2 x) dx$

$= \int \tan^2 x \sec^2 x \, dx - \int (\sec^2 x - 1) dx$

$= \frac{1}{3} \tan^3 x - \tan x + x + C$

19. $\tan^3 x = \int (\tan x)(\tan^2 x) dx$

$= \int (\tan x)(\sec^2 x - 1) dx$

$= \frac{1}{2} \tan^2 x + \ln|\cos x| + C$

21. $\int \tan^5 \left(\frac{\theta}{2} \right) d\theta$

$u = \left(\frac{\theta}{2} \right); \ du = \frac{d\theta}{2}$

$\int \tan^5 \left(\frac{\theta}{2} \right) d\theta = 2 \int \tan^5 u \, du$

$= 2 \int (\tan^3 u)(\sec^2 u - 1) du$

$= 2 \int \tan^3 u \sec^2 u \, du - 2 \int \tan^3 u \, du$

$$= 2\int \tan^3 u \sec^2 u \; du - 2\int \tan u \left(\sec^2 u - 1\right) du$$

$$= 2\int \tan^3 u \sec^2 u \; du - 2\int \tan u \sec^2 u \; du + 2\int \tan u \; du$$

$$= \frac{1}{2}\tan^4\left(\frac{\theta}{2}\right) - \tan^2\left(\frac{\theta}{2}\right) - 2\ln\left|\cos\frac{\theta}{2}\right| + C$$

23. $\displaystyle\int \tan^{-3} x \sec^4 x \, dx = \int \left(\tan^{-3} x\right)\left(\sec^2 x\right)\left(\sec^2 x\right) dx$

$$= \int \left(\tan^{-3} x\right)\left(1 + \tan^2 x\right)\left(\sec^2 x\right) dx$$

$$= \int \tan^{-3} x \sec^2 x \; dx + \int \left(\tan x\right)^{-1} \sec^2 x \; dx$$

$$= -\frac{1}{2}\tan^{-2} x + \ln\left|\tan x\right| + C$$

25. $\displaystyle\int \tan^3 x \sec^2 x \, dx = \int \left(\tan^2 x\right)\left(\sec x\right)\left(\sec x \tan x\right) dx$

$$= \int \left(\sec^2 x - 1\right)\left(\sec x\right)\left(\sec x \tan x\right) dx$$

$$= \int \sec^3 x \sec x \tan x \; dx - \int \sec x \left(\sec x \tan x\right) dx$$

$$= \frac{1}{4}\sec^4 x - \frac{1}{2}\sec^2 x + C$$

27. $\displaystyle\int_{-\pi}^{\pi} \cos mx \cos nx \, dx = \frac{1}{2}\int_{-\pi}^{\pi}\left(\cos[(m+n)x] + \cos[(m-n)x]\right)dx = \frac{1}{2}\left[\frac{1}{m+n}\sin[(m+n)x] + \frac{1}{m-n}\sin[(m-n)x]\right]_{-\pi}^{\pi}$

$= 0$ for $m \neq n$, since $\sin k\pi = 0$ for all integers k.

29. $\displaystyle\int_0^\pi \pi(x + \sin x)^2 \, dx = \pi\int_0^\pi \left(x^2 + 2x\sin x + \sin^2 x\right) dx = \pi\int_0^\pi x^2 \, dx + 2\pi\int_0^\pi x\sin x \, dx + \frac{\pi}{2}\int_0^\pi (1 - \cos 2x) dx$

$$= \pi\left[\frac{1}{3}x^3\right]_0^\pi + 2\pi\left[\sin x - x\cos x\right]_0^\pi + \frac{\pi}{2}\left[x - \frac{1}{2}\sin 2x\right]_0^\pi = \frac{1}{3}\pi^4 + 2\pi(0 + \pi - 0) + \frac{\pi}{2}(\pi - 0 - 0) = \frac{1}{3}\pi^4 + \frac{5}{2}\pi^2 \approx 57.1437$$

Use Formula 40 with $u = x$ for $\displaystyle\int x\sin x \, dx$

31. a. $\displaystyle\frac{1}{\pi}\int_{-\pi}^{\pi} f(x)\sin(mx)dx = \frac{1}{\pi}\int_{-\pi}^{\pi}\left(\sum_{n=1}^{N} a_n \sin(nx)\right)\sin(mx)dx = \frac{1}{\pi}\sum_{n=1}^{N} a_n \int_{-\pi}^{\pi} \sin(nx)\sin(mx)dx$

From Example 6,

$$\int_{-\pi}^{\pi} \sin(nx)\sin(mx)dx = \begin{cases} 0 \text{ if } n \neq m \\ \pi \text{ if } n = m \end{cases} \text{ so every term in the sum is 0 except for when } n = m.$$

If $m > N$, there is no term where $n = m$, while if $m \leq N$, then $n = m$ occurs. When $n = m$

$a_n \displaystyle\int_{-\pi}^{\pi} \sin(nx)\sin(mx) \, dx = a_m\pi$ so when $m \leq N$,

$$\frac{1}{\pi}\int_{-\pi}^{\pi} f(x)\sin(mx) \, dx = \frac{1}{\pi} \cdot a_m \cdot \pi = a_m.$$

b. $\displaystyle\frac{1}{\pi}\int_{-\pi}^{\pi} f^2(x)dx = \frac{1}{\pi}\int_{-\pi}^{\pi}\left(\sum_{n=1}^{N} a_n \sin(nx)\right)\left(\sum_{m=1}^{N} a_m \sin(mx)\right)dx = \frac{1}{\pi}\sum_{n=1}^{N}\sum_{m=1}^{N} a_n a_m \int_{-\pi}^{\pi} \sin(nx)\sin(mx) \, dx$

From Example 6, the integral is 0 except when $m = n$. When $m = n$, we obtain

$$\frac{1}{\pi}\sum_{n=1}^{N} a_n(a_n\pi) = \sum_{n=1}^{N} a_n^2.$$

33. Using the half-angle identity $\cos\dfrac{x}{2} = \sqrt{\dfrac{1+\cos x}{2}}$, we see that since

$$\cos\frac{\pi}{4} = \cos\frac{\frac{\pi}{2}}{2} = \frac{\sqrt{2}}{2}$$

$$\cos\frac{\pi}{8} = \cos\frac{\frac{\pi}{2}}{4} = \sqrt{\frac{1+\frac{\sqrt{2}}{2}}{2}} = \sqrt{\frac{2+\sqrt{2}}{2}},$$

$$\cos\frac{\pi}{16} = \cos\frac{\frac{\pi}{2}}{8} = \sqrt{\frac{1+\frac{\sqrt{2+\sqrt{2}}}{2}}{2}} = \frac{\sqrt{2+\sqrt{2+\sqrt{2}}}}{2}, \text{ etc.}$$

Thus, $\dfrac{\sqrt{2}}{2} \cdot \dfrac{\sqrt{2+\sqrt{2}}}{2} \cdot \dfrac{\sqrt{2+\sqrt{2+\sqrt{2}}}}{2} \cdots = \cos\left(\dfrac{\frac{\pi}{2}}{2}\right)\cos\left(\dfrac{\frac{\pi}{2}}{4}\right)\cos\left(\dfrac{\frac{\pi}{2}}{8}\right)\cdots$

$$= \lim_{n\to\infty} \cos\left(\frac{\frac{\pi}{2}}{2}\right)\cos\left(\frac{\frac{\pi}{2}}{4}\right)\cdots\cos\left(\frac{\frac{\pi}{2}}{2^n}\right) = \frac{\sin\left(\frac{\pi}{2}\right)}{\frac{\pi}{2}} = \frac{2}{\pi}$$

8.3 Concepts Review

1. $\sqrt{x-3}$ **3.** $2\tan t$

Problem Set 8.3

1. $u = \sqrt{x+1}, u^2 = x+1, 2u\,du = dx$

$$\int x\sqrt{x+1}\,dx = \int (u^2-1)u(2u\,du)$$

$$= \int (2u^4 - 2u^2)\,du = \frac{2}{5}u^5 - \frac{2}{3}u^3 + C$$

$$= \frac{2}{5}(x+1)^{5/2} - \frac{2}{3}(x+1)^{3/2} + C$$

3. $u = \sqrt{3t+4}, u^2 = 3t+4, \ 2u\,du = 3\,dt$

$$\int \frac{t\,dt}{\sqrt{3t+4}} = \int \frac{\frac{1}{3}(u^2-4)\frac{2}{3}u\,du}{u} = \frac{2}{9}\int (u^2-4)\,du$$

$$= \frac{2}{27}u^3 - \frac{8}{9}u + C$$

$$= \frac{2}{27}(3t+4)^{3/2} - \frac{8}{9}(3t+4)^{1/2} + C$$

5. $u = \sqrt{t}, u^2 = t, \ 2u\,du = dt$

$$\int_1^2 \frac{dt}{\sqrt{t}+e} = \int_1^{\sqrt{2}} \frac{2u\,du}{u+e} = 2\int_1^{\sqrt{2}} \frac{u+e-e}{u+e}\,du$$

$$= 2\int_1^{\sqrt{2}} du - 2\int_1^{\sqrt{2}} \frac{e}{u+e}\,du$$

$$= 2[u]_1^{\sqrt{2}} - 2e\left[\ln|u+e|\right]_1^{\sqrt{2}}$$

$$= 2(\sqrt{2}-1) - 2e[\ln(\sqrt{2}+e) - \ln(1+e)]$$

$$= 2\sqrt{2} - 2 - 2e\ln\left(\frac{\sqrt{2}+e}{1+e}\right)$$

7. $u = (3t+2)^{1/2}, u^2 = 3t+2, 2u\,du = 3dt$

$$\int t(3t+2)^{3/2}\,dt = \int \frac{1}{3}(u^2-2)u^3\left(\frac{2}{3}u\,du\right)$$

$$= \frac{2}{9}\int (u^6 - 2u^4)\,du = \frac{2}{63}u^7 - \frac{4}{45}u^5 + C$$

$$= \frac{2}{63}(3t+2)^{7/2} - \frac{4}{45}(3t+2)^{5/2} + C$$

9. $x = 2\sin t, dx = 2\cos t\,dt$

$$\int \frac{\sqrt{4-x^2}}{x}\,dx = \int \frac{2\cos t}{2\sin t}(2\cos t\,dt)$$

$$= 2\int \frac{1-\sin^2 t}{\sin t}\,dt = 2\int \csc t\,dt - 2\int \sin t\,dt$$

$$= 2\ln|\csc t - \cot t| + 2\cos t + C$$

$$= 2\ln\left|\frac{2-\sqrt{4-x^2}}{x}\right| + \sqrt{4-x^2} + C$$

11. $x = 2\tan t, dx = 2\sec^2 t\,dt$

$$\int \frac{dx}{(x^2+4)^{3/2}} = \int \frac{2\sec^2 t\,dt}{(4\sec^2 t)^{3/2}} = \frac{1}{4}\int \cos t\,dt$$

$$= \frac{1}{4}\sin t + C = \frac{x}{4\sqrt{x^2+4}} + C$$

13. $t = \sec x$, $dt = \sec x \tan x \, dx$

Note that $\dfrac{\pi}{2} < x \le \pi$.

$\sqrt{t^2 - 1} = |\tan x| = -\tan x$

$\displaystyle \int_{-2}^{-3} \frac{\sqrt{t^2-1}}{t^3} \, dt = \int_{2\pi/3}^{\sec^{-1}(-3)} \frac{-\tan x}{\sec^3 x} \sec x \tan x \, dx$

$\displaystyle = \int_{2\pi/3}^{\sec^{-1}(-3)} -\sin^2 x \, dx = \int_{2\pi/3}^{\sec^{-1}(-3)} \left(\frac{1}{2} \cos 2x - \frac{1}{2} \right) dx$

$\displaystyle = \left[\frac{1}{4} \sin 2x - \frac{1}{2} x \right]_{2\pi/3}^{\sec^{-1}(-3)}$

$\displaystyle = \left[\frac{1}{2} \sin x \cos x - \frac{1}{2} x \right]_{2\pi/3}^{\sec^{-1}(-3)}$

$\displaystyle = -\frac{\sqrt{2}}{9} - \frac{1}{2} \sec^{-1}(-3) + \frac{\sqrt{3}}{8} + \frac{\pi}{3} \approx 0.151252$

15. $z = \sin t$, $dz = \cos t \, dt$

$\displaystyle \int \frac{2z - 3}{\sqrt{1 - z^2}} \, dz = \int (2\sin t - 3) dt$

$= -2\cos t - 3t + C$

$= -2\sqrt{1 - z^2} - 3\sin^{-1} z + C$

17. $x^2 + 2x + 5 = x^2 + 2x + 1 + 4 = (x+1)^2 + 4$

$u = x + 1$, $du = dx$

$\displaystyle \int \frac{dx}{\sqrt{x^2 + 2x + 5}} = \int \frac{du}{\sqrt{u^2 + 4}}$

$u = 2\tan t$, $du = 2\sec^2 t \, dt$

$\displaystyle \int \frac{du}{\sqrt{u^2 + 4}} = \int \sec t \, dt = \ln|\sec t + \tan t| + C$

$\displaystyle \ln\left| \frac{\sqrt{u^2 + 4}}{2} + \frac{u}{2} \right| + C_1$

$\displaystyle = \ln\left| \frac{\sqrt{x^2 + 2x + 5} + x + 1}{2} \right| + C_1$

$= \ln\left| \sqrt{x^2 + 2x + 5} + x + 1 \right| + C$

19. $x^2 + 2x + 5 = x^2 + 2x + 1 + 4 = (x+1)^2 + 4$

$u = x + 1$, $du = dx$

$\displaystyle \int \frac{3x}{\sqrt{x^2 + 2x + 5}} \, dx = \int \frac{3u - 3}{\sqrt{u^2 + 4}} \, du$

$\displaystyle = 3\int \frac{u}{\sqrt{u^2 + 4}} \, du - 3\int \frac{du}{\sqrt{u^2 + 4}}$

(Use the result of Problem 17.)

$= 3\sqrt{u^2 + 4} - 3\ln\left| \sqrt{u^2 + 4} + u \right| + C$

$= 3\sqrt{x^2 + 2x + 5} - 3\ln\left| \sqrt{x^2 + 2x + 5} + x + 1 \right| + C$

21. $5 - 4x - x^2 = 9 - (4 + 4x + 4x^2) = 9 - (x+2)^2$

$u = x + 2$, $du = dx$

$\displaystyle \int \sqrt{5 - 4x - x^2} \, dx = \int \sqrt{9 - u^2} \, du$

$u = 3\sin t$, $du = 3\cos t \, dt$

$\displaystyle \int \sqrt{9 - u^2} \, du = 9\int \cos^2 t \, dt = \frac{9}{2} \int (1 + \cos 2t) dt$

$\displaystyle = \frac{9}{2}\left(t + \frac{1}{2} \sin 2t \right) + C = \frac{9}{2}(t + \sin t \cos t) + C$

$\displaystyle = \frac{9}{2} \sin^{-1}\left(\frac{u}{3} \right) + \frac{1}{2} u\sqrt{9 - u^2} + C$

$\displaystyle = \frac{9}{2} \sin^{-1}\left(\frac{x+2}{3} \right) + \frac{x+2}{2} \sqrt{5 - 4x - x^2} + C$

23. $4x - x^2 = 4 - (4 - 4x + x^2) = 4 - (x-2)^2$

$u = x - 2$, $du = dx$

$\displaystyle \int \frac{dx}{\sqrt{4x - x^2}} = \int \frac{du}{\sqrt{4 - u^2}}$

$u = 2\sin t$, $du = 2\cos t \, dt$

$\displaystyle \int \frac{du}{\sqrt{4 - u^2}} = \int dt = t + C = \sin^{-1}\left(\frac{u}{2} \right) + C$

$\displaystyle = \sin^{-1}\left(\frac{x-2}{2} \right) + C$

25. $x^2 + 2x + 2 = x^2 + 2x + 1 + 1 = (x+1)^2 + 1$

$u = x + 1$, $du = dx$

$\displaystyle \int \frac{2x + 1}{x^2 + 2x + 2} \, dx = \int \frac{2u - 1}{u^2 + 1} \, du$

$\displaystyle = \int \frac{2u}{u^2 + 1} \, du - \int \frac{du}{u^2 + 1}$

$= \ln|u^2 + 1| - \tan^{-1} u + C$

$= \ln|x^2 + 2x + 2| - \tan^{-1}(x+1) + C$

27. $\displaystyle V = \pi \int_0^1 \left(\frac{1}{x^2 + 2x + 5} \right)^2 dx$

$\displaystyle = \pi \int_0^1 \left[\frac{1}{(x+1)^2 + 4} \right]^2 dx$

$x + 1 = 2\tan t$, $dx = 2\sec^2 t \, dt$

$\displaystyle V = \pi \int_{\tan^{-1}(1/2)}^{\pi/4} \left(\frac{1}{4\sec^2 t} \right)^2 2\sec^2 t \, dt$

$$= \frac{\pi}{8} \int_{\tan^{-1}(1/2)}^{\pi/4} \frac{1}{\sec^2 t} dt = \frac{\pi}{8} \int_{\tan^{-1}(1/2)}^{\pi/4} \cos^2 t \, dt$$

$$= \frac{\pi}{8} \int_{\tan^{-1}(1/2)}^{\pi/4} \left(\frac{1}{2} + \frac{1}{2} \cos 2t \right) dt$$

$$= \frac{\pi}{8} \left[\frac{1}{2} t + \frac{1}{4} \sin 2t \right]_{\tan^{-1}(1/2)}^{\pi/4}$$

$$= \frac{\pi}{8} \left[\frac{1}{2} t + \frac{1}{2} \sin t \cos t \right]_{\tan^{-1}(1/2)}^{\pi/4}$$

$$= \frac{\pi}{8} \left[\left(\frac{\pi}{8} + \frac{1}{4} \right) - \left(\frac{1}{2} \tan^{-1} \frac{1}{2} + \frac{1}{5} \right) \right]$$

$$= \frac{\pi}{16} \left(\frac{1}{10} + \frac{\pi}{4} - \tan^{-1} \frac{1}{2} \right) \approx 0.082811$$

29. a. $u = x^2 + 9, du = 2x \, dx$

$$\int \frac{x \, dx}{x^2 + 9} = \frac{1}{2} \int \frac{du}{u} = \frac{1}{2} \ln |u| + C$$

$$= \frac{1}{2} \ln |x^2 + 9| + C$$

b. $x = 3 \tan t, \ dx = 3 \sec^2 t \, dt$

$$\int \frac{x \, dx}{x^2 + 9} = \int \tan t \, dt = -\ln |\cos t| + C$$

$$= -\ln \left| \frac{3}{\sqrt{x^2 + 9}} \right| + C_1$$

$$= \ln \left| \sqrt{x^2 + 9} \right| - \ln |3| + C_1$$

$$= \ln \left| (x^2 + 9)^{1/2} \right| + C = \frac{1}{2} \ln |x^2 + 9| + C$$

31. a. $u = \sqrt{4 - x^2}, u^2 = 4 - x^2, \ 2u \, du = -2x \, dx$

$$\int \frac{\sqrt{4 - x^2}}{x} dx = \int \frac{\sqrt{4 - x^2}}{x^2} x \, dx = -\int \frac{u^2 \, du}{4 - u^2}$$

$$= \int \frac{-4 + 4 - u^2}{4 - u^2} du = -4 \int \frac{1}{4 - u^2} du + \int du$$

$$= -4 \cdot \frac{1}{4} \ln \left| \frac{u + 2}{u - 2} \right| + u + C$$

$$= -\ln \left| \frac{\sqrt{4 - x^2} + 2}{\sqrt{4 - x^2} - 2} \right| + \sqrt{4 - x^2} + C$$

b. $x = 2 \sin t, \ dx = 2 \cos t \, dt$

$$\int \frac{\sqrt{4 - x^2}}{x} dx = 2 \int \frac{\cos^2 t}{\sin t} dt$$

$$= 2 \int \frac{(1 - \sin^2 t)}{\sin t} dt$$

$$= 2 \int \csc t \, dt - 2 \int \sin t \, dt$$

$$= 2 \ln |\csc t - \cot t| + 2 \cos t + C$$

$$= 2 \ln \left| \frac{2}{x} - \frac{\sqrt{4 - x^2}}{x} \right| + \sqrt{4 - x^2} + C$$

$$= 2 \ln \left| \frac{2 - \sqrt{4 - x^2}}{x} \right| + \sqrt{4 - x^2} + C$$

To reconcile the answers, note that

$$-\ln \left| \frac{\sqrt{4 - x^2} + 2}{\sqrt{4 - x^2} - 2} \right| = \ln \left| \frac{\sqrt{4 - x^2} - 2}{\sqrt{4 - x^2} + 2} \right|$$

$$= \ln \left| \frac{(\sqrt{4 - x^2} - 2)^2}{(\sqrt{4 - x^2} + 2)(\sqrt{4 - x^2} - 2)} \right|$$

$$= \ln \left| \frac{(2 - \sqrt{4 - x^2})^2}{4 - x^2 - 4} \right| = \ln \left| \frac{(2 - \sqrt{4 - x^2})^2}{-x^2} \right|$$

$$= \ln \left| \left(\frac{2 - \sqrt{4 - x^2}}{x} \right)^2 \right| = 2 \ln \left| \frac{2 - \sqrt{4 - x^2}}{x} \right|$$

33. a. The coordinate of C is $(0, -a)$. The lower arc of the lune lies on the circle given by the equation $x^2 + (y + a)^2 = 2a^2$ or

$y = \pm \sqrt{2a^2 - x^2} - a$. The upper arc of the lune lies on the circle given by the equation $x^2 + y^2 = a^2$ or $y = \pm \sqrt{a^2 - x^2}$.

$$A = \int_{-a}^{a} \sqrt{a^2 - x^2} dx - \int_{-a}^{a} \left(\sqrt{2a^2 - x^2} - a \right) dx$$

$$= \int_{-a}^{a} \sqrt{a^2 - x^2} dx - \int_{-a}^{a} \sqrt{2a^2 - x^2} dx + 2a^2$$

Note that $\int_{-a}^{a} \sqrt{a^2 - x^2} dx$ is the area of a semicircle with radius a, so

$$\int_{-a}^{a} \sqrt{a^2 - x^2} dx = \frac{\pi a^2}{2}.$$

For $\int_{-a}^{a} \sqrt{2a^2 - x^2} dx$, let

$x = \sqrt{2} a \sin t, dx = \sqrt{2} a \cos t \, dt$

$$\int_{-a}^{a} \sqrt{2a^2 - x^2} dx = \int_{-\pi/4}^{\pi/4} 2a^2 \cos^2 t \, dt$$

$$= a^2 \int_{-\pi/4}^{\pi/4} (1 + \cos 2t) dt = a^2 \left[t + \frac{1}{2} \sin 2t \right]_{-\pi/4}^{\pi/4}$$

$$= \frac{\pi a^2}{2} + a^2$$

$$A = \frac{\pi a^2}{2} - \left(\frac{\pi a^2}{2} + a^2 \right) + 2a^2 = a^2$$

Thus, the area of the lune is equal to the area of the square.

b. Without using calculus, consider the following labels on the figure.

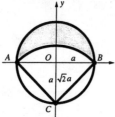

Area of the lune = Area of the semicircle of radius a at O + Area ($\triangle ABC$) – Area of the sector ABC.

$$A = \frac{1}{2}\pi a^2 + a^2 - \frac{1}{2}\left(\frac{\pi}{2}\right)(\sqrt{2}a)^2$$

$$= \frac{1}{2}\pi a^2 + a^2 - \frac{1}{2}\pi a^2 = a^2$$

Note that since BC has length $\sqrt{2}a$, the measure of angle OCB is $\frac{\pi}{4}$, so the measure of angle ACB is $\frac{\pi}{2}$.

35. $\dfrac{dy}{dx} = -\dfrac{\sqrt{a^2 - x^2}}{x}$

$y = \displaystyle\int -\frac{\sqrt{a^2 - x^2}}{x}\,dx$

$x = a\sin t,\ dx = a\cos t\,dt$

$$y = \int -\frac{a\cos t}{a\sin t}a\cos t\,dt = -a\int\frac{\cos^2 t}{\sin t}\,dt$$

$$= -a\int\frac{1-\sin^2 t}{\sin t}\,dt = a\int(\sin t - \csc t)\,dt$$

$$= a\left(-\cos t - \ln|\csc t - \cot t|\right) + C$$

$\cos t = \dfrac{\sqrt{a^2 - x^2}}{a},\ \csc t = \dfrac{a}{x},\ \cot t = \dfrac{\sqrt{a^2 - x^2}}{x}$

$$y = a\left(-\frac{\sqrt{a^2 - x^2}}{a} - \ln\left|\frac{a}{x} - \frac{\sqrt{a^2 - x^2}}{x}\right|\right) + C$$

$$= -\sqrt{a^2 - x^2} - a\ln\left|\frac{a - \sqrt{a^2 - x^2}}{x}\right| + C$$

Since $y = 0$ when $x = a$,
$0 = 0 - a\ln 1 + C$, so $C = 0$.

$$y = -\sqrt{a^2 - x^2} - a\ln\left|\frac{a - \sqrt{a^2 - x^2}}{x}\right|$$

8.4 Concepts Review

1. $uv - \displaystyle\int v\,du$

3. 1

Problem Set 8.4

1. $u = x$ $dv = e^x\,dx$
$du = dx$ $v = e^x$
$\displaystyle\int xe^x\,dx = xe^x - \int e^x\,dx = xe^x - e^x + C$

3. $u = t$ $dv = e^{5t+\pi}\,dt$
$du = dt$ $v = \dfrac{1}{5}e^{5t+\pi}$

$\displaystyle\int te^{5t+\pi}\,dt = \frac{1}{5}te^{5t+\pi} - \int\frac{1}{5}e^{5t+\pi}\,dt$

$= \dfrac{1}{5}te^{5t+\pi} - \dfrac{1}{25}e^{5t+\pi} + C$

5. $u = x$ $dv = \cos x\,dx$
$du = dx$ $v = \sin x$

$\displaystyle\int x\cos x\,dx = x\sin x - \int\sin x\,dx$
$= x\sin x + \cos x + C$

7. $u = t - 3$ $dv = \cos(t - 3)\,dt$
$du = dt$ $v = \sin(t - 3)$
$\displaystyle\int(t - 3)\cos(t - 3)\,dt = (t - 3)\sin(t - 3) - \int\sin(t - 3)\,dt$
$= (t - 3)\sin(t - 3) + \cos(t - 3) + C$

9. $u = t$ $dv = \sqrt{t+1}\,dt$
$du = dt$ $v = \dfrac{2}{3}(t+1)^{3/2}$

$\displaystyle\int t\sqrt{t+1}\,dt = \frac{2}{3}t(t+1)^{3/2} - \int\frac{2}{3}(t+1)^{3/2}\,dt$

$= \dfrac{2}{3}t(t+1)^{3/2} - \dfrac{4}{15}(t+1)^{5/2} + C$

11. $u = \ln 3x$ $dv = dx$
$du = \dfrac{1}{x}\,dx$ $v = x$

$\displaystyle\int\ln 3x\,dx = x\ln 3x - \int x\frac{1}{x}\,dx = x\ln 3x - x + C$

13. $u = \arctan x \qquad dv = dx$

$\qquad du = \dfrac{1}{1+x^2}dx \qquad v = x$

$\qquad \int \arctan x = x\arctan x - \int \dfrac{x}{1+x^2}dx$

$\qquad = x\arctan x - \dfrac{1}{2}\int \dfrac{2x}{1+x^2}dx$

$\qquad = x\arctan x - \dfrac{1}{2}\ln(1+x^2) + C$

15. $u = \ln x \qquad dv = \dfrac{dx}{x^2}$

$\qquad du = \dfrac{1}{x}dx \qquad v = -\dfrac{1}{x}$

$\qquad \int \dfrac{\ln x}{x^2}dx = -\dfrac{\ln x}{x} - \int -\dfrac{1}{x}\left(\dfrac{1}{x}\right)dx$

$\qquad = -\dfrac{\ln x}{x} - \dfrac{1}{x} + C$

17. $u = \ln t \qquad dv = \sqrt{t}\,dt$

$\qquad du = \dfrac{1}{t}dt \qquad v = \dfrac{2}{3}t^{3/2}$

$\qquad \int_1^e \sqrt{t}\,\ln t\,dt = \left[\dfrac{2}{3}t^{3/2}\ln t\right]_1^e - \int_1^e \dfrac{2}{3}t^{1/2}dt$

$\qquad = \dfrac{2}{3}e^{3/2}\ln e - \dfrac{2}{3}\cdot 1\ln 1 - \left[\dfrac{4}{9}t^{3/2}\right]_1^e$

$\qquad = \dfrac{2}{3}e^{3/2} - 0 - \dfrac{4}{9}e^{3/2} + \dfrac{4}{9} = \dfrac{2}{9}e^{3/2} + \dfrac{4}{9}$

19. $u = \ln z \qquad dv = z^3\,dz$

$\qquad du = \dfrac{1}{z}dz \qquad v = \dfrac{1}{4}z^4$

$\qquad \int z^3 \ln z\,dz = \dfrac{1}{4}z^4 \ln z - \int \dfrac{1}{4}z^4 \cdot \dfrac{1}{z}dz$

$\qquad = \dfrac{1}{4}z^4 \ln z - \dfrac{1}{4}\int z^3\,dz$

$\qquad = \dfrac{1}{4}z^4 \ln z - \dfrac{1}{16}z^4 + C$

21. $u = \arctan\left(\dfrac{1}{t}\right) \qquad dv = dt$

$\qquad du = -\dfrac{1}{1+t^2}dt \qquad v = t$

$\qquad \int \arctan\left(\dfrac{1}{t}\right)dt = t\arctan\left(\dfrac{1}{t}\right) + \int \dfrac{t}{1+t^2}dt$

23. $u = x\cos^2 x \qquad\qquad dv = \sin x\,dx$

$\quad du = (\cos^2 x - 2x\cos x\sin x)dx \qquad v = -\cos x$

$\quad \int x\cos^2 x\sin x\,dx = -x\cos^3 x + \int(\cos^3 x - 2x\cos^2 x\sin x)dx = -x\cos^3 x + \int \cos^3 x\,dx - 2\int x\cos^2 x\sin x\,dx$

$\quad 3\int x\cos^2 x\sin x\,dx = -x\cos^3 x + \int \cos x(1 - \sin^2 x)dx = -x\cos^3 x + \sin x - \dfrac{1}{3}\sin^3 x + C$

$\quad \int x\cos^2 x\sin x\,dx = -\dfrac{x}{3}\cos^3 x + \dfrac{1}{3}\sin x - \dfrac{1}{9}\sin^3 x + C$

25. $u = x \qquad dv = \csc^2 x\,dx$

$\quad du = dx \qquad v = -\cot x$

$\quad \int_{\pi/6}^{\pi/2} x\csc^2 x\,dx = \left[-x\cot x\right]_{\pi/6}^{\pi/2} + \int_{\pi/6}^{\pi/2}\cot x\,dx = \left[-x\cot x + \ln|\sin x|\right]_{\pi/6}^{\pi/2}$

$\quad = -\dfrac{\pi}{2}\cdot 0 + \ln 1 + \dfrac{\pi}{6}\sqrt{3} - \ln\dfrac{1}{2} = \dfrac{\pi}{2\sqrt{3}} + \ln 2 \approx 1.60$

27. $u = x \qquad dv = \sec^2 x\,dx$

$\quad du = dx \qquad v = \tan x$

$$\int_{\pi/6}^{\pi/4} x\sec^2 x\,dx = \left[x\tan x\right]_{\pi/6}^{\pi/4} - \int_{\pi/6}^{\pi/4}\tan x\,dx = \left[x\tan x + \ln|\cos x|\right]_{\pi/6}^{\pi/4} = \frac{\pi}{4} + \ln\frac{\sqrt{2}}{2} - \left(\frac{\pi}{6\sqrt{3}} + \ln\frac{\sqrt{3}}{2}\right)$$

$$= \frac{\pi}{4} + \ln\frac{\sqrt{2}}{2} - \frac{\pi}{6\sqrt{3}} - \ln\frac{\sqrt{3}}{2} = \frac{\pi}{4} - \frac{\pi}{6\sqrt{3}} + \frac{1}{2}\ln\frac{2}{3} \approx 0.28$$

29. $u = x^3 \qquad dv = x^2\sqrt{x^3+4}\,dx$

$du = 3x^2\,dx \quad v = \frac{2}{9}(x^3+4)^{3/2}$

$\int x^5\sqrt{x^3+4}\,dx = \frac{2}{9}x^3(x^3+4)^{3/2} - \int\frac{2}{3}x^2(x^3+4)^{3/2}\,dx = \frac{2}{9}x^3(x^3+4)^{3/2} - \frac{4}{45}(x^3+4)^{5/2} + C$

31. $u = t^4 \qquad dv = \dfrac{t^3}{(7-3t^4)^{3/2}}\,dt$

$du = 4t^3\,dt \quad v = \dfrac{1}{6(7-3t^4)^{1/2}}$

$\int\dfrac{t^7}{(7-3t^4)^{3/2}}\,dt = \dfrac{t^4}{6(7-3t^4)^{1/2}} - \dfrac{2}{3}\int\dfrac{t^3}{(7-3t^4)^{1/2}}\,dt = \dfrac{t^4}{6(7-3t^4)^{1/2}} + \dfrac{1}{9}(7-3t^4)^{1/2} + C$

33. $u = z^4 \qquad dv = \dfrac{z^3}{(4-z^4)^2}\,dz$

$du = 4z^3\,dz \quad v = \dfrac{1}{4(4-z^4)}$

$\int\dfrac{z^7}{(4-z^4)^2}\,dz = \dfrac{z^4}{4(4-z^4)} - \int\dfrac{z^3}{4-z^4}\,dz = \dfrac{z^4}{4(4-z^4)} + \dfrac{1}{4}\ln\left|4-z^4\right| + C$

35. $u = x \qquad dv = \sinh x\,dx$
$du = dx \qquad v = \cosh x$
$\int x\sinh x\,dx = x\cosh x - \int\cosh x\,dx = x\cosh x - \sinh x + C$

37. $u = \ln x \qquad dv = x^{-1/2}dx$

$du = \dfrac{1}{x}dx \quad v = 2x^{1/2}$

$\int\dfrac{\ln x}{\sqrt{x}}\,dx = 2\sqrt{x}\ln x - 2\int\dfrac{1}{x^{1/2}}\,dx = 2\sqrt{x}\ln x - 4\sqrt{x} + C$

39. $u = x \qquad dv = 2^x\,dx$

$du = dx \qquad v = \dfrac{1}{\ln 2}2^x$

$\int x2^x\,dx = \dfrac{x}{\ln 2}2^x - \dfrac{1}{\ln 2}\int 2^x\,dx$

$= \dfrac{x}{\ln 2}2^x - \dfrac{1}{(\ln 2)^2}2^x + C$

$\int x^2 e^x\,dx = x^2 e^x - \int 2xe^x\,dx$

$u = x \qquad dv = e^x\,dx$
$du = dx \qquad v = e^x$
$\int x^2 e^x\,dx = x^2 e^x - 2\left(xe^x - \int e^x\,dx\right)$
$= x^2 e^x - 2xe^x + 2e^x + C$

41. $u = x^2 \qquad dv = e^x\,dx$
$du = 2x\,dx \quad v = e^x$

43. $u = \ln^2 z \qquad dv = dz$

$du = \dfrac{2\ln z}{z}\,dz \qquad v = z$

$$\int \ln^2 z \, dz = z \ln^2 z - 2 \int \ln z \, dz$$
$$u = \ln z \qquad dv = dz$$
$$du = \frac{1}{z} dz \qquad v = z$$
$$\int \ln^2 z \, dz = z \ln^2 z - 2 \left(z \ln z - \int dz \right)$$
$$= z \ln^2 z - 2z \ln z + 2z + C$$

45. $\quad u = e^t \qquad dv = \cos t \, dt$

$\qquad du = e^t dt \qquad v = \sin t$

$$\int e^t \cos t \, dt = e^t \sin t - \int e^t \sin t \, dt$$

$\qquad u = e^t \qquad dv = \sin t \, dt$

$\qquad du = e^t dt \qquad v = -\cos t$

$$\int e^t \cos t \, dt = e^t \sin t - \left[-e^t \cos t + \int e^t \cos t \, dt \right]$$

$$\int e^t \cos t \, dt = e^t \sin t + e^t \cos t - \int e^t \cos t \, dt$$
$$2 \int e^t \cos t \, dt = e^t \sin t + e^t \cos t + C$$
$$\int e^t \cos t \, dt = \frac{1}{2} e^t (\sin t + \cos t) + C$$

47. $\quad u = x^2 \qquad dv = \cos x \, dx$

$\qquad du = 2x \, dx \qquad v = \sin x$

$$\int x^2 \cos x \, dx = x^2 \sin x - \int 2x \sin x \, dx$$

$\qquad u = 2x \qquad dv = \sin x \, dx$

$\qquad du = 2dx \qquad v = -\cos x$

$$\int x^2 \cos x \, dx = x^2 \sin x - \left(-2x \cos x + \int 2 \cos x \, dx \right)$$

$$= x^2 \sin x + 2x \cos x - 2 \sin x + C$$

49. $\quad u = \sin(\ln x) \qquad dv = dx$

$\qquad du = \cos(\ln x) \cdot \frac{1}{x} dx \qquad v = x$

$$\int \sin(\ln x) dx = x \sin(\ln x) - \int \cos(\ln x) \, dx$$

$\qquad u = \cos(\ln x) \qquad dv = dx$

$\qquad du = -\sin(\ln x) \cdot \frac{1}{x} dx \qquad v = x$

$$\int \sin(\ln x) dx = x \sin(\ln x) - \left[x \cos(\ln x) - \int -\sin(\ln x) dx \right]$$

$$\int \sin(\ln x) dx = x \sin(\ln x) - x \cos(\ln x) - \int \sin(\ln x) dx$$

$$2 \int \sin(\ln x) dx = x \sin(\ln x) - x \cos(\ln x) + C$$

$$\int \sin(\ln x) dx = \frac{x}{2} [\sin(\ln x) - \cos(\ln x)] + C$$

51. $\quad u = (\ln x)^3 \qquad dv = dx$

$\qquad du = \frac{3 \ln^2 x}{x} dx \qquad v = x$

$$\int (\ln x)^3 \, dx = x(\ln x)^3 - 3 \int \ln^2 x \, dx$$

$$= x \ln^3 x - 3(x \ln^2 x - 2x \ln x + 2x + C)$$

$$= x \ln^3 x - 3x \ln^2 x + 6x \ln x - 6x + C$$

53. $\quad u = \sin x \qquad dv = \sin(3x) dx$

$\qquad du = \cos x \, dx \qquad v = -\frac{1}{3} \cos(3x)$

$$\int \sin x \sin(3x) dx = -\frac{1}{3} \sin x \cos(3x) + \frac{1}{3} \int \cos x \cos(3x) dx$$

$\qquad u = \cos x \qquad dv = \cos(3x) dx$

$\qquad du = -\sin x \, dx \qquad v = \frac{1}{3} \sin(3x)$

$$\int \sin x \sin(3x) dx = -\frac{1}{3} \sin x \cos(3x) + \frac{1}{3} \left[\frac{1}{3} \cos x \sin(3x) + \frac{1}{3} \int \sin x \sin(3x) dx \right]$$

$$= -\frac{1}{3}\sin x \cos(3x) + \frac{1}{9}\cos x \sin(3x) + \frac{1}{9}\int \sin x \sin(3x)dx$$

$$\frac{8}{9}\int \sin x \sin(3x)dx = -\frac{1}{3}\sin x \cos(3x) + \frac{1}{9}\cos x \sin(3x) + C$$

$$\int \sin x \sin(3x)dx = -\frac{3}{8}\sin x \cos(3x) + \frac{1}{8}\cos x \sin(3x) + C$$

55. $\quad u = e^{\alpha z} \qquad\qquad dv = \sin \beta z \, dz$

$\qquad du = \alpha e^{\alpha z} dz \qquad v = -\frac{1}{\beta}\cos \beta z$

$$\int e^{\alpha z}\sin \beta z \, dz = -\frac{1}{\beta}e^{\alpha z}\cos \beta z + \frac{\alpha}{\beta}\int e^{\alpha z}\cos \beta z \, dz$$

$\qquad u = e^{\alpha z} \qquad\qquad dv = \cos \beta z \, dz$

$\qquad du = \alpha e^{\alpha z} dz \qquad v = \frac{1}{\beta}\sin \beta z$

$$\int e^{\alpha z}\sin \beta z \, dz = -\frac{1}{\beta}e^{\alpha z}\cos \beta z + \frac{\alpha}{\beta}\left[\frac{1}{\beta}e^{\alpha z}\sin \beta z - \frac{\alpha}{\beta}\int e^{\alpha z}\sin \beta z \, dz\right]$$

$$= -\frac{1}{\beta}e^{\alpha z}\cos \beta z + \frac{\alpha}{\beta^2}e^{\alpha z}\sin \beta z - \frac{\alpha^2}{\beta^2}\int e^{\alpha z}\sin \beta z \, dz$$

$$\frac{\beta^2 + \alpha^2}{\beta^2}\int e^{\alpha z}\sin \beta z \, dz = -\frac{1}{\beta}e^{\alpha z}\cos \beta z + \frac{\alpha}{\beta^2}e^{\alpha z}\sin \beta z + C$$

$$\int e^{\alpha z}\sin \beta z \, dz = \frac{-\beta}{\alpha^2 + \beta^2}e^{\alpha z}\cos \beta z + \frac{\alpha}{\alpha^2 + \beta^2}e^{\alpha z}\sin \beta z + C = \frac{e^{\alpha z}(\alpha \sin \beta z - \beta \cos \beta z)}{\alpha^2 + \beta^2} + C$$

57. $\quad u = \ln x \qquad dv = x^\alpha dx$

$\qquad du = \frac{1}{x}dx \qquad v = \frac{x^{\alpha+1}}{\alpha+1}, \, \alpha \neq -1$

$$\int x^\alpha \ln x \, dx = \frac{x^{\alpha+1}}{\alpha+1}\ln x - \frac{1}{\alpha+1}\int x^\alpha dx = \frac{x^{\alpha+1}}{\alpha+1}\ln x - \frac{x^{\alpha+1}}{(\alpha+1)^2} + C, \alpha \neq -1$$

59. $\quad u = x^\alpha \qquad\qquad dv = e^{\beta x} dx$

$\qquad du = \alpha x^{\alpha-1} dx \qquad v = \frac{1}{\beta}e^{\beta x}$

$$\int x^\alpha e^{\beta x} dx = \frac{x^\alpha e^{\beta x}}{\beta} - \frac{\alpha}{\beta}\int x^{\alpha-1}e^{\beta x} dx$$

61. $\quad u = x^\alpha \qquad\qquad dv = \cos \beta x \, dx$

$\qquad du = \alpha x^{\alpha-1} dx \qquad v = \frac{1}{\beta}\sin \beta x$

$$\int x^\alpha \cos \beta x \, dx = \frac{x^\alpha \sin \beta x}{\beta} - \frac{\alpha}{\beta}\int x^{\alpha-1}\sin \beta x \, dx$$

63. $\quad u = (a^2 - x^2)^\alpha \qquad\qquad dv = dx$

$\qquad du = -2\alpha x(a^2 - x^2)^{\alpha-1} dx \quad v = x$

$$\int (a^2 - x^2)^\alpha dx = x(a^2 - x^2)^\alpha + 2\alpha \int x^2(a^2 - x^2)^{\alpha-1} dx$$

65. $u = \cos^{\alpha-1} \beta x$ $\qquad\qquad dv = \cos \beta x\, dx$

$du = -\beta(\alpha-1)\cos^{\alpha-2} \beta x \sin \beta x\, dx \quad v = \dfrac{1}{\beta}\sin \beta x$

$\displaystyle\int \cos^{\alpha} \beta x\, dx = \frac{\cos^{\alpha-1} \beta x \sin \beta x}{\beta} + (\alpha-1)\int \cos^{\alpha-2} \beta x \sin^2 \beta x\, dx$

$\displaystyle = \frac{\cos^{\alpha-1} \beta x \sin \beta x}{\beta} + (\alpha-1)\int \cos^{\alpha-2} \beta x(1 - \cos^2 \beta x)\, dx$

$\displaystyle = \frac{\cos^{\alpha-1} \beta x \sin \beta x}{\beta} + (\alpha-1)\int \cos^{\alpha-2} \beta x\, dx - (\alpha-1)\int \cos^{\alpha} \beta x\, dx$

$\displaystyle \alpha \int \cos^{\alpha} \beta x = \frac{\cos^{\alpha-1} \beta x \sin \beta x}{\beta} + (\alpha-1)\int \cos^{\alpha-2} \beta x\, dx$

$\displaystyle \int \cos^{\alpha} \beta x\, dx = \frac{\cos^{\alpha-1} \beta x \sin \beta x}{\alpha\beta} + \frac{\alpha-1}{\alpha}\int \cos^{\alpha-2} \beta x\, dx$

67. $\displaystyle\int x^4 \cos 3x\, dx = \frac{1}{3}x^4 \sin 3x - \frac{4}{3}\int x^3 \sin 3x\, dx = \frac{1}{3}x^4 \sin 3x - \frac{4}{3}\left[-\frac{1}{3}x^3 \cos 3x + \int x^2 \cos 3x\, dx\right]$

$\displaystyle = \frac{1}{3}x^4 \sin 3x + \frac{4}{9}x^3 \cos 3x - \frac{4}{3}\left[\frac{1}{3}x^2 \sin 3x - \frac{2}{3}\int x \sin 3x\, dx\right]$

$\displaystyle = \frac{1}{3}x^4 \sin 3x + \frac{4}{9}x^3 \cos 3x - \frac{4}{9}x^2 \sin 3x + \frac{8}{9}\left[-\frac{1}{3}x \cos 3x + \frac{1}{3}\int \cos 3x\, dx\right]$

$\displaystyle = \frac{1}{3}x^4 \sin 3x + \frac{4}{9}x^3 \cos 3x - \frac{4}{9}x^2 \sin 3x - \frac{8}{27}x \cos 3x + \frac{8}{81}\sin 3x + C$

69. First make a sketch.

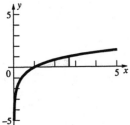

From the sketch, the area is given by

$\displaystyle\int_1^e \ln x\, dx$

$u = \ln x \qquad dv = dx$

$du = \dfrac{1}{x}dx \qquad v = x$

$\displaystyle\int_1^e \ln x\, dx = \left[x \ln x\right]_1^e - \int_1^e dx = [x \ln x - x]_1^e = (e - e) - (1 \cdot 0 - 1) = 1$

71.

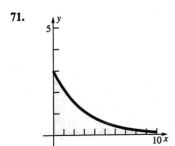

$$\int_0^9 3e^{-x/3}\,dx = -9\int_0^9 e^{-x/3}\left(-\frac{1}{3}\,dx\right) = -9[e^{-x/3}]_0^9 = -\frac{9}{e^3}+9 \approx 8.55$$

73. $\displaystyle\int_0^{\pi/4}(x\cos x - x\sin x)\,dx = \int_0^{\pi/4} x\cos x\,dx - \int_0^{\pi/4} x\sin x\,dx$

$$= \left([x\sin x]_0^{\pi/4} - \int_0^{\pi/4}\sin x\,dx\right) - \left([-x\cos x]_0^{\pi/4} + \int_0^{\pi/4}\cos x\,dx\right)$$

$$= [x\sin x + \cos x + x\cos x - \sin x]_0^{\pi/4} = \frac{\sqrt{2}\pi}{4}-1 \approx 0.11$$

Use Problems 60 and 61 for $\int x\sin x\,dx$ and $\int x\cos x\,dx$.

75. $\displaystyle\int_1^e \ln x^2\,dx = 2\int_1^e \ln x\,dx$

$u = \ln x \qquad dv = dx$

$du = \dfrac{1}{x}\,dx \qquad v = x$

$2\displaystyle\int_1^e \ln x\,dx = 2\left([x\ln x]_1^e - \int_1^e dx\right) = 2\left(e - [x]_1^e\right) = 2$

$\displaystyle\int_1^e x\ln x^2\,dx = 2\int_1^e x\ln x\,dx$

$u = \ln x \qquad dv = x\,dx$

$du = \dfrac{1}{x}\,dx \qquad v = \dfrac{1}{2}x^2$

$2\displaystyle\int_1^e x\ln x\,dx = 2\left(\left[\frac{1}{2}x^2\ln x\right]_1^e - \int_1^e \frac{1}{2}x\,dx\right) = 2\left(\frac{1}{2}e^2 - \left[\frac{1}{4}x^2\right]_1^e\right) = \frac{1}{2}(e^2+1)$

$\dfrac{1}{2}\displaystyle\int_1^e (\ln x)^2\,dx$

$u = (\ln x)^2 \qquad\qquad dv = dx$

$du = \dfrac{2\ln x}{x}\,dx \qquad v = x$

$\dfrac{1}{2}\displaystyle\int_1^e (\ln x)^2\,dx = \frac{1}{2}\left([x(\ln x)^2]_1^e - 2\int_1^e \ln x\,dx\right) = \frac{1}{2}(e-2)$

$\bar{x} = \dfrac{\frac{1}{2}(e^2+1)}{2} = \dfrac{e^2+1}{4}$

$\bar{y} = \dfrac{\frac{1}{2}(e-2)}{2} = \dfrac{e-2}{4}$

77. a. $p(x) = x^3 - 2x$

$\quad g(x) = e^x$

All antiderivatives of $g(x) = e^x$

$\displaystyle\int(x^3 - 2x)e^x\,dx = (x^3-2x)e^x - (3x^2-2)e^x + 6xe^x - 6e^x + C$

b. $p(x) = x^2 - 3x + 1$

$\quad g(x) = \sin x$

$\quad G_1(x) = -\cos x$

$\quad G_2(x) = -\sin x$

$$G_3(x) = \cos x$$

$$\int (x^2 - 3x + 1) \sin x \, dx = (x^2 - 3x + 1)(-\cos x) - (2x - 3)(-\sin x) + 2\cos x + C$$

79. $\displaystyle \int_0^{2\pi} \sin^n x \, dx = \begin{cases} 0 & \text{if } n \text{ is odd} \\ 4\int_0^{\pi/2} \sin^n x \, dx & \text{if } n \text{ is even} \end{cases}$

From Formula 113,

$$4\int_0^{\pi/2} \sin^n x \, dx = 4\left(\frac{1 \cdot 3 \cdot 5 \cdot \ldots \cdot (n-1)}{2 \cdot 4 \cdot 6 \cdot \ldots \cdot n} \frac{\pi}{2} \right) \text{ if } n \text{ is even}$$

$$\int_0^{2\pi} \sin^n x \, dx = \begin{cases} 0 & \text{if } n \text{ is odd} \\ \dfrac{1 \cdot 3 \cdot 5 \cdot \ldots \cdot (n-1)}{2 \cdot 4 \cdot 6 \cdot \ldots \cdot n} 2\pi & \text{if } n \text{ is even} \end{cases}$$

81. $\quad u = f(x) \qquad\qquad dv = \sin nx \, dx$

$\quad du = f'(x)dx \qquad v = -\dfrac{1}{n}\cos nx$

$$a_n = \frac{1}{\pi}\left[\underbrace{\left[-\frac{1}{n}\cos(nx)f(x) \right]_{-\pi}^{\pi}}_{\text{Term 1}} + \underbrace{\frac{1}{n}\int_{-\pi}^{\pi} \cos(nx)f'(x)dx}_{\text{Term 2}} \right]$$

$$\text{Term 1} = \frac{1}{n}\cos(n\pi)(f(-\pi) - f(\pi)) = \pm\frac{1}{n}(f(-\pi) - f(\pi))$$

Since $f'(x)$ is continuous on $[-\infty, \infty]$, it is bounded. Thus, $\int_{-\pi}^{\pi} \cos(nx)f'(x)dx$ is bounded so

$$\lim_{n\to\infty} a_n = \lim_{n\to\infty} \frac{1}{\pi n}\left[\pm(f(-\pi) - f(\pi)) + \int_{-\pi}^{\pi} \cos(nx)f'(x)\,dx \right] = 0.$$

83. The proof fails to consider the constants when integrating $\dfrac{1}{t}$.

$$\int \frac{1}{t}dt = 1 + \int \frac{1}{t}dt$$

$$\ln t + C_1 = 1 + \ln t + C_2$$

Thus $C_1 - C_2 = 1$.

85. $\quad u = f(x) \qquad\qquad dv = dx$

$\quad du = f'(x)dx \qquad v = x$

$$\int_a^b f(x)dx = \left[xf(x) \right]_a^b - \int_a^b xf'(x)dx$$

Starting with the same integral,

$u = f(x) \qquad\qquad dv = dx$

$du = f'(x)dx \qquad v = x - a$

$$\int_a^b f(x)\,dx = \left[(x-a)f(x) \right]_a^b - \int_a^b (x-a)f'(x)dx$$

87. Use proof by induction.

$$n = 1: \; f(a) + f'(a)(t-a) + \int_a^t (t-x)f''(x)dx = f(a) + f'(a)(t-a) + \left[f'(x)(t-x) \right]_a^t + \int_a^t f'(x)dx$$

$$= f(a) + f'(a)(t-a) - f'(a)(t-a) + \left[f(x) \right]_a^t = f(t)$$

Thus, the statement is true for $n = 1$. Note that integration by parts was used with $u = (t-x)$, $dv = f''(x)dx$.

Suppose the statement is true for n.

$$f(t) = f(a) + \sum_{i=1}^{n} \frac{f^{(i)}(a)}{i!}(t-a)^i + \int_a^t \frac{(t-x)^n}{n!} f^{(n+1)}(x)\,dx$$

Integrate $\int_a^t \frac{(t-x)^n}{n!} f^{(n+1)}(x)\,dx$ by parts.

$$u = f^{(n+1)}(x) \qquad dv = \frac{(t-x)^n}{n!}\,dx$$

$$du = f^{(n+2)}(x) \qquad v = -\frac{(t-x)^{n+1}}{(n+1)!}$$

$$\int_a^t \frac{(t-x)^n}{n!} f^{(n+1)}(x)\,dx = \left[-\frac{(t-x)^{n+1}}{(n+1)!} f^{(n+1)}(x) \right]_a^t + \int_a^t \frac{(t-x)^{n+1}}{(n+1)!} f^{(n+2)}(x)\,dx$$

$$= \frac{(t-a)^{n+1}}{(n+1)!} f^{(n+1)}(a) + \int_a^t \frac{(t-x)^{n+1}}{(n+1)!} f^{(n+2)}(x)\,dx$$

Thus $f(t) = f(a) + \sum_{i=1}^{n} \frac{f^{(i)}(a)}{i!}(t-a)^i + \frac{(t-a)^{n+1}}{(n+1)!} f^{(n+1)}(a) + \int_a^t \frac{(t-x)^{n+1}}{(n+1)!} f^{(n+2)}(x)\,dx$

$$= f(a) + \sum_{i=1}^{n+1} \frac{f^{(i)}(a)}{i!}(t-a)^i + \int_a^t \frac{(t-x)^{n+1}}{(n+1)!} f^{(n+2)}(x)\,dx$$

Thus, the statement is true for $n+1$.

89. $u = f(t) \qquad\qquad dv = f''(t)\,dt$

$\quad\;\; du = f'(t)\,dt \qquad\quad v = f'(t)$

$$\int_a^b f''(t)f(t)\,dt = \left[f(t)f'(t) \right]_a^b - \int_a^b [f'(t)]^2\,dt$$

$$= f(b)f'(b) - f(a)f'(a) - \int_a^b [f'(t)]^2\,dt = -\int_a^b [f'(t)]^2\,dt$$

$[f'(t)]^2 \geq 0$, so $-\int_a^b [f'(t)]^2 \leq 0$.

91. Let $I = \int_0^x \int_0^{t_1} \cdots \int_0^{t_{n-1}} f(t_n)\,dt_n \ldots dt_2 dt_1$ be the iterated integral. Note that for $i \geq 2$, the limits of integration of the integral with respect to t_i are 0 to t_{i-1} so that any change of variables in an outer integral affects the limits, and hence the variables in all interior integrals.

$$I = \int_0^x \left[\int_0^{t_1} \int_0^{t_2} \cdots \int_0^{t_{n-1}} f(t_n)\,dt_n \ldots dt_3 dt_2 \right] dt_1$$

$$u = \int_0^{t_1} \int_0^{t_2} \cdots \int_0^{t_{n-1}} f(t_n)\,dt_n \ldots dt_3 dt_2 \qquad dv = dt_1$$

$$du = \left(\int_0^{t_2} \cdots \int_0^{t_{n-1}} f(t_n)\,dt_n \ldots dt_3 \right) dt_1 \qquad v = t_1$$

$$I = \left[t_1 \int_0^{t_1} \int_0^{t_2} \cdots \int_0^{t_{n-1}} f(t_n)\,dt_n \ldots dt_3 dt_2 \right]_{t_1=0}^{t_1=x} - \int_0^x t_1 \left(\int_0^{t_2} \cdots \int_0^{t_{n-1}} f(t_n)\,dt_n \ldots dt_3 \right) dt_1$$

$$= x \int_0^x \int_0^{t_2} \cdots \int_0^{t_{n-1}} f(t_n)\,dt_n \ldots dt_3 dt_2 - \int_0^x t_1 \left(\int_0^{t_2} \cdots \int_0^{t_{n-1}} f(t_n)\,dt_n \ldots dt_3 \right) dt_1$$

Make the change of variables $t_i = t_{i-1}$, so $dt_i = dt_{i-1}$ for $i = 2$ to n.

Hence, $I = \int_0^x x \int_0^{t_1} \cdots \int_0^{t_{n-2}} f(t_{n-1})\,dt_{n-1} \ldots dt_2 dt_1 - \int_0^x t_1 \left(\int_0^{t_1} \cdots \int_0^{t_{n-2}} f(t_{n-1})\,dt_{n-1} \ldots dt_2 \right) dt_1$

$$= \int_0^x (x - t_1) \left(\int_0^{t_1} \int_0^{t_2} \cdots \int_0^{t_{n-2}} f(t_{n-1})\,dt_{n-1} \ldots dt_3 dt_2 \right) dt_1$$

Use integration by parts again.

$$u = \int_0^{t_1} \int_0^{t_2} \cdots \int_0^{t_{n-2}} f(t_{n-1})\, dt_{n-1}...dt_3 dt_2 \qquad dv = (x - t_1)\, dt_1$$

$$du = \left(\int_0^{t_2} \cdots \int_0^{t_{n-2}} f(t_{n-1})\, dt_{n-1}...dt_3 \right) dt_1 \qquad v = -\frac{1}{2}(x - t_1)^2$$

$$I = \left[-\frac{1}{2}(x - t_1)^2 \int_0^{t_1} \int_0^{t_2} \cdots \int_0^{t_{n-2}} f(t_{n-1})\, dt_{n-1}...dt_3 dt_2 \right]_{t_1=0}^{t_1=x} + \int_0^x \frac{1}{2}(x - t_1)^2 \left(\int_0^{t_2} \cdots \int_0^{t_{n-2}} f(t_{n-1})\, dt_{n-1}...dt_3 \right) dt_1$$

$$= \int_0^x \frac{1}{2}(x - t_1)^2 \left(\int_0^{t_2} \cdots \int_0^{t_{n-2}} f(t_{n-1})\, dt_{n-1}...dt_3 \right) dt_1$$

Since $(x - t_1)^2 = 0$ when $t_1 = x$ while $\int_0^{t_1} g(t_2)\, dt_2 = 0$ when $t_1 = 0$ for any integrand $g(t_2)$.

Again change variables so that $t_i = t_{i-1}$ for $i = 2$ to $n - 1$.

$$I = \int_0^x \frac{1}{2}(x - t_1)^2 \left(\int_0^{t_1} \int_0^{t_2} \cdots \int_0^{t_{n-3}} f(t_{n-2})\, dt_{n-2}...dt_3 dt_2 \right) dt_1$$

$$u = \int_0^{t_1} \int_0^{t_2} \cdots \int_0^{t_{n-3}} f(t_{n-2})\, dt_{n-2}...dt_3 dt_2 \qquad dv = \frac{1}{2}(x - t_1)^2\, dt_1$$

$$du = \left(\int_0^{t_2} \cdots \int_0^{t_{n-3}} f(t_{n-2})\, dt_{n-2}...dt_3 \right) dt_1 \qquad v = -\frac{1}{3!}(x - t_1)^3$$

$$I = \left[-\frac{1}{3!}(x - t_1)^3 \int_0^{t_1} \int_0^{t_2} \cdots \int_0^{t_{n-3}} f(t_{n-2})\, dt_{n-2}...dt_3 dt_2 \right]_{t_1=0}^{t_1=x} + \int_0^x \frac{1}{3!}(x - t_1)^3 \left(\int_0^{t_2} \cdots \int_0^{t_{n-3}} f(t_{n-2})\, dt_{n-2}...dt_3 \right) dt_1$$

Changing variables and using integration by parts as before, then changing variables again yields

$$I = \int_0^x \frac{1}{4!}(x - t_1)^4 \left(\int_0^{t_1} \cdots \int_0^{t_{n-5}} f(t_{n-4})\, dt_{n-4}...dt_2 \right) dt_1$$

Reducing the n-fold iterated integral to a single integral requires $n - 1$ integrations by parts, each one contributing a factor of $x - t_1$. The integral is multiplied by $\frac{1}{k!}$ after k integrations by parts, hence

$$I = \frac{1}{(n-1)!} \int_0^x f(t_1)(x - t_1)^{n-1}\, dt_1 \ .$$

93. $\int (3x^4 + 2x^2)e^x\, dx = e^x \sum_{j=0}^{4} (-1)^j \frac{d^j(3x^4 + 2x^2)}{dx^j}$

$$= e^x[3x^4 + 2x^2 - 12x^3 - 4x + 36x^2 + 4 - 72x + 72]$$

$$= e^x(3x^4 - 12x^3 + 38x^2 - 76x + 76)$$

8.5 Concepts Review

1. proper

3. $2x^2 + 3x - 1 = ax^2 + bx + c$
   $a = 2; b = 3; c = -1$

Problem Set 8.5

1. $\dfrac{1}{x(x+1)} = \dfrac{A}{x} + \dfrac{B}{x+1}$
   $1 = A(x+1) + Bx$
   $A = 1, B = -1$

$$\int \frac{1}{x(x+1)}\, dx = \int \frac{1}{x}\, dx - \int \frac{1}{x+1}\, dx$$
$$= \ln|x| - \ln|x+1| + C$$

3. $\dfrac{3}{x^2 - 1} = \dfrac{3}{(x+1)(x-1)} = \dfrac{A}{x+1} + \dfrac{B}{x-1}$
   $3 = A(x - 1) + B(x + 1)$
   $A = -\dfrac{3}{2}, B = \dfrac{3}{2}$

$$\int \frac{3}{x^2 - 1}\, dx = -\frac{3}{2} \int \frac{1}{x+1}\, dx + \frac{3}{2} \int \frac{1}{x-1}\, dx$$
$$= -\frac{3}{2} \ln|x+1| + \frac{3}{2} \ln|x-1| + C$$

5. $\dfrac{x-11}{x^2+3x-4}=\dfrac{x-11}{(x+4)(x-1)}=\dfrac{A}{x+4}+\dfrac{B}{x-1}$

$x-11=A(x-1)+B(x+4)$

$A=3,\,B=-2$

$\displaystyle\int\dfrac{x-11}{x^2+3x-4}\,dx=3\int\dfrac{1}{x+4}\,dx-2\int\dfrac{1}{x-1}\,dx$

$=3\ln|x+4|-2\ln|x-1|+C$

7. $\dfrac{3x-13}{x^2+3x-10}=\dfrac{3x-13}{(x+5)(x-2)}=\dfrac{A}{x+5}+\dfrac{B}{x-2}$

$3x-13=A(x-2)+B(x+5)$

$A=4,\,B=-1$

$\displaystyle\int\dfrac{3x-13}{x^2+3x-10}\,dx=4\int\dfrac{1}{x+5}\,dx-\int\dfrac{1}{x-2}\,dx$

$=4\ln|x+5|-\ln|x-2|+C$

9. $\dfrac{2x+21}{2x^2+9x-5}=\dfrac{2x+21}{(2x-1)(x+5)}=\dfrac{A}{2x-1}+\dfrac{B}{x+5}$

$2x+21=A(x+5)+B(2x-1)$

$A=4,\,B=-1$

$\displaystyle\int\dfrac{2x+21}{2x^2+9x-5}\,dx=\int\dfrac{4}{2x-1}\,dx-\int\dfrac{1}{x+5}\,dx$

$=2\ln|2x-1|-\ln|x+5|+C$

11. $\dfrac{17x-3}{3x^2+x-2}=\dfrac{17x-3}{(3x-2)(x+1)}=\dfrac{A}{3x-2}+\dfrac{B}{x+1}$

$17x-3=A(x+1)+B(3x-2)$

$A=5,\,B=4$

$\displaystyle\int\dfrac{17x-3}{3x^2+x-2}\,dx=\int\dfrac{5}{3x-2}\,dx+\int\dfrac{4}{x+1}\,dx=\dfrac{5}{3}\ln|3x-2|+4\ln|x+1|+C$

13. $\dfrac{2x^2+x-4}{x^3-x^2-2x}=\dfrac{2x^2+x-4}{x(x+1)(x-2)}=\dfrac{A}{x}+\dfrac{B}{x+1}+\dfrac{C}{x-2}$

$2x^2+x-4=A(x+1)(x-2)+Bx(x-2)+Cx(x+1)$

$A=2,\,B=-1,\,C=1$

$\displaystyle\int\dfrac{2x^2+x-4}{x^3-x^2-2x}\,dx=\int\dfrac{2}{x}\,dx-\int\dfrac{1}{x+1}\,dx+\int\dfrac{1}{x-2}\,dx=2\ln|x|-\ln|x+1|+\ln|x-2|+C$

15. $\dfrac{6x^2+22x-23}{(2x-1)(x^2+x-6)}=\dfrac{6x^2+22x-23}{(2x-1)(x+3)(x-2)}=\dfrac{A}{2x-1}+\dfrac{B}{x+3}+\dfrac{C}{x-2}$

$6x^2+22x-23=A(x+3)(x-2)+B(2x-1)(x-2)+C(2x-1)(x+3)\quad A=2,\,B=-1,\,C=3$

$\displaystyle\int\dfrac{6x^2+22x-23}{(2x-1)(x^2+x-6)}\,dx=\int\dfrac{2}{2x-1}\,dx-\int\dfrac{1}{x+3}\,dx+\int\dfrac{3}{x-2}\,dx=\ln|2x-1|-\ln|x+3|+3\ln|x-2|+C$

17. $\dfrac{x^3}{x^2+x-2}=x-1+\dfrac{3x-2}{x^2+x-2}$

$\dfrac{3x-2}{x^2+x-2}=\dfrac{3x-2}{(x+2)(x-1)}=\dfrac{A}{x+2}+\dfrac{B}{x-1}$

$3x-2=A(x-1)+B(x+2)$

$A=\dfrac{8}{3},\,B=\dfrac{1}{3}$

$\displaystyle\int\dfrac{x^3}{x^2+x-2}\,dx=\int(x-1)\,dx+\dfrac{8}{3}\int\dfrac{1}{x+2}\,dx+\dfrac{1}{3}\int\dfrac{1}{x-1}\,dx=\dfrac{1}{2}x^2-x+\dfrac{8}{3}\ln|x+2|+\dfrac{1}{3}\ln|x-1|+C$

19. $\dfrac{x^4+8x^2+8}{x^3-4x}=x+\dfrac{12x^2+8}{x(x+2)(x-2)}$

$\dfrac{12x^2+8}{x(x+2)(x-2)}=\dfrac{A}{x}+\dfrac{B}{x+2}+\dfrac{C}{x-2}$

$12x^2 + 8 = A(x+2)(x-2) + Bx(x-2) + Cx(x+2)$

$A = -2, B = 7, C = 7$

$\int \dfrac{x^4 + 8x^2 + 8}{x^3 - 4x}\,dx = \int x\,dx - 2\int\dfrac{1}{x}\,dx + 7\int\dfrac{1}{x+2}\,dx + 7\int\dfrac{1}{x-2}\,dx = \dfrac{1}{2}x^2 - 2\ln|x| + 7\ln|x+2| + 7\ln|x-2| + C$

21. $\dfrac{x+1}{(x-3)^2} = \dfrac{A}{x-3} + \dfrac{B}{(x-3)^2}$

$x + 1 = A(x-3) + B$

$A = 1, B = 4$

$\int\dfrac{x+1}{(x-3)^2}\,dx = \int\dfrac{1}{x-3}\,dx + \int\dfrac{4}{(x-3)^2}\,dx = \ln|x-3| - \dfrac{4}{x-3} + C$

23. $\dfrac{3x+2}{x^3 + 3x^2 + 3x + 1} = \dfrac{3x+2}{(x+1)^3} = \dfrac{A}{x+1} + \dfrac{B}{(x+1)^2} + \dfrac{C}{(x+1)^3}$

$3x + 2 = A(x+1)^2 + B(x+1) + C$

$A = 0, B = 3, C = -1$

$\int\dfrac{3x+2}{x^3 + 3x^2 + 3x + 1}\,dx = \int\dfrac{3}{(x+1)^2}\,dx - \int\dfrac{1}{(x+1)^3}\,dx = -\dfrac{3}{x+1} + \dfrac{1}{2(x+1)^2} + C$

25. $\dfrac{3x^2 - 21x + 32}{x^3 - 8x^2 + 16x} = \dfrac{3x^2 - 21x + 32}{x(x-4)^2} = \dfrac{A}{x} + \dfrac{B}{x-4} + \dfrac{C}{(x-4)^2}$

$3x^2 - 21x + 32 = A(x-4)^2 + Bx(x-4) + Cx$

$A = 2, B = 1, C = -1$

$\int\dfrac{3x^2 - 21x + 32}{x^3 - 8x^2 + 16}\,dx = \int\dfrac{2}{x}\,dx + \int\dfrac{1}{x-4}\,dx - \int\dfrac{1}{(x-4)^2}\,dx = 2\ln|x| + \ln|x-4| + \dfrac{1}{x-4} + C$

27. $\dfrac{2x^2 + x - 8}{x^3 + 4x} = \dfrac{2x^2 + x - 8}{x(x^2 + 4)} = \dfrac{A}{x} + \dfrac{Bx + C}{x^2 + 4}$

$A = -2, B = 4, C = 1$

$\int\dfrac{2x^2 + x - 8}{x^3 + 4x}\,dx = -2\int\dfrac{1}{x}\,dx + \int\dfrac{4x+1}{x^2 + 4}\,dx = -2\int\dfrac{1}{x}\,dx + 2\int\dfrac{2x}{x^2 + 4}\,dx + \int\dfrac{1}{x^2 + 4}\,dx$

$= -2\ln|x| + 2\ln\left|x^2 + 4\right| + \dfrac{1}{2}\tan^{-1}\left(\dfrac{x}{2}\right) + C$

29. $\dfrac{2x^2 - 3x - 36}{(2x-1)(x^2 + 9)} = \dfrac{A}{2x-1} + \dfrac{Bx + C}{x^2 + 9}$

$A = -4, B = 3, C = 0$

$\int\dfrac{2x^2 - 3x - 36}{(2x-1)(x^2 + 9)}\,dx = -4\int\dfrac{1}{2x-1}\,dx + \int\dfrac{3x}{x^2 + 9}\,dx = -2\ln|2x-1| + \dfrac{3}{2}\ln\left|x^2 + 9\right| + C$

31. $\dfrac{1}{(x-1)^2(x+4)^2} = \dfrac{A}{x-1} + \dfrac{B}{(x-1)^2} + \dfrac{C}{x+4} + \dfrac{D}{(x+4)^2}$

$A = -\dfrac{2}{125}, B = \dfrac{1}{25}, C = \dfrac{2}{125}, D = \dfrac{1}{25}$

$$\int \frac{1}{(x-1)^2(x+4)^2}\,dx = -\frac{2}{125}\int \frac{1}{x-1}\,dx + \frac{1}{25}\int \frac{1}{(x-1)^2}\,dx + \frac{2}{125}\int \frac{1}{x+4}\,dx + \frac{1}{25}\int \frac{1}{(x+4)^2}\,dx$$

$$= -\frac{2}{125}\ln|x-1| - \frac{1}{25(x-1)} + \frac{2}{125}\ln|x+4| - \frac{1}{25(x+4)} + C$$

33. $x = \sin t,\ dx = \cos t\,dt$

$$\int \frac{(\sin^3 t - 8\sin^2 t - 1)\cos t}{(\sin t + 3)(\sin^2 t - 4\sin t + 5)}\,dt = \int \frac{x^3 - 8x^2 - 1}{(x+3)(x^2 - 4x + 5)}\,dx$$

$$= x - \frac{50}{13}\ln|x+3| - \frac{68}{13}\tan^{-1}(x-2) - \frac{41}{26}\ln\left|x^2 - 4x + 5\right| + C$$

which is the result of Problem 32.

$$\int \frac{(\sin^3 t - 8\sin^2 t - 1)\cos t}{(\sin t + 3)(\sin^2 t - 4\sin t + 5)}\,dt = \sin t - \frac{50}{13}\ln|\sin t + 3| - \frac{68}{13}\tan^{-1}(\sin t - 2) - \frac{41}{26}\ln\left|\sin^2 t - 4\sin t + 5\right| + C$$

35. $\dfrac{x^3 - 4x}{(x^2+1)^2} = \dfrac{Ax+B}{x^2+1} + \dfrac{Cx+D}{(x^2+1)^2}$

$A = 1, B = 0, C = -5, D = 0$

$$\int \frac{x^3 - 4x}{(x^2+1)^2}\,dx = \int \frac{x}{x^2+1}\,dx - 5\int \frac{x}{(x^2+1)^2}\,dx = \frac{1}{2}\ln\left|x^2 + 1\right| + \frac{5}{2(x^2+1)} + C$$

37. $\dfrac{2x^3 + 5x^2 + 16x}{x^5 + 8x^3 + 16x} = \dfrac{x(2x^2 + 5x + 16)}{x(x^4 + 8x^2 + 16)} = \dfrac{2x^2 + 5x + 16}{(x^2+4)^2} = \dfrac{Ax+B}{x^2+4} + \dfrac{Cx+D}{(x^2+4)^2}$

$A = 0, B = 2, C = 5, D = 8$

$$\int \frac{2x^3 + 5x^2 + 16x}{x^5 + 8x^3 + 16x}\,dx = \int \frac{2}{x^2+4}\,dx + \int \frac{5x+8}{(x^2+4)^2}\,dx = \int \frac{2}{x^2+4}\,dx + \int \frac{5x}{(x^2+4)^2}\,dx + \int \frac{8}{(x^2+4)^2}\,dx$$

To integrate $\displaystyle\int \frac{8}{(x^2+4)^2}\,dx$, let $x = 2\tan\theta,\ dx = 2\sec^2\theta\,d\theta$.

$$\int \frac{8}{(x^2+4)^2}\,dx = \int \frac{16\sec^2\theta}{16\sec^4\theta}\,d\theta = \int \cos^2\theta\,d\theta = \int \left(\frac{1}{2} + \frac{1}{2}\cos 2\theta\right)d\theta$$

$$= \frac{1}{2}\theta + \frac{1}{4}\sin 2\theta + C = \frac{1}{2}\theta + \frac{1}{2}\sin\theta\cos\theta + C = \frac{1}{2}\tan^{-1}\frac{x}{2} + \frac{x}{x^2+4} + C$$

$$\int \frac{2x^3 + 5x^2 + 16x}{x^5 + 8x^3 + 16x}\,dx = \tan^{-1}\frac{x}{2} - \frac{5}{2(x^2+4)} + \frac{1}{2}\tan^{-1}\frac{x}{2} + \frac{x}{x^2+4} + C = \frac{3}{2}\tan^{-1}\frac{x}{2} + \frac{2x-5}{2(x^2+4)} + C$$

39. $u = \sin\theta,\ du = \cos\theta\,d\theta$

$$\int_0^{\pi/4} \frac{\cos\theta}{(1-\sin^2\theta)(\sin^2\theta+1)^2}\,d\theta = \int_0^{1/\sqrt{2}} \frac{1}{(1-u^2)(u^2+1)^2}\,du = \int_0^{1/\sqrt{2}} \frac{1}{(1-u)(1+u)(u^2+1)^2}\,du$$

$$\frac{1}{(1-u^2)(u^2+1)^2} = \frac{A}{1-u} + \frac{B}{1+u} + \frac{Cu+D}{u^2+1} + \frac{Eu+F}{(u^2+1)^2}$$

$A = \dfrac{1}{8}, B = \dfrac{1}{8}, C = 0, D = \dfrac{1}{4}, E = 0, F = \dfrac{1}{2}$

$$\int_0^{1/\sqrt{2}} \frac{1}{(1-u^2)(u^2+1)^2}\,du = \frac{1}{8}\int_0^{1/\sqrt{2}} \frac{1}{1-u}\,du + \frac{1}{8}\int_0^{1/\sqrt{2}} \frac{1}{1+u}\,du + \frac{1}{4}\int_0^{1/\sqrt{2}} \frac{1}{u^2+1}\,du + \frac{1}{2}\int_0^{1/\sqrt{2}} \frac{1}{(u^2+1)^2}\,du$$

$$= \left[-\frac{1}{8}\ln|1-u| + \frac{1}{8}\ln|1+u| + \frac{1}{4}\tan^{-1}u + \frac{1}{4}\left(\tan^{-1}u + \frac{u}{u^2+1}\right) \right]_0^{1/\sqrt{2}} = \left[\frac{1}{8}\ln\left|\frac{1+u}{1-u}\right| + \frac{1}{2}\tan^{-1}u + \frac{u}{4(u^2+1)} \right]_0^{1/\sqrt{2}}$$

$$= \frac{1}{8}\ln\left|\frac{\sqrt{2}+1}{\sqrt{2}-1}\right| + \frac{1}{2}\tan^{-1}\frac{1}{\sqrt{2}} + \frac{1}{6\sqrt{2}} \approx 0.65$$

(To integrate $\int \frac{1}{(u^2+1)^2}\,du$, let $u = \tan t$.)

41. a. Separating variables, we obtain

$$\frac{dx}{(a-x)(b-x)} = k\,dt$$

$$\frac{1}{(a-x)(b-x)} = \frac{A}{a-x} + \frac{B}{b-x}$$

$$A = -\frac{1}{a-b},\ B = \frac{1}{a-b}$$

$$\int \frac{dx}{(a-x)(b-x)}$$

$$= \frac{1}{a-b}\int\left(-\frac{1}{a-x} + \frac{1}{b-x}\right)dx = \int k\,dt$$

$$\frac{\ln|a-x| - \ln|b-x|}{a-b} = kt + C$$

$$\frac{1}{a-b}\ln\left|\frac{a-x}{b-x}\right| = kt + C$$

$$\frac{a-x}{b-x} = Ce^{(a-b)kt}$$

Since $x = 0$ when $t = 0$, $C = \dfrac{a}{b}$, so

$$a - x = (b-x)\frac{a}{b}e^{(a-b)kt}$$

$$a\left(1 - e^{(a-b)kt}\right) = x\left(1 - \frac{a}{b}e^{(a-b)kt}\right)$$

$$x(t) = \frac{a(1 - e^{(a-b)kt})}{1 - \frac{a}{b}e^{(a-b)kt}} = \frac{ab(1 - e^{(a-b)kt})}{b - ae^{(a-b)kt}}$$

b. Since $b > a$ and $k > 0$, $e^{(a-b)kt} \to 0$ as $t \to \infty$. Thus,

$$x \to \frac{ab(1)}{b-0} = a.$$

c. $x(t) = \dfrac{8(1 - e^{-2kt})}{4 - 2e^{-2kt}}$

$x(20) = 1$, so $4 - 2e^{-40k} = 8 - 8e^{-40k}$

$6e^{-40k} = 4$

$k = -\dfrac{1}{40}\ln\dfrac{2}{3}$

$$e^{-2kt} = e^{t/20\ln 2/3} = e^{\ln(2/3)^{t/20}} = \left(\frac{2}{3}\right)^{t/20}$$

$$x(t) = \frac{4\left(1 - \left(\frac{2}{3}\right)^{t/20}\right)}{2 - \left(\frac{2}{3}\right)^{t/20}}$$

$$x(60) = \frac{4\left(1 - \left(\frac{2}{3}\right)^3\right)}{2 - \left(\frac{2}{3}\right)^3} = \frac{38}{23} \approx 1.65 \text{ grams}$$

d. If $a = b$, the differential equation is, after separating variables

$$\frac{dx}{(a-x)^2} = k\,dt$$

$$\int \frac{dx}{(a-x)^2} = \int k\,dt$$

$$\frac{1}{a-x} = kt + C$$

$$\frac{1}{kt + C} = a - x$$

$$x(t) = a - \frac{1}{kt + C}$$

Since $x = 0$ when $t = 0$, $C = \dfrac{1}{a}$, so

$$x(t) = a - \frac{1}{kt + \frac{1}{a}} = a - \frac{a}{akt+1}$$

$$= a\left(1 - \frac{1}{akt+1}\right) = a\left(\frac{akt}{akt+1}\right).$$

43. a. $\dfrac{dy}{dt} = ky(10 - y)$

$$\frac{dy}{y(10-y)} = k\,dt$$

$$\frac{1}{10}\int\left(\frac{1}{y} + \frac{1}{10-y}\right)dy = \int k\,dt$$

$$\ln\left|\frac{y}{10-y}\right| = 10kt + C$$

$$\frac{y}{10-y} = Ce^{10kt}$$

$$y(0) = 2: \frac{1}{4} = C;\ \frac{y}{10-y} = \frac{1}{4}e^{10kt}$$

$$y(50) = 4: \frac{2}{3} = \frac{1}{4}e^{500k},\ k = \frac{1}{500}\ln\frac{8}{3}$$

$$\frac{y}{10-y} = \frac{1}{4}e^{\left(\frac{1}{50}\ln\frac{8}{3}\right)t}$$

$$4y = 10e^{\left(\frac{1}{50}\ln\frac{8}{3}\right)t} - ye^{\left(\frac{1}{50}\ln\frac{8}{3}\right)t}$$

$$y = \frac{10e^{\left(\frac{1}{50}\ln\frac{8}{3}\right)t}}{4+e^{\left(\frac{1}{50}\ln\frac{8}{3}\right)t}} = \frac{10}{1+4e^{-\left(\frac{1}{50}\ln\frac{8}{3}\right)t}}$$

b. $y(90) = \dfrac{10}{1+4e^{-\left(\frac{1}{50}\ln\frac{8}{3}\right)90}} \approx 5.94$ billion

c. $9 = \dfrac{10}{1+4e^{-\left(\frac{1}{50}\ln\frac{8}{3}\right)t}}$

$$4e^{-\left(\frac{1}{50}\ln\frac{8}{3}\right)t} = \frac{10}{9} - 1$$

$$e^{-\left(\frac{1}{50}\ln\frac{8}{3}\right)t} = \frac{1}{36}$$

$$-\left(\frac{1}{50}\ln\frac{8}{3}\right)t = \ln\frac{1}{36}$$

$$t = -50\left(\frac{\ln\frac{1}{36}}{\ln\frac{8}{3}}\right) \approx 182.68$$

The population will be 9 billion in 2108.

8.6 Chapter Review

Concepts Test

1. True: The resulting integrand will be of the form sin u.

3. False: Try the substitution
$u = x^4, du = 4x^3\,dx$

5. True: The resulting integrand will be of the form $\dfrac{1}{a^2+u^2}$.

7. True: This integral is most easily solved with a partial fraction decomposition.

9. True: Because both exponents are even positive integers, half-angle formulas are used.

11. False: Use the substitution
$u = -x^2 - 4x, du = (-2x-4)dx$

45. Separating variables, we obtain

$$\frac{dy}{(A-y)(B+y)} = k\,dt$$

$$\frac{1}{(A-y)(B+y)} = \frac{C}{A-y} + \frac{D}{B+y}$$

$$C = \frac{1}{A+B}, D = \frac{1}{A+B}$$

$$\int\frac{dy}{(A-y)(B+y)} = \frac{1}{A+B}\int\left(\frac{1}{A-y} + \frac{1}{B+y}\right)dy$$

$$= \int k\,dt$$

$$\frac{-\ln(A-y) + \ln(B+y)}{A+B} = kt + C$$

$$\frac{1}{A+B}\ln\left|\frac{B+y}{A-y}\right| = kt + C$$

$$\frac{B+y}{A-y} = Ce^{(A+B)kt}$$

$$B + y = (A-y)Ce^{(A+B)kt}$$

$$y(1+Ce^{(A+B)kt}) = ACe^{(A+B)kt} - B$$

$$y(t) = \frac{ACe^{(A+B)kt} - B}{1+Ce^{(A+B)kt}}$$

13. True: Then expand and use the substitution
$u = \sin x, du = \cos x\,dx$

15. True: Let $u = \ln x$ $dv = x^2dx$
$du = \dfrac{1}{x}dx$ $v = \dfrac{1}{3}x^3$

17. False: $\dfrac{x^2}{x^2-1} = 1 + \dfrac{1}{2(x-1)} - \dfrac{1}{2(x+1)}$

19. True: $\dfrac{x^2+2}{x(x^2+1)} = \dfrac{2}{x} + \dfrac{-x}{x^2+1}$

21. False: To complete the square, add $\left(\dfrac{b}{2a}\right)^2$.

23. True: Polynomials with the same values for all x will have identical coefficients for like degree terms.

Sample Test Problems

1. $\int_0^4 \frac{t}{\sqrt{9+t^2}}\,dt = \left[\sqrt{9+t^2}\right]_0^4 = 5-3 = 2$

3. $\int_0^{\pi/2} e^{\cos x} \sin x\,dx = \left[-e^{\cos x}\right]_0^{\pi/2} = e-1 \approx 1.718$

5. $\int \frac{y^3+y}{y+1}\,dy = \int\left(y^2-y+2-\frac{2}{1+y}\right)dy$

$= \frac{1}{3}y^3 - \frac{1}{2}y^2 + 2y - 2\ln|1+y| + C$

7. $\int \frac{y-2}{y^2-4y+2}\,dy = \frac{1}{2}\int \frac{2y-4}{y^2-4y+2}\,dy$

$= \frac{1}{2}\ln\left|y^2-4y+2\right| + C$

9. $\int \frac{e^{2t}}{e^t-2}\,dt = e^t + 2\ln\left|e^t-2\right| + C$

(Use the substitution $u = e^t - 2$,

$du = e^t\,dt$

which gives the integral $\int \frac{u+2}{u}\,du$.)

11. $\int \frac{dx}{\sqrt{16+4x-2x^2}} = \frac{1}{\sqrt{2}}\sin^{-1}\left(\frac{x-1}{3}\right) + C$

(Complete the square.)

13. $y = \sqrt{\frac{2}{3}}\tan t,\ dy = \sqrt{\frac{2}{3}}\sec^2 t\,dt$

$\int \frac{dy}{\sqrt{2+3y^2}} = \int \frac{\sqrt{\frac{2}{3}}\sec^2 t}{\sqrt{2}\sec t}\,dt$

$= \frac{1}{\sqrt{3}}\int \sec t\,dt = \frac{1}{\sqrt{3}}\ln|\sec t + \tan t| + C_1$

$= \frac{1}{\sqrt{3}}\ln\left|\frac{\sqrt{y^2+\frac{2}{3}}}{\sqrt{\frac{2}{3}}} + \frac{y}{\sqrt{\frac{2}{3}}}\right| + C_1$

$= \frac{1}{\sqrt{3}}\ln\left|\frac{\sqrt{y^2+\frac{2}{3}}+y}{\sqrt{\frac{2}{3}}}\right| + C_1$

$= \frac{1}{\sqrt{3}}\ln\left|\sqrt{y^2+\frac{2}{3}}+y\right| + C$

Note that $\tan t = \frac{y}{\sqrt{\frac{2}{3}}}$, so $\sec t = \frac{\sqrt{y^2+\frac{2}{3}}}{\sqrt{\frac{2}{3}}}$.

15. $\int \frac{\tan x}{\ln|\cos x|}\,dx = -\ln\left|\ln|\cos x|\right| + C$

Use the substitution $u = \ln|\cos x|$.

17. $\int \sinh x\,dx = \cosh x + C$

19. $u = x \qquad dv = \cot^2 x\,dx$

$du = dx \qquad v = -\cot x - x$

$\int x\cot^2 x\,dx = -x\cot x - x^2 - \int(-\cot x - x)dx$

$= -x\cot x - \frac{1}{2}x^2 + \ln|\sin x| + C$

Use $\cot^2 x = \csc^2 x - 1$ for $\int \cot^2 x\,dx$.

21. $u = \ln t^2,\ du = \frac{2}{t}\,dt$

$\int \frac{\ln t^2}{t}\,dt = \frac{[\ln(t^2)]^2}{4} + C$

23. $\int e^{t/3}\sin 3t\,dt = \frac{-3e^{t/3}(9\cos 3t - \sin 3t)}{82} + C$

Use integration by parts twice.

25. $\int \sin\frac{3x}{2}\cos\frac{x}{2}\,dx = -\frac{\cos x}{2} - \frac{\cos 2x}{4} + C$

Use a product identity.

27. $\int \tan^3 2x\sec 2x\,dx = \frac{1}{2}\int(\sec^2 2x-1)d(\sec 2x)$

$= \frac{1}{6}\sec^3(2x) - \frac{1}{2}\sec(2x) + C$

29. $\int \tan^{3/2} x\sec^4 x\,dx = \int \tan^{3/2} x(1+\tan^2 x)\sec^2 x\,dx = \int \tan^{3/2} x\sec^2 x\,dx + \int \tan^{7/2} x\sec^2 x\,dx$

$= \frac{2}{5}\tan^{5/2} x + \frac{2}{9}\tan^{9/2} x + C$

31. $u = 9 - e^{2y},\ du = -2e^{2y}\,dy$

$\int \frac{e^{2y}}{\sqrt{9-e^{2y}}}\,dy = -\frac{1}{2}\int u^{-1/2}\,du = -\sqrt{u} + C = -\sqrt{9-e^{2y}} + C$

33. $\int e^{\ln(3\cos x)}\,dx = \int 3\cos x\,dx = 3\sin x + C$

35. $u = e^{4x},\ du = 4e^{4x}\,dx$

$$\int \frac{e^{4x}}{1+e^{8x}}\,dx = \frac{1}{4}\int \frac{du}{1+u^2} = \frac{1}{4}\tan^{-1}(e^{4x}) + C$$

37. $u = \sqrt{w+5},\ u^2 = w+5,\ 2u\,du = dw$

$$\int \frac{w}{\sqrt{w+5}}\,dw = 2\int (u^2-5)\,du = \frac{2}{3}u^3 - 10u + C$$

$$= \frac{2}{3}(w+5)^{3/2} - 10(w+5)^{1/2} + C$$

39. $u = \cos^2 y,\ du = -2\cos y \sin y\,dy$

$$\int \frac{\sin y \cos y}{9+\cos^4 y}\,dy = -\frac{1}{2}\int \frac{du}{9+u^2}$$

$$= -\frac{1}{6}\tan^{-1}\left(\frac{\cos^2 y}{3}\right) + C$$

41. $\dfrac{4x^2+3x+6}{x^2(x^2+3)} = \dfrac{A}{x} + \dfrac{B}{x^2} + \dfrac{Cx+D}{x^2+3}$

$A = 1,\ B = 2,\ C = -1,\ D = 2$

$$\int \frac{4x^2+3x+6}{x^2(x^2+3)}\,dx = \int \frac{1}{x}\,dx + 2\int \frac{1}{x^2}\,dx + \int \frac{-x+2}{x^2+3}\,dx$$

$$= \int \frac{1}{x}\,dx + 2\int \frac{1}{x^2}\,dx - \frac{1}{2}\int \frac{2x}{x^2+3}\,dx + 2\int \frac{1}{x^2+3}\,dx$$

$$= \ln|x| - \frac{2}{x} - \frac{1}{2}\ln\left|x^2+3\right| + \frac{2}{\sqrt{3}}\tan^{-1}\left(\frac{x}{\sqrt{3}}\right) + C$$

43. a. $\dfrac{3-4x^2}{(2x+1)^3} = \dfrac{A}{2x+1} + \dfrac{B}{(2x+1)^2} + \dfrac{C}{(2x+1)^3}$

b. $\dfrac{7x-41}{(x-1)^2(2-x)^3} = \dfrac{A}{x-1} + \dfrac{B}{(x-1)^2} + \dfrac{C}{2-x} + \dfrac{D}{(2-x)^2} + \dfrac{E}{(2-x)^3}$

c. $\dfrac{3x+1}{(x^2+x+10)^2} = \dfrac{Ax+B}{x^2+x+10} + \dfrac{Cx+D}{(x^2+x+10)^2}$

d. $\dfrac{(x+1)^2}{(x^2-x+10)^2(1-x^2)^2} = \dfrac{A}{1-x} + \dfrac{B}{(1-x)^2} + \dfrac{C}{1+x} + \dfrac{D}{(1+x)^2} + \dfrac{Ex+F}{x^2-x+10} + \dfrac{Gx+H}{(x^2-x+10)^2}$

e. $\dfrac{x^5}{(x+3)^4(x^2+2x+10)^2} = \dfrac{A}{x+3} + \dfrac{B}{(x+3)^2} + \dfrac{C}{(x+3)^3} + \dfrac{D}{(x+3)^4} + \dfrac{Ex+F}{x^2+2x+10} + \dfrac{Gx+H}{(x^2+2x+10)^2}$

f. $\dfrac{(3x^2+2x-1)^2}{(2x^2+x+10)^3} = \dfrac{Ax+B}{2x^2+x+10} + \dfrac{Cx+D}{(2x^2+x+10)^2} + \dfrac{Ex+F}{(2x^2+x+10)^3}$

45. $y = \dfrac{x^2}{16},\ y' = \dfrac{x}{8}$

$$L = \int_0^4 \sqrt{1+\left(\frac{x}{8}\right)^2}\,dx = \int_0^4 \sqrt{1+\frac{x^2}{64}}\,dx$$

$x = 8\tan t,\ dx = 8\sec^2 t$

$$L = \int_0^{\tan^{-1}\frac{1}{2}} \sec t \cdot 8\sec^2 t\,dt = 8\int_0^{\tan^{-1}\frac{1}{2}} \sec^3 t\,dt = 4\left[\sec t \tan t + \ln|\sec t + \tan t|\right]_0^{\tan^{-1}\frac{1}{2}}$$

$$= 4\left[\left(\frac{\sqrt{5}}{2}\right)\left(\frac{1}{2}\right) + \ln\left|\frac{1}{2} + \frac{\sqrt{5}}{2}\right|\right] = \sqrt{5} + 4\ln\left(\frac{1+\sqrt{5}}{2}\right) \approx 4.1609$$

Use Formula 28 for $\int \sec^3 t\, dt$.

47. $V = 2\pi \int_0^3 \frac{x}{x^2 + 5x + 6}\, dx$

$$\frac{x}{x^2 + 5x + 6} = \frac{A}{x+2} + \frac{B}{x+3}$$

$A = -2, B = 3$

$$V = 2\pi \int_0^3 \left[-\frac{2}{x+2} + \frac{3}{x+3}\right] dx = 2\pi\left[-2\ln(x+2) + 3\ln(x+3)\right]_0^3$$

$$= 2\pi[(-2\ln 5 + 3\ln 6) - (-2\ln 2 + 3\ln 3)] = 2\pi\left(3\ln 2 + 2\ln\frac{2}{5}\right) = 2\pi\ln\frac{32}{25} \approx 1.5511$$

49. $V = 2\pi \int_0^{\ln 3} 2(e^x - 1)(\ln 3 - x)\, dx = 4\pi \int_0^{\ln 3} [(\ln 3)e^x - xe^x - \ln 3 + x]\, dx$

Note that $\int xe^x dx = xe^x - \int e^x dx = xe^x - e^x + C$ by using integration by parts.

$$V = 4\pi\left[(\ln 3)e^x - xe^x + e^x - (\ln 3)x + \frac{1}{2}x^2\right]_0^{\ln 3} = 4\pi\left[\left(3\ln 3 - 3\ln 3 + 3 - (\ln 3)^2 + \frac{1}{2}(\ln 3)^2\right) - (\ln 3 + 1)\right]$$

$$= 4\pi\left[2 - \ln 3 - \frac{1}{2}(\ln 3)^2\right] \approx 3.7437$$

51. $A = -\int_{-6}^0 \frac{t}{(t-1)^2}\, dt$

$$\frac{t}{(t-1)^2} = \frac{A}{(t-1)} + \frac{B}{(t-1)^2}$$

$A = 1, B = 1$

$$A = -\int_{-6}^0 \left[\frac{1}{t-1} + \frac{1}{(t-1)^2}\right] dt = -\left[\ln|t-1| - \frac{1}{t-1}\right]_{-6}^0 = -\left[(0+1) - \left(\ln 7 + \frac{1}{7}\right)\right] = \ln 7 - \frac{6}{7} \approx 1.0888$$

53. The length is given by

$$\int_{\pi/6}^{\pi/3} \sqrt{1 + [f'(x)]^2}\, dx = \int_{\pi/6}^{\pi/3} \sqrt{1 + \frac{\cos^2 x}{\sin^2 x}}\, dx = \int_{\pi/6}^{\pi/3} \sqrt{\frac{\sin^2 x + \cos^2 x}{\sin^2 x}}\, dx = \int_{\pi/6}^{\pi/3} \frac{1}{\sin x}\, dx = \int_{\pi/6}^{\pi/3} \csc x\, dx$$

$$= \left[\ln|\csc x - \cot x|\right]_{\pi/6}^{\pi/3} = \ln\left|\frac{2}{\sqrt{3}} - \frac{1}{\sqrt{3}}\right| - \ln\left|2 - \sqrt{3}\right| = \ln\left(\frac{1}{\sqrt{3}}\right) - \ln(2 - \sqrt{3}) = \ln\left(\frac{2\sqrt{3}+3}{3}\right) \approx 0.768$$

CHAPTER 9

Indeterminate Forms and Improper Integrals

9.1 Concepts Review

1. $\lim_{x \to a} f(x); \lim_{x \to a} g(x)$

3. $\sec^2 x; 1; \lim_{x \to 0} \cos x \neq 0$

Problem Set 9.1

1. The limit is of the form $\frac{0}{0}$.

$$\lim_{x \to 0} \frac{2x - \sin x}{x} = \lim_{x \to 0} \frac{2 - \cos x}{1} = 1$$

3. The limit is of the form $\frac{0}{0}$.

$$\lim_{x \to 0} \frac{x - \sin 2x}{\tan x} = \lim_{x \to 0} \frac{1 - 2\cos 2x}{\sec^2 x} = \frac{1-2}{1} = -1$$

5. The limit is of the form $\frac{0}{0}$.

$$\lim_{x \to -2} \frac{x^2 + 6x + 8}{x^2 - 3x - 10} = \lim_{x \to -2} \frac{2x + 6}{2x - 3}$$
$$= \frac{2}{-7} = -\frac{2}{7}$$

7. The limit is not of the form $\frac{0}{0}$.

As $x \to 1^-, x^2 - 2x + 2 \to 1$, and $x^2 - 1 \to 0^-$ so

$$\lim_{x \to 1^-} \frac{x^2 - 2x + 2}{x^2 + 1} = -\infty$$

9. The limit is of the form $\frac{0}{0}$.

$$\lim_{x \to \pi/2} \frac{\ln(\sin x)^3}{\pi/2 - x} = \lim_{x \to \pi/2} \frac{\frac{1}{\sin^3 x} 3\sin^2 x \cos x}{-1}$$
$$= \frac{0}{-1} = 0$$

11. The limit is of the form $\frac{0}{0}$.

$$\lim_{t \to 1} \frac{\sqrt{t} - t^2}{\ln t} = \lim_{t \to 1} \frac{\frac{1}{2\sqrt{t}} - 2t}{\frac{1}{t}} = \frac{-\frac{3}{2}}{1} = -\frac{3}{2}$$

13. The limit is of the form $\frac{0}{0}$. (Apply l'Hôpital's Rule twice.)

$$\lim_{x \to 0} \frac{\ln \cos 2x}{7x^2} = \lim_{x \to 0} \frac{\frac{-2\sin 2x}{\cos 2x}}{14x} = \lim_{x \to 0} \frac{-2\sin 2x}{14x \cos 2x}$$
$$= \lim_{x \to 0} \frac{-4\cos 2x}{14\cos 2x - 28x\sin 2x} = \frac{-4}{14 - 0} = -\frac{2}{7}$$

15. The limit is of the form $\frac{0}{0}$. (Apply l'Hôpital's Rule three times.)

$$\lim_{x \to 0} \frac{\tan x - x}{\sin 2x - 2x} = \lim_{x \to 0} \frac{\sec^2 x - 1}{2\cos 2x - 2}$$
$$= \lim_{x \to 0} \frac{2\sec^2 x \tan x}{-4\sin 2x} = \lim_{x \to 0} \frac{2\sec^4 x + 4\sec^2 x \tan^2 x}{-8\cos 2x}$$
$$= \frac{2 + 0}{-8} = -\frac{1}{4}$$

17. The limit is of the form $\frac{0}{0}$. (Apply l'Hôpital's Rule twice.)

$$\lim_{x \to 0^+} \frac{x^2}{\sin x - x} = \lim_{x \to 0^+} \frac{2x}{\cos x - 1} = \lim_{x \to 0^+} \frac{2}{-\sin x}$$

This limit is not of the form $\frac{0}{0}$. As $x \to 0^+, 2 \to 2$,

and $-\sin x \to 0^-$, so $\lim_{x \to 0^+} \frac{2}{\sin x} = -\infty$.

19. The limit is of the form $\frac{0}{0}$. (Apply l'Hôpital's Rule twice.)

$$\lim_{x \to 0} \frac{\tan^{-1} x - x}{8x^3} = \lim_{x \to 0} \frac{\frac{1}{1+x^2} - 1}{24x^2} = \lim_{x \to 0} \frac{\frac{-2x}{(1+x^2)^2}}{48x}$$
$$= \lim_{x \to 0} -\frac{1}{24(1+x^2)^2} = -\frac{1}{24}$$

21. The limit is of the form $\frac{0}{0}$. (Apply l'Hôpital's Rule twice.)

$$\lim_{x\to 0^+}\frac{1-\cos x-x\sin x}{2-2\cos x-\sin^2 x}=\lim_{x\to 0^+}\frac{-x\cos x}{2\sin x-2\cos x\sin x}$$

$$=\lim_{x\to 0^+}\frac{x\sin x-\cos x}{2\cos x-2\cos^2 x+2\sin^2 x}$$

This limit is not of the form $\frac{0}{0}$.

As $x\to 0^+$, $x\sin x-\cos x\to -1$ and $2\cos x-2\cos^2 x+2\sin^2 x\to 0^+$, so

$$\lim_{x\to 0^+}\frac{x\sin x-\cos x}{2\cos x-2\cos^2 x+2\sin^2 x}=-\infty$$

23. The limit is of the form $\frac{0}{0}$.

$$\lim_{x\to 0}\frac{\int_0^x \sqrt{1+\sin t}\,dt}{x}=\lim_{x\to 0}\sqrt{1+\sin x}=1$$

25. It would not have helped us because we proved

$$\lim_{x\to 0}\frac{\sin x}{x}=1$$ in order to find the derivative of $\sin x$.

27. a. $\overline{OB}=\cos t,\ \overline{BC}=\sin t$ and $\overline{AB}=1-\cos t$, so the area of triangle ABC is $\frac{1}{2}\sin t(1-\cos t)$.

The area of the sector COA is $\frac{1}{2}t$ while the area of triangle COB is $\frac{1}{2}\cos t\sin t$, thus the area of the curved region ABC is $\frac{1}{2}(t-\cos t\sin t)$.

$$\lim_{t\to 0^+}\frac{\text{area of triangle }ABC}{\text{area of curved region }ABC}=\lim_{t\to 0^+}\frac{\frac{1}{2}\sin t(1-\cos t)}{\frac{1}{2}(t-\cos t\sin t)}$$

$$=\lim_{t\to 0^+}\frac{\sin t(1-\cos t)}{t-\cos t\sin t}=\lim_{t\to 0^+}\frac{\cos t-\cos^2 t+\sin^2 t}{1-\cos^2 t+\sin^2 t}=\lim_{t\to 0^+}\frac{4\sin t\cos t-\sin t}{4\cos t\sin t}=\lim_{t\to 0^+}\frac{4\cos t-1}{4\cos t}=\frac{3}{4}$$

(L'Hôpital's Rule was applied twice.)

b. The area of the sector BOD is $\frac{1}{2}t\cos^2 t$, so the area of the curved region BCD is $\frac{1}{2}\cos t\sin t-\frac{1}{2}t\cos^2 t$.

$$\lim_{t\to 0^+}\frac{\text{area of curved region }BCD}{\text{area of curved region }ABC}=\lim_{t\to 0^+}\frac{\frac{1}{2}\cos t(\sin t-t\cos t)}{\frac{1}{2}(t-\cos t\sin t)}$$

$$=\lim_{t\to 0^+}\frac{\cos t(\sin t-t\cos t)}{t-\sin t\cos t}=\lim_{t\to 0^+}\frac{\sin t(2t\cos t-\sin t)}{1-\cos^2 t+\sin^2 t}=\lim_{t\to 0^+}\frac{2t(\cos^2 t-\sin^2 t)}{4\cos t\sin t}=\lim_{t\to 0^+}\frac{t(\cos^2 t-\sin^2 t)}{2\cos t\sin t}$$

$$=\lim_{t\to 0^+}\frac{\cos^2 t-4t\cos t\sin t-\sin^2 t}{2\cos^2 t-2\sin^2 t}=\frac{1-0-0}{2-0}=\frac{1}{2}$$

(L'Hôpital's Rule was applied three times.)

29. A should approach $4\pi b^2$, the surface area of a sphere of radius b.

$$\lim_{a\to b^+}\left[2\pi b^2+\frac{2\pi a^2 b\arcsin\frac{\sqrt{a^2-b^2}}{a}}{\sqrt{a^2-b^2}}\right]=2\pi b^2+2\pi b\lim_{a\to b^+}\frac{a^2\arcsin\frac{\sqrt{a^2-b^2}}{a}}{\sqrt{a^2-b^2}}$$

Focusing on the limit, we have

$$\lim_{a \to b^+} \frac{a^2 \arcsin \frac{\sqrt{a^2-b^2}}{a}}{\sqrt{a^2-b^2}} = \lim_{a \to b^+} \frac{2a \arcsin \frac{\sqrt{a^2-b^2}}{a} + a^2\left(\frac{b}{a\sqrt{a^2-b^2}}\right)}{\frac{a}{\sqrt{a^2-b^2}}} = \lim_{a \to b^+} \left(2\sqrt{a^2-b^2}\arcsin \frac{\sqrt{a^2-b^2}}{a} + b\right) = b.$$

Thus, $\lim\limits_{a \to b^+} A = 2\pi b^2 + 2\pi b(b) = 4\pi b^2$.

31. If $f'(a)$ and $g'(a)$ both exist, then f and g are both continuous at a. Thus, $\lim\limits_{x \to a} f(x) = 0 = f(a)$ and $\lim\limits_{x \to a} g(x) = 0 = g(a)$.

$$\lim_{x \to a} \frac{f(x)}{g(x)} = \lim_{x \to a} \frac{f(x) - f(a)}{g(x) - g(a)}$$

$$\lim_{x \to a} \frac{\frac{f(x)-f(a)}{x-a}}{\frac{g(x)-g(a)}{x-a}} = \frac{\lim\limits_{x \to a} \frac{f(x)-f(a)}{x-a}}{\lim\limits_{x \to a} \frac{g(x)-g(a)}{x-a}} = \frac{f'(a)}{g'(a)}$$

33. $\lim\limits_{x \to 0} \dfrac{e^x - 1 - x - \frac{x^2}{2} - \frac{x^3}{6}}{x^4} = \dfrac{1}{24}$

35. $\lim\limits_{x \to 0} \dfrac{\tan x - x}{\arcsin x - x} = 2$

37.

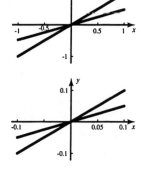

39.

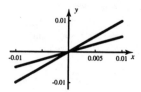

The slopes are approximately $0.005 / 0.01 = 1/2$ and $0.01 / 0.01 = 1$. The ratio of the slopes is therefore $1/2$, indicating that the limit of the ratio should be about $1/2$. An application of l'Hopital's Rule confirms this.

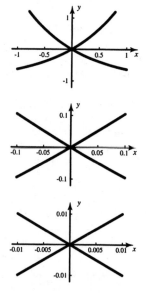

The slopes are approximately $0.01 / 0.01 = 1$ and $-0.01 / 0.01 = 1$. The ratio of the slopes is therefore $-1/1 = -1$, indicating that the limit of the ratio should be about -1. An application of l'Hopital's Rule confirms this.

9.2 Concepts Review

1. $\dfrac{f'(x)}{g'(x)}$

3. $\infty - \infty, 0°, \infty°, 1^\infty$

Problem Set 9.2

1. The limit is of the form $\dfrac{\infty}{\infty}$.

$$\lim_{x\to\infty} \frac{\ln x^{1000}}{x} = \lim_{x\to\infty} \frac{\frac{1}{x^{1000}}1000x^{999}}{1}$$

$$= \lim_{x\to\infty} \frac{1000}{x} = 0$$

3. $\lim\limits_{x\to\infty} \dfrac{x^{10000}}{e^x} = 0$ (See Example 2).

5. The limit is of the form $\dfrac{\infty}{\infty}$.

$$\lim_{x\to\frac{\pi}{2}} \frac{3\sec x+5}{\tan x} = \lim_{x\to\frac{\pi}{2}} \frac{3\sec x\tan x}{\sec^2 x}$$

$$= \lim_{x\to\frac{\pi}{2}} \frac{3\tan x}{\sec x} = \lim_{x\to\frac{\pi}{2}} 3\sin x = 3$$

7. The limit is of the form $\dfrac{\infty}{\infty}$.

$$\lim_{x\to\infty} \frac{\ln(\ln x^{1000})}{\ln x} = \lim_{x\to\infty} \frac{\frac{1}{\ln x^{1000}}\left(\frac{1}{x^{1000}}1000x^{999}\right)}{\frac{1}{x}}$$

$$= \lim_{x\to\infty} \frac{1000}{x\ln x^{1000}} = 0$$

9. The limit is of the form $\dfrac{\infty}{\infty}$.

$$\lim_{x\to 0^+} \frac{\cot x}{\sqrt{-\ln x}} = \lim_{x\to 0^+} \frac{-\csc^2 x}{-\frac{1}{2x\sqrt{-\ln x}}}$$

$$= \lim_{x\to 0^+} \frac{2x\sqrt{-\ln x}}{\sin^2 x}$$

$$= \lim_{x\to 0^+} \left[\frac{2x}{\sin x}\csc x\sqrt{-\ln x}\right] = \infty$$

since $\lim\limits_{x\to 0^+} \dfrac{x}{\sin x} = 1$ while $\lim\limits_{x\to 0^+} \csc x = \infty$ and $\lim\limits_{x\to 0^+} \sqrt{-\ln x} = \infty$.

11. $\lim\limits_{x\to 0}(x\ln x^{1000}) = \lim\limits_{x\to 0} \dfrac{\ln x^{1000}}{\frac{1}{x}}$

The limit is of the form $\dfrac{\infty}{\infty}$.

$$\lim_{x\to 0} \frac{\ln x^{1000}}{\frac{1}{x}} = \lim_{x\to 0} \frac{\frac{1}{x^{1000}}1000x^{999}}{-\frac{1}{x^2}}$$

$$= \lim_{x\to 0} -1000x = 0$$

13. $\lim\limits_{x\to 0}(\csc^2 x - \cot^2 x) = \lim\limits_{x\to 0} \dfrac{1-\cos^2 x}{\sin^2 x}$

$$= \lim_{x\to 0} \frac{\sin^2 x}{\sin^2 x} = 1$$

15. The limit is of the form 0^0.

Let $y = (3x)^{x^2}$, then $\ln y = x^2\ln 3x$

$$\lim_{x\to 0^+} x^2\ln 3x = \lim_{x\to 0^+} \frac{\ln 3x}{\frac{1}{x^2}}$$

The limit is of the form $\dfrac{\infty}{\infty}$.

$$\lim_{x\to 0^+} \frac{\ln 3x}{\frac{1}{x^2}} = \lim_{x\to 0^+} \frac{\frac{1}{3x}\cdot 3}{-\frac{2}{x^3}} = \lim_{x\to 0^+} -\frac{x^2}{2} = 0$$

$$\lim_{x\to 0^+}(3x)^{x^2} = \lim_{x\to 0^+} e^{\ln y} = 1$$

17. The limit is of the form 0^∞, which is not an indeterminate form. $\lim\limits_{x\to(\pi/2)^-}(5\cos x)^{\tan x} = 0$

19. The limit is of the form 1^∞.

Let $y = (x+e^{x/3})^{3/x}$, then $\ln y = \dfrac{3}{x}\ln(x+e^{x/3})$.

$$\lim_{x\to 0} \frac{3}{x}\ln(x+e^{x/3}) = \lim_{x\to 0} \frac{3\ln(x+e^{x/3})}{x}$$

The limit is of the form $\dfrac{0}{0}$.

$$\lim_{x\to 0} \frac{3\ln(x+e^{x/3})}{x} = \lim_{x\to 0} \frac{\frac{3}{x+e^{x/3}}\left(1+\frac{1}{3}e^{x/3}\right)}{1}$$

$$= \lim_{x\to 0} \frac{3+e^{x/3}}{x+e^{x/3}} = \frac{4}{1} = 4$$

$$\lim_{x\to 0}(x+e^{x/3})^{3/x} = \lim_{x\to 0} e^{\ln y} = e^4$$

21. The limit is of the form 1^0, which is not an indeterminate form.

$$\lim_{x\to\frac{\pi}{2}}(\sin x)^{\cos x} = 1$$

23. The limit is of the form ∞^0. Let

$y = x^{1/x}$, then $\ln y = \dfrac{1}{x}\ln x$.

$\lim\limits_{x\to\infty} \dfrac{1}{x}\ln x = \lim\limits_{x\to\infty}\dfrac{\ln x}{x}$

The limit is of the form $\dfrac{-\infty}{\infty}$.

$\lim\limits_{x\to\infty}\dfrac{\ln x}{x} = \lim\limits_{x\to\infty}\dfrac{\frac{1}{x}}{1} = \lim\limits_{x\to\infty}\dfrac{1}{x} = 0$

$\lim\limits_{x\to\infty} x^{1/x} = \lim\limits_{x\to\infty} e^{\ln y} = 1$

25. The limit is of the form 0^∞, which is not an indeterminate form.

$\lim\limits_{x\to 0^+}(\tan x)^{2/x} = 0$

27. The limit is of the form 0^0. Let

$y = (\sin x)^x$, then $\ln y = x\ln(\sin x)$.

$\lim\limits_{x\to 0^+} x\ln(\sin x) = \lim\limits_{x\to 0^+}\dfrac{\ln(\sin x)}{\frac{1}{x}}$

The limit is of the form $\dfrac{-\infty}{\infty}$.

$\lim\limits_{x\to 0^+}\dfrac{\ln(\sin x)}{\frac{1}{x}} = \lim\limits_{x\to 0^+}\dfrac{\frac{1}{\sin x}\cos x}{-\frac{1}{x^2}}$

$= \lim\limits_{x\to 0^+}\left[\dfrac{x}{\sin x}(-x\cos x)\right] = 1\cdot 0 = 0$

$\lim\limits_{x\to 0^+}(\sin x)^x = \lim\limits_{x\to 0^+} e^{\ln y} = 1$

29. The limit is of the form $\infty - \infty$.

$\lim\limits_{x\to 0}\left(\csc x - \dfrac{1}{x}\right) = \lim\limits_{x\to 0}\left(\dfrac{1}{\sin x} - \dfrac{1}{x}\right) = \lim\limits_{x\to 0}\dfrac{x - \sin x}{x\sin x}$

The limit is of the form $\dfrac{0}{0}$. (Apply l'Hôpital's Rule twice.)

$\lim\limits_{x\to 0}\dfrac{x - \sin x}{x\sin x} = \lim\limits_{x\to 0}\dfrac{1 - \cos x}{\sin x + x\cos x}$

$= \lim\limits_{x\to 0}\dfrac{\sin x}{2\cos x - x\sin x} = \dfrac{0}{2} = 0$

31. The limit is of the form 3^∞, which is not an indeterminate form.

$\lim\limits_{x\to 0^+}(1 + 2e^x)^{1/x} = \infty$

33. The limit is of the form 1^∞.

Let $y = (\cos x)^{1/x}$, then $\ln y = \dfrac{1}{x}\ln(\cos x)$.

$\lim\limits_{x\to 0}\dfrac{1}{x}\ln(\cos x) = \lim\limits_{x\to 0}\dfrac{\ln(\cos x)}{x}$

The limit is of the form $\dfrac{0}{0}$.

$\lim\limits_{x\to 0}\dfrac{\ln(\cos x)}{x} = \lim\limits_{x\to 0}\dfrac{\frac{1}{\cos x}(-\sin x)}{1} = \lim\limits_{x\to 0}-\dfrac{\sin x}{\cos x} = 0$

$\lim\limits_{x\to 0}(\cos x)^{1/x} = \lim\limits_{x\to 0} e^{\ln y} = 1$

35. Since $\cos x$ oscillates between -1 and 1 as $x \to \infty$, this limit is not of an indeterminate form previously seen.

Let $y = e^{\cos x}$, then $\ln y = (\cos x)\ln e = \cos x$

$\lim\limits_{x\to\infty}\cos x$ does not exist, so $\lim\limits_{x\to\infty} e^{\cos x}$ does not exist.

37. The limit is of the form $\dfrac{0}{-\infty}$, which is not an indeterminate form.

$\lim\limits_{x\to 0^+}\dfrac{x}{\ln x} = 0$

39. $\sqrt{1 + e^{-t}} > 1$ for all t, so

$\displaystyle\int_1^x \sqrt{1 + e^{-t}}\,dt > \int_1^x dt = x - 1$.

The limit is of the form $\dfrac{\infty}{\infty}$.

$\lim\limits_{x\to\infty}\dfrac{\displaystyle\int_1^x \sqrt{1 + e^{-t}}\,dt}{x} = \lim\limits_{x\to\infty}\dfrac{\sqrt{1 + e^{-x}}}{1} = 1$

41. a. Let $y = \sqrt[n]{a}$, then $\ln y = \dfrac{1}{n}\ln a$.

$\lim\limits_{n\to\infty}\dfrac{1}{n}\ln a = 0$

$\lim\limits_{n\to\infty}\sqrt[n]{a} = \lim\limits_{n\to\infty} e^{\ln y} = 1$

b. The limit is of the form ∞^0.

Let $y = \sqrt[n]{n}$, then $\ln y = \dfrac{1}{n}\ln n$.

$\lim\limits_{n\to\infty}\dfrac{1}{n}\ln n = \lim\limits_{n\to\infty}\dfrac{\ln n}{n}$

This limit is of the form $\dfrac{\infty}{\infty}$.

$\lim\limits_{n\to\infty}\dfrac{\ln n}{n} = \lim\limits_{n\to\infty}\dfrac{\frac{1}{n}}{1} = 0$

$\lim\limits_{n\to\infty}\sqrt[n]{n} = \lim\limits_{n\to\infty} e^{\ln y} = 1$

c. $\lim_{n\to\infty} n\left(\sqrt[n]{a}-1\right) = \lim_{n\to\infty} \dfrac{\sqrt[n]{a}-1}{\frac{1}{n}}$

This limit is of the form $\dfrac{0}{0}$,

since $\lim_{n\to\infty} \sqrt[n]{a} = 1$ by part a.

$\lim_{n\to\infty} \dfrac{\sqrt[n]{a}-1}{\frac{1}{n}} = \lim_{n\to\infty} \dfrac{-\frac{1}{n^2}\sqrt[n]{a}\ln a}{-\frac{1}{n^2}}$

$= \lim_{n\to\infty} \sqrt[n]{a}\ln a = \ln a$

d. $\lim_{n\to\infty} n\left(\sqrt[n]{n}-1\right) = \lim_{n\to\infty} \dfrac{\sqrt[n]{n}-1}{\frac{1}{n}}$

This limit is of the form $\dfrac{0}{0}$,

since $\lim_{n\to\infty} \sqrt[n]{n} = 1$ by part b.

$\lim_{n\to\infty} \dfrac{\sqrt[n]{n}-1}{\frac{1}{n}} = \lim_{n\to\infty} \dfrac{\sqrt[n]{n}\left(\frac{1}{n^2}\right)(1-\ln n)}{-\frac{1}{n^2}}$

$= \lim_{n\to\infty} \sqrt[n]{n}(\ln n - 1) = \infty$

43.

$\ln y = \dfrac{\ln x}{x}$

$\lim_{x\to 0^+} \dfrac{\ln x}{x} = -\infty$, so $\lim_{x\to 0^+} x^{1/x} = \lim_{x\to 0^+} e^{\ln y} = 0$

$\lim_{x\to\infty} \dfrac{\ln x}{x} = \lim_{x\to\infty} \dfrac{\frac{1}{x}}{1} = 0$, so $\lim_{x\to\infty} x^{1/x} = \lim_{x\to\infty} e^{\ln y} = 1$

$y = x^{1/x} = e^{\frac{1}{x}\ln x}$

9.3 Concepts Review

1. converge

3. $\displaystyle\int_{-\infty}^{0} f(x)\,dx;\ \int_{0}^{\infty} f(x)\,dx$

$y' = \left(\dfrac{1}{x^2} - \dfrac{\ln x}{x^2}\right)e^{\frac{1}{x}\ln x}$

$y' = 0$ when $x = e$.

y is maximum at $x = e$ since $y' > 0$ on $(0, e)$ and

$y' < 0$ on (e, ∞). When $x = e$, $y = e^{1/e}$.

45. $\lim_{n\to\infty} \dfrac{1^k + 2^k + \ldots + n^k}{n^{k+1}}$

$= \lim_{n\to\infty} \dfrac{1}{n}\left[\left(\dfrac{1}{n}\right)^k + \left(\dfrac{2}{n}\right)^k + \ldots + \left(\dfrac{n}{n}\right)^k\right]$

$= \lim_{n\to\infty} \dfrac{1}{n}\left[\dfrac{1}{n^k} + \dfrac{2^k}{n^k} + \ldots + 1\right] = 0$

47. a. $\lim_{t\to 0^+} \left(\dfrac{1}{2}2^t + \dfrac{1}{2}5^t\right)^{1/t} = \sqrt{2}\sqrt{5} \approx 3.162$

b. $\lim_{t\to 0^+} \left(\dfrac{1}{5}2^t + \dfrac{4}{5}5^t\right)^{1/t} = \sqrt[5]{2}\cdot\sqrt[5]{5^4} \approx 4.163$

c. $\lim_{t\to 0^+} \left(\dfrac{1}{10}2^t + \dfrac{9}{10}5^t\right)^{1/t} = \sqrt[10]{2}\cdot\sqrt[10]{5^9} \approx 4.562$

49. Note $f(x) > 0$ on $[0, \infty)$.

$\lim_{x\to\infty} f(x) = \lim_{x\to\infty} \left(\dfrac{x^{25}}{e^x} + \dfrac{x^3}{e^x} + \left(\dfrac{2}{e}\right)^x\right) = 0$

Therefore there is no absolute minimum.

$f'(x) = (25x^{24} + 3x^2 + 2^x \ln 2)e^{-x}$

$\qquad -(x^{25} + x^3 + 2^x)e^{-x}$

$= (-x^{25} + 25x^{24} - x^3 + 3x^2 - 2^x + 2^x \ln 2)e^{-x}$

Solve for x when $f'(x) = 0$. Using a numerical method, $x \approx 25$.

A graph using a computer algebra system verifies that an absolute maximum occurs at about $x = 25$.

Problem Set 9.3

In this section and the chapter review, it is understood that $[g(x)]_a^\infty$ means $\lim_{b\to\infty} [g(x)]_a^b$ and likewise for similar expressions.

1. $\int_{100}^{\infty} e^x \, dx = \left[e^x \right]_{100}^{\infty} = \infty - e^{100} = \infty$

The integral diverges.

3. $\int_1^{\infty} 2xe^{-x^2} \, dx = \left[-e^{-x^2} \right]_1^{\infty} = 0 - (-e^{-1}) = \dfrac{1}{e}$

5. $\int_9^{\infty} \dfrac{x \, dx}{\sqrt{1+x^2}} = \left[\sqrt{1+x^2} \right]_9^{\infty} = \infty - \sqrt{82} = \infty$

The integral diverges.

7. $\int_1^{\infty} \dfrac{dx}{x^{1.00001}} = \left[-\dfrac{1}{0.00001 x^{0.00001}} \right]_1^{\infty}$

$= 0 - \left(-\dfrac{1}{0.00001} \right) = \dfrac{1}{0.00001} = 100,000$

9. $\int_1^{\infty} \dfrac{dx}{x^{0.99999}} = \left[\dfrac{x^{0.00001}}{0.00001} \right]_1^{\infty} = \infty - 100,000 = \infty$

The integral diverges.

11. $\int_e^{\infty} \dfrac{1}{x \ln x} \, dx = [\ln(\ln x)]_e^{\infty} = \infty - 0 = \infty$

The integral diverges.

13. Let $u = \ln x, \ du = \dfrac{1}{x} dx, dv = \dfrac{1}{x^2} dx, v = -\dfrac{1}{x}.$

$\int_2^{\infty} \dfrac{\ln x}{x^2} dx = \left[-\dfrac{\ln x}{x} \right]_2^{\infty} + \int_2^{\infty} \dfrac{1}{x^2} dx$

$= \left[-\dfrac{\ln x}{x} - \dfrac{1}{x} \right]_2^{\infty} = \lim_{b \to \infty} \left(-\dfrac{\ln x + 1}{x} \right) + \dfrac{\ln 2 + 1}{2}$

$= \dfrac{\ln 2 + 1}{2}$

15. $\int_{-\infty}^1 \dfrac{dx}{(2x-3)^3} = \left[-\dfrac{1}{4(2x-3)^2} \right]_{-\infty}^1$

$= -\dfrac{1}{4} - (-0) = -\dfrac{1}{4}$

17. $\int_{-\infty}^{\infty} \dfrac{x}{\sqrt{x^2+9}} dx = \int_{-\infty}^0 \dfrac{x}{\sqrt{x^2+9}} dx + \int_0^{\infty} \dfrac{x}{\sqrt{x^2+9}} dx = \left[\sqrt{x^2+9} \right]_{-\infty}^0 + \left[\sqrt{x^2+9} \right]_0^{\infty} = (3-\infty) + (\infty - 3)$

The integral diverges since both $\int_{-\infty}^0 \dfrac{x}{\sqrt{x^2+9}} dx$ and $\int_0^{\infty} \dfrac{x}{\sqrt{x^2+9}} dx$ diverge.

19. $\int_{-\infty}^{\infty} \dfrac{1}{x^2+2x+10} dx = \int_{-\infty}^{\infty} \dfrac{1}{(x+1)^2+9} dx = \int_{-\infty}^0 \dfrac{1}{(x+1)^2+9} dx + \int_0^{\infty} \dfrac{1}{(x+1)^2+9} dx$

$\int \dfrac{1}{(x+1)^2+9} dx = \dfrac{1}{3} \tan^{-1} \dfrac{x+1}{3}$ by using the substitution $x + 1 = 3 \tan \theta.$

$\int_{-\infty}^0 \dfrac{1}{(x+1)^2+9} dx = \left[\dfrac{1}{3} \tan^{-1} \dfrac{x+1}{3} \right]_{-\infty}^0 = \dfrac{1}{3} \tan^{-1} \dfrac{1}{3} - \dfrac{1}{3}\left(-\dfrac{\pi}{2} \right) = \dfrac{1}{6}\left(\pi + 2\tan^{-1} \dfrac{1}{3} \right)$

$\int_0^{\infty} \dfrac{1}{(x+1)^2+9} dx = \left[\dfrac{1}{3} \tan^{-1} \dfrac{x+1}{3} \right]_0^{\infty} = \dfrac{1}{3}\left(\dfrac{\pi}{2} \right) - \dfrac{1}{3}\tan^{-1} \dfrac{1}{3} = \dfrac{1}{6}\left(\pi - 2\tan^{-1} \dfrac{1}{3} \right)$

$\int_{-\infty}^{\infty} \dfrac{1}{x^2+2x+10} dx = \dfrac{1}{6}\left(\pi + 2\tan^{-1} \dfrac{1}{3} \right) + \dfrac{1}{6}\left(\pi - 2\tan^{-1} \dfrac{1}{3} \right) = \dfrac{\pi}{3}$

21. $\int_{-\infty}^{\infty} \text{sech } x \, dx = \int_{-\infty}^0 \text{sech } x \, dx = \int_0^{\infty} \text{sech } x \, dx$

$= [\tan^{-1}(\sinh x)]_{-\infty}^0 + [\tan^{-1}(\sinh x)]_0^{\infty}$

$= \left[0 - \left(-\dfrac{\pi}{2} \right) \right] + \left[\dfrac{\pi}{2} - 0 \right] = \pi$

23. $\int_0^{\infty} e^{-x} \cos x \, dx = \left[\dfrac{1}{2e^x}(\sin x - \cos x) \right]_0^{\infty}$

$= 0 - \dfrac{1}{2}(0-1) = \dfrac{1}{2}$

(Use Formula 68 with $a = -1$ and $b = 1$.)

25. The area is given by

$$\int_1^\infty \frac{2}{4x^2-1}\,dx = \int_1^\infty \left(\frac{1}{2x-1}-\frac{1}{2x+1}\right)dx$$

$$=\frac{1}{2}\Big[\ln|2x-1|-\ln|2x+1|\Big]_1^\infty = \frac{1}{2}\left[\ln\left|\frac{2x-1}{2x+1}\right|\right]_1^\infty$$

$$=\frac{1}{2}\left(0-\ln\left(\frac{1}{3}\right)\right)=\frac{1}{2}\ln 3$$

Note:. $\lim\limits_{x\to\infty}\ln=\left|\dfrac{2x-1}{2x+1}\right|=0$ since

$$\lim_{x\to\infty}\left(\frac{2x-1}{2x+1}\right)=1\cdot$$

31. **a.**

$$\int_{-\infty}^\infty f(x)\,dx = \int_{-\infty}^a 0\,dx + \int_a^b \frac{1}{b-a}\,dx + \int_b^\infty 0\,dx$$

$$=0+\frac{1}{b-a}[x]_a^b+0$$

$$=\frac{1}{b-a}(b-a)$$

b.

$$\mu=\int_{-\infty}^\infty x f(x)\,dx$$

$$=\int_{-\infty}^a x\cdot 0\,dx + \int_a^b x\frac{1}{b-a}\,dx + \int_b^\infty x\cdot 0\,dx$$

$$=0+\frac{1}{b-a}\left[\frac{x^2}{2}\right]_a^b+0$$

$$=\frac{b^2-a^2}{2(b-a)}$$

$$=\frac{(b+a)(b-a)}{2(b-a)}$$

$$=\frac{a+b}{2}$$

$$\sigma^2=\int_{-\infty}^\infty (x-\mu)^2\,dx$$

$$=\int_{-\infty}^a (x-\mu)^2\cdot 0\,dx + \int_a^b (x-\mu)^2\frac{1}{b-a}\,dx + \int_b^\infty (x-\mu)^2\cdot 0\,dx$$

$$=0+\frac{1}{b-a}\left[\frac{(x-\mu)^3}{3}\right]_a^b+0$$

$$=\frac{1}{b-a}\frac{(b-\mu)^3-(a-\mu)^3}{3}$$

$$=\frac{1}{b-a}\frac{b^3-3b^2\mu+3b\mu^2-a^3+3a^2\mu-3a\mu^2}{3}$$

Next, substitute $\mu=(a+b)/2$ to obtain

$$\sigma^2=\frac{1}{3(b-a)}\left[\tfrac{1}{4}b^3-\tfrac{3}{4}b^2a+\tfrac{3}{4}ba^2-\tfrac{1}{4}a^3\right]$$

$$=\frac{1}{12(b-a)}(b-a)^3$$

27. The integral would take the form

$$k\int_{3960}^\infty \frac{1}{x}\,dx = [k\ln x]_{3960}^\infty = \infty$$

which would make it impossible to send anything out of the earth's gravitational field.

29. $FP=\int_0^\infty e^{-rt}f(t)\,dt = \int_0^\infty 100{,}000e^{-0.08t}$

$$=\left[-\frac{1}{0.08}100{,}000e^{-0.08t}\right]_0^\infty = 1{,}250{,}000$$

The present value is $1,250,000.

$$= \frac{(b-a)^2}{12}$$

c.
$$P(X < 2) = \int_{-\infty}^{2} f(x)\,dx$$

$$= \int_{-\infty}^{0} 0\,dx + \int_{0}^{2} \frac{1}{10-0}\,dx$$

$$= \frac{2}{10} = \frac{1}{5}$$

33. a.
$$\int_{-\infty}^{\infty} f(x)\,dx = \int_{-\infty}^{0} 0\,dx + \int_{0}^{\infty} \alpha e^{-\alpha x}\,dx$$

$$= 0 + \left[-e^{-\alpha x} \right]_{0}^{\infty} = 1$$

b.
$$\mu = \int_{-\infty}^{\infty} x f(x)\,dx = \int_{-\infty}^{0} x \cdot 0\,dx + \int_{0}^{\infty} x \alpha e^{-\alpha x}\,dx$$

$$= \int_{0}^{\infty} x \alpha e^{-\alpha x}\,dx$$

Integrate by parts: Let $u = x$, $dv = \alpha e^{-\alpha x}$. Then
$du = dx$, $v = -e^{\alpha x}$. Thus,

$$\mu = \left[-x e^{-\alpha x} \right]_{0}^{\infty} - \int_{0}^{\infty} \left(-e^{-\alpha x} \right) dx$$

$$= [0-0] + \int_{0}^{\infty} e^{-\alpha x}\,dx$$

$$= \left[-\frac{1}{\alpha} e^{-\alpha x} \right]_{0}^{\infty} = -0 + \frac{1}{\alpha} = \frac{1}{\alpha}$$

$$\sigma^2 = \int_{-\infty}^{\infty} (x - \mu)^2 f(x)\,dx$$

$$= \int_{-\infty}^{0} (x-\mu)^2 \cdot 0\,dx + \int_{0}^{\infty} (x-\mu)^2 \alpha e^{-\alpha x}\,dx$$

$$= 0 + \int_{0}^{\infty} \left(x^2 - 2x\mu + \mu^2 \right) \alpha e^{-\alpha x}\,dx$$

$$= \int_{0}^{\infty} x^2 \alpha e^{-\alpha x}\,dx - 2\mu \int_{0}^{\infty} x \alpha e^{-\alpha x}\,dx$$
$$+ \mu^2 \int_{0}^{\infty} \alpha e^{-\alpha x}\,dx$$

$$= \int_{0}^{\infty} x^2 \alpha e^{-\alpha x}\,dx - 2\frac{1}{\alpha}\left(\frac{1}{\alpha}\right) + \left(\frac{1}{\alpha}\right)^2$$

On this last line, we have used the results that
$\mu = 1/\alpha$, $\int_{0}^{\infty} \alpha e^{-\alpha x}\,dx = 1$, and

$\int_{0}^{\infty} x \alpha e^{-\alpha x}\,dx = 1/\alpha$. To evaluate the integral,
use integration by parts. Let
$u = x^2, dv = \alpha e^{-\alpha x}$. Then, $du = 2x\,dx$ and
$v = -e^{-\alpha x}$. Thus

$$\sigma^2 = \left[x^2 e^{-\alpha x} \right]_{0}^{\infty} - \int_{0}^{\infty} \left(-e^{-\alpha x} \right) 2x\,dx$$

$$- 2\frac{1}{\alpha}\left(\frac{1}{\alpha}\right) + \left(\frac{1}{\alpha}\right)^2$$

$$= 0 + \frac{2}{\alpha} \int_0^\infty x\alpha e^{-\alpha x}\, dx - 2\frac{1}{\alpha^2} + \frac{1}{\alpha^2}$$

$$= 2\frac{1}{\alpha^2} - 2\frac{1}{\alpha^2} + \frac{1}{\alpha^2} = \frac{1}{\alpha^2}$$

35. a. $\displaystyle\int_{-\infty}^\infty \sin x\, dx = \int_{-\infty}^0 \sin x\, dx + \int_0^\infty \sin x\, dx$

$$= \lim_{a\to\infty} \left[-\cos x\right]_0^a + \lim_{a\to-\infty} \left[-\cos x\right]_a^0$$

Both do not converge since $-\cos x$ is oscillating between -1 and 1, so the integral diverges.

b. $\displaystyle\lim_{a\to\infty} \int_{-a}^a \sin x\, dx = \lim_{a\to\infty} \left[-\cos x\right]_{-a}^a$

$$= \lim_{a\to\infty} \left[-\cos a + \cos(-a)\right]$$

$$= \lim_{a\to\infty} \left[-\cos a + \cos a\right] = \lim_{a\to\infty} 0 = 0$$

37. For example, the region under the curve $y = \dfrac{1}{x}$

to the right of $x = 1$.
Rotated about the x-axis the volume is

$\pi \displaystyle\int_1^\infty \frac{1}{x^2}\, dx = \pi$. Rotated about the y-axis, the

volume is $2\pi \displaystyle\int_1^\infty \frac{1}{x}\, dx$ which diverges.

39. $\displaystyle\int_1^{100} \frac{1}{x^2}\, dx = \left[-\frac{1}{x}\right]_1^{100} = 0.99$

$$\int_1^{100} \frac{1}{x^{1.1}}\, dx = \left[-\frac{1}{0.1x^{0.1}}\right]_1^{100} \approx 3.69$$

$$\int_1^{100} \frac{1}{x^{1.01}}\, dx = \left[-\frac{1}{0.01x^{0.01}}\right]_1^{100} \approx 4.50$$

$$\int_1^{100} \frac{1}{x}\, dx = \left[\ln x\right]_1^{100} = \ln 100 \approx 4.61$$

$$\int_1^{100} \frac{1}{x^{0.99}}\, dx = \left[\frac{x^{0.01}}{0.01}\right]_1^{100} \approx 4.71$$

41. $\displaystyle\int_{-1}^1 \frac{1}{\sqrt{2\pi}} \exp(-0.5x^2)\, dx \approx 0.6827$

$$\int_{-2}^2 \frac{1}{\sqrt{2\pi}} \exp(-0.5x^2)\, dx \approx 0.9545$$

$$\int_{-3}^3 \frac{1}{\sqrt{2\pi}} \exp(-0.5x^2)\, dx \approx 0.9973$$

$$\int_{-4}^4 \frac{1}{\sqrt{2\pi}} \exp(-0.5x^2)\, dx \approx 0.9999$$

9.4 Concepts Review

1. unbounded

3. $\displaystyle\lim_{x\to4^-} \int_0^b \frac{1}{\sqrt{4-x}}\, dx$

Problem Set 9.4

1. $\displaystyle\int_1^3 \frac{dx}{(x-1)^{1/3}} = \lim_{b\to1^+} \left[\frac{3(x-1)^{2/3}}{2}\right]_b^3$

$$= \frac{3}{2}\sqrt[3]{2^2} - \lim_{b\to1^+} \frac{3(b-1)^{2/3}}{2} = \frac{3}{\sqrt[3]{2}} - 0 = \frac{3}{\sqrt[3]{2}}$$

3. $\displaystyle\int_3^{10} \frac{dx}{\sqrt{x-3}} = \lim_{b\to3^+} \left[2\sqrt{x-3}\right]_b^{10}$

$$= 2\sqrt{7} - \lim_{b\to3^+} 2\sqrt{b-3} = 2\sqrt{7}$$

5. $\displaystyle\int_0^1 \frac{dx}{\sqrt{1-x^2}} = \lim_{b\to1^-} \left[\sin^{-1} x\right]_0^b$

$$= \lim_{b\to1^-} \sin^{-1} b - \sin^{-1} 0 = \frac{\pi}{2} - 0 = \frac{\pi}{2}$$

7. $\displaystyle\int_{-1}^3 \frac{1}{x^3}\, dx = \lim_{b\to0^-} \int_{-1}^b \frac{1}{x^3}\, dx + \lim_{b\to0^+} \int_b^3 \frac{1}{x^3}\, dx$

$$= \lim_{b\to0^-} \left[-\frac{1}{2x^2}\right]_{-1}^b + \lim_{b\to0^+} \left[-\frac{1}{2x^2}\right]_b^3$$

$$= \left(\lim_{b\to0^-} -\frac{1}{2b^2} + \frac{1}{2}\right) + \left(-\frac{1}{18} + \lim_{b\to0^+} \frac{1}{2b^2}\right)$$

$$= \left(-\infty + \frac{1}{2}\right) + \left(-\frac{1}{8} + \infty\right)$$

The integral diverges.

9. $\displaystyle\int_{-1}^{128} x^{-5/7}\,dx = \lim_{b\to 0^-}\int_{-1}^{b} x^{-5/7}\,dx + \lim_{b\to 0^+}\int_{b}^{128} x^{-5/7}\,dx = \lim_{b\to 0^-}\left[\frac{7}{2}x^{2/7}\right]_{-1}^{b} + \lim_{b\to 0^+}\left[\frac{7}{2}x^{2/7}\right]_{b}^{128}$

$\displaystyle = \lim_{b\to 0^-}\frac{7}{2}b^{2/7} - \frac{7}{2}(-1)^{2/7} + \frac{7}{2}(128)^{2/7} - \lim_{b\to 0^+}\frac{7}{2}b^{2/7} = 0 - \frac{7}{2} + \frac{7}{2}(4) - 0 = \frac{21}{2}$

11. $\displaystyle\int_{0}^{4}\frac{dx}{(2-3x)^{1/3}} = \lim_{b\to\frac{2}{3}^-}\int_{0}^{b}\frac{dx}{(2-3x)^{1/3}} + \lim_{b\to\frac{2}{3}^+}\int_{b}^{4}\frac{dx}{(2-3x)^{1/3}} = \lim_{b\to\frac{2}{3}^-}\left[-\frac{1}{2}(2-3x)^{2/3}\right]_{0}^{b} + \lim_{b\to\frac{2}{3}^+}\left[-\frac{1}{2}(2-3x)^{2/3}\right]_{b}^{4}$

$\displaystyle = \lim_{b\to\frac{2}{3}^-}-\frac{1}{2}(2-3b)^{2/3} + \frac{1}{2}(2)^{2/3} - \frac{1}{2}(-10)^{2/3} + \lim_{b\to\frac{2}{3}^+}\frac{1}{2}(2-3b)^{2/3}$

$\displaystyle = 0 + \frac{1}{2}2^{2/3} - \frac{1}{2}10^{2/3} + 0 = \frac{1}{2}(2^{2/3} - 10^{2/3})$

13. $\displaystyle\int_{0}^{4}\frac{x}{16-2x^2}\,dx = \lim_{b\to-\sqrt{8}^+}\int_{0}^{b}\frac{x}{16-2x^2}\,dx + \lim_{b\to-\sqrt{8}^-}\int_{b}^{4}\frac{x}{16-2x^2}\,dx$

$\displaystyle = \lim_{b\to-\sqrt{8}^+}\left[-\frac{1}{4}\ln\left|16-2x^2\right|\right]_{0}^{b} + \lim_{b\to-\sqrt{8}^-}\left[-\frac{1}{4}\ln\left|16-2x^2\right|\right]_{b}^{-4}$

$\displaystyle = \lim_{b\to-\sqrt{8}^+}-\frac{1}{4}\ln\left|16-2b^2\right| + \frac{1}{4}\ln 16 - \frac{1}{4}\ln 16 + \lim_{b\to-\sqrt{8}^-}\frac{1}{4}\ln\left|16-2b^2\right|$

$\displaystyle = \left[-(-\infty) + \frac{1}{4}\ln 16\right] + \left[-\frac{1}{4}\ln 16 + (-\infty)\right]$

The integral diverges.

15. $\displaystyle\int_{-2}^{-1}\frac{dx}{(x+1)^{4/3}} = \lim_{b\to-1^-}\left[-\frac{3}{(x+1)^{1/3}}\right]_{-2}^{b} = \lim_{b\to-1^-}-\frac{3}{(b+1)^{1/3}} + \frac{3}{(-1)^{1/3}} = -(-\infty) - 3$

The integral diverges.

17. Note that $\displaystyle\frac{1}{x^3-x^2-x+1} = \frac{1}{2(x-1)^2} - \frac{1}{4(x-1)} + \frac{1}{4(x+1)}$

$\displaystyle\int_{0}^{3}\frac{dx}{x^3-x^2-x+1} = \lim_{b\to 1^-}\int_{0}^{b}\frac{dx}{x^3-x^2-x+1} + \lim_{b\to 1^+}\int_{b}^{3}\frac{dx}{x^3-x^2-x+1}$

$\displaystyle = \lim_{b\to 1^-}\left[-\frac{1}{2(x-1)} - \frac{1}{4}\ln|x-1| + \frac{1}{4}\ln|x+1|\right]_{0}^{b} + \lim_{b\to 1^+}\left[-\frac{1}{2(x-1)} - \frac{1}{4}\ln|x-1| + \frac{1}{4}\ln|x+1|\right]_{b}^{3}$

$\displaystyle \lim_{b\to 1^-}\left\{\left(-\frac{1}{2(b-1)} + \frac{1}{4}\ln\left|\frac{b+1}{b-1}\right|\right) + \left(-\frac{1}{2} + 0\right) + \left[-\frac{1}{4} + \frac{1}{4}\ln 2 - \left(-\frac{1}{2(b-1)} + \frac{1}{4}\ln\left|\frac{b+1}{b-1}\right|\right)\right]\right\}$

$\displaystyle = \left(\infty + \infty - \frac{1}{2}\right) + \left(-\frac{1}{4} + \frac{1}{4}\ln 2 + \infty - \infty\right)$

The integral diverges.

19. $\displaystyle\int_{0}^{\pi/4}\tan 2x\,dx = \lim_{b\to\frac{\pi}{4}^-}\left[-\frac{1}{2}\ln|\cos 2x|\right]_{0}^{b}$

$\displaystyle = \lim_{b\to\frac{\pi}{4}^-}-\frac{1}{2}\ln|\cos 2b| + \frac{1}{2}\ln 1 = -(-\infty) + 0$

The integral diverges.

21. $\displaystyle\int_{0}^{\pi/2}\frac{\sin x}{1-\cos x}\,dx = \lim_{b\to 0^+}\left[\ln|1-\cos x|\right]_{b}^{\pi/2}$

$\displaystyle = \ln 1 - \lim_{b\to 0^+}\ln|1-\cos b| = 0 - (-\infty)$

The integral diverges.

23. $\displaystyle\int_0^{\pi/2} \tan^2 x \sec^2 x\, dx = \lim_{b\to\frac{\pi}{2}^-}\left[\frac{1}{3}\tan^3 x\right]_0^b$

$\displaystyle = \lim_{b\to\frac{\pi}{2}^-}\frac{1}{3}\tan^3 b - \frac{1}{3}(0)^3 = \infty$

The integral diverges.

25. Since $\dfrac{1-\cos x}{2} = \sin^2\dfrac{x}{2}$,

$\dfrac{1}{\cos x - 1} = -\dfrac{1}{2}\csc^2\dfrac{x}{2}.$

$\displaystyle\int_0^\pi \frac{dx}{\cos x - 1} = \lim_{b\to 0^+}\left[\cot\frac{x}{2}\right]_b^\pi$

$\displaystyle = \cot\frac{\pi}{2} - \lim_{b\to 0^+}\cot\frac{b}{2} = 0 - \infty$

The integral diverges.

27. $\displaystyle\int_0^{\ln 3}\frac{e^x\,dx}{\sqrt{e^x-1}} = \lim_{b\to 0^+}\left[2\sqrt{e^x-1}\right]_b^{\ln 3}$

$\displaystyle = 2\sqrt{3-1} - \lim_{b\to 0^+}2\sqrt{e^b-1} = 2\sqrt{2}-0 = 2\sqrt{2}$

29. $\displaystyle\int_1^e \frac{dx}{x\ln x} = \lim_{b\to 1^+}[\ln(\ln x)]_b^e = \ln(\ln e) - \lim_{b\to 1^+}\ln(\ln b) = \ln 1 - \ln 0 = 0 + \infty$

The integral diverges.

31. $\displaystyle\int_{2c}^{4c}\frac{dx}{\sqrt{x^2-4c^2}} = \lim_{b\to 2c^+}\left[\ln\left|x+\sqrt{x^2-4c^2}\right|\right]_b^{4c} = \ln\left[(4+2\sqrt{3})c\right] - \lim_{b\to 2c^+}\ln\left|b+\sqrt{b^2-4c^2}\right|$

$\displaystyle = \ln\left[(4+2\sqrt{3})c\right] - \ln 2c = \ln(2+\sqrt{3})$

33. For $0 < c < 1$, $\dfrac{1}{\sqrt{x}(1+x)}$ is continuous. Let $u = \dfrac{1}{1+x}$, $du = -\dfrac{1}{(1+x)^2}dx$.

$dv = \dfrac{1}{\sqrt{x}}dx, v = 2\sqrt{x}$.

$\displaystyle\int_c^1 \frac{1}{\sqrt{x}(1+x)}dx = \left[\frac{2\sqrt{x}}{1+x}\right]_c^1 + 2\int_c^1\frac{\sqrt{x}\,dx}{(1+x)^2} = \frac{2}{2} - \frac{2\sqrt{c}}{1+c} + 2\int_c^1\frac{\sqrt{x}\,dx}{(1+x)^2} = 1 - \frac{2\sqrt{c}}{1+c} + 2\int_c^1\frac{\sqrt{x}\,dx}{(1+x)^2}$

Thus, $\displaystyle\lim_{c\to 0}\int_c^1\frac{1}{\sqrt{x}(1+x)}dx = \lim_{c\to 0}\left[1 - \frac{2\sqrt{c}}{1+c} + 2\int_c^1\frac{\sqrt{x}\,dx}{(1+x)^2}\right] = 1 - 0 + 2\int_0^1\frac{\sqrt{x}\,dx}{(1+x)^2}$

This last integral is a proper integral.

35. $\displaystyle\int_{-3}^3\frac{x}{\sqrt{9-x^2}}dx = \int_{-3}^0\frac{x}{\sqrt{9-x^2}}dx + \int_0^3\frac{x}{\sqrt{9-x^2}}dx = \lim_{b\to -3^+}\left[-\sqrt{9-x^2}\right]_b^0 + \lim_{b\to 3^-}\left[-\sqrt{9-x^2}\right]_0^b$

$\displaystyle = -\sqrt{9} + \lim_{b\to -3^+}\sqrt{9-b^2} - \lim_{b\to 3^-}\sqrt{9-b^2} + \sqrt{9} = -3 + 0 - 0 + 3 = 0$

37. $\displaystyle\int_{-4}^4\frac{1}{16-x^2}dx = \int_{-4}^0\frac{1}{16-x^2}dx + \int_0^4\frac{1}{16-x^2}dx = \lim_{b\to -4^+}\left[\frac{1}{8}\ln\left|\frac{x+4}{x-4}\right|\right]_b^0 + \lim_{b\to 4^-}\left[\frac{1}{8}\ln\left|\frac{x+4}{x-4}\right|\right]_0^b$

$\displaystyle = \frac{1}{8}\ln 1 - \lim_{b\to -4^+}\frac{1}{8}\ln\left|\frac{b+4}{b-4}\right| + \lim_{b\to 4^-}\frac{1}{8}\ln\left|\frac{b+4}{b-4}\right| - \frac{1}{8}\ln 1 = (0+\infty) + (\infty - 0)$

The integral diverges.

39. $\displaystyle\int_0^\infty\frac{1}{x^p}dx = \int_0^1\frac{1}{x^p}dx + \int_1^\infty\frac{1}{x^p}dx$

If $p > 1$, $\displaystyle\int_0^1\frac{1}{x^p}dx = \left[\frac{1}{-p+1}x^{-p+1}\right]_0^1$ diverges

since $\displaystyle\lim_{x\to 0^+}x^{-p+1} = \infty$.

If $p < 1$ and $p \geq 0$, $\displaystyle\int_1^\infty\frac{1}{x^p}dx = \left[\frac{1}{-p+1}x^{-p+1}\right]_1^\infty$

diverges since $\lim\limits_{x\to\infty} x^{-p+1} = \infty$.

If $p = 0$, $\int_0^\infty dx = \infty$.

If $p = 1$, both $\int_0^1 \frac{1}{x}\,dx$ and $\int_1^\infty \frac{1}{x}\,dx$ diverge.

41. $\int_0^8 (x-8)^{-2/3}\,dx = \lim\limits_{b\to 8^-}\left[3(x-8)^{1/3}\right]_0^b$

$= 3(0) - 3(-2) = 6$

43. a. $\int_0^1 x^{-2/3}\,dx = \lim\limits_{b\to 0^+}\left[3x^{1/3}\right]_b^1 = 3$

b. $V = \pi\int_0^1 x^{-4/3}\,dx = \lim\limits_{b\to 0^+}\pi\left[-3x^{-1/3}\right]_b^1$

$= -3\pi + 3\pi\lim\limits_{b\to 0} b^{-1/3}$

The limit tends to infinity as $b \to 0$, so the volume is infinite.

45. $\int_0^1 \frac{\sin x}{x}\,dx$ is not an improper integral since

$\dfrac{\sin x}{x}$ is bounded in the interval $0 \le x \le 1$.

47. For $x \ge 1$, $x^2 \ge x$ so $-x^2 \le -x$, thus

$e^{-x^2} \le e^{-x}$.

$\int_1^\infty e^{-x}\,dx = \lim\limits_{b\to\infty}[-e^{-x}]_1^b = -\lim\limits_{b\to\infty}\dfrac{1}{e^b} + e^{-1}$

$= -0 + \dfrac{1}{e} = \dfrac{1}{e}$

Thus, by the Comparison Test, $\int_1^\infty e^{-x^2}\,dx$ converges.

49. Since $x^2 \ln(x+1) \ge x^2$, we know that

$\dfrac{1}{x^2 \ln(x+1)} \le \dfrac{1}{x^2}$. Since $\int_1^\infty \dfrac{1}{x^2}\,dx = \left[-\dfrac{1}{x}\right]_1^\infty = 1$

we can apply the Comparison Test of Problem 46

to conclude that $\int_1^\infty \dfrac{1}{x^2 \ln(x+1)}\,dx$ converges.

51. a. From Example 2 of Section 9.2, $\lim\limits_{x\to\infty}\dfrac{x^a}{e^x} = 0$

for a any positive real number.

Thus $\lim\limits_{x\to\infty}\dfrac{x^{n+1}}{e^x} = 0$ for any positive real

number n, hence there is a number M such

that $0 < \dfrac{x^{n+1}}{e^x} \le 1$ for $x \ge M$. Divide the

inequality by x^2 to get that $0 < \dfrac{x^{n-1}}{e^x} \le \dfrac{1}{x^2}$

for $x \ge M$.

b. $\int_1^\infty \dfrac{1}{x^2}\,dx = \lim\limits_{b\to\infty}\left[-\dfrac{1}{x}\right]_1^b = -\lim\limits_{b\to\infty}\dfrac{1}{b} + \dfrac{1}{1}$

$= -0 + 1 = 1$

$\int_1^\infty x^{n-1}e^{-x}\,dx = \int_1^M x^{n-1}e^{-x}\,dx + \int_M^\infty x^{n-1}e^{-x}\,dx$

$\le \int_1^M x^{n-1}e^{-x}\,dx + \int_1^\infty \dfrac{1}{x^2}\,dx$

$= 1 + \int_1^M x^{n-1}e^{-x}\,dx$

by part a and Problem 46. The remaining

integral is finite, so $\int_1^\infty x^{n-1}e^{-x}\,dx$

converges.

53. a. $\Gamma(1) = \int_0^\infty x^0 e^{-x}\,dx = \left[-e^{-x}\right]_0^\infty = 1$

b. $\Gamma(n+1) = \int_0^\infty x^n e^{-x}\,dx$

Let $u = x^n$, $dv = e^{-x}\,dx$,

$du = nx^{n-1}\,dx$, $v = -e^{-x}$.

$\Gamma(n+1) = [-x^n e^{-x}]_0^\infty + \int_0^\infty nx^{n-1}e^{-x}\,dx$

$= 0 + n\int_0^\infty x^{n-1}e^{-x}\,dx = n\Gamma(n)$

c. From parts a and b,
$\Gamma(1) = 1, \Gamma(2) = 1\cdot\Gamma(1) = 1$,
$\Gamma(3) = 2\cdot\Gamma(2) = 2\cdot 1 = 2!$.
Suppose $\Gamma(n) = (n-1)!$, then by part b,
$\Gamma(n+1) = n\Gamma(n) = n[(n-1)!] = n!$.

55. a. The integral is the area between the curve

$y^2 = \dfrac{1-x}{x}$ and the x-axis from $x = 0$ to $x = 1$.

$y^2 = \dfrac{1-x}{x}$; $xy^2 = 1-x$; $x(y^2+1) = 1$

$x = \dfrac{1}{y^2+1}$

As $x \to 0$, $y = \sqrt{\dfrac{1-x}{x}} \to \infty$, while

when $x = 1$, $y = \sqrt{\dfrac{1-1}{1}} = 0$, thus the area is

$$\int_0^\infty \frac{1}{y^2+1}\,dy = \lim_{b\to\infty}\left[\tan^{-1} y\right]_0^b$$

$$= \lim_{b\to\infty}\tan^{-1} b - \tan^{-1} 0 = \frac{\pi}{2}$$

b. The integral is the area between the curve $y^2 = \dfrac{1+x}{1-x}$ and the x-axis from $x=-1$ to $x=1$.

$$y^2 = \frac{1+x}{1-x};\ y^2 - xy^2 = 1+x;\ y^2 - 1 = x(y^2+1);$$

$$x = \frac{y^2-1}{y^2+1}$$

When $x = -1,\ y = \sqrt{\dfrac{1+(-1)}{1-(-1)}} = \sqrt{\dfrac{0}{2}} = 0$, while

as $x\to 1,\ y = \sqrt{\dfrac{1+x}{1-x}} \to \infty$.

The area in question is the area to the right of the curve $y = \sqrt{\dfrac{1+x}{1-x}}$ and to the left of the line $x=1$. Thus, the area is

$$\int_0^\infty \left(1 - \frac{y^2-1}{y^2+1}\right)dy = \int_0^\infty \frac{2}{y^2+1}\,dy$$

$$= \lim_{b\to\infty}\left[2\tan^{-1} y\right]_0^b$$

$$\lim_{b\to\infty} 2\tan^{-1} b - 2\tan^{-1} 0 = 2\left(\frac{\pi}{2}\right) = \pi$$

9.5 Chapter Review

Concepts Test

1. True: See Example 2 of Section 9.2.

3. False: $\displaystyle\lim_{x\to\infty}\frac{1000x^4+1000}{0.001x^4+1} = \frac{1000}{0.001} = 10^6$

5. False: For example, if $f(x) = x$ and $g(x) = e^x$,

$$\lim_{x\to\infty}\frac{x}{e^x} = 0.$$

7. True: Take the inner limit first.

9. True: Since $\displaystyle\lim_{x\to a} f(x) = -1 \neq 0$, it serves only to affect the sign of the limit of the product.

11. False: Consider $f(x) = 3x^2$ and $g(x) = x^2 + 1$, then

$$\lim_{x\to\infty}\frac{f(x)}{g(x)} = \lim_{x\to\infty}\frac{3x^2}{x^2+1}$$

$$= \lim_{x\to\infty}\frac{3}{1+\frac{1}{x^2}} = 3,\ \text{but}$$

$$\lim_{x\to\infty}[f(x) - 3g(x)]$$

$$= \lim_{x\to\infty}[3x^2 - 3(x^2+1)]$$

$$= \lim_{x\to\infty}[-3] = -3$$

13. True: See Example 7 of Section 9.2.

15. True: Use repeated applications of l'Hôpital's Rule.

17. False: Consider $f(x) = 3x^2 + x + 1$ and $g(x) = 4x^3 + 2x + 3;\ f'(x) = 6x + 1$ $g'(x) = 12x^2 + 2$, and so

$$\lim_{x\to 0}\frac{f'(x)}{g'(x)} = \lim_{x\to 0}\frac{6x+1}{12x^2+2} = \frac{1}{2}\ \text{while}$$

$$\lim_{x\to 0}\frac{f(x)}{g(x)} = \lim_{x\to 0}\frac{3x^2+x+1}{4x^3+2x+3} = \frac{1}{3}$$

19. True: $\displaystyle\int_0^\infty \frac{1}{x^p}\,dx = \int_0^1 \frac{1}{x^p}\,dx + \int_1^\infty \frac{1}{x^p}\,dx;$

$\displaystyle\int_0^1 \frac{1}{x^p}\,dx$ diverges for $p \geq 1$ and

$\displaystyle\int_1^\infty \frac{1}{x^p}\,dx$ diverges for $p \leq 1$.

21. True: $\displaystyle\int_{-\infty}^\infty f(x)\,dx = \int_{-\infty}^0 f(x)\,dx + \int_0^\infty f(x)\,dx$

If f is an even function, then $f(-x) = f(x)$ so

$$\int_{-\infty}^0 f(x)\,dx = \int_0^\infty f(x)\,dx.$$

Thus, both integrals making up $\displaystyle\int_{-\infty}^\infty f(x)\,dx$ converge so their sum converges.

23. True: $\displaystyle\int_0^\infty f'(x)\,dx = \lim_{b\to\infty}\int_0^b f'(x)\,dx$

$$= \lim_{b\to\infty}[f(x)]_0^b = \lim_{b\to\infty} f(b) - f(0)$$

$$= 0 - f(0) = -f(0).$$

$f(0)$ must exist and be finite since $f'(x)$ is continuous on $[0, \infty)$.

25. **False:** The integrand is bounded on the interval $\left[0, \dfrac{\pi}{4}\right]$.

Sample Test Problems

1. The limit is of the form $\dfrac{0}{0}$.

$$\lim_{x\to 0} \frac{4x}{\tan x} = \lim_{x\to 0} \frac{4}{\sec^2 x} = 4$$

3. The limit is of the form $\dfrac{0}{0}$. (Apply l'Hôpital's Rule twice.)

$$\lim_{x\to 0} \frac{\sin x - \tan x}{\frac{1}{3}x^2} = \lim_{x\to 0} \frac{\cos x - \sec^2 x}{\frac{2}{3}x}$$

$$= \lim_{x\to 0} \frac{-\sin x - 2\sec x(\sec x \tan x)}{\frac{2}{3}} = 0$$

5. $\lim\limits_{x\to 0} 2x\cot x = \lim\limits_{x\to 0} \dfrac{2x\cos x}{\sin x}$

The limit is of the form $\dfrac{0}{0}$.

$$\lim_{x\to 0} \frac{2x\cos x}{\sin x} = \lim_{x\to 0} \frac{2\cos x - 2x\sin x}{\cos x}$$

$$= \frac{2-0}{1} = 2$$

7. The limit is of the form $\dfrac{\infty}{\infty}$.

$$\lim_{t\to\infty} \frac{\ln t}{t^2} = \lim_{t\to\infty} \frac{\frac{1}{t}}{2t} = \lim_{t\to\infty} \frac{1}{2t^2} = 0$$

9. As $x\to 0$, $\sin x \to 0$, and $\dfrac{1}{x} \to \infty$. A number less than 1, raised to a large power, is a very small number $\left(\left(\dfrac{1}{2}\right)^{32} = 2.328\times 10^{-10}\right)$ so

$$\lim_{x\to 0^+} (\sin x)^{1/x} = 0.$$

11. The limit is of the form 0^0.

Let $y = x^x$, then $\ln y = x \ln x$.

$$\lim_{x\to 0^+} x\ln x = \lim_{x\to 0^+} \frac{\ln x}{\frac{1}{x}}$$

The limit is of the form $\dfrac{\infty}{\infty}$.

$$\lim_{x\to 0^+} \frac{\ln x}{\frac{1}{x}} = \lim_{x\to 0^+} \frac{\frac{1}{x}}{-\frac{1}{x^2}} = \lim_{x\to 0^+} -x = 0$$

$$\lim_{x\to 0^+} x^x = \lim_{x\to 0^+} e^{\ln y} = 1$$

13. $\lim\limits_{x\to 0^+} \sqrt{x}\ln x = \lim\limits_{x\to 0^+} \dfrac{\ln x}{\frac{1}{\sqrt{x}}}$

The limit is of the form $\dfrac{\infty}{\infty}$.

$$\lim_{x\to 0^+} \frac{\ln x}{\frac{1}{\sqrt{x}}} = \lim_{x\to 0^+} \frac{\frac{1}{x}}{-\frac{1}{2x^{3/2}}} = \lim_{x\to 0^+} -2\sqrt{x} = 0$$

15. $\lim\limits_{x\to 0^+} \left(\dfrac{1}{\sin x} - \dfrac{1}{x}\right) = \lim\limits_{x\to 0^+} \dfrac{x - \sin x}{x\sin x}$

The limit is of the form $\dfrac{0}{0}$. (Apply l'Hôpital's Rule twice.)

$$\lim_{x\to 0^+} \frac{x - \sin x}{x\sin x} = \lim_{x\to 0^+} \frac{1 - \cos x}{\sin x + x\cos x}$$

$$= \lim_{x\to 0^+} \frac{\sin x}{2\cos x - x\sin x} = \frac{0}{2} = 0$$

17. The limit is of the form 1^∞.

Let $y = (\sin x)^{\tan x}$, then $\ln y = \tan x \ln(\sin x)$.

$$\lim_{x\to\frac{\pi}{2}} \tan x \ln(\sin x) = \lim_{x\to\frac{\pi}{2}} \frac{\sin x \ln(\sin x)}{\cos x}$$

The limit is of the form $\dfrac{0}{0}$.

$$\lim_{x\to\frac{\pi}{2}} \frac{\sin x \ln(\sin x)}{\cos x} = \lim_{x\to\frac{\pi}{2}} \frac{\cos x \ln(\sin x) + \frac{\sin x}{\sin x}\cos x}{\sin x}$$

$$= \lim_{x\to\frac{\pi}{2}} \frac{\cos x(1 + \ln(\sin x))}{\sin x} = \frac{0}{1} = 0$$

$$\lim_{x\to\frac{\pi}{2}} (\sin x)^{\tan x} = \lim_{x\to\frac{\pi}{2}} e^{\ln y} = 1$$

19. $\displaystyle\int_0^\infty \frac{dx}{(x+1)^2} = \left[-\frac{1}{x+1}\right]_0^\infty = 0 + 1 = 1$

21. $\displaystyle\int_{-\infty}^1 e^{2x}dx = \left[\frac{1}{2}e^{2x}\right]_{-\infty}^1 = \frac{1}{2}e^2 - 0 = \frac{1}{2}e^2$

23. $\displaystyle\int_0^\infty \frac{dx}{x+1} = [\ln(x+1)]_0^\infty = \infty - 0 = \infty$

The integral diverges.

25. $\displaystyle\int_1^\infty \frac{dx}{x^2+x^4} = \int_1^\infty\left(\frac{1}{x^2}-\frac{1}{1+x^2}\right)dx = \left[-\frac{1}{x}-\tan^{-1}x\right]_1^\infty = 0-\frac{\pi}{2}+1+\tan^{-1}1 = 1+\frac{\pi}{4}-\frac{\pi}{2} = 1-\frac{\pi}{4}$

27. $\displaystyle\int_{-2}^0 \frac{dx}{2x+3} = \lim_{b\to-\frac{3}{2}^-}\int_{-2}^b \frac{dx}{2x+3} + \lim_{b\to-\frac{3}{2}^+}\int_b^0 \frac{dx}{2x+3} = \lim_{b\to-\frac{3}{2}^-}\left[\frac{1}{2}\ln|2x+3|\right]_{-2}^b + \lim_{b\to-\frac{3}{2}^+}\left[\frac{1}{2}\ln|2x+3|\right]_b^0$

$\displaystyle = \left(\lim_{b\to-\frac{3}{2}^-}\frac{1}{2}\ln|2b+3|-\frac{1}{2}(0)\right)+\left(\frac{1}{2}\ln3 - \lim_{b\to-\frac{3}{2}^+}\frac{1}{2}\ln|2b+3|\right) = (-\infty)+\left(\frac{1}{2}\ln3+\infty\right)$

The integral diverges.

29. $\displaystyle\int_2^\infty \frac{dx}{x(\ln x)^2} = \left[-\frac{1}{\ln x}\right]_2^\infty = -0+\frac{1}{\ln2} = \frac{1}{\ln2}$

31. $\displaystyle\int_3^5 \frac{dx}{(4-x)^{2/3}} = \lim_{b\to4^-}\int_3^b \frac{dx}{(4-x)^{2/3}} + \lim_{b\to4^+}\int_b^5 \frac{dx}{(4-x)^{2/3}} = \lim_{b\to4^-}\left[-3(4-x)^{1/3}\right]_3^b + \lim_{b\to4^+}\left[-3(4-x)^{1/3}\right]_b^5$

$\displaystyle = \lim_{b\to4^-}-3(4-b)^{1/3}+3(1)^{1/3} - 3(-1)^{1/3} + \lim_{b\to4^+}3(4-b)^{1/3} = 0+3+3+0 = 6$

33. $\displaystyle\int_{-\infty}^\infty \frac{x}{x^2+1}dx = \int_{-\infty}^0 \frac{x}{x^2+1}dx + \int_0^\infty \frac{x}{x^2+1}dx$

$\displaystyle = \frac{1}{2}\left[\ln(x^2+1)\right]_{-\infty}^0 + \frac{1}{2}\left[\ln(x^2+1)\right]_0^\infty = (0+\infty)+(\infty-0)$

The integral diverges.

35. $\displaystyle\frac{e^x}{e^{2x}+1} = \frac{e^x}{(e^x)^2+1}$

Let $u = e^x$, $du = e^x dx$

$\displaystyle\int_0^\infty \frac{e^x}{e^{2x}+1}dx = \int_1^\infty \frac{1}{u^2+1}du = \left[\tan^{-1}u\right]_1^\infty$

$\displaystyle = \frac{\pi}{2}-\tan^{-1}1 = \frac{\pi}{2}-\frac{\pi}{4} = \frac{\pi}{4}$

37. $\displaystyle\int_{-3}^3 \frac{x}{\sqrt{9-x^2}}dx = 0$

See Problem 35 in Section 9.4.

39. For $p\neq1, p\neq0$, $\displaystyle\int_1^\infty \frac{1}{x^p}dx = \left[-\frac{1}{(p-1)x^{p-1}}\right]_1^\infty$

$\displaystyle = \lim_{b\to\infty}\frac{1}{(1-p)b^{p-1}}+\frac{1}{p-1}$

$\displaystyle\lim_{b\to\infty}\frac{1}{b^{p-1}} = 0 \text{ when } p-1>0 \text{ or } p>1,$

and $\displaystyle\lim_{b\to\infty}\frac{1}{b^{p-1}} = \infty \text{ when } p<1, p\neq0.$

When $p=1$, $\displaystyle\int_1^\infty \frac{1}{x}dx = [\ln x]_1^\infty = \infty-0$

The integral diverges.

When $p=0$, $\displaystyle\int_1^\infty 1dx = [x]_1^\infty = \infty-1$. The integral diverges.

$\displaystyle\int_1^\infty \frac{1}{x^p}dx$ converges when $p>1$ and diverges when $p\leq1$.

41. For $x\geq1$, $x^6+x>x^6$, so $\sqrt{x^6+x}>\sqrt{x^6} = x^3$

and $\displaystyle\frac{1}{\sqrt{x^6+x}}<\frac{1}{x^3}$. Hence,

$\displaystyle\int_1^\infty \frac{1}{\sqrt{x^6+x}}dx < \int_1^\infty \frac{1}{x^3}dx$ which converges

since $3>1$ (see Problem 39). Thus

$\displaystyle\int_1^\infty \frac{1}{\sqrt{x^6+x}}dx$ converges.

43. For $x>3$, $\ln x>1$, so $\displaystyle\frac{\ln x}{x}>\frac{1}{x}$. Hence,

$\displaystyle\int_3^\infty \frac{\ln x}{x}dx > \int_3^\infty \frac{1}{x}dx = [\ln x]_3^\infty = \infty-\ln3.$

The integral diverges, thus $\displaystyle\int_3^\infty \frac{\ln x}{x}dx$ also

diverges.

9.6 Additional Problem Set

1. a. For $1 < x < c$, $\dfrac{1}{x}$ is a differentiable function.

Let $u = \dfrac{1}{x}$, so $x = \dfrac{1}{u}$; $du = -\dfrac{1}{x^2}dx$ so

$$dx = -x^2 du = -\frac{1}{u^2}du \; .$$

When $x = 1$, $u = 1$, while when $x = c$, $u = \dfrac{1}{c}$.

$$\int_1^c \frac{1}{1+x^2}dx = \int_1^{\frac{1}{c}} \frac{1}{1+\frac{1}{u^2}}\left(-\frac{1}{u^2}du\right)$$

$$= -\int_1^{\frac{1}{c}} \frac{1}{u^2+1}du = \int_{\frac{1}{c}}^1 \frac{1}{u^2+1}du$$

b. $\displaystyle\int_1^\infty \frac{1}{1+x^2}dx = \lim_{c\to\infty}\int_1^c \frac{1}{1+x^2}dx$

$$= \lim_{c\to\infty}\int_{\frac{1}{c}}^1 \frac{1}{u^2+1}du = \int_0^1 \frac{1}{u^2+1}du \text{ by using}$$

part a and $\displaystyle\lim_{c\to\infty}\frac{1}{c} = 0$.

$$\int_0^1 \frac{1}{u^2+1}du = [\tan^{-1}u]_0^1 = \tan^{-1}1 - \tan^{-1}0$$

$$= \frac{\pi}{4} - 0 = \frac{\pi}{4}$$

3. a. $f(x)$ is an even function, so $\displaystyle\int_{-\infty}^0 f(x)dx = \int_0^\infty f(x)dx$, thus $\displaystyle\int_{-\infty}^\infty f(x)dx = 2\int_0^\infty f(x)dx$.

$$2\int_0^\infty C|x|e^{-kx^2}dx = \left[-\frac{C}{k}e^{-kx^2}\right]_0^\infty = -0 + \frac{C}{k}$$

$$\frac{C}{k} = 1 \text{ when } C = k$$

b. $\displaystyle\int_{-\infty}^\infty kx|x|e^{-kx^2}dx = \int_{-\infty}^0 -kx^2e^{-kx^2}dx + \int_0^\infty kx^2e^{-kx^2}dx$

In both integrals, use $u = x$, $du = dx$, $dv = kxe^{-kx^2}dx$, $v = -\dfrac{1}{2}e^{-kx^2}$

$$\int_{-\infty}^0 -kx^2e^{-kx^2}dx + \int_0^\infty kx^2e^{-kx^2}dx = -\left(\left[-\frac{1}{2}xe^{-kx^2}\right]_{-\infty}^0 + \int_{-\infty}^0 \frac{1}{2}e^{-kx^2}dx\right) + \left(\left[-\frac{1}{2}xe^{-kx^2}\right]_0^\infty + \int_0^\infty \frac{1}{2}e^{-kx^2}dx\right)$$

$$= -(0+0) - \frac{1}{2}\int_{-\infty}^0 e^{-kx^2}dx + (0-0) + \frac{1}{2}\int_0^\infty e^{-kx^2}dx = -\frac{1}{2}\int_{-\infty}^0 e^{-kx^2}dx + \frac{1}{2}\int_0^\infty e^{-kx^2}dx$$

Now let $u = \sqrt{2k}x$, so that $kx^2 = \dfrac{u^2}{2}$ and $du = \sqrt{2k}dx$. Then the integrals become

$$-\frac{1}{2\sqrt{2k}}\int_{-\infty}^0 e^{-u^2/2}du + \frac{1}{2\sqrt{2k}}\int_0^\infty e^{-u^2/2}du = -\frac{1}{2\sqrt{2k}}\left(\frac{1}{2}\sqrt{2\pi}\right) + \frac{1}{2\sqrt{2k}}\left(\frac{1}{2}\sqrt{2\pi}\right) = 0.$$

$$\int_0^\infty e^{-u^2/2}du = \int_{-\infty}^0 e^{-u^2/2}du = \frac{1}{2}\sqrt{2\pi} \text{ from Example 5 in Section 9.3.}$$

5. a. $\displaystyle\int_{-\infty}^\infty f(x)dx = \int_M^\infty \frac{CM^k}{x^{k+1}}dx = CM^k\left[-\frac{1}{kx^k}\right]_M^\infty = CM^k\left(0 + \frac{1}{kM^k}\right) = \frac{C}{k}$. Thus, $\dfrac{C}{k} = 1$ when $C = k$.

b. $\mu = \displaystyle\int_{-\infty}^\infty xf(x)dx = \int_M^\infty x\frac{kM^k}{x^{k+1}}dx = kM^k\int_M^\infty \frac{1}{x^k}dx = kM^k\left(\lim_{b\to\infty}\int_M^b \frac{1}{x^k}dx\right)$

This integral converges when $k > 1$.

When $k > 1$, $\mu = kM^k\left(\lim_{b\to\infty}\left[-\frac{1}{(k-1)x^{k-1}}\right]_M^b\right) = kM^k\left(-0 + \frac{1}{(k-1)M^{k-1}}\right) = \frac{kM}{k-1}$

The mean is finite only when $k > 1$.

c. Since the mean is finite only when $k > 1$, the variance is only defined when $k > 1$.

$$\sigma^2 = \int_{-\infty}^{\infty} (x-\mu)^2 f(x)dx = \int_{M}^{\infty} \left(x - \frac{kM}{k-1}\right)^2 \frac{kM^k}{x^{k+1}} dx = kM^k \int_{M}^{\infty} \left(x^2 - \frac{2kM}{k-1}x + \frac{k^2M^2}{(k-1)^2}\right)\frac{1}{x^{k+1}} dx$$

$$= kM^k \int_{M}^{\infty} \frac{1}{x^{k-1}} dx - \frac{2k^2M^{k+1}}{k-1}\int_{M}^{\infty} \frac{1}{x^k} dx + \frac{k^3M^{k+2}}{(k-1)^2}\int_{M}^{\infty} \frac{1}{x^{k+1}} dx$$

The first integral converges only when $k - 1 > 1$ or $k > 2$. The second integral converges only when $k > 1$, which is taken care of by requiring $k > 2$.

$$\sigma^2 = kM^k \left[-\frac{1}{(k-2)x^{k-2}}\right]_{M}^{\infty} - \frac{2k^2M^{k+1}}{k-1}\left[-\frac{1}{(k-1)x^{k-1}}\right]_{M}^{\infty} + \frac{k^3M^{k+2}}{(k-1)^2}\left[-\frac{1}{kx^k}\right]_{M}^{\infty}$$

$$= kM^k \left(-0 + \frac{1}{(k-2)M^{k-2}}\right) - \frac{2k^2M^{k+1}}{k-1}\left(-0 + \frac{1}{(k-1)M^{k-1}}\right) + \frac{k^3M^{k+2}}{(k-1)^2}\left(-0 + \frac{1}{kM^k}\right)$$

$$= \frac{kM^2}{k-2} - \frac{2k^2M^2}{(k-1)^2} + \frac{k^2M^2}{(k-1)^2}$$

$$= kM^2\left(\frac{1}{k-2} - \frac{k}{(k-1)^2}\right) = kM^2\left(\frac{k^2 - 2k + 1 - k^2 + 2k}{(k-2)(k-1)^2}\right) = \frac{kM^2}{(k-2)(k-1)^2}$$

7. a.

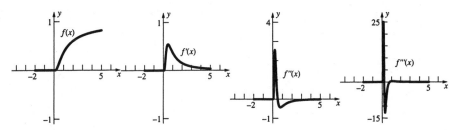

b.
$$f^{(1)}(x) = \begin{cases} \dfrac{1}{x^2}e^{-1/x} & \text{if } x > 0 \\ 0 & \text{if } x \le 0 \end{cases}$$

$$f^{(2)}(x) = \begin{cases} \left(\dfrac{1}{x^4} - \dfrac{2}{x^3}\right)e^{-1/x} & \text{if } x > 0 \\ 0 & \text{if } x \le 0 \end{cases}$$

$$f^{(3)}(x) = \begin{cases} \left(\dfrac{1}{x^6} - \dfrac{6}{x^5} + \dfrac{6}{x^4}\right)e^{-1/x} & \text{if } x > 0 \\ 0 & \text{if } x \le 0 \end{cases}$$

$$\lim_{x\to 0^+} f^{(0)}(x) = \lim_{x\to 0^+} e^{-1/x} = \lim_{x\to 0^+} \frac{1}{e^{1/x}}$$

Let $v = \dfrac{1}{x}$, then as $x \to 0^+, v \to \infty$ and

$$\lim_{x\to 0^+} \frac{1}{e^{1/x}} = \lim_{v\to\infty} \frac{1}{e^v} = 0.$$

Use the same change of variables.

$$\lim_{x\to 0^+} f^{(1)}(x) = \lim_{x\to 0^+} \frac{1}{x^2}e^{-1/x} = \lim_{v\to\infty} v^2 e^{-v} = \lim_{v\to\infty} \frac{v^2}{e^v} = 0$$

$$\lim_{x\to 0^+} f^{(2)}(x) = \lim_{x\to 0^+}\left[\left(\frac{1}{x^4} - \frac{2}{x^3}\right)e^{-1/x}\right] = \lim_{v\to\infty}[(v^4 - 2v^3)e^{-v}] = \lim_{v\to\infty}\left(\frac{v^4}{e^v} - \frac{2v^3}{e^v}\right) = 0$$

$$\lim_{x \to 0^+} f^{(3)}(x) = \lim_{x \to 0^+} \left[\left(\frac{1}{x^6} - \frac{6}{x^5} + \frac{6}{x^4} \right) e^{-1/x} \right] = \lim_{v \to \infty} [(v^6 - 6v^5 + 6v^4)e^{-v}] = \lim_{v \to \infty} \left(\frac{v^6}{e^v} - \frac{6v^5}{e^v} + \frac{6v^4}{e^v} \right) = 0$$

(Recall from Example 2 of Section 9.2 that $\displaystyle\lim_{x \to \infty} \frac{x^a}{e^x} = 0$ when a is any positive real number.)

For $x > 0$ and $x < 0$, $f^{(i)}(x)$ is continuous and since $\displaystyle\lim_{x \to 0^+} f^{(i)}(x) = 0 = f^{(i)}(0)$ for $i = 0, 1, 2, 3$, these are continuous for all x.

c. For $x \le 0$, $f^{(n)}(x) = 0$ for all n, while for $x > 0$, $f^{(n)}(x)$ will be a sum of terms of the form $\dfrac{b}{x^a} e^{-1/x}$. Using

the same change of variables as in part b, this is $\dfrac{bv^a}{e^v}$ which approaches 0 as $v \to \infty$. Thus

$\displaystyle\lim_{x \to 0^+} f^{(n)}(x) = 0 = f^{(n)}(0)$ so $f^{(n)}(x)$ will be continuous.

Infinite Series

10.1 Concepts Review

1. a sequence

3. bounded above

Problem Set 10.1

1. $a_1 = \dfrac{1}{2}, a_2 = \dfrac{2}{5}, a_3 = \dfrac{3}{8}, a_4 = \dfrac{4}{11}, a_5 = \dfrac{5}{14}$

$$\lim_{n \to \infty} \frac{n}{3n-1} = \lim_{n \to \infty} \frac{1}{3 - \frac{1}{n}} = \frac{1}{3}; \text{ converges}$$

3. $a_1 = \dfrac{6}{3} = 2, a_2 = \dfrac{18}{9} = 2, a_3 = \dfrac{38}{17},$

$a_4 = \dfrac{66}{27} = \dfrac{22}{9}, a_5 = \dfrac{102}{39} = \dfrac{34}{13}$

$$\lim_{n \to \infty} \frac{4n^2 + 2}{n^2 + 3n - 1} = \lim_{n \to \infty} \frac{4 + \frac{2}{n^2}}{1 + \frac{3}{n} - \frac{1}{n^2}} = 4; \text{ converges}$$

5. $a_1 = \dfrac{7}{8}, a_2 = \dfrac{26}{27}, a_3 = \dfrac{63}{64}, \ a_4 = \dfrac{124}{125}, a_5 = \dfrac{215}{216}$

$$\lim_{n \to \infty} \frac{n^3 + 3n^2 + 3n}{(n+1)^3} = \lim_{n \to \infty} \frac{n^3 + 3n^2 + 3n}{n^3 + 3n^2 + 3n + 1}$$

$$= \lim_{n \to \infty} \frac{1 + \frac{3}{n} + \frac{3}{n^2}}{1 + \frac{3}{n} + \frac{3}{n^2} + \frac{1}{n^3}} = 1$$

7. $a_1 = -\dfrac{1}{3}, a_2 = \dfrac{2}{4} = \dfrac{1}{2}, a_3 = -\dfrac{3}{5}, a_4 = \dfrac{4}{6} = \dfrac{2}{3},$

$a_5 = -\dfrac{5}{7}$

$$\lim_{n \to \infty} \frac{n}{n+2} = \lim_{n \to \infty} \frac{1}{1 + \frac{2}{n}} = 1, \text{ but since it alternates}$$

between positive and negative, the sequence diverges.

9. $a_1 = -1, a_2 = \dfrac{1}{2}, a_3 = -\dfrac{1}{3}, a_4 = \dfrac{1}{4}, a_5 = -\dfrac{1}{5}$

$-1 \le \cos(n\pi) \le 1$ for all n, so

$$-\frac{1}{n} \le \frac{\cos(n\pi)}{n} \le \frac{1}{n}.$$

$\lim_{n \to \infty} -\dfrac{1}{n} = \lim_{n \to \infty} \dfrac{1}{n} = 0,$ so by the Squeeze Theorem, the sequence converges to 0.

11. $a_1 = \dfrac{e^2}{3} \approx 2.4630, a_2 = \dfrac{e^4}{9} \approx 6.0665,$

$a_3 = \dfrac{e^6}{17} \approx 23.7311, a_4 = \dfrac{e^8}{27} \approx 110.4059,$

$a_5 = \dfrac{e^{10}}{39} \approx 564.7812$

Consider

$$\lim_{x \to \infty} \frac{e^{2x}}{x^2 + 3x - 1} = \lim_{x \to \infty} \frac{2e^{2x}}{2x + 3} = \lim_{x \to \infty} \frac{4e^{2x}}{2} = \infty$$

by using l'Hôpital's Rule twice. The sequence diverges.

13. $a_1 = -\dfrac{\pi}{5} \approx -0.6283, a_2 = \dfrac{\pi^2}{25} \approx 0.3948,$

$a_3 = -\dfrac{\pi^3}{125} \approx -0.2481, a_4 = \dfrac{\pi^4}{625} \approx 0.1559,$

$a_5 = -\dfrac{\pi^5}{3125} \approx -0.0979$

$\dfrac{(-\pi)^n}{5^n} = \left(-\dfrac{\pi}{5}\right)^n, -1 < -\dfrac{\pi}{5} < 1,$ thus the sequence converges to 0.

15. $a_1 = 2.99, a_2 = 2.9801, a_3 \approx 2.9703,$
$a_4 \approx 2.9606, a_5 \approx 2.9510$

$(0.99)^n$ converges to 0 since $-1 < 0.99 < 1,$ thus $2 + (0.99)^n$ converges to 2.

17. $a_1 = \dfrac{\ln 1}{\sqrt{1}} = 0, a_2 = \dfrac{\ln 2}{\sqrt{2}} \approx 0.4901,$

$a_3 = \dfrac{\ln 3}{\sqrt{3}} \approx 0.6343, a_4 = \dfrac{\ln 4}{2} \approx 0.6931,$

$a_5 = \dfrac{\ln 5}{\sqrt{5}} \approx 0.7198$

Consider $\lim\limits_{x\to\infty}\dfrac{\ln x}{\sqrt{x}}=\lim\limits_{x\to\infty}\dfrac{\frac{1}{x}}{\frac{1}{2\sqrt{x}}}=\lim\limits_{x\to\infty}\dfrac{2}{\sqrt{x}}=0$ by

using l'Hôpital's Rule. Thus, $\lim\limits_{n\to\infty}\dfrac{\ln n}{\sqrt{n}}=0$;
converges.

19. $a_1=\left(1+\dfrac{2}{1}\right)^{1/2}=\sqrt{3}\approx 1.7321,$

$a_2=\left(1+\dfrac{2}{2}\right)^{2/2}=2,$

$a_3=\left(1+\dfrac{2}{3}\right)^{3/2}=\left(\dfrac{5}{3}\right)^{3/2}\approx 2.1517,$

$a_4=\left(1+\dfrac{2}{4}\right)^{4/2}=\left(\dfrac{3}{2}\right)^{2}=\dfrac{9}{4},$

$a_5=\left(1+\dfrac{2}{5}\right)^{5/2}=\left(\dfrac{7}{5}\right)^{5/2}\approx 2.3191$

Let $\dfrac{2}{n}=h$, then as $n\to\infty, h\to 0$ and

$\lim\limits_{n\to\infty}\left(1+\dfrac{2}{n}\right)^{n/2}=\lim\limits_{h\to 0}(1+h)^{1/h}=e$ by
Theorem 7.5A; converges

21. $a_n=\dfrac{n}{n+1}$ or $a_n=1-\dfrac{1}{n+1}$;

$\lim\limits_{n\to\infty}\left(1-\dfrac{1}{n+1}\right)=1-\lim\limits_{n\to\infty}\dfrac{1}{n+1}=1$; converges

23. $a_n=(-1)^n\dfrac{n}{2n-1}$; $\lim\limits_{n\to\infty}\dfrac{n}{2n-1}$

$=\lim\limits_{n\to\infty}\dfrac{1}{2-\frac{1}{n}}=\dfrac{1}{2}$, but due to $(-1)^n$, the terms of

the sequence alternate between positive and
negative, so the sequence diverges.

25. $a_n=\dfrac{n}{n^2-(n-1)^2}=\dfrac{n}{n^2-(n^2-2n+1)}=\dfrac{n}{2n-1}$;

$\lim\limits_{n\to\infty}\dfrac{n}{2n-1}=\lim\limits_{n\to\infty}\dfrac{1}{2-\frac{1}{n}}=\dfrac{1}{2}$; converges

27. $a_n=n\sin\dfrac{1}{n}$; $\lim\limits_{n\to\infty}n\sin\dfrac{1}{n}=\lim\limits_{n\to\infty}\dfrac{\sin\frac{1}{n}}{\frac{1}{n}}=1$ since

$\lim\limits_{x\to 0}\dfrac{\sin x}{x}=1$; converges

29. $a_n=\dfrac{2^n}{n^2}$;

$\lim\limits_{n\to\infty}\dfrac{2^n}{n^2}=\lim\limits_{n\to\infty}\dfrac{2^n\ln 2}{2n}=\lim\limits_{n\to\infty}\dfrac{2^n(\ln 2)^2}{2}=\infty$;
diverges

31. $a_1=\dfrac{1}{2}, a_2=\dfrac{5}{4}, a_3=\dfrac{9}{8}, a_4=\dfrac{13}{16}$

a_n is positive for all n, and $a_{n+1}<a_n$ for all

$n\ge 2$ since $a_{n+1}-a_n=-\dfrac{4n-7}{2^{n+1}}$, so $\{a_n\}$

converges to a limit $L\ge 0$.

33. $a_2=\dfrac{3}{4}; a_3=\left(\dfrac{3}{4}\right)\left(\dfrac{8}{9}\right)=\dfrac{2}{3}$;

$a_4=\left(\dfrac{3}{4}\right)\left(\dfrac{8}{9}\right)\left(\dfrac{15}{16}\right)=\dfrac{5}{8}$;

$a_5=\left(\dfrac{3}{4}\right)\left(\dfrac{8}{9}\right)\left(\dfrac{15}{16}\right)\left(\dfrac{24}{25}\right)=\dfrac{3}{5}$

$a_n>0$ for all n and $a_{n+1}<a_n$ since

$a_{n+1}=a_n\left(1-\dfrac{1}{(n+1)^2}\right)$ and $1-\dfrac{1}{(n+1)^2}<1$, so

$\{a_n\}$ converges to a limit $L\ge 0$.

35. $a_1=1, a_2=1+\dfrac{1}{2}(1)=\dfrac{3}{2}, a_3=1+\dfrac{1}{2}\left(\dfrac{3}{2}\right)=\dfrac{7}{4},$

$a_4=1+\dfrac{1}{2}\left(\dfrac{7}{4}\right)=\dfrac{15}{8}$

Suppose that $1<a_n<2$, then $\dfrac{1}{2}<\dfrac{1}{2}a_n<1$, so

$\dfrac{3}{2}<1+\dfrac{1}{2}a_n<2$, or $\dfrac{3}{2}<a_{n+1}<2$. Thus, since

$1<a_2<2$, every subsequent term is between $\dfrac{3}{2}$

and 2.

$a_n<2$ thus $\dfrac{1}{2}a_n<1$, so $a_n<1+\dfrac{1}{2}a_n=a_{n+1}$

and the sequence is nondecreasing, so $\{a_n\}$
converges to a limit $L\le 2$.

37.

n	u_n
1	1.73205
2	2.17533
3	2.27493
4	2.29672
5	2.30146
6	2.30249
7	2.30271
8	2.30276

9	2.30277
10	2.30278
11	2.30278

$\lim\limits_{n\to\infty} u_n \approx 2.3028$

39. If $u = \lim\limits_{n\to\infty} u_n$, then $u = \sqrt{3+u}$ or $u^2 = 3+u$;

$u^2 - u - 3 = 0$ when $u = \dfrac{1}{2}\left(1 \pm \sqrt{13}\right)$ so

$u = \dfrac{1}{2}\left(1 + \sqrt{13}\right) \approx 2.3028$ since $u > 0$ and

$\dfrac{1}{2}\left(1 - \sqrt{13}\right) < 0$.

41.

n	u_n
1	0
2	1
3	1.1
4	1.11053
5	1.11165
6	1.11177
7	1.11178
8	1.11178

$\lim\limits_{n\to\infty} u_n \approx 1.1118$

43. As $n \to \infty$, $\dfrac{k}{n} \to 0$; using $\Delta x = \dfrac{1}{n}$, an equivalent

definite integral is

$\int_0^1 \sin x\, dx = [-\cos x]_0^1 = -\cos 1 + \cos 0 = 1 - \cos 1$

≈ 0.4597

45. $\left|\dfrac{n}{n+1} - 1\right| = \left|\dfrac{n-(n+1)}{n+1}\right| = \left|\dfrac{-1}{n+1}\right| = \dfrac{1}{n+1}$;

$\dfrac{1}{n+1} < \varepsilon$ is the same as $\dfrac{1}{\varepsilon} < n+1$. For whatever

ε is given, choose $N > \dfrac{1}{\varepsilon} - 1$ then

$n \ge N \Rightarrow \left|\dfrac{n}{n+1} - 1\right| < \varepsilon$.

47. Recall from Section 1.2 that every rational
number can be written as either a terminating or a
repeating decimal.
Thus if the sequence 1, 1.4, 1.41, 1.414, ... has a
limit within the rational numbers, the terms of the
sequence would eventually either repeat or
terminate, which they do not since they are the
decimal approximations to $\sqrt{2}$, which is
irrational. Within the real numbers, the least
upper bound is $\sqrt{2}$.

49. If $\{b_n\}$ is bounded, there are numbers N and M
with $N \le |b_n| \le M$ for all n. Then

$|a_n N| \le |a_n b_n| \le |a_n M|$.

$\lim\limits_{n\to\infty} |a_n N| = |N| \lim\limits_{n\to\infty} |a_n| = 0$ and

$\lim\limits_{n\to\infty} |a_n M| = |M| \lim\limits_{n\to\infty} |a_n| = 0$, so $\lim\limits_{n\to\infty} |a_n b_n| = 0$

by the Squeeze Theorem, and by Theorem C,

$\lim\limits_{n\to\infty} a_n b_n = 0$.

51. No. Consider $a_n = (-1)^n$ and $b_n = (-1)^{n+1}$. Both
$\{a_n\}$ and $\{b_n\}$ diverge, but

$a_n + b_n = (-1)^n + (-1)^{n+1} = (-1)^n(1+(-1)) = 0$ so
$\{a_n + b_n\}$ converges.

53.

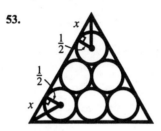

From the figure shown, the sides of the triangle
have length $n - 1 + 2x$. The small right triangles

marked are 30-60-90 right triangles, so $x = \dfrac{\sqrt{3}}{2}$;

thus the sides of the large triangle have lengths

$n - 1 + \sqrt{3}$ and $B_n = \dfrac{\sqrt{3}}{4}\left(n - 1 + \sqrt{3}\right)^2$

$= \dfrac{\sqrt{3}}{4}\left(n^2 + 2\sqrt{3}n - 2n - 2\sqrt{3} + 4\right)$ while

$A_n = \dfrac{n(n+1)}{2}\pi\left(\dfrac{1}{2}\right)^2 = \dfrac{\pi}{8}(n^2 + n)$

$\lim\limits_{n\to\infty} \dfrac{A_n}{B_n} = \lim\limits_{n\to\infty} \dfrac{\frac{\pi}{8}(n^2+n)}{\frac{\sqrt{3}}{4}(n^2 + 2\sqrt{3}n - 2n - 2\sqrt{3} + 4)}$

$= \lim\limits_{n\to\infty} \dfrac{\pi\left(1 + \frac{1}{n}\right)}{2\sqrt{3}\left(1 + \frac{2\sqrt{3}}{n} - \frac{2}{n} - \frac{2\sqrt{3}}{n^2} + \frac{4}{n^2}\right)} = \dfrac{\pi}{2\sqrt{3}}$

55. Let $f(x) = \left(1 + \dfrac{1}{2x}\right)^x$.

$\lim\limits_{x\to\infty}\left(1 + \dfrac{1}{2x}\right)^x = \lim\limits_{x\to 0^+}\left(1 + \dfrac{x}{2}\right)^{1/x}$

$$= \lim_{x\to 0^+}\left[\left(1+\frac{x}{2}\right)^{2/x}\right]^{1/2} = e^{1/2}, \text{ so}$$

$$\lim_{n\to\infty}\left(1+\frac{1}{2n}\right)^n = e^{1/2}.$$

57. Let $f(x)=\left(\dfrac{x-1}{x+1}\right)^x$.

$$\lim_{x\to\infty}\left(\frac{x-1}{x+1}\right)^x = \lim_{x\to 0^+}\left(\frac{\frac{1}{x}-1}{\frac{1}{x}+1}\right)^{1/x} = \lim_{x\to 0^+}\left(\frac{\frac{1-x}{x}}{\frac{1+x}{x}}\right)^{1/x}$$

$$= \lim_{x\to 0^+}\left(\frac{1-x}{1+x}\right)^{1/x} = e^{-2}, \text{ so}$$

$$\lim_{n\to\infty}\left(\frac{n-1}{n+1}\right)^n = e^{-2}.$$

59. Let $f(x)=\left(\dfrac{2+x^2}{3+x^2}\right)^{x^2}$.

$$\lim_{x\to\infty}\left(\frac{2+x^2}{3+x^2}\right)^{x^2} = \lim_{x\to 0^+}\left(\frac{2+\frac{1}{x^2}}{3+\frac{1}{x^2}}\right)^{1/x^2}$$

$$= \lim_{x\to 0^+}\left(\frac{\frac{2x^2+1}{x^2}}{\frac{3x^2+1}{x^2}}\right)^{1/x^2} = \lim_{x\to 0^+}\left(\frac{2x^2+1}{3x^2+1}\right)^{1/x^2} = e^{-1},$$

$$\text{so } \lim_{n\to\infty}\left(\frac{2+n^2}{3+n^2}\right)^{n^2} = e^{-1}.$$

10.2 Concepts Review

1. an infinite series

3. $|r|<1; \dfrac{a}{1-r}$

Problem Set 10.2

1. $\displaystyle\sum_{k=1}^{\infty}\left(\frac{1}{7}\right)^k = \frac{1}{7}+\frac{1}{7}\cdot\frac{1}{7}+\frac{1}{7}\left(\frac{1}{7}\right)^2+\ldots;$ a geometric

series with $a=\dfrac{1}{7}, r=\dfrac{1}{7}; \ S=\dfrac{\frac{1}{7}}{1-\frac{1}{7}}=\dfrac{\frac{1}{7}}{\frac{6}{7}}=\dfrac{1}{6}$

3. $\displaystyle\sum_{k=0}^{\infty}2\left(\frac{1}{4}\right)^k = 2+2\cdot\frac{1}{4}+2\left(\frac{1}{4}\right)^2+\ldots;$ a geometric

series with $a=2, \ r=\dfrac{1}{4}; \ S=\dfrac{2}{1-\frac{1}{4}}=\dfrac{2}{\frac{3}{4}}=\dfrac{8}{3}.$

7. $\displaystyle\sum_{k=2}^{\infty}\left(\frac{1}{k}-\frac{1}{k-1}\right) = \left(\frac{1}{2}-\frac{1}{1}\right)+\left(\frac{1}{3}-\frac{1}{2}\right)+\left(\frac{1}{4}-\frac{1}{3}\right)+\ldots;$

$$S_n = \left(\frac{1}{2}-1\right)+\left(\frac{1}{3}-\frac{1}{2}\right)+\ldots+\left(\frac{1}{n-1}-\frac{1}{n-2}\right)+\left(\frac{1}{n}-\frac{1}{n-1}\right) = -1+\frac{1}{n};$$

$$\lim_{n\to\infty} S_n = \lim_{n\to\infty} -1+\frac{1}{n} = -1, \text{ so } \sum_{k=2}^{\infty}\left(\frac{1}{k}-\frac{1}{k-1}\right) = -1$$

9. $\displaystyle\sum_{k=1}^{\infty}\frac{k!}{100^k} = \frac{1}{100}+\frac{2}{10,000}+\frac{6}{1,000,000}+\ldots$

Consider $\{a_n\}$, where $a_{n+1}=\dfrac{n+1}{100}a_n, a_1=\dfrac{1}{100}$. $a_n>0$ for all n, and for $n>99$, $a_{n+1}>a_n$, so the

$$\sum_{k=0}^{\infty}3\left(-\frac{1}{5}\right)^k = 3-3\cdot\frac{1}{5}+3\left(\frac{1}{5}\right)^2 -\ldots;$$ a geometric

series with $a=3, \ r=-\dfrac{1}{5};$

$$S = \frac{3}{1-\left(-\frac{1}{5}\right)} = \frac{3}{\frac{6}{5}} = \frac{5}{2}$$

Thus, by Theorem B,

$$\sum_{k=0}^{\infty}\left[2\left(\frac{1}{4}\right)^k+3\left(-\frac{1}{5}\right)^k\right] = \frac{8}{3}+\frac{5}{2} = \frac{31}{6}$$

5. $\displaystyle\sum_{k=1}^{\infty}\frac{k-5}{k+2} = -\frac{4}{3}-\frac{3}{4}-\frac{2}{5}-\frac{1}{6}+0+\frac{1}{8}+\frac{2}{9}+\ldots;$

$$\lim_{k\to\infty}\frac{k-5}{k+2} = \lim_{k\to\infty}\frac{1-\frac{5}{k}}{1+\frac{2}{k}} = 1\neq 0;$$ the series

diverges.

sequence is eventually an increasing sequence, hence $\lim\limits_{n\to\infty} a_n \neq 0$. The sequence can also be described by

$a_n = \dfrac{n!}{100^n}$, hence $\sum\limits_{k=1}^{\infty} \dfrac{k!}{100^k}$ diverges.

11. $\sum\limits_{k=1}^{\infty}\left(\dfrac{e}{\pi}\right)^{k+1} = \left(\dfrac{e}{\pi}\right)^2 + \left(\dfrac{e}{\pi}\right)^2 \cdot \dfrac{e}{\pi} + \left(\dfrac{e}{\pi}\right)^2 \left(\dfrac{e}{\pi}\right)^2 + \ldots;$ a geometric series with $a = \left(\dfrac{e}{\pi}\right)^2, r = \dfrac{e}{\pi} < 1;$

$S = \dfrac{\left(\dfrac{e}{\pi}\right)^2}{1-\dfrac{e}{\pi}} = \dfrac{\left(\dfrac{e}{\pi}\right)^2}{\dfrac{\pi-e}{\pi}} = \dfrac{e^2}{\pi(\pi-e)} \approx 5.5562$

13. $\sum\limits_{k=2}^{\infty}\left(\dfrac{3}{(k-1)^2} - \dfrac{3}{k^2}\right) = \left(\dfrac{3}{1} - \dfrac{3}{4}\right) + \left(\dfrac{3}{4} - \dfrac{3}{9}\right) + \left(\dfrac{3}{9} - \dfrac{3}{16}\right) + \ldots;$

$S_n = \left(3 - \dfrac{3}{4}\right) + \left(\dfrac{3}{4} - \dfrac{1}{3}\right) + \left(\dfrac{1}{3} - \dfrac{3}{16}\right) + \ldots + \left(\dfrac{3}{(n-2)^2} - \dfrac{3}{(n-1)^2}\right) + \left(\dfrac{3}{(n-1)^2} - \dfrac{3}{n^2}\right)$

$= 3 - \dfrac{3}{n^2}; \lim\limits_{n\to\infty} S_n = 3 - \lim\limits_{n\to\infty} \dfrac{3}{n^2} = 3$, so

$\sum\limits_{k=2}^{\infty}\left(\dfrac{3}{(k-1)^2} - \dfrac{3}{k^2}\right) = 3.$

15. $0.22222\ldots = \sum\limits_{k=1}^{\infty} \dfrac{2}{10}\left(\dfrac{1}{10}\right)^{k-1}$

$= \dfrac{\dfrac{2}{10}}{1-\dfrac{1}{10}} = \dfrac{2}{9}$

19. $0.4999\ldots = \dfrac{4}{10} + \sum\limits_{k=1}^{\infty} \dfrac{9}{100}\left(\dfrac{1}{10}\right)^{k-1}$

$= \dfrac{4}{10} + \dfrac{\dfrac{9}{100}}{1-\dfrac{1}{10}} = \dfrac{1}{2}$

17. $0.013013013\ldots = \sum\limits_{k=1}^{\infty} \dfrac{13}{1000}\left(\dfrac{1}{1000}\right)^{k-1}$

$= \dfrac{\dfrac{13}{1000}}{1-\dfrac{1}{1000}} = \dfrac{13}{999}$

21. Let $s = 1 - r$, so $r = 1 - s$. Since $0 < r < 2$, $-1 < 1 - r < 1$, so

$|s| < 1$, and $\sum\limits_{k=0}^{\infty} r(1-r)^k = \sum\limits_{k=0}^{\infty} (1-s)s^k$

$= \sum\limits_{k=1}^{\infty} (1-s)s^{k-1} = \dfrac{1-s}{1-s} = 1$

23. $\ln\dfrac{k}{k+1} = \ln k - \ln(k+1)$

$S_n = (\ln 1 - \ln 2) + (\ln 2 - \ln 3) + (\ln 3 - \ln 4) + \ldots + (\ln(n-1) - \ln n) + (\ln n - \ln(n+1)) = \ln 1 - \ln(n+1) = -\ln(n+1)$

$\lim\limits_{n\to\infty} S_n = \lim\limits_{n\to\infty} -\ln(n+1) = -\infty$, thus $\sum\limits_{k=1}^{\infty} \ln\dfrac{k}{k+1}$ diverges.

25. The ball drops 100 feet, rebounds up $100\left(\dfrac{2}{3}\right)$ feet, drops $100\left(\dfrac{2}{3}\right)$ feet, rebounds up $100\left(\dfrac{2}{3}\right)^2$ feet, drops $100\left(\dfrac{2}{3}\right)^2$, etc. The total distance it travels is

$$100 + 200\left(\frac{2}{3}\right) + 200\left(\frac{2}{3}\right)^2 + 200\left(\frac{2}{3}\right)^3 + \ldots = -100 + 200 + 200\left(\frac{2}{3}\right) + 200\left(\frac{2}{3}\right)^2 + 200\left(\frac{2}{3}\right)^3 + \ldots$$

$$= -100 + \sum_{k=1}^{\infty} 200\left(\frac{2}{3}\right)^{k-1} = -100 + \frac{200}{1-\frac{2}{3}} = 500 \text{ feet}$$

27. \$1 billion + 75% of \$1 billion + 75% of 75% of \$1 billion + $\ldots = \sum_{k=1}^{\infty}(\$1 \text{ billion})0.75^{k-1} = \frac{\$1 \text{ billion}}{1-0.75} = \$4$ billion

29. As the midpoints of the sides of a square are connected, a new square is formed. The new square has sides $\frac{1}{\sqrt{2}}$ times the sides of the old square. Thus, the new square has area $\frac{1}{2}$ the area of the old square. Then in the next step, $\frac{1}{8}$ of each new square is shaded.

$$\text{Area} = \frac{1}{8} \cdot 1 + \frac{1}{8} \cdot \frac{1}{2} + \frac{1}{8} \cdot \frac{1}{4} + \ldots = \sum_{k=1}^{\infty} \frac{1}{8}\left(\frac{1}{2}\right)^{k-1} = \frac{\frac{1}{8}}{1-\frac{1}{2}} = \frac{1}{4}$$

The area will be $\frac{1}{4}$.

31. $\dfrac{3}{4} + \dfrac{3}{4}\left(\dfrac{1}{4} \cdot \dfrac{1}{4}\right) + \dfrac{3}{4}\left(\dfrac{1}{4} \cdot \dfrac{1}{4}\right)\left(\dfrac{1}{4} \cdot \dfrac{1}{4}\right) + \ldots = \sum_{k=1}^{\infty} \dfrac{3}{4}\left(\dfrac{1}{16}\right)^{k-1} = \dfrac{\frac{3}{4}}{1-\frac{1}{16}} = \dfrac{4}{5}$

The original does not need to be equilateral since each smaller triangle will have $\frac{1}{4}$ area of the previous larger triangle.

33. Both Achilles and the tortoise will have moved.

$$100 + 10 + 1 + \frac{1}{10} + \frac{1}{100} + \ldots = \sum_{k=1}^{\infty} 100\left(\frac{1}{10}\right)^{k-1}$$

$$= \frac{100}{1-\frac{1}{10}} = 111\frac{1}{9} \text{ yards}$$

Also, one can see this by the following reasoning. In the time it takes the tortoise to run $\frac{d}{10}$ yards, Achilles will run d yards. Solve

$$d = 100 + \frac{d}{10}. \; d = \frac{1000}{9} = 111\frac{1}{9} \text{ yards}$$

35. (Proof by contradiction) Assume $\displaystyle\sum_{k=1}^{\infty} ca_k$ converges, and $c \neq 0$. Then $\frac{1}{c}$ is defined, so

$$\sum_{k=1}^{\infty} a_k = \sum_{k=1}^{\infty} \frac{1}{c} ca_k = \frac{1}{c}\sum_{k=1}^{\infty} ca_k \text{ would also}$$

converge, by Theorem B(i).

37. a. The top block is supported *exactly* at its center of mass. The location of the center of mass of the top n blocks is the average of the locations of their individual centers of mass, so the nth block moves the center of mass left by $\frac{1}{n}$ of the location of its center of mass, that is, $\frac{1}{n} \cdot \frac{1}{2}$ or $\frac{1}{2n}$ to the left. But this is exactly how far the $(n+1)$st block underneath it is offset.

b. Since $\dfrac{1}{2} + \dfrac{1}{4} + \dfrac{1}{6} + \ldots = \dfrac{1}{2}\displaystyle\sum_{k=1}^{\infty} \dfrac{1}{k}$, which diverges, there is no limit to how far the top block can protrude.

39. (Proof by contradiction) Assume $\displaystyle\sum_{k=1}^{\infty}(a_k + b_k)$ converges. Since $\displaystyle\sum_{k=1}^{\infty} b_k$ converges, so would

$\sum_{k=1}^{\infty} a_k = \sum_{k=1}^{\infty} (a_k + b_k) + (-1) \sum_{k=1}^{\infty} b_k$, by Theorem B(ii).

41. Taking vertical strips, the area is

$$1 \cdot 1 + 1 \cdot \frac{1}{2} + 1 \cdot \frac{1}{4} + 1 \cdot \frac{1}{8} + \cdots = \sum_{k=1}^{\infty} \left(\frac{1}{2}\right)^{k-1}$$

Taking horizontal strips, the area is

$$\frac{1}{2} \cdot 1 + \frac{1}{4} \cdot 2 + \frac{1}{8} \cdot 3 + \frac{1}{16} \cdot 4 + \cdots = \sum_{k=1}^{\infty} \frac{k}{2^k}.$$

a. $\displaystyle\sum_{k=1}^{\infty} \frac{k}{2^k} = \sum_{k=1}^{\infty} \left(\frac{1}{2}\right)^{k-1} = \frac{1}{1 - \frac{1}{2}} = 2$

b. The moment about $x = 0$ is

$$\sum_{k=0}^{\infty} \left(\frac{1}{2}\right)^k \cdot (1) k = \sum_{k=0}^{\infty} \frac{k}{2^k} = \sum_{k=1}^{\infty} \frac{k}{2^k} = 2.$$

$$\bar{x} = \frac{\text{moment}}{\text{area}} = \frac{2}{2} = 1$$

43. a. $A = \displaystyle\sum_{n=0}^{\infty} Ce^{-nkt} = \sum_{n=1}^{\infty} C\left(\frac{1}{e^{kt}}\right)^{n-1}$

$$= \frac{C}{1 - \frac{1}{e^{kt}}} = \frac{Ce^{kt}}{e^{kt} - 1}$$

b.

$$\frac{1}{2} = e^{-kt} = e^{-6k} \Rightarrow k = \frac{\ln 2}{6} \Rightarrow A = \frac{4}{3}C;$$

if $C = 2$ mg, then $A = \dfrac{8}{3}$ mg.

45. $\dfrac{1}{f_k f_{k+1}} - \dfrac{1}{f_{k+1} f_{k+2}} = \dfrac{f_{k+2} - f_k}{f_k f_{k+1} f_{k+2}} = \dfrac{1}{f_k f_{k+2}}$

since $f_{k+2} = f_{k+1} + f_k$. Thus,

$$\sum_{k=1}^{\infty} \frac{1}{f_k f_{k+2}} = \sum_{k=1}^{\infty} \left(\frac{1}{f_k f_{k+1}} - \frac{1}{f_{k+1} f_{k+2}}\right) \text{ and}$$

$$S_n = \left(\frac{1}{f_1 f_2} - \frac{1}{f_2 f_3}\right) + \left(\frac{1}{f_2 f_3} - \frac{1}{f_3 f_4}\right) + \cdots + \left(\frac{1}{f_{n-1} f_n} - \frac{1}{f_n f_{n+1}}\right) + \left(\frac{1}{f_n f_{n+1}} - \frac{1}{f_{n+1} f_{n+2}}\right)$$

$$= \frac{1}{f_1 f_2} - \frac{1}{f_{n+1} f_{n+2}} = \frac{1}{1 \cdot 1} - \frac{1}{f_{n+1} f_{n+2}} = 1 - \frac{1}{f_{n+1} f_{n+2}}$$

The terms of the Fibonacci sequence increase without bound, so

$$\lim_{n \to \infty} S_n = 1 - \lim_{n \to \infty} \frac{1}{f_{n+1} f_{n+2}} = 1 - 0 = 1$$

10.3 Concepts Review

1. bounded above

3. convergence or divergence

Problem Set 10.3

1. $\dfrac{1}{x+3}$ is continuous, positive, and nonincreasing on $[0, \infty)$.

$$\int_0^{\infty} \frac{1}{x+3} dx = \left[\ln|x+3|\right]_0^{\infty} = \infty - \ln 3 = \infty$$

The series diverges.

3. $\dfrac{x}{x^2+3}$ is continuous, positive, and nonincreasing on $[2, \infty)$.

$$\int_2^{\infty} \frac{x}{x^2+3} dx = \left[\frac{1}{2} \ln|x^2+3|\right]_2^{\infty} = \infty - \frac{1}{2} \ln 7 = \infty$$

The series diverges.

5. $\dfrac{2}{\sqrt{x+2}}$ is continuous, positive, and nonincreasing on $[1, \infty)$.

$$\int_1^{\infty} \frac{2}{\sqrt{x+2}} dx = \left[4\sqrt{x+2}\right]_1^{\infty} = \infty - 4\sqrt{3} = \infty$$

Thus $\displaystyle\sum_{k=1}^{\infty} \frac{2}{\sqrt{k+2}}$ diverges, hence

$$\sum_{k=1}^{\infty} \frac{-2}{\sqrt{k+2}} = -\sum_{k=1}^{\infty} \frac{2}{\sqrt{k+2}} \text{ also diverges.}$$

7. $\dfrac{7}{4x+2}$ is continuous, positive, and nonincreasing on. $[2, \infty)$

$$\int_2^\infty \frac{7}{4x+2}\,dx = \left[\frac{7}{4}\ln|4x+2|\right]_2^\infty = \infty - \frac{7}{4}\ln 10 = \infty$$

The series diverges.

9. $\dfrac{3}{(4+3x)^{7/6}}$ is continuous, positive, and nonincreasing on $[1, \infty)$.

$$\int_1^\infty \frac{3}{(4+3x)^{7/6}}\,dx = \left[-\frac{6}{(4+3x)^{1/6}}\right]_1^\infty$$

$$= 0 + \frac{6}{7^{1/6}} = 6 \cdot 7^{-1/6} < \infty$$

The series converges.

11. xe^{-3x^2} is continuous, positive, and nonincreasing on $[1, \infty)$.

$$\int_1^\infty xe^{-3x^2}\,dx = \left[-\frac{1}{6}e^{-3x^2}\right]_1^\infty = 0 + \frac{1}{6}e^{-3}$$

$$= \frac{1}{6e^3} < \infty$$

The series converges.

13. $\displaystyle\lim_{k\to\infty} \frac{k^2+1}{k^2+5} = \lim_{k\to\infty} \frac{1+\frac{1}{k^2}}{1+\frac{5}{k^2}} = 1 \neq 0$, so the series

diverges.

15. $\displaystyle\sum_{k=1}^\infty \left(\frac{1}{2}\right)^k$ is a geometric series with $r = \dfrac{1}{2}; \left|\dfrac{1}{2}\right| < 1$

so the series converges.

In $\displaystyle\sum_{k=1}^\infty \frac{k-1}{2k+1}$, $\displaystyle\lim_{k\to\infty} \frac{k-1}{2k+1} = \lim_{k\to\infty} \frac{1-\frac{1}{k}}{2+\frac{1}{k}} = \frac{1}{2} \neq 0$, so

the series diverges. Thus, the sum of the series diverges.

17. $\sin\left(\dfrac{k\pi}{2}\right) = \begin{cases} 1 & k = 4j+1 \\ -1 & k = 4j+3, \\ 0 & k \text{ is even} \end{cases}$

where j is any nonnegative integer.

Thus $\displaystyle\lim_{k\to\infty} \left|\sin\left(\frac{k\pi}{2}\right)\right|$ does not exist, hence

$\displaystyle\lim_{k\to\infty} \left|\sin\left(\frac{k\pi}{2}\right)\right| \neq 0$ and the series diverges.

19. $x^2e^{-x^3}$ is continuous, positive, and nonincreasing on $[1, \infty)$.

$$\int_1^\infty x^2e^{-x^3}\,dx = \left[-\frac{1}{3}e^{-x^3}\right]_1^\infty = 0 + \frac{1}{3}e^{-1} < \infty, \text{ so}$$

the series converges.

21. $\dfrac{\tan^{-1}x}{1+x^2}$ is continuous, positive, and nonincreasing on $[1, \infty)$.

$$\int_1^\infty \frac{\tan^{-1}x}{1+x^2}\,dx = \left[\frac{1}{2}(\tan^{-1}x)^2\right]_1^\infty$$

$$= \frac{1}{2}\left(\frac{\pi}{2}\right)^2 - \frac{1}{2}\left(\frac{\pi}{4}\right)^2 = \frac{3\pi^2}{32} < \infty, \text{ so the series}$$

converges.

23. $\dfrac{x}{e^x}$ is continuous, positive, and nonincreasing on $[5, \infty)$.

$$E = \sum_{k=6}^\infty \frac{k}{e^k} \leq \int_5^\infty \frac{x}{e^x}\,dx = [-xe^{-x}]_5^\infty + \int_5^\infty e^{-x}dx$$

$$= [-xe^{-x} - e^{-x}]_5^\infty = 0 + 5e^{-5} + e^{-5} = 6e^{-5}$$

$$\approx 0.0404$$

25. $\dfrac{1}{1+x^2}$ is continuous, positive, and nonincreasing on $[5, \infty)$.

$$E = \sum_{k=6}^\infty \frac{1}{1+k^2} \leq \int_5^\infty \frac{1}{1+x^2}\,dx = [\tan^{-1}x]_5^\infty$$

$$= \frac{\pi}{2} - \tan^{-1}5 \approx 0.1974$$

27. Consider $\displaystyle\int_2^\infty \frac{1}{x(\ln x)^p}\,dx$. Let $u = \ln x$,

$du = \dfrac{1}{x}dx$.

$$\int_2^\infty \frac{1}{x(\ln x)^p}\,dx = \int_{\ln 2}^\infty \frac{1}{u^p}\,du \text{ which converges for}$$

$p > 1$.

29.

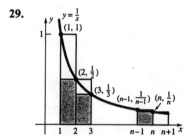

The upper rectangles, which extend to $n + 1$ on the right, have area $1 + \frac{1}{2} + \frac{1}{3} + \ldots + \frac{1}{n}$. These rectangles are above the curve $y = \frac{1}{x}$ from $x = 1$ to $x = n + 1$. Thus,

$$\int_1^{n+1} \frac{1}{x}\,dx = [\ln x]_1^{n+1} = \ln(n+1) - \ln 1 = \ln(n+1)$$

$$< 1 + \frac{1}{2} + \frac{1}{3} + \ldots + \frac{1}{n}.$$

The lower (shaded) rectangles have area $\frac{1}{2} + \frac{1}{3} + \ldots + \frac{1}{n}$. These rectangles lie below the curve $y = \frac{1}{x}$ from $x = 1$ to $x = n$. Thus

$$\frac{1}{2} + \frac{1}{3} + \ldots + \frac{1}{n} < \int_1^n \frac{1}{x}\,dx = \ln n, \text{ so}$$

$$1 + \frac{1}{2} + \frac{1}{3} + \ldots + \frac{1}{n} < 1 + \ln n.$$

31. $\{B_n\}$ is a nondecreasing sequence that is bounded above, thus by the Monotone Sequence Theorem (Theorem D of Section 10.1), $\lim_{n \to \infty} B_n$ exists.

The rationality of γ is a famous unsolved problem.

33. $\gamma + \ln(n+1) > 20 \Rightarrow \ln(n+1) > 20 - \gamma \approx 19.4228$

$\Rightarrow n + 1 > e^{19.4228} \approx 272,404,867$

$\Rightarrow n > 272,404,866$

35. Every time n is incremented by 1, a positive amount of area is added, thus $\{A_n\}$ is an increasing sequence.

Each curved region has horizontal width 1, and can be moved into the heavily outlined triangle without any overlap. This can be done by shifting the nth shaded region, which goes from $(n, f(n))$ to $(n + 1, f(n + 1))$, as follows:

shift $(n + 1, f(n + 1))$ to $(2, f(2))$ and $(n, f(n))$ to $(1, f(2)-[f(n + 1) - f(n)])$.

The slope of the line forming the bottom of the shaded region between $x = n$ and $x = n + 1$ is

$$\frac{f(n+1) - f(n)}{(n+1) - n} = f(n+1) - f(n) > 0$$

since f is increasing. By the Mean Value Theorem, $f(n+1) - f(n) = f'(c)$ for some c in $(n, n + 1)$. Since f is concave down, $n < c < n + 1$ means that $f'(c) < f'(b)$ for all b in $[1, n]$. Thus, the nth shaded region will not overlap any other shaded region when shifted into the heavily outlined triangle. Thus, the area of all of the shaded regions is less than or equal to the area of the heavily outlined triangle, so $\lim_{n \to \infty} A_n$ exists.

10.4 Concepts Review

1. $0 \le a_k \le b_k$

3. $\rho < 1; \rho > 1; \rho = 1$

Problem Set 10.4

1. $a_n = \frac{n}{n^2 + 2n + 3}; b_n = \frac{1}{n}$

$$\lim_{n \to \infty} \frac{a_n}{b_n} = \lim_{n \to \infty} \frac{n^2}{n^2 + 2n + 3} = \lim_{n \to \infty} \frac{1}{1 + \frac{2}{n} + \frac{3}{n^2}} = 1;$$

$0 < 1 < \infty$

$\sum_{n=1}^{\infty} b_n$ diverges $\Rightarrow \sum_{n=1}^{\infty} a_n$ diverges.

3. $a_n = \frac{1}{n\sqrt{n+1}} = \frac{1}{\sqrt{n^3 + n^2}}; \quad b_n = \frac{1}{n^{3/2}}$

$$\lim_{n \to \infty} \frac{a_n}{b_n} = \lim_{n \to \infty} \frac{n^{3/2}}{\sqrt{n^3 + n^2}} = \lim_{n \to \infty} \sqrt{\frac{n^3}{n^3 + n^2}}$$

$$= \lim_{n \to \infty} \sqrt{\frac{1}{1 + \frac{1}{n}}} = 1; \ 0 < 1 < \infty$$

$\sum_{n=1}^{\infty} b_n$ converges $\Rightarrow \sum_{n=1}^{\infty} a_n$ converges

5. $\lim_{n \to \infty} \frac{a_{n+1}}{a_n} = \lim_{n \to \infty} \frac{8^{n+1} n!}{(n+1)! 8^n} = \lim_{n \to \infty} \frac{8}{n+1} = 0 < 1$

The series converges.

7. $\lim_{n \to \infty} \frac{a_{n+1}}{a_n} = \lim_{n \to \infty} \frac{(n+1)! n^{100}}{(n+1)^{100} n!} = \lim_{n \to \infty} \frac{n^{100}}{(n+1)^{99}}$

$$= \lim_{n \to \infty} \frac{n}{\left(\frac{n+1}{n}\right)^{99}} = \infty \text{ since } \lim_{n \to \infty} \left(\frac{n+1}{n}\right)^{99} = 1$$

The series diverges.

9. $\lim_{n \to \infty} \frac{a_{n+1}}{a_n} = \lim_{n \to \infty} \frac{(n+1)^3 (2n)!}{(2n+2)! n^3}$

$$= \lim_{n \to \infty} \frac{(n+1)^3}{(2n+2)(2n+1)n^3} = \lim_{n \to \infty} \frac{n^3 + 3n^2 + 3n + 1}{4n^5 + 6n^4 + 2n^3}$$

$$= \lim_{n \to \infty} \frac{\frac{1}{n^2} + \frac{3}{n^3} + \frac{3}{n^4} + \frac{1}{n^5}}{4 + \frac{6}{n} + \frac{2}{n^2}} = 0 < 1$$

The series converges.

11. $\lim_{n \to \infty} \frac{n}{n + 200} = \lim_{n \to \infty} \frac{1}{1 + \frac{200}{n}} = 1 \neq 0$

The series diverges; nth-Term Test

13. $a_n = \frac{n+3}{n^2 \sqrt{n}}; b_n = \frac{1}{n^{3/2}}$

$\lim_{n \to \infty} \frac{a_n}{b_n} = \lim_{n \to \infty} \frac{n^{5/2} + 3n^{3/2}}{n^{5/2}} = \lim_{n \to \infty} \frac{1 + \frac{3}{n}}{1} = 1;$

$0 < 1 < \infty. \sum_{n=1}^{\infty} b_n$ converges $\Rightarrow \sum_{n=1}^{\infty} a_n$

converges; Limit Comparison Test

15. $\lim_{n \to \infty} \frac{a_{n+1}}{a_n} = \lim_{n \to \infty} \frac{(n+1)^2 n!}{(n+1)! n^2} = \lim_{n \to \infty} \frac{n^2 + 2n + 1}{(n+1)n^2}$

$= \lim_{n \to \infty} \frac{n^2 + 2n + 1}{n^3 + n^2} = \lim_{n \to \infty} \frac{\frac{1}{n} + \frac{2}{n^2} + \frac{1}{n^3}}{1 + \frac{1}{n}} = 0 < 1$

The series converges; Ratio Test

17. $a_n = \frac{4n^3 + 3n}{n^5 - 4n^2 + 1}; b_n = \frac{1}{n^2}$

$\lim_{n \to \infty} \frac{a_n}{b_n} = \lim_{n \to \infty} \frac{4n^5 + 3n^3}{n^5 - 4n^2 + 1} = \lim_{n \to \infty} \frac{4 + \frac{3}{n^2}}{1 - \frac{4}{n^3} + \frac{1}{n^5}} = 4;$

$0 < 4 < \infty$

$\sum_{n=1}^{\infty} b_n$ converges $\Rightarrow \sum_{n=1}^{\infty} a_n$ converges; Limit

Comparison Test

19. $a_n = \frac{1}{n(n+1)} = \frac{1}{n^2 + n}; b_n = \frac{1}{n^2}$

$\lim_{n \to \infty} \frac{a_n}{b_n} = \lim_{n \to \infty} \frac{n^2}{n^2 + n} = \lim_{n \to \infty} \frac{1}{1 + \frac{1}{n}} = 1;$

$0 < 1 < \infty$

$\sum_{n=1}^{\infty} b_n$ converges $\Rightarrow \sum_{n=1}^{\infty} a_n$ converges;

Limit Comparison Test

21. $a_n = \frac{n+1}{n(n+2)(n+3)} = \frac{n+1}{n^3 + 5n^2 + 6n}; b_n = \frac{1}{n^2}$

$\lim_{n \to \infty} \frac{a_n}{b_n} = \lim_{n \to \infty} \frac{n^3 + n^2}{n^3 + 5n^2 + 6n} = \lim_{n \to \infty} \frac{1 + \frac{1}{n}}{1 + \frac{5}{n} + \frac{6}{n^2}} = 1;$

$0 < 1 < \infty$

$\sum_{n=1}^{\infty} b_n$ converges $\Rightarrow \sum_{n=1}^{\infty} a_n$ converges;

Limit Comparison Test

23. $a_n = \frac{n}{3^n}; \lim_{n \to \infty} \frac{a_{n+1}}{a_n} = \lim_{n \to \infty} \frac{(n+1)3^n}{3^{n+1} n}$

$= \lim_{n \to \infty} \frac{n+1}{3n} = \lim_{n \to \infty} \frac{1 + \frac{1}{n}}{3} = \frac{1}{3} < 1$

The series converges; Ratio Test

25. $a_n = \frac{1}{n\sqrt{n}} = \frac{1}{n^{3/2}}; \frac{1}{x^{3/2}}$ is continuous, positive,

and nonincreasing on $[1, \infty)$.

$\int_1^{\infty} \frac{1}{x^{3/2}} dx = \left[-\frac{2}{\sqrt{x}} \right]_1^{\infty} = 0 + 2 = 2 < \infty$

The series converges; Integral Test

27. $0 \leq \sin^2 n \leq 1$ for all n, so

$2 \leq 2 + \sin^2 n \leq 3 \Rightarrow \frac{1}{2} \geq \frac{1}{2 + \sin^2 n} \geq \frac{1}{3}$ for all n.

Thus, $\lim_{n \to \infty} \frac{1}{2 + \sin^2 n} \neq 0$ and the series diverges;

nth-Term Test

29. $-1 \leq \cos n \leq 1$ for all n, so

$3 \leq 4 + \cos n \leq 5 \Rightarrow \frac{3}{n^3} \leq \frac{4 + \cos n}{n^3} \leq \frac{5}{n^3}$ for all n.

$\sum_{n=1}^{\infty} \frac{5}{n^3}$ converges $\Rightarrow \sum_{n=1}^{\infty} \frac{4 + \cos n}{n^3}$ converges;

Comparison Test

31. $\lim_{n \to \infty} \frac{a_{n+1}}{a_n} = \lim_{n \to \infty} \frac{(n+1)^{n+1}(2n)!}{(2n+2)! n^n}$

$= \lim_{n \to \infty} \frac{(n+1)^{n+1}}{(2n+2)(2n+1)n^n} = \lim_{n \to \infty} \frac{(n+1)^{n+1}}{2(n+1)(2n+1)n^n}$

$= \lim_{n \to \infty} \frac{(n+1)^n}{2(2n+1)n^n} = \lim_{n \to \infty} \left[\frac{1}{4n+2} \left(\frac{n+1}{n} \right)^n \right]$

$= \left[\lim_{n \to \infty} \frac{1}{4n+2} \right] \left[\lim_{n \to \infty} \left(\frac{n+1}{n} \right)^n \right] = 0 \cdot e = 0 < 1$

(The limits can be separated since both limits
exist.) The series converges; Ratio Test

33. $\lim_{n \to \infty} \frac{a_{n+1}}{a_n} = \lim_{n \to \infty} \frac{(4^{n+1} + n + 1)n!}{(n+1)!(4^n + n)}$

$= \lim_{n \to \infty} \frac{4^{n+1} + n + 1}{(n+1)(4^n + n)} = \lim_{n \to \infty} \frac{4 + \frac{n}{4^n} + \frac{1}{4^n}}{(n+1)\left(1 + \frac{n}{4^n}\right)}$

$$= \lim_{n \to \infty} \frac{4 + \frac{n}{4^n} + \frac{1}{4^n}}{1 + n + \frac{n}{4^n} + \frac{n^2}{4^n}} = 0$$

since $\lim_{n \to \infty} \frac{n^2}{4^n} = 0$, $\lim_{n \to \infty} \frac{n}{4^n} = 0$, and

$\lim_{n \to \infty} \frac{1}{4^n} = 0$. The series converges; Ratio Test

35. Since $\sum a_n$ converges, $\lim_{n \to \infty} a_n = 0$. Thus, there

is some positive integer N such that $0 < a_n < 1$

for all $n \geq N$. $a_n < 1 \Rightarrow a_n^2 < a_n$, thus

$$\sum_{n=N}^{\infty} a_n^2 < \sum_{n=N}^{\infty} a_n. \text{ Hence } \sum_{n=N}^{\infty} a_n^2 \text{ converges,}$$

and $\sum a_n^2$ also converges, since adding a finite number of terms does not affect the convergence or divergence of a series.

37. If $\lim_{n \to \infty} \frac{a_n}{b_n} = 0$ then there is some positive integer

N such that $0 \leq \frac{a_n}{b_n} < 1$ for all $n \geq N$. Thus, for

$n \geq N$, $a_n < b_n$. By the Comparison Test, since

$$\sum_{n=N}^{\infty} b_n \text{ converges, } \sum_{n=N}^{\infty} a_n \text{ also converges. Thus,}$$

$\sum a_n$ converges since adding a finite number of terms will not affect the convergence or divergence of a series.

39. If $\lim_{n \to \infty} n a_n = 1$ then there is some positive

integer N such that $a_n \geq 0$ for all $n \geq N$, Let

$b_n = \frac{1}{n}$, so $\lim_{n \to \infty} \frac{a_n}{b_n} = \lim_{n \to \infty} n a_n = 1 < \infty.$

Since $\sum_{n=N}^{\infty} \frac{1}{n}$ diverges, $\sum_{n=N}^{\infty} a_n$ diverges by the

Limit Comparison Test.

Thus $\sum a_n$ diverges since adding a finite number of terms will not affect the convergence or divergence of a series.

41. Suppose that $\lim_{n \to \infty} (a_n)^{1/n} = R$ where $a_n > 0$.

If $R < 1$, there is some number r with $R < r < 1$ and some positive integer N such that

$\left| (a_n)^{1/n} - R \right| < r - R$ for all $n \geq N$. Thus,

$R - r < (a_n)^{1/n} - R < r - R$ or

$-r < (a_n)^{1/n} < r < 1$. Since $a_n > 0$,

$0 < (a_n)^{1/n} < r$ and $0 < a_n < r^n$ for all $n \geq N$.

Thus, $\sum_{n=N}^{\infty} a_n < \sum_{n=N}^{\infty} r^n$, which converges since

$|r| < 1$. Thus, $\sum_{n=N}^{\infty} a_n$ converges so $\sum a_n$ also

converges.

If $R > 1$, there is some number r with $1 < r < R$ and some positive integer N such that

$\left| (a_n)^{1/n} - R \right| < R - r$ for all $n \geq N$. Thus,

$r - R < (a_n)^{1/n} - R < R - r$ or

$r < (a_n)^{1/n} < 2R - r$ for all $n \geq N$. Hence

$r^n < a_n$ for all $n \geq N$, so $\sum_{n=N}^{\infty} r^n < \sum_{n=N}^{\infty} a_n$, and

since $\sum_{n=N}^{\infty} r^n$ diverges $(r > 1)$, $\sum_{n=N}^{\infty} a_n$ also

diverges, so $\sum a_n$ diverges.

43. a. $\ln\left(1 + \frac{1}{n}\right) = \ln\left(\frac{n+1}{n}\right) = \ln(n+1) - \ln n$

$S_n = (\ln 2 - \ln 1) + (\ln 3 - \ln 2) + \dots$
$\quad + (\ln n - \ln(n-1)) + (\ln(n+1) - \ln n)$
$\quad = -\ln 1 + \ln(n+1) = \ln(n+1)$

$\lim_{n \to \infty} S_n = \lim_{n \to \infty} \ln(n+1) = \infty$

Since the partial sums are unbounded, the series diverges.

b. $\ln \frac{(n+1)^2}{n(n+2)} = 2 \ln(n+1) - \ln n - \ln(n+2)$

$S_n = (2 \ln 2 - \ln 1 - \ln 3) + (2 \ln 3 - \ln 2 - \ln 4)$
$\quad + (2 \ln 4 - \ln 3 - \ln 5) + \dots$
$\quad + (2 \ln n - \ln(n-1) - \ln(n+1))$
$\quad + (2 \ln(n+1) - \ln n - \ln(n+2))$
$\quad = \ln 2 - \ln 1 + \ln(n+1) - \ln(n+2)$
$\quad = \ln 2 + \ln \frac{n+1}{n+2}$

$\lim_{n \to \infty} S_n = \ln 2 + \lim_{n \to \infty} \ln \frac{n+1}{n+2} = \ln 2$

Since the partial sums converge, the series converges.

c. $\left(\frac{1}{\ln x}\right)^{\ln x}$ is continuous, positive, and

nonincreasing on $[2, \infty)$, thus $\sum_{n=2}^{\infty} \left(\frac{1}{\ln n}\right)^{\ln n}$

converges if and only if $\displaystyle\int_2^\infty \left(\frac{1}{\ln x}\right)^{\ln x} dx$ converges.

Let $u = \ln x$, so $x = e^u$ and $dx = e^u\, du$.

$$\int_2^\infty \left(\frac{1}{\ln x}\right)^{\ln x} dx = \int_{\ln 2}^\infty \left(\frac{1}{u}\right)^u e^u\, du = \int_{\ln 2}^\infty \left(\frac{e}{u}\right)^u du$$

This integral converges if and only if the associated series, $\displaystyle\sum_{n=1}^\infty \left(\frac{e}{n}\right)^n$ converges. With $a_n = \left(\dfrac{e}{n}\right)^n$, the Root Test (Problem 41) gives

$$\lim_{n\to\infty} (a_n)^{1/n} = \lim_{n\to\infty}\left[\left(\frac{e}{n}\right)^n\right]^{1/n}$$

$$= \lim_{n\to\infty}\frac{e}{n} = 0 < 1$$

Thus, $\displaystyle\sum_{n=1}^\infty \left(\frac{e}{n}\right)^n$ converges, so $\displaystyle\int_{\ln 2}^\infty \left(\frac{e}{u}\right)^u du$ converges, whereby $\displaystyle\sum_{n=2}^\infty \frac{1}{(\ln n)^{\ln n}}$ converges.

d. $\left(\dfrac{1}{\ln(\ln x)}\right)^{\ln x}$ is continuous, positive, and nonincreasing on $[3, \infty)$, thus

$\displaystyle\sum_{n=3}^\infty \left(\frac{1}{\ln(\ln n)}\right)^{\ln n}$ converges if and only if

$\displaystyle\int_3^\infty \left(\frac{1}{\ln(\ln x)}\right)^{\ln x} dx$ converges.

Let $u = \ln x$, so $x = e^u$ and $dx = e^u\, du$.

$$\int_3^\infty \left(\frac{1}{\ln(\ln x)}\right)^{\ln x} dx$$

$$= \int_{\ln 3}^\infty \left(\frac{1}{\ln u}\right)^u e^u\, du = \int_{\ln 3}^\infty \left(\frac{e}{\ln u}\right)^u du.$$

This integral converges if and only if the associated series, $\displaystyle\sum_{n=2}^\infty \left(\frac{e}{\ln n}\right)^n$ converges.

With $a_n = \left(\dfrac{e}{\ln n}\right)^n$, the Root Test (Problem 41) gives

$$\lim_{n\to\infty} (a_n)^{1/n} = \lim_{n\to\infty}\left[\left(\frac{e}{\ln n}\right)^n\right]^{1/n}$$

$$= \lim_{n\to\infty}\frac{e}{\ln n} = 0 < 1$$

Thus, $\displaystyle\sum_{n=2}^\infty \left(\frac{e}{\ln n}\right)^n$ converges, so $\displaystyle\int_{\ln 3}^\infty \left(\frac{e}{\ln u}\right)^u du$ converges, whereby $\displaystyle\sum_{n=3}^\infty \frac{1}{(\ln(\ln n))^{\ln n}}$ converges.

e. $a_n = 1/n$; $b_n = 1/(\ln n)^4$

$$\lim_{n\to\infty}\frac{a_n}{b_n} = \lim_{n\to\infty}\frac{1/n}{1/(\ln n)^4} = \lim_{n\to\infty}\frac{(\ln n)^4}{n}$$

$$= \lim_{n\to\infty}\frac{4(\ln n)^3(1/n)}{1} = \lim_{n\to\infty}\frac{4(\ln n)^3}{n}$$

$$= \lim_{n\to\infty}\frac{12(\ln n)^2(1/n)}{1} = \lim_{n\to\infty}\frac{12(\ln n)^2}{n}$$

$$= \lim_{n\to\infty}\frac{24(\ln n)}{n} = \lim_{n\to\infty}\frac{24(1/n)}{1}$$

$$= \lim_{n\to\infty}\frac{24}{n} = 0$$

$\displaystyle\sum_{n=2}^\infty \frac{1}{n}$ diverges $\Rightarrow \displaystyle\sum_{n=2}^\infty \frac{1}{(\ln n)^4}$ diverges

f. $\left(\dfrac{\ln x}{x}\right)^2$ is continuous, positive, and nonincreasing on $[3, \infty)$. Using integration by parts twice,

$$\int_3^\infty \left(\frac{\ln x}{x}\right)^2 dx = \left[-\frac{(\ln x)^2}{x}\right]_3^\infty + \int_3^\infty \frac{2\ln x}{x^2} dx$$

$$= \left[-\frac{(\ln x)^2}{x}\right]_3^\infty + \left[-\frac{2\ln x}{x}\right]_3^\infty + \int_3^\infty \frac{2}{x^2} dx$$

$$= \left[-\frac{(\ln x)^2}{x} - \frac{2\ln x}{x} - \frac{2}{x}\right]_3^\infty \approx 1.8 < \infty$$

Thus, $\displaystyle\sum_{n=3}^\infty \left(\frac{\ln x}{x}\right)^2$ converges.

45. Let $a_n = \dfrac{1}{n^p}\left(1 + \dfrac{1}{2^p} + \dfrac{1}{3^p} + \ldots + \dfrac{1}{n^p}\right)$ and $b_n = \dfrac{1}{n^p}$. Then

$$\lim_{n\to\infty}\frac{a_n}{b_n}=\lim_{n\to\infty}\left(1+\frac{1}{2^P}+\frac{1}{3^P}+...+\frac{1}{n^P}\right)=\sum_{n=1}^{\infty}\frac{1}{n^P}$$

which converges if $p>1$. Thus, by the Limit

Comparison Test, if $\sum_{n=1}^{\infty}b_n$ converges for $p>1$,

so does $\sum_{n=1}^{\infty}a_n$. Since $\sum_{n=1}^{\infty}b_n=\sum_{n=1}^{\infty}\frac{1}{n^P}$ converges

for $p>1$, $\sum_{n=1}^{\infty}\frac{1}{n^P}\left(1+\frac{1}{2^P}+\frac{1}{3^P}+\cdots+\frac{1}{n^P}\right)$ also

converges. For $p\le 1$, since $1+\frac{1}{2^P}+...+\frac{1}{n^P}>1,$

$\frac{1}{n^P}\left(1+\frac{1}{2^P}+...+\frac{1}{n^P}\right)>\frac{1}{n^P}$. Hence, since

$\sum_{n=1}^{\infty}\frac{1}{n^P}$ diverges for $p\le 1,$

$\sum_{n=1}^{\infty}\frac{1}{n^P}\left(1+\frac{1}{2^P}+...+\frac{1}{n^P}\right)$ also diverges. The

series converges for $p>1$ and diverges for $p\le 1$.

10.5 Concepts Review

1. $\displaystyle\lim_{n\to\infty}a_n=0$

3. the alternating harmonic series

Problem Set 10.5

1. $a_n=\dfrac{2}{3n+1};\dfrac{2}{3n+1}>\dfrac{2}{3n+4}$, so $a_n>a_{n+1}$;

   $\displaystyle\lim_{n\to\infty}\frac{2}{3n+1}=0.$ $S_9\approx 0.363.$ The error made by
   using S_9 is not more than $a_{10}\approx 0.065$.

3. $a_n=\dfrac{1}{\ln(n+1)};\dfrac{1}{\ln(n+1)}>\dfrac{1}{\ln(n+2)}$, so

   $a_n>a_{n+1}$;

   $\displaystyle\lim_{n\to\infty}\frac{1}{\ln(n+1)}=0.$ $S_9\approx 1.137.$ The error made by
   using S_9 is not more than $a_{10}\approx 0.417.$

5. $a_n=\dfrac{\ln n}{n};\dfrac{\ln n}{n}>\dfrac{\ln(n+1)}{n+1}$ is equivalent to

   $\ln\dfrac{n^{n+1}}{(n+1)^n}>0$ or $\dfrac{n^{n+1}}{(n+1)^n}>1$ which is true for

   $n>2$. $S_9\approx -0.041$. The error made by using S_9
   is not more than $a_{10}\approx 0.230.$

7. $\dfrac{|u_{n+1}|}{|u_n|}=\dfrac{\left|\left(-\frac{3}{4}\right)^{n+1}\right|}{\left|\left(-\frac{3}{4}\right)^n\right|}=\dfrac{3}{4}<1$, so the series

   converges absolutely.

9. $\dfrac{|u_{n+1}|}{|u_n|}=\dfrac{\frac{n+1}{2^{n+1}}}{\frac{n}{2^n}}=\dfrac{n+1}{2n};\displaystyle\lim_{n\to\infty}\frac{n+1}{2n}=\frac{1}{2}<1$, so the

   series converges absolutely.

11. $n(n+1)=n^2+n>n^2$ for all $n>0$, thus

    $\dfrac{1}{n(n+1)}<\dfrac{1}{n^2}$, so $\displaystyle\sum_{n=1}^{\infty}|u_n|=\sum_{n=1}^{\infty}\frac{1}{n(n+1)}<\sum_{n=1}^{\infty}\frac{1}{n^2}$

    which converges since $2>1$, thus

    $\displaystyle\sum_{n=1}^{\infty}(-1)^{n+1}\frac{1}{n(n+1)}$ converges absolutely.

13. $\displaystyle\sum_{n=1}^{\infty}(-1)^{n+1}\frac{1}{5n}=\frac{1}{5}\sum_{n=1}^{\infty}\frac{(-1)^{n+1}}{n}$ which converges

    since $\displaystyle\sum_{n=1}^{\infty}\frac{(-1)^{n+1}}{n}$ converges. The series is

    conditionally convergent since $\dfrac{1}{5}\displaystyle\sum_{n=1}^{\infty}\frac{1}{n}$ diverges.

15. $\displaystyle\lim_{n\to\infty}\frac{n}{10n+1}=\frac{1}{10}\ne 0.$ Thus the sequence of
    partial sums does not converge; the series
    diverges.

17. $\displaystyle\lim_{n\to\infty}\frac{1}{n\ln n}=0;\dfrac{1}{n\ln n}>\dfrac{1}{(n+1)\ln(n+1)}$ is

    equivalent to $(n+1)^{n+1}>n^n$ which is true for all
    $n>0$ so $a_n>a_{n+1}$. The alternating series
    converges.

    $\displaystyle\sum_{n=2}^{\infty}|u_n|=\sum_{n=2}^{\infty}\frac{1}{n\ln n};\dfrac{1}{x\ln x}$ is continuous,
    positive, and nonincreasing on $[2,\infty)$.

    Using $u=\ln x,\ du=\dfrac{1}{x}dx,$

$\int_2^\infty \frac{1}{x \ln x} dx = \int_{\ln 2}^\infty \frac{1}{u} du = \left[\ln |u| \right]_{\ln 2}^\infty = \infty.$ Thus,

$\sum_{n=2}^\infty \frac{1}{n \ln n}$ diverges and $\sum_{n=2}^\infty (-1)^n \frac{1}{n \ln n}$ is conditionally convergent.

19. $\frac{|u_{n+1}|}{|u_n|} = \frac{\frac{(n+1)^4}{2^{n+1}}}{\frac{n^4}{2^n}} = \frac{n^4}{2(n+1)^4};$

$\lim_{n \to \infty} \frac{n^4}{2(n+1)^4} = \frac{1}{2} < 1.$

The series is absolutely convergent.

21. $a_n = \frac{n}{n^2+1}; \frac{n}{n^2+1} > \frac{n+1}{(n+1)^2+1}$ is equivalent to

$n^2 + n - 1 > 0,$ which is true for $n > 1$, so

$a_n > a_{n+1}; \lim_{n \to \infty} \frac{n}{n^2+1} = 0,$ hence the alternating

series converges. Let $b_n = \frac{1}{n}$, then

$\lim_{n \to \infty} \frac{a_n}{b_n} = \lim_{n \to \infty} \frac{n^2}{n^2+1} = 1; 0 < 1 < \infty.$ Thus, since

$\sum_{n=1}^\infty b_n = \sum_{n=1}^\infty \frac{1}{n}$ diverges, $\sum_{n=1}^\infty a_n = \sum_{n=1}^\infty \frac{n}{n^2+1}$ also

diverges. The series is conditionally convergent.

23. $\cos n\pi = (-1)^n = \frac{1}{(-1)} (-1)^{n+1}$ so the series is

$-1 \sum_{n=1}^\infty \frac{(-1)^{n+1}}{n},$ -1 times the alternating

harmonic series. The series is conditionally convergent.

25. $|\sin n| \le 1$ for all n, so

$\sum_{n=1}^\infty |u_n| = \sum_{n=1}^\infty \frac{|\sin n|}{n\sqrt{n}} \le \sum_{n=1}^\infty \frac{1}{n^{3/2}}$ which converges

since $\frac{3}{2} > 1$. Thus the series is absolutely convergent.

27. $a_n = \frac{1}{\sqrt{n(n+1)}}; \frac{1}{\sqrt{n(n+1)}} > \frac{1}{\sqrt{(n+1)(n+2)}}$ and

$\lim_{n \to \infty} \frac{1}{\sqrt{n(n+1)}} = 0$ so the alternating series converges.

Let $b_n = \frac{1}{n}$, then

$\lim_{n \to \infty} \frac{a_n}{b_n} = \lim_{n \to \infty} \frac{n}{\sqrt{n^2+n}} = \lim_{n \to \infty} \frac{1}{\sqrt{1+\frac{1}{n}}} = 1;$

$0 < 1 < \infty.$

Thus, since $\sum_{n=1}^\infty b_n = \sum_{n=1}^\infty \frac{1}{n}$ diverges,

$\sum_{n=1}^\infty a_n = \sum_{n=1}^\infty \frac{1}{\sqrt{n(n+1)}}$ also diverges.

The series is conditionally convergent.

29. $\sum_{n=1}^\infty \frac{(-3)^{n+1}}{n^2} = \sum_{n=1}^\infty (-1)^{n+1} \frac{3^{n+1}}{n^2}; \lim_{n \to \infty} \frac{3^{n+1}}{n^2} \ne 0,$ so

the series diverges.

31. Suppose $\sum |a_n|$ converges. Thus, $\sum 2|a_n|$

converges, so $\sum (|a_n| + a_n)$ converges since

$0 \le |a_n| + a_n \le 2|a_n|.$ By the linearity of

convergent series, $\sum a_n = \sum (|a_n| + a_n) - \sum |a_n|$

converges, which is a contradiction.

33. The positive-term series is

$1 + \frac{1}{3} + \frac{1}{5} + \frac{1}{7} + ... = \sum_{n=1}^\infty \frac{1}{2n-1}.$

$\sum_{n=1}^\infty \frac{1}{2n-1} > \frac{1}{2} \sum_{n=1}^\infty \frac{1}{n}$ which diverges since the

harmonic series diverges.

Thus, $\sum_{n=1}^\infty \frac{1}{2n-1}$ diverges.

The negative-term series is

$-\frac{1}{2} - \frac{1}{4} - \frac{1}{6} - \frac{1}{8} - ... = -\frac{1}{2} \sum_{n=1}^\infty \frac{1}{n}$ which diverges,

since the harmonic series diverges.

35. a. $1 + \frac{1}{3} \approx 1.33$

b. $1 + \frac{1}{3} - \frac{1}{2} \approx 0.833$

c. $1 + \frac{1}{3} - \frac{1}{2} + \frac{1}{5} + \frac{1}{7} + \frac{1}{9} + \frac{1}{11} \approx 1.38$

$1 + \frac{1}{3} - \frac{1}{2} + \frac{1}{5} + \frac{1}{7} + \frac{1}{9} + \frac{1}{11} - \frac{1}{4} \approx 1.13$

37. Written response

39. Consider $1-1+\dfrac{1}{2}-\dfrac{1}{4}+\dfrac{1}{3}-\dfrac{1}{9}+\ldots$

It is clear that $\lim\limits_{n\to\infty} a_n = 0$. Pairing successive

terms, we obtain $\dfrac{1}{n}-\dfrac{1}{n^2}=\dfrac{n-1}{n^2}>0$ for $n>1$.

Let $a_n=\dfrac{n-1}{n^2}$ and $b_n=\dfrac{1}{n}$. Then

$\lim\limits_{n\to\infty}\dfrac{a_n}{b_n}=\lim\limits_{n\to\infty}\dfrac{n^2-n}{n^2}=1;\ 0<1<\infty.$

Thus, since $\displaystyle\sum_{n=1}^{\infty} b_n=\sum_{n=1}^{\infty}\dfrac{1}{n}$ diverges,

$\displaystyle\sum_{n=1}^{\infty} a_n=\sum_{n=1}^{\infty}\left(\dfrac{1}{n}-\dfrac{1}{n^2}\right)$ also diverges.

41. Note that $(a_k+b_k)^2\geq 0$ and $(a_k-b_k)^2\geq 0$ for

all k. Thus, $a_k^2\pm 2a_kb_k+b_k^2\geq 0$, or

$a_k^2+b_k^2\geq\pm 2a_kb_k$ for all k, and

$a_k^2+b_k^2\geq 2|a_kb_k|$. Since $\displaystyle\sum_{k=1}^{\infty} a_k^2$ and $\displaystyle\sum_{k=1}^{\infty} b_k^2$ both

converge, $\displaystyle\sum_{k=1}^{\infty}(a_k^2+b_k^2)$ also converges, and by

the Comparison Test, $\displaystyle\sum_{k=1}^{\infty} 2|a_kb_k|$ converges.

Hence, $\displaystyle\sum_{k=1}^{\infty}|a_kb_k|=\dfrac{1}{2}\sum_{k=1}^{\infty} 2|a_kb_k|$ converges, i.e.,

$\displaystyle\sum_{k=1}^{\infty} a_kb_k$ converges absolutely.

43. Consider the graph of $\dfrac{|\sin x|}{x}$ on the interval $[k\pi,(k+1)\,\pi]$.

Note that for $k\pi+\dfrac{\pi}{6}\leq x\leq k\pi+\dfrac{5\pi}{6}$, $\dfrac{1}{2}\leq|\sin x|$ while $\dfrac{1}{\left(k+\frac{5}{6}\right)\pi}\leq\dfrac{1}{x}$. Thus on $\left[\left(k+\dfrac{1}{6}\right)\pi,\left(k+\dfrac{5}{6}\right)\pi\right]$

$\dfrac{1}{2\left(k+\frac{5}{6}\right)\pi}=\dfrac{1}{\left(2k+\frac{5}{3}\right)\pi}\leq\dfrac{|\sin x|}{x}$, so $\displaystyle\int_{k\pi}^{(k+1)\pi}\dfrac{|\sin x|}{x}dx\geq\int_{(k+1/6)\pi}^{(k+5/6)\pi}\dfrac{|\sin x|}{x}dx\geq\dfrac{1}{\left(2k+\frac{5}{3}\right)\pi}\int_{(k+1/6)\pi}^{(k+5/6)\pi}dx=\dfrac{1}{3k+\frac{5}{2}}.$

Hence, $\displaystyle\int_{\pi}^{\infty}\dfrac{|\sin x|}{x}dx\geq\sum_{k=1}^{\infty}\dfrac{1}{3k+\frac{5}{2}}$. Let $a_k=\dfrac{1}{3k+\frac{5}{2}}$ and $b_k=\dfrac{1}{k}$.

$\lim\limits_{k\to\infty}\dfrac{a_k}{b_k}=\lim\limits_{k\to\infty}\dfrac{k}{3k+\frac{5}{2}}=\lim\limits_{k\to\infty}\dfrac{1}{3+\frac{5}{2k}}=\dfrac{1}{3};\ 0<\dfrac{1}{3}<\infty.$ Thus, since $\displaystyle\sum_{k=1}^{\infty} b_k=\sum_{k=1}^{\infty}\dfrac{1}{k}$ diverges, $\displaystyle\sum_{k=1}^{\infty} a_k=\sum_{k=1}^{\infty}\dfrac{1}{3k+\frac{5}{2}}$ also

diverges. Hence, $\displaystyle\int_{\pi}^{\infty}\dfrac{|\sin x|}{x}dx$ also diverges and adding $\displaystyle\int_{0}^{\pi}\dfrac{|\sin x|}{x}dx$ will not affect its divergence.

45. $\dfrac{1}{n+1}+\dfrac{1}{n+2}+\cdots+\dfrac{1}{2n}=\left[\dfrac{1}{1+\frac{1}{n}}+\dfrac{1}{1+\frac{2}{n}}+\cdots+\dfrac{1}{1+\frac{n}{n}}\right]\left(\dfrac{1}{n}\right)$

This is a Riemann sum for the function $f(x)=\dfrac{1}{x}$ from $x=1$ to 2 where $\Delta x=\dfrac{1}{n}$.

$\lim\limits_{n\to\infty}\sum_{k=1}^{n}\left[\dfrac{1}{1+\frac{k}{n}}\left(\dfrac{1}{n}\right)\right]=\int_{1}^{2}\dfrac{1}{x}dx=\ln 2$

10.6 Concepts Review

1. power series   **3.** interval; endpoints

Problem Set 10.6

1. $\displaystyle\sum_{n=1}^{\infty} \frac{(-1)^{n+1} x^n}{n(n+1)}; \rho = \lim_{n\to\infty} \left| \frac{x^{n+1}}{(n+1)(n+2)} \div \frac{x^n}{n(n+1)} \right| = \lim_{n\to\infty} |x| \left| \frac{n}{n+2} \right| = |x|$

When $x = 1$, the series is $\displaystyle\sum_{n=1}^{\infty} (-1)^{n+1} \frac{1}{n(n+1)}$ which converges absolutely by comparison with the series $\displaystyle\sum_{n=1}^{\infty} \frac{1}{n^2}$.

When $x = -1$, the series is $\displaystyle\sum_{n=1}^{\infty} (-1)^{n+1} \frac{(-1)^n}{n(n+1)} = \sum_{n=1}^{\infty} (-1)^{2n+1} \frac{1}{n(n+1)}$

$\displaystyle = \sum_{n=1}^{\infty} (-1) \frac{1}{n(n+1)} = (-1) \sum_{n=1}^{\infty} \frac{1}{n(n+1)}$ which converges since $\displaystyle\sum_{n=1}^{\infty} \frac{1}{n(n+1)}$ converges.

The series converges on $-1 \le x \le 1$.

3. $\displaystyle\sum_{n=1}^{\infty} \frac{(-1)^{n+1} x^{2n-1}}{(2n-1)!}; \rho = \lim_{n\to\infty} \left| \frac{x^{2n+1}}{(2n+1)!} \div \frac{x^{2n-1}}{(2n-1)!} \right| = \lim_{n\to\infty} |x^2| \left| \frac{1}{2n(2n+1)} \right| = 0$

The series converges for all x.

5. $\displaystyle\sum_{n=1}^{\infty} nx^n; \rho = \lim_{n\to\infty} \left| \frac{(n+1)x^{n+1}}{nx^n} \right|$

$\displaystyle = \lim_{n\to\infty} |x| \left| \frac{n+1}{n} \right| = |x|$

When $x = 1$, the series is $\displaystyle\sum_{n=1}^{\infty} n$ which clearly diverges.

When $x = -1$, the series is $\displaystyle\sum_{n=1}^{\infty} n(-1)^n; a_n = n$;

$\displaystyle\lim_{n\to\infty} a_n \ne 0$, thus the series diverges.

The series converges on $-1 < x < 1$.

7. $\displaystyle 1 + \sum_{n=1}^{\infty} \frac{(-1)^n x^n}{n}; \rho = \lim_{n\to\infty} \left| \frac{x^{n+1}}{n+1} \div \frac{x^n}{n} \right|$

$\displaystyle = \lim_{n\to\infty} |x| \left| \frac{n}{n+1} \right| = |x|$

When $x = 1$, the series is $\displaystyle 1 + \sum_{n=1}^{\infty} (-1)^n \frac{1}{n}$, which is

1 added to the alternating harmonic series multiplied by -1, which converges.
When $x = -1$, the series is

$\displaystyle 1 + \sum_{n=1}^{\infty} (-1)^n \frac{(-1)^n}{n} = 1 + \sum_{n=1}^{\infty} \frac{1}{n}$, which diverges.

The series converges on $-1 < x \le 1$.

9. $\displaystyle 1 + \sum_{n=1}^{\infty} \frac{(-1)^n x^n}{n(n+2)}$;

$\displaystyle \rho = \lim_{n\to\infty} \left| \frac{x^{n+1}}{(n+1)(n+3)} \div \frac{x^n}{n(n+2)} \right|$

$\displaystyle = \lim_{n\to\infty} |x| \left| \frac{n^2 + 2n}{n^2 + 4n + 3} \right| = |x|$

When $x = 1$ the series is $\displaystyle 1 + \sum_{n=1}^{\infty} (-1)^n \frac{1}{n(n+2)}$

which converges absolutely by comparison with

the series $\displaystyle\sum_{n=1}^{\infty} \frac{1}{n^2}$.

When $x = -1$, the series is

$\displaystyle 1 + \sum_{n=1}^{\infty} (-1)^n \frac{(-1)^n}{n(n+2)} = 1 + \sum_{n=1}^{\infty} \frac{1}{n(n+2)}$ which

converges by comparison with $\displaystyle\sum_{n=1}^{\infty} \frac{1}{n^2}$.

The series converges on $-1 \le x \le 1$.

11. $\displaystyle\sum_{n=0}^{\infty} \frac{(-1)^n x^n}{2^n}; \rho = \lim_{n\to\infty} \left| \frac{x^{n+1}}{2^{n+1}} \div \frac{x^n}{2^n} \right| = \lim_{n\to\infty} \left| \frac{x}{2} \right|$

$\displaystyle = \left| \frac{x}{2} \right|; \left| \frac{x}{2} \right| < 1$ when $-2 < x < 2$.

When $x = 2$, the series is $\displaystyle\sum_{n=0}^{\infty} (-1)^n \frac{2^n}{2^n} = \sum_{n=0}^{\infty} (-1)^n$ which diverges.

When $x = -2$, the series is

$\displaystyle\sum_{n=0}^{\infty} (-1)^n \frac{(-2)^n}{2^n} = \sum_{n=0}^{\infty} (-1)^n (-1)^n = \sum_{n=0}^{\infty} 1$ which

diverges. The series converges on $-2 < x < 2$.

13. $\displaystyle\sum_{n=0}^{\infty} \frac{2^n x^n}{n!}; \rho = \lim_{n \to \infty} \left| \frac{2^{n+1} x^{n+1}}{(n+1)!} \div \frac{2^n x^n}{n!} \right|$

$= \displaystyle\lim_{n \to \infty} |2x| \left| \frac{1}{n+1} \right| = 0$.

The series converges for all x.

15. $\displaystyle\sum_{n=1}^{\infty} \frac{(x-1)^n}{n}; \rho = \lim_{n \to \infty} \left| \frac{(x-1)^{n+1}}{n+1} \div \frac{(x-1)^n}{n} \right|$

$= \displaystyle\lim_{n \to \infty} |x-1| \left| \frac{n}{n+1} \right| = |x-1|; |x-1| < 1$ when
$0 < x < 2$.

When $x = 0$, the series is $\displaystyle\sum_{n=1}^{\infty} \frac{(-1)^n}{n}$ which

converges.

When $x = 2$, the series is $\displaystyle\sum_{n=1}^{\infty} \frac{1}{n}$ which diverges.

The series converges on $0 \leq x < 2$.

17. $\displaystyle\sum_{n=0}^{\infty} \frac{(x+1)^n}{2^n}; \rho = \lim_{n \to \infty} \left| \frac{(x+1)^{n+1}}{2^{n+1}} \div \frac{(x+1)^n}{2^n} \right|$

$= \displaystyle\lim_{n \to \infty} \left| \frac{x+1}{2} \right| = \left| \frac{x+1}{2} \right|; \left| \frac{x+1}{2} \right| < 1$ when
$-3 < x < 1$.

When $x = -3$, the series is $\displaystyle\sum_{n=0}^{\infty} \frac{(-2)^n}{2^n} = \sum_{n=0}^{\infty} (-1)^n$

which diverges.

When $x = 1$, the series is $\displaystyle\sum_{n=0}^{\infty} \frac{2^n}{2^n} = \sum_{n=0}^{\infty} 1$ which

diverges.
The series converges on $-3 < x < 1$.

19. $\displaystyle\sum_{n=1}^{\infty} \frac{(x+5)^n}{n(n+1)}; \rho = \lim_{n \to \infty} \left| \frac{(x+5)^{n+1}}{(n+1)(n+2)} \div \frac{(x+5)^n}{n(n+1)} \right|$

$= \displaystyle\lim_{n \to \infty} |x+5| \left| \frac{n}{n+2} \right| = |x+5|; |x+5| < 1$ when
$-6 < x < -4$.

When $x = -4$, the series is $\displaystyle\sum_{n=1}^{\infty} \frac{1}{n(n+1)}$ which

converges by comparison with $\displaystyle\sum_{n=1}^{\infty} \frac{1}{n^2}$.

When $x = -6$, the series is $\displaystyle\sum_{n=1}^{\infty} \frac{(-1)^n}{n(n+1)}$ which

converges absolutely since $\displaystyle\sum_{n=1}^{\infty} \frac{1}{n(n+1)}$

converges.
The series converges on $-6 \leq x \leq -4$.

21. If for some x_0, $\displaystyle\lim_{n \to \infty} \frac{x_0^n}{n!} \neq 0$, then $\displaystyle\sum \frac{x_0^n}{n!}$ could

not converge.

23. The Absolute Ratio Test gives

$\rho = \displaystyle\lim_{n \to \infty} \left| \frac{(n+1)! x^{2n+3}}{1 \cdot 3 \cdot 5 \cdots (2n+1)} \div \frac{n! x^{2n+1}}{1 \cdot 3 \cdot 5 \cdots (2n-1)} \right|$

$= \displaystyle\lim_{n \to \infty} |x^2| \left| \frac{n+1}{2n+1} \right| = \left| \frac{x^2}{2} \right|; \left| \frac{x^2}{2} \right| < 1$ when
$|x| < \sqrt{2}$.

The radius of convergence is $\sqrt{2}$.

25. This is a geometric series, so it converges for
$|x-3| < 1$, $2 < x < 4$. For these values of x, the

series converges to $\dfrac{1}{1-(x-3)} = \dfrac{1}{4-x}$.

27. a. $\rho = \displaystyle\lim_{n \to \infty} \left| \frac{(3x+1)^{n+1}}{(n+1) \cdot 2^{n+1}} \div \frac{(3x+1)^n}{n \cdot 2^n} \right| = \lim_{n \to \infty} |3x+1| \left| \frac{n}{2n+2} \right| = \frac{1}{2}|3x+1|; \frac{1}{2}|3x+1| < 1$ when $-1 < x < \frac{1}{3}$.

When $x = -1$, the series is $\displaystyle\sum_{n=1}^{\infty} \frac{(-2)^n}{n \cdot 2^n} = \sum_{n=1}^{\infty} (-1)^n \frac{1}{n}$, which converges.

When $x = \frac{1}{3}$, the series is $\displaystyle\sum_{n=1}^{\infty} \frac{2^n}{n \cdot 2^n} = \sum_{n=1}^{\infty} \frac{1}{n}$, which diverges. The series converges on $-1 \leq x < \frac{1}{3}$.

b. $\rho = \lim\limits_{n\to\infty} \left| \dfrac{(-1)^{n+1}(2x-3)^{n+1}}{4^{n+1}\sqrt{n+1}} \div \dfrac{(-1)^n(2x-3)^n}{4^n\sqrt{n}} \right| = \lim\limits_{n\to\infty}|2x-3|\left|\dfrac{\sqrt{n}}{4\sqrt{n+1}}\right| = \dfrac{1}{4}|2x-3|;$

$\dfrac{1}{4}|2x-3| < 1$ when $-\dfrac{1}{2} < x < \dfrac{7}{2}.$

When $x = -\dfrac{1}{2}$, the series is $\sum\limits_{n=1}^{\infty}(-1)^n\dfrac{(-4)^n}{4^n\sqrt{n}} = \sum\limits_{n=1}^{\infty}\dfrac{1}{\sqrt{n}}$ which diverges since $\dfrac{1}{2} < 1.$

When $x = \dfrac{7}{2}$, the series is $\sum\limits_{n=1}^{\infty}(-1)^n\dfrac{4^n}{4^n\sqrt{n}} = \sum\limits_{n=1}^{\infty}(-1)^n\dfrac{1}{\sqrt{n}};$

$a_n = \dfrac{1}{\sqrt{n}}; \dfrac{1}{\sqrt{n}} > \dfrac{1}{\sqrt{n+1}},$ so $a_n > a_{n+1};$

$\lim\limits_{n\to\infty}\dfrac{1}{\sqrt{n}} = 0,$ so $\sum\limits_{n=1}^{\infty}(-1)^n\dfrac{1}{\sqrt{n}}$ converges.

The series converges on $-\dfrac{1}{2} < x \le \dfrac{7}{2}.$

29. If $a_{n+3} = a_n$, then $a_0 = a_3 = a_6 = a_{3n}, a_1 = a_4 = a_7 = a_{3n+1}$, and $a_2 = a_5 = a_8 = a_{3n+2}$. Thus,

$\sum\limits_{n=0}^{\infty}a_n x^n = a_0 + a_1 x + a_2 x^2 + a_0 x^3 + a_1 x^4 + a_2 x^5 + \cdots = (a_0 + a_1 x + a_2 x^2)(1 + x^3 + x^6 + \cdots)$

$= (a_0 + a_1 x + a_2 x^2)\sum\limits_{n=0}^{\infty}x^{3n} = (a_0 + a_1 x + a_2 x^2)\sum\limits_{n=0}^{\infty}(x^3)^n.$

$a_0 + a_1 x + a_2 x^2$ is a polynomial, which will converge for all x.

$\sum\limits_{n=0}^{\infty}(x^3)^n$ is a geometric series which, converges for $|x^3| < 1$, or, equivalently, $|x| < 1$.

Since $\sum\limits_{n=0}^{\infty}(x^3)^n = \dfrac{1}{1-x^3}$ for $|x| < 1$, $S(x) = \dfrac{a_0 + a_1 x + a_2 x^2}{1-x^3}$ for $|x| < 1$.

10.7 Concepts Review

1. integrated; interior

3. $1 + x^2 + \dfrac{x^4}{2} + \dfrac{x^6}{6}$

Problem Set 10.7

1. From the geometric series for $\dfrac{1}{1-x}$ with x replaced by $-x$, we get

$\dfrac{1}{1+x} = 1 - x + x^2 - x^3 + x^4 - x^5 + \cdots,$

radius of convergence 1.

3. $\dfrac{d}{dx}\left(\dfrac{1}{1-x}\right) = \dfrac{1}{(1-x)^2}; \dfrac{d}{dx}\left(\dfrac{1}{(1-x)^2}\right) = \dfrac{2}{(1-x)^3},$

so $\dfrac{1}{(1-x)^3}$ is $\dfrac{1}{2}$ of the second derivative of

$\dfrac{1}{1-x}.$ Thus,

$\dfrac{1}{(1-x)^3} = 1 + 3x + 6x^2 + 10x^3 + \cdots;$

radius of convergence 1.

5. From the geometric series for $\dfrac{1}{1-x}$ with x

replaced by $\dfrac{3}{2}x$, we get

$$\frac{1}{2-3x} = \frac{1}{2} + \frac{3x}{4} + \frac{9x^2}{8} + \frac{27x^3}{16} + \cdots;$$

radius of convergence $\frac{2}{3}$.

7. From the geometric series for $\frac{1}{1-x}$ with x replaced by x^4, we get

$$\frac{x^2}{1-x^4} = x^2 + x^6 + x^{10} + x^{14} + \cdots;$$

radius of convergence 1.

9. From the geometric series for $\ln(1+x)$ with x replaced by t, we get

$$\int_0^x \ln(1+t)dt = \frac{x^2}{2} - \frac{x^3}{6} + \frac{x^4}{12} - \frac{x^5}{20} + \cdots;$$

radius of convergence 1.

11. $\ln(1+x) = x - \frac{x^2}{2} + \frac{x^3}{3} - \frac{x^4}{4} + \cdots, -1 < x \le 1$

$\ln(1-x) = -x - \frac{x^2}{2} - \frac{x^3}{3} - \frac{x^4}{4} + \cdots, -1 \le x < 1$

$\ln\frac{1+x}{1-x} = \ln(1+x) - \ln(1-x)$

$= 2x + \frac{2x^3}{3} + \frac{2x^5}{5} + \cdots;$ radius of convergence 1.

13. Substitute $-x$ for x in the series for e^x to get:

$$e^{-x} = 1 - x + \frac{x^2}{2!} - \frac{x^3}{3!} + \frac{x^4}{4!} - \frac{x^5}{5!} + \cdots.$$

15. Add the result of Problem 13 to the series for e^x to get:

$$e^x + e^{-x} = 2 + \frac{2x^2}{2!} + \frac{2x^4}{4!} + \frac{2x^6}{6!} + \cdots.$$

17. $e^{-x} \cdot \frac{1}{1-x} = \left(1 - x + \frac{x^2}{2!} - \frac{x^3}{3!} + \cdots\right)(1 + x + x^2 + \cdots) = 1 + \frac{x^2}{2} + \frac{x^3}{3} + \frac{3x^4}{8} + \frac{11x^5}{30} + \cdots$

19. $\frac{\tan^{-1} x}{e^x} = e^{-x}\tan^{-1} x = \left(1 - x + \frac{x^2}{2!} - \frac{x^3}{3!} + \cdots\right)\left(x - \frac{x^3}{3} + \frac{x^5}{5} - \frac{x^7}{7} + \cdots\right) = x - x^2 + \frac{x^3}{6} + \frac{x^4}{6} + \frac{3x^5}{40} + \cdots$

21. $(\tan^{-1} x)(1 + x^2 + x^4) = \left(x - \frac{x^3}{3} + \frac{x^5}{5} - \frac{x^7}{7} + \cdots\right)(1 + x^2 + x^4) = x + \frac{2x^3}{3} + \frac{13x^5}{15} - \frac{29x^7}{105} + \cdots$

23. The series representation of $\frac{e^x}{1+x}$ is $1 + \frac{x^2}{2} - \frac{x^3}{3} + \frac{3x^4}{8} - \frac{11x^5}{30} + \cdots$, so $\int_0^x \frac{e^t}{1+t}dt = x + \frac{1}{6}x^3 - \frac{1}{12}x^4 + \frac{3}{40}x^5 - \ldots.$

25. **a.** $\frac{1}{1+x} = 1 - x + x^2 - x^3 + x^4 - x^5 + \cdots$, so $\frac{x}{1+x} = x - x^2 + x^3 - x^4 + x^5 - \cdots.$

b. $e^x = 1 + x + \frac{x^2}{2!} + \frac{x^3}{3!} + \frac{x^4}{4!} + \frac{x^5}{5!} + \cdots$, so $\frac{e^x - (1+x)}{x^2} = \frac{1}{2!} + \frac{x}{3!} + \frac{x^2}{4!} + \frac{x^3}{5!} + \cdots.$

c. $-\ln(1-x) = x + \frac{x^2}{2} + \frac{x^3}{3} + \frac{x^4}{4} + \frac{x^5}{5} + \cdots$, so $-\ln(1-2x) = 2x + \frac{4x^2}{2} + \frac{8x^3}{3} + \frac{16x^4}{4} + \cdots.$

27. Differentiating the series for $\dfrac{1}{1-x}$ yields $\dfrac{1}{(1-x)^2} = 1 + 2x + 3x^2 + 4x^3 + \cdots$ multiplying this series by x gives

$$\dfrac{x}{(1-x)^2} = x + 2x^2 + 3x^3 + 4x^4 + \cdots, \text{ hence } \sum_{n=1}^{\infty} nx^n = \dfrac{x}{(1-x)^2} \text{ for } -1 < x < 1.$$

29. a. $\tan^{-1}(e^x - 1) = (e^x - 1) - \dfrac{(e^x - 1)^3}{3} + \dfrac{(e^x - 1)^5}{5} - \cdots = \left(x + \dfrac{x^2}{2!} + \dfrac{x^3}{3!} + \cdots \right) - \dfrac{1}{3}\left(x + \dfrac{x^2}{2!} + \cdots \right)^3 + \cdots$

$$= x + \dfrac{x^2}{2} - \dfrac{x^3}{6} - \cdots$$

b. $e^{e^x - 1} = 1 + (e^x - 1) + \dfrac{(e^x - 1)^2}{2!} + \dfrac{(e^x - 1)^3}{3!} + \cdots$

$$= 1 + \left(x + \dfrac{x^2}{2!} + \dfrac{x^3}{3!} + \cdots \right) + \dfrac{1}{2!}\left(x + \dfrac{x^2}{2!} + \cdots \right)^2 + \dfrac{1}{3!}\left(x + \dfrac{x^2}{2!} + \cdots \right)^3$$

$$= 1 + \left(x + \dfrac{x^2}{2!} + \dfrac{x^3}{3!} + \cdots \right) + \dfrac{1}{2!}\left(x^2 + 2\dfrac{x^3}{2!} + \cdots \right) + \dfrac{1}{3!}\left(x^3 + 3\dfrac{x^4}{2!} + \cdots \right) = 1 + x + x^2 + \dfrac{5x^3}{6} + \cdots$$

31. $\dfrac{x}{x^2 - 3x + 2} = \dfrac{x}{(x-2)(x-1)} = \dfrac{2}{x-2} - \dfrac{1}{x-1} = -\dfrac{1}{1 - \frac{x}{2}} + \dfrac{1}{1-x} = -\left(1 + \dfrac{x}{2} + \dfrac{x^2}{4} + \dfrac{x^3}{8} + \cdots \right) + \left(1 + x + x^2 + x^3 + \cdots \right)$

$$= \dfrac{x}{2} + \dfrac{3x^2}{4} + \dfrac{7x^3}{8} + \cdots = \sum_{n=1}^{\infty} \dfrac{(2^n - 1)x^n}{2^n}$$

33. $F(x) - xF(x) - x^2 F(x) = (f_0 + f_1 x + f_2 x^2 + f_3 x^3 + \cdots) - (f_0 x + f_1 x^2 + f_2 x^3 + \cdots) - (f_0 x^2 + f_1 x^3 + f_2 x^4 + \cdots)$

$$= f_0 + (f_1 - f_0)x + (f_2 - f_1 - f_0)x^2 + (f_3 - f_2 - f_1)x^3 + \cdots$$

$$= f_0 + (f_1 - f_0)x + \sum_{n=2}^{\infty} (f_n - f_{n-1} - f_{n-2})x^n = 0 + x + \sum_{n=0}^{\infty} (f_{n+2} - f_{n+1} - f_n)x^{n+2}$$

Since $f_{n+2} = f_{n+1} + f_n$, $f_{n+2} - f_{n+1} - f_n = 0$. Thus $F(x) - xF(x) - x^2 F(x) = x$.

$$F(x) = \dfrac{x}{1 - x - x^2}$$

35. $\pi \approx 16\left(\dfrac{1}{5} - \dfrac{1}{375} + \dfrac{1}{15,625} - \dfrac{1}{546,875} + \dfrac{1}{17,578,125} \right) - 4\left(\dfrac{1}{239} \right) \approx 3.14159$

10.8 Concepts Review

1. $\dfrac{f^{(k)}(0)}{k!}$ **3.** $-\infty$; ∞

Problem Set 10.8

1. $\tan x = \dfrac{\sin x}{\cos x} = \dfrac{x - \frac{x^3}{3!} + \frac{x^5}{5!} - \frac{x^7}{7!} + \cdots}{1 - \frac{x^2}{2!} + \frac{x^4}{4!} - \frac{x^6}{6!} + \cdots} = x + \dfrac{x^3}{3} + \dfrac{2x^5}{15} + \cdots$

3. $e^x \sin x = \left(1+x+\dfrac{x^2}{2!}+\dfrac{x^3}{3!}+\dfrac{x^4}{4!}+\cdots\right)\left(x-\dfrac{x^3}{3!}+\dfrac{x^5}{5!}-\cdots\right) = x+x^2+\dfrac{x^3}{3}-\dfrac{x^5}{30}-\cdots$

5. $\cos x \ln(1+x) = \left(1-\dfrac{x^2}{2!}+\dfrac{x^4}{4!}-\cdots\right)\left(x-\dfrac{x^2}{2}+\dfrac{x^3}{3}-\dfrac{x^4}{4}+\cdots\right) = x-\dfrac{x^2}{2}-\dfrac{x^3}{6}+\dfrac{3x^5}{40}-\cdots$

7. $e^x+x+\sin x = x+\left(1+x+\dfrac{x^2}{2!}+\cdots\right)+\left(x-\dfrac{x^3}{3!}+\dfrac{x^5}{5!}-\cdots\right) = 1+3x+\dfrac{x^2}{2!}+\dfrac{x^4}{4!}+\dfrac{2x^5}{5!}+\cdots$

9. $\dfrac{1}{1-x}\cosh x = (1+x+x^2+x^3+\cdots)\left(1+\dfrac{x^2}{2!}+\dfrac{x^4}{4!}+\cdots\right) = 1+x+\dfrac{3x^2}{2}+\dfrac{3x^3}{2}+\dfrac{37x^4}{24}+\dfrac{37x^5}{24}+\cdots, \quad -1<x<1$

11. $\dfrac{1}{1+x+x^2} = \dfrac{1}{1-(-x^2-x)} = 1+(-x^2-x)+(-x^2-x)^2+(-x^2-x)^3+\cdots$

   $= 1-x+x^3-x^4+\cdots, \quad -1<-x^2-x<1 \text{ or } -1<x^2+x<1$

13. $\sin^3 x = \left(x-\dfrac{x^3}{3!}+\dfrac{x^5}{5!}-\cdots\right)^2\left(x-\dfrac{x}{3!}+\dfrac{x^5}{5!}-\cdots\right) = \left(x^2-2\dfrac{x^4}{3!}+\cdots\right)\left(x-\dfrac{x^3}{3!}+\dfrac{x^5}{5!}-\cdots\right) = x^3-\dfrac{x^5}{2}+\cdots$

15. $x\sec(x^2)+\sin x = \dfrac{x}{\cos(x^2)}+\sin x = \dfrac{x}{1-\frac{x^4}{2!}+\frac{x^8}{4!}-\cdots}+\left(x-\dfrac{x^3}{3!}+\dfrac{x^5}{5!}-\cdots\right)$

   $= \left(x+\dfrac{x^5}{2}+\cdots\right)+\left(x-\dfrac{x^3}{3!}+\dfrac{x^5}{5!}-\cdots\right) = 2x-\dfrac{x^3}{3!}+\dfrac{61x^5}{120}+\cdots$

17. $(1+x)^{3/2} = 1+\dfrac{3x}{2}+\dfrac{3x^2}{8}-\dfrac{x^3}{16}+\dfrac{3x^4}{128}-\dfrac{3x^5}{256}+\cdots, \quad -1<x<1$

19. $f^{(n)}(x) = e^x$ for all n. $f(1) = f'(1) = f''(1) = f'''(1) = e$

   $e^x \approx e+e(x-1)+\dfrac{e}{2}(x-1)^2+\dfrac{e}{6}(x-1)^3$

21. $f\left(\dfrac{\pi}{3}\right) = \dfrac{1}{2}; f'\left(\dfrac{\pi}{3}\right) = -\dfrac{\sqrt{3}}{2}; f''\left(\dfrac{\pi}{3}\right) = -\dfrac{1}{2}; f'''\left(\dfrac{\pi}{3}\right) = \dfrac{\sqrt{3}}{2}; \cos x \approx \dfrac{1}{2}-\dfrac{\sqrt{3}}{2}\left(x-\dfrac{\pi}{3}\right)-\dfrac{1}{4}\left(x-\dfrac{\pi}{3}\right)^2+\dfrac{\sqrt{3}}{12}\left(x-\dfrac{\pi}{3}\right)^3$

23. $f(1) = 3; f'(1) = 2+3 = 5;$

   $f''(1) = 2+6 = 8; f'''(1) = 6$

   $1+x^2+x^3 = 3+5(x-1)+4(x-1)^2+(x-1)^3$

   This is exact since $f^{(n)}(x) = 0$ for $n \geq 4$.

25. The derivative of an even function is an odd function and the derivative of an odd function is an even function. (Problem 50 of Section 3.2).

   Since $f(x) = \sum a_n x^n$ is an even function, $f'(x)$ is an odd function, so $f''(x)$ is an even function, hence $f'''(x)$ is an odd function, etc.

Thus $f^{(n)}(x)$ is an even function when n is even and an odd function when n is odd.

By the Uniqueness Theorem, if $f(x) = \sum a_n x^n$, then $a_n = \dfrac{f^{(n)}(0)}{n!}$. If $g(x)$ is an odd function, $g(0) = 0$, thence $a_n = 0$ for all odd n since $f^{(n)}(x)$ is an odd function for odd n.

27. $\dfrac{1}{\sqrt{1-t^2}} = [1+(-t^2)]^{-1/2}$

$= 1 - \dfrac{1}{2}(-t^2) + \dfrac{3}{8}(-t^2)^2 - \dfrac{5}{16}(-t^2)^3 + \cdots$

$= 1 + \dfrac{t^2}{2} + \dfrac{3t^4}{8} + \dfrac{5t^6}{16} + \cdots$

Thus, $\sin^{-1} x = \displaystyle\int_0^x \dfrac{1}{\sqrt{1-t^2}}\, dt$

$= \displaystyle\int_0^x \left(1 + \dfrac{t^2}{2} + \dfrac{3t^4}{8} + \dfrac{5t^6}{16} + \cdots\right) dt$

$= \left[t + \dfrac{t^3}{6} + \dfrac{3t^5}{40} + \dfrac{5t^7}{112} + \cdots\right]_0^x$

$= x + \dfrac{x^3}{6} + \dfrac{3x^5}{40} + \dfrac{5x^7}{112} + \cdots$

29. $\cos(x^2) = 1 - \dfrac{x^4}{2!} + \dfrac{x^8}{4!} - \dfrac{x^{12}}{6!} + \dfrac{x^{16}}{8!} - \cdots$

$\displaystyle\int_0^1 \cos(x^2)\, dx = \int_0^1 \left(1 - \dfrac{x^4}{2!} + \dfrac{x^8}{4!} - \dfrac{x^{12}}{6!} + \dfrac{x^{16}}{8!} - \cdots\right) dx$

$= \left[x - \dfrac{x^5}{10} + \dfrac{x^9}{216} - \dfrac{x^{13}}{9360} + \dfrac{x^{17}}{685,440} - \cdots\right]_0^1$

$= 1 - \dfrac{1}{10} + \dfrac{1}{216} - \dfrac{1}{9360} + \dfrac{1}{685,440} - \cdots \approx 0.90452$

31. $\dfrac{1}{x} = \dfrac{1}{1-(1-x)} = 1 + (1-x) + (1-x)^2 + (1-x)^3 + \cdots = 1 - (x-1) + (x-1)^2 - (x-1)^3 + \cdots$

for $-1 < 1-x < 1$, or $0 < x < 2$.

33. a. $f(x) = 1 + (x+x^2) + \dfrac{(x+x^2)^2}{2!} + \dfrac{(x+x^2)^3}{3!} + \dfrac{(x+x^2)^4}{4!} + \cdots$

$= 1 + (x+x^2) + \dfrac{1}{2}(x^2 + 2x^3 + x^4) + \dfrac{1}{6}(x^3 + 3x^4 + 3x^5 + x^6) + \dfrac{1}{24}(x^4 + 4x^5 + 6x^6 + 4x^7 + x^8) + \cdots$

$= 1 + x + \dfrac{3x^2}{2} + \dfrac{7x^3}{6} + \dfrac{25x^4}{24} + \cdots = \displaystyle\sum_{n=0}^{\infty} \dfrac{f^{(n)}(0)}{n!} x^n$

Thus $\dfrac{f^{(4)}(0)}{4!} = \dfrac{25}{24}$ so $f^{(4)}(0) = \dfrac{25}{24} 4! = 25$.

b. $f(x) = 1 + \sin x + \dfrac{\sin^2 x}{2!} + \dfrac{\sin^3 x}{3!} + \dfrac{\sin^4 x}{4!} + \cdots$

$= 1 + \left(x - \dfrac{x^3}{3!} + \cdots\right) + \dfrac{1}{2}\left(x - \dfrac{x^3}{3!} + \cdots\right)^2 + \dfrac{1}{6}\left(x - \dfrac{x^3}{3!} + \cdots\right)^3 + \dfrac{1}{24}\left(x - \dfrac{x^3}{3!} + \cdots\right)^4$

$= 1 + \left(x - \dfrac{x^3}{3!} + \cdots\right) + \dfrac{1}{2}\left(x^2 - 2\dfrac{x^4}{3!} + \cdots\right) + \dfrac{1}{6}(x^3 - \cdots) + \dfrac{1}{24}(x^4 - \cdots) = 1 + x + \dfrac{x^2}{2} - \dfrac{x^4}{8} - \cdots = \displaystyle\sum_{n=0}^{\infty} \dfrac{f^{(n)}(0)}{n!} x^n$

Thus, $\dfrac{f^{(4)}(0)}{4!} = -\dfrac{1}{8}$ so $f^{(4)}(0) = -\dfrac{1}{8} 4! = -3$.

c. $e^{t^2} - 1 = -1 + \left(1 + t^2 + \dfrac{t^4}{2!} + \dfrac{t^6}{3!} + \dfrac{t^8}{4!} + \cdots\right) = t^2 + \dfrac{t^4}{2!} + \dfrac{t^6}{3!} + \dfrac{t^8}{4!} + \cdots$

so $\dfrac{e^{t^2} - 1}{t^2} = 1 + \dfrac{t^2}{2} + \dfrac{t^4}{6} + \dfrac{t^6}{24} + \cdots$

$$f(x) = \int_0^x \left(1 + \frac{t^2}{2} + \frac{t^4}{6} + \frac{t^6}{24} + \ldots\right) dt = \left[t + \frac{t^3}{6} + \frac{t^5}{30} + \ldots\right]_0^x = x + \frac{x^3}{6} + \frac{x^5}{30} + \ldots = \sum_{n=0}^{\infty} \frac{f^{(n)}(0)}{n!} x^n$$

Thus, $\dfrac{f^{(4)}(0)}{4!} = 0$ so $f^{(4)}(0) = 0$.

d. $e^{\cos x - 1} = 1 + (\cos x - 1) + \dfrac{(\cos x - 1)^2}{2!} + \dfrac{(\cos x - 1)^3}{3!} + \ldots$

$$= 1 + \left(-\frac{x^2}{2!} + \frac{x^4}{4!} - \ldots\right) + \frac{1}{2}\left(-\frac{x^2}{2!} + \frac{x^4}{4!} - \ldots\right)^2 + \frac{1}{6}\left(-\frac{x^2}{2!} + \frac{x^4}{4!} - \ldots\right)^3 + \ldots$$

$$= 1 + \left(-\frac{x^2}{2} + \frac{x^4}{24} - \ldots\right) + \frac{1}{2}\left(\frac{x^4}{4} - \ldots\right) + \frac{1}{6}\left(-\frac{x^6}{8} + \ldots\right) = 1 - \frac{x^2}{2} + \frac{x^4}{6} - \ldots$$

Hence $f(x) = e - \dfrac{e}{2}x^2 + \dfrac{e}{6}x^4 - \ldots = \displaystyle\sum_{n=0}^{\infty} \dfrac{f^{(n)}(0)}{n!} x^n$

Thus, $\dfrac{f^{(4)}(0)}{4!} = \dfrac{e}{6}$ so $f^{(4)}(0) = \dfrac{e}{6}4! = 4e$.

e. Observe that $\ln(\cos^2 x) = \ln(1 - \sin^2 x)$.

$$\sin^2 x = \left(x - \frac{x^3}{3!} + \frac{x^5}{5!} - \cdots\right)^2 = x^2 - \frac{x^4}{3} + \frac{2x^6}{45} - \cdots$$

$$\ln(1 - \sin^2 x) = -\sin^2 x - \frac{\sin^4 x}{2} - \frac{\sin^6 x}{3} - \cdots$$

$$= -\left(x^2 - \frac{x^4}{3} + \cdots\right) - \frac{1}{2}\left(x^2 - \frac{x^4}{3} + \cdots\right)^2 - \frac{1}{3}\left(x^2 - \frac{x^4}{3} + \cdots\right)^3$$

$$= -\left(x^2 - \frac{x^4}{3} + \frac{2x^6}{45} - \cdots\right) - \frac{1}{2}\left(x^4 - \frac{2x^6}{3} + \cdots\right) - \frac{1}{3}(x^6 - \cdots) = -x^2 - \frac{x^4}{6} - \frac{2x^6}{45} - \cdots$$

Hence $f(x) = -x^2 - \dfrac{1}{6}x^4 - \dfrac{2}{45}x^6 - \cdots = \displaystyle\sum_{n=0}^{\infty} \dfrac{f^{(n)}(0)}{n!} x^n$.

Thus, $\dfrac{f^{(4)}(0)}{4!} = -\dfrac{1}{6}$ so $f^{(4)}(0) = -\dfrac{1}{6}4! = -4$.

35. $\tanh x = \dfrac{\sinh x}{\cosh x} = a_0 + a_1 x + a_2 x^2 + \ldots$

so $\sinh x = \cosh x(a_0 + a_1 x + a_2 x^2 + \ldots)$

or $x + \dfrac{x^3}{6} + \dfrac{x^5}{120} + \ldots = \left(1 + \dfrac{x^2}{2} + \dfrac{x^4}{24} + \ldots\right)(a_0 + a_1 x + a_2 x^2 + \ldots)$

$$= a_0 + a_1 x + \left(a_2 + \frac{a_0}{2}\right)x^2 + \left(a_3 + \frac{a_1}{2}\right)x^3 + \left(a_4 + \frac{a_2}{2} + \frac{a_0}{24}\right)x^4 + \left(a_5 + \frac{a_3}{2} + \frac{a_1}{24}\right)x^5 + \ldots$$

Thus $a_0 = 0$, $a_1 = 1$, $a_2 + \dfrac{a_0}{2} = 0$, $a_3 + \dfrac{a_1}{2} = \dfrac{1}{6}$,

$a_4 + \dfrac{a_2}{2} + \dfrac{a_0}{24} = 0$, $a_5 + \dfrac{a_3}{2} + \dfrac{a_1}{24} = \dfrac{1}{120}$, so

$a_0 = 0, a_1 = 1, a_2 = 0, a_3 = -\dfrac{1}{3}, a_4 = 0, a_5 = \dfrac{2}{15}$ and therefore

$$\tanh x = x - \frac{1}{3}x^3 + \frac{2}{15}x^5 - \cdots$$

37.   **a.**   First define $R_3(x)$ by

$$R_3(x) = f(x) - f(a) - f'(a)(x-a) - \frac{f''(a)}{2!}(x-a)^2 - \frac{f'''(a)}{3!}(x-a)^3$$

For any t in the interval $[a, x]$ we define

$$g(t) = f(x) - f(t) - f'(t)(x-t) - \frac{f''(t)}{2!}(x-t)^2 - \frac{f'''(t)}{3!}(x-t)^3 - R_3(x)\frac{(x-t)^4}{(x-a)^4}$$

Next we differentiate with respect to t using the Product and Power Rules:

$$g'(t) = 0 - f'(t) - \left[-f'(t) + f''(t)(x-t)\right] - \frac{1}{2!}\left[-2f''(t)(x-t) + f'''(t)(x-t)^2\right]$$

$$- \frac{1}{3!}\left[-3f'''(t)(x-t)^2 + f^{(4)}(t)(x-t)^3\right] + R_3(x)\frac{4(x-t)^3}{(x-a)^4}$$

$$= -\frac{f^{(4)}(t)(x-t)^3}{3!} + 4R_3(x)\frac{(x-t)^3}{(x-a)^4}$$

Since $g(x) = 0$, $g(a) = R_3(x) - R_3(x) = 0$, and $g(t)$ is continuous on $[a, x]$, we can apply the Mean Value Theorem for Derivatives. There exists, therefore, a number c between a and x such that $g'(c) = 0$. Thus,

$$0 = g'(c) = -\frac{f^{(4)}(c)(x-c)^3}{3!} + 4R_3(x)\frac{(x-c)^3}{(x-a)^4}$$

which leads to:

$$R_3(x) = \frac{f^{(4)}(c)}{4!}(x-a)^4$$

b.   Like the previous part, first define $R_n(x)$ by

$$R_n(x) = f(x) - f(a) - f'(a)(x-a) - \frac{f''(a)}{2!}(x-a)^2 - \cdots - \frac{f^{(n)}(a)}{n!}(x-a)^n$$

For any t in the interval $[a, x]$ we define

$$g(t) = f(x) - f(t) - f'(t)(x-t) - \frac{f''(t)}{2!}(x-t)^2 - \cdots - \frac{f^{(n)}(t)}{n!}(x-t)^n - R_n(x)\frac{(x-t)^{n+1}}{(x-a)^{n+1}}$$

Next we differentiate with respect to t using the Product and Power Rules:

$$g'(t) = 0 - f'(t) - \left[-f'(t) + f''(t)(x-t)\right] - \frac{1}{2!}\left[-2f''(t)(x-t) + f'''(t)(x-t)^2\right] - \cdots$$

$$- \frac{1}{n!}\left[-nf^{(n)}(t)(x-t)^{n-1} + f^{(n+1)}(t)(x-t)^n\right] + R_n(x)\frac{(n+1)(x-t)^n}{(x-a)^{n+1}}$$

$$= -\frac{f^{(n+1)}(t)(x-t)^n}{n!} + (n+1)R_n(x)\frac{(x-t)^n}{(x-a)^{n+1}}$$

Since $g(x) = 0$, $g(a) = R_n(x) - R_n(x) = 0$, and $g(t)$ is continuous on $[a, x]$, we can apply the Mean Value Theorem for Derivatives. There exists, therefore, a number c between a and x such that $g'(c) = 0$. Thus,

$$0 = g'(c) = -\frac{f^{(n+1)}(c)(x-c)^n}{n!} + (n+1)R_n(x)\frac{(x-c)^n}{(x-a)^{n+1}}$$

which leads to:

$$R_n(x) = \frac{f^{(n+1)}(c)}{(n+1)!}(x-a)^{n+1}$$

39. $f'(t) = \begin{cases} 0 & \text{if } t < 0 \\ 4t^3 & \text{if } t \geq 0 \end{cases}$,

$f''(t) = \begin{cases} 0 & \text{if } t < 0 \\ 12t^2 & \text{if } t \geq 0 \end{cases}$,

$f'''(t) = \begin{cases} 0 & \text{if } t < 0 \\ 24t & \text{if } t \geq 0 \end{cases}$,

$f^{(4)}(t) = \begin{cases} 0 & \text{if } t < 0 \\ 24 & \text{if } t \geq 0 \end{cases}$

$\lim_{t \to 0^+} f^{(4)}(t) = 24$ while $\lim_{t \to 0^-} f^{(4)}(t) = 0$, thus

$f^{(4)}(0)$ does not exist, and $f(t)$ cannot be represented by a Maclaurin series.

Suppose that $g(t)$ as described in the text is represented by a Maclaurin series, so

$g(t) = a_0 + a_1 t + a_2 t^2 + \ldots = \sum_{n=0}^{\infty} \frac{g^{(n)}(0)}{n!} t^n$ for all

t in $(-R, R)$ for some $R > 0$. It is clear that, for $t \leq 0$, $g(t)$ is represented by

$g(t) = 0 + 0t + 0t^2 + \ldots$. However, this will not represent $g(t)$ for any $t > 0$ since the car is moving for $t > 0$. Similarly, any series that represents $g(t)$ for $t > 0$ cannot be 0 everywhere, so it will not represent $g(t)$ for $t < 0$. Thus, $g(t)$ cannot be represented by a Maclaurin series.

41. $\sin x = x - \dfrac{x^3}{6} + \dfrac{x^5}{120} - \dfrac{x^7}{5040} + \cdots$

43. $3\sin x - 2\exp x = -2 + x - x^2 - \dfrac{5x^3}{6} - \cdots$

$3\sin x = 3x - \dfrac{x^3}{2} + \dfrac{x^5}{40} - \dfrac{x^7}{1680} + \cdots$

$-2\exp x = -2 - 2x - x^2 - \dfrac{x^3}{3} - \cdots$

Thus, $3\sin x - 2\exp x = -2 + x - x^2 - \dfrac{5x^3}{6} - \cdots$

45. $\sin(\exp x - 1) = x + \dfrac{x^2}{2} - \dfrac{5x^4}{24} - \dfrac{23x^5}{120} - \cdots$

$\exp x - 1 = x + \dfrac{x^2}{2} + \dfrac{x^3}{6} + \dfrac{x^4}{24} + \cdots$

$\sin(\exp x - 1) = \left(x + \dfrac{x^2}{2} + \dfrac{x^3}{6} + \dfrac{x^4}{24} + \dfrac{x^5}{120} + \cdots \right) - \dfrac{1}{6}\left(x + \dfrac{x^2}{2} + \dfrac{x^3}{6} + \cdots \right)^3 + \dfrac{1}{120}\left(x + \dfrac{x^2}{2} + \cdots \right)^5 - \cdots$

$= \left(x + \dfrac{x^2}{2} + \dfrac{x^3}{6} + \dfrac{x^4}{24} + \dfrac{x^5}{120} + \cdots \right) - \dfrac{1}{6}\left(x^3 + \dfrac{3x^4}{2} + \dfrac{5x^5}{4} + \cdots \right) + \dfrac{1}{120}(x^5 + \cdots) - \cdots$

$= \left(x + \dfrac{x^2}{2} + \dfrac{x^3}{6} + \dfrac{x^4}{24} + \dfrac{x^5}{120} + \cdots \right) - \left(\dfrac{x^3}{6} + \dfrac{x^4}{4} + \dfrac{5x^5}{24} + \cdots \right) + \left(\dfrac{x^5}{120} + \cdots \right) - \cdots = x + \dfrac{x^2}{2} - \dfrac{5x^4}{24} - \dfrac{23x^5}{120} - \cdots$

47. $(\sin x)(\exp x) = x + x^2 + \dfrac{x^3}{3} - \dfrac{x^5}{30} - \cdots$

$(\sin x)(\exp x) = \left(x - \dfrac{x^3}{6} + \dfrac{x^5}{120} - \cdots \right)\left(1 + x + \dfrac{x^2}{2} + \dfrac{x^3}{6} + \dfrac{x^4}{24} + \cdots \right)$

$= \left(x - \dfrac{x^3}{6} + \dfrac{x^5}{120} + \cdots \right) + \left(x^2 - \dfrac{x^4}{6} + \dfrac{x^6}{120} - \cdots \right) + \left(\dfrac{x^3}{2} - \dfrac{x^5}{12} + \cdots \right) + \left(\dfrac{x^4}{6} - \dfrac{x^6}{36} + \cdots \right) + \left(\dfrac{x^5}{24} - \dfrac{x^7}{144} + \cdots \right) + \cdots$

$= x + x^2 + \dfrac{x^3}{3} - \dfrac{x^5}{30} - \cdots$

10.9 Chapter Review

Concepts Test

1. **False:** If $b_n = 100$ and $a_n = 50 + (-1)^n$ then

   since $a_n = \begin{cases} 51 & \text{if } n \text{ is even} \\ 49 & \text{if } n \text{ is odd} \end{cases}$,

   $0 \le a_n \le b_n$ for all n and

   $\lim_{n \to \infty} b_n = 100$ while $\lim_{n \to \infty} a_n$ does not exist.

3. **True:** If $\lim_{n \to \infty} a_n = L$ then for any $\varepsilon > 0$ there is a number $M > 0$ such that $|a_n - L| < \varepsilon$ for all $n \ge M$. Thus, for the same ε, $|a_{3n+4} - L| < \varepsilon$ for $3n + 4 \ge M$ or $n \ge \dfrac{M-4}{3}$. Since ε was arbitrary, $\lim_{n \to \infty} a_{3n+4} = L$.

5. **False:** Let a_n be given by

   $a_n = \begin{cases} 1 & \text{if } n \text{ is prime} \\ 0 & \text{if } n \text{ is composite} \end{cases}$

   Then $a_{mn} = 0$ for all mn since $m \ge 2$, hence $\lim_{n \to \infty} a_{mn} = 0$ for $m \ge 2$.

   $\lim_{n \to \infty} a_n$ does not exist since for any $M > 0$ there will be a_n's with $a_n = 1$ since there are infinitely many prime numbers.

7. **False:** Let $a_n = 1 + \dfrac{1}{2} + \cdots + \dfrac{1}{n}$. Then

   $a_n - a_{n+1} = -\dfrac{1}{n+1}$ so

   $\lim_{n \to \infty} (a_n - a_{n+1}) = 0$ but $\lim_{n \to \infty} a_n$ is

   not finite since $\lim_{n \to \infty} a_n = \sum_{k=1}^{\infty} \dfrac{1}{k}$,

   which diverges.

9. **True:** If $\{a_n\}$ converges, then for some N, there are numbers m and M with $m \le a_n \le M$ for all $n \ge N$. Thus

   $\dfrac{m}{n} \le \dfrac{a_n}{n} \le \dfrac{M}{n}$ for all $n \ge N$. Since

   $\left\{\dfrac{m}{n}\right\}$ and $\left\{\dfrac{M}{n}\right\}$ both converge to 0,

   $\left\{\dfrac{a_n}{n}\right\}$ must also converge to 0.

11. **True:** The series converges by the Alternating Series Test. $S_1 = a_1, S_2 = a_1 - a_2, S_3 = a_1 - a_2 + a_3,$ $S_4 = a_1 - a_2 + a_3 - a_4,$ etc. $0 < a_2 < a_1 \Rightarrow 0 < a_1 - a_2 = S_2 < a_1;$ $0 < a_3 < a_2 \Rightarrow -a_2 < -a_2 + a_3 < 0$ so $0 < a_1 - a_2 < a_1 - a_2 + a_3 = S_3 < a_1;$ $0 < a_4 < a_3 \Rightarrow 0 < a_3 - a_4 < a_3,$ so $-a_2 < -a_2 + a_3 - a_4 < -a_2 + a_3 < 0,$ hence $0 < a_1 - a_2 < a_1 - a_2 + a_3 - a_4 = S_4$ $< a_1 - a_2 + a_3 < a_1;$ etc. For each even n, $0 < S_{n-1} - a_n$ while for each odd n, $n > 1$, $S_{n-1} + a_n < a_1$.

13. **False:** $\sum_{n=1}^{\infty} (-1)^n$ diverges but the partial sums are bounded ($S_n = -1$ for odd n and $S_n = 0$ for even n.)

15. **True:** $\rho = \lim_{n \to \infty} \dfrac{a_{n+1}}{a_n} = 1$, Ratio Test is inconclusive. (See the discussion before Example 5 in Section 10.4.)

17. **False:** $\lim_{n \to \infty} \left(1 - \dfrac{1}{n}\right)^n = \dfrac{1}{e} \ne 0$ so the series cannot converge.

19. **True:** $\displaystyle\sum_{n=2}^{\infty} \dfrac{n+1}{(n \ln n)^2}$

   $= \displaystyle\sum_{n=2}^{\infty} \left[\dfrac{n}{(n \ln n)^2} + \dfrac{1}{(n \ln n)^2} \right]$

   $= \displaystyle\sum_{n=2}^{\infty} \left[\dfrac{1}{n(\ln n)^2} + \dfrac{1}{n^2 (\ln n)^2} \right]$

   $\dfrac{1}{x(\ln x)^2}$ is continuous, positive, and nonincreasing on $[2, \infty)$. Using $u = \ln x$, $du = \dfrac{1}{x} dx$,

   $\displaystyle\int_2^{\infty} \dfrac{1}{x(\ln x)^2} dx = \int_{\ln 2}^{\infty} \dfrac{1}{u^2} du$

$$=\left[-\frac{1}{u}\right]_{\ln 2}^{\infty}=0+\frac{1}{\ln 2}<\infty \text{ so}$$

$$\sum_{n=2}^{\infty}\frac{1}{n(\ln n)^2} \text{ converges.}$$

For $n \geq 3$, $\ln n > 1$, so $(\ln n)^2 > 1$ and

$$\frac{1}{n^2(\ln n)^2}<\frac{1}{n^2}. \text{ Thus}$$

$$\sum_{n=3}^{\infty}\frac{1}{n^2(\ln n)^2}<\sum_{n=3}^{\infty}\frac{1}{n^2} \text{ so}$$

$$\sum_{n=3}^{\infty}\frac{1}{n^2(\ln n)^2} \text{ converges by the}$$

Comparison Test. Since both series converge, so does their sum.

21. True: If $0 \leq a_{n+100} \leq b_n$ for all n in N, then

$$\sum_{n=101}^{\infty} a_n \leq \sum_{n=1}^{\infty} b_n \text{ so } \sum_{n=1}^{\infty} a_n \text{ also}$$

converges, since adding a finite number of terms does not affect the convergence or divergence of a series.

23. True: $\frac{1}{3}+\left(\frac{1}{3}\right)^2+\left(\frac{1}{3}\right)^3+\ldots=\sum_{n=1}^{\infty}\left(\frac{1}{3}\right)^n$

$$=\sum_{n=1}^{\infty}\frac{1}{3}\left(\frac{1}{3}\right)^{n-1}=\frac{\frac{1}{3}}{1-\frac{1}{3}}=\frac{\frac{1}{3}}{\frac{2}{3}}=\frac{1}{2}, \text{ so the}$$

sum of the first thousand terms is less than $\frac{1}{2}$.

25. True: If $b_n \leq a_n \leq 0$ for all n in N then $0 \leq -a_n \leq -b_n$ for all n in N.

$$\sum_{n=1}^{\infty}-b_n=(-1)\sum_{n=1}^{\infty}b_n \text{ which converges}$$

since $\sum_{n=1}^{\infty}b_n$ converges.

Thus, by the Comparison Test,

$$\sum_{n=1}^{\infty}-a_n \text{ converges, hence}$$

$$\sum_{n=1}^{\infty}a_n=(-1)\sum_{n=1}^{\infty}(-a_n) \text{ also}$$

converges.

27. True: $\left|\sum_{n=1}^{\infty}(-1)^{n+1}\frac{1}{n}-\sum_{n=1}^{99}(-1)^{n+1}\frac{1}{n}\right|$

$$=\left|-\frac{1}{100}+\frac{1}{101}-\frac{1}{102}-\cdots\right|<\frac{1}{100}=0.01$$

29. True: $|3-(-1.1)|=4.1$, so the radius of convergence of the series is at least 4.1.

$|3-7|=4<4.1$ so $x=7$ is within the interval of convergence.

31. True: The radius of convergence is at least 1.5, so 1 is within the interval of convergence.

Thus $\int_0^1 f(x)dx=\left[\sum_{n=0}^{\infty}\frac{a_n x^{n+1}}{n+1}\right]_0^1$

$$=\sum_{n=0}^{\infty}\frac{a_n}{n+1}.$$

33. False: Consider the function

$$f(x)=\begin{cases} e^{-\frac{1}{x^2}} & x \neq 0 \\ 0 & x=0 \end{cases}.$$

The Maclaurin series for this function represents the function only at $x=0$. (See Problem 40 of Section 10.8.)

35. True: $\sum_{n=0}^{\infty}\frac{(-1)^n x^n}{n!}=e^{-x}, \frac{d}{dx}e^{-x}+e^{-x}=0$

Sample Test Problems

1. $\lim_{n\to\infty}\frac{9n}{\sqrt{9n^2+1}}=\lim_{n\to\infty}\frac{9}{\sqrt{9+\frac{1}{n^2}}}=3$

The sequence converges to 3.

3. $\lim_{n\to\infty}\left(1+\frac{4}{n}\right)^n=\lim_{n\to\infty}\left(\left(1+\frac{4}{n}\right)^{n/4}\right)^4=e^4$

The sequence converges to e^4.

5. Let $y=\sqrt[n]{n}=n^{1/n}$ then $\ln y=\frac{1}{n}\ln n$.

$\lim_{n\to\infty}\frac{1}{n}\ln n=\lim_{n\to\infty}\frac{\ln n}{n}=\lim_{n\to\infty}\frac{\frac{1}{n}}{1}=\lim_{n\to\infty}\frac{1}{n}=0$ by using l'Hôpital's Rule. Thus,

$\lim_{n\to\infty}\sqrt[n]{n}=\lim_{n\to\infty}e^{\ln y}=1$. The sequence converges to 1.

7. $a_n \geq 0; \lim_{n\to\infty}\frac{\sin^2 n}{\sqrt{n}}\leq\lim_{n\to\infty}\frac{1}{\sqrt{n}}=0$

The sequence converges to 0.

9. $S_n = \left(\dfrac{1}{\sqrt{1}} - \dfrac{1}{\sqrt{2}}\right) + \left(\dfrac{1}{\sqrt{2}} - \dfrac{1}{\sqrt{3}}\right) + \ldots + \left(\dfrac{1}{\sqrt{n-1}} - \dfrac{1}{\sqrt{n}}\right) + \left(\dfrac{1}{\sqrt{n}} - \dfrac{1}{\sqrt{n+1}}\right) = 1 - \dfrac{1}{\sqrt{n+1}}$, so

$\lim\limits_{n\to\infty} S_n = \lim\limits_{n\to\infty}\left(1 - \dfrac{1}{\sqrt{n+1}}\right) = 1$. The series converges to 1.

11. $\ln\dfrac{1}{2} + \ln\dfrac{2}{3} + \ln\dfrac{3}{4} + \ldots = \displaystyle\sum_{n=1}^{\infty} \ln\dfrac{n}{n+1} = \sum_{n=1}^{\infty} [\ln n - \ln(n+1)]$

$S_n = (\ln 1 - \ln 2) + (\ln 2 - \ln 3) + \ldots + (\ln(n-1) - \ln n) + (\ln n - \ln(n+1)) = \ln 1 - \ln(n+1) = \ln\dfrac{1}{n+1}$

As $n \to \infty$, $\dfrac{1}{n+1} \to 0$ so $\lim\limits_{n\to\infty} S_n = \lim\limits_{n\to\infty} \ln\dfrac{1}{n+1} = -\infty$.
The series diverges.

13. $\displaystyle\sum_{k=0}^{\infty} e^{-2k} = \sum_{k=0}^{\infty} \left(\dfrac{1}{e^2}\right)^k = \dfrac{1}{1 - \frac{1}{e^2}} = \dfrac{e^2}{e^2 - 1} \approx 1.1565$

since $\dfrac{1}{e^2} < 1$.

15. $\displaystyle\sum_{k=1}^{\infty} 91\left(\dfrac{1}{100}\right)^k = \dfrac{91}{1 - \frac{1}{100}} - 91 = \dfrac{9100}{99} - 91 = \dfrac{91}{99}$

The series converges since $\left|\dfrac{1}{100}\right| < 1$.

17. $\cos x = 1 - \dfrac{x^2}{2!} + \dfrac{x^4}{4!} - \dfrac{x^6}{6!} + \cdots$, so

$1 - \dfrac{2^2}{2!} + \dfrac{2^4}{4!} - \dfrac{2^6}{6!} + \cdots$ converges to
$\cos 2 \approx -0.41615$.

19. Let $a_n = \dfrac{n}{1+n^2}$ and $b_n = \dfrac{1}{n}$.

$\lim\limits_{n\to\infty} \dfrac{a_n}{b_n} = \lim\limits_{n\to\infty} \dfrac{n^2}{1+n^2} = \lim\limits_{n\to\infty} \dfrac{1}{\frac{1}{n^2}+1} = 1$;

$0 < 1 < \infty$.
By the Limit Comparison Test, since

$\displaystyle\sum_{n=1}^{\infty} b_n = \sum_{n=1}^{\infty} \dfrac{1}{n}$ diverges, $\displaystyle\sum_{n=1}^{\infty} a_n = \sum_{n=1}^{\infty} \dfrac{n}{1+n^2}$ also
diverges.

21. Since the series alternates, $\dfrac{1}{\sqrt[3]{n}} > \dfrac{1}{\sqrt[3]{n+1}} > 0$, and

$\lim\limits_{n\to\infty} \dfrac{1}{\sqrt[3]{n}} = 0$, the series converges by the
Alternating Series Test.

23. $\displaystyle\sum_{n=1}^{\infty} \dfrac{2^n + 3^n}{4^n} = \sum_{n=1}^{\infty} \left(\left(\dfrac{1}{2}\right)^n + \left(\dfrac{3}{4}\right)^n\right)$

$= \left(\dfrac{1}{1 - \frac{1}{2}} - 1\right) + \left(\dfrac{1}{1 - \frac{3}{4}} - 1\right) = 1 + 3 = 4$

The series converges to 4. The 1's must be
subtracted since the index starts with $n = 1$.

25. $\lim\limits_{n\to\infty} \dfrac{n+1}{10n+12} = \dfrac{1}{10} \neq 0$, so the series diverges.

27. $\rho = \lim\limits_{n\to\infty} \left|\dfrac{(n+1)^2}{(n+1)!} \div \dfrac{n^2}{n!}\right| = \lim\limits_{n\to\infty} \left|\dfrac{n+1}{n^2}\right| = 0 < 1$, so

the series converges.

29. $\rho = \lim\limits_{n\to\infty} \left|\dfrac{2^{n+1}(n+1)!}{(n+3)!} \div \dfrac{2^n n!}{(n+2)!}\right|$

$= \lim\limits_{n\to\infty} \left|\dfrac{2(n+1)}{n+3}\right| = 2 > 1$
The series diverges.

31. $\rho = \lim\limits_{n\to\infty} \left|\dfrac{(n+1)^2\left(\frac{2}{3}\right)^{n+1}}{n^2\left(\frac{2}{3}\right)^n}\right| = \lim\limits_{n\to\infty} \left|\dfrac{2}{3}\right|\left|\dfrac{(n+1)^2}{n^2}\right|$

$= \dfrac{2}{3} < 1$, so the series converges.

33. $a_n = \dfrac{1}{3n-1}$; $\dfrac{1}{3n-1} > \dfrac{1}{3n+2}$ so $a_n > a_{n+1}$;

$\lim\limits_{n\to\infty} a_n = \lim\limits_{n\to\infty} \dfrac{1}{3n-1} = 0$, so the series

$\displaystyle\sum_{n=1}^{\infty} (-1)^n \dfrac{1}{3n-1}$ converges by the Alternating
Series Test.

Let $b_n = \dfrac{1}{n}$, then

$$\lim_{n\to\infty} \frac{a_n}{b_n} = \lim_{n\to\infty} \frac{n}{3n-1} = \lim_{n\to\infty} \frac{1}{3-\frac{1}{n}} = \frac{1}{3};$$

$0 < \dfrac{1}{3} < \infty$. By the Limit Comparison Test, since

$$\sum_{n=1}^{\infty} b_n = \sum_{n=1}^{\infty} \frac{1}{n} \text{ diverges, } \sum_{n=1}^{\infty} a_n = \sum_{n=1}^{\infty} \frac{1}{3n-1} \text{ also}$$

diverges.
The series is conditionally convergent.

35. $\dfrac{3^n}{2^{n+8}} = \dfrac{1}{2^8}\left(\dfrac{3}{2}\right)^n$;

$$\lim_{n\to\infty} \frac{1}{2^8}\left(\frac{3}{2}\right)^n = \frac{1}{2^8} \lim_{n\to\infty} \left(\frac{3}{2}\right)^n = \infty \text{ since } \frac{3}{2} > 1.$$

The series is divergent.

37. $\rho = \lim_{n\to\infty} \left| \dfrac{x^{n+1}}{(n+1)^3+1} \div \dfrac{x^n}{n^3+1} \right|$

$$= \lim_{n\to\infty} |x| \left| \frac{n^3+1}{(n+1)^3+1} \right| = |x|$$

When $x = 1$, the series is

$$\sum_{n=0}^{\infty} \frac{1}{n^3+1} = 1 + \sum_{n=1}^{\infty} \frac{1}{n^3+1} \le 1 + \sum_{n=1}^{\infty} \frac{1}{n^3}, \text{ which}$$

converges.

When $x = -1$, the series is $\sum_{n=0}^{\infty} \dfrac{(-1)^n}{n^3+1}$ which

converges absolutely since $\sum_{n=0}^{\infty} \dfrac{1}{n^3+1}$ converges.

The series converges on $-1 \le x \le 1$.

39. $\rho = \lim_{n\to\infty} \left| \dfrac{(x-4)^{n+1}}{n+2} \div \dfrac{(x-4)^n}{n+1} \right|$

$$= \lim_{n\to\infty} |x-4| \left| \frac{n+1}{n+2} \right| = |x-4|; \; |x-4| < 1 \text{ when}$$

$3 < x < 5$.
When $x = 5$, the series is

$$\sum_{n=0}^{\infty} \frac{(-1)^n (1)^n}{n+1} = \sum_{n=0}^{\infty} \frac{(-1)^n}{n+1}.$$

$a_n = \dfrac{1}{n+1}; \dfrac{1}{n+1} > \dfrac{1}{n+2}$, so $a_n > a_{n+1}$;

$$\lim_{n\to\infty} \frac{1}{n+1} = 0 \text{ so } \sum_{n=0}^{\infty} \frac{(-1)^n}{n+1} \text{ converges by the}$$

Alternating Series Test.
When $x = 3$, the series is

$$\sum_{n=0}^{\infty} \frac{(-1)^n(-1)^n}{n+1} = \sum_{n=0}^{\infty} \frac{1}{n+1}. \; a_n = \frac{1}{n+1}, \text{ let}$$

$b_n = \dfrac{1}{n}$ then $\lim_{n\to\infty} \dfrac{a_n}{b_n} = \lim_{n\to\infty} \dfrac{n}{n+1} = 1; \; 0 < 1 < \infty$

hence since $\sum_{n=1}^{\infty} b_n = \sum_{n=1}^{\infty} \dfrac{1}{n}$ diverges, $\sum_{n=0}^{\infty} \dfrac{1}{n+1}$

also diverges.
The series converges on $3 < x \le 5$.

41. $\rho = \lim_{n\to\infty} \left| \dfrac{(x-3)^{n+1}}{2^{n+1}+1} \div \dfrac{(x-3)^n}{2^n+1} \right|$

$$= \lim_{n\to\infty} |x-3| \left| \frac{1+\frac{1}{2^n}}{2+\frac{1}{2^n}} \right| = \frac{|x-3|}{2}; \; \frac{|x-3|}{2} < 1$$

when $1 < x < 5$.
When $x = 5$, the series is

$$\sum_{n=0}^{\infty} \frac{2^n}{2^n+1} = \sum_{n=0}^{\infty} \frac{1}{1+\left(\frac{1}{2}\right)^n}; \; \lim_{n\to\infty} \frac{1}{1+\left(\frac{1}{2}\right)^n} = 1 \ne 0$$

so the series diverges.
When $x = 1$, the series is

$$\sum_{n=0}^{\infty} \frac{(-2)^n}{2^n+1} = \sum_{n=0}^{\infty} \frac{(-1)^n}{1+\left(\frac{1}{2}\right)^n}; \; \lim_{n\to\infty} \frac{1}{1+\left(\frac{1}{2}\right)^n} = 1 \ne 0$$

so the series diverges.
The series converges on $1 < x < 5$.

43. $\dfrac{1}{1+x} = 1 - x + x^2 - x^3 + \dots$ for $-1 < x < 1$.

If $f(x) = \dfrac{1}{1+x}$, then $f'(x) = -\dfrac{1}{(1+x)^2}$. Thus,

differentiating the series for $\dfrac{1}{1+x}$ and

multiplying by -1 yields

$\dfrac{1}{(1+x)^2} = 1 - 2x + 3x^2 - 4x^3 + \dots$. The series

converges on $-1 < x < 1$.

45. $\sin^2 x = \left(x - \dfrac{x^3}{3!} + \dfrac{x^5}{5!} - \dfrac{x^7}{7!} + \dots \right)^2$

$$= x^2 - \frac{x^4}{3} + \frac{2x^6}{45} - \frac{x^8}{315} + \dots$$

Since the series for $\sin x$ converges for all x, so
does the series for $\sin^2 x$.

47. $\sin x + \cos x = 1 + x - \dfrac{x^2}{2!} - \dfrac{x^3}{3!} + \dfrac{x^4}{4!} + \dfrac{x^5}{5!} - \cdots$

Since the series for $\sin x$ and $\cos x$ converge for all x, so does the series for $\sin x + \cos x$.

49. $\dfrac{e^x - 1}{x} = 1 + \dfrac{x}{2!} + \dfrac{x^2}{3!} + \dfrac{x^3}{4!} + \cdots$

Thus, $\displaystyle\int_0^{0.2} \dfrac{e^x - 1}{x}\, dx = \left[x + \dfrac{x^2}{2 \cdot 2!} + \dfrac{x^3}{3 \cdot 3!} + \dfrac{x^4}{4 \cdot 4!} + \cdots \right]_0^{0.2} = 0.2 + \dfrac{0.2^2}{2 \cdot 2!} + \dfrac{0.2^3}{3 \cdot 3!} + \dfrac{0.2^4}{4 \cdot 4!} + \cdots \approx 0.21046.$

51. $1 - \dfrac{x^2}{2}$ is the Maclaurin polynomial of order 3 for $\cos x$, so $|R_3(x)| = \left| \dfrac{\cos c}{4!} x^4 \right| \le \dfrac{0.1^4}{4!} \approx 0.000004167.$

11

Numerical Methods, Approximations

11.1 Concepts Review

1. $f(1);\ f'(1);\ f''(1)$

3. error of the method; error of calculation

Problem Set 11.1

1. $f(x) = e^{2x}$ $f(0) = 1$

$f'(x) = 2e^{2x}$ $f'(0) = 2$

$f''(x) = 4e^{2x}$ $f''(0) = 4$

$f^{(3)}(x) = 8e^{2x}$ $f^{(3)}(0) = 8$

$f^{(4)}(x) = 16e^{2x}$ $f^{(4)}(0) = 16$

$f(x) \approx 1 + 2x + \dfrac{4}{2!}x^2 + \dfrac{8}{3!}x^3 + \dfrac{16}{4!}x^4 = 1 + 2x + 2x^2 + \dfrac{4}{3}x^3 + \dfrac{2}{3}x^4$

$f(0.12) \approx 1 + 2(0.12) + 2(0.12)^2 + \dfrac{4}{3}(0.12)^3 + \dfrac{2}{3}(0.12)^4 \approx 1.2712$

3. $f(x) = \sin 2x$ $f(0) = 0$

$f'(x) = 2\cos 2x$ $f'(0) = 2$

$f''(x) = -4\sin 2x$ $f''(0) = 0$

$f^{(3)}(x) = -8\cos 2x$ $f^{(3)}(0) = -8$

$f^{(4)}(x) = 16\sin 2x$ $f^{(4)}(0) = 0$

$f(x) \approx 2x - \dfrac{8}{3!}x^3 = 2x - \dfrac{4}{3}x^3$

$f(0.12) \approx 2(0.12) - \dfrac{4}{3}(0.12)^3 \approx 0.2377$

5. $f(x) = \ln(1 + x)$ $f(0) = 0$

$f'(x) = \dfrac{1}{1+x}$ $f'(0) = 1$

$f''(x) = -\dfrac{1}{(1+x)^2}$ $f''(0) = -1$

$f^{(3)}(x) = \dfrac{2}{(1+x)^3}$ $f^{(3)}(0) = 2$

$f^{(4)}(x) = -\dfrac{6}{(1+x)^4}$ $f^{(4)}(0) = -6$

$f(x) \approx x - \dfrac{1}{2!}x^2 + \dfrac{2}{3!}x^3 - \dfrac{6}{4!}x^4$

$= x - \dfrac{1}{2}x^2 + \dfrac{1}{3}x^3 - \dfrac{1}{4}x^4$

$f(0.12) \approx 0.12 - \dfrac{1}{2}(0.12)^2 + \dfrac{1}{3}(0.12)^3 - \dfrac{1}{4}(0.12)^4$

≈ 0.1133

7. $f(x) = \tan^{-1} x$ $f(0) = 0$

$f'(x) = \dfrac{1}{1+x^2}$ $f'(0) = 1$

$f''(x) = -\dfrac{2x}{(1+x^2)^2}$ $f''(0) = 0$

$f'''(x) = \dfrac{6x^2 - 2}{(1+x^2)^3}$ $f'''(0) = -2$

$f^{(4)}(x) = \dfrac{-24x^3 + 24x}{(1+x^2)^4}$ $f^{(4)}(0) = 0$

$f(x) \approx x - \dfrac{2}{3!}x^3 = x - \dfrac{1}{3}x^3$

$f(0.12) \approx 0.12 - \dfrac{1}{3}(0.12)^3 \approx 0.1194$

9. $f(x) = e^x$ $\qquad$ $f(1) = e$

$f'(x) = e^x$ $\qquad$ $f'(1) = e$

$f''(x) = e^x$ $\qquad$ $f''(1) = e$

$f'''(x) = e^x$ $\qquad$ $f'''(1) = e$

$P_3(x) = e + e(x-1) + \dfrac{e}{2}(x-1)^2 + \dfrac{e}{6}(x-1)^3$

11. $f(x) = \tan x$; $\quad f\left(\dfrac{\pi}{6}\right) = \dfrac{\sqrt{3}}{3}$

$f'(x) = \sec^2 x$; $\quad f'\left(\dfrac{\pi}{6}\right) = \dfrac{4}{3}$

$f''(x) = 2\sec^2 x \tan x$; $\quad f''\left(\dfrac{\pi}{6}\right) = \dfrac{8\sqrt{3}}{9}$

$f'''(x) = 2\sec^4 x + 4\sec^2 x \tan^2 x$; $\quad f'''\left(\dfrac{\pi}{6}\right) = \dfrac{16}{3}$

$P_3(x) = \dfrac{\sqrt{3}}{3} + \dfrac{4}{3}\left(x - \dfrac{\pi}{6}\right) + \dfrac{4\sqrt{3}}{9}\left(x - \dfrac{\pi}{6}\right)^2$

$\qquad + \dfrac{8}{9}\left(x - \dfrac{\pi}{6}\right)^3$

13. $f(x) = \cot^{-1} x$; $\quad f(1) = \dfrac{\pi}{4}$

$f'(x) = -\dfrac{1}{1+x^2}$; $\quad f'(1) = -\dfrac{1}{2}$

$f''(x) = \dfrac{2x}{(1+x^2)^2}$; $\quad f''(1) = \dfrac{1}{2}$

$f'''(x) = \dfrac{-6x^2 + 2}{(1+x^2)^3}$; $\quad f'''(1) = -\dfrac{1}{2}$

$P_3(x) = \dfrac{\pi}{4} - \dfrac{1}{2}(x-1) + \dfrac{1}{4}(x-1)^2 - \dfrac{1}{12}(x-1)^3$

15. $f(x) = x^3 - 2x^2 + 3x + 5$; $\quad f(1) = 7$

$f'(x) = 3x^2 - 4x + 3$; $\quad f'(1) = 2$

$f''(x) = 6x - 4$; $\quad f''(1) = 2$

$f^{(3)}(x) = 6$; $\quad f^{(3)}(1) = 6$

$P_3(x) = 7 + 2(x-1) + (x-1)^2 + (x-1)^3$

$\qquad = 5 + 3x - 2x^2 + x^3 = f(x)$

17. $f(x) = \dfrac{1}{1-x}$; $\quad f(0) = 1$

$f'(x) = \dfrac{1}{(1-x)^2}$; $\quad f'(0) = 1$

$f''(x) = \dfrac{2}{(1-x)^3}$; $\quad f''(0) = 2$

$f^{(3)}(x) = \dfrac{6}{(1-x)^4}$; $\quad f^{(3)}(0) = 6$

$f^{(4)}(x) = \dfrac{24}{(1-x)^5}$; $\quad f^{(4)}(0) = 24$

$f^{(n)}(x) = \dfrac{n!}{(1-x)^{n+1}}$; $\quad f^{(n)}(0) = n!$

$f(x) \approx 1 + x + \dfrac{2}{2!}x^2 + \dfrac{6}{3!}x^3 + \ldots + \dfrac{n!}{n!}x^n$

$\qquad = 1 + x + x^2 + x^3 + \ldots + x^n$

Using $n = 4$, $f(x) \approx 1 + x + x^2 + x^3 + x^4$

a. $f(0.1) \approx 1.1111$

b. $f(0.5) \approx 1.9375$

c. $f(0.9) \approx 4.0951$

d. $f(2) \approx 31$

19. The area of the sector with angle t is $\dfrac{1}{2}tr^2$. The area of the triangle is

$\dfrac{1}{2}\left(r\sin\dfrac{t}{2}\right)\left(2r\cos\dfrac{t}{2}\right) = r^2 \sin\dfrac{t}{2}\cos\dfrac{t}{2} = \dfrac{1}{2}r^2 \sin t$

$A = \dfrac{1}{2}tr^2 - \dfrac{1}{2}r^2 \sin t$

Using $n = 3$, $\sin t \approx t - \dfrac{1}{6}t^3$.

$A \approx \dfrac{1}{2}tr^2 - \dfrac{1}{2}r^2\left(t - \dfrac{1}{6}t^3\right) = \dfrac{1}{12}r^2 t^3$

21. a. $\ln\left(1 + \dfrac{r}{12}\right)^{12n} = \ln 2$

$12n \ln\left(1 + \dfrac{r}{12}\right) = \ln 2$

$n = \dfrac{\ln 2}{12\ln\left(1 + \dfrac{r}{12}\right)}$

b. $f(x) = \ln(1 + x)$; $\quad f(0) = 0$

$f'(x) = \dfrac{1}{1+x}$; $\quad f'(0) = 1$

$f''(x) = -\dfrac{1}{(1+x)^2}$; $\quad f''(0) = -1$

$\ln(1 + x) \approx x - \dfrac{x^2}{2}$

$n \approx \dfrac{\ln 2}{r - \frac{r^2}{24}} = \left[\dfrac{24}{r(24-r)}\right]\ln 2$

$$= \frac{\ln 2}{r} + \frac{\ln 2}{24 - r}$$

$$\approx \frac{\ln 2}{r} + \frac{\ln 2}{24} \approx \frac{0.693}{r} + 0.029$$

We let $24 - r \approx 24$ since the interest rate r is going to be close to 0.

c.

r	n (exact)	n (approx.)	n (rule 72)
0.05	13.8918	13.889	14.4
0.10	6.9603	6.959	7.2
0.15	4.6498	4.649	4.8
0.20	3.4945	3.494	3.6

23. a.

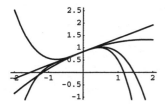

b.

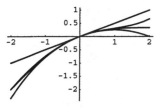

25. $\left| e^{2c} + e^{-2c} \right| \le \left| e^{2c} \right| + \left| \frac{1}{e^{2c}} \right| \le e^6 + 1$

27. $\left| \frac{4c}{\sin c} \right| = \frac{|4c|}{|\sin c|} \le \frac{2\pi}{\frac{1}{\sqrt{2}}} = 2\sqrt{2}\pi$

29. $\left| \frac{e^c}{c+5} \right| = \frac{\left| e^c \right|}{|c+5|} \le \frac{e^4}{3}$

31. $\left| \frac{c^2 + \sin c}{10 \ln c} \right| = \frac{\left| c^2 + \sin c \right|}{|10 \ln c|} \le \frac{\left| c^2 \right| + |\sin c|}{|10 \ln c|}$

$\le \frac{16 + 1}{10 \ln 2} = \frac{17}{10 \ln 2}$

33. $f(x) = \ln(2 + x); \ f'(x) = \frac{1}{2 + x};$

$f''(x) = -\frac{1}{(2+x)^2}; \ f^{(3)}(x) = \frac{2}{(2+x)^3};$

$f^{(4)}(x) = -\frac{6}{(2+x)^4}; \ f^{(5)}(x) = \frac{24}{(2+x)^5};$

$f^{(6)}(x) = -\frac{120}{(2+x)^6}; \ f^{(7)}(x) = \frac{720}{(2+x)^7}$

$R_6(x) = \frac{1}{7!} \cdot \frac{720}{(2+c)^7} x^7 = \frac{x^7}{7(2+c)^7}$

$\left| R_6(0.5) \right| \le \left| \frac{0.5^7}{7 \cdot 2^7} \right| \approx 8.719 \times 10^{-6}$

35. $f(x) = \sin x; \ f^{(7)}(x) = -\cos x$

$R_6(x) = \frac{-\cos c}{7!} \left(x - \frac{\pi}{4} \right)^7 = \frac{-\cos c \left(x - \frac{\pi}{4} \right)^7}{5040}$

$\left| R_6(0.5) \right| \le \left| \frac{\cos 0.5 \left(0.5 - \frac{\pi}{4} \right)^7}{5040} \right| \approx 2.685 \times 10^{-8}$

37. $f(x) = e^x \qquad\qquad f(0) = 1$

$f'(x) = e^x \qquad\qquad f'(0) = 1$

$f''(x) = e^x \qquad\qquad f''(0) = 1$

$f^{(3)}(x) = e^x \qquad\qquad f^{(3)}(0) = 1$

$f^{(4)}(x) = e^x \qquad\qquad f^{(4)}(c) = e^c$

$e^x \approx 1 + x + \frac{1}{2}x^2 + \frac{1}{6}x^3$

$R_3(x) = \frac{e^c}{4!} x^4$

$R_3(-0.1) = \frac{e^c}{240,000} \le \frac{1}{240,000} \approx 4.17 \times 10^{-6}$

$$R_3(-0.1) \geq \frac{e^{-0.1}}{240,000} \approx 3.77 \times 10^{-6}$$

$$e^{-0.1} - \left[1 + (-0.1) + \frac{1}{2}(-0.1)^2 + \frac{1}{6}(-0.1)^3 \right] \approx 4.08 \times 10^{-6}$$

39. $R_n(x) = \frac{e^c}{(n+1)!} x^{n+1}$

Note that $e^1 < 3$.

$|R_n(1)| < \frac{3}{(n+1)!}$

$\frac{3}{(n+1)!} < 0.000005$ or $600000 < (n+1)!$ when

$n \geq 9$.

41. $f(x) = (1+x)^{-1/2}$ $\qquad f(0) = 1$

$f'(x) = -\frac{1}{2}(1+x)^{-3/2}$ $\qquad f'(0) = -\frac{1}{2}$

$f''(x) = \frac{3}{4}(1+x)^{-5/2}$ $\qquad f''(0) = \frac{3}{4}$

$f^{(3)}(x) = -\frac{15}{8}(1+x)^{-7/2}$ $\qquad f^{(3)}(0) = -\frac{15}{8}$

$f^{(4)}(x) = \frac{105}{16}(1+x)^{-9/2}$ $\qquad f^{(4)}(c) = \frac{105}{16}(1+c)^{-9/2}$

$(1+x)^{-1/2} \approx 1 - \frac{1}{2}x + \frac{3}{8}x^2 - \frac{5}{16}x^3$

$R_3(x) = \frac{35}{128}(1+c)^{-9/2} x^4$

$|R_3(x)| \leq \left| \frac{35}{128}(0.95)^{-9/2}(0.05)^4 \right| \approx 2.15 \times 10^{-6}$

43. $R_4(x) = \frac{\cos c}{5!} x^5$

$|R_4(x)| \leq \frac{(0.5)^5}{5!} \approx 0.00026042 \leq 0.0002605$

$\int_0^{0.5} \sin x\, dx \approx \int_0^{0.5} \left(x - \frac{1}{6}x^3 \right) dx$

$= \left[\frac{1}{2}x^2 - \frac{1}{24}x^4 \right]_0^{0.5} \approx 0.1224$

Error $\leq 0.0002605(0.5 - 0) = 0.00013025$

45. $f(x) = x^4 - 3x^3 + 2x^2 + x - 2$; $f(1) = -1$

$f'(x) = 4x^3 - 9x^2 + 4x + 1$; $f'(1) = 0$

$f''(x) = 12x^2 - 18x + 4$; $f''(1) = -2$

$f^{(3)}(x) = 24x - 18$; $f^{(3)}(1) = 6$

$f^{(4)}(x) = 24$; $f^{(4)}(1) = 24$

$f^{(5)}(x) = 0$

Since $f^{(5)}(x) = 0$, $R_5(x) = 0$.

$x^4 - 3x^3 + 2x^2 + x - 2$

$= -1 - (x-1)^2 + (x-1)^3 + (x-1)^4$

47. $f(x) = \sin x$; $f\left(\dfrac{\pi}{4}\right) = \dfrac{\sqrt{2}}{2}$

$f'(x) = \cos x$; $f'\left(\dfrac{\pi}{4}\right) = \dfrac{\sqrt{2}}{2}$

$f''(x) = -\sin x$; $f''\left(\dfrac{\pi}{4}\right) = -\dfrac{\sqrt{2}}{2}$

$f^{(3)}(x) = -\cos x$; $f^{(3)}\left(\dfrac{\pi}{4}\right) = -\dfrac{\sqrt{2}}{2}$

$f^{(4)}(x) = \sin x$; $f^{(4)}(c) = \sin c$

$43° = \dfrac{\pi}{4} - \dfrac{\pi}{90}$ radians

$\sin x = \dfrac{\sqrt{2}}{2} + \dfrac{\sqrt{2}}{2}\left(x - \dfrac{\pi}{4}\right) - \dfrac{\sqrt{2}}{4}\left(x - \dfrac{\pi}{4}\right)^2$

$\qquad - \dfrac{\sqrt{2}}{12}\left(x - \dfrac{\pi}{4}\right)^3 + R_3(x)$

$\sin\left(\dfrac{\pi}{4} - \dfrac{\pi}{90}\right) = \dfrac{\sqrt{2}}{2} + \dfrac{\sqrt{2}}{2}\left(-\dfrac{\pi}{90}\right) - \dfrac{\sqrt{2}}{4}\left(-\dfrac{\pi}{90}\right)^2 \approx 0.681998 + R_3$

$\qquad\qquad - \dfrac{\sqrt{2}}{12}\left(-\dfrac{\pi}{90}\right)^3 + R_3\left(\dfrac{\pi}{4} - \dfrac{\pi}{90}\right)$

$|R_3| = \left|\dfrac{\sin c}{4!}\left(-\dfrac{\pi}{90}\right)^4\right| < \dfrac{1}{24}\left(\dfrac{\pi}{90}\right)^4 \approx 6.19 \times 10^{-8}$

49. $|R_9(x)| \leq \dfrac{1}{10!}x^{10} \leq \dfrac{1}{10!}\left(\dfrac{\pi}{2}\right)^{10} \approx 2.5202 \times 10^{-5}$

51. The kth derivative of $h(x)f(x)$ is $\displaystyle\sum_{i=0}^{k}\binom{k}{i}h^{(i)}(x)f^{(k-i)}(x)$. If $h(x) = x^{n+1}$, $h^{(i)}(x) = \dfrac{(n+1)!}{(n+1-i)!}x^{n+1-i}$.

Thus for $i \leq n + 1$, $h^{(i)}(0) = 0$. Let $q(x) = x^{n+1}f(x)$. Then $q^{(k)}(0) = \displaystyle\sum_{i=0}^{k}\binom{k}{i}h^{(i)}(0)f^{(k-i)}(0) = 0$

for $k \leq n + 1$.

$g^{(k)}(x) = p^{(k)}(x) + q^{(k)}(x)$, so $g^{(k)}(0) = p^{(k)}(0) + q^{(k)}(0) = p^{(k)}(0)$

for $k \leq n + 1$.

The Maclaurin polynomial of order n for g is $p(0) + p'(0)x + \dfrac{p''(0)}{2!}x^2 + \ldots + \dfrac{p^{(n)}(0)}{n!}x^n$ which is the Maclaurin polynomial of order n for $p(x)$. Since $p(x)$ is a polynomial of degree at most n, the remainder $R_n(x)$ of Maclaurin's Formula for $p(x)$ is 0, so the Maclaurin polynomial of order n for $g(x)$ is $p(x)$.

53. **a.** $L_{51}(x)$ is of degree 4 since it is the product of four linear factors.

$$L_{51}(x_1) = \frac{(x_1 - x_2)(x_1 - x_3)(x_1 - x_4)(x_1 - x_5)}{(x_1 - x_2)(x_1 - x_3)(x_1 - x_4)(x_1 - x_5)} = 1$$

$L_{51}(x_j) = 0$ for $j = 2, 3, 4, 5$ since $x - x_j$ is a linear factor of $L_{51}(x)$.

b. $L_{52}(x) = \dfrac{(x - x_1)(x - x_3)(x - x_4)(x - x_5)}{(x_2 - x_1)(x_2 - x_3)(x_2 - x_4)(x_2 - x_5)}$

$L_{53}(x) = \dfrac{(x - x_1)(x - x_2)(x - x_4)(x - x_5)}{(x_3 - x_1)(x_3 - x_2)(x_3 - x_4)(x_3 - x_5)}$

$L_{54}(x) = \dfrac{(x - x_1)(x - x_2)(x - x_3)(x - x_5)}{(x_4 - x_1)(x_4 - x_2)(x_4 - x_3)(x_4 - x_5)}$

$L_{55}(x) = \dfrac{(x - x_1)(x - x_2)(x - x_3)(x - x_4)}{(x_5 - x_1)(x_5 - x_2)(x_5 - x_3)(x_5 - x_4)}$

c. Since $L_{51}, L_{52}, L_{53}, L_{54},$ and L_{55} are of degree 4, L_5 is of degree less than or equal to 4. Since $L_{5j}(x_i) = 0$ for $i \neq j$ and $L_{5i}(x_i) = 1, L_5(x_i) = y_i$.

d. $L_{31}(x) = \dfrac{(x-2)(x-0)}{(1-2)(1-0)} = -(x-2)(x) = -x^2 + 2x$

$L_{32}(x) = \dfrac{(x-1)(x-0)}{(2-1)(2-0)} = \dfrac{1}{2}(x-1)(x) = \dfrac{1}{2}x^2 - \dfrac{1}{2}x$

$L_{33}(x) = \dfrac{(x-1)(x-2)}{(0-1)(0-2)} = \dfrac{1}{2}(x-1)(x-2) = \dfrac{1}{2}x^2 - \dfrac{3}{2}x + 1$

$L_3(x) = (-x^2 + 2x)(2) + \left(\dfrac{1}{2}x^2 - \dfrac{1}{2}x\right)(2.5) + \left(\dfrac{1}{2}x^2 - \dfrac{3}{2}x + 1\right)(0) = -0.75x^2 + 2.75x$

55. $x_1 = 1; \quad y_1 = 0$

$x_2 = 3; \quad y_2 = 1.099$

$x_3 = 5; \quad y_3 = 1.609$

$L_{31}(x) = \dfrac{(x-3)(x-5)}{(1-3)(1-5)} = \dfrac{1}{8}(x-3)(x-5)$

$L_{32}(x) = \dfrac{(x-1)(x-5)}{(3-1)(3-5)} = -\dfrac{1}{4}(x-1)(x-5)$

$L_{33}(x) = \dfrac{(x-1)(x-3)}{(5-1)(5-3)} = \dfrac{1}{8}(x-1)(x-3)$

$L_3(x) = L_{31}(x) \cdot 0 + L_{32}(x) \cdot 1.099 + L_{33}(x) \cdot 1.609$

$= -\dfrac{1.099}{4}(x-1)(x-5) + \dfrac{1.609}{8}(x-1)(x-3)$

$= \dfrac{1}{4}(x-1)\left[-1.099(x-5) + 0.8045(x-3)\right]$

$= \dfrac{1}{4}(x-1)(-0.2945x + 3.0815)$

Thus,

$\ln 2 \approx \dfrac{1}{4}(2-1)(-0.2945x + 3.0815) \approx 0.623$

The estimate for the maximum error is

$R_2(x) = \dfrac{(x-1)(x-3)(x-5)}{3!}f^{(3)}(\alpha)$

$f(x) = \ln x; \qquad f'(x) = \dfrac{1}{x}$

$f''(x) = -\dfrac{1}{x^2}; \qquad f'''(x) = \dfrac{2}{x^3}$

$|R_2(2)| = \left|\dfrac{(2-1)(2-3)(2-5)}{3!}\right|\left|f^{(3)}(\alpha)\right|$

$= \left|\dfrac{1 \cdot (-1) \cdot (-3)}{6}\right|\dfrac{2}{\alpha^3} = \dfrac{1}{\alpha^3} \leq 1$ for $\alpha \in [1,5]$.

A calculator gives $\ln 2 \approx 0.693$, so the actual error is approximately $0.693 - 0.623 \approx 0.07$.

57. The second order Maclauring polynomial is

$$P_2(x) = 1 + x + \frac{x^2}{2}$$

From Problem 56, we know that the interpolating polynomial is $L(x) = \frac{50}{3}(x-0.2)(x-0.3)$

$$-61.05x(x-0.3)$$

$$+45x(x-0.2)$$

For the Maclaurin polynomial

$$R_2(x) = \frac{f^{(3)}(c)}{3!}x^3 = \frac{1}{6}e^c x^3$$

Thus,

$$\left|R_2(x)\right| = \frac{1}{6}e^c\left|x^3\right| \le \frac{1}{6}e^{0.3}0.3^3 \approx 0.0060744$$

For the interpolating polynomial,

$$\left|R_2(x)\right| = \left|\frac{x(x-0.2)(x-0.3)}{6}\right|e^c$$

The expression in absolute values reaches a maximum of 0.0003521 when $x = 0.078475$. On the interval $[0, 0.3]$, the expression e^c reaches a maximum of $e^{0.3} \approx 1.350$. Thus, the largest possible error for the interpolating polynomial on $[0, 0.3]$ is

$$R_2(x) \le 0.0003521 \cdot 1.350 \approx 0.0004753$$

A calculator gives $e^{0.1} \approx 1.105$. When $x = 0.1$, the error for the Maclaurin polynomial is

$$\left|1.105 - P_2(0.1)\right| = \left|1.105 - 1.105\right| = 0$$

and the error for the interpolating polynomial is

$$\left|1.105 - L_2(0.1)\right| = \left|1.105 - 1.104\right| = 0.001$$

11.2 Concepts Review

1. $1, 2, 2, 2, ..., 2, 1$ **3.** n^4

Problem Set 11.2

1. $f(x) = \frac{1}{x^2}; h = \frac{3-1}{8} = 0.25$

$x_0 = 1.00$	$f(x_0) = 1$	$x_5 = 2.25$	$f(x_5) \approx 0.1975$
$x_1 = 1.25$	$f(x_1) = 0.64$	$x_6 = 2.50$	$f(x_6) = 0.16$
$x_2 = 1.50$	$f(x_2) \approx 0.4444$	$x_7 = 2.75$	$f(x_7) \approx 0.1322$
$x_3 = 1.75$	$f(x_3) \approx 0.3265$	$x_8 = 3.00$	$f(x_8) \approx 0.1111$
$x_4 = 2.00$	$f(x_4) = 0.25$		

Left Riemann Sum: $\int_1^3 \frac{1}{x^2}dx \approx 0.25[f(x_0) + f(x_1) + ... + f(x_7)] \approx 0.7846$

Trapezoidal Rule: $\int_1^3 \frac{1}{x^2}dx \approx \frac{0.25}{2}[f(x_0) + 2f(x_1) + ... + 2f(x_7) + f(x_8)] \approx 0.6766$

Parabolic Rule: $\int_1^3 \frac{1}{x^2}dx \approx \frac{0.25}{3}[f(x_0) + 4f(x_1) + 2f(x_2) + ... + 4f(x_7) + f(x_8)] \approx 0.6671$

Fundamental Theorem of Calculus: $\int_1^3 \frac{1}{x^2}dx = \left[-\frac{1}{x}\right]_1^3 = -\frac{1}{3} + 1 = \frac{2}{3} \approx 0.6667$

3. $f(x) = \sqrt{x}; h = \frac{2-0}{8} = 0.25$

$x_0 = 0.00$	$f(x_0) = 0$	$x_5 = 1.25$	$f(x_5) \approx 1.1180$
$x_1 = 0.25$	$f(x_1) = 0.5$	$x_6 = 1.50$	$f(x_6) \approx 1.2247$
$x_2 = 0.50$	$f(x_2) \approx 0.7071$	$x_7 = 1.75$	$f(x_7) \approx 1.3229$
$x_3 = 0.75$	$f(x_3) \approx 0.8660$	$x_8 = 2.00$	$f(x_8) \approx 1.4142$
$x_4 = 1.00$	$f(x_4) = 1$		

Left Riemann Sum: $\int_0^2 \sqrt{x}\, dx \approx 0.25[f(x_0) + f(x_1) + \ldots + f(x_7)] \approx 1.6847$

Trapezoidal Rule: $\int_0^2 \sqrt{x}\,dx \approx \frac{0.25}{2}[f(x_0) + 2f(x_1) + \ldots + 2f(x_7) + f(x_8)] \approx 1.8615$

Parabolic Rule: $\int_0^2 \sqrt{x}\,dx \approx \frac{0.25}{3}[f(x_1) + 4f(x_2) + 2f(x_3) + \ldots + 4f(x_7) + f(x_8)] \approx 1.8755$

Fundamental Theorem of Calculus: $\int_0^2 \sqrt{x}\,dx = \left[\frac{2}{3} x^{3/2}\right]_0^2 = \frac{4\sqrt{2}}{3} \approx 1.8856$

5. $f(x) = \sin x$

$n = 2: \ h = \frac{\pi}{2}$

$x_0 = 0 \qquad\qquad\qquad f(x_0) = 0$

$x_1 = \frac{\pi}{2} \qquad\qquad\quad f(x_1) = 1$

$x_2 = \pi \qquad\qquad\quad f(x_2) = 0$

$\int_0^\pi \sin x\, dx \approx \frac{\pi}{4}[f(x_0) + 2f(x_1) + f(x_0)] = \frac{\pi}{2} \approx 1.5708$

$n = 6: \ h = \frac{\pi}{6}$

$x_0 = 0 \qquad\qquad\quad f(x_0) = 0 \qquad\qquad\qquad x_4 = \frac{2\pi}{3} \qquad\qquad f(x_4) = \frac{\sqrt{3}}{2}$

$x_1 = \frac{\pi}{6} \qquad\qquad\quad f(x_1) = \frac{1}{2} \qquad\qquad\qquad x_5 = \frac{5\pi}{6} \qquad\qquad f(x_5) = \frac{1}{2}$

$x_2 = \frac{\pi}{3} \qquad\qquad\quad f(x_2) = \frac{\sqrt{3}}{2} \qquad\qquad\qquad x_6 = \pi \qquad\qquad\quad f(x_6) = 0$

$x_3 = \frac{\pi}{2} \qquad\qquad\quad f(x_3) = 1$

$\int_0^\pi \sin x\, dx \approx \frac{\pi}{12}[f(x_0) + 2f(x_1) + \ldots + 2f(x_5) + f(x_6)] = \frac{\pi}{12}\left(4 + 2\sqrt{3}\right) \approx 1.9541$

$n = 12: \ h = \frac{\pi}{12}$

$x_0 = 0 \qquad\qquad\qquad f(x_0) = 0 \qquad\qquad\qquad\quad x_7 = \frac{7\pi}{12} \qquad\qquad f(x_7) = \frac{\sqrt{2}}{4}\left(\sqrt{3} + 1\right)$

$x_1 = \frac{\pi}{12} \qquad\qquad\quad f(x_1) = \frac{\sqrt{2}}{4}\left(\sqrt{3} - 1\right) \qquad\quad x_8 = \frac{2\pi}{3} \qquad\qquad f(x_8) = \frac{\sqrt{3}}{2}$

$x_2 = \frac{\pi}{6} \qquad\qquad\quad f(x_2) = \frac{1}{2} \qquad\qquad\qquad\quad x_9 = \frac{3\pi}{4} \qquad\qquad f(x_9) = \frac{\sqrt{2}}{2}$

$x_3 = \frac{\pi}{4} \qquad\qquad\quad f(x_3) = \frac{\sqrt{2}}{2} \qquad\qquad\qquad x_{10} = \frac{5\pi}{6} \qquad\qquad f(x_{10}) = \frac{1}{2}$

$x_4 = \frac{\pi}{3} \qquad\qquad\quad f(x_4) = \frac{\sqrt{3}}{2} \qquad\qquad\qquad x_{11} = \frac{11\pi}{12} \qquad\quad f(x_{11}) = \frac{\sqrt{2}}{4}\left(\sqrt{3} - 1\right)$

$x_5 = \frac{5\pi}{12} \qquad\qquad f(x_5) = \frac{\sqrt{2}}{4}\left(\sqrt{3} + 1\right) \qquad\quad x_{12} = \pi \qquad\qquad\quad f(x_{12}) = 0$

$x_6 = \frac{\pi}{2} \qquad\qquad\quad f(x_6) = 1$

$\int_0^\pi \sin x\, dx \approx \frac{\pi}{24}[f(x_0) + 2f(x_1) + \ldots + 2f(x_{11}) + f(x_{12})] \approx \frac{\pi}{24}\left(4 + 2\sqrt{3} + 2\sqrt{2} + 2\sqrt{6}\right) \approx 1.9886$

7. $f(x) = \dfrac{4}{1+x^2}; h = \dfrac{1}{10}$

$x_0 = 0.0$	$f(x_0) = 4$	$x_6 = 0.6$	$f(x_6) \approx 2.9412$
$x_1 = 0.1$	$f(x_1) \approx 3.9604$	$x_7 = 0.7$	$f(x_7) \approx 2.6846$
$x_2 = 0.2$	$f(x_2) \approx 3.8462$	$x_8 = 0.8$	$f(x_8) \approx 2.4390$
$x_3 = 0.3$	$f(x_3) \approx 3.6697$	$x_9 = 0.9$	$f(x_9) \approx 2.2099$
$x_4 = 0.4$	$f(x_4) \approx 3.4483$	$x_{10} = 1.0$	$f(x_{10}) = 2$
$x_5 = 0.5$	$f(x_5) = 3.2$		

$$\int_0^1 \frac{4}{1+x^2}\,dx \approx \frac{1}{30}[f(x_0)+4f(x_1)+2f(x_2)+\ldots+4f(x_9)+f(x_{10})] \approx 3.1416$$

9. $f(x) = e^{-x^2}$

$$f'(x) = -2xe^{-x^2}$$

$$f''(x) = -2e^{-x^2} + 4x^2 e^{-x^2} = e^{-x^2}(4x^2 - 2)$$

$$|E_n| = \frac{1}{12n^2}\left|e^{-c^2}(4c^2 - 2)\right| \leq \frac{1}{12n^2}\cdot 2 = \frac{1}{6n^2}$$

$\dfrac{1}{6n^2} \leq 0.01$ when $n \geq 5$.

$h = \dfrac{1}{5} = 0.2$

$x_0 = 0.0$	$f(x_0) = 1$	$x_3 = 0.6$	$f(x_3) \approx 0.6977$
$x_1 = 0.2$	$f(x_1) \approx 0.9608$	$x_4 = 0.8$	$f(x_4) \approx 0.5273$
$x_2 = 0.4$	$f(x_2) \approx 0.8521$	$x_5 = 1.0$	$f(x_5) \approx 0.3679$

$$\int_0^1 e^{-x^2}\,dx \approx \frac{0.2}{2}[f(x_0)+2f(x_1)+\ldots+2f(x_4)+f(x_5)] \approx 0.74$$

11. $f(x) = \sqrt{\cos x}$

$$f'(x) = -\frac{\sin x}{2\sqrt{\cos x}}$$

$$f''(x) = -\frac{2\cos^2 x + \sin^2 x}{4\cos^{3/2} x}$$

$$|E_n| = \frac{(0.5)^3}{12n^2}\left|\frac{2\cos^2 x + \sin^2 x}{4\cos^{3/2} x}\right|$$

$$\leq \frac{(0.5)^3}{12n^2}\left|\frac{2\cos^2 1 + \sin^2 1.5}{4\cos^{3/2} 1.5}\right| \leq \frac{2.6225}{12n^2}$$

$\dfrac{2.6225}{12n^2} \leq 0.01$ when $n \geq 5$.

$h = \dfrac{0.5}{5} = 0.1$

$x_0 = 1.0$	$f(x_0) \approx 0.7351$
$x_1 = 1.1$	$f(x_1) \approx 0.6735$
$x_2 = 1.2$	$f(x_2) \approx 0.6020$
$x_3 = 1.3$	$f(x_3) \approx 0.5172$

$x_4 = 1.4$	$f(x_4) \approx 0.4123$
$x_5 = 1.5$	$f(x_5) \approx 0.2660$

$$\int_1^{1.5} \sqrt{\cos x}\,dx \approx \frac{0.1}{2}[f(x_0)+2f(x_1)+2f(x_2)$$
$$+2f(x_3)+2f(x_4)+f(x_5)] \approx 0.27$$

13. $f(x) = \dfrac{1+x}{1-x}$

$$f'(x) = \frac{2}{(1-x)^2}$$

$$f''(x) = \frac{4}{(1-x)^3}$$

$$f^{(3)}(x) = \frac{12}{(1-x)^4}$$

$$f^{(4)}(x) = \frac{48}{(1-x)^5}$$

$$|E_n| = \frac{4^5}{180n^4}\left|\frac{48}{(1-c)^5}\right| \le \frac{4^5}{180n^4}\left(\frac{48}{1}\right) = \frac{4096}{15n^4}$$

$$\frac{4096}{15n^4} \le 0.005 \quad \text{when } n \ge 16.$$

$$h = \frac{4}{16} = 0.25$$

$$\int_2^6 \frac{1+x}{1-x}\,dx \approx \frac{0.25}{3}[f(x_0) + 4f(x_1) + 2f(x_2) + \ldots$$
$$+ 4f(x_{15}) + f(x_{16})] \approx -7.219$$

15. $\int_{m-h}^{m+h}(ax^2 + bx + c)\,dx = \left[\frac{a}{3}x^3 + \frac{b}{2}x^2 + cx\right]_{m-h}^{m+h}$

$\quad = \frac{a}{3}(m+h)^3 + \frac{b}{2}(m+h)^2 + c(m+h) - \frac{a}{3}(m-h)^3 - \frac{b}{2}(m-h)^2 - c(m-h)$

$\quad = \frac{a}{3}(6m^2 h + 2h^3) + \frac{b}{2}(4mh) + c(2h) = \frac{h}{3}[a(6m^2 + 2h^2) + b(6m) + 6]$

$\quad = \frac{h}{3}[f(m-h) + 4f(m) + f(m+h)]$

$\quad = \frac{h}{3}[a(m-h)^2 + b(m-h) + c + 4am^2 + 4bm + 4c + a(m+h)^2 + b(m+h) + c]$

$\quad = \frac{h}{3}[a(6m^2 + 2h^2) + b(6m) + 6c]$

17. $f(x) = \dfrac{1}{x}$

$\quad f'(x) = -\dfrac{1}{x^2}$

$\quad f''(x) = \dfrac{2}{x^3}$

$|E_n| = \dfrac{1}{12n^2}\left(\dfrac{2}{c^3}\right) \le \dfrac{1}{6n^2}$

$\dfrac{1}{6n^2} \le 10^{-10} \quad \text{when } n \ge 40825.$

19. Let $n = 2$.

$\quad f(x) = x^k; \quad h = a$

$\quad x_0 = -a \qquad f(x_0) = -a^k$

$\quad x_1 = 0 \qquad f(x_1) = 0$

$\quad x_2 = a \qquad f(x_2) = a^k$

$\quad \int_{-a}^a x^k\,dx \approx \dfrac{a}{2}[-a^k + 2 \cdot 0 + a^k] = 0$

$\quad \int_{-a}^a x^k\,dx = \left[\dfrac{1}{k+1}x^{k+1}\right]_{-a}^a = \dfrac{1}{k+1}[a^{k+1} - (-a)^{k+1}] = \dfrac{1}{k+1}[a^{k+1} - a^{k+1}] = 0$

21. $A \approx \dfrac{10}{2}[75 + 2 \cdot 71 + 2 \cdot 60 + 2 \cdot 45 + 2 \cdot 45 + 2 \cdot 52 + 2 \cdot 57 + 2 \cdot 60 + 59] = 4570 \text{ ft}^2$

23. $A \approx \dfrac{20}{3}[0 + 4 \cdot 7 + 2 \cdot 12 + 4 \cdot 18 + 2 \cdot 20 + 4 \cdot 20 + 2 \cdot 17 + 4 \cdot 10 + 0] = 2120 \text{ ft}^2$

$\quad 4 \text{ mi/h} = 21{,}120 \text{ ft/h}$

$\quad (2120)(21{,}120)(24) = 1{,}074{,}585{,}600 \text{ ft}^3$

25. $h = \dfrac{24-0}{8} = 3 \text{ minutes} = \dfrac{1}{20} \text{ hour}$

Suppose t is time measured in hours and $v(t)$ is the velocity at time t.

a. Using the Trapezoidal Rule, the distance traveled is

$$\int_0^{0.4} v(t)\,dt \approx \frac{1/20}{2}\left[0+2\cdot31+2\cdot54+2\cdot53+2\cdot52+2\cdot35+2\cdot31+2\cdot28+0\right]$$

$$=\frac{464}{40}=11.6\,\text{miles}$$

b. Using the Parabolic Rule, the distance traveled is

$$\int_0^{0.4} v(t)\,dt \approx \frac{1/20}{3}\left[1\cdot0+4\cdot31+2\cdot54+4\cdot53+2\cdot52+4\cdot35+2\cdot31+4\cdot28+1\cdot0\right]$$

$$=\frac{568}{40}=14.2\,\text{miles.}$$

27. **a.** Lay the part with the long flat side along the x-axis, with the upper left corner (as shown in Figure 22) at the origin. Let $h(x)$ denote the height of the lamina at the value of x. Then, using the Trapezoidal Rule,

$$m=\int_0^{40} h(x)\,dx \approx \frac{5}{2}\left[5+2(6.5+8+9+10+10.5+10.5+10)+8\right]=355$$

$$M_y=\int_0^{40} x\,h(x)\,dx \approx \frac{5}{2}\left[0\cdot5+2(5\cdot6.5+10\cdot8+15\cdot9+20\cdot10+25\cdot10.5+30\cdot10.5+35\cdot10)+40\cdot8\right]=7675$$

$$\bar{x}=\frac{M_y}{m}\approx\frac{7675}{355}\approx21.62$$

$$M_x=\frac{1}{2}\int_0^{40} h^2(x)\,dx \approx \frac{1}{2}\frac{5}{2}\left[5^2+2\left(6.5^2+8^2+9^2+10^2+10.5^2+10.5^2+10^2\right)+8^2\right]=1630.625$$

$$\bar{y}=\frac{M_x}{m}\approx\frac{1630.625}{355}=4.59$$

b. Using the Parabolic Rule,

$$m=\int_0^{40} h(x)\,dx \approx \frac{5}{3}\left[5+4\cdot6.5+2\cdot8+4\cdot9+2\cdot10+4\cdot10.5+2\cdot10.5+4\cdot10+8\right]=\frac{5}{3}214\approx356.67$$

$$M_y=\int_0^{40} x\,h(x)\,dx$$

$$\approx\frac{5}{3}\left[0\cdot5+4(5\cdot6.5)+2(10\cdot8)+4(15\cdot9)+2(20\cdot10)+4(25\cdot10.5)+2(30\cdot10.5)+4(35\cdot10)+40\cdot8\right]$$

$$=\frac{5}{3}4630\approx7716.67$$

$$\bar{x}=\frac{M_y}{m}\approx\frac{7716.67}{356.67}\approx21.64$$

$$M_x=\frac{1}{2}\int_0^{40} h^2(x)\,dx$$

$$\approx\frac{1}{2}\frac{5}{3}\left[5^2+4\left(6.5^2\right)+2\left(8^2\right)+4\left(9^2\right)+2\left(10^2\right)+4\left(10.5^2\right)+2\left(10.5^2\right)+4\left(10^2\right)+8^2\right]$$

$$=\frac{5}{6}1971.5\approx1642.9$$

$$\bar{y}=\frac{M_x}{m}\approx\frac{1642.9}{356.67}=4.61$$

29. Rotate the map $90°$ clockwise and put the x-axis along the bottom, with the origin below the southernmost tip of Illinois. Let $f(x)$ denote the "upper" function, and $g(x)$ the "lower" function. Then the east/west dimensions given on the map are $f(x)-g(x)$, and $g(x)$ is equal to 85, 50, 30, 25, 15, and 10 at the southern end of the state, and 10 and 13 at the northern end.

$$m=\int_0^{380}\left[f(x)-g(x)\right]dx \approx \frac{20}{2}\left[0+2(58+79+\cdots+139+132)+140\right]=57,000$$

$$M_y = \int_0^{380} x[f(x) - g(x)]\,dx \approx \frac{20}{2}\left[0 \cdot 0 + 2(20 \cdot 58 + 40 \cdot 79 + \cdots + 340 \cdot 139 + 360 \cdot 132) + 380 \cdot 140\right] \approx 12{,}002{,}800$$

$$M_x = \frac{1}{2}\int_0^{380}\left[f^2(x) - g^2(x)\right]dx \approx \frac{1}{2}\frac{20}{2}\left\{\left(85^2 - 85^2\right) + 2\left[\left(108^2 - 50^2\right) + \left(109^2 - 30^2\right) + \cdots + \left(142^2 - 10^2\right)\right]\right\}$$

$$\approx 4{,}994{,}480$$

$$\bar{x} = \frac{M_y}{m} \approx \frac{12{,}002{,}800}{57{,}000} \approx 210.6$$

$$\bar{y} = \frac{M_x}{m} \approx \frac{4{,}994{,}480}{57{,}000} \approx 87.6$$

This is a point just southeast of Lincoln, IL and it is about 30 miles northeast of Springfield.

31. a. Since $x > X > 0$, $x^2 > Xx$. Thus $e^{-x^2} < e^{-Xx}$.

$$\int_X^\infty e^{-x^2}\,dx < \int_X^\infty e^{-Xx}\,dx = \left[-\frac{1}{X}e^{-Xx}\right]_X^\infty = \frac{1}{X}e^{-X^2}$$

b. $\displaystyle\int_5^\infty e^{-x^2}\,dx < \int_5^\infty e^{-5x}\,dx = \frac{1}{5}e^{-25} \approx 2.78 \times 10^{-12} < 10^{-11}$

c. $n = 10$, $h = \dfrac{5 - 0}{10} = 0.5$

$x_0 = 0.0$	$f(x_0) = 1$	$x_6 = 3.0$	$f(x_6) \approx 1.234 \times 10^{-4}$
$x_1 = 0.5$	$f(x_1) \approx 0.7788$	$x_7 = 3.5$	$f(x_7) \approx 4.785 \times 10^{-6}$
$x_2 = 1.0$	$f(x_2) \approx 0.3679$	$x_8 = 4.0$	$f(x_8) \approx 1.125 \times 10^{-7}$
$x_3 = 1.5$	$f(x_3) \approx 0.1054$	$x_9 = 4.5$	$f(x_9) \approx 1.605 \times 10^{-9}$
$x_4 = 2.0$	$f(x_4) \approx 0.0183$	$x_{10} = 5.0$	$f(x_{10}) \approx 1.389 \times 10^{-11}$
$x_5 = 2.5$	$f(x_5) \approx 0.0019$		

$$\int_0^5 e^{-x^2}\,dx \approx \frac{0.5}{3}[f(x_0) + 4f(x_1) + 2f(x_2) + \cdots + 4f(x_9) + f(x_{10})] \approx 0.8862$$

$$\frac{2}{\sqrt{\pi}}\int_0^\infty e^{-x^2}\,dx \approx \frac{2}{\sqrt{\pi}}(0.8862) \approx 0.99997$$

$$f(x) = e^{-x^2}$$

$$f'(x) = -2xe^{-x^2} \qquad\qquad f^{(3)}(x) = xe^{-x^2}(12 - 8x^2)$$

$$f''(x) = e^{-x^2}(4x^2 - 2) \qquad\qquad f^{(4)}(x) = e^{-x^2}(16x^4 - 48x^2 + 12)$$

$$|E_{10}| = \frac{5^5}{180 \times 10^4}\left|\frac{16c^4 - 48c^2 + 12}{e^{c^2}}\right| \le \frac{5^5}{1{,}800{,}000}\cdot 12$$

A graphing utility or computer shows that

$$\left|\frac{16c^4 - 48c^2 + 12}{e^{c^2}}\right| \le 12 \text{ on } [0, 5].$$

$$|E_{10}| \le 0.0209$$

Thus, the error in computing

$$\frac{2}{\sqrt{\pi}}\int_0^\infty e^{-x^2}\,dx \text{ is less than } \frac{2}{\sqrt{\pi}}(0.0209 + 10^{-11}) \approx 0.0236.$$

33. a. $u = \dfrac{2}{\sqrt{4+x}}$ $dv = \dfrac{1}{2\sqrt{x}}dx$

   $du = -\dfrac{1}{(4+x)^{3/2}}dx$ $v = \sqrt{x}$

   $$\int_0^2 \frac{dx}{\sqrt{4x+x^2}} = \left[\frac{2\sqrt{x}}{\sqrt{4+x}}\right]_0^2 + \int_0^2 \frac{\sqrt{x}}{(4+x)^{3/2}}dx = \frac{2}{\sqrt{3}} + \int_0^2 \frac{\sqrt{x}}{(4+x)^{3/2}}dx$$

b. $u = \dfrac{1}{x}$ $dv = \sin x\, dx$

   $du = -\dfrac{1}{x^2}dx$ $v = -\cos x$

   $$\int_1^\infty \frac{\sin x}{x}dx = \left[-\frac{\cos x}{x}\right]_1^\infty - \int_1^\infty \frac{\cos x}{x^2}dx = \cos 1 - \int_1^\infty \frac{\cos x}{x^2}dx$$

   $|\cos x| \le 1$ for all x, so

   $$\int_1^\infty \frac{\cos x}{x^2}dx \le \int_1^\infty \frac{1}{x^2}dx = \left[-\frac{1}{x}\right]_1^\infty = 1.$$

   Therefore, $\displaystyle\int_1^\infty \frac{\sin x}{x}dx$ exists.

c. $u = \dfrac{1}{4+x}$ $dv = \dfrac{1}{\sqrt{x}}dx$

   $du = -\dfrac{1}{(4+x)^2}$ $v = 2\sqrt{x}$

   $$\int_0^1 \frac{1}{\sqrt{x}}\frac{1}{4+x}dx = \left[\frac{2\sqrt{x}}{4+x}\right]_0^1 + \int_0^1 \frac{2\sqrt{x}}{(4+x)^2}dx = \frac{2}{5} + \int_0^1 \frac{2\sqrt{x}}{(4+x)^2}dx$$

35. $\displaystyle\int_0^{\pi/2} \ln(\sin x)dx = \int_0^{\pi/2} \ln x\, dx + \int_0^{\pi/2} \ln\left(\frac{\sin x}{x}\right)dx$

   $\displaystyle\int_0^{\pi/2} \ln x\, dx = \lim_{a\to 0}\int_a^{\pi/2} \ln x\, dx = \lim_{a\to 0}[x\ln x - x]_a^{\pi/2} = \left(\frac{\pi}{2}\ln\frac{\pi}{2} - \frac{\pi}{2}\right) - \lim_{a\to 0} a\ln a = \frac{\pi}{2}\ln\frac{\pi}{2} - \frac{\pi}{2} \approx -0.8615$

   $f(x) = \ln\left(\dfrac{\sin x}{x}\right), h = \dfrac{\pi}{8} \approx 0.3927$

   Note that $\displaystyle\lim_{x\to 0}\ln\left(\frac{\sin x}{x}\right) = \ln 1 = 0.$

i	x_i	$f(x_i)$
0	0.0000	0
1	0.3927	−0.02584
2	0.7854	−0.10501
3	1.1781	−0.24307
4	1.5708	−0.45158

   $\displaystyle\int_0^{\pi/2} \ln\left(\frac{\sin x}{x}\right)dx \approx \frac{\pi}{24}[f(x_0) + 4f(x_1) + 2f(x_2) + 4f(x_3) + f(x_4)] \approx -0.2274$

   $\displaystyle\int_0^{\pi/2} \ln(\sin x)dx \approx -1.0889$

11.3 Concepts Review

1. slowness of convergence

3. algorithms

Problem Set 11.3

1. Let $f(x) = x^3 + 2x - 6$.

$f(1) = -3, f(2) = 6$

n	h_n	m_n	$f(m_n)$
1	0.5	1.5	0.375
2	0.25	1.25	−1.546875
3	0.125	1.375	−0.650391
4	0.0625	1.4375	−0.154541
5	0.03125	1.46875	0.105927
6	0.015625	1.45312	−0.0253716
7	0.0078125	1.46094	0.04001
8	0.00390625	1.45703	0.00725670
9	0.00195312	1.45508	−0.00907617

$r \approx 1.46$

3. Let $f(x) = 2\cos x - e^{-x}$.

$f(1) \approx 0.712725, f(2) \approx -0.967629$

n	h_n	m_n	$f(m_n)$
1	0.5	1.5	−0.0816558
2	0.25	1.25	0.34414
3	0.125	1.375	0.136256
4	0.0625	1.4375	0.0282831
5	0.03125	1.46875	−0.0264745
6	0.015625	1.45313	0.000961516
7	0.0078125	1.46094	−0.0127427
8	0.00390625	1.45703	−0.00588708
9	0.00195312	1.45508	−0.0024619
10	0.000976562	1.4541	−0.000749968
11	0.000488281	1.45361	0.00010583

$r \approx 1.45$

5. Let $f(x) = x^3 + 6x^2 + 9x + 1 = 0$.

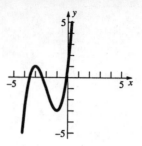

$f'(x) = 3x^2 + 12x + 9$

n	x_n
1	0
2	−0.1111111
3	−0.1205484
4	−0.1206148
5	−0.1206148

$r \approx -0.12061$

7. Let $f(x) = x - 2 + 2\ln x$.

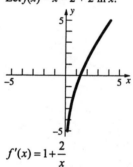

$f'(x) = 1 + \dfrac{2}{x}$

n	x_n
1	1.5
2	1.366744
3	1.370151
4	1.370154
5	1.370154

$r \approx 1.37015$

9. Let $f(x) = \cos x - 2x$.

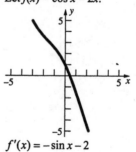

$f'(x) = -\sin x - 2$

n	x_n
1	0.5
2	0.4506267
3	0.4501836
4	0.4501836

$r \approx 0.45018$

11. Let $f(x) = x^4 - 8x^3 + 22x^2 - 24x + 8$.

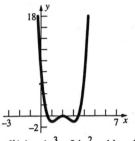

$f'(x) = 4x^3 - 24x^2 + 44x - 24$

Note that $f(2) = 0$.

n	x_n
1	0.5
2	0.575
3	0.585586
4	0.585786

n	x_n
1	3.5
2	3.425
3	3.414414
4	3.414214
5	3.414214

$r = 2,\, r \approx 0.58579,\, r \approx 3.41421$

13. Let $f(x) = 2x^2 - \sin^{-1} x$.

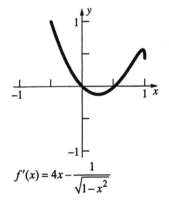

$f'(x) = 4x - \dfrac{1}{\sqrt{1-x^2}}$

n	x_n
1	0.5
2	0.527918
3	0.526583
4	0.526580
5	0.526580

$r \approx 0.52658$

15. Let $f(x) = x^3 - 6$.

$f'(x) = 3x^2$

n	x_n
1	1.5
2	1.888889
3	1.819813
4	1.817125
5	1.817121
6	1.817121

$\sqrt[3]{6} \approx 1.81712$

17. Let $g(x) = \dfrac{\sin x}{x}$.

$f(x) = g'(x) = \dfrac{\cos x}{x} - \dfrac{\sin x}{x^2}$

$f'(x) = -\dfrac{\sin x}{x} - \dfrac{2\cos x}{x^2} + \dfrac{2\sin x}{x^3}$

n	x_n
1	4.5
2	4.4934
3	4.49341
4	4.49341

Minimum at $x \approx 4.49341$
Minimum value is ≈ -0.21723.

19. Let $f(x) = x^2 - 2$.

$f'(x) = 2x,\, f''(x) = 2$

$|f'(x)| \geq 2$ on $[1, 2]$, so $m = 2$.

$|f''(x)| \leq 2$ on $[1, 2]$, so $M = 2$.

$\left| x_6 - \sqrt{2} \right| \leq \dfrac{2(2)}{2}\left(\dfrac{2}{2(2)} |1.5 - r| \right)^{2^{6-1}}$

$\leq 2(0.25)^{32} \approx 1.08 \times 10^{-19}$

21. Let $f(x) = \dfrac{1 + \ln x}{x}$.

$$f'(x) = -\frac{\ln x}{x^2}$$

$$x_{n+1} = x_n - \frac{f(x_n)}{f'(x_n)} = x_n + \frac{x_n}{\ln x_n} + x_n$$

$$= 2x_n + \frac{x_n}{\ln x_n}$$

Suppose $x_1 = 1.2$.

n	x_n
1	1.2
2	8.981778
3	22.05511
4	51.23963
5	115.4958
6	255.3103
7	556.6849
8	1201.425

Suppose $x_1 = 0.5$.

n	x_n
1	0.5
2	0.2786525
3	0.3392312
4	0.3646713
5	0.3678377
6	0.3678794
7	0.3678794

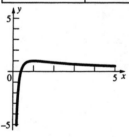

23. a. For Tom's car, $P = 2000$, $R = 100$, and $k = 24$, thus

$$2000 = \frac{100}{i}\left[1 - \frac{1}{(1+i)^{24}}\right] \text{ or}$$

$$20i = 1 - \frac{1}{(1+i)^{24}}, \text{ which is equivalent}$$

to $20i(1+i)^{24} - (1+i)^{24} + 1 = 0$.

b. Let

$$f(i) = 20i(1+i)^{24} - (1+i)^{24} + 1$$
$$= (1+i)^{24}(20i - 1) + 1.$$

Then

$$f'(i) = 20(1+i)^{24} + 480i(1+i)^{23} - 24(1+i)^{23}$$
$$= (1+i)^{23}(500i - 4), \text{ so}$$

$$i_{n+1} = i_n - \frac{f(i_n)}{f'(i_n)} = i_n - \frac{(1+i_n)^{24}(20i_n - 1) + 1}{(1+i_n)^{23}(500i_n - 4)}$$

$$= i_n - \left[\frac{20i_n^2 + 19i_n - 1 + (1+i_n)^{-23}}{500i_n - 4}\right].$$

c.

n	i_n
1	0.012
2	0.0165297
3	0.0152651
4	0.0151323
5	0.0151308
6	0.0151308

$i = 0.0151308$
$r = 18.157\%$

25. a. The algorithm computes the root of $\dfrac{1}{x} - a = 0$ for x_1 close to $\dfrac{1}{a}$.

b. Let $f(x) = \dfrac{1}{x} - a$.

$$f'(x) = -\frac{1}{x^2}$$

$$\frac{f(x)}{f'(x)} = -x + ax^2$$

The recursion formula is

$$x_{n+1} = x_n - \frac{f(x_n)}{f'(x_n)} = 2x_n - ax_n^2.$$

27. $r \approx -1.87939$, $r \approx 0.34730$, $r \approx 1.53209$

29. $r \approx -2.08204$, $r \approx 0.09251$, $r \approx 0.91314$, $r \approx 1.62015$, $r \approx 1.85411$

11.4 Concepts Review

1. fixed point

3. $|g'(x)| \le M < 1$

Problem Set 11.4

1. $x_{n+1} = \dfrac{1}{9}e^{-2x_n}$

n	x_n
1	1
2	0.015037
3	0.107819
4	0.089559
5	0.092890
6	0.092273
7	0.092387
8	0.092366
9	0.092370
10	0.092369
11	0.092369

$x \approx 0.09237$

3. $x_{n+1} = \sqrt{2.7 + x_n}$

n	x_n
1	1
2	1.923538
3	2.150241
4	2.202326
5	2.214120
6	2.216781
7	2.217382
8	2.217517
9	2.217548
10	2.217554
11	2.217556
12	2.217556

$x \approx 2.21756$

5. a.

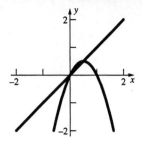

$x \approx 0.5$

b. $x_{n+1} = 2(x_n - x_n^2)$

n	x_n
1	0.7
2	0.42
3	0.4872
4	0.4996723
5	0.4999998
6	0.5
7	0.5

c. $x = 2(x - x^2)$

$2x^2 - x = 0$

$x(2x - 1) = 0$

$x = 0, \ x = \dfrac{1}{2}$

d. $g'(x) = 2 - 4x$

$g'\left(\dfrac{1}{2}\right) = 0$

7. a.

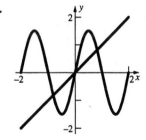

b. The algorithm does not yield a convergent sequence.

c. $g'(x) = \dfrac{3\pi}{2}\cos \pi x$

$|g'(x)| > 1$ in a neighborhood of the fixed points.

9. a. $x = \dfrac{3}{2}\sin \pi x$

$$6x = 5x + \dfrac{3}{2}\sin \pi x$$

$$x = \dfrac{5}{6}x + \dfrac{1}{4}\sin \pi x$$

b. Let $x_1 = 0.8$.

n	x_n
1	0.8
2	0.813613
3	0.816176
4	0.816629
5	0.816709
6	0.816723
7	0.816725
8	0.816726
9	0.816726

Let $x_1 = -0.8$.

n	x_n
1	−0.8
2	−0.813613
3	−0.816176
4	−0.816629
5	−0.816709
6	−0.816723
7	−0.816725
8	−0.816726
9	−0.816726

c. $g'(x) = \dfrac{5}{6} + \dfrac{\pi}{4}\cos \pi x$

$g'(0.81673) \approx 0.17456$

$g'(-0.81673) \approx 0.17456$

11. Graph $y = x^3 - x^2 - x - 1$.

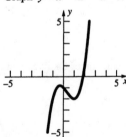

The positive root is near 2.

Rewrite the equation as $x = 1 + \dfrac{1}{x} + \dfrac{1}{x^2} = g(x)$.

Let $x_1 = 1.8$

n	x_n
1	1.8
2	1.8642
3	1.8242
4	1.8487
5	1.8335
6	1.8429
7	1.8371
8	1.8406
9	1.8384
10	1.8398
11	1.8390
12	1.8395
13	1.8392
14	1.8394
15	1.8392
16	1.8393
17	1.8393

$r \approx 1.839$

13. a. $x_1 = 0$

$x_2 = \sqrt{1} = 1$

$x_3 = \sqrt{1 + \sqrt{1}} = \sqrt{2} \approx 1.4142136$

$x_4 = \sqrt{1 + \sqrt{1 + \sqrt{1}}} \approx 1.553774$

$x_5 = \sqrt{1 + \sqrt{1 + \sqrt{1 + \sqrt{1}}}} \approx 1.5980532$

b. $x = \sqrt{1 + x}$

$x^2 = 1 + x$

$x^2 - x - 1 = 0$

$$x = \dfrac{1 \pm \sqrt{1 + 4 \cdot 1 \cdot 1}}{2} = \dfrac{1 \pm \sqrt{5}}{2}$$

Taking the minus sign gives a negative solution for x, violating the requirement that $x \geq 0$.

Hence, $x = \dfrac{1 + \sqrt{5}}{2} \approx 1.618034$.

c. Let $x = \sqrt{5 + \sqrt{5 + \sqrt{5 + \dots}}}$. Then x satisfies the equation $x = \sqrt{5 + x}$. From part b we know that x must equal $\left(1 + \sqrt{5}\right)/2 \approx 1.618034$.

15. Let $a = \pi$ and $x_1 = 2$.

n	x_n
1	2
2	1.785398
3	1.772501
4	1.772454
5	1.772454

$\sqrt{\pi} \approx 1.77245$

$g'(x) = \dfrac{1}{2}\left(1 - \dfrac{a}{x^2}\right)$

$g'(a) = \dfrac{1}{2}\left(1 - \dfrac{1}{a}\right) < 1$ for $a > 0$.

17. a. $10,000 = \dfrac{R}{0.015}[1-(1.015)^{-48}]$

$R = \dfrac{150}{1-(1.015)^{-48}} \approx 293.75$

b. $10,000 = \dfrac{300}{i}[1-(1+i)^{-48}]$

$i = \dfrac{3}{100}[1-(1+i)^{-48}]$

Let $i_1 = 0.015$.

n	i_n
1	0.015
2	0.015319
3	0.015539
4	0.015689
5	0.015789
6	0.015857

7	0.015902
8	0.015932
9	0.015952
10	0.15965
11	0.015974
12	0.015980
13	0.015983
14	0.015986
15	0.015988
16	0.015989
17	0.015989

$i \approx 0.01599$

19. a. Suppose r is a root. Then $r = r - \dfrac{f(r)}{f'(r)}$.

$\dfrac{f(r)}{f'(r)} = 0$, so $f(r) = 0$.

Suppose $f(r) = 0$. Then $r - \dfrac{f(r)}{f'(r)} = r - 0 = r$, so r is

a root of $x = x - \dfrac{f(x)}{f'(x)}$.

b. If we want to solve $f(x) = 0$ and $f'(x) \neq 0$ in $[a, b]$,

then $\dfrac{f(x)}{f'(x)} = 0$ or $x = x - \dfrac{f(x)}{f'(x)} = g(x)$.

$g'(x) = 1 - \dfrac{f'(x)}{f'(x)} + \dfrac{f(x)}{[f'(x)]^2}f''(x)$

$= \dfrac{f(x)f''(x)}{[f'(x)]^2}$

and $g'(r) = \dfrac{f(r)f''(r)}{[f'(r)]^2} = 0$.

11.5 Concepts Review

1. slope field

3. $y_{n-1} + hf(x_{n-1}, y_{n-1})$

Problem Set 11.5

1.

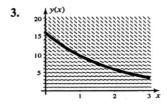

$\lim_{t \to \infty} y(t) = 12$ and $y(2) \approx 10.5$

3.

$\lim_{t \to \infty} y(t) = 0$ and $y(2) \approx 6$

5.

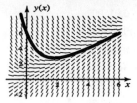

The oblique asymptote is $y = x$.

7.

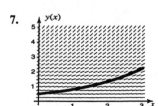

$$\frac{dy}{dx} = \frac{1}{2}y; \quad y(0) = \frac{1}{2}$$

$$\frac{dy}{y} = \frac{1}{2}dx$$

$$\ln y = \frac{x}{2} + C$$

$$y = C_1 e^{x/2}$$

To find C_1, apply the initial condition:

$$\frac{1}{2} = y(0) = C_1 e^0 = C_1$$

$$y = \frac{1}{2}e^{x/2}$$

9.

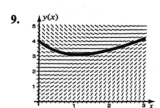

$$y' + y = x + 2$$

The integrating factor is $e^{\int 1\,dx} = e^x$.

$$e^x y' + y e^x = e^x(x+2)$$

$$\frac{d}{dx}\left(e^x y\right) = (x+2)e^x$$

$$e^x y = \int (x+2)e^x\, dx$$

Integrate by parts: let $u = x+2$, $dv = e^x\,dx$.

Then $du = dx$ and $v = e^x$. Thus

$$e^x y = (x+2)e^x - \int e^x\,dx$$

$$e^x y = (x+2)e^x - e^x + C$$

$$y = x + 2 - 1 + Ce^{-x}$$

To find C, apply the initial condition:

$$4 = y(0) = 0 + 1 + Ce^{-0} = 1 + C$$

Thus, $y = x + 1 + e^{-x}$.

Note: Solutions to Problems 17-25 are given along with the corresponding solutions to 11-15.

11., 17.

x_n	Euler's Method y_n	Improved Euler Method y_n
0.0	3.0	3.0
0.2	4.2	4.44
0.4	5.88	6.5712
0.6	8.232	9.72538
0.8	11.5248	14.39356
1.0	16.1347	21.30246

13., 19.

x_n	Euler's Method y_n	Improved Euler Method y_n
0.0	0.0	0.0
0.2	0.0	0.2
0.4	0.04	0.8
0.6	0.12	0.18
0.8	0.24	0.032
1.0	0.40	0.05

15., 21.

x_n	Euler's Method y_n	Improved Euler Method y_n
1.0	1.0	1.0
1.2	1.2	1.244
1.4	1.488	1.60924
1.6	1.90464	2.16410
1.8	2.51412	3.02455
2.0	3.41921	4.391765

23.

h	Error from Euler's Method	Error from Improved Euler Method
0.2	0.229962	0.015574
0.1	0.124539	0.004201
0.05	0.064984	0.001091
0.01	0.013468	0.000045
0.005	0.006765	0.000011

For Euler's method, the error is halved as the step size h is halved. Thus, the error is proportional to h. For the improved Euler method, when h is halved, the error decreases to approximately one-fourth of what is was.

Hence, for the improved Euler method, the error is proportional to h^2.

25. a. $y_0 = 1$

$y_1 = y_0 + hf(x_0, y_0)$
$\quad = y_0 + hy_0 = (1+h)y_0$

$y_2 = y_1 + hf(x_1, y_1) = y_1 + hy_1$
$\quad = (1+h)y_1 = (1+h)^2 y_0$

$y_3 = y_2 + hf(x_2, y_2) = y_2 + hy_2$
$\quad = (1+h)y_2 = (1+h)^3 y_0$

$\vdots$

$y_n = y_{n-1} + hf(x_{n-1}, y_{n-1}) = y_{n-1} + hy_{n-1}$
$\quad = (1+h)y_{n-1} = (1+h)^n y_0$

b. Let $N = 1/h$. Then y_N is an approximation to the solution at $x = Nh = (1/h)h = 1$. The exact solution is $y(1) = e$. Thus, $(1 + 1/N)^N \approx e$ for large N. From Chapter 7, we know that

$$\lim_{N \to \infty} (1 + 1/N)^N = e.$$

27. a. $\int_{x_0}^{x_1} y'(x)dx = \int_{x_0}^{x_1} \sin x^2 dx$

$y(x_1) - y(x_0) \approx (x_1 - x_0)\sin x_0^2$

$y(x_1) - y(0) = h \sin x_0^2$

$y(x_1) - 0 \approx 0.1 \sin 0^2$

$y(x_1) \approx 0$

b. $\int_{x_0}^{x_2} y'(x)dx = \int_{x_0}^{x_2} \sin x^2 dx$

$y(x_2) - y(x_0) \approx (x_1 - x_0)\sin x_0^2$
$\qquad\qquad + (x_2 - x_1)\sin x_1^2$

$y(x_2) - y(0) = h \sin x_0^2 + h \sin x_1^2$

$y(x_2) - 0 \approx 0.1 \sin 0^2 + 0.1 \sin 0.1^2$

$y(x_2) \approx 0.00099998$

c. $\int_{x_0}^{x_3} y'(x)dx = \int_{x_0}^{x_3} \sin x^2 dx$

$y(x_3) - y(x_0) \approx (x_1 - x_0)\sin x_0^2$
$\qquad\qquad + (x_2 - x_1)\sin x_1^2 + (x_3 - x_2)\sin x_1^2$

$y(x_3) - y(0) = h \sin x_0^2 + h \sin x_1^2 + h \sin x_2^2$

$y(x_3) - 0 \approx 0.1 \sin 0^2 + 0.1 \sin 0.1^2$
$\qquad\qquad + 0.1 \sin 0.2^2$

$y(x_3) \approx 0.004999$

Continuing in this fashion, we have

$$\int_{x_0}^{x_n} y'(x)dx = \int_{x_0}^{x_n} \sin x^2 dx$$

$$y(x_n) - y(x_0) \approx \sum_{i=0}^{n-1}(x_{i+1} - x_i)\sin x_i^2$$

$$y(x_n) \approx h \sum_{i=0}^{n-1} f(x_{i-1})$$

When $n = 10$, this becomes

$y(x_{10}) = y(1) \approx 0.269097$

d. The result $y(x_n) \approx h \sum_{i=0}^{n-1} f(x_{i-1})$ is the same as that given in Problem 26. Thus, when $f(x, y)$ depends only on x, then the two methods (1) Euler's method for approximating the solution to $y' = f(x)$ at x_n, and (2) the left-endpoint Riemann sum for approximating $\int_0^{x_n} f(x)dx$, are equivalent.

11.6 Chapter Review

Concepts Test

1. True: $P(x) = f(0) + f'(0)x + \dfrac{f''(0)}{2}x^2$

3. True: $f(0) = f'(0) = f''(0) = 0$, its second order Maclaurin polynomial is 0.

5. True: Any Maclaurin polynomial for $\cos x$ involves only even powers of x.

7. True: Taylor's Formula with Remainder for $n = 0$ is $f(x) = f(a) + f'(c)(x - a)$ which is equivalent to the Mean Value Theorem.

9. False: For example $\int \dfrac{\sin x}{x}dx$ cannot be expressed in terms of elementary functions.

11. True: $E_{10} = 0$ since the fourth derivative of x^3 is 0.

13. True:

$$\left| e^{-x^2} + x^2 + \sin(x+1) \right|$$

$$\le \left| e^{-x^2} \right| + \left| x^2 \right| + \left| \sin(x+1) \right|$$

$$\le 1 + 4 + 1 = 6$$

15. True: Intermediate Value Theorem

17. False: $x_{n+1} = x_n - \dfrac{f(x_n)}{f'(x_n)} = -2x_n$. (See Problem 22 of Problem Set 10.4.)

19. False: $g(x) = 5(x - x^2) + 0.01$, $g'(x) = 5 - 10x$; $|g'(x)| > 1$ in a neighborhood of the fixed point.

21. True: At $(2,1)$ the slope is $y' = 2 \cdot 1 = 2$

Sample Test Problems

1. $f(0) = 0$

$$f'(x) = \cos^2 x - 2x^2 \sin^2 x$$
$$f'(0) = 1$$
$$p(x) = x;\ p(0.2) = 0.2;\ f(0.2) = 0.1998$$

3. $g(x) = x^3 - 2x^2 + 5x - 7 \qquad g(2) = 3$

$\quad g'(x) = 3x^2 - 4x + 5 \qquad g'(2) = 9$

$\quad g''(x) = 6x - 4 \qquad\qquad g''(2) = 8$

$\quad g^{(3)}(x) = 6 \qquad\qquad\quad g^{(3)}(2) = 6$

Since $g^{(4)}(x) = 0$, $R_3(x) = 0$, so the Taylor polynomial of order 3 based at 2 is an exact representation.

$$g(x) = P_4(x) = 3 + 9(x-2) + 4(x-2)^2 + (x-2)^3$$

5. $f(x) = \dfrac{1}{x+1} \qquad\qquad f(1) = \dfrac{1}{2}$

$\quad f'(x) = -\dfrac{1}{(x+1)^2} \qquad\quad f'(1) = -\dfrac{1}{4}$

$\quad f''(x) = \dfrac{2}{(x+1)^3} \qquad\quad f''(1) = \dfrac{1}{4}$

$\quad f^{(3)}(x) = -\dfrac{6}{(x+1)^4} \qquad f^{(3)}(1) = -\dfrac{3}{8}$

$\quad f^{(4)}(x) = \dfrac{24}{(x+1)^5} \qquad\ f^{(4)}(1) = \dfrac{3}{4}$

$$f(x) \approx \frac{1}{2} - \frac{1}{4}(x-1) + \frac{1}{8}(x-1)^2$$
$$- \frac{1}{16}(x-1)^3 + \frac{1}{32}(x-1)^4$$

7. $f(x) = \dfrac{1}{2}(1 - \cos 2x) \qquad f(0) = 0$

$\quad f'(x) = \sin 2x \qquad\qquad\quad f'(0) = 0$

$\quad f''(x) = 2\cos 2x \qquad\qquad f''(0) = 2$

$\quad f^{(3)}(x) = -4\sin 2x \qquad\quad f^{(3)}(0) = 0$

$\quad f^{(4)}(x) = -8\cos 2x \qquad\ f^{(4)}(0) = -8$

$\quad f^{(5)}(x) = 16\sin 2x \qquad\quad f^{(5)}(0) = 0$

$\quad f^{(6)}(x) = 32\cos 2x \qquad\ f^{(6)}(c) = 32\cos 2c$

$$\sin^2 x \approx \frac{2}{2!}x^2 - \frac{8}{4!}x^4 = x^2 - \frac{1}{3}x^4$$

$$|R_4(x)| = |R_5(x)| = \left| \frac{32}{6!}(\cos 2c)x^6 \right| \le \frac{2}{45}(0.2)^6$$

$$< 2.85 \times 10^{-6}$$

9. From Problem 8,

$$\ln x \approx (x-1) - \frac{1}{2}(x-1)^2 + \frac{1}{3}(x-1)^3$$

$$- \frac{1}{4}(x-1)^4 + \frac{1}{5}(x-1)^5.$$

$$|R_5(x)| = \left| \frac{1}{6c^6}(x-1)^6 \right| \le \frac{0.2^6}{6 \cdot 0.8^6} < 4.07 \times 10^{-5}$$

$$\int_{0.8}^{1.2} \ln x\, dx \approx \int_{0.8}^{1.2} \left[(x-1) - \frac{1}{2}(x-1)^2 + \frac{1}{3}(x-1)^3 \right.$$

$$\left. - \frac{1}{4}(x-1)^4 + \frac{1}{5}(x-1)^5 \right] dx$$

$$= \left[\frac{1}{2}(x-1)^2 - \frac{1}{6}(x-1)^3 + \frac{1}{12}(x-1)^4 \right.$$

$$\left. - \frac{1}{20}(x-1)^5 + \frac{1}{30}(x-1)^6 \right]_{0.8}^{1.2}$$

$$\approx -0.00269867$$

$$\text{Error} \le (1.2 - 0.8)4.07 \times 10^{-5} < 1.63 \times 10^{-5}$$

11. $f(x) = \ln x,\ h = 0.05$

n	x_n	$f(x_n)$
0	0.80	−0.22314
1	0.85	−0.16252
2	0.90	−0.10536
3	0.95	−0.051293
4	1.00	0
5	1.05	0.04879
6	1.10	0.09531
7	`1.15	0.13976
8	1.20	0.18232

$$\int_{0.8}^{1.2} \ln x \, dx \approx \frac{0.05}{3} \left[f(x_0) + 4f(x_1) + 2f(x_2) + \ldots \right.$$
$$\left. + 4f(x_7) + f(x_8) \right]$$

$$\approx -0.00269939$$

$$|E_8| = \left| \frac{0.4^5}{180 \cdot 8^4} \frac{6}{c^4} \right| \le \frac{0.4^5 \cdot 6}{180 \cdot 8^4 \cdot 0.8^4} < 2.04 \times 10^{-7}$$

13. $f(x) = 3x - \cos 2x$, $f'(x) = 3 + 2\sin 2x$

Let $x_1 = 0.5$.

n	x_n
1	0.5
2	0.2950652
3	0.2818563
4	0.2817846
5	0.2817846

$x \approx 0.281785$

15. $y = x$ and $y = \tan x$

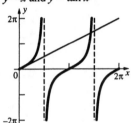

Let $x_1 = \dfrac{11\pi}{8}$.

$f(x) = x - \tan x$, $f'(x) = 1 - \sec^2 x$.

n	x_n
1	$\dfrac{11\pi}{8}$
2	4.64661795
3	4.60091050
4	4.54662258
5	4.50658016
6	4.49422443
7	4.49341259
8	4.49340946

$x \approx 4.4934$

17. Graph $y = e^x$ and $y = \sin x$

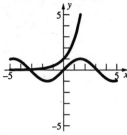

Let $x_1 = -3$.

$$f(x) = e^x - \sin x, \; f'(x) = e^x - \cos x$$

$$x_{n+1} = x_n - \frac{e^{x_n} - \sin x_n}{e^{x_n} - \cos x_n}$$

n	x_n
1	-3
2	-3.183603
3	-3.183063
4	-3.183063

$x \approx -3.18306$

19. Improved Euler Method

x_n	y_n
0	2
0.2	3.56
0.4	6.3368
0.6	11.2795
0.8	20.07752
1.0	35.73798
1.2	63.61361
1.4	113.2322
1.6	201.5533
1.8	358.7650
2.0	638.6016

11.7 Additional Problem Set

1. Using a computer, the table is as follows:

i	(x_i, y_i)
0	$(\pi, 1)$
1	$(3.4558, 0.9511)$
2	$(3.7699, 0.809)$
3	$(4.0841, 0.5878)$
4	$(4.3982, 0.309)$
5	$(4.7124, 0)$
6	$(5.0265, -0.309)$
7	$(5.3407, -0.5878)$
8	$(5.6549, -0.809)$
9	$(5.9690, -0.9511)$
10	$(6.2832, -1)$

3. a. $y_{n+1}^{\text{predicted}} = y_n + 0.25(y_n) = 1.25 y_n$

$y_{n+1} = y_n + \dfrac{0.25}{2}(y_n + 1.25 y_n) = 1.28125 y_n;$

$x_0 = 0,\ y_0 = 1$

n	x_n	y_n
1	0.25	1.28125
2	0.50	1.641602
3	0.75	2.103302
4	1.00	2.694856

$y_{n+1}^{\text{predicted}} = y_n + 0.1(y_n) = 1.1 y_n$

$y_{n+1} = y_n + \dfrac{0.1}{2}(y_n + 1.1 y_n) = 1.105 y_n;$

$x_0 = 0,\ y_0 = 1$

n	x_n	y_n
1	0.1	1.105
2	0.2	1.221025
3	0.3	1.349233
4	0.4	1.490902
5	0.5	1.647447
6	0.6	1.820429
7	0.7	2.011574
8	0.8	2.222789
9	0.9	2.456182
10	1.0	2.714081

b. $y_{n+1} = y_n + 0.05(x_n + y_n)$

$= 0.05 x_n + 1.05 y_n;$

$x_0 = 0,\ y_0 = 1$

n	x_n	y_n
1	0.05	1.05
2	0.10	1.105
3	0.15	1.16525
4	0.20	1.23101
5	0.25	1.30256
6	0.30	1.38019
7	0.35	1.46420
8	0.40	1.55491
9	0.45	1.65266
10	0.50	1.75779
11	0.55	1.87068
12	0.60	1.99171
13	0.65	2.12130
14	0.70	2.25986
15	0.75	2.40786
16	0.80	2.56575
17	0.85	2.73404
18	0.90	2.91324
19	0.95	3.10390
20	1.00	3.30660

$y_{n+1}^{\text{predicted}} = y_n + 0.1(x_n + y_n)$

$= 0.1 x_n + 1.1 y_n$

$y_{n+1} = y_n + \dfrac{0.1}{2}(x_n + y_n + x_{n+1} + 0.1 x_n + 1.1 y_n)$

$= y_n + 0.05(1.1 x_n + 2.1 y_n + x_n + 0.1)$

$= 0.105 x_n + 1.105 y_n + 0.005;$

$x_0 = 0,\ y_0 = 1$

n	x_n	y_n
1	0.1	1.11
2	0.2	1.24205
3	0.3	1.39847
4	0.4	1.58180
5	0.5	1.79489
6	0.6	2.04086
7	0.7	2.32315
8	0.8	2.64558
9	0.9	3.01236
10	1.0	3.42816

$y(1) = 2e - 1 - 1 = 3.4365$

12.1 Concepts Review

1. $e < 1; e = 1; e > 1$

3. $(0, 1); y = -1$

Problem Set 12.1

1. $y^2 = 4(1)x$
   Focus at $(1, 0)$
   Directrix: $x = -1$

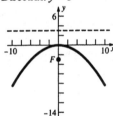

3. $x^2 = -4(3)y$
   Focus at $(0, -3)$
   Directrix: $y = 3$

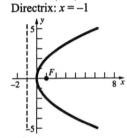

5. $y^2 = 4\left(\frac{1}{4}\right)x$
   Focus at $\left(\frac{1}{4}, 0\right)$

   Directrix: $x = -\frac{1}{4}$

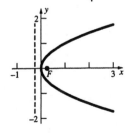

7. $2x^2 = 6y$
   $x^2 = 4\left(\frac{3}{4}\right)y$

   Focus at $\left(0, \frac{3}{4}\right)$

   Directrix: $y = -\frac{3}{4}$

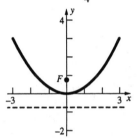

9. The parabola opens to the right, and $p = 2$.
   $y^2 = 8x$

11. The parabola opens downward, and $p = 2$.
    $x^2 = -8y$

13. The parabola opens to the left, and $p = 4$.
    $y^2 = -16x$

15. The equation has the form $y^2 = cx$, so
    $(-1)^2 = 3c$.

    $c = \frac{1}{3}$

    $y^2 = \frac{1}{3}x$

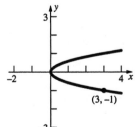

17. The equation has the form $x^2 = cy$, so

$$(6)^2 = -5c.$$

$$c = -\frac{36}{5}$$

$$x^2 = -\frac{36}{5}y$$

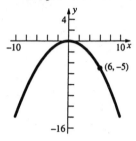

(6, -5)

19. $y^2 = 16x$

$$2yy' = 16$$

$$y' = \frac{16}{2y}$$

At $(1, -4)$, $y' = -2$.

Tangent: $y + 4 = -2(x - 1)$ or $y = -2x - 2$

Normal: $y + 4 = \frac{1}{2}(x - 1)$ or $y = \frac{1}{2}x - \frac{9}{2}$

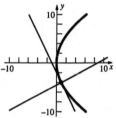

21. $x^2 = 2y$

$$2x = 2y'$$

$$y' = x$$

At $(4, 8)$, $y' = 4$.

Tangent: $y - 8 = 4(x - 4)$ or $y = 4x - 8$

Normal: $y - 8 = -\frac{1}{4}(x - 4)$ or $y = -\frac{1}{4}x + 9$

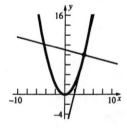

23. $y^2 = -15x$

$$2yy' = -15$$

$$y' = -\frac{15}{2y}$$

At $\left(-3, -3\sqrt{5}\right)$, $y' = \frac{\sqrt{5}}{2}$.

Tangent: $y + 3\sqrt{5} = \frac{\sqrt{5}}{2}(x + 3)$ or

$$y = \frac{\sqrt{5}}{2}x - \frac{3\sqrt{5}}{2}$$

Normal: $y + 3\sqrt{5} = -\frac{2\sqrt{5}}{5}(x + 3)$ or

$$y = -\frac{2\sqrt{5}}{5}x - \frac{21\sqrt{5}}{5}$$

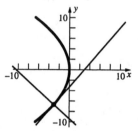

25. $x^2 = -6y$

$$2x = -6y'$$

$$y' = -\frac{x}{3}$$

At $\left(3\sqrt{2}, -3\right)$, $y' = -\sqrt{2}$.

Tangent: $y + 3 = -\sqrt{2}\left(x - 3\sqrt{2}\right)$ or $y = -\sqrt{2}x + 3$

Normal: $y + 3 = \frac{\sqrt{2}}{2}\left(x - 3\sqrt{2}\right)$ or $y = \frac{\sqrt{2}}{2}x - 6$

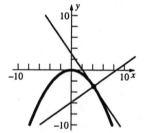

27. $y^2 = 5x$

$2yy' = 5$

$y' = \dfrac{5}{2y}$

$y' = \dfrac{\sqrt{5}}{4}$ when $y = 2\sqrt{5}$, so $x = 4$.

The point is $\left(4, 2\sqrt{5}\right)$.

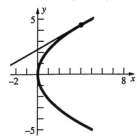

29. The slope of the line is $\dfrac{3}{2}$.

$y^2 = -18x;\ 2yy' = -18$

$2y\left(\dfrac{3}{2}\right) = -18;\ y = -6$

$(-6)^2 = -18x;\ x = -2$

The equation of the tangent line is

$y + 6 = \dfrac{3}{2}(x + 2)$ or $y = \dfrac{3}{2}x - 3$.

31. From Problem 30, if the parabola is described by the equation $y^2 = 4px$, the slopes of the tangent

lines are $\dfrac{2p}{y_0}$ and $-\dfrac{y_0}{2p}$. Since they are negative reciprocals, the tangent lines are perpendicular.

33. Assume that the x- and y-axes are positioned such that the axis of the parabola is the y-axis with the vertex at the origin and opening upward. Then the equation of the parabola is $x^2 = 4py$ and $(0, p)$ is the focus. Let D be the distance from a point on the parabola to the focus.

$D = \sqrt{(x-0)^2 + (y-p)^2} = \sqrt{x^2 + \left(\dfrac{x^2}{4p} - p\right)^2}$

$= \sqrt{\dfrac{x^4}{16p^2} + \dfrac{x^2}{2} + p^2} = \dfrac{x^2}{4p} + p$

$D' = \dfrac{x}{2p};\ \dfrac{x}{2p} = 0,\ x = 0$

$D'' > 0$ so at $x = 0$, D is minimum. $y = 0$
Therefore, the vertex $(0, 0)$ is the point closest to the focus.

35. Let the y-axis be the axis of the parabola, so the Earth's coordinates are $(0, p)$ and the equation of the path is $x^2 = 4py$, where the coordinates are in millions of miles. When the line from Earth to the asteroid makes an angle of $75°$ with the axis of the parabola, the asteroid is at $(40\sin 75°, p + 40\cos 75°)$. (See figure.)

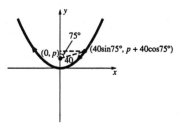

$(40\sin 75°)^2 = 4p(p + 40\cos 75°)$

$p^2 + 40p\cos 75° - 400\sin^2 75° = 0$

$p = \dfrac{-40\cos 75° \pm \sqrt{1600\cos^2 75° + 1600\sin^2 75°}}{2}$

$= -20\cos 75° \pm 20$

$p = 20 - 20\cos 75° \approx 14.8\ (p > 0)$

The closest point to Earth is $(0, 0)$, so the asteroid will come within 14.8 million miles of Earth.

37. Let $|RL|$ be the distance from R to the directrix. Observe that the distance from the latus rectum to the directrix is $2p$ so $|RG| = 2p - |RL|$. From the definition of a parabola, $|RL| = |FR|$. Thus,

$|FR| + |RG| = |RL| + 2p - |RL| = 2p$.

39. Let P_1 and P_2 denote (x_1, y_1) and (x_2, y_2), respectively. $|P_1P_2| = |P_1F| + |P_2F|$ since the focal chord passes through the focus. By definition of a parabola, $|P_1F| = p + x_1$ and $|P_2F| = p + x_2$. Thus, the length of the chord is

$|P_1P_2| = x_1 + x_2 + 2p$.

$L = p + p + 2p = 4p$

41. $\dfrac{dy}{dx} = \dfrac{\delta x}{H}$

$y = \dfrac{\delta x^2}{2H} + C$

$y(0) = 0$ implies that $C = 0$. $y = \dfrac{\delta x^2}{2H}$

This is an equation for a parabola.

43.

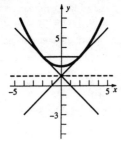

12.2 Concepts Review

1. $\dfrac{x^2}{a^2} + \dfrac{y^2}{b^2} = 1$

3. $\dfrac{x^2}{a^2} - \dfrac{y^2}{b^2} = 1$

Problem Set 12.2

1. Horizontal ellipse

3. Vertical hyperbola

5. Vertical parabola

7. Vertical ellipse

9. $\dfrac{x^2}{16} + \dfrac{y^2}{4} = 1$; horizontal ellipse

   $a = 4, b = 2, c = 2\sqrt{3}$
   Vertices: $(\pm 4, 0)$
   Foci: $\left(\pm 2\sqrt{3}, 0\right)$

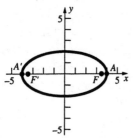

11. $\dfrac{y^2}{4} - \dfrac{x^2}{9} = 1$; vertical hyperbola

    $a = 2, b = 3, c = \sqrt{13}$
    Vertices: $(0, \pm 2)$
    Foci: $\left(0, \pm \sqrt{13}\right)$

Asymptotes: $y = \pm \dfrac{2}{3}x$

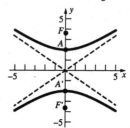

13. $\dfrac{x^2}{2} + \dfrac{y^2}{8} = 1$; vertical ellipse

    $a = 2\sqrt{2}, b = \sqrt{2}, c = \sqrt{6}$
    Vertices: $(0, \pm 2\sqrt{2})$
    Foci: $\left(0, \pm \sqrt{6}\right)$

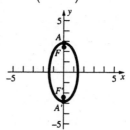

15. $\dfrac{x^2}{10} - \dfrac{y^2}{4} = 1$; horizontal hyperbola

    $a = \sqrt{10}, b = 2, c = \sqrt{14}$
    Vertices: $\left(\pm\sqrt{10}, 0\right)$
    Foci: $\left(\pm\sqrt{14}, 0\right)$

Asymptotes: $y = \pm \dfrac{2}{\sqrt{10}} x$

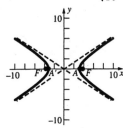

17. This is a horizontal ellipse with $a = 6$ and $c = 3$.
$$b = \sqrt{36-9} = \sqrt{27}$$
$$\frac{x^2}{36} + \frac{y^2}{27} = 1$$

19. This is a vertical ellipse with $c = 5$.
$$a = \frac{c}{e} = \frac{5}{\frac{1}{3}} = 15, b = \sqrt{225-25} = \sqrt{200}$$
$$\frac{x^2}{200} + \frac{y^2}{225} = 1$$

21. This is a horizontal ellipse with $a = 5$.
$$\frac{x^2}{25} + \frac{y^2}{b^2} = 1$$
$$\frac{4}{25} + \frac{9}{b^2} = 1$$
$$b^2 = \frac{225}{21}$$
$$\frac{x^2}{25} + \frac{y^2}{\frac{225}{21}} = 1$$

23. This is a vertical hyperbola with $a = 4$ and $c = 5$.
$$b = \sqrt{25-16} = 3$$
$$\frac{y^2}{16} - \frac{x^2}{9} = 1$$

25. This is a horizontal hyperbola with $a = 8$.
The asymptotes are $y = \pm \dfrac{1}{2} x$, so $\dfrac{b}{8} = \dfrac{1}{2}$ or $b = 4$.
$$\frac{x^2}{64} - \frac{y^2}{16} = 1$$

27. This is a horizontal ellipse with $c = 2$.
$$8 = \frac{a}{e}, 8 = \frac{a}{\frac{c}{a}}, \text{ so } a^2 = 8c = 16.$$
$$b = \sqrt{16-4} = \sqrt{12}$$
$$\frac{x^2}{16} + \frac{y^2}{12} = 1$$

29. The asymptotes are $y = \pm \dfrac{1}{2} x$. If the hyperbola is
horizontal $\dfrac{b}{a} = \dfrac{1}{2}$ or $a = 2b$. If the hyperbola is
vertical, $\dfrac{a}{b} = \dfrac{1}{2}$ or $b = 2a$.
Suppose the hyperbola is horizontal.
$$\frac{x^2}{4b^2} - \frac{y^2}{b^2} = 1$$
$$\frac{16}{4b^2} - \frac{9}{b^2} = 1$$
$$b^2 = -5$$
This is not possible.
Suppose the hyperbola is vertical.
$$\frac{y^2}{a^2} - \frac{x^2}{4a^2} = 1$$
$$\frac{9}{a^2} - \frac{16}{4a^2} = 1$$
$$a^2 = 5$$
$$\frac{y^2}{5} - \frac{x^2}{20} = 1$$

31. Let the y-axis run through the center of the arch
and the x-axis lie on the floor. Thus $a = 5$ and
$b = 4$ and the equation of the arch is $\dfrac{x^2}{25} + \dfrac{y^2}{16} = 1$.
When $y = 2$, $\dfrac{x^2}{25} + \dfrac{(2)^2}{16} = 1$, so $x = \pm \dfrac{5\sqrt{3}}{2}$.
The width of the box can at most be
$5\sqrt{3} \approx 8.66$ ft.

33. The foci are at $(\pm c, 0)$.
$$c = \sqrt{a^2 - b^2}$$
$$\frac{a^2 - b^2}{a^2} + \frac{y^2}{b^2} = 1$$
$$y^2 = \frac{b^4}{a^2}, y = \pm \frac{b^2}{a}$$
Thus, the length of the latus rectum is $\dfrac{2b^2}{a}$.

35. $a = 18.09, b = 4.56,$
$$c = \sqrt{(18.09)^2 - (4.56)^2} \approx 17.51$$
The comet's minimum distance from the sun is
$18.09 - 17.51 \approx 0.58$ AU.

37. $a - c = 4132; a + c = 4583$
$2a = 8715; a = 4357.5$
$c = 4357.5 - 4132 = 225.5$
$$e = \frac{c}{a} = \frac{225.5}{4357.5} \approx 0.05175$$

39. $2a = p + q,\ 2c = |p - q|$

$$b^2 = a^2 - c^2 = \frac{(p+q)^2}{4} - \frac{(p-q)^2}{4} = pq$$

$$b = \sqrt{pq}$$

41. Let (x, y) be the coordinates of P as the ladder slides. Using a property of similar triangles,

$$\frac{x}{a} = \frac{\sqrt{b^2 - y^2}}{b}.$$

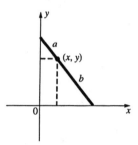

Square both sides to get

$$\frac{x^2}{a^2} = \frac{b^2 - y^2}{b^2} \text{ or } b^2 x^2 + a^2 y^2 = a^2 b^2 \text{ or}$$

$$\frac{x^2}{a^2} + \frac{y^2}{b^2} = 1.$$

43. The equations of the hyperbolas are $\dfrac{x^2}{a^2} - \dfrac{y^2}{b^2} = 1$

and $\dfrac{y^2}{b^2} - \dfrac{x^2}{a^2} = 1.$

$$e = \frac{c}{a} = \frac{\sqrt{a^2 + b^2}}{a}$$

$$E = \frac{c}{b} = \frac{\sqrt{a^2 + b^2}}{b}$$

$$e^{-2} + E^{-2} = \frac{a^2}{a^2 + b^2} + \frac{b^2}{a^2 + b^2} = 1$$

45.

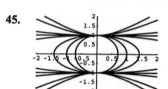

When $a < 0$, the conic is an ellipse. When $a > 0$, the conic is a hyperbola. When $a = 0$, the graph is two parallel lines.

12.3 Concepts Review

1. foci

3. to the other focus

Problem Set 12.3

1. This is a vertical ellipse with $a = 13$ and $c = 9$.
$$b = \sqrt{169 - 81} = 2\sqrt{22}$$
$$\frac{x^2}{88} + \frac{y^2}{169} = 1$$

3. This is a horizontal hyperbola with $a = 6$ and $c = 7$.
$$b = \sqrt{49 - 36} = \sqrt{13}$$
$$\frac{x^2}{36} - \frac{y^2}{13} = 1$$

5. $a^2 = 27, b^2 = 9$
$$\frac{3x}{27} + \frac{\sqrt{6}y}{9} = 1$$
$$x + \sqrt{6}y = 9$$

7. $a^2 = 27, b^2 = 9$
$$\frac{3x}{27} + \frac{-\sqrt{6}y}{9} = 1$$
$$x - \sqrt{6}y = 9$$

9. $a^2 = 169, b^2 = 169$
$$\frac{5x}{169} + \frac{12y}{169} = 1$$
$$5x + 12y = 169$$

11. $a^2 = 169, b^2 = 88$
$$\frac{13y}{169} = 1$$
$$y = 13$$

13. $\dfrac{x^2}{4} + \dfrac{y^2}{9} = 1$

Equation of tangent line at (x_0, y_0):
$$\frac{xx_0}{4} + \frac{yy_0}{9} = 1$$

At $(0, 6)$, $y_0 = \dfrac{3}{2}.$

When $y = \dfrac{3}{2}$, $\dfrac{x^2}{4} + \dfrac{1}{4} = 1$, $x = \pm\sqrt{3}$.

The points of tangency are $\left(\sqrt{3}, \dfrac{3}{2}\right)$ and

$\left(-\sqrt{3}, \dfrac{3}{2}\right)$

15. $2x^2 - 7y^2 - 35 = 0$; $4x - 14yy' = 0$

$y' = \dfrac{2x}{7y}$; $-\dfrac{2}{3} = \dfrac{2x}{7y}$; $x = -\dfrac{7y}{3}$

Substitute $x = -\dfrac{7y}{3}$ into the equation of the hyperbola.

$\dfrac{98}{9}y^2 - 7y^2 - 35 = 0$, $y = \pm 3$

The coordinates of the points of tangency are $(-7, 3)$ and $(7, -3)$.

17. $y = \pm b\sqrt{1 - \dfrac{x^2}{a^2}}$

$A = 4b \displaystyle\int_0^a \sqrt{1 - \dfrac{x^2}{a^2}}\, dx$

Let $x = a\sin t$ then $dx = a\cos t\, dt$. Then the limits are 0 and $\dfrac{\pi}{2}$.

$A = 4b \displaystyle\int_0^a \sqrt{1 - \dfrac{x^2}{a^2}}\, dx = 4ab \displaystyle\int_0^{\pi/2} \cos^2 t\, dt$

$= 2ab \displaystyle\int_0^{\pi/2} (1 + \cos 2t)\, dt = 2ab\left[t + \dfrac{\sin 2t}{2}\right]_0^{\pi/2}$

$= \pi ab$

19. $y = \pm b\sqrt{\dfrac{x^2}{a^2} - 1}$

The vertical line at one focus is $x = \sqrt{a^2 + b^2}$.

$V = \pi \displaystyle\int_a^{\sqrt{a^2+b^2}} \left(b\sqrt{\dfrac{x^2}{a^2} - 1}\right)^2 dx$

$= b^2\pi \displaystyle\int_a^{\sqrt{a^2+b^2}} \left(\dfrac{x^2}{a^2} - 1\right) dx$

$= b^2\pi\left[\dfrac{x^3}{3a^2} - x\right]_a^{\sqrt{a^2+b^2}}$

$= b^2\pi\left[\dfrac{(a^2 + b^2)^{3/2}}{3a^2} - \sqrt{a^2 + b^2} + \dfrac{2}{3}a\right]$

$= \dfrac{\pi b^2}{3a^2}\left[(a^2 + b^2)^{3/2} - 3a^2\sqrt{a^2 + b^2} + 2a^3\right]$

21. If one corner of the rectangle is at (x, y) the sides have length $2x$ and $2y$.

$x = \pm a\sqrt{1 - \dfrac{y^2}{b^2}}$

$A = 4xy = 4ya\sqrt{1 - \dfrac{y^2}{b^2}} = 4a\sqrt{y^2 - \dfrac{y^4}{b^2}}$

$\dfrac{dA}{dy} = \dfrac{2a\left(2y - \dfrac{4y^3}{b^2}\right)}{\sqrt{y^2 - \dfrac{y^4}{b^2}}}$; $\dfrac{dA}{dy} = 0$ when

$y - \dfrac{2y^3}{b^2} = 0$

$y\left(1 - \dfrac{2y^2}{b^2}\right) = 0$

$y = 0$ or $y = \pm\dfrac{b}{\sqrt{2}}$

The Second Derivative Test shows that $y = \dfrac{b}{\sqrt{2}}$ is a maximum.

$x = a\sqrt{1 - \dfrac{\left(\dfrac{b}{\sqrt{2}}\right)^2}{b^2}} = \dfrac{a}{\sqrt{2}}$

Therefore, the rectangle is $a\sqrt{2}$ by $b\sqrt{2}$.

23. Add the two equations to get $9y^2 = 675$.

$y = \pm 5\sqrt{3}$

Substitute $y = 5\sqrt{3}$ into either of the two equations and solve for x.

$x = \pm 6$

The point in the first quadrant is $\left(6, 5\sqrt{3}\right)$.

25.

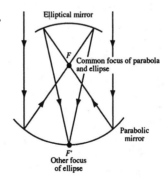

27. Written response. Possible answer: the ball will follow a path that does not go between the foci.

29. Possible answer: Attach one end of a string to F and attach one end of another string to F'. Place a spool at a vertex. Tightly wrap both strings in the same direction around the spool. Insert a pencil through the spool. Then trace out a branch of the hyperbola by unspooling the strings while keeping both strings taut.

12.4 Concepts Review

1. $\dfrac{a^2}{4}$

3. a line, parallel lines, intersecting lines, a point, the empty set

Problem Set 12.4

1. $x^2 + y^2 - 2x + 2y + 1 = 0$

$(x^2 - 2x + 1) + (y^2 + 2y + 1) = -1 + 1 + 1$

$(x-1)^2 + (y+1)^2 = 1$

This is a circle.

3. $9x^2 + 4y^2 + 72x - 16y + 124 = 0$

$9(x^2 + 8x + 16) + 4(y^2 - 4y + 4) = -124 + 144 + 16$

$9(x+4)^2 + 4(y-2)^2 = 36$

This is an ellipse.

5. $9x^2 + 4y^2 + 72x - 16y + 160 = 0$

$9(x^2 + 8x + 16) + 4(y^2 - 4y + 4) = -160 + 144 + 16$

$9(x+4)^2 + 4(y-2)^2 = 0$

This is a point.

7. $y^2 - 5x - 4y - 6 = 0$

$(y^2 - 4y + 4) = 5x + 6 + 4$

$(y-2)^2 = 5(x+2)$

This is a parabola.

31. Let $P(x, y)$ be the location of the explosion.

$3|AP| = 3|BP| + 12$

$|AP| - |BP| = 4$

Thus, P lies on the right branch of the horizontal hyperbola with $a = 2$ and $c = 8$, so $b = 2\sqrt{15}$.

$\dfrac{x^2}{4} - \dfrac{y^2}{60} = 1$

Since $|BP| = |CP|$, the y-coordinate of P is 5.

$\dfrac{x^2}{4} - \dfrac{25}{60} = 1, x = \pm\sqrt{\dfrac{17}{3}}$

P is at $\left(\sqrt{\dfrac{17}{3}}, 5\right)$.

9. $3x^2 + 3y^2 - 6x + 12y + 60 = 0$

$3(x^2 - 2x + 1) + 3(y^2 + 4y + 4) = -60 + 3 + 12$

$3(x-1)^2 + 3(y+2)^2 = -45$

This is the empty set.

11. $4x^2 - 4y^2 + 8x + 12y - 5 = 0$

$4(x^2 + 2x + 1) - 4\left(y^2 - 3y + \dfrac{9}{4}\right) = 5 + 4 - 9$

$4(x+1)^2 - 4\left(y - \dfrac{3}{2}\right)^2 = 0$

This is two intersecting lines.

13. $4x^2 - 24x + 36 = 0$

$4(x^2 - 6x + 9) = -36 + 36$

$4(x-3)^2 = 0$

This is a line.

15. $25x^2 + 4y^2 + 150x - 8y + 129 = 0$

$25(x^2 + 6x + 9) + 4(y^2 - 2y + 1) = -129 + 225 + 4$

$25(x+3)^2 + 4(y-1)^2 = 100$

This is an ellipse.

17. $\dfrac{(x+3)^2}{4} + \dfrac{(y+2)^2}{16} = 1$

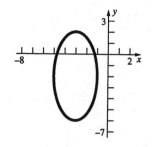

19. $\dfrac{(x+3)^2}{4} - \dfrac{(y+2)^2}{16} = 1$

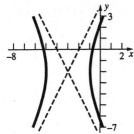

21. $(x+2)^2 = 8(y-1)$

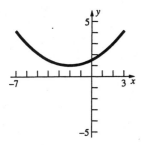

23. $(y-1)^2 = 16$

$y - 1 = \pm 4$

$y = 5,\ y = -3$

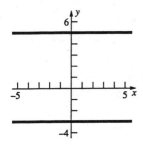

25. $x^2 + 4y^2 - 2x + 16y + 1 = 0$

$(x^2 - 2x + 1) + 4(y^2 + 4y + 4) = -1 + 1 + 16$

$(x-1)^2 + 4(y+2)^2 = 16$

$\dfrac{(x-1)^2}{16} + \dfrac{(y+2)^2}{4} = 1$

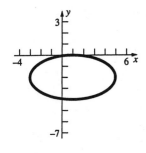

27. $9x^2 - 16y^2 + 54x + 64y - 127 = 0$

$9(x^2 + 6x + 9) - 16(y^2 - 4y + 4) = 127 + 81 - 64$

$9(x+3)^2 - 16(y-2)^2 = 144$

$\dfrac{(x+3)^2}{16} - \dfrac{(y-2)^2}{9} = 1$

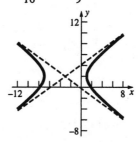

29. $4x^2 + 16x - 16y + 32 = 0$

$4(x^2 + 4x + 4) = 16y - 32 + 16$

$4(x+2)^2 = 16(y-1)$

$(x+2)^2 = 4(y-1)$

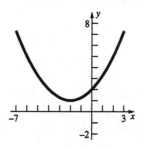

31. $2y^2 - 4y - 10x = 0$

$2(y^2 - 2y + 1) = 10x + 2$

$(y-1)^2 = 5\left(x + \dfrac{1}{5}\right)$

$(y-1)^2 = 4\left(\dfrac{5}{4}\right)\left(x + \dfrac{1}{5}\right)$

Horizontal parabola, $p = \dfrac{5}{4}$

Vertex $\left(-\dfrac{1}{5}, 1\right)$

Focus is at $\left(\dfrac{21}{20}, 1\right)$ and directrix is at $x = -\dfrac{29}{20}$.

33. $16(x-1)^2 + 25(y+2)^2 = 400$

$\dfrac{(x-1)^2}{25} + \dfrac{(y+2)^2}{16} = 1$

Horizontal ellipse, center $(1, -2)$, $a = 5$, $b = 4$,

$c = \sqrt{25 - 16} = 3$

Foci are at $(-2, -2)$ and $(4, -2)$.

35. $a = 5, b = 4$

$$\frac{(x-5)^2}{25} + \frac{(y-1)^2}{16} = 1$$

37. Vertical parabola, opens upward, $p = 5 - 3 = 2$

$$(x-2)^2 = 4(2)(y-3)$$

$$(x-2)^2 = 8(y-3)$$

39. Vertical hyperbola, center $(0, 3)$, $2a = 6$, $a = 3$,

$c = 5$, $b = \sqrt{25-9} = 4$

$$\frac{(y-3)^2}{9} - \frac{x^2}{16} = 1$$

41. Horizontal parabola, opens to the left

Vertex $(6, 5)$, $p = \dfrac{10-2}{2} = 4$

$$(y-5)^2 = -4(4)(x-6)$$

$$(y-5)^2 = -16(x-6)$$

43. Horizontal ellipse, center $(0, 2)$, $c = 2$

Since it passes through the origin and center is at $(0, 2)$, $b = 2$.

$$a = \sqrt{4+4} = \sqrt{8}$$

$$\frac{x^2}{8} + \frac{(y-2)^2}{4} = 1$$

45. a. If C is a vertical parabola, the equation for C can be written in the form $y = ax^2 + bx + c$.

Substitute the three points into the equation.

$2 = a - b + c$

$0 = c$

$6 = 9a + 3b + c$

Solve the system to get $a = 1, b = -1, c = 0$.

$$y = x^2 - x$$

b. If C is a horizontal parabola, an equation for C can be written in the form $x = ay^2 + by + c$.

Substitute the three points into the equation.

$-1 = 4a + 2b + c$

$0 = c$

$3 = 36a + 6b + c$

Solve the system to get $a = \dfrac{1}{4}, b = -1, c = 0$.

$$x = \frac{1}{4}y^2 - y$$

c. If C is a circle, an equation for C can be written in the form $(x-h)^2 + (y-k)^2 = r^2$.

Substitute the three points into the equation.

$$(-1-h)^2 + (2-k)^2 = r^2$$

$$h^2 + k^2 = r^2$$

$$(3-h)^2 + (6-k)^2 = r^2$$

Solve the system to get $h = \dfrac{5}{2}, k = \dfrac{5}{2}$, and

$$r^2 = \frac{25}{2}.$$

$$\left(x - \frac{5}{2}\right)^2 + \left(y - \frac{5}{2}\right)^2 = \frac{25}{2}$$

47. $y^2 = Lx + Kx^2$

$$K\left(x^2 + \frac{L}{K}x + \frac{L^2}{4K^2}\right) - y^2 = \frac{L^2}{4K}$$

$$K\left(x + \frac{L}{2K}\right)^2 - y^2 = \frac{L^2}{4K}$$

$$\frac{(x + \frac{L}{2K})^2}{\frac{L^2}{4K^2}} - \frac{y^2}{\frac{L^2}{4K}} = 1$$

If $K < -1$, the conic is a vertical ellipse. If $K = -1$, the conic is a circle. If $-1 < K < 0$, the conic is a horizontal ellipse. If $K = 0$, the original equation is $y^2 = Lx$, so the conic is a horizontal parabola. If $K > 0$, the conic is a horizontal hyperbola.

If $-1 < K < 0$ (a horizontal ellipse) the length of the latus rectum is (see problem 33, Section 12.2)

$$\frac{2b^2}{a} = 2\frac{\frac{L^2}{4|K|}}{\frac{|L|}{2|K|}} = |L|$$

From general considerations, the result for a vertical ellipse is the same as the one just obtained. For $K = -1$ (a circle) we have

$$\left(x - \frac{L}{2}\right)^2 + y^2 = \frac{L^2}{4} \Rightarrow 2\frac{|L|}{2} = |L|$$

If $K = 0$ (a horizontal parabola) we have

$$y^2 = Lx; y^2 = 4\frac{L}{4}x; p = \frac{L}{4}$$, and the latus rectum is

$$2\sqrt{Lp} = 2\sqrt{L\frac{L}{4}} = |L|.$$

If $K > 0$ (a horizontal hyperbola) we can use the result of Problem 34, Section 12.2. The length of the latus rectum is $\dfrac{2b^2}{a}$, which is equal to $|L|$.

12.5 Concepts Test

1. $Ax^2 + Bxy + Cy^2 + Dx + Ey + F = 0$

3. hyperbola

Problem Set 12.5

1. $x^2 + xy + y^2 = 6$

$\cot 2\theta = 0$

$2\theta = \dfrac{\pi}{2}, \theta = \dfrac{\pi}{4}$

$x = u\dfrac{\sqrt{2}}{2} - v\dfrac{\sqrt{2}}{2} = \dfrac{\sqrt{2}}{2}(u - v)$

$y = u\dfrac{\sqrt{2}}{2} + v\dfrac{\sqrt{2}}{2} = \dfrac{\sqrt{2}}{2}(u + v)$

$\dfrac{1}{2}(u - v)^2 + \dfrac{1}{2}(u - v)(u + v) + \dfrac{1}{2}(u + v)^2 = 6$

$\dfrac{3}{2}u^2 + \dfrac{1}{2}v^2 = 6$

$\dfrac{u^2}{4} + \dfrac{v^2}{12} = 1$

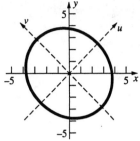

3. $4x^2 + xy + 4y^2 = 56$

$\cot 2\theta = 0, \ 2\theta = \dfrac{\pi}{2}, \theta = \dfrac{\pi}{4}$

$x = \dfrac{\sqrt{2}}{2}(u - v)$

$y = \dfrac{\sqrt{2}}{2}(u + v)$

$2(u - v)^2 + \dfrac{1}{2}(u - v)(u + v) + 2(u + v)^2 = 56$

$\dfrac{9}{2}u^2 + \dfrac{7}{2}v^2 = 56$

7. $-\dfrac{1}{2}x^2 + 7xy - \dfrac{1}{2}y^2 - 6\sqrt{2}x - 6\sqrt{2}y = 0$

$\cot 2\theta = 0, \ 2\theta = \dfrac{\pi}{2}, \theta = \dfrac{\pi}{4}$

$x = \dfrac{\sqrt{2}}{2}(u - v)$

$\dfrac{u^2}{\frac{112}{9}} + \dfrac{v^2}{16} = 1$

5. $4x^2 - 3xy = 18$

$\cot 2\theta = -\dfrac{4}{3}, \ r = 5$

$\cos 2\theta = -\dfrac{4}{5}$

$\cos\theta = \sqrt{\dfrac{1 - \frac{4}{5}}{2}} = \dfrac{1}{\sqrt{10}}$

$\sin\theta = \sqrt{\dfrac{1 + \frac{4}{5}}{2}} = \dfrac{3}{\sqrt{10}}$

$x = \dfrac{1}{\sqrt{10}}(u - 3v)$

$y = \dfrac{1}{\sqrt{10}}(3u + v)$

$\dfrac{2}{5}(u - 3v)^2 - \dfrac{3}{10}(u - 3v)(3u + v) = 18$

$-\dfrac{u^2}{2} + \dfrac{9v^2}{2} = 18$

$\dfrac{v^2}{4} - \dfrac{u^2}{36} = 1$

$$y = \frac{\sqrt{2}}{2}(u+v)$$

$$-\frac{1}{4}(u-v)^2 + \frac{7}{2}(u-v)(u+v) - \frac{1}{4}(u+v)^2 - 6(u-v) - 6(u+v) = 0$$

$$3u^2 - 4v^2 - 12u = 0$$

$$3(u^2 - 4u + 4) - 4v^2 = 12$$

$$\frac{(u-2)^2}{4} - \frac{v^2}{3} = 1$$

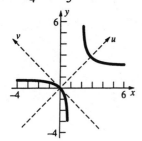

9. $34x^2 + 24xy + 41y^2 + 250y = -325$

$$\cot 2\theta = -\frac{7}{24}, r = 25$$

$$\cos 2\theta = -\frac{7}{25}$$

$$\cos\theta = \sqrt{\frac{1 - \frac{7}{25}}{2}} = \frac{3}{5}$$

$$\sin\theta = \sqrt{\frac{1 + \frac{7}{25}}{2}} = \frac{4}{5}$$

$$x = \frac{1}{5}(3u - 4v)$$

$$y = \frac{1}{5}(4u + 3v)$$

$$\frac{34}{25}(3u - 4v)^2 + \frac{24}{25}(3u - 4v)(4u + 3v) + \frac{41}{25}(4u + 3v)^2 + 50(4u + 3v) = -325$$

$$50u^2 + 25v^2 + 200u + 150v = -325$$

$$2u^2 + v^2 + 8u + 6v = -13$$

$$2(u^2 + 4u + 4) + (v^2 + 6v + 9) = -13 + 8 + 9$$

$$2(u+2)^2 + (v+3)^2 = 4$$

$$\frac{(u+2)^2}{2} + \frac{(v+3)^2}{4} = 1$$

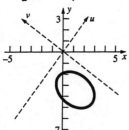

11. $5x^2 - 3xy + y^2 + 65x - 25y + 203 = 0$

$$\cot 2\theta = -\frac{4}{3}, \ r = 5$$

$$\cos 2\theta = -\frac{4}{5}$$

$$\cos\theta = \sqrt{\frac{1-\frac{4}{5}}{2}} = \frac{1}{\sqrt{10}}$$

$$\sin\theta = \sqrt{\frac{1+\frac{4}{5}}{2}} = \frac{3}{\sqrt{10}}$$

$$x = \frac{1}{\sqrt{10}}(u - 3v)$$

$$y = \frac{1}{\sqrt{10}}(3u + v)$$

$$\frac{1}{2}(u-3v)^2 - \frac{3}{10}(u-3v)(3u+v) + \frac{1}{10}(3u+v)^2 + \frac{65}{\sqrt{10}}(u-3v) - \frac{25}{\sqrt{10}}(3u+v) + 203 = 0$$

$$\frac{1}{2}u^2 + \frac{11}{2}v^2 - \sqrt{10}u - 22\sqrt{10}v + 203 = 0$$

$$\frac{1}{2}\left(u^2 - 2\sqrt{10}u + 10\right) + \frac{11}{2}\left(v^2 - 4\sqrt{10}v + 40\right) = -203 + 5 + 220$$

$$\frac{1}{2}\left(u - \sqrt{10}\right)^2 + \frac{11}{2}\left(v - 2\sqrt{10}\right)^2 = 22$$

$$\frac{\left(u - \sqrt{10}\right)^2}{44} + \frac{\left(v - 2\sqrt{10}\right)^2}{4} = 1$$

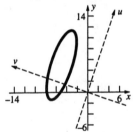

13. $x = u \cos\alpha - v \sin\alpha$
$y = u \sin\alpha + v \cos\alpha$
$(u \cos\alpha - v \sin\alpha) \cos\alpha + (u \sin\alpha + v \cos\alpha) \sin\alpha = d$
$u(\cos^2\alpha + \sin^2\alpha) = d$
$u = d$
Thus, the perpendicular distance from the origin is d.

15. $x = u \cos\theta - v \sin\theta$
$y = u \sin\theta + v \cos\theta$
$x(\cos\theta) + y(\sin\theta) = (u\cos^2\theta - v\cos\theta\sin\theta) + (u\sin^2\theta + v\cos\theta\sin\theta) = u$
$x(-\sin\theta) + y(\cos\theta) = (-u\cos\theta\sin\theta + v\sin^2\theta) + (u\cos\theta\sin\theta + v\cos^2\theta) = v$
Thus, $u = x \cos\theta + y \sin\theta$ and $v = -x \sin\theta + y \cos\theta$.

17. Rotate to eliminate the xy-term.
$x^2 + 14xy + 49y^2 = 100$

$$\cot 2\theta = -\frac{24}{7}$$

$$\cos 2\theta = -\frac{24}{25}$$

$$\cos\theta = \sqrt{\frac{1-\frac{24}{25}}{2}} = \frac{1}{5\sqrt{2}}$$

$$\sin\theta = \sqrt{\frac{1+\frac{24}{25}}{2}} = \frac{7}{5\sqrt{2}}$$

$$x = \frac{1}{5\sqrt{2}}(u-7v)$$

$$y = \frac{1}{5\sqrt{2}}(7u+v)$$

$$\frac{1}{50}(u-7v)^2 + \frac{14}{50}(u-7v)(7u+v) + \frac{49}{50}(7u+v)^2 = 100$$

$$50u^2 = 100$$

$$u^2 = 2$$

$$u = \pm\sqrt{2}$$

Thus the points closest to the origin in uv-coordinates are $\left(\sqrt{2},0\right)$ and $\left(-\sqrt{2},0\right)$.

$$x = \frac{1}{5\sqrt{2}}\left(\sqrt{2}\right) = \frac{1}{5} \text{ or } x = \frac{1}{5\sqrt{2}}\left(-\sqrt{2}\right) = -\frac{1}{5}$$

$$y = \frac{1}{5\sqrt{2}}\left(7\sqrt{2}\right) = \frac{7}{5} \text{ or } y = \frac{1}{5\sqrt{2}}\left(-7\sqrt{2}\right) = -\frac{7}{5}$$

The points closest to the origin in xy-coordinates are $\left(\frac{1}{5},\frac{7}{5}\right)$ and $\left(-\frac{1}{5},-\frac{7}{5}\right)$.

19. From Problem 18 $a = A\cos^2\theta + B\cos\theta\sin\theta + C\sin^2\theta$,

$b = -2A\cos\theta\sin\theta + B(\cos^2\theta - \sin^2\theta) + 2C\cos\theta\sin\theta$, and

$c = A\sin^2\theta - B\cos\theta\sin\theta + C\cos^2\theta$.

$b^2 = B^2\cos^4\theta + 4(-AB+BC)\cos^3\theta\sin\theta + 2(2A^2 - B^2 - 4AC + 2C^2)\cos^2\theta\sin^2\theta$

$\qquad +4(AB-BC)\cos\theta\sin^3\theta + B^2\sin^4\theta$

$4ac = 4AC\cos^4\theta + 4(-AB+BC)\cos^3\theta\sin\theta + 4(A^2 - B^2 + C^2)\cos^2\theta\sin^2\theta + 4(AB-BC)\cos\theta\sin^3\theta + 4AC\sin^4\theta$

$b^2 - 4ac = (B^2 - 4AC)\cos^4\theta + 2(B^2 - 4AC)\cos^2\theta\sin^2\theta + (B^2 - 4AC)\sin^4\theta$

$= (B^2 - 4AC)(\cos^2\theta)(\cos^2\theta + \sin^2\theta) + (B^2 - 4AC)(\sin^2\theta)(\cos^2\theta + \sin^2\theta)$

$= (B^2 - 4AC)(\cos^2\theta + \sin^2\theta) = B^2 - 4AC$

21. **a.** From Problem 19, $-4ac = B^2 - 4AC = -\Delta$ or

$$\frac{1}{ac} = \frac{4}{\Delta}.$$

b. From Problem 18, $a + c = A + C$.

$$\frac{1}{a} + \frac{1}{c} = \frac{a+c}{ac} = \frac{4(A+C)}{\Delta}$$

c. $\dfrac{2}{\Delta}\left(A+C\pm\sqrt{(A-C)^2+B^2}\right)$

$$= \frac{2}{\Delta}\left[\frac{\Delta}{4}\left(\frac{1}{a}+\frac{1}{c}\right)\pm\sqrt{A^2+2AC+C^2+B^2-4AC}\right]$$

$$= \frac{1}{2}\left(\frac{1}{a}+\frac{1}{c}\right)\pm\frac{2}{\Delta}\sqrt{(A+C)^2-\Delta}$$

$$= \frac{1}{2}\left(\frac{1}{a}+\frac{1}{c}\right) \pm \frac{2}{\Delta}\sqrt{\frac{\Delta^2}{16}\left(\frac{1}{a}+\frac{1}{c}\right)^2 - \Delta}$$

$$= \frac{1}{2}\left(\frac{1}{a}+\frac{1}{c}\right) \pm \frac{1}{2}\sqrt{\left(\frac{1}{a}+\frac{1}{c}\right)^2 - 4\left(\frac{4}{\Delta}\right)}$$

$$= \frac{1}{2}\left(\frac{1}{a}+\frac{1}{c}\right) \pm \sqrt{\frac{1}{a^2}+\frac{2}{ac}+\frac{1}{c^2} - 4\left(\frac{1}{ac}\right)}$$

$$= \frac{1}{2}\left(\frac{1}{a}+\frac{1}{c}\right) \pm \frac{1}{2}\sqrt{\left(\frac{1}{a}-\frac{1}{c}\right)^2} = \frac{1}{2}\left(\frac{1}{a}+\frac{1}{c} \pm \left|\frac{1}{a}-\frac{1}{c}\right|\right)$$

The two values of this expression are $\dfrac{1}{a}$ and $\dfrac{1}{c}$.

23. $\cot 2\theta = 0,\ \theta = \dfrac{\pi}{4}$

$$x = \frac{\sqrt{2}}{2}(u-v)$$

$$y = \frac{\sqrt{2}}{2}(u+v)$$

$$\frac{1}{2}(u-v)^2 + \frac{B}{2}(u-v)(u+v) + \frac{1}{2}(u+v)^2 = 1$$

$$\frac{2+B}{2}u^2 + \frac{2-B}{2}v^2 = 1$$

a. The graph is an ellipse if $\dfrac{2+B}{2} > 0$ and

$\dfrac{2-B}{2} > 0$, so $-2 < B < 2$.

b. The graph is a circle if $\dfrac{2+B}{2} = \dfrac{2-B}{2}$, so

$B = 0$.

c. The graph is a hyperbola if $\dfrac{2+B}{2} > 0$ and

$\dfrac{2-B}{2} < 0$ or if $\dfrac{2+B}{2} < 0$ and $\dfrac{2-B}{2} > 0$, so

$B < -2$ or $B > 2$.

d. The graph is two parallel lines if $\dfrac{2+B}{2} = 0$ or

$\dfrac{2-B}{2} = 0$, so $B = \pm 2$.

25. From Figure 1 it is clear that
$v = r\sin\phi$ and $u = r\cos\phi$.
Also noting that $y = r\sin(\theta + \phi)$ leads us to
$y = r\sin(\theta + \phi) = r\sin\theta\cos\phi + r\cos\theta\sin\phi$
$= (r\cos\phi)(\sin\theta) + (r\sin\phi)(\cos\theta)$
$= u\sin\theta + v\cos\theta$

12.6 Concepts Review

1. infinitely many

3. circle; line

Problem Set 12.6

1.

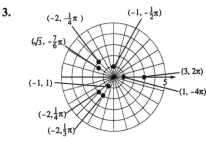

3.

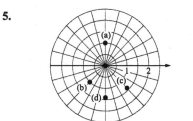

5.

a. $\left(1, -\dfrac{3}{2}\pi\right), \left(1, \dfrac{5}{2}\pi\right), \left(-1, -\dfrac{1}{2}\pi\right), \left(-1, \dfrac{3}{2}\pi\right)$

b. $\left(1, -\dfrac{3}{4}\pi\right), \left(1, \dfrac{5}{4}\pi\right), \left(-1, -\dfrac{7}{4}\pi\right), \left(-1, \dfrac{9}{4}\pi\right)$

c. $\left(\sqrt{2}, -\frac{7}{3}\pi\right), \left(\sqrt{2}, \frac{5}{3}\pi\right), \left(-\sqrt{2}, -\frac{4}{3}\pi\right),$

$\left(-\sqrt{2}, \frac{2}{3}\pi\right)$

d. $\left(\sqrt{2}, -\frac{1}{2}\pi\right), \left(\sqrt{2}, \frac{3}{2}\pi\right), \left(-\sqrt{2}, -\frac{3}{2}\pi\right),$

$\left(-\sqrt{2}, \frac{1}{2}\pi\right)$

7. a. $x = 1\cos\frac{1}{2}\pi = 0$

$y = 1\sin\frac{1}{2}\pi = 1$

$(0, 1)$

b. $x = -1\cos\frac{1}{4}\pi = -\frac{\sqrt{2}}{2}$

$y = -1\sin\frac{1}{4}\pi = -\frac{\sqrt{2}}{2}$

$\left(-\frac{\sqrt{2}}{2}, -\frac{\sqrt{2}}{2}\right)$

c. $x = \sqrt{2}\cos\left(-\frac{1}{3}\pi\right) = \frac{\sqrt{2}}{2}$

$y = \sqrt{2}\sin\left(-\frac{1}{3}\pi\right) = -\frac{\sqrt{6}}{2}$

$\left(\frac{\sqrt{2}}{2}, -\frac{\sqrt{6}}{2}\right)$

d. $x = -\sqrt{2}\cos\frac{5}{2}\pi = 0$

$y = -\sqrt{2}\sin\frac{5}{2}\pi = -\sqrt{2}$

$\left(0, -\sqrt{2}\right)$

9. a. $r^2 = \left(3\sqrt{3}\right)^2 + 3^2 = 36, \ r = 6$

$\tan\theta = \frac{1}{\sqrt{3}}, \theta = \frac{\pi}{6}$

$\left(6, \frac{\pi}{6}\right)$

b. $r^2 = \left(-2\sqrt{3}\right)^2 + 2^2 = 16, \ r = 4$

$\tan\theta = \frac{2}{-2\sqrt{3}}, \theta = \frac{5\pi}{6}$

$\left(4, \frac{5\pi}{6}\right)$

c. $r^2 = \left(-\sqrt{2}\right)^2 + \left(-\sqrt{2}\right)^2 = 4, \ r = 2$

$\tan\theta = \frac{-\sqrt{2}}{-\sqrt{2}}, \theta = \frac{5\pi}{4}$

$\left(2, \frac{5\pi}{4}\right)$

d. $r^2 = 0^2 + 0^2 = 0, \ r = 0$

$\tan\theta = 0, \ \theta = 0$

$(0, 0)$

11. $x - 3y + 2 = 0$

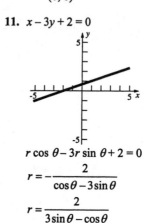

$r\cos\theta - 3r\sin\theta + 2 = 0$

$r = -\frac{2}{\cos\theta - 3\sin\theta}$

$r = \frac{2}{3\sin\theta - \cos\theta}$

13. $y = -2$

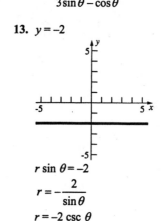

$r\sin\theta = -2$

$r = -\frac{2}{\sin\theta}$

$r = -2\csc\theta$

15. $x^2 + y^2 = 4$

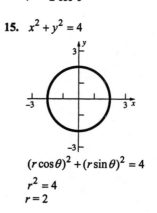

$(r\cos\theta)^2 + (r\sin\theta)^2 = 4$

$r^2 = 4$

$r = 2$

17. $\theta = \dfrac{\pi}{2}$

$\cot \theta = 0$

$\dfrac{x}{y} = 0$

$x = 0$

19. $r \cos \theta + 3 = 0$

$x + 3 = 0$

$x = -3$

21. $r \sin \theta - 1 = 0$

$y - 1 = 0$

$y = 1$

23. $r = 6$, circle

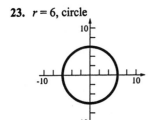

25. $r = \dfrac{3}{\sin \theta}$

$r = \dfrac{3}{\cos\left(\theta - \frac{\pi}{2}\right)}$, line

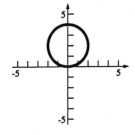

27. $r = 4 \sin \theta$

$r = 2(2) \cos\left(\theta - \dfrac{\pi}{2}\right)$, circle

29. $r = \dfrac{4}{1 + \cos \theta}$

$r = \dfrac{(1)(4)}{1 + (1)\cos \theta}$, parabola

$e = 1$

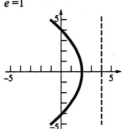

31. $r = \dfrac{6}{2 + \sin \theta}$

$r = \dfrac{\left(\frac{1}{2}\right)6}{1 + \left(\frac{1}{2}\right)\cos\left(\theta - \frac{\pi}{2}\right)}$, ellipse

$e = \dfrac{1}{2}$

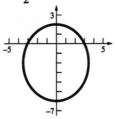

33. $r = \dfrac{4}{2 + 2 \cos \theta}$

$r = \dfrac{(1)(2)}{1 + (1)\cos \theta}$, parabola

$e = 1$

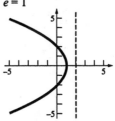

35. $r = \dfrac{4}{\frac{1}{2} + \cos(\theta - \pi)}$

$r = \dfrac{4(2)}{1 + 2 \cos(\theta - \pi)}$,

hyperbola

$e = 2$

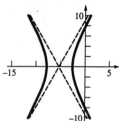

$$x = \frac{4e}{1+e\cos t}\cos t, \quad y = \frac{4e}{1+e\cos t}\sin t$$

$e = 0.1$

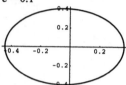

37. By the Law of Cosines,
$a^2 = r^2 + c^2 - 2rc\cos(\theta - \alpha)$ (see figure below).

$e = 0.5$

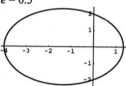

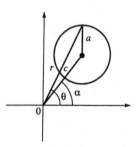

$e = 0.9$

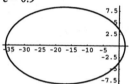

39. Recall that the latus rectum is perpendicular to the axis of the conic through a focus.

$$r\left(\theta_0 + \frac{\pi}{2}\right) = \frac{ed}{1+e\cos\frac{\pi}{2}} = ed$$

Thus the length of the latus rectum is $2ed$.

$e = 1$

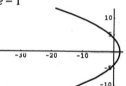

41. $a + c = 183, \; a - c = 17$
$2a = 200, \; a = 100$
$2c = 166, \; c = 83$

$$e = \frac{c}{a} = 0.83$$

43. Let sun lie at the pole and the axis of the parabola lie on the pole so that the parabola opens to the left. Then the path is described by the equation

$r = \dfrac{d}{1+\cos\theta}$. Substitute $(100, 120°)$ into the

equation and solve for d.

$$100 = \frac{d}{1+\cos 120°}$$

$d = 50$

The closest distance occurs when $\theta = 0°$.

$$r = \frac{50}{1+\cos 0°} = 25 \text{ million miles}$$

$e = 1.1$

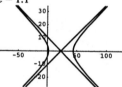

$e = 1.3$

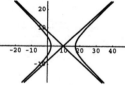

12.7 Concepts Review

1. limaçon   3. rose; odd; even

Problem Set 12.7

1. $\theta^2 - \dfrac{\pi^2}{16} = 0$; $\theta = \pm\dfrac{\pi}{4}$

Changing $\theta \to -\theta$ or $r \to -r$ yields an equivalent set of equations. Therefore all 3 tests are passed.

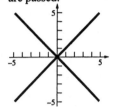

3. $r \sin\theta + 4 = 0$; $r = -\dfrac{4}{\sin\theta}$

Since $\sin(-\theta) = -\sin\theta$, test 2 is passed. The other two tests fail so the graph has only y-axis symmetry.

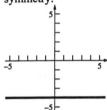

5. $r = 2\cos\theta$

Since $\cos(-\theta) = \cos\theta$, the graph is symmetric about the x-axis. The other symmetry tests fail.

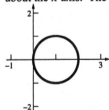

7. $r = \dfrac{2}{1 - \cos\theta}$

Since $\cos(-\theta) = \cos\theta$, the graph is symmetric about the x-axis. The other symmetry tests fail.

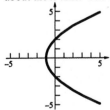

9. $r = 3 - 3\cos\theta$ (cardioid)

Since $\cos(-\theta) = \cos\theta$, the graph is symmetric about the x-axis. The other symmetry tests fail.

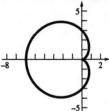

11. $r = 1 - 1\sin\theta$ (cardioid)

Since $\sin(\pi - \theta) = \sin\theta$, the graph is symmetric about the y-axis. The other symmetry tests fail.

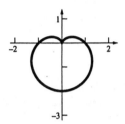

13. $r = 1 - 2\sin\theta$ (limaçon)

Since $\sin(\pi - \theta) = \sin\theta$, the graph is symmetric about the y-axis. The other symmetry tests fail.

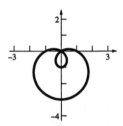

15. $r = 2 - 3\sin\theta$ (limaçon)

Since $\sin(\pi - \theta) = \sin\theta$, the graph is symmetric about the y-axis. The other symmetry tests fail.

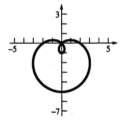

17. $r^2 = 4\cos 2\theta$ (lemniscate)

$r = \pm 2\sqrt{\cos 2\theta}$

Since $\cos(-2\theta) = \cos 2\theta$ and

$\cos(2(\pi - \theta)) = \cos(2\pi - 2\theta) = \cos(-2\theta) = \cos 2\theta$

the graph is symmetric about both axes and the origin.

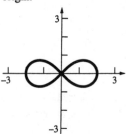

19. $r^2 = -9\cos 2\theta$ (lemniscate)

$r = \pm 3\sqrt{-\cos 2\theta}$

Since $\cos(-2\theta) = \cos 2\theta$ and

$\cos(2(\pi - \theta)) = \cos(2\pi - 2\theta) = \cos(-2\theta) = \cos 2\theta$

the graph is symmetric about both axes and the origin.

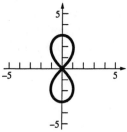

21. $r = 5\cos 3\theta$ (three-leaved rose)

Since $\cos(-3\theta) = \cos(3\theta)$, the graph is symmetric about the x-axis. The other symmetry tests fail.

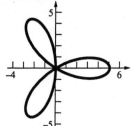

23. $r = 6\sin 2\theta$ (four-leaved rose)

Since

$\sin(2(\pi - \theta)) = \sin(2\pi - 2\theta)$

$= \sin(-2\theta) = -\sin(2\theta)$

and $\sin(-2\theta) = -\sin(2\theta)$, the graph is symmetric about both axes and the origin.

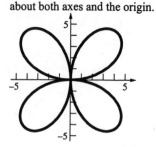

25. $r = 7\cos 5\theta$ (five-leaved rose)

Since $\cos(-5\theta) = \cos 5\theta$, the graph is symmetric about the x-axis. The other symmetry tests fail.

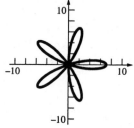

27. $r = \frac{1}{2}\theta, \theta \ge 0$ (spiral of Archimedes)

No symmetry. All three tests fail.

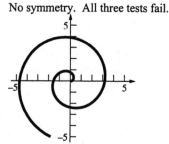

29. $r = e^\theta, \theta \ge 0$ (logarithmic spiral)

No symmetry. All three tests fail.

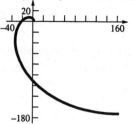

31. $r = \dfrac{2}{\theta}, \theta > 0$ (reciprocal spiral)

No symmetry. All three tests fail.

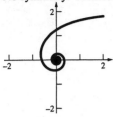

33. $r = 6, \; r = 4 + 4\cos\theta$

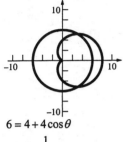

$6 = 4 + 4\cos\theta$

$\cos\theta = \dfrac{1}{2}$

$\theta = \dfrac{\pi}{3}, \theta = \dfrac{5\pi}{3}$

$\left(6, \dfrac{\pi}{3}\right), \left(6, \dfrac{5\pi}{3}\right)$

35. $r = 3\sqrt{3}\cos\theta, r = 3\sin\theta$

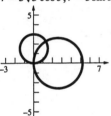

$3\sqrt{3}\cos\theta = 3\sin\theta$

$\tan\theta = \sqrt{3}$

$\theta = \dfrac{\pi}{3}, \theta = \dfrac{4\pi}{3}$

$\left(\dfrac{3\sqrt{3}}{2}, \dfrac{\pi}{3}\right) = \left(-\dfrac{3\sqrt{3}}{2}, \dfrac{4\pi}{3}\right)$

(0, 0) is also a solution since both graphs include the pole.

37. $r = 6\sin\theta, r = \dfrac{6}{1 + 2\sin\theta}$

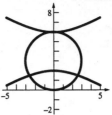

$6\sin\theta = \dfrac{6}{1 + 2\sin\theta}$

$12\sin^2\theta + 6\sin\theta - 6 = 0$

$6(2\sin\theta - 1)(\sin\theta + 1) = 0$

$\sin\theta = \dfrac{1}{2}, \sin\theta = -1$

$\theta = \dfrac{\pi}{6}, \theta = \dfrac{5\pi}{6}, \theta = \dfrac{3\pi}{2}$

$\left(3, \dfrac{\pi}{6}\right), \left(3, \dfrac{5\pi}{6}\right), \left(-6, \dfrac{3\pi}{2}\right)$ or $\left(6, \dfrac{\pi}{2}\right)$

39. Consider $r = \cos\dfrac{1}{2}\theta$.

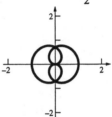

The graph is clearly symmetric with respect to the y-axis.
Substitute (r, θ) by $(-r, -\theta)$.

$-r = \cos\left(-\dfrac{1}{2}\theta\right) = \cos\dfrac{1}{2}\theta$

$r = -\cos\dfrac{1}{2}\theta$

Substitute (r, θ) by $(r, \pi - \theta)$

$r = \cos\dfrac{1}{2}(\pi - \theta) = \cos\dfrac{1}{2}\pi\cos\dfrac{1}{2}\theta + \sin\dfrac{1}{2}\pi\sin\dfrac{1}{2}\theta$

$= \sin\dfrac{1}{2}\theta$

$r = \sin\dfrac{1}{2}\theta$

41.

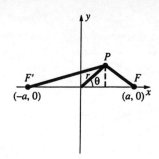

$$|PF| = \sqrt{(a - r\cos\theta)^2 + (r\sin\theta)^2}$$
$$= \sqrt{r^2 + a^2 - 2ar\cos\theta}$$
$$|PF'| = \sqrt{(a + r\cos\theta)^2 + (r\sin\theta)^2}$$
$$= \sqrt{r^2 + a^2 + 2ar\cos\theta}$$
$$|PF||PF'| = \sqrt{(r^2 + a^2)^2 - 4a^2r^2\cos^2\theta} = a^2$$
$$(r^4 + 2a^2r^2 + a^4) - 4a^2r^2\cos^2\theta = a^4$$
$$r^4 - 4a^2r^2\cos^2\theta + 2a^2r^2 = 0$$
$$r^2(r^2 - 2a^2(2\cos^2\theta - 1)) = 0$$
$$r^2 - 2a^2(2\cos^2\theta - 1) = 0$$
$$r^2 = 2a^2(1 + \cos 2\theta - 1)$$
$$r^2 = 2a^2\cos 2\theta$$

This is the equation of a lemniscate.

43. a. $y = 45$
$$r\sin\theta = 45$$
$$r = \frac{45}{\sin\theta}$$

b. $x^2 + y^2 = 36$
$$r^2 = 36$$
$$r = 6$$

c. $x^2 - y^2 = 1$
$$r^2\cos^2\theta - r^2\sin^2\theta = 1$$
$$r^2 = \frac{1}{\cos 2\theta}$$
$$r = \pm\frac{1}{\sqrt{\cos 2\theta}}$$

d. $4xy = 1$
$$4r^2\cos\theta\sin\theta = 1$$
$$r^2 = \frac{1}{2\sin 2\theta}$$
$$r = \pm\frac{1}{\sqrt{2\sin 2\theta}}$$

e. $y = 3x + 2$
$$r\sin\theta = 3r\cos\theta + 2$$
$$r(\sin\theta - 3\cos\theta) = 2$$
$$r = \frac{2}{\sin\theta - 3\cos\theta}$$

f. $3x^2 + 4y = 2$
$$3r^2\cos^2\theta + 4r\sin\theta = 2$$
$$(3\cos^2\theta)r^2 + (4\sin\theta)r - 2 = 0$$
$$r = \frac{-4\sin\theta \pm \sqrt{16\sin^2\theta + 24\cos^2\theta}}{6\cos^2\theta}$$
$$r = \frac{-2\sin\theta \pm \sqrt{4\sin^2\theta + 6\cos^2\theta}}{3\cos^2\theta}$$

g. $x^2 + 2x + y^2 - 4y - 25 = 0$
$$r^2 + 2r\cos\theta - 4r\sin\theta - 25 = 0$$
$$r^2 + (2\cos\theta - 4\sin\theta)r - 25 = 0$$
$$r = \frac{-2\cos\theta + 4\sin\theta \pm \sqrt{(2\cos\theta - 4\sin\theta)^2 + 100}}{2}$$
$$r = -\cos\theta + 2\sin\theta \pm \sqrt{(\cos\theta - 2\sin\theta)^2 + 25}$$

45. a. VII

b. I

c. VIII

d. III

e. V

f. II

g. VI

h. IV

47. $r = \cos\left(\dfrac{13\theta}{5}\right)$

49. $r = 1 + 3\cos\left(\dfrac{\theta}{3}\right)$

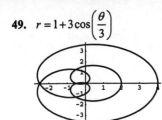

51. a. The graph for $\phi = 0$ is the graph for $\phi \neq 0$ rotated by ϕ counterclockwise about the pole.

    b. As n increases, the number of "leaves" increases.

    c. If $|a| > |b|$, the graph will not pass through the pole and will not "loop." If $|b| < |a|$, the graph will pass through the pole and will have $2n$ "loops" (n small "loops" and n large

"loops"). If $|a| = |b|$, the graph passes through the pole and will have n "loops." If $ab \neq 0, n > 1$, and $\phi = 0$, the graph will be symmetric about $\theta = \dfrac{\pi}{n}k$, where $k = 0, n-1$.

53. The spiral will unwind clockwise for $c < 0$. The spiral will unwind counter-clockwise for $c > 0$.

55. a. III

    b. IV

    c. I

    d. II

    e. VI

    f. V

12.8 Concepts Review

1. $\dfrac{1}{2}r^2\theta$

3. $\dfrac{1}{2}\displaystyle\int_0^{2\pi}(2+2\cos\theta)^2\,d\theta$

Problem Set 12.8

1. $r = a,\ a > 0$

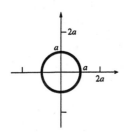

$$A = \frac{1}{2}\int_0^{2\pi}a^2\,d\theta = \pi a^2$$

3. $r = 2 + \cos\theta$

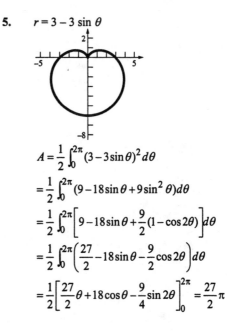

$$A = \frac{1}{2}\int_0^{2\pi}(2+\cos\theta)^2\,d\theta$$

$$= \frac{1}{2}\int_0^{2\pi}(4+4\cos\theta+\cos^2\theta)\,d\theta$$

$$= \frac{1}{2}\int_0^{2\pi}\left[4+4\cos\theta+\frac{1}{2}(1+\cos 2\theta)\right]d\theta$$

$$= \frac{1}{2}\int_0^{2\pi}\left(\frac{9}{2}+4\cos\theta+\frac{1}{2}\cos 2\theta\right)d\theta$$

$$= \frac{1}{2}\left[\frac{9}{2}\theta+4\sin\theta+\frac{1}{4}\sin 2\theta\right]_0^{2\pi} = \frac{9}{2}\pi$$

5. $r = 3 - 3\sin\theta$

$$A = \frac{1}{2}\int_0^{2\pi}(3-3\sin\theta)^2\,d\theta$$

$$= \frac{1}{2}\int_0^{2\pi}(9-18\sin\theta+9\sin^2\theta)\,d\theta$$

$$= \frac{1}{2}\int_0^{2\pi}\left[9-18\sin\theta+\frac{9}{2}(1-\cos 2\theta)\right]d\theta$$

$$= \frac{1}{2}\int_0^{2\pi}\left(\frac{27}{2}-18\sin\theta-\frac{9}{2}\cos 2\theta\right)d\theta$$

$$= \frac{1}{2}\left[\frac{27}{2}\theta+18\cos\theta-\frac{9}{4}\sin 2\theta\right]_0^{2\pi} = \frac{27}{2}\pi$$

7. $r = a(1 + \cos \theta)$

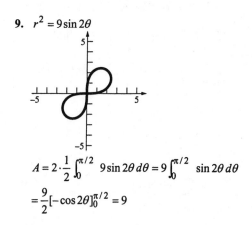

$$A = \frac{1}{2} \int_0^{2\pi} [a(1 + \cos \theta)]^2 \, d\theta$$

$$= \frac{a^2}{2} \int_0^{2\pi} (1 + 2 \cos \theta + \cos^2 \theta) \, d\theta$$

$$= \frac{a^2}{2} \int_0^{2\pi} \left[1 + 2 \cos \theta + \frac{1}{2}(1 + \cos 2\theta) \right] d\theta$$

$$= \frac{a^2}{2} \int_0^{2\pi} \left(\frac{3}{2} + 2 \cos \theta + \frac{1}{2} \cos 2\theta \right) d\theta$$

$$= \frac{a^2}{2} \left[\frac{3}{2}\theta + 2 \sin \theta + \frac{1}{4} \sin 2\theta \right]_0^{2\pi} = \frac{3\pi a^2}{2}$$

9. $r^2 = 9 \sin 2\theta$

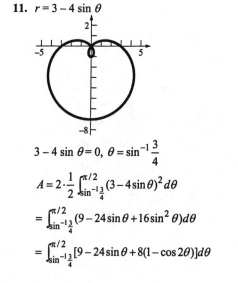

$$A = 2 \cdot \frac{1}{2} \int_0^{\pi/2} 9 \sin 2\theta \, d\theta = 9 \int_0^{\pi/2} \sin 2\theta \, d\theta$$

$$= \frac{9}{2} [-\cos 2\theta]_0^{\pi/2} = 9$$

11. $r = 3 - 4 \sin \theta$

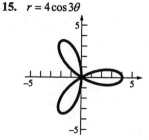

$3 - 4 \sin \theta = 0, \quad \theta = \sin^{-1} \frac{3}{4}$

$$A = 2 \cdot \frac{1}{2} \int_{\sin^{-1}\frac{3}{4}}^{\pi/2} (3 - 4 \sin \theta)^2 \, d\theta$$

$$= \int_{\sin^{-1}\frac{3}{4}}^{\pi/2} (9 - 24 \sin \theta + 16 \sin^2 \theta) d\theta$$

$$= \int_{\sin^{-1}\frac{3}{4}}^{\pi/2} [9 - 24 \sin \theta + 8(1 - \cos 2\theta)] d\theta$$

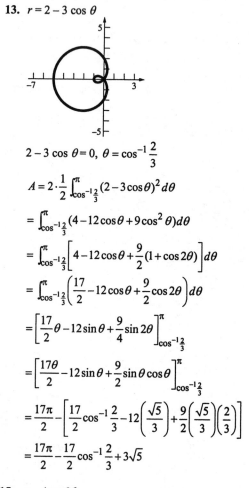

$$= \int_{\sin^{-1}\frac{3}{4}}^{\pi/2} (17 - 24 \sin \theta - 8 \cos 2\theta) d\theta$$

$$= [17\theta + 24 \cos \theta - 4 \sin 2\theta]_{\sin^{-1}\frac{3}{4}}^{\pi/2}$$

$$= [17\theta + 24 \cos \theta - 8 \sin \theta \cos \theta]_{\sin^{-1}\frac{3}{4}}^{\pi/2}$$

$$= \frac{17\pi}{2} - \left[17 \sin^{-1} \frac{3}{4} + 24 \left(\frac{\sqrt{7}}{4} \right) - 8 \left(\frac{3}{4} \right) \left(\frac{\sqrt{7}}{4} \right) \right]$$

$$= \frac{17\pi}{2} - 17 \sin^{-1} \frac{3}{4} - \frac{9\sqrt{7}}{2}$$

13. $r = 2 - 3 \cos \theta$

$2 - 3 \cos \theta = 0, \quad \theta = \cos^{-1} \frac{2}{3}$

$$A = 2 \cdot \frac{1}{2} \int_{\cos^{-1}\frac{2}{3}}^{\pi} (2 - 3 \cos \theta)^2 \, d\theta$$

$$= \int_{\cos^{-1}\frac{2}{3}}^{\pi} (4 - 12 \cos \theta + 9 \cos^2 \theta) d\theta$$

$$= \int_{\cos^{-1}\frac{2}{3}}^{\pi} \left[4 - 12 \cos \theta + \frac{9}{2}(1 + \cos 2\theta) \right] d\theta$$

$$= \int_{\cos^{-1}\frac{2}{3}}^{\pi} \left(\frac{17}{2} - 12 \cos \theta + \frac{9}{2} \cos 2\theta \right) d\theta$$

$$= \left[\frac{17}{2}\theta - 12 \sin \theta + \frac{9}{4} \sin 2\theta \right]_{\cos^{-1}\frac{2}{3}}^{\pi}$$

$$= \left[\frac{17\theta}{2} - 12 \sin \theta + \frac{9}{2} \sin \theta \cos \theta \right]_{\cos^{-1}\frac{2}{3}}^{\pi}$$

$$= \frac{17\pi}{2} - \left[\frac{17}{2} \cos^{-1} \frac{2}{3} - 12 \left(\frac{\sqrt{5}}{3} \right) + \frac{9}{2} \left(\frac{\sqrt{5}}{3} \right) \left(\frac{2}{3} \right) \right]$$

$$= \frac{17\pi}{2} - \frac{17}{2} \cos^{-1} \frac{2}{3} + 3\sqrt{5}$$

15. $r = 4 \cos 3\theta$

$$A = 6 \cdot \frac{1}{2} \int_0^{\pi/6} (4\cos 3\theta)^2 \, d\theta = 48 \int_0^{\pi/6} \cos^2 3\theta \, d\theta$$

$$= 24 \int_0^{\pi/6} (1 + \cos 6\theta) d\theta = 24 \left[\theta + \frac{1}{6} \sin 6\theta \right]_0^{\pi/6} = 4\pi$$

$$= \frac{1}{2} \int_0^{\pi/4} (18\cos\theta + 9\cos^2\theta - 18\sin\theta - 9\sin^2\theta) d\theta$$

$$= \frac{1}{2} \int_0^{\pi/4} (18\cos\theta - 18\sin\theta + 9\cos 2\theta) d\theta$$

$$= \frac{1}{2} \left[18\sin\theta + 18\cos\theta + \frac{9}{2}\sin 2\theta \right]_0^{\pi/4}$$

$$= 9\sqrt{2} - \frac{27}{4}$$

17. $A = \dfrac{1}{2} \displaystyle\int_0^{2\pi} 100 \, d\theta - \dfrac{1}{2} \displaystyle\int_0^{2\pi} 49 \, d\theta = 51\pi$

19. $r = 2, r^2 = 8\cos 2\theta$

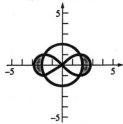

Solve for the θ-coordinate of the first
intersection point.

$$4 = 8\cos 2\theta$$

$$\cos 2\theta = \frac{1}{2}$$

$$2\theta = \frac{\pi}{3}$$

$$\theta = \frac{\pi}{6}$$

$$A = 4 \cdot \frac{1}{2} \int_0^{\pi/6} (8\cos 2\theta - 4) d\theta$$

$$= 2[4\sin 2\theta - 4\theta]_0^{\pi/6}$$

$$= 4\sqrt{3} - \frac{4\pi}{3}$$

21. $r = 3 + 3\cos\theta, r = 3 + 3\sin\theta$

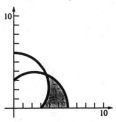

Solve for the θ-coordinate of the intersection
point.

$$3 + 3\cos\theta = 3 + 3\sin\theta$$

$$\tan\theta = 1$$

$$\theta = \frac{\pi}{4}$$

$$A = \frac{1}{2} \int_0^{\pi/4} [(3 + 3\cos\theta)^2 - (3 + 3\sin\theta)^2] d\theta$$

23. a. $f(\theta) = 2\cos\theta, f'(\theta) = -2\sin\theta$

$$m = \frac{(2\cos\theta)\cos\theta + (-2\sin\theta)\sin\theta}{-(2\cos\theta)\sin\theta + (-2\sin\theta)\cos\theta}$$

$$= \frac{2\cos^2\theta - 2\sin^2\theta}{-4\cos\theta\sin\theta} = \frac{\cos 2\theta}{-\sin 2\theta}$$

At $\theta = \dfrac{\pi}{3}, m = \dfrac{-\frac{1}{2}}{-\frac{\sqrt{3}}{2}} = \dfrac{1}{\sqrt{3}}$.

b. $f(\theta) = 1 + \sin\theta, f'(\theta) = \cos\theta$

$$m = \frac{(1 + \sin\theta)\cos\theta + (\cos\theta)\sin\theta}{-(1 + \sin\theta)\sin\theta + (\cos\theta)\cos\theta}$$

$$= \frac{\cos\theta + 2\sin\theta\cos\theta}{\cos^2\theta - \sin^2\theta - \sin\theta} = \frac{\cos\theta + \sin 2\theta}{\cos 2\theta - \sin\theta}$$

At $\theta = \dfrac{\pi}{3}, m = \dfrac{\frac{1}{2} + \frac{\sqrt{3}}{2}}{-\frac{1}{2} - \frac{\sqrt{3}}{2}} = -1$.

c. $f(\theta) = \sin 2\theta, f'(\theta) = 2\cos 2\theta$

$$m = \frac{(\sin 2\theta)\cos\theta + (2\cos 2\theta)\sin\theta}{-(\sin 2\theta)\sin\theta + (2\cos 2\theta)\cos\theta}$$

At $\theta = \dfrac{\pi}{3}$, .

$$m = \frac{\left(\frac{\sqrt{3}}{2}\right)\left(\frac{1}{2}\right) + (-1)\left(\frac{\sqrt{3}}{2}\right)}{-\left(\frac{\sqrt{3}}{2}\right)\left(\frac{\sqrt{3}}{2}\right) + (-1)\left(\frac{1}{2}\right)} = \frac{-\frac{\sqrt{3}}{4}}{-\frac{5}{4}} = \frac{\sqrt{3}}{5}$$

d. $f(\theta) = 4 - 3\cos\theta, f'(\theta) = 3\sin\theta$

$$m = \frac{(4 - 3\cos\theta)\cos\theta + (3\sin\theta)\sin\theta}{-(4 - 3\cos\theta)\sin\theta + (3\sin\theta)\cos\theta}$$

$$= \frac{4\cos\theta - 3\cos^2\theta + 3\sin^2\theta}{-4\sin\theta + 6\sin\theta\cos\theta}$$

$$= \frac{4\cos\theta - 3\cos 2\theta}{-4\sin\theta + 3\sin 2\theta}$$

At $\theta = \dfrac{\pi}{3}$,

$$m = \frac{4\left(\frac{1}{2}\right) - 3\left(-\frac{1}{2}\right)}{-4\left(\frac{\sqrt{3}}{2}\right) + 3\left(\frac{\sqrt{3}}{2}\right)} = \frac{\frac{7}{2}}{-\frac{\sqrt{3}}{2}} = -\frac{7}{\sqrt{3}}.$$

25. $f(\theta) = 1 - 2\sin\theta, \, f'(\theta) = -2\cos\theta$

$$m = \frac{(1 - 2\sin\theta)\cos\theta + (-2\cos\theta)\sin\theta}{-(1 - 2\sin\theta)\sin\theta + (-2\cos\theta)\cos\theta}$$

$$= \frac{\cos\theta - 4\sin\theta\cos\theta}{-\sin\theta + 2\sin^2\theta - 2\cos^2\theta}$$

$$= \frac{\cos\theta(1 - 4\sin\theta)}{-\sin\theta + 2\sin^2\theta - 2\cos^2\theta}$$

$m = 0$ when $\cos\theta(1 - 4\sin\theta) = 0$

$\cos\theta = 0, 1 - 4\sin\theta = 0$

$\theta = \dfrac{\pi}{2}, \theta = \dfrac{3\pi}{2}, \theta = \sin^{-1}\left(\dfrac{1}{4}\right) \approx 0.25,$

$\theta = \pi - \sin^{-1}\left(\dfrac{1}{4}\right) \approx 2.89$

$f\left(\dfrac{\pi}{2}\right) = -1, \, f\left(\dfrac{3\pi}{2}\right) = 3, \, f\left(\sin^{-1}\left(\dfrac{1}{4}\right)\right) = \dfrac{1}{2},$

$f\left(\pi - \sin^{-1}\left(\dfrac{1}{4}\right)\right) = \dfrac{1}{2}$

$\left(-1, \dfrac{\pi}{2}\right), \left(3, \dfrac{3\pi}{2}\right), \left(\dfrac{1}{2}, 0.25\right), \left(\dfrac{1}{2}, 2.89\right)$

27. $f(\theta) = a(1 + \cos\theta), \, f'(\theta) = -a\sin\theta$

$$L = \int_0^{2\pi} \sqrt{[a(1 + \cos\theta)]^2 + [-a\sin\theta]^2}\, d\theta = a\int_0^{2\pi} \sqrt{2 + 2\cos\theta}\, d\theta = 2a\int_0^{2\pi} \sqrt{\frac{1 + \cos\theta}{2}}\, d\theta = 2a\int_0^{2\pi} \left|\cos\frac{\theta}{2}\right| d\theta$$

$$= 2a\left[\int_0^{\pi} \cos\frac{\theta}{2}\, d\theta - \int_{\pi}^{2\pi} \cos\frac{\theta}{2}\, d\theta\right] = 2a\left(\left[2\sin\frac{\theta}{2}\right]_0^{\pi} - \left[2\sin\frac{\theta}{2}\right]_{\pi}^{2\pi}\right) = 8a$$

29. If n is even, there are $2n$ leaves.

$$A = 2n\frac{1}{2}\int_{-\pi/2n}^{\pi/2n} (a\cos n\theta)^2\, d\theta = na^2\int_{-\pi/2n}^{\pi/2n} \cos^2 n\theta\, d\theta = na^2\int_{-\pi/2n}^{\pi/2n} \frac{1 + \cos 2n\theta}{2}\, d\theta$$

$$= na^2\left[\frac{1}{2}\theta + \frac{\sin 2n\theta}{4n}\right]_{-\pi/2n}^{\pi/2n} = \frac{1}{2}a^2\pi$$

If n is odd, there are n leaves.

$$A = n\cdot\frac{1}{2}\int_{-\pi/2n}^{\pi/2n} (a\cos n\theta)^2\, d\theta = \frac{na^2}{2}\int_{-\pi/2n}^{\pi/2n} \cos^2 n\theta\, d\theta = \frac{na^2}{2}\left[\frac{1}{2}\theta + \frac{\sin 2n\theta}{4n}\right]_{-\pi/2n}^{\pi/2n} = \frac{1}{4}a^2\pi$$

31. a. Sketch the graph.

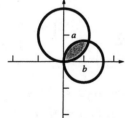

Solve for the θ-coordinate of the intersection.

$2a\sin\theta = 2b\cos\theta$

$\tan\theta = \dfrac{b}{a}$

$\theta = \tan^{-1}\left(\dfrac{b}{a}\right)$

Let $\theta_0 = \tan^{-1}\left(\dfrac{b}{a}\right)$.

$$A = \frac{1}{2}\int_0^{\theta_0} (2a\sin\theta)^2\, d\theta + \frac{1}{2}\int_{\theta_0}^{\pi/2} (2b\cos\theta)^2\, d\theta$$

$$= 2a^2\int_0^{\theta_0} \sin^2\theta\, d\theta + 2b^2\int_{\theta_0}^{\pi/2} \cos^2\theta\, d\theta$$

$$= a^2\int_0^{\theta_0} (1 - \cos 2\theta)\, d\theta + b^2\int_{\theta_0}^{\pi/2} (1 + \cos 2\theta)\, d\theta$$

$$= a^2\left[\theta - \frac{\sin 2\theta}{2}\right]_0^{\theta_0} + b^2\left[\theta + \frac{\sin 2\theta}{2}\right]_{\theta_0}^{\pi/2}$$

$$= a^2\theta_0 + b^2\left(\frac{\pi}{2} - \theta_0\right) - \frac{a^2 + b^2}{2}\sin 2\theta_0$$

$$= a^2\theta_0 + b^2\left(\frac{\pi}{2} - \theta_0\right) - (a^2 + b^2)\sin\theta_0\cos\theta_0$$

$$= a^2\tan^{-1}\left(\frac{b}{a}\right) + b^2\left(\frac{\pi}{2} - \tan^{-1}\left(\frac{b}{a}\right)\right) - ab.$$

Note that since $\tan\theta = \dfrac{b}{a}, \cos\theta = \dfrac{a}{\sqrt{a^2 + b^2}}$

and $\sin\theta = \dfrac{b}{\sqrt{a^2 + b^2}}$.

b. Let m_1 be the slope of $r = 2a\sin\theta$.

$$m_1 = \frac{2a\sin\theta\cos\theta + 2a\cos\theta\sin\theta}{-2a\sin\theta\sin\theta + 2a\cos\theta\cos\theta}$$

$$= \frac{2\sin\theta\cos\theta}{\cos^2\theta - \sin^2\theta}$$

At $\theta = \tan^{-1}\left(\dfrac{b}{a}\right)$, $m_1 = \dfrac{2ab}{a^2 - b^2}$.

At $\theta = 0$ (the pole), $m_1 = 0$.

Let m_2 be the slope of $r = 2b\cos\theta$.

$$m_2 = \frac{2b\cos\theta\cos\theta - 2b\sin\theta\sin\theta}{-2b\cos\theta\sin\theta - 2b\sin\theta\cos\theta}$$

$$= \frac{\cos^2\theta - \sin^2\theta}{-2\sin\theta\cos\theta}$$

At $\theta = \tan^{-1}\left(\dfrac{b}{a}\right)$, $m_2 = -\dfrac{a^2 - b^2}{2ab}$.

At $\theta = \dfrac{\pi}{2}$ (the pole), m_2 is undefined.

Therefore the two circles intersect at right angles.

33. The edge of the pond is described by the equation $r = 2a\cos\theta$.

Solve for intersection points of the circles $r = ak$ and $r = 2a\cos\theta$.

$$ak = 2a\cos\theta$$

$$\cos\theta = \frac{k}{2}, \theta = \cos^{-1}\left(\frac{k}{2}\right)$$

Let A be the grazing area.

$$A = \frac{1}{2}\pi(ka)^2 + 2\cdot\frac{1}{2}\int_{\cos^{-1}\left(\frac{k}{2}\right)}^{\pi/2} [(ka)^2 - (2a\cos\theta)^2]d\theta = \frac{1}{2}k^2a^2\pi + a^2\int_{\cos^{-1}\left(\frac{k}{2}\right)}^{\pi/2} (k^2 - 4\cos^2\theta)d\theta$$

$$= \frac{1}{2}k^2a^2\pi + a^2\int_{\cos^{-1}\left(\frac{k}{2}\right)}^{\pi/2} ((k^2 - 2) - 2\cos 2\theta)d\theta = \frac{1}{2}k^2a^2\pi + a^2\left[(k^2 - 2)\theta - \sin 2\theta\right]_{\cos^{-1}\left(\frac{k}{2}\right)}^{\pi/2}$$

$$= \frac{1}{2}k^2a^2\pi + a^2\left[k^2\theta - 2\theta - 2\sin\theta\cos\theta\right]_{\cos^{-1}\left(\frac{k}{2}\right)}^{\pi/2}$$

$$= \frac{1}{2}k^2a^2\pi + a^2\left[\frac{k^2\pi}{2} - \pi - k^2\cos^{-1}\left(\frac{k}{2}\right) + 2\cos^{-1}\left(\frac{k}{2}\right) + \frac{k\sqrt{4 - k^2}}{2}\right]$$

$$= a^2\left[(k^2 - 1)\pi + (2 - k^2)\cos^{-1}\left(\frac{k}{2}\right) + \frac{k\sqrt{4 - k^2}}{2}\right]$$

35. The untethered goat has a grazing area of πa^2.

From Problem 34, the tethered goat has a grazing area of $a^2\left(\dfrac{\pi k^2}{2} + \dfrac{k^3}{3}\right)$.

$$\pi a^2 = a^2\left(\frac{\pi k^2}{2} + \frac{k^3}{3}\right)$$

$$\pi = \frac{\pi k^2}{2} + \frac{k^3}{3}$$

$$2k^3 + 3\pi k^2 - 6\pi = 0$$

Using a numerical method or graphing calculator, $k \approx 1.26$. The length of the rope is approximately $1.26a$.

37. $A = 3\cdot\dfrac{1}{2}\displaystyle\int_0^{\pi/3} (4\sin 3\theta)^2 \, d\theta = 24\int_0^{\pi/3} \sin^2 3\theta \, d\theta$

$$= 12\int_0^{\pi/3} (1 - \cos 6\theta) \, d\theta$$

$$= 12\left[\theta - \frac{\sin 6\theta}{6}\right]_0^{\pi/3} = 4\pi$$

$$f(\theta) = 4\sin 3\theta, f'(\theta) = 12\cos 3\theta$$

$$L = 3\int_0^{\pi/3} \sqrt{(4\sin 3\theta)^2 + (12\cos 3\theta)^2} \, d\theta$$

$$= 3\int_0^{\pi/3} \sqrt{16\sin^2 3\theta + 144\cos^2 3\theta} \, d\theta$$

$$= 12\int_0^{\pi/3} \sqrt{1 + 8\cos^2 3\theta} \, d\theta \approx 26.73$$

39. $r = 4\sin\left(\dfrac{3\theta}{2}\right), 0 \le \theta \le 4\pi$

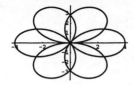

$$f(\theta) = 4\sin\left(\frac{3\theta}{2}\right), f'(\theta) = 6\cos\left(\frac{3\theta}{2}\right)$$

$$L = \int_0^{4\pi} \sqrt{\left[4\sin\left(\frac{3\theta}{2}\right)\right]^2 + \left[6\cos\left(\frac{3\theta}{2}\right)\right]^2}\, d\theta$$

$$= \int_0^{4\pi} \sqrt{16\sin^2\left(\frac{3\theta}{2}\right) + 36\cos^2\left(\frac{3\theta}{2}\right)}\, d\theta$$

$$= \int_0^{4\pi} \sqrt{16 + 20\cos^2\left(\frac{3\theta}{2}\right)}\, d\theta \approx 63.46$$

12.9 Chapter Review

Concepts Test

1. **False:** If $a = 0$, the graph is a line.

3. **False:** The defining condition of an ellipse is $|PF| = e|PL|$ where $0 < e < 1$. Hence the distance from the vertex to a directrix is a greater than the distance to a focus.

5. **True:** The asymptotes for both hyperbolas are $y = \pm\dfrac{b}{a}x$.

7. **True:** As e approaches 0, the ellipse becomes more circular.

9. **False:** The equation $x^2 - y^2 = 0$ represents the two lines $y = \pm x$.

11. **True:** If $k > 0$, the equation is a horizontal hyperbola; if $k < 0$, the equation is a vertical hyperbola.

13. **False:** If $b > a$, the distance is $2\sqrt{b^2 - a^2}$.

15. **True:** Since light from one focus reflects to the other focus, light emanating from a point between a focus and the nearest vertex will reflect beyond the other focus.

17. **True:** Since angular momentum $mr^2\dfrac{d\theta}{dt}$ is assumed to be constant, $\dfrac{d\theta}{dt}$ is greatest at the vertex nearest the sun.

19. **True:** The equation is equivalent to
$$\left(x + \frac{C}{2}\right)^2 + \left(y + \frac{D}{2}\right)^2 = -F + \frac{C^2}{4} + \frac{D^2}{4}.$$

Thus, the graph is a circle if
$$-F + \frac{C^2}{4} + \frac{D^2}{4} > 0, \text{ a point if}$$
$$-F + \frac{C^2}{4} + \frac{D^2}{4} = 0, \text{ or the empty set if}$$
$$-F^2 + \frac{C^2}{4} + \frac{D^2}{4} < 0.$$

21. **False:** The limiting forms of two parallel lines and the empty set cannot be formed in such a manner.

23. **False:** For example, $xy = 1$ is a hyperbola with coordinates only in the first and third quadrants.

25. **True:** The graph of $r = 4\cos\theta$ is a circle of radius 2 centered at $(2, 0)$. The graph $r = 4\cos\left(\theta - \dfrac{\pi}{3}\right)$ is the graph of $r = 4\cos\theta$ rotated $\dfrac{\pi}{3}$ counter-clockwise about the pole.

27. **False:** For example, if $f(\theta) = \cos\theta$ and $g(\theta) = \sin\theta$, solving the two equations simultaneously does not give the pole (which is $\left(0, \dfrac{\pi}{2}\right)$ for $f(\theta)$ and $(0, 0)$ for $g(\theta)$).

29. **True:** Since f is even $f(-\theta) = f(\theta)$. Thus, if we replace (r, θ) by $(r, -\theta)$, the equation $r = f(-\theta)$ is $r = f(\theta)$. Therefore, the graph is symmetric about the x-axis.

Sample Test Problems

1. **a.** $x^2 - 4y^2 = 0; y = \pm\dfrac{x}{2}$
    (5) Two intersecting lines

b. $x^2 - 4y^2 = 0.01; \dfrac{x^2}{0.01} - \dfrac{y^2}{0.0025} = 1$

(9) A hyperbola

c. $x^2 - 4 = 0; x = \pm 2$

(4) Two parallel lines

d. $x^2 - 4x + 4 = 0; x = 2$

(3) A single line

e. $x^2 + 4y^2 = 0; (0, 0)$

(2) A single point

f. $x^2 + 4y^2 = x; x^2 - x + \dfrac{1}{4} + 4y^2 = \dfrac{1}{4};$

$\dfrac{\left(x - \frac{1}{2}\right)^2}{\frac{1}{4}} + \dfrac{y^2}{\frac{1}{16}} = 1$

(8) An ellipse

g. $x^2 + 4y^2 = -x; x^2 + x + \dfrac{1}{4} + 4y^2 = \dfrac{1}{4};$

$\dfrac{\left(x + \frac{1}{2}\right)^2}{\frac{1}{4}} + \dfrac{y^2}{\frac{1}{16}} = 1$

(8) An ellipse

h. $x^2 + 4y^2 = -1$

(1) No graph

i. $(x^2 + 4y - 1)^2 = 0; x^2 + 4y - 1 = 0$

(7) A parabola

j. $3x^2 + 4y^2 = -x^2 + 1; x^2 + y^2 = \dfrac{1}{4}$

(6) A circle

3. $9x^2 + 4y^2 - 36 = 0; \dfrac{x^2}{4} + \dfrac{y^2}{9} = 1$

Vertical ellipse; $a = 3, b = 2, c = \sqrt{5}$

Foci are at $\left(0, \pm\sqrt{5}\right)$ and vertices are at $(0, \pm 3)$.

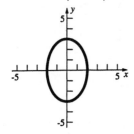

5. $x^2 + 9y = 0; x^2 = -9y; x^2 = -4\left(\dfrac{9}{4}\right)y$

Vertical parabola; opens downward; $p = \dfrac{9}{4}$

Focus at $\left(0, -\dfrac{9}{4}\right)$ and vertex at $(0, 0)$.

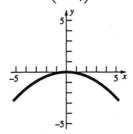

7. $9x^2 + 25y^2 - 225 = 0; \dfrac{x^2}{25} + \dfrac{y^2}{9} = 1$

Horizontal ellipse, $a = 5, b = 3, c = 4$

Foci are at $(\pm 4, 0)$ and vertices are at $(\pm 5, 0)$.

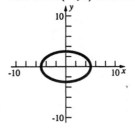

9. $r = \dfrac{5}{2 + 2\sin\theta} = \dfrac{\left(\frac{5}{2}\right)(1)}{1 + (1)\cos\left(\theta - \frac{\pi}{2}\right)}$

$e = 1$; parabola

Focus is at $(0, 0)$ and vertex is at $\left(0, \dfrac{5}{4}\right)$ (in Cartesian coordinates).

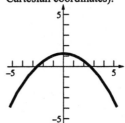

11. Horizontal ellipse; center at $(0, 0)$, $a = 4$,

$e = \dfrac{c}{a} = \dfrac{1}{2}, c = 2, b = \sqrt{16 - 4} = 2\sqrt{3}$

$\dfrac{x^2}{16} + \dfrac{y^2}{12} = 1$

13. Horizontal parabola;

$y^2 = ax, (3)^2 = a(-1), a = -9$

$y^2 = -9x$

15. Horizontal hyperbola, $a = 2$,

$x = \pm 2y, \dfrac{a}{b} = 2, b = 1$

$\dfrac{x^2}{4} - \dfrac{y^2}{1} = 1$

17. Horizontal ellipse; $2a = 10, a = 5, c = 4 - 1 = 3$,

$b = \sqrt{25 - 9} = 4$

$\dfrac{(x-1)^2}{25} + \dfrac{(y-2)^2}{16} = 1$

19. $4x^2 + 4y^2 - 24x + 36y + 81 = 0$

$4(x^2 - 6x + 9) + 4\left(y^2 + 9y + \dfrac{81}{4}\right) = -81 + 36 + 81$

$4(x-3)^2 + 4\left(y + \dfrac{9}{2}\right)^2 = 36$

$(x-3)^2 + \left(y + \dfrac{9}{2}\right)^2 = 9;$ circle

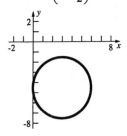

21. $x^2 + 8x + 6y + 28 = 0$

$(x^2 + 8x + 16) = -6y - 28 + 16$

$(x+4)^2 = -6(y+2);$ parabola

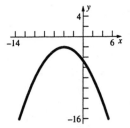

23. $x = \dfrac{\sqrt{2}}{2}(u - v)$

$y = \dfrac{\sqrt{2}}{2}(u + v)$

$\dfrac{1}{2}(u-v)^2 + \dfrac{3}{2}(u-v)(u+v) + \dfrac{1}{2}(u+v)^2 = 10$

$\dfrac{5}{2}u^2 - \dfrac{1}{2}v^2 = 10$

$r = \dfrac{5}{2}, s = -\dfrac{1}{2}$

$\dfrac{u^2}{4} - \dfrac{v^2}{20} = 1;$ hyperbola

$a = 2, \ b = 2\sqrt{5}, c = \sqrt{4 + 20} = 2\sqrt{6}$

The distance between foci is $4\sqrt{6}$.

25. $r = 6\cos\theta$

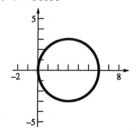

27. $r = \cos 2\theta$

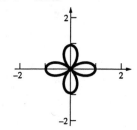

29. $r = 4$

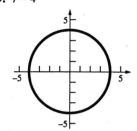

31. $r = 4 - 3\cos\theta$

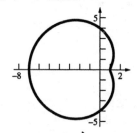

33. $\theta = \dfrac{2}{3}\pi$

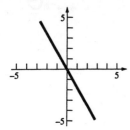

35. $r^2 = 16\sin 2\theta$

$r = \pm 4\sqrt{\sin 2\theta}$

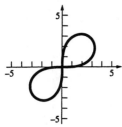

37. $r^2 - 6r(\cos\theta + \sin\theta) + 9 = 0$

$x^2 + y^2 - 6x - 6y + 9 = 0$

$(x^2 - 6x + 9) + (y^2 - 6y + 9) = -9 + 9 + 9$

$(x-3)^2 + (y-3)^2 = 9$

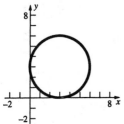

39. $f(\theta) = 3 + 3\cos\theta, \; f'(\theta) = -3\sin\theta$

$m = \dfrac{(3 + 3\cos\theta)\cos\theta + (-3\sin\theta)\sin\theta}{-(3 + 3\cos\theta)\sin\theta + (-3\sin\theta)\cos\theta}$

$= \dfrac{\cos\theta + \cos^2\theta - \sin^2\theta}{-\sin\theta - 2\cos\theta\sin\theta} = \dfrac{\cos\theta + \cos 2\theta}{-\sin\theta - \sin 2\theta}$

At $\theta = \dfrac{\pi}{6}, \; m = \dfrac{\cos\frac{\pi}{6} + \cos\frac{\pi}{3}}{-\sin\frac{\pi}{6} - \sin\frac{\pi}{3}} = -1$.

$5\sin\theta = 2 + \sin\theta$

$\sin\theta = \dfrac{1}{2}$

$\theta = \dfrac{\pi}{6}, \dfrac{5\pi}{6}$

$\left(\dfrac{5}{2}, \dfrac{\pi}{6}\right), \left(\dfrac{5}{2}, \dfrac{5\pi}{6}\right)$

41. $A = 2 \cdot \dfrac{1}{2}\displaystyle\int_0^\pi (5 - 5\cos\theta)^2 \, d\theta$

$= 25\displaystyle\int_0^\pi (1 - 2\cos\theta + \cos^2\theta)\, d\theta$

$= 25\displaystyle\int_0^\pi \left(\dfrac{3}{2} - 2\cos\theta + \dfrac{1}{2}\cos 2\theta\right) d\theta$

$= 25\left[\dfrac{3}{2}\theta - 2\sin\theta + \dfrac{1}{4}\sin 2\theta\right]_0^\pi = \dfrac{75\pi}{2}$

43. $\dfrac{x^2}{400} + \dfrac{y^2}{100} = 1; \; \dfrac{x}{200} + \dfrac{yy'}{50} = 0$

$y' = -\dfrac{x}{4y}; \; y' = -\dfrac{2}{3}$ at $(16, 6)$

Tangent line: $y - 6 = -\dfrac{2}{3}(x - 16)$

When $x = 14, \; y = -\dfrac{2}{3}(14 - 16) + 6 = \dfrac{22}{3}$.

$k = \dfrac{22}{3}$

45. a. I

b. IV

c. III

d. II

e. V

13.1 Concepts Review

1. simple; closed; simple

3. cycloid

Problem Set 13.1

1. a.

t	x	y
-2	-6	-4
-1	-3	-2
0	0	0
1	3	2
2	6	4

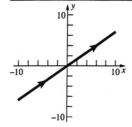

b. Simple; not closed

c. $t = \dfrac{x}{3}$

$y = \dfrac{2}{3}x$

3. a.

t	x	y
0	-1	0
1	2	1
2	5	2
3	8	3
4	11	4

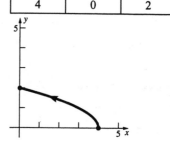

b. Simple; not closed

c. $t = \dfrac{1}{3}(x+1)$

$y = \dfrac{1}{3}(x+1)$

5. a.

t	x	y
0	4	0
1	3	1
2	2	$\sqrt{2}$
3	1	$\sqrt{3}$
4	0	2

b. Simple; not closed

c. $t = 4 - x$

$y = \sqrt{4-x}$

7. a.

s	x	y
1	1	1
3	$\dfrac{1}{3}$	3
5	$\dfrac{1}{5}$	5
7	$\dfrac{1}{7}$	7
9	$\dfrac{1}{9}$	9

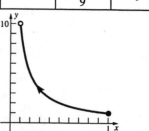

b. Simple; not closed

c. $s = \dfrac{1}{x}$

$y = \dfrac{1}{x}$

9. a.

t	x	y
-3	-15	5
-2	0	0
-1	3	-3
0	0	-4
1	-3	-3
2	0	0
3	15	5

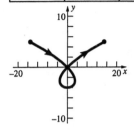

b. Not simple; not closed

c. $x^2 = t^6 - 8t^4 + 16t^2$

$t^2 = y + 4$

$x^2 = (y+4)^3 - 8(y+4)^2 + 16(y+4)$

$x^2 = y^3 + 4y^2$

11. a.

t	x	y
2	0	$3\sqrt{2}$
3	2	3
4	$2\sqrt{2}$	0

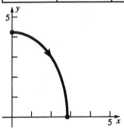

b. Simple; not closed

c. $t = \dfrac{1}{4}x^2 + 2$

$t = 4 - \dfrac{1}{9}y^2$

$\dfrac{1}{4}x^2 + 2 = 4 - \dfrac{1}{9}y^2$

$\dfrac{x^2}{8} + \dfrac{y^2}{18} = 1$

13. a.

t	x	y
0	0	3
$\dfrac{\pi}{2}$	2	0
π	0	-3
$\dfrac{3\pi}{2}$	-2	0
2π	0	3

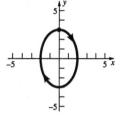

b. Simple; closed

c. $\sin^2 t = \dfrac{x^2}{4}$; $\cos^2 t = \dfrac{y^2}{9}$

$\dfrac{x^2}{4} + \dfrac{y^2}{9} = 1$

15. a.

r	x	y
0	0	-3
$\dfrac{\pi}{2}$	-2	0
π	0	3
$\dfrac{3\pi}{2}$	2	0
2π	0	-3
$\dfrac{5\pi}{2}$	-2	0
3π	0	3
$\dfrac{7\pi}{2}$	2	0
4π	0	-3

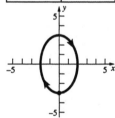

b. Not simple; closed

c. $\sin^2 r = \dfrac{x^2}{4}$

$\cos^2 r = \dfrac{y^2}{9}$

$\dfrac{x^2}{4} + \dfrac{y^2}{9} = 1$

17. a.

θ	x	y
0	0	9
$\dfrac{\pi}{4}$	$\dfrac{9}{2}$	$\dfrac{9}{2}$
$\dfrac{\pi}{2}$	0	9
$\dfrac{3\pi}{4}$	$\dfrac{9}{2}$	$\dfrac{9}{2}$
π	0	9

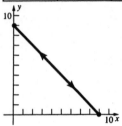

b. Not simple; closed

c. $\sin^2\theta = \dfrac{x}{9}$

$\cos^2\theta = \dfrac{y}{9}$

$\dfrac{x}{9} + \dfrac{y}{9} = 1$

$x + y = 9$

19. a.

θ	x	y
$0 + 2\pi n$	1	0
$\dfrac{\pi}{3} + 2\pi n$	$\dfrac{1}{2}$	$-\dfrac{3}{2}$
$\dfrac{\pi}{2} + 2\pi n$	0	0
$\dfrac{2\pi}{3} + 2\pi n$	$-\dfrac{1}{2}$	$-\dfrac{3}{2}$
$\pi + 2\pi n$	-1	0
$\dfrac{4\pi}{3} + 2\pi n$	$-\dfrac{1}{2}$	$-\dfrac{3}{2}$
$\dfrac{3\pi}{2} + 2\pi n$	0	0
$\dfrac{5\pi}{3} + 2\pi n$	$\dfrac{1}{2}$	$-\dfrac{3}{2}$

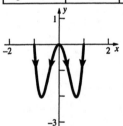

b. Not simple; not closed

c. $\cos\theta = x$

$\sin\theta = \sqrt{1-x^2}$

$y = -8\sin^2\theta\cos^2\theta$

$y = -8x^2(1-x^2)$

21. $\dfrac{dx}{d\tau} = 6\tau$

$\dfrac{dy}{d\tau} = 12\tau^2$

$\dfrac{dy}{dx} = 2\tau$

$\dfrac{dy'}{d\tau} = 2$

$\dfrac{d^2y}{dx^2} = \dfrac{1}{3\tau}$

23. $\dfrac{dx}{d\theta} = 4\theta$

$\dfrac{dy}{d\theta} = 3\sqrt{5}\theta^2$

$\dfrac{dy}{dx} = \dfrac{3\sqrt{5}}{4}\theta$

$\dfrac{dy'}{d\theta} = \dfrac{3\sqrt{5}}{4}$

$\dfrac{d^2y}{dx^2} = \dfrac{3\sqrt{5}}{16\theta}$

25. $\dfrac{dx}{dt} = \sin t$

$\dfrac{dy}{dt} = \cos t$

$\dfrac{dy}{dx} = \cot t$

$\dfrac{dy'}{dt} = -\csc^2 t$

$\dfrac{d^2y}{dx^2} = -\csc^3 t$

27. $\dfrac{dx}{dt} = 3\sec^2 t$

$\dfrac{dy}{dt} = 5\sec t\tan t$

$\dfrac{dy}{dx} = \dfrac{5}{3}\sin t$

$\dfrac{dy'}{dt} = \dfrac{5}{3}\cos t$

$\dfrac{d^2y}{dx^2} = \dfrac{5}{9}\cos^3 t$

29. $\dfrac{dx}{dt} = -\dfrac{2t}{(1+t^2)^2}$

$\dfrac{dy}{dt} = \dfrac{2t-1}{t^2(1-t)^2}$

$\dfrac{dy}{dx} = \dfrac{(1-2t)(1+t^2)^2}{2t^3(1-t)^2}$

$\dfrac{dy'}{dt} = -\dfrac{3t^5 + 7t^4 - 6t^3 + 10t^2 - 9t + 3}{2t^4(1-t)^3}$

$\dfrac{d^2y}{dx^2} = \dfrac{(3t^5 + 7t^4 - 6t^3 + 10t^2 - 9t + 3)(1+t^2)^2}{4t^5(1-t)^3}$

31. $\dfrac{dx}{dt} = 2t, \dfrac{dy}{dt} = 3t^2$

$\dfrac{dy}{dx} = \dfrac{3}{2}t$

At $t = 2$, $x = 4$, $y = 8$, and $\dfrac{dy}{dx} = 3$.

Tangent line: $y - 8 = 3(x - 4)$ or $3x - y - 4 = 0$

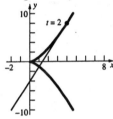

33. $\dfrac{dx}{dt} = 2\sec t \tan t, \dfrac{dy}{dt} = 2\sec^2 t$

$\dfrac{dy}{dx} = \csc t$

At $t = -\dfrac{\pi}{6}$, $x = \dfrac{4}{\sqrt{3}}$, $y = -\dfrac{2}{\sqrt{3}}$, and $\dfrac{dy}{dx} = -2$.

Tangent line: $y + \dfrac{2}{\sqrt{3}} = -2\left(x - \dfrac{4}{\sqrt{3}}\right)$ or

$2\sqrt{3}x + \sqrt{3}y - 6 = 0$

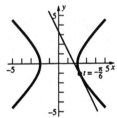

35. $\dfrac{dx}{dt} = 2, \dfrac{dy}{dt} = 3$

$L = \int_0^3 \sqrt{4+9}\,dt = \sqrt{13}\int_0^3 dt = \sqrt{13}[t]_0^3 = 3\sqrt{13}$

37. $\dfrac{dx}{dt} = 1, \dfrac{dy}{dt} = \dfrac{3}{2}t^{1/2}$

$L = \int_0^3 \sqrt{1 + \dfrac{9}{4}t}\,dt$

$= \dfrac{1}{2}\int_0^3 \sqrt{4+9t}\,dt$

$= \dfrac{1}{18}\left[\dfrac{2}{3}(4+9t)^{3/2}\right]_0^3$

$= \dfrac{1}{27}(31^{3/2} - 8) = \dfrac{1}{27}(31\sqrt{31} - 8)$

39. $\dfrac{dx}{dt} = 6t, \dfrac{dy}{dt} = 3t^2$

$L = \int_0^2 \sqrt{36t^2 + 9t^4}\,dt = 3\int_0^2 t\sqrt{4+t^2}\,dt$

$3\left[\dfrac{1}{3}(4+t^2)^{3/2}\right]_0^2 = 16\sqrt{2} - 8$

41. $\dfrac{dx}{dt} = 2e^t, \dfrac{dy}{dt} = \dfrac{9}{2}e^{3t/2}$

$L = \int_{\ln 3}^{2\ln 3} \sqrt{4e^{2t} - \dfrac{81}{4}e^{3t}}\,dt = \int_{\ln 3}^{2\ln 3} e^t\sqrt{4 - \dfrac{81}{4}e^t}\,dt$

$= \left[-\dfrac{8}{243}\left(4 - \dfrac{81}{4}e^t\right)^{3/2}\right]_{\ln 3}^{2\ln 3}$

$= \dfrac{713\sqrt{713} - 227\sqrt{227}}{243}$

43. $\dfrac{dx}{dt} = \dfrac{2}{\sqrt{t}}, \dfrac{dy}{dt} = 2t - \dfrac{1}{2t^2}$

$L = \int_{1/4}^1 \sqrt{\dfrac{4}{t} + \left(4t^2 - \dfrac{2}{t} + \dfrac{1}{4t^4}\right)}\,dt$

$= \int_{1/4}^1 \sqrt{4t^2 + \dfrac{2}{t} + \dfrac{1}{4t^4}}\,dt$

$= \int_{1/4}^1 \sqrt{\left(2t + \dfrac{1}{2t^2}\right)^2}\,dt$

$= \int_{1/4}^1 \left(2t + \dfrac{1}{2t^2}\right)\,dt = \left[t^2 - \dfrac{1}{2t}\right]_{1/4}^1 = \dfrac{39}{16}$

45. $\dfrac{dx}{dt} = -\sin t,$

$\dfrac{dy}{dt} = \dfrac{\sec t \tan t + \sec^2 t}{\sec t + \tan t} - \cos t = \sec t - \cos t$

$L = \int_0^{\pi/4} \sqrt{\sin^2 t + (\sec^2 t - 2 + \cos^2 t)}\,dt$

$$= \int_0^{\pi/4} \tan t\, dt$$

$$= \left[-\ln|\cos t| \right]_0^{\pi/4} = -\ln \frac{1}{\sqrt{2}} = \frac{1}{2}\ln 2$$

47. a. $\dfrac{dx}{d\theta} = \cos\theta, \dfrac{dy}{d\theta} = -\sin\theta$

$$L = \int_0^{2\pi} \sqrt{\cos^2\theta + \sin^2\theta}\, d\theta = \int_0^{2\pi} d\theta$$

$$= [\theta]_0^{2\pi} = 2\pi$$

b. $\dfrac{dx}{d\theta} = 3\cos 3\theta, \dfrac{dy}{d\theta} = -3\sin 3\theta$

$$L = \int_0^{2\pi} \sqrt{9\cos^2 3\theta + 9\sin^2 3\theta}\, d\theta$$

$$= 3\int_0^{2\pi} d\theta = 3[\theta]_0^{2\pi} = 6\pi$$

c. The curve in part a goes around the unit circle once, while the curve in part b goes around the unit circle three times.

49. $\dfrac{dx}{dt} = -\sin t, \dfrac{dy}{dt} = \cos t$

$$S = \int_0^{2\pi} 2\pi(1+\cos t)\sqrt{\sin^2 t + \cos^2 t}\, dt$$

$$= 2\pi \int_0^{2\pi} (1+\cos t)\, dt = 2\pi[t + \sin t]_0^{2\pi} = 4\pi^2$$

51. $\dfrac{dx}{dt} = -\sin t, \dfrac{dy}{dt} = \cos t$

$$S = \int_0^{2\pi} 2\pi(1+\sin t)\sqrt{\sin^2 t + \cos^2 t}\, dt$$

$$= 2\pi \int_0^{2\pi} (1+\sin t)\, dt = 2\pi[t - \cos t]_0^{2\pi} = 4\pi^2$$

53. $\dfrac{dx}{dt} = 1, \dfrac{dy}{dt} = t + \sqrt{7}$

$$S = \int_{-\sqrt{7}}^{\sqrt{7}} 2\pi\left(t + \sqrt{7}\right)\sqrt{1 + \left(t + \sqrt{7}\right)^2}\, dt$$

$$= 2\pi \left[\frac{1}{3}\left(1 + \left(t + \sqrt{7}\right)^2\right)^{3/2} \right]_{-\sqrt{7}}^{\sqrt{7}}$$

$$= \frac{2\pi}{3}\left(29\sqrt{29} - 1\right)$$

55. $dx = dt$; when $x = 0$, $t = -1$; when $x = 1$, $t = 0$.

$$\int_0^1 (x^2 - 4y)dx = \int_{-1}^0 [(t+1)^2 - 4(t^3 + 4)]dt$$

$$= \int_{-1}^0 (-4t^3 + t^2 + 2t - 15)dt$$

$$= \left[-t^4 + \frac{1}{3}t^3 + t^2 - 15t \right]_{-1}^0 = -\frac{44}{3}$$

57. $dx = 2e^{2t}\, dt$

$$A = \int_1^{25} y\, dx = \int_0^{\ln 5} 2e^t\, dt = [2e^t]_0^{\ln 5} = 8$$

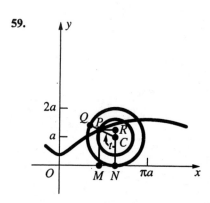

59.

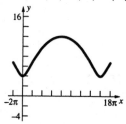

Let the wheel roll along the x-axis with P initially at $(0, a - b)$.

$|ON| = \text{arc } NQ = at$

$x = |OM| = |ON| - |MN| = at - b\sin t$

$y = |MP| = |RN| = |NC| + |CR| = a - b\cos t$

61. The x- and y-coordinates of the center of the circle of radius b are $(a - b)\cos t$ and $(a - b)\sin t$, respectively. The angle measure (in a clockwise direction) of arc BP is $\dfrac{a}{b}t$. The horizontal change from the center of the circle of radius b to P is $b\cos\left(-\left(\dfrac{a}{b}t - t\right)\right) = b\cos\left(\dfrac{a - b}{b}t\right)$ and the vertical change is

$b\sin\left(-\left(\dfrac{a}{b}t - t\right)\right) = -b\sin\left(\dfrac{a - b}{b}t\right)$. Therefore,

$$x = (a-b)\cos t + b\cos\left(\frac{a-b}{b}t\right) \text{ and}$$

$$y = (a-b)\sin t - b\sin\left(\frac{a-b}{b}t\right).$$

63. Consider the following figure similar to the one in the text for Problem 61.

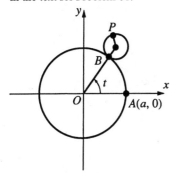

The x- and y-coordinates of the center of the circle of radius b are $(a+b)\cos t$ and $(a+b)\sin t$ respectively. The angle measure (in a counterclockwise direction) of arc PB is $\frac{a}{b}t$. The horizontal change from the center of the circle of radius b to P is

$$b\cos\left(\frac{a}{b}t + t + \pi\right) = -b\cos\left(\frac{a+b}{b}t\right) \text{ and the}$$

vertical change is

$$b\sin\left(\frac{a}{b}t + t + \pi\right) = -b\sin\left(\frac{a+b}{b}t\right). \text{ Therefore,}$$

$$x = (a+b)\cos t - b\cos\left(\frac{a+b}{b}t\right) \text{ and}$$

$$y = (a+b)\sin t - b\sin\left(\frac{a+b}{b}t\right).$$

65. $\dfrac{dx}{dt} = \left(\dfrac{a}{3}\right)(-2\sin t - 2\sin 2t),$

$$\frac{dy}{dt} = \left(\frac{a}{3}\right)(2\cos t - 2\cos 2t)$$

$$\left(\frac{dx}{dt}\right)^2 = \left(\frac{a}{3}\right)^2 (4\sin^2 t + 8\sin t \sin 2t + 4\sin^2 2t)$$

$$\left(\frac{dy}{dt}\right)^2 = \left(\frac{a}{3}\right)^2 (4\cos^2 t - 8\cos t \cos 2t + 4\cos^2 2t)$$

$$\left(\frac{dx}{dt}\right)^2 + \left(\frac{dy}{dt}\right)^2 = \left(\frac{a}{3}\right)^2 (8 + 8\sin t \sin 2t - 8\cos t \cos 2t)$$

$$= \left(\frac{a}{3}\right)^2 (8 + 16\sin^2 t \cos t - 8\cos^3 t + 8\sin^2 t \cos t)$$

$$= \left(\frac{a}{3}\right)^2 (8 + 24\cos t \sin^2 t - 8\cos^3 t)$$

$$= \left(\frac{a}{3}\right)^2 (8 + 24\cos t - 32\cos^3 t)$$

$$L = 3\int_0^{2\pi/3} \sqrt{\left(\frac{dx}{dt}\right)^2 + \left(\frac{dy}{dt}\right)^2} \, dt$$

$$= a\int_0^{2\pi/3} \sqrt{8 + 24\cos t - 32\cos^3 t} \, dt$$

Using a CAS to evaluate the length, $L = \dfrac{16a}{3}$.

67. a.

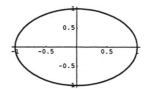

The curve touches a horizontal border once and touches a vertical border once.

b.

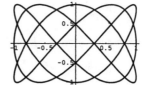

The curve touches a horizontal border five times and touches a vertical border three times.

c.

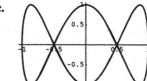

The curve touches a horizontal border three times and touches a vertical border once. Note that the curve is traced out three times.

d.

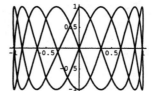

The curve touches a horizontal border nine times and touches a vertical border twice.

69. a. $0 \leq t \leq 2$

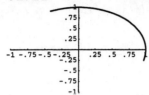

b. $0 \leq t \leq 1$

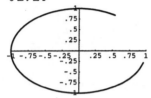

c. $0.25 \leq t \leq 2$

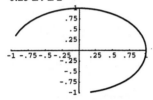

d. $0 \leq t \leq 2\pi$

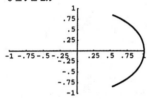

Given a parameterization of the form
$x = \cos f(t)$ and $y = \sin f(t)$,
the point moves around the curve (which is a
circle of radius 1) at a speed of $|f'(t)|$. The
point travels clockwise around the circle when
$f(t)$ is decreasing and counterclockwise when
$f(t)$ is increasing. Note that in part d, only part
of the circle will be traced out since the range
of $f(t) = \sin t$ is $[-1, 1]$.

71. a. $0 \leq t \leq 2\pi$

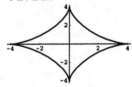

b. $0 \leq t \leq 2\pi$

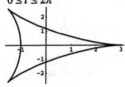

c. $0 \leq t \leq 4\pi$

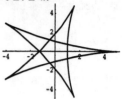

d. $0 \leq t \leq 8\pi$

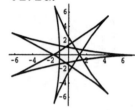

Let $\dfrac{p}{q} = \dfrac{a}{b}$ where $\dfrac{p}{q}$ is the reduced fraction

of $\dfrac{a}{b}$. The length of the t-interval is $2q\pi$.

The number of times the graph would touch
the circle of radius a during the t-interval is p.

73. $x = \dfrac{3t}{t^3 + 1}, y = \dfrac{3t^2}{t^3 + 1}$

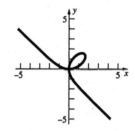

When $x > 0$, $t > 0$ or $t < -1$.
When $x < 0$, $-1 < t < 0$.
When $y > 0$, $t > -1$. When $y < 0$, $t < -1$.
Therefore the graph is in quadrant I for $t > 0$,
quadrant II for $-1 < t < 0$,
quadrant III for no t, and quadrant IV for $t < -1$.

13.2 Concepts Review

1. magnitude; direction

3. the tail of **u**; the head of **v**

Problem Set 13.2

1.

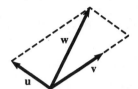

3.

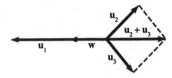

5. $\mathbf{w} = \dfrac{1}{2}(\mathbf{u} + \mathbf{v}) = \dfrac{1}{2}\mathbf{u} + \dfrac{1}{2}\mathbf{v}$

7. $|\mathbf{w}| = |\mathbf{u}|\cos 60° + |\mathbf{v}|\cos 60° = \dfrac{1}{2} + \dfrac{1}{2} = 1$

9. Let θ be the angle of **w** measured clockwise from south.

$|\mathbf{w}|\cos\theta = |\mathbf{u}|\cos 30° + |\mathbf{v}|\cos 45° = 25\sqrt{3} + 25\sqrt{2}$

$= 25\left(\sqrt{3} + \sqrt{2}\right)$

$|\mathbf{w}|\sin\theta = |\mathbf{v}|\sin 45° - |\mathbf{u}|\sin 30° = 25\sqrt{2} - 25$

$= 25\left(\sqrt{2} - 1\right)$

$|\mathbf{w}|^2 = |\mathbf{w}|^2 \cos^2\theta + |\mathbf{w}|^2 \sin^2\theta$

$= 625\left(\sqrt{3} + \sqrt{2}\right)^2 + 625\left(\sqrt{2} - 1\right)^2$

$= 625\left(8 - 2\sqrt{2} + 2\sqrt{6}\right)$

$|\mathbf{w}| = \sqrt{625\left(8 - 2\sqrt{2} + 2\sqrt{6}\right)} = 25\sqrt{8 - 2\sqrt{2} + 2\sqrt{6}}$

≈ 79.34

$\tan\theta = \dfrac{|\mathbf{w}|\sin\theta}{|\mathbf{w}|\cos\theta} = \dfrac{\sqrt{2} - 1}{\sqrt{3} + \sqrt{2}}$

$\theta = \tan^{-1}\left(\dfrac{\sqrt{2} - 1}{\sqrt{3} + \sqrt{2}}\right) = 7.5°$

w has magnitude 79.34 lb in the direction S 7.5° W.

11. The force of 300 N parallel to the plane has magnitude 300 sin 30° = 150 N. Thus, a force of

13. Let θ be the angle the plane makes from north, measured clockwise.

425 sin θ = 45 sin 20°

$\sin\theta = \dfrac{9}{85}\sin 20°$

$\theta = \sin^{-1}\left(\dfrac{9}{85}\sin 20°\right) \approx 2.08°$

Let x be the speed of airplane with respect to the ground.

$x = 45\cos 20° + 425\cos\theta \approx 467$

The plane flies in the direction N 2.08° E, flying 467 mi/h with respect to the ground.

15. Let x be the air speed.

$x\cos 60° = 40$

$x = \dfrac{40}{\cos 60°} = 80$

The air speed of the plane is 80 mi/hr

17. Given triangle ABC, let D be the midpoint of AB and E be the midpoint of BC. $\mathbf{u} = \overline{AB}$, $\mathbf{v} = \overline{BC}$, $\mathbf{w} = \overline{AC}$, $\mathbf{z} = \overline{DE}$

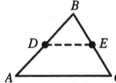

$\mathbf{u} + \mathbf{v} = \mathbf{w}$

$\mathbf{z} = \dfrac{1}{2}\mathbf{u} + \dfrac{1}{2}\mathbf{v} = \dfrac{1}{2}(\mathbf{u} + \mathbf{v}) = \dfrac{1}{2}\mathbf{w}$

Thus, DE is parallel to AC.

19. Let P_i be the tail of $\mathbf{v}_i$. Then

$\mathbf{v}_1 + \mathbf{v}_2 + \ldots + \mathbf{v}_n = \overrightarrow{P_1 P_2} + \overrightarrow{P_2 P_3} + \cdots + \overrightarrow{P_n P_1}$

$= \overrightarrow{P_1 P_1} = \mathbf{0}.$

21. The components of the forces along the lines containing AP, BP, and CP are in equilibrium; that is,

$W = W\cos\alpha + W\cos\beta$

$W = W\cos\beta + W\cos\gamma$

$W = W\cos\alpha + W\cos\gamma$

Thus, $\cos\alpha + \cos\beta = 1$, $\cos\beta + \cos\gamma = 1$, and $\cos\alpha + \cos\gamma = 1$. Solving this system of equations results in $\cos\alpha = \cos\beta = \cos\gamma = \dfrac{1}{2}$.

Hence $\alpha = \beta = \gamma = 60°$.

Therefore, $\alpha + \beta = \alpha + \gamma = \beta + \gamma = 120°$.

23. The components of the forces along the lines containing AP, BP, and CP are in equilibrium; that is,

$5w \cos \alpha + 4w \cos \beta = 3w$
$3w \cos \beta + 5w \cos \gamma = 4w$
$3w \cos \alpha + 4w \cos \gamma = 5w$

Thus, $5 \cos \alpha + 4 \cos \beta = 3$, $3 \cos \beta + 5 \cos \gamma = 4$, and $3 \cos \alpha + 4 \cos \gamma = 5$. Solving this system of equations results in

$\cos \alpha = \dfrac{3}{5}$, $\cos \beta = 0$, $\cos \gamma = \dfrac{4}{5}$, from which it

follows that $\sin \alpha = \dfrac{4}{5}$, $\sin \beta = 1$, $\sin \gamma = \dfrac{3}{5}$.

Therefore, $\cos(\alpha + \beta) = -\dfrac{4}{5}$, $\cos(\alpha + \gamma) = 0$,

$\cos(\beta + \gamma) = -\dfrac{3}{5}$, so

$\alpha + \beta = \cos^{-1}\left(-\dfrac{4}{5}\right) \approx 143.13°$, $\alpha + \gamma = 90°$,

$\beta + \gamma = \cos^{-1}\left(-\dfrac{3}{5}\right) \approx 126.87°$.

This problem can be modeled with three strings going through A, four strings through B, and five strings through C, with equal weights attached to the twelve strings. Then the quantity to be minimized is $3|AP| + 4|BP| + 5|CP|$.

13.3 Concepts Review

1. $\langle u_1 + v_1, u_2 + v_2 \rangle$; $\langle cu_1, cu_2 \rangle$; $\sqrt{u_1^2 + u_2^2}$

3. basis vectors

Problem Set 13.3

1. a. $2\mathbf{a} - 4\mathbf{b} = (-4\mathbf{i} + 6\mathbf{j}) + (-8\mathbf{i} + 12\mathbf{j})$
   $= -12\mathbf{i} + 18\mathbf{j}$

   b. $\mathbf{a} \cdot \mathbf{b} = (-2)(2) + (3)(-3) = -13$

   c. $\mathbf{a} \cdot (\mathbf{b} + \mathbf{c}) = (-2\mathbf{i} + 3\mathbf{j}) \cdot (2\mathbf{i} - 8\mathbf{j})$
   $= (-2)(2) + (3)(-8) = -28$

   d. $(-2\mathbf{a} + 3\mathbf{b}) \cdot 5\mathbf{c} = 5[(10\mathbf{i} - 15\mathbf{j}) \cdot (-5\mathbf{j})]$
   $= 5[(10)(0) + (-15)(-5)] = 375$

   e. $|\mathbf{a}|\mathbf{c} \cdot \mathbf{a} = \sqrt{4+9}[(0)(-2) + (-5)(3)] = -15\sqrt{13}$

   f. $\mathbf{b} \cdot \mathbf{b} - |\mathbf{b}| = (2)(2) + (-3)(-3) - \sqrt{4+9}$
   $= 13 - \sqrt{13}$

3. a. $\cos \theta = \dfrac{\mathbf{a} \cdot \mathbf{b}}{|\mathbf{a}||\mathbf{b}|} = \dfrac{(1)(-1) + (-3)(2)}{\left(\sqrt{10}\right)\left(\sqrt{5}\right)} = -\dfrac{7}{\sqrt{50}}$

   $= -\dfrac{7}{5\sqrt{2}} \approx -0.9899$

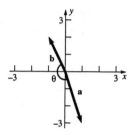

b. $\cos \theta = \dfrac{\mathbf{a} \cdot \mathbf{b}}{|\mathbf{a}||\mathbf{b}|} = \dfrac{(-1)(6) + (-2)(0)}{\left(\sqrt{5}\right)(6)} = -\dfrac{6}{6\sqrt{5}}$

   $= -\dfrac{1}{\sqrt{5}} \approx -0.4472$

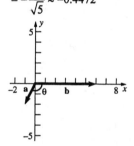

c. $\cos \theta = \dfrac{\mathbf{a} \cdot \mathbf{b}}{|\mathbf{a}||\mathbf{b}|} = \dfrac{(2)(-2) + (-1)(-4)}{\left(\sqrt{5}\right)\left(2\sqrt{5}\right)} = \dfrac{0}{10} = 0$

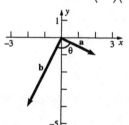

d. $\cos\theta = \dfrac{\mathbf{a}\cdot\mathbf{b}}{|\mathbf{a}||\mathbf{b}|} = \dfrac{(4)(-8)+(-7)(10)}{\left(\sqrt{65}\right)\left(2\sqrt{41}\right)}$

$= \dfrac{-102}{2\sqrt{2665}} = -\dfrac{51}{\sqrt{2665}} \approx -0.9879$

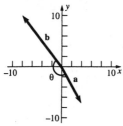

5. a. $\mathbf{a} = (-3-2)\mathbf{i} + (4-2)\mathbf{j} = -5\mathbf{i} + 2\mathbf{j}$

b. $\mathbf{a} = (-6-0)\mathbf{i} + (0-4)\mathbf{j} = -6\mathbf{i} - 4\mathbf{j}$

c. $\mathbf{a} = \left(0-\sqrt{2}\right)\mathbf{i} + (0+\pi)\mathbf{j} = -\sqrt{2}\mathbf{i} + \pi\mathbf{j}$

d. $\mathbf{a} = (-4+7)\mathbf{i} + \left(\dfrac{1}{3}-e\right)\mathbf{j} = 3\mathbf{i} + \left(\dfrac{1}{3}-e\right)\mathbf{j}$

7. $(\mathbf{u}+\mathbf{v})\cdot(\mathbf{u}-\mathbf{v}) = \mathbf{u}\cdot\mathbf{u} - \mathbf{u}\cdot\mathbf{v} + \mathbf{v}\cdot\mathbf{u} - \mathbf{v}\cdot\mathbf{v}$

$= |\mathbf{u}|^2 - |\mathbf{v}|^2 = 0$

Thus, $|\mathbf{u}|^2 = |\mathbf{v}|^2$ or $|\mathbf{u}| = |\mathbf{v}|$.

9. $a\mathbf{u} + b\mathbf{u} = a\langle u_1, u_2\rangle + b\langle u_1, u_2\rangle$

$= \langle au_1, au_2\rangle + \langle bu_1, bu_2\rangle = \langle au_1 + bu_1, au_2 + bu_2\rangle$

$\langle (a+b)u_1, (a+b)u_2\rangle = (a+b)\langle u_1, u_2\rangle$

$= (a+b)\mathbf{u}$

11. $c(\mathbf{u}\cdot\mathbf{v}) = c(\langle u_1, u_2\rangle \cdot \langle v_1, v_2\rangle)$

$= c(u_1 v_1 + u_2 v_2) = c(u_1 v_1) + c(u_2 v_2)$

$= (cu_1)v_1 + (cu_2)v_2 = \langle cu_1, cu_2\rangle \cdot \langle v_1, v_2\rangle$

$= (c\langle u_1, u_2\rangle)\cdot\langle v_1, v_2\rangle = (c\mathbf{u})\cdot\mathbf{v}$

13. $3(6\mathbf{i} - 8\mathbf{j}) = 18\mathbf{i} - 24\mathbf{j}$

15. $\langle 6, 3\rangle \cdot \langle -1, 2\rangle = (6)(-1) + (3)(2) = -6 + 6 = 0$

By Theorem C, they are perpendicular.

17. $\langle c, 6\rangle \cdot \langle c, -4\rangle = (c)(c) + (6)(-4) = c^2 - 24 = 0$

$c^2 = 24$

$c = \pm 2\sqrt{6}$

19. $\mathbf{r} = k\mathbf{a} + m\mathbf{b} \Rightarrow 7 = k(3) + m(-3)$ and

$-8 = k(-2) + m(4)$

$3k - 3m = 7$

$-2k + 4m = -8$

Solve the system of equations to get

$k = \dfrac{2}{3}, m = -\dfrac{5}{3}.$

21. $r_1\mathbf{i} + r_2\mathbf{j} = k(a_1\mathbf{i} + a_2\mathbf{j}) + m(b_1\mathbf{i} + b_2\mathbf{j})$

$= (ka_1 + mb_1)\mathbf{i} + (ka_2 + mb_2)\mathbf{j}$

$\Leftrightarrow r_1 = ka_1 + mb_1$ and $r_2 = ka_2 + mb_2$.

Solve these two equations simultaneously (noting that $a_1 b_2 - a_2 b_1 \neq 0$ since $\mathbf{a}$ and $\mathbf{b}$ are noncollinear) and obtain

$k = \dfrac{b_2 r_1 - b_1 r_2}{a_1 b_2 - a_2 b_1}$ and $m = \dfrac{a_1 r_2 - a_2 r_1}{a_1 b_2 - a_2 b_1}.$

23. Work $= \mathbf{F}\cdot\mathbf{D} = (3\mathbf{i} + 10\mathbf{j})\cdot(10\mathbf{j})$

$= 0 + 100 = 100$ joules

25. $\mathbf{D} = 5\mathbf{i} + 8\mathbf{j}$

Work $= \mathbf{F}\cdot\mathbf{D} = (6)(5) + (8)(8) = 94$ ft-lb

27. $|\mathbf{u}\cdot\mathbf{v}| = |\cos\theta||\mathbf{u}||\mathbf{v}| \leq |\mathbf{u}||\mathbf{v}|$ since $|\cos\theta| \leq 1$.

29. a. $|\mathbf{u}|^2 + |\mathbf{v}|^2 = 2\mathbf{u}\cdot\mathbf{v} \Rightarrow \mathbf{u}\cdot\mathbf{u} - 2\mathbf{u}\cdot\mathbf{v} + \mathbf{v}\cdot\mathbf{v} = 0$

$\Rightarrow (\mathbf{u}-\mathbf{v})\cdot(\mathbf{u}-\mathbf{v}) = 0 \Rightarrow |\mathbf{u}-\mathbf{v}|^2 = 0$

$\Rightarrow \mathbf{u} - \mathbf{v} = \mathbf{0} \Rightarrow \mathbf{u} = \mathbf{v}$

b. $|\mathbf{u}|^2 = |\mathbf{v}|^2 = 2\mathbf{u}\cdot\mathbf{v} \Rightarrow |\mathbf{u}| = |\mathbf{v}|$ and

$|\mathbf{u}|^2 = 2|\mathbf{u}||\mathbf{v}|\cos\theta \Rightarrow |\mathbf{u}|^2 = 2|\mathbf{u}|^2\cos\theta$

$\Rightarrow \cos\theta = \dfrac{1}{2} \Rightarrow \theta = \dfrac{\pi}{3}$

Thus, $\mathbf{u}$ and $\mathbf{v}$ have the same length and the angle between them is $\dfrac{\pi}{3}$.

31. Let $x^2 + y^2 = r^2$ be the equation of the circle and let $A(-r, 0)$, $B(r, 0)$, and $C(x, y)$. Then

$\overrightarrow{AC}\cdot\overrightarrow{BC} = \langle x+r, y\rangle \cdot \langle x-r, y\rangle$

$= (x+r)(x-r) + y^2 = (x^2 + y^2) - r^2$

$= r^2 - r^2 = 0.$

33. a. $\mathrm{pr_v}\mathbf{u} = \left(\dfrac{\langle 0,5\rangle \cdot \langle 3,4\rangle}{|\langle 3,4\rangle|^2}\right)\langle 3,4\rangle$

$= \dfrac{20}{9+16}\langle 3,4\rangle = \left\langle \dfrac{12}{5}, \dfrac{16}{5}\right\rangle$

b. $\mathrm{pr_v}\mathbf{u} = \left(\dfrac{\langle -3,2\rangle \cdot \langle 3,4\rangle}{|\langle 3,4\rangle|}\right)\langle 3,4\rangle$

$= \dfrac{-9+8}{9+16}\langle 3,4\rangle = \left\langle -\dfrac{3}{25}, -\dfrac{4}{25}\right\rangle$

35. a. a and b cannot both be zero. If $a = 0$, then the line $ax + by + c = 0$, is horizontal and $\mathbf{n} = b\mathbf{j}$ is vertical, so $\mathbf{n}$ is perpendicular to the line. Use a similar argument if $b = 0$.

If $a \neq 0$ and $b \neq 0$, then $\left(0, -\dfrac{c}{b}\right)$ and

$\left(-\dfrac{c}{a}, 0\right)$ are on the line. Therefore,

$\left\langle -\dfrac{c}{a}, \dfrac{c}{b}\right\rangle$ is a vector in the direction of the

line. $\left\langle -\dfrac{c}{a}, \dfrac{c}{b}\right\rangle \cdot \langle a, b\rangle = -c + c = 0$, so

$\left\langle -\dfrac{c}{a}, \dfrac{c}{b}\right\rangle$ is perpendicular to $\langle a, b\rangle$. Hence,

$\langle a, b\rangle$ is perpendicular to the line.

b. Consider the following figure.

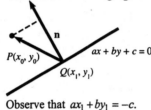

$ax + by + c = 0$

$P(x_0, y_0)$

$Q(x_1, y_1)$

$\mathbf{n}$

Observe that $ax_1 + by_1 = -c$.

$$\text{distance} = \left| \text{pr}_{\mathbf{n}}\ \overrightarrow{QP} \right| = \left| \frac{\overrightarrow{QP} \cdot \mathbf{n}}{|\mathbf{n}|^2} \mathbf{n}\right|$$

$$= \frac{\left|\overrightarrow{QP} \cdot \mathbf{n}\right|}{|\mathbf{n}|^2}|\mathbf{n}| = \frac{\left|\overrightarrow{QP} \cdot \mathbf{n}\right|}{|\mathbf{n}|}$$

$$= \frac{\left|\langle x_0 - x_1, y_0 - y_1\rangle \cdot \langle a, b\rangle\right|}{\sqrt{a^2 + b^2}}$$

$$= \frac{\left|(ax_0 + by_0) - (ax_1 + by_1)\right|}{\sqrt{a^2 + b^2}}$$

$$= \frac{\left|ax_0 + by_0 + c\right|}{\sqrt{a^2 + b^2}}$$

13.4 Concepts Review

1. a vector-valued function of a real variable

3. position

Problem Set 13.4

1. $\lim\limits_{t\to 1}[2t\mathbf{i} - t^2\mathbf{j}] = \lim\limits_{t\to 1}(2t)\mathbf{i} - \lim\limits_{t\to 1}(t^2)\mathbf{j} = 2\mathbf{i} - \mathbf{j}$

3.
$$\lim\limits_{t\to 1}\left[\frac{t-1}{t^2-1}\mathbf{i} - \frac{t^2+2t-3}{t-1}\mathbf{j}\right]$$

$$= \lim\limits_{t\to 1}\left(\frac{t-1}{t^2-1}\right)\mathbf{i} - \lim\limits_{t\to 1}\left(\frac{t^2+2t-3}{t-1}\right)\mathbf{j}$$

$$= \lim\limits_{t\to 1}\left(\frac{1}{t+1}\right)\mathbf{i} - \lim\limits_{t\to 1}(t+3)\mathbf{j} = \frac{1}{2}\mathbf{i} - 4\mathbf{j}$$

5.
$$\lim\limits_{t\to 0}\left[\frac{\sin t \cos t}{t}\mathbf{i} - \frac{7t^3}{e^t}\mathbf{j}\right]$$

$$= \lim\limits_{t\to 0}\left(\frac{\sin t \cos t}{t}\right)\mathbf{i} - \lim\limits_{t\to 0}\left(\frac{7t^3}{e^t}\right)\mathbf{j} = \mathbf{i}$$

7. $\lim\limits_{t\to 0^+}\left\langle \ln(t^3), t^2 \ln t\right\rangle$ does not exist because

$\lim\limits_{t\to 0^+} \ln(t^3) = -\infty$.

9. a. The domain of $f(t) = \dfrac{2}{t-4}$ is

$(-\infty, 4) \cup (4, \infty)$. The domain of

$g(t) = \sqrt{3-t}$ is $(-\infty, 3]$. Thus, the domain

of $\mathbf{r}$ is $(-\infty, 3]$ or $\{t \in \mathbb{R} : t \le 3\}$.

b. The domain of $f(t) = \left[\!\left[t^2 \right]\!\right]$ is $(-\infty, \infty)$. The

domain of $g(t) = \sqrt{20-t}$ is $(-\infty, 20]$. Thus,

the domain of $\mathbf{r}$ is $(-\infty, 20]$ or

$\{t \in \mathbb{R} : t \le 20\}$.

11. a. $f(t) = \dfrac{2}{t-4}$ is continuous on

$(-\infty, 4) \cup (4, \infty)$. $g(t) = \sqrt{3-t}$ is continuous on $(-\infty, 3)$. Thus, $\mathbf{r}$ is continuous on $(-\infty, 3)$ or $\{t \in \mathbb{R} : t < 3\}$.

b. $f(t) = \left[\!\left[t^2 \right]\!\right]$ is continuous on

$\left(-\sqrt{n+1}, -\sqrt{n}\right) \cup \left(\sqrt{n}, \sqrt{n+1}\right)$ where n is a non-negative integer. $g(t) = \sqrt{20-t}$ is continuous on $(-\infty, 20)$ or $\{t \in \mathbb{R} : t < 20\}$. Thus, $\mathbf{r}$ is continuous on

$\left(-\sqrt{n+1}, -\sqrt{n}\right) \cup \left(\sqrt{k}, \sqrt{k+1}\right)$ where n and k are non-negative integers and $k < 400$ or $\{t \in \mathbb{R} : t < 20, t^2 \text{ not an integer}\}$.

13. a. $D_t \mathbf{r}(t) = 9(3t+4)^2 \mathbf{i} + 2te^{t^2} \mathbf{j}$

$D_t^2 \mathbf{r}(t) = 54(3t+4)\mathbf{i} + 2(2t^2+1)e^{t^2} \mathbf{j}$

b. $D_t \mathbf{r}(t) = \sin 2t \mathbf{i} - 3\sin 3t \mathbf{j}$

$D_t^2 \mathbf{r}(t) = 2\cos 2t \mathbf{i} - 9\cos 3t \mathbf{j}$

15. $\mathbf{r}'(t) = -e^{-t}\mathbf{i} - \dfrac{2}{t}\mathbf{j}$

$\mathbf{r}(t) \cdot \mathbf{r}'(t) = -e^{-2t} + \dfrac{2}{t}\ln(t^2)$

$D_t[\mathbf{r}(t) \cdot \mathbf{r}'(t)] = 2e^{-2t} + \dfrac{4}{t^2} - \dfrac{2}{t^2}\ln(t^2)$

17. $h(t)\mathbf{r}(t) = e^{-3t}\sqrt{t-1}\mathbf{i} + e^{-3t}\ln(2t^2)\mathbf{j}$

$D_t[h(t)\mathbf{r}(t)] = -\dfrac{e^{-3t}}{2}\left(\dfrac{6t-7}{\sqrt{t-1}}\right)\mathbf{i}$

$+ e^{-3t}\left(\dfrac{2}{t} - 3\ln(2t^2)\right)\mathbf{j}$

19. $\mathbf{f}'(u) = -\sin u \mathbf{i} + 3e^{3u}\mathbf{j}; \; g'(t) = 6t$

$\mathbf{F}'(t) = \mathbf{f}'(g(t))g'(t)$

$= -6t\sin(3t^2-4)\mathbf{i} + 18te^{9t^2-12}\mathbf{j}$

21. $\displaystyle\int_0^1 (e^t\mathbf{i} + e^{-t}\mathbf{j})\,dt = [e^t\mathbf{i} - e^{-t}\mathbf{j}]_0^1$

$= (e-1)\mathbf{i} + (1-e^{-1})\mathbf{j}$

23. $\mathbf{r}'(t) = \mathbf{v}(t) = -e^{-t}\mathbf{i} + e^t\mathbf{j}$

$\mathbf{r}''(t) = \mathbf{a}(t) = e^{-t}\mathbf{i} + e^t\mathbf{j}$

$|\mathbf{v}(t)| = \sqrt{e^{-2t} + e^{2t}}$

$\mathbf{r}(1) = e^{-1}\mathbf{i} + e\mathbf{j}$

$\mathbf{v}(1) = -e^{-1}\mathbf{i} + e\mathbf{j}$

$\mathbf{a}(1) = e^{-1}\mathbf{i} + e\mathbf{j}$

$|\mathbf{v}(1)| = \sqrt{e^{-2} + e^2}$

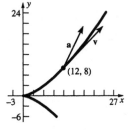

25. $\mathbf{r}'(t) = \mathbf{v}(t) = -2\sin t \mathbf{i} - 3\sin 2t \mathbf{j}$

$\mathbf{r}''(t) = \mathbf{a}(t) = -2\cos t \mathbf{i} - 6\cos 2t \mathbf{j}$

$\mathbf{r}\left(\dfrac{\pi}{3}\right) = \mathbf{i} - \dfrac{9}{4}\mathbf{j}$

$\mathbf{v}\left(\dfrac{\pi}{3}\right) = -\sqrt{3}\mathbf{i} - \dfrac{3\sqrt{3}}{2}\mathbf{j}$

$\mathbf{a}\left(\dfrac{\pi}{3}\right) = -\mathbf{i} + 3\mathbf{j}$

$\left|\mathbf{v}\left(\dfrac{\pi}{3}\right)\right| = \sqrt{3 + \dfrac{27}{4}} = \dfrac{\sqrt{39}}{2}$

27. $\mathbf{r}'(t) = \mathbf{v}(t) = 6t\mathbf{i} + 3t^2\mathbf{j}$

$\mathbf{r}''(t) = \mathbf{a}(t) = 6\mathbf{i} + 6t\mathbf{j}$

$\mathbf{r}(2) = 12\mathbf{i} + 8\mathbf{j}$

$\mathbf{v}(2) = 12\mathbf{i} + 12\mathbf{j}$

$\mathbf{a}(2) = 6\mathbf{i} + 12\mathbf{j}$

$|\mathbf{v}(2)| = \sqrt{144 + 144} = 12\sqrt{2}$

29. $\mathbf{r}'(t) = \mathbf{v}(t) = -\sin t \mathbf{i} - 2\sec^2 t \mathbf{j}$

$\mathbf{r}''(t) = \mathbf{a}(t) = -\cos t \mathbf{i} - 4\sec^2 t \tan t \mathbf{j}$

$$r\left(-\frac{\pi}{4}\right) = \frac{1}{\sqrt{2}}\mathbf{i} + 2\mathbf{j}$$

$$v\left(-\frac{\pi}{4}\right) = \frac{1}{\sqrt{2}}\mathbf{i} - 4\mathbf{j}$$

$$a\left(-\frac{\pi}{4}\right) = -\frac{1}{\sqrt{2}}\mathbf{i} + 8\mathbf{j}$$

$$\left|v\left(-\frac{\pi}{4}\right)\right| = \sqrt{\frac{1}{2} + 16} = \sqrt{\frac{33}{2}}$$

31. $v(t) = \int a(t)\,dt = C_1\mathbf{i} + (-32t + C_2)\mathbf{j}$

$v(0) = \mathbf{0} = C_1\mathbf{i} + C_2\mathbf{j}$

$C_1 = C_2 = 0$

$v(t) = -32t\mathbf{j}$

$r(t) = \int v(t)\,dt = C_1\mathbf{i} + (-16t^2 + C_2)\mathbf{j}$

$r(0) = \mathbf{0} = C_1\mathbf{i} + C_2\mathbf{j}$

$C_1 = C_2 = 0$

$r(t) = -16t^2\mathbf{j}$

33. $v(t) = \int a(t)\,dt = (t + C_1)\mathbf{i} + (-e^{-t} + C_2)\mathbf{j}$

$v(0) = 2\mathbf{i} + \mathbf{j} = C_1\mathbf{i} + (-1 + C_2)\mathbf{j}$

$C_1 = C_2 = 2$

$v(t) = (t + 2)\mathbf{i} + (-e^{-t} + 2)\mathbf{j}$

$r(t) = \int v(t)\,dt = \left(\frac{1}{2}t^2 + 2t + C_1\right)\mathbf{i} + (e^{-t} + 2t + C_2)\mathbf{j}$

$r(0) = \mathbf{i} + \mathbf{j} = C_1\mathbf{i} + (1 + C_2)\mathbf{j}$

$C_1 = 1,\ C_2 = 0$

$r(t) = \left(\frac{1}{2}t^2 + 2t + 1\right)\mathbf{i} + (e^{-t} + 2t)\mathbf{j}$

35. $r(t) = 5\cos 6t\mathbf{i} + 5\sin 6t\mathbf{j}$

$v(t) = -30\sin 6t\mathbf{i} + 30\cos 6t\mathbf{j}$

$|v(t)| = \sqrt{900\sin^2 6t + 900\cos^2 6t} = 30$

$a(t) = -180\cos 6t\mathbf{i} - 180\sin 6t\mathbf{j}$

37. Substitute $\theta = 30°$ and $v_0 = 96$ ft/s into the equations for $r(t)$ and $v(t)$ in Example 5.

$r(t) = 48\sqrt{3}t\mathbf{i} + (48t - 16t^2)\mathbf{j}$,

$v(t) = 48\sqrt{3}\mathbf{i} + (48 - 32t)\mathbf{j}$

The projectile is at ground level when

$48t - 16t^2 = 0$, so $t = 0$ or $t = 3$. Thus, the

projectile hits the ground after 3 seconds. Its speed is $|v(3)| = |48\sqrt{3}\mathbf{i} - 48\mathbf{j}| = 96$ ft/s. It will land $48\sqrt{3}(3) = 144\sqrt{3}$ ft from the origin.

39. $r'(t) = v(t) = \omega\sinh \omega t\mathbf{i} + \omega\cosh \omega t\mathbf{j}$

$r''(t) = a(t) = \omega^2\cosh \omega t\mathbf{i} + \omega^2\sinh \omega t\mathbf{j}$

$= \omega^2(\cosh \omega t\mathbf{i} + \sinh \omega t\mathbf{j}) = \omega^2 r(t)$

Thus, $a(t) = cr(t)$ where $c = \omega^2$.

41. Substitute $\theta = 45°$ into the equation for $r(t)$ in Example 5.

$$r(t) = \frac{v_0}{\sqrt{2}}t\mathbf{i} + \left(\frac{v_0}{\sqrt{2}}t - 16t^2\right)\mathbf{j},$$

If t_1 is the time that the ball is caught,

$\frac{v_0}{\sqrt{2}}t_1 = 300$ and $\frac{v_0}{\sqrt{2}}t_1 - 16t_1^2 = 0$. Thus, from

the first equation $t_1 = \frac{300\sqrt{2}}{v_0}$. Substitute into the

second equation and solve for v_0.

$v_0 = \sqrt{9600} = 40\sqrt{6} \approx 97.98$ ft/s

43. From Example 3, the acceleration due to rotation is $a(t) = -60\omega^2\cos \omega t\mathbf{i} - 60\omega^2\sin \omega t\mathbf{j}$. Since it must be at least greater than gravity,

$980 \le |a(t)| = 60\omega^2$. Thus, $\omega^2 \ge \frac{49}{3}$ or

$\omega \ge \frac{7}{\sqrt{3}} \approx 4.04$ rad/s. The angular speed must be at least 4.04 rad/s.

45. a. From Problem 44a and c, $v^2 = \frac{k}{r} = \frac{gR^2}{r}$.

Convert gravity into mi/h² so

$g \approx (32.17)\left(\frac{1}{5280}\right)\left(\frac{3600}{1}\right)^2 \approx 78{,}963$ mi/h²

$v^2 \approx \frac{(78{,}963)(3960)^2}{(4160)}$

$\approx 297{,}660{,}140$ mi²/h²

$v \approx 17{,}253$ mi/h

b. From Problem 44b and c,

$r^3 = \left(\frac{k}{4\pi^2}\right)T^2 = \left(\frac{gR^2}{4\pi^2}\right)T^2$. Since the

satellite is in synchronous orbit, $T = 24$ h.

$r^3 \approx \left[\frac{(78{,}963)(3960)^2}{4\pi^2}\right](24)^2$

$\approx 1.80666 \times 10^{13}$ mi³

$r \approx 26{,}240$ mi

47. $\mathbf{r}'(t) = -4\sin t\mathbf{i} + 3\cos t\mathbf{j}$, $\mathbf{r}''(t) = -4\cos t\mathbf{i} - 3\sin t\mathbf{j}$

Observe that the graphs of $\mathbf{r}(t)$, $\mathbf{r}'(t)$, and $\mathbf{r}''(t)$ are the same.

a. $\mathbf{r}\left(\dfrac{\pi}{6}\right) = 2\sqrt{3}\mathbf{i} + \dfrac{3}{2}\mathbf{j}$, $\mathbf{r}'\left(\dfrac{\pi}{6}\right) = -2\mathbf{i} + \dfrac{3\sqrt{3}}{2}\mathbf{j}$,

$\mathbf{r}''\left(\dfrac{\pi}{6}\right) = -2\sqrt{3}\mathbf{i} - \dfrac{3}{2}\mathbf{j}$

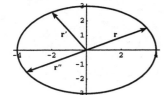

b. $\mathbf{r}\left(\dfrac{\pi}{4}\right) = 2\sqrt{2}\mathbf{i} + \dfrac{3\sqrt{2}}{2}\mathbf{j}$,

$\mathbf{r}'\left(\dfrac{\pi}{4}\right) = -2\sqrt{2}\mathbf{i} + \dfrac{3\sqrt{2}}{2}\mathbf{j}$,

$\mathbf{r}''\left(\dfrac{\pi}{4}\right) = -2\sqrt{2}\mathbf{i} - \dfrac{3\sqrt{2}}{2}\mathbf{j}$

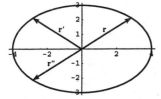

c. $\mathbf{r}\left(\dfrac{\pi}{3}\right) = 2\mathbf{i} + \dfrac{3\sqrt{3}}{2}\mathbf{j}$, $\mathbf{r}'\left(\dfrac{\pi}{3}\right) = -2\sqrt{3}\mathbf{i} + \dfrac{3}{2}\mathbf{j}$,

$\mathbf{r}''\left(\dfrac{\pi}{3}\right) = -2\mathbf{i} - \dfrac{3\sqrt{3}}{2}\mathbf{j}$

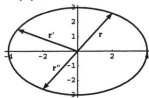
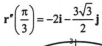

d. $\mathbf{r}\left(\dfrac{\pi}{2}\right) = 3\mathbf{j}$, $\mathbf{r}'\left(\dfrac{\pi}{2}\right) = -4\mathbf{i}$, $\mathbf{r}''\left(\dfrac{\pi}{2}\right) = -3\mathbf{j}$

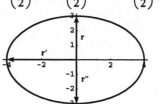

49. $\mathbf{r}(t)$:

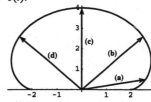

$\mathbf{r}'(t)$:

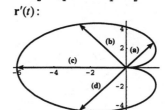

$\mathbf{r}''(t)$:

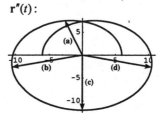

13.5 Concepts Review

1. $\left|\dfrac{d\mathbf{T}}{ds}\right|$

3. $\kappa = \dfrac{1}{R}$

Problem Set 13.5

1. $\mathbf{r}'(t) = 3\mathbf{i} + 6t\mathbf{j}$

$|\mathbf{r}'(t)| = 3\sqrt{1 + 4t^2}$

$\mathbf{T}(t) = \dfrac{\mathbf{r}'(t)}{|\mathbf{r}'(t)|} = \dfrac{1}{\sqrt{1 + 4t^2}}\mathbf{i} + \dfrac{2t}{\sqrt{1 + 4t^2}}\mathbf{j}$

$\mathbf{T}\left(\dfrac{1}{3}\right) = \dfrac{3}{\sqrt{13}}\mathbf{i} + \dfrac{2}{\sqrt{13}}\mathbf{j}$

$$x'(t) = 3 \qquad\qquad y'(t) = 6t$$
$$x''(t) = 0 \qquad\qquad y''(t) = 6$$
$$\kappa(t) = \frac{|x'y'' - y'x''|}{\left[x'^2 + y'^2\right]^{3/2}} = \frac{18}{27(1+4t^2)^{3/2}}$$
$$= \frac{2}{3(1+4t^2)^{3/2}}$$
$$\kappa\left(\frac{1}{3}\right) = \frac{2}{3\left(\frac{13}{9}\right)^{3/2}} = \frac{18}{13\sqrt{13}}$$

3. $\mathbf{u}'(t) = 8t\mathbf{i} + 4\mathbf{j}$

$$|\mathbf{u}'(t)| = 4\sqrt{4t^2 + 1}$$
$$\mathbf{T}(t) = \frac{\mathbf{u}'(t)}{|\mathbf{u}'(t)|} = \frac{2t}{\sqrt{4t^2 + 1}}\mathbf{i} + \frac{1}{\sqrt{4t^2 + 1}}\mathbf{j}$$
$$\mathbf{T}\left(\frac{1}{2}\right) = \frac{1}{\sqrt{2}}\mathbf{i} + \frac{1}{\sqrt{2}}\mathbf{j}$$
$$x'(t) = 8t \qquad\qquad y'(t) = 4$$
$$x''(t) = 8 \qquad\qquad y''(t) = 0$$
$$\kappa(t) = \frac{|x'y'' - y'x''|}{\left[x'^2 + y'^2\right]^{3/2}} = \frac{32}{64(4t^2+1)^{3/2}}$$
$$= \frac{1}{2(4t^2+1)^{3/2}}$$
$$\kappa\left(\frac{1}{2}\right) = \frac{1}{2(2)^{3/2}} = \frac{1}{4\sqrt{2}}$$

5. $\mathbf{z}'(t) = -3\sin t\mathbf{i} + 4\cos t\mathbf{j}$

$$|\mathbf{z}'(t)| = \sqrt{9 + 7\cos^2 t}$$
$$\mathbf{T}(t) = \frac{\mathbf{r}'(t)}{|\mathbf{r}'(t)|} = -\frac{3\sin t}{\sqrt{9 + 7\cos^2 t}}\mathbf{i} + \frac{4\cos t}{\sqrt{9 + 7\cos^2 t}}\mathbf{j}$$
$$\mathbf{T}\left(\frac{\pi}{4}\right) = -\frac{3}{5}\mathbf{i} + \frac{4}{5}\mathbf{j}$$
$$x'(t) = -3\sin t \qquad\qquad y'(t) = 4\cos t$$
$$x''(t) = -3\cos t \qquad\qquad y''(t) = -4\sin t$$

$$\kappa(t) = \frac{|x'y'' - y'x''|}{\left[x'^2 + y'^2\right]^{3/2}} = \frac{12}{(9 + 7\cos^2 t)^{3/2}}$$
$$\kappa\left(\frac{\pi}{4}\right) = \frac{12}{\left(\frac{25}{2}\right)^{3/2}} = \frac{24\sqrt{2}}{125}$$

7. $\mathbf{r}'(t) = (\cos t - \sin t)e^t\mathbf{i} + (\sin t + \cos t)e^t\mathbf{j}$

$$|\mathbf{r}'(t)| = \sqrt{2}e^t$$
$$\mathbf{T}(t) = \frac{\mathbf{r}'(t)}{|\mathbf{r}'(t)|} = \frac{\cos t - \sin t}{\sqrt{2}}\mathbf{i} + \frac{\sin t + \cos t}{\sqrt{2}}\mathbf{j}$$
$$\mathbf{T}\left(\frac{\pi}{2}\right) = -\frac{1}{\sqrt{2}}\mathbf{i} + \frac{1}{\sqrt{2}}\mathbf{j}$$
$$x'(t) = (\cos t - \sin t)e^t \qquad y'(t) = (\sin t + \cos t)e^t$$
$$x''(t) = (-2\sin t)e^t \qquad\quad y''(t) = (2\cos t)e^t$$
$$\kappa(t) = \frac{|x'y'' - y'x''|}{\left[x'^2 + y'^2\right]^{3/2}} = \frac{2e^{2t}}{2\sqrt{2}e^{3t}} = \frac{1}{\sqrt{2}e^t}$$
$$\kappa\left(\frac{\pi}{2}\right) = \frac{1}{\sqrt{2}e^{\pi/2}}$$

9. $\mathbf{r}'(t) = -2t\mathbf{i} - 3t^2\mathbf{j}$

$$|\mathbf{r}'(t)| = t\sqrt{4 + 9t^2}$$
$$\mathbf{T}(t) = \frac{\mathbf{r}'(t)}{|\mathbf{r}'(t)|} = -\frac{2}{\sqrt{4 + 9t^2}}\mathbf{i} - \frac{3t}{\sqrt{4 + 9t^2}}\mathbf{j}$$
$$\mathbf{T}(1) = -\frac{2}{\sqrt{13}}\mathbf{i} - \frac{3}{\sqrt{13}}\mathbf{j}$$
$$x'(t) = -2t \qquad\qquad y'(t) = -3t^2$$
$$x''(t) = -2 \qquad\qquad y''(t) = -6t$$
$$\kappa(t) = \frac{|x'y'' - y'x''|}{\left[x'^2 + y'^2\right]^{3/2}} = \frac{6t^2}{t^3(4 + 9t^2)^{3/2}}$$
$$= \frac{6}{t(4 + 9t^2)^{3/2}}$$
$$\kappa(1) = \frac{6}{13\sqrt{13}}$$

11. $\mathbf{r}'(t) = -(\cos t + \sin t)e^{-t}\mathbf{i} + (\cos t - \sin t)e^t\mathbf{j}$

$$|\mathbf{r}'(t)| = \sqrt{2}e^{-t}$$
$$\mathbf{T}(t) = \frac{\mathbf{r}'(t)}{|\mathbf{r}'(t)|} = -\frac{\cos t + \sin t}{\sqrt{2}}\mathbf{i} + \frac{\cos t - \sin t}{\sqrt{2}}\mathbf{j}$$
$$\mathbf{T}(0) = -\frac{1}{\sqrt{2}}\mathbf{i} + \frac{1}{\sqrt{2}}\mathbf{j}$$
$$x'(t) = -(\cos t - \sin t)e^{-t} \quad y'(t) = (\cos t - \sin t)e^{-t}$$

$$x''(t) = (2\sin t)e^{-t} \qquad y''(t) = (-2\cos t)e^{-t}$$

$$\kappa(t) = \frac{|x'y'' - y'x''|}{\left[x'^2 + y'^2\right]^{3/2}} = \frac{2e^{-2t}}{2\sqrt{2}e^{-3t}} = \frac{e^t}{\sqrt{2}}$$

$$\kappa(0) = \frac{1}{\sqrt{2}}$$

13.

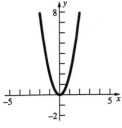

$$y' = 4x, \; y'' = 4$$

$$\kappa = \frac{4}{(1+16x^2)^{3/2}}$$

At $(1, 2)$, $\kappa = \dfrac{4}{17\sqrt{17}}$ and $R = \dfrac{17\sqrt{17}}{4}$.

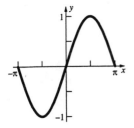

15.

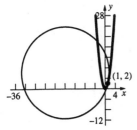

$$y' = \cos x, \; y'' = -\sin x$$

$$\kappa = \frac{|\sin x|}{(1+\cos^2 x)^{3/2}}$$

At $\left(\dfrac{\pi}{4}, \dfrac{\sqrt{2}}{2}\right)$, $\kappa = \dfrac{\frac{1}{\sqrt{2}}}{\left(\frac{3}{2}\right)^{3/2}} = \dfrac{2}{3\sqrt{3}}$ and $R = \dfrac{3\sqrt{3}}{2}$.

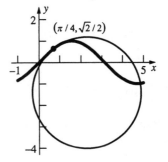

17.

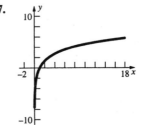

$$y' = \frac{2}{x}, \; y'' = -\frac{2}{x^2}$$

$$\kappa = \frac{\frac{2}{x^2}}{\left(1+\frac{4}{x^2}\right)^{3/2}} = \frac{2|x|}{(x^2+4)^{3/2}}$$

At $(1, 0)$, $\kappa = \dfrac{2}{5\sqrt{5}}$ and $R = \dfrac{5\sqrt{5}}{2}$.

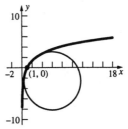

19.

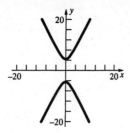

$$2yy' - 8x = 0$$

$$y' = \frac{4x}{y}, \quad y'' = \frac{4(y - xy')}{y^2} = \frac{4(y^2 - 4x^2)}{y^3}$$

$$\kappa = \frac{\frac{4y^2 - 16x^2}{y^3}}{\left(1 + \frac{16x^2}{y^2}\right)^{3/2}} = \frac{4(y^2 - 4x^2)}{(y^2 + 16x^2)^{3/2}}$$

At $(2, 6)$, $\kappa = \dfrac{80}{1000} = \dfrac{2}{25}$ and $R = \dfrac{25}{2}$.

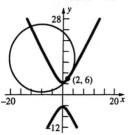

21.

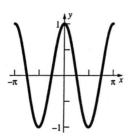

$$y' = -2\sin 2x, \quad y'' = -4\cos 2x$$

$$\kappa = \frac{|4\cos 2x|}{(1 + 4\sin^2 2x)^{3/2}}$$

At $\left(\dfrac{\pi}{6}, \dfrac{1}{2}\right)$, $\kappa = \dfrac{2}{8} = \dfrac{1}{4}$ and $R = 4$.

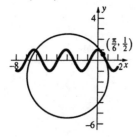

23.

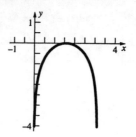

$$y' = \cot x, \quad y'' = -\csc^2 x$$

$$\kappa = \frac{|\csc^2 x|}{(1 + \cot^2 x)^{3/2}} = \frac{1}{|\csc x|}$$

At $\left(\dfrac{\pi}{4}, -\ln\sqrt{2}\right)$, $\kappa = \dfrac{1}{\sqrt{2}}$ and $R = \sqrt{2}$.

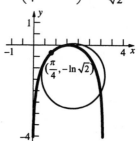

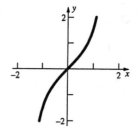

25.

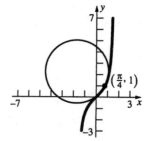

$$y' = \sec^2 x, \quad y'' = 2\sec^2 x \tan x$$

$$\kappa = \frac{|2\sec^2 x \tan x|}{(1 + \sec^4 x)^{3/2}}$$

At $\left(\dfrac{\pi}{4}, 1\right)$, $\kappa = \dfrac{4}{5\sqrt{5}}$ and $R = \dfrac{5\sqrt{5}}{4}$.

27.

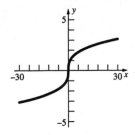

At $(1, 1)$, $\kappa = \dfrac{6}{10\sqrt{10}} = \dfrac{3}{5\sqrt{10}}$ and $R = \dfrac{5\sqrt{10}}{3}$.

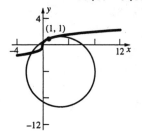

$y' = \dfrac{1}{3x^{2/3}}$, $y'' = -\dfrac{2}{9x^{5/3}}$

$\kappa = \dfrac{\left|\dfrac{2}{9x^{5/3}}\right|}{\left(1 + \dfrac{1}{9x^{4/3}}\right)^{3/2}} = \dfrac{6x^{1/3}}{(9x^{4/3} + 1)^{3/2}}$

29. $y' = \dfrac{1}{x}$, $y'' = -\dfrac{1}{x^2}$

$\kappa = \dfrac{\left|\dfrac{1}{x^2}\right|}{\left(1 + \dfrac{1}{x^2}\right)^{3/2}} = \dfrac{|x|}{(x^2 + 1)^{3/2}}$

Since $0 < x < \infty$, $\kappa = \dfrac{x}{(x^2 + 1)^{3/2}}$.

$\kappa' = \dfrac{(x^2 + 1)^{3/2} - 3x^2(x^2 + 1)^{1/2}}{(x^2 + 1)^3} = \dfrac{-2x^2 + 1}{(x^2 + 1)^{5/2}}$

$\kappa' = 0$ when $x = \dfrac{1}{\sqrt{2}}$. Since $\kappa' > 0$ on $\left(0, \dfrac{1}{\sqrt{2}}\right)$ and $\kappa' < 0$ on $\left(\dfrac{1}{\sqrt{2}}, \infty\right)$, so κ is maximum when

$x = \dfrac{1}{\sqrt{2}}$, $y = \ln\dfrac{1}{\sqrt{2}} = -\dfrac{\ln 2}{2}$. The point of maximum curvature is $\left(\dfrac{1}{\sqrt{2}}, -\dfrac{\ln 2}{2}\right)$.

31. $y' = \sinh x$, $y'' = \cosh x$

$\kappa = \dfrac{\cosh x}{(1 + \sinh^2 x)^{3/2}} = \operatorname{sech}^2 x$

$\kappa' = -2\operatorname{sech}^2 x \tanh x$

$\kappa' = 0$ when $x = 0$. Since $\kappa' > 0$ on $(-\infty, 0)$ and $\kappa' < 0$ on $(0, \infty)$, so κ is maximum when $x = 0$, $y = 1$. The point of maximum curvature is $(0, 1)$.

33. $y' = e^x$, $y'' = e^x$

$\kappa = \dfrac{e^x}{(1 + e^{2x})^{3/2}}$

$\kappa' = \dfrac{e^x(1 + e^{2x})^{3/2} - 3e^{3x}(1 + e^{2x})^{1/2}}{(1 + e^{2x})^3}$

$= \dfrac{e^x(1 - 2e^{2x})}{(1 + e^{2x})^{5/2}}$

$\kappa' = 0$ when $x = -\dfrac{1}{2}\ln 2$. Since $\kappa' > 0$ on

$\left(-\infty, -\dfrac{1}{2}\ln 2\right)$ and $\kappa' < 0$ on $\left(-\dfrac{1}{2}\ln 2, \infty\right)$, so

κ is maximum when $x = -\dfrac{1}{2}\ln 2$, $y = \dfrac{1}{\sqrt{2}}$. The

point of maximum curvature is $\left(-\dfrac{1}{2}\ln 2, \dfrac{1}{\sqrt{2}}\right)$.

35. $\mathbf{r}'(t) = 3\mathbf{i} + 6t\mathbf{j}$

$\mathbf{r}''(t) = 6\mathbf{j}$

$\dfrac{ds}{dt} = |\mathbf{r}'(t)| = 3\sqrt{1+4t^2}$

$a_T = \dfrac{d^2s}{dt^2} = \dfrac{12t}{\sqrt{1+4t^2}}$

$a_N^2 = |\mathbf{r}''(t)|^2 - a_T^2 = 36 - \dfrac{144t^2}{1+4t^2} = \dfrac{36}{1+4t^2}$

$a_N = \dfrac{6}{\sqrt{1+4t^2}}$

At $t_1 = \dfrac{1}{3}$, $a_T = \dfrac{12}{\sqrt{13}}$ and $a_N = \dfrac{18}{\sqrt{13}}$.

37. $\mathbf{r}'(t) = 2\mathbf{i} + 2t\mathbf{j}$

$\mathbf{r}''(t) = 2\mathbf{j}$

$\dfrac{ds}{dt} = |\mathbf{r}'(t)| = 2\sqrt{1+t^2}$

$a_T = \dfrac{d^2s}{dt^2} = \dfrac{2t}{\sqrt{1+t^2}}$

$a_N^2 = |\mathbf{r}''(t)|^2 - a_T^2 = 4 - \dfrac{4t^2}{1+t^2} = \dfrac{4}{1+t^2}$

$a_N = \dfrac{2}{\sqrt{1+t^2}}$

At $t_1 = -1$, $a_T = -\sqrt{2}$ and $a_N = \sqrt{2}$.

39. $\mathbf{r}'(t) = a\sinh t\,\mathbf{i} + a\cosh t\,\mathbf{j}$

$\mathbf{r}''(t) = a\cosh t\,\mathbf{i} + a\sinh t\,\mathbf{j}$

$\dfrac{ds}{dt} = |\mathbf{r}'(t)| = a\sqrt{\sinh^2 t + \cosh^2 t}$

$a_T = \dfrac{d^2s}{dt^2} = \dfrac{2a\cosh t\sinh t}{\sqrt{\sinh^2 t + \cosh^2 t}}$

$a_N^2 = |\mathbf{r}''(t)|^2 - a_T^2 = a^2(\cosh^2 t + \sinh^2 t) - \dfrac{4a^2\cosh^2 t\sinh^2 t}{\sinh^2 t + \cosh^2 t} = \dfrac{a^2(\cosh^2 t - \sinh t^2)^2}{\sinh^2 t + \cosh^2 t}$

$a_N = \dfrac{a(\cosh^2 t - \sinh^2 t)}{\sqrt{\sinh^2 t + \cosh^2 t}}$

At $t_1 = \ln 3$, $\cosh t = \dfrac{5}{3}$, $\sinh t = \dfrac{4}{3}$, and $\sqrt{\sinh^2 t + \cosh^2 t} = \dfrac{\sqrt{41}}{3}$, so $a_T = \dfrac{40a}{3\sqrt{41}}$ and $a_N = \dfrac{3a}{\sqrt{41}}$.

41. $\mathbf{r}'(t) = e^t\mathbf{i} - e^{-t}\mathbf{j}$; $\mathbf{r}''(t) = e^t\mathbf{i} + e^{-t}\mathbf{j}$

$\dfrac{ds}{dt} = |\mathbf{r}'(t)| = \sqrt{e^{2t} + e^{-2t}}$

$a_T = \dfrac{d^2s}{dt^2} = \dfrac{e^{2t} - e^{-2t}}{\sqrt{e^{2t} + e^{-2t}}}$

$a_N^2 = |\mathbf{r}''(t)|^2 - a_T^2$

$= e^{2t} + e^{-2t} - \dfrac{e^{4t} - 2 + e^{-4t}}{e^{2t} + e^{-2t}} = \dfrac{4}{e^{2t} + e^{-2t}}$

$a_N = \dfrac{2}{\sqrt{e^{2t} + e^{-2t}}}$

At $t_1 = 0$, $a_T = 0$ and $a_N = \sqrt{2}$.

43. $x'(t) = (\sin t + \cos t)e^t$ $y'(t) = (\cos t - \sin t)e^t$

$x''(t) = (2\cos t)e^t$ $y''(t) = (-2\sin t)e^t$

$\dfrac{ds}{dt} = \sqrt{x'(t)^2 + y'(t)^2} = \sqrt{2}e^t$

$a_T = \dfrac{d^2s}{dt^2} = \sqrt{2}e^t$

$a_N^2 = [x''(t)^2 + y''(t)^2] - a_T^2 = 4e^{2t} - 2e^{2t} = 2e^{2t}$

$a_N = \sqrt{2}e^t$

At $t_1 = \dfrac{\pi}{3}$, $a_T = \sqrt{2}e^{\pi/3}$ and $a_N = \sqrt{2}e^{\pi/3}$.

45. See the discussion dealing with s and a parameter t in the text. The important ingredients are that s increases as t increases and t can be expressed as a function of s. Use the Chain Rule.

$\mathbf{N} = \dfrac{\frac{d\mathbf{T}}{ds}}{\left|\frac{d\mathbf{T}}{ds}\right|} = \dfrac{\frac{\mathbf{T}'(t)}{|\mathbf{v}(t)|}}{\left|\frac{\mathbf{T}'(t)}{|\mathbf{v}(t)|}\right|} = \dfrac{\mathbf{T}'(t)}{|\mathbf{T}'(t)|} = \dfrac{\frac{d\mathbf{T}}{dt}}{\left|\frac{d\mathbf{T}}{dt}\right|}$

47.

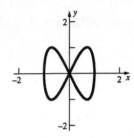

$\mathbf{v}(t) = \langle \cos t, 2\cos 2t \rangle$, $\mathbf{a}(t) = \langle -\sin t, -4\sin 2t \rangle$

$\mathbf{a}(t) = 0$ if and only if $-\sin t = 0$ and $-4\sin 2t = 0$, which occurs if and only if $t = 0$, π, 2π, so it occurs only at the origin.

$\mathbf{a}(t)$ points to the origin if and only if $\mathbf{a}(t) = -k\mathbf{r}(t)$ for some k and $\mathbf{r}(t)$ is not $\mathbf{0}$. This occurs if and only if $t = \dfrac{\pi}{2}, \dfrac{3\pi}{2}$, so it occurs only at $(1, 0)$ and $(-1, 0)$.

49. $s''(t) = a_T = 0 \Rightarrow$ speed $= s'(t) = c$ (a constant)

$$\kappa \left(\frac{ds}{dt} \right)^2 = a_N = 0 \Rightarrow \kappa = 0 \text{ or } \frac{ds}{dt} = 0$$
$$\Rightarrow \kappa = 0$$

51. $\mathbf{T} \cdot \mathbf{T} = 1 \Rightarrow \mathbf{T} \cdot \left(\dfrac{d\mathbf{T}}{ds} \right) + \mathbf{T} \cdot \left(\dfrac{d\mathbf{T}}{ds} \right) = 0$

$$\Rightarrow \mathbf{T} \cdot \left(\frac{d\mathbf{T}}{ds} \right) = 0 = \mathbf{T} \cdot \kappa \mathbf{N} = 0$$

Therefore, $\mathbf{T}$ is perpendicular to $\kappa \mathbf{N}$, and hence to $\mathbf{N}$.

53. It is given that at $(-12, 16)$, $s'(t) = 10$ ft/s and $s''(t) = 5$ ft/s^2. From Example 2, $\kappa = \dfrac{1}{20}$.

Therefore, $a_T = 5$ and $a_N = \left(\dfrac{1}{20} \right)(10)^2 = 5$, so $\mathbf{a} = 5\mathbf{T} + 5\mathbf{N}$. Let $\mathbf{r}(t) = \langle 20\cos t, 20\sin t \rangle$ describe the circle.

$\mathbf{r}(t) = \langle -12, 16 \rangle \Rightarrow \cos t = -\dfrac{3}{5}$ and $\sin t = \dfrac{4}{5}$.

$\mathbf{v}(t) = \langle -20\sin t, 20\cos t \rangle$, so $|\mathbf{v}(t)| = 20$.

Then $\mathbf{T}(t) = \dfrac{\mathbf{v}(t)}{|\mathbf{v}(t)|} = \langle -\sin t, \cos t \rangle$.

Thus, at $(-12, 16)$, $\mathbf{T} = \left\langle -\dfrac{4}{5}, -\dfrac{3}{5} \right\rangle$ and

$\mathbf{N} = \left\langle \dfrac{3}{5}, -\dfrac{4}{5} \right\rangle$ since $\mathbf{N}$ is a unit vector perpendicular to $\mathbf{T}$ and pointing to the concave side of the curve.

Therefore,

$$\mathbf{a} = 5 \left\langle -\frac{4}{5}, -\frac{3}{5} \right\rangle + 5 \left\langle \frac{3}{5}, -\frac{4}{5} \right\rangle = -\mathbf{i} - 7\mathbf{j}.$$

55. Let $\mu mg = \dfrac{mv_R^2}{R}$. Then $v_R = \sqrt{\mu g R}$. At the values given,

$v_R = \sqrt{(0.4)(32)(400)} = \sqrt{5120} \approx 71.55$ ft/s (about 49.79 mi/h).

57. Let the polar coordinate equation of the curve be $r = f(\theta)$. Then the curve is parametrized by $x = r \cos \theta$ and $y = r \sin \theta$.

$x' = -r\sin\theta + r'\cos\theta$ $\qquad$ $y' = r\cos\theta + r'\sin\theta$

$x'' = -r\cos\theta - 2r'\sin\theta + r''\cos\theta$ $\qquad$ $y'' = -r\sin\theta + 2r'\cos\theta + r''\sin\theta$

By Theorem A, the curvature is

$$\kappa = \frac{|x'y'' - y'x''|}{|x'^2 + y'^2|^{3/2}}$$

$$= \frac{|(-r\sin\theta + r'\cos\theta)(-r\sin\theta + 2r'\cos\theta + r''\sin\theta) - (r\cos\theta + r'\sin\theta)(-r\cos\theta - 2r'\sin\theta + r''\cos\theta)|}{[(-r\sin\theta + r'\cos\theta)^2 + (r\cos\theta + r'\sin\theta)^2]^{3/2}}$$

$$= \frac{|r^2 + 2r'^2 - rr''|}{(r^2 + r'^2)^{3/2}}.$$

59. $r' = -\sin\theta, r'' = -\cos\theta$

At $\theta = 0, r = 2, r' = 0,$ and $r'' = -1.$

$$\kappa = \frac{|4+0+2|}{(4+0)^{3/2}} = \frac{6}{8} = \frac{3}{4}$$

61. $r' = -4\sin\theta, r'' = -4\cos\theta$

At $\theta = \dfrac{\pi}{2}, r = 4, r' = -4,$ and $r'' = 0.$

$$\kappa = \frac{|16+32-0|}{(16+16)^{3/2}} = \frac{48}{128\sqrt{2}} = \frac{3}{8\sqrt{2}}$$

63. $r' = 4\cos\theta, r'' = -4\sin\theta$

At $\theta = \dfrac{\pi}{2}, r = 8, r' = 0, r'' = -4.$

$$\kappa = \frac{|64+0+32|}{(64+0)^{3/2}} = \frac{96}{512} = \frac{3}{16}$$

65. $r = \sqrt{\cos 2\theta}$

$$r' = -\frac{\sin 2\theta}{\sqrt{\cos 2\theta}}, r'' = -\frac{\cos^2 2\theta + 1}{(\cos 2\theta)^{3/2}}$$

$$\kappa = \frac{\left|\cos 2\theta + \dfrac{2\sin^2 2\theta}{\cos 2\theta} + \dfrac{\cos^2 2\theta + 1}{\cos 2\theta}\right|}{\left(\cos 2\theta + \dfrac{\sin^2 2\theta}{\cos 2\theta}\right)^{3/2}} = \frac{\left|\dfrac{3}{\cos 2\theta}\right|}{\left(\dfrac{1}{\cos 2\theta}\right)^{3/2}}$$

$$= 3\left(\sqrt{\cos 2\theta}\right) = 3r$$

67.

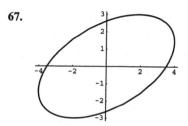

Maximum curvature $\approx 0.7606,$
minimum curvature ≈ 0.1248

69. If $a > b$, then the graph is an ellipse centered at the origin with the length of the major axis $2a$ and the length of the minor axis $2b$. The angle of the major axis with the positive x-axis is θ.

13.6 Chapter Review

Concepts Test

1. False: For example, $x = 0$, $y = t$, and $x = 0$, $y = -t$ both represent the line $x = 0$.

3. False: For example, the graph of $x = t^2$, $y = t$ does not represent y as a function of x. $y = \pm\sqrt{x}$, but $h(x) = \pm\sqrt{x}$ is not a function.

5. False: For example, if $x = t^3$, $y = t^3$ then

$$y = x \text{ so } \frac{d^2 y}{dx^2} = 0, \text{ but } \frac{g''(t)}{f''(t)} = 1.$$

7. True: $(2\mathbf{i} - 3\mathbf{j}) \cdot (6\mathbf{i} + 4\mathbf{j}) = 0$ if and only if the vectors are perpendicular.

9. False: The dot product for three vectors $(\mathbf{a} \cdot \mathbf{b}) \cdot \mathbf{c}$ is not defined.

11. True: If $|\mathbf{u} \cdot \mathbf{v}| = |\mathbf{u}||\mathbf{v}|$, then $|\cos\theta| = 1$ since $|\mathbf{u} \cdot \mathbf{v}| = |\mathbf{u}||\mathbf{v}||\cos\theta|$. Thus, $\mathbf{u}$ is a scalar multiple of $\mathbf{v}$. If $\mathbf{u}$ is a scalar multiple of $\mathbf{v}$,

$$|\mathbf{u} \cdot \mathbf{v}| = |\mathbf{u} \cdot k\mathbf{u}| = k|\mathbf{u}|^2 = |\mathbf{u}||k\mathbf{u}| = |\mathbf{u}||\mathbf{v}|.$$

13. True: $(\mathbf{u} + \mathbf{v}) \cdot (\mathbf{u} - \mathbf{v}) = |\mathbf{u}|^2 - |\mathbf{v}|^2 = 0,$ so

$$|\mathbf{u}|^2 = |\mathbf{v}|^2 \text{ or } |\mathbf{u}| = |\mathbf{v}|.$$

15. True: Theorem 13.4A

17. True: $\begin{aligned} x' &= 3 & y' &= 2 \\ x'' &= 0 & y'' &= 0 \end{aligned}$

Thus, $\kappa = \dfrac{|x'y'' - y'x''|}{\left[x'^2 + y'^2\right]^{3/2}} = 0.$

19. True: $\mathbf{T}(t) = \dfrac{1}{|\mathbf{r}'(t)|}\mathbf{r}'(t);$

$$\mathbf{T}'(t) = -\frac{\mathbf{r}'(t) \cdot \mathbf{r}''(t)}{|\mathbf{r}'(t)|^3}\mathbf{r}'(t) + \frac{1}{|\mathbf{r}'(t)|}\mathbf{r}''(t);$$

$$\mathbf{T}(t) \cdot \mathbf{T}'(t)$$

$$= \frac{1}{|\mathbf{r}'(t)|}\mathbf{r}'(t) \cdot$$

$$\left[-\frac{\mathbf{r}'(t) \cdot \mathbf{r}''(t)}{|\mathbf{r}'(t)|^3}\mathbf{r}'(t) + \frac{1}{|\mathbf{r}'(t)|}\mathbf{r}''(t)\right]$$

$$= -\frac{\mathbf{r}'(t) \cdot \mathbf{r}''(t)}{|\mathbf{r}'(t)|^2} + \frac{\mathbf{r}'(t) \cdot \mathbf{r}''(t)}{|\mathbf{r}'(t)|^2} = 0$$

21. True: $\kappa = \dfrac{|y''|}{[1+y'^2]^{3/2}} = 0$

23. False: For example, if $\mathbf{u} = \mathbf{i}$ and $\mathbf{v} = \mathbf{j}$, then $\mathbf{u} \cdot \mathbf{v} = 0$.

25. True: If $\mathbf{v} \cdot \mathbf{v} = $ constant, differentiate both sides to get $\mathbf{v} \cdot \mathbf{v}' + \mathbf{v} \cdot \mathbf{v}' = 2\mathbf{v} \cdot \mathbf{v}' = 0$, so $\mathbf{v} \cdot \mathbf{v}' = 0$.

Sample Test Problems

1. $t = \dfrac{1}{6}(x-2)$

$y = \dfrac{1}{3}(x-2)$

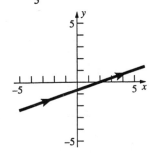

3. $\sin t = \dfrac{x+2}{4}, \cos t = \dfrac{y-1}{3}$

$\dfrac{(x+2)^2}{16} + \dfrac{(y-1)^2}{9} = 1$

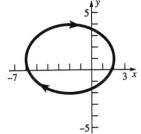

5. $\dfrac{dx}{dt} = 6t^2 - 4, \dfrac{dy}{dt} = 1 + \dfrac{1}{t+1} = \dfrac{t+2}{t+1}$

$\dfrac{dy}{dx} = \dfrac{\frac{t+2}{t+1}}{6t^2 - 4} = \dfrac{t+2}{(t+1)(6t^2-4)}$

At $t = 0, x = 7, y = 0$, and $\dfrac{dy}{dx} = -\dfrac{1}{2}$.

Tangent line: $y = -\dfrac{1}{2}(x-7)$ or $x + 2y - 7 = 0$

Normal line: $y = 2(x-7)$ or $2x - y - 14 = 0$.

7. $\dfrac{dx}{dt} = -\sin t + \sin t + t \cos t = t \cos t$

$\dfrac{dy}{dt} = \cos t - \cos t + t \sin t = t \sin t$

$L = \int_0^{2\pi} \sqrt{(t \cos t)^2 + (t \sin t)^2}\, dt = \int_0^{2\pi} t\, dt$

$= \left[\dfrac{1}{2}t^2\right]_0^{2\pi} = 2\pi^2$

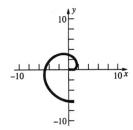

9. a. $3\langle 2, -5\rangle - 2\langle 1,1\rangle = \langle 6, -15\rangle - \langle 2, 2\rangle = \langle 4, -17\rangle$

b. $\langle 2, -5\rangle \cdot \langle 1, 1\rangle = 2 + (-5) = -3$

c. $\langle 2, -5\rangle \cdot (\langle 1, 1\rangle + \langle -6, 0\rangle) = \langle 2, -5\rangle \cdot \langle -5, 1\rangle$
$= -10 + (-5) = -15$

d. $(4\langle 2, -5\rangle + 5\langle 1, 1\rangle) \cdot 3\langle -6, 0\rangle$
$= \langle 13, -15\rangle \cdot \langle -18, 0\rangle$
$= -234 + 0 = -234$

e. $\sqrt{36+0}\,\langle -6, 0\rangle \cdot \langle 1, -1\rangle = 6(-6+0) = -36$

f. $\langle -6, 0\rangle \cdot \langle -6, 0\rangle - \sqrt{36+0}$

$= (36+0) - 6 = 30$

11. $5\mathbf{i} - 4\mathbf{j} = k(-2\mathbf{i}) + m(3\mathbf{i} - 2\mathbf{j})$
$= (-2k + 3m)\mathbf{i} - (2m)\mathbf{j}$, so $5 = -2k + 3m$ and
$4 = 2m$. Therefore, $m = 2$ and $k = \dfrac{1}{2}$.

13. One vector that makes such an angle is
$\langle \cos 150°, \sin 150°\rangle$. The required vector is then

$10\dfrac{\langle \cos 150°, \sin 150°\rangle}{|\langle \cos 150°, \sin 150°\rangle|} = 5\langle -\sqrt{3}, 1\rangle$.

15. Let the wind vector be
$\mathbf{w} = \langle 100\cos 30°, 100\sin 30°\rangle = \langle 50\sqrt{3}, 50\rangle$.
Let $\mathbf{p} = \langle p_1, p_2\rangle$ be the plane's air velocity vector.
We want $\mathbf{w} + \mathbf{p} = 450\,\mathbf{j}$

$$= \langle 0, 450 \rangle \cdot \langle 50\sqrt{3}, 50 \rangle + \langle p_1, p_2 \rangle = \langle 0, 450 \rangle$$
$$\Rightarrow 50\sqrt{3} + p_1 = 0, 50 + p_2 = 450$$
$$\Rightarrow p_1 = -50\sqrt{3}, p_2 = 400$$

Therefore, $\mathbf{p} = \langle -50\sqrt{3}, 400 \rangle$. The angle β formed with the vertical satisfies

$$\cos \beta = \frac{\mathbf{p} \cdot \mathbf{j}}{|\mathbf{p}||\mathbf{j}|} = \frac{400}{\sqrt{167,500}}; \ \beta \approx 12.22°. \text{ Thus, the}$$

heading is N12.22°W. The air speed is $|\mathbf{p}| = \sqrt{167,500} \approx 409.27$ mi/h.

17. **a.** $\mathbf{r}'(t) = \left\langle \frac{1}{t}, -6t \right\rangle; \mathbf{r}''(t) = \left\langle -t^{-2}, -6 \right\rangle$

b. $\mathbf{r}'(t) = \langle \cos t, -2 \sin 2t \rangle;$
$\mathbf{r}''(t) = \langle -\sin t, -4 \cos 2t \rangle$

c. $\mathbf{r}'(t) = \left\langle \sec^2 t, -4t^3 \right\rangle;$
$\mathbf{r}''(t) = \left\langle 2\sec^2 t \tan t, -12t^2 \right\rangle$

19. $\mathbf{v}(t) = \langle 4t, 4 \rangle; \ \mathbf{a}(t) = \langle 4, 0 \rangle; \ \mathbf{v}(-1) = \langle -4, 4 \rangle;$
$|\mathbf{v}(-1)| = 4\sqrt{2}; \ \mathbf{a}(-1) = \langle 4, 0 \rangle$

21. **a.** $y = x^2 - x; y' = 2x - 1; y'' = 2;$
$y'(1) = 1; y''(1) = 2;$
$$\kappa(1) = \frac{|2|}{(1+1)^{3/2}} = \frac{1}{\sqrt{2}} \approx 0.7071$$

b. Let $x = t + t^3; x' = 1 + 3t^2; x'' = 6t;$
$y = t + t^2; y' = 1 + 2t; y'' = 2$.
At $(2, 2)$, $t = 1$, so $x' = 4, x'' = 6$,
$y' = 3, y'' = 2$.
Therefore, $\kappa = \dfrac{|(4)(2) - (3)(6)|}{(16+9)^{3/2}} = 0.08$.

c. $y = a \cosh\left(\dfrac{x}{a}\right); y' = \sinh\left(\dfrac{x}{a}\right);$
$y'' = \left(\dfrac{1}{a}\right) \cosh\left(\dfrac{x}{a}\right)$.
At $(a, a \cosh 1)$,
$y' = \sinh 1, y'' = \left(\dfrac{1}{a}\right)(\cosh 1)$.
Therefore,
$$\kappa = \frac{\left(\frac{1}{a}\right)(\cosh 1)}{(1 + \sinh^2 1)^{3/2}} = \frac{\cosh 1}{a(\cosh^2 1)^{3/2}}$$
$$= \frac{1}{a \cosh^2 1} \approx \frac{0.4120}{a}.$$

23. $\mathbf{v}(t) = -2t\mathbf{i} + 2\mathbf{j}; \ |\mathbf{v}(t)| = 2(t^2 + 1)^{1/2}; \ \mathbf{a}(t) = -2\mathbf{i};$
$|\mathbf{a}(t)| = 2$
$$a_T = \frac{d^2 s}{dt^2} = \left(\frac{d}{dt}\right)[2(t^2 + 1)^{1/2}] = 2t(t^2 + 1)^{-1/2}$$
At $(0, 2)$:
$t = 1$
$a_T = \sqrt{2}$
$a_N = |a|^2 - a_T = 4 - 2 = 2$, so $a_N = \sqrt{2}$.

14

Geometry in Space, Vectors

14.1 Concepts Review

1. coordinates

3. $(-1, 3, 5)$; 4

Problem Set 14.1

1. $A(1, 2, 3)$, $B(2, 0, 1)$, $C(-2, 4, 5)$, $D(0, 3, 0)$, $E(-1, -2, -3)$

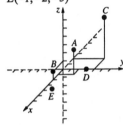

3. $x = 0$ in the yz-plane. $x = 0$ and $y = 0$ on the z-axis.

5. **a.** $\sqrt{(6-1)^2 + (-1-2)^2 + (0-3)^2}$
$= \sqrt{25+9+9} = \sqrt{43}$

b. $\sqrt{(-2-2)^2 + (-2+2)^2 + (0+3)^2}$
$= \sqrt{16+0+9} = 5$

c. $\sqrt{(e+\pi)^2 + (\pi+4)^2 + \left(0-\sqrt{3}\right)^2}$
$\sqrt{(e+\pi)^2 + (\pi+4)^2 + 3} \approx 9.399$

7. $P(2, 1, 6)$, $Q(4, 7, 9)$, $R(8, 5, -6)$
$|PQ| = \sqrt{(2-4)^2 + (1-7)^2 + (6-9)^2}$

$= \sqrt{4+36+9} = 7$

$|PR| = \sqrt{(2-8)^2 + (1-5)^2 + (6+6)^2}$
$= \sqrt{36+16+144} = 14$

$|QR| = \sqrt{(4-8)^2 + (7-5)^2 + (9+6)^2}$
$= \sqrt{16+4+225} = \sqrt{245}$

$|PQ|^2 + |PR|^2 = 49+196 = 245 = |QR|^2$, so the triangle formed by joining P, Q, and R is a right triangle, since it satisfies the Pythagorean Theorem.

9. Since the faces are parallel to the coordinate planes, the sides of the box are in the planes $x = 2$, $y = 3$, $z = 4$, $x = 6$, $y = -1$, and $z = 0$ and the vertices are at the points where 3 of these planes intersect. Thus, the vertices are $(2, 3, 4)$, $(2, 3, 0)$, $(2, -1, 4)$, $(2, -1, 0)$, $(6, 3, 4)$, $(6, 3, 0)$, $(6, -1, 4)$, and $(6, -1, 0)$.

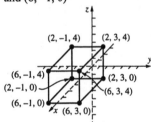

11. **a.** $(x-1)^2 + (y-2)^2 + (z-3)^2 = 25$

b. $(x+2)^2 + (y+3)^2 + (z+6)^2 = 5$

c. $(x-\pi)^2 + (y-e)^2 + \left(z-\sqrt{2}\right)^2 = \pi$

13. $(x^2 - 12x + 36) + (y^2 + 14y + 49) + (z^2 - 8z + 16) = -1+36+49+16$
$(x-6)^2 + (y+7)^2 + (z-4)^2 = 100$
Center: $(6, -7, 4)$; radius 10

15. $x^2 + y^2 + z^2 - x + 2y + 4z = \dfrac{13}{4}$

$\left(x^2 - x + \dfrac{1}{4}\right) + (y^2 + 2y + 1) + (z^2 + 4z + 4) = \dfrac{13}{4} + \dfrac{1}{4} + 1 + 4$

$\left(x - \dfrac{1}{2}\right)^2 + (y+1)^2 + (z+2)^2 = \dfrac{17}{2}$

Center: $\left(\dfrac{1}{2}, -1, -2\right)$; radius $\sqrt{\dfrac{17}{2}} \approx 2.92$

17. x-intercept: $y = z = 0 \Rightarrow 2x = 12, x = 6$
y-intercept: $x = z = 0 \Rightarrow 6y = 12, y = 2$
z-intercept: $x = y = 0 \Rightarrow 3z = 12, z = 4$

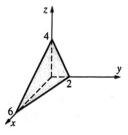

19. x-intercept: $y = z = 0 \Rightarrow x = 6$
y-intercept: $x = z = 0 \Rightarrow 3y = 6, y = 2$
z-intercept: $x = y = 0 \Rightarrow -z = 6, z = -6$

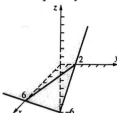

21. x and y cannot both be zero, so the plane is parallel to the z-axis.
x-intercept: $y = z = 0 \Rightarrow x = 8$

y-intercept: $x = z = 0 \Rightarrow 3y = 8, \ y = \dfrac{8}{3}$

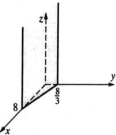

23. This is a sphere with center $(0, 0, 0)$ and radius 3.

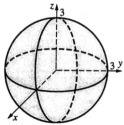

25. The center of the sphere is the midpoint of the diameter, so it is
$\left(\dfrac{-2+4}{2}, \dfrac{3-1}{2}, \dfrac{6+5}{2}\right) = \left(1, 1, \dfrac{11}{2}\right)$. The radius is
$\dfrac{1}{2}\sqrt{(-2-4)^2 + (3+1)^2 + (6-5)^2} = \dfrac{\sqrt{53}}{2}$. The
equation is $(x-1)^2 + (y-1)^2 + \left(z - \dfrac{11}{2}\right)^2 = \dfrac{53}{4}$.

27. The center must be 6 units from each coordinate plane. Since it is in the first octant, the center is $(6, 6, 6)$. The equation is
$(x-6)^2 + (y-6)^2 + (z-6)^2 = 36$.

29. a. Plane parallel to and two units above the xy-plane

   b. Plane perpendicular to the xy-plane whose trace in the xy-plane is the line $x = y$.

   c. Union of the yz-plane ($x = 0$) and the xz-plane ($y = 0$)

   d. Union of the three coordinate planes

   e. Cylinder of radius 2, parallel to the z-axis

   f. Top half of the sphere with center $(0, 0, 0)$ and radius 3

31. If $P(x, y, z)$ denotes the moving point,
$$\sqrt{(x-1)^2+(y-2)^2+(z+3)^2} = 2\sqrt{(x-1)^2+(y-2)^2+(z-3)^2},$$
which simplifies to $(x-1)^2+(y-2)^2+(z-5)^2 = 16$, is a sphere with radius 4 and center $(1, 2, 5)$.

33. Note that the volume of a segment of height h in a hemisphere of radius r is $\pi h^2\left[r-\left(\dfrac{h}{3}\right)\right]$.

The resulting solid is the union of two segments, one for each ball. Since the two balls have the same radius, each segment will have the same value for h. h is the radius minus half the distance between the centers of the two balls.
$$h = 2 - \frac{1}{2}\sqrt{(2-1)^2+(4-2)^2+(3-1)^2} = 2 - \frac{3}{2} = \frac{1}{2}$$
$$V = 2\left[\pi\left(\frac{1}{2}\right)^2\left(2-\frac{1}{6}\right)\right] = \frac{11\pi}{12}$$

14.2 Concepts Review

1. $\sqrt{2^2+(-3)^2+\left(\sqrt{3}\right)^2} = \sqrt{4+9+3} = 4$;
$2\cdot 3 + (-3)\cdot 2 + \sqrt{3}\cdot\left(-2\sqrt{3}\right) = 6-6-6 = -6$

3. $(2, -1, 0)$; $\langle 3, -2, 4\rangle$

Problem Set 14.2

1. a. $(4-1)\mathbf{i} + (5-2)\mathbf{j} + (6-4)\mathbf{k} = 3\mathbf{i} + 3\mathbf{j} + 2\mathbf{k}$

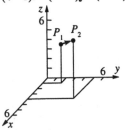

b. $(-14+1)\mathbf{i} + (52+3)\mathbf{j} + (26-204)\mathbf{k}$
$= -13\mathbf{i} + 55\mathbf{j} - 178\mathbf{k}$

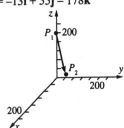

3. a. $\text{length} = \sqrt{4^2+1^2+2^2} = \sqrt{21} \approx 4.58$
$$\cos\alpha = \frac{4}{\sqrt{21}} \approx 0.87,\ \cos\beta = \frac{1}{\sqrt{21}} \approx 0.22,$$
$$\cos\gamma = \frac{2}{\sqrt{21}} \approx 0.44$$

b. $\text{length} = \sqrt{(-2)^2+(-3)^2+7^2} = \sqrt{62} \approx 7.87$
$$\cos\alpha = -\frac{2}{\sqrt{62}} \approx -0.25,$$
$$\cos\beta = -\frac{3}{\sqrt{62}} \approx -0.38,\ \cos\gamma = \frac{7}{\sqrt{62}} \approx 0.89$$

5. $|\langle 3, -4, 5\rangle| = \sqrt{3^2+(-4)^2+5^2} = 5\sqrt{2}$
$$\frac{1}{5\sqrt{2}}\langle 3, -4, 5\rangle = \left\langle\frac{3}{5\sqrt{2}}, -\frac{4}{5\sqrt{2}}, \frac{1}{\sqrt{2}}\right\rangle \text{ is the unit}$$
vector with the same direction as $\langle 3, -4, 5\rangle$.
$$-5\left\langle\frac{3}{5\sqrt{2}}, -\frac{4}{5\sqrt{2}}, \frac{1}{\sqrt{2}}\right\rangle = \left\langle-\frac{3}{\sqrt{2}}, \frac{4}{\sqrt{2}}, -\frac{5}{\sqrt{2}}\right\rangle \text{ is}$$
a vector of length 5 oriented in the opposite direction.

7. $\cos\theta = \dfrac{\langle 4, -3, -1\rangle\cdot\langle -2, -3, 5\rangle}{|\langle 4, -3, -1\rangle||\langle -2, -3, 5\rangle|}$
$$= \frac{(4)(-2)+(-3)(-3)+(-1)(5)}{\sqrt{16+9+1}\sqrt{4+9+25}} = \frac{-4}{\sqrt{26}\sqrt{38}}$$
$$= -\frac{2}{\sqrt{247}}$$
$$\theta = \cos^{-1}\left(-\frac{2}{\sqrt{247}}\right) \approx 97°$$

9. If $x\mathbf{i} + y\mathbf{j} + z\mathbf{k}$ is perpendicular to $-4\mathbf{i} + 5\mathbf{j} + \mathbf{k}$ and $4\mathbf{i} + \mathbf{j}$, then $-4x + 5y + z = 0$ and $4x + y = 0$ since the dot product of perpendicular vectors is 0. Solving these equations yields $y = -4x$ and

$z = 24x$. Hence, for any x, $x\mathbf{i} - 4x\mathbf{j} + 24x\mathbf{k}$ is perpendicular to the given vectors.

$|x\mathbf{i} - 4x\mathbf{j} + 24x\mathbf{k}| = \sqrt{x^2 + 16x^2 + 576x^2}$
$= |x|\sqrt{593}$

This length is 10 when $x = \pm\dfrac{10}{\sqrt{593}}$. The vectors

are $\dfrac{10}{\sqrt{593}}\mathbf{i} - \dfrac{40}{\sqrt{593}}\mathbf{j} + \dfrac{240}{\sqrt{593}}\mathbf{k}$ and

$-\dfrac{10}{\sqrt{593}}\mathbf{i} + \dfrac{40}{\sqrt{593}}\mathbf{j} - \dfrac{240}{\sqrt{593}}\mathbf{k}$.

11. A vector equivalent to $\overrightarrow{BA}$ is
$\mathbf{u} = \langle 1+4, 2-5, 3-6 \rangle = \langle 5, -3, -3 \rangle$.

A vector equivalent to $\overrightarrow{BC}$ is
$\mathbf{v} = \langle 1+4, 0-5, 1-6 \rangle = \langle 5, -5, -5 \rangle$.

$\cos\theta = \dfrac{\mathbf{u} \cdot \mathbf{v}}{|\mathbf{u}||\mathbf{v}|} = \dfrac{5\cdot 5 + (-3)(-5) + (-3)(-5)}{\sqrt{25+9+9}\sqrt{25+25+25}}$

$= \dfrac{55}{\sqrt{43}\sqrt{75}} = \dfrac{11}{\sqrt{129}}$, so $\theta = \cos^{-1}\dfrac{11}{\sqrt{129}} \approx 14.4°$.

13. $\mathbf{u} \cdot \mathbf{v} = (-1)(-1) + 5\cdot 1 + 3(-1) = 3$

$|\mathbf{v}| = \sqrt{1+1+1} = \sqrt{3}$

$\dfrac{\mathbf{u} \cdot \mathbf{v}}{|\mathbf{v}|} = \dfrac{3}{\sqrt{3}} = \sqrt{3}$

15. $\mathbf{m} = \mathrm{pr_v}\,\mathbf{u} = \dfrac{\mathbf{u} \cdot \mathbf{v}}{|\mathbf{v}|^2}\mathbf{v}$

$= \dfrac{(-3)(-3) + 2\cdot 5 + 1\cdot(-3)}{9+25+9}(-3\mathbf{i} + 5\mathbf{j} - 3\mathbf{k})$

$= \dfrac{16}{43}(-3\mathbf{i} + 5\mathbf{j} - 3\mathbf{k}) = -\dfrac{48}{43}\mathbf{i} + \dfrac{80}{43}\mathbf{j} - \dfrac{48}{43}\mathbf{k}$

$\mathbf{n} = \mathbf{u} - \mathbf{m} = \left(-3 + \dfrac{48}{43}\right)\mathbf{i} + \left(2 - \dfrac{80}{43}\right)\mathbf{j} + \left(1 + \dfrac{48}{43}\right)\mathbf{k}$

$= -\dfrac{81}{43}\mathbf{i} + \dfrac{6}{43}\mathbf{j} + \dfrac{91}{43}\mathbf{k}$

17. a. $|\mathbf{u}| = \sqrt{9+4+1} = \sqrt{14}$

$\cos\alpha = -\dfrac{3}{\sqrt{14}}, \alpha = \cos^{-1}\left(-\dfrac{3}{\sqrt{14}}\right) \approx 143°$

$\cos\beta = \dfrac{2}{\sqrt{14}}, \beta = \cos^{-1}\left(\dfrac{2}{\sqrt{14}}\right) \approx 58°$

$\cos\gamma = \dfrac{1}{\sqrt{14}}, \gamma = \cos^{-1}\left(\dfrac{1}{\sqrt{14}}\right) \approx 74°$

b. $|\mathbf{u}| = \sqrt{9+36+1} = \sqrt{46}$

$\cos\alpha = \dfrac{3}{\sqrt{46}}, \alpha = \cos^{-1}\left(\dfrac{3}{\sqrt{46}}\right) \approx 64°$

$\cos\beta = \dfrac{6}{\sqrt{46}}, \beta = \cos^{-1}\left(\dfrac{6}{\sqrt{46}}\right) \approx 28°$

$\cos\gamma = -\dfrac{1}{\sqrt{46}}, \gamma = \cos^{-1}\left(-\dfrac{1}{\sqrt{46}}\right) \approx 98°$

19. $|\mathbf{u}| = (4 + 9 + z^2)^{1/2} = 5$ and $z > 0$, so
$z = 2\sqrt{3} \approx 3.4641$.

21. There are infinitely many such pairs. Note that
$\langle -4, 2, 5\rangle \cdot \langle 1, 2, 0\rangle = -4 + 4 + 0 = 0$, so
$\mathbf{u} = \langle 1, 2, 0\rangle$ is perpendicular to $\langle -4, 2, 5\rangle$. For
any c, $\langle -2, 1, c\rangle \cdot \langle 1, 2, 0\rangle = -2 + 2 + 0 = 0$ so
$\mathbf{v} = \langle -2, 1, c\rangle$ is a candidate.
$\langle -4, 2, 5\rangle \cdot \langle -2, 1, c\rangle = 8 + 2 + 5c$
$8 + 2 + 5c = 0 \Rightarrow c = -2$, so one pair is
$\mathbf{u} = \langle 1, 2, 0\rangle$, $\mathbf{v} = \langle -2, 1, -2\rangle$.

23. The following do not make sense.

a. $\mathbf{v} \cdot \mathbf{w}$ is not a vector.

b. $\mathbf{u} \cdot \mathbf{w}$ is not a vector.

f. $|\mathbf{u}|$ is not a vector.

25. a. $2(x-1) - 4(y-2) + 3(z+3) = 0$ or
$2x - 4y + 3z = -15$.

b. $3(x+2) - 2(y+3) - 1(z-4) = 0$ or
$3x - 2y - z = -4$.

27. The planes are $2x - 4y + 3z = -15$ and
$3x - 2y - z = 4$. The normals to the planes are
$\mathbf{u} = \langle 2, -4, 3\rangle$ and $\mathbf{v} = \langle 3, -2, -1\rangle$. If θ is the angle
between the planes,

$\cos\theta = \dfrac{\mathbf{u} \cdot \mathbf{v}}{|\mathbf{u}||\mathbf{v}|} = \dfrac{6 + 8 - 3}{\sqrt{29}\sqrt{14}} = \dfrac{11}{\sqrt{406}}$, so

$\theta = \cos^{-1}\dfrac{11}{\sqrt{406}} \approx 56.91°$.

29. a. Planes parallel to the xy-plane may be
expressed as $z = D$, so $z = 2$ is an equation of
the plane.

b. An equation of the plane is
$2(x+4) - 3(y+1) - 4(z-2) = 0$ or
$2x - 3y - 4z = -13$.

31. The distance is 0 since the point is in the plane.

$$(|(-3)(2)+2(6)+(3)-9|=0)$$

33. $(1, 0, 0)$ is on $5x - 3y - 2z = 5$. The distance from $(1, 0, 0)$ to $-5x + 3y + 2z = 7$ is

$$\frac{|-5(1)+3(0)+2(0)-7|}{\sqrt{25+9+4}} = \frac{12}{\sqrt{38}} \approx 1.9467.$$

35. $|\mathbf{u}+\mathbf{v}|^2 + |\mathbf{u}-\mathbf{v}|^2 = (\mathbf{u}+\mathbf{v})\cdot(\mathbf{u}+\mathbf{v}) + (\mathbf{u}-\mathbf{v})\cdot(\mathbf{u}-\mathbf{v}) = [\mathbf{u}\cdot\mathbf{u}+2(\mathbf{u}\cdot\mathbf{v})+\mathbf{v}\cdot\mathbf{v}] + [\mathbf{u}\cdot\mathbf{u}-2(\mathbf{u}\cdot\mathbf{v})+\mathbf{v}\cdot\mathbf{v}]$

$= 2(\mathbf{u}\cdot\mathbf{u}) + 2(\mathbf{v}\cdot\mathbf{v}) = 2|\mathbf{u}|^2 + 2|\mathbf{v}|^2$

37. Orient and scale the axes so that the cube has one end of a main diagonal at $(0, 0, 0)$ and the other end at $(1, 1, 1)$. Thus $\langle 1, 1, 1 \rangle$ is along the diagonal. The angle between the main diagonal and one of its faces is the angle between $\langle 1, 1, 1 \rangle$ and $\langle 1, 1, 0 \rangle$. Let θ be the angle.

$$\cos\theta = \frac{\langle 1, 1, 1 \rangle \cdot \langle 1, 1, 0 \rangle}{\sqrt{3}\cdot\sqrt{2}} = \frac{2}{\sqrt{6}},$$

$$\theta = \cos^{-1}\frac{2}{\sqrt{6}} \approx 35.26°$$

39. Place the box so that its corners are at the points $(0, 0, 0)$, $(4, 0, 0)$, $(0, 6, 0)$, $(4, 6, 0)$, $(0, 0, 10)$, $(4, 0, 10)$, $(0, 6, 10)$, and $(4, 6, 10)$. The main diagonals are $(0, 0, 0)$ to $(4, 6, 10)$, $(4, 0, 0)$ to $(0, 6, 10)$, $(0, 6, 0)$ to $(4, 0, 10)$, and $(0, 0, 10)$ to $(4, 6, 0)$. The corresponding vectors are $\langle 4, 6, 10 \rangle$, $\langle -4, 6, 10 \rangle$, $\langle 4, -6, 10 \rangle$, and $\langle 4, 6, -10 \rangle$.

All of these vectors have length $\sqrt{16+36+100} = \sqrt{152}$. Thus, the smallest angle θ between any pair, $\mathbf{u}$ and $\mathbf{v}$ of the diagonals is found from the largest value of $\mathbf{u}\cdot\mathbf{v}$, since

$$\cos\theta = \frac{\mathbf{u}\cdot\mathbf{v}}{|\mathbf{u}||\mathbf{v}|} = \frac{\mathbf{u}\cdot\mathbf{v}}{152}.$$

There are six ways of pairing the four vectors. The largest value of $\mathbf{u}\cdot\mathbf{v}$ is 120 which occurs with

$\mathbf{u} = \langle 4, 6, 10 \rangle$ and $\mathbf{v} = \langle -4, 6, 10 \rangle$. Thus,

$$\cos\theta = \frac{120}{152} = \frac{15}{19} \text{ so } \theta = \cos^{-1}\frac{15}{19} \approx 37.86°.$$

41. $\mathbf{D} = (4-0)\mathbf{i} + (4-0)\mathbf{j} + (0-8)\mathbf{k} = 4\mathbf{i} + 4\mathbf{j} - 8\mathbf{k}$

Thus, $W = \mathbf{F}\cdot\mathbf{D} = 0(4) + 0(4) - 4(-8) = 32$ joules.

43. $\mathbf{D} = (3-0)\mathbf{i} + (5-1)\mathbf{j} + (7-2)\mathbf{k} = 3\mathbf{i} + 4\mathbf{j} + 5\mathbf{k}$

$$\mathbf{F} = 5\frac{(2\mathbf{i}+2\mathbf{j}-\mathbf{k})}{\sqrt{4+4+1}} = \frac{10}{3}\mathbf{i} + \frac{10}{3}\mathbf{j} - \frac{5}{3}\mathbf{k}$$

$$W = \mathbf{F}\cdot\mathbf{D} = \frac{10}{3}(3) + \frac{10}{3}(4) - \frac{5}{3}(5) = 15 \text{ joules.}$$

45. $\langle x, y, z \rangle = \langle 2, 3, -1 \rangle + \frac{1}{5}\langle 7-2, -2-3, 9+1 \rangle$

$= \langle 2+1, 3-1, -1+2 \rangle = \langle 3, 2, 1 \rangle$,

so $(3, 2, 1)$ is the point.

47. The equation of the sphere in standard form is $(x+1)^2 + (y+3)^2 + (z-4)^2 = 26$, so its center is $(-1, -3, 4)$ and radius is $\sqrt{26}$. The distance from the sphere to the plane is the distance from the center to the plane minus the radius of the sphere or

$$\frac{|3(-1)+4(-3)+1(4)-15|}{\sqrt{9+16+1}} - \sqrt{26} = \sqrt{26} - \sqrt{26} = 0,$$

so the sphere is tangent to the plane.

49. Let $\mathbf{x} = \langle x, y, z \rangle$, so

$(\mathbf{x}-\mathbf{a})\cdot(\mathbf{x}-\mathbf{b}) = \langle x-a_1, y-a_2, z-a_3 \rangle \cdot \langle x-b_1, y-b_2, z-b_3 \rangle$

$= x^2 - (a_1+b_1)x + a_1b_1 + y^2 - (a_2+b_2)y + a_2b_2 + z^2 - (a_3+b_3)z + a_3b_3$

Setting this equal to 0 and completing the squares yields

$$\left(x - \frac{a_1+b_1}{2}\right)^2 + \left(y - \frac{a_2+b_2}{2}\right)^2 + \left(z - \frac{a_3+b_3}{2}\right)^2 = \frac{1}{4}\left[(a_1-b_1)^2 + (a_2-b_2)^2 + (a_3-b_3)^2\right].$$

A sphere with center $\left(\frac{a_1+b_1}{2}, \frac{a_2+b_2}{2}, \frac{a_3+b_3}{2}\right)$ and radius $\frac{1}{4}\left[(a_1-b_1)^2 + (a_2-b_2)^2 + (a_3-b_3)^2\right] = \frac{1}{4}|\mathbf{a}-\mathbf{b}|^2$.

51. If **a**, **b**, and **c** are the position vectors of the vertices labeled A, B, and C, respectively, then the side BC is

represented by the vector $\mathbf{c} - \mathbf{b}$. The position vector of the midpoint of BC is $\mathbf{b} + \frac{1}{2}(\mathbf{c} - \mathbf{b}) = \frac{1}{2}(\mathbf{b} + \mathbf{c})$. The segment

from A to the midpoint of BC is $\frac{1}{2}(\mathbf{b} + \mathbf{c}) - \mathbf{a}$. Thus, the position vector of P is $\mathbf{a} + \frac{2}{3}\left[\frac{1}{2}(\mathbf{b} + \mathbf{c}) - \mathbf{a}\right] = \frac{\mathbf{a} + \mathbf{b} + \mathbf{c}}{3}$

If the vertices are $(2, 6, 5)$, $(4, -1, 2)$, and $(6, 1, 2)$, the corresponding position vectors are $\langle 2, 6, 5 \rangle$, $\langle 4, -1, 2 \rangle$, and

$\langle 6, 1, 2 \rangle$. The position vector of P is $\frac{1}{3}\langle 2 + 4 + 6, \ 6 - 1 + 1, \ 5 + 2 + 2 \rangle = \frac{1}{3}\langle 12, \ 6, \ 9 \rangle = \langle 4, 2, 3 \rangle$. Thus P is $(4, 2, 3)$.

14.3 Concepts Review

1.
$$\begin{vmatrix} \mathbf{i} & \mathbf{j} & \mathbf{k} \\ -1 & 2 & 1 \\ 3 & 1 & -1 \end{vmatrix} = \begin{vmatrix} 2 & 1 \\ 1 & -1 \end{vmatrix}\mathbf{i} - \begin{vmatrix} -1 & 1 \\ 3 & -1 \end{vmatrix}\mathbf{j} + \begin{vmatrix} -1 & 2 \\ 3 & 1 \end{vmatrix}\mathbf{k}$$
$$= (-2 - 1)\mathbf{i} - (1 - 3)\mathbf{j} + (-1 - 6)\mathbf{k}$$
$$= -3\mathbf{i} + 2\mathbf{j} - 7\mathbf{k} = \langle -3, 2, -7 \rangle$$

3. $-(\mathbf{v} \times \mathbf{u})$

Problem Set 14.3

1. a. $\mathbf{a} \times \mathbf{b} = \begin{vmatrix} \mathbf{i} & \mathbf{j} & \mathbf{k} \\ -3 & 2 & -2 \\ -1 & 2 & -4 \end{vmatrix}$

$$= \begin{vmatrix} 2 & -2 \\ 2 & -4 \end{vmatrix}\mathbf{i} - \begin{vmatrix} -3 & -2 \\ -1 & -4 \end{vmatrix}\mathbf{j} + \begin{vmatrix} -3 & 2 \\ -1 & 2 \end{vmatrix}\mathbf{k}$$
$$= (-8 + 4)\mathbf{i} - (12 - 2)\mathbf{j} + (-6 + 2)\mathbf{k}$$
$$= -4\mathbf{i} - 10\mathbf{j} - 4\mathbf{k}$$

b. $\mathbf{b} + \mathbf{c} = 6\mathbf{i} + 5\mathbf{j} - 8\mathbf{k}$, so
$$\mathbf{a} \times (\mathbf{b} + \mathbf{c}) = \begin{vmatrix} \mathbf{i} & \mathbf{j} & \mathbf{k} \\ -3 & 2 & -2 \\ 6 & 5 & -8 \end{vmatrix}$$

$$= \begin{vmatrix} 2 & -2 \\ 5 & -8 \end{vmatrix}\mathbf{i} - \begin{vmatrix} -3 & -2 \\ 6 & -8 \end{vmatrix}\mathbf{j} + \begin{vmatrix} -3 & 2 \\ 6 & 5 \end{vmatrix}\mathbf{k}$$
$$= (-16 + 10)\mathbf{i} - (24 + 12)\mathbf{j} + (-15 - 12)\mathbf{k}$$
$$= -6\mathbf{i} - 36\mathbf{j} - 27\mathbf{k}$$

c. $\mathbf{a} \cdot (\mathbf{b} + \mathbf{c}) = -3(6) + 2(5) - 2(-8) = 8$

d. $\mathbf{b} \times \mathbf{c} = \begin{vmatrix} \mathbf{i} & \mathbf{j} & \mathbf{k} \\ -1 & 2 & -4 \\ 7 & 3 & -4 \end{vmatrix}$

$$= \begin{vmatrix} 2 & -4 \\ 3 & -4 \end{vmatrix}\mathbf{i} - \begin{vmatrix} -1 & -4 \\ 7 & -4 \end{vmatrix}\mathbf{j} + \begin{vmatrix} -1 & 2 \\ 7 & 3 \end{vmatrix}\mathbf{k}$$
$$= (-8 + 12)\mathbf{i} - (4 + 28)\mathbf{j} + (-3 - 14)\mathbf{k}$$
$$= 4\mathbf{i} - 32\mathbf{j} - 17\mathbf{k}$$

$$\mathbf{a} \times (\mathbf{b} \times \mathbf{c}) = \begin{vmatrix} \mathbf{i} & \mathbf{j} & \mathbf{k} \\ -3 & 2 & -2 \\ 4 & -32 & -17 \end{vmatrix}$$

$$= \begin{vmatrix} 2 & -2 \\ -32 & -17 \end{vmatrix}\mathbf{i} - \begin{vmatrix} -3 & -2 \\ 4 & -17 \end{vmatrix}\mathbf{j} + \begin{vmatrix} -3 & 2 \\ 4 & -32 \end{vmatrix}\mathbf{k}$$
$$= (-34 - 64)\mathbf{i} - (51 + 8)\mathbf{j} + (96 - 8)\mathbf{k}$$
$$= -98\mathbf{i} - 59\mathbf{j} + 88\mathbf{k}$$

3. $\mathbf{a} \times \mathbf{b} = \begin{vmatrix} \mathbf{i} & \mathbf{j} & \mathbf{k} \\ 1 & 2 & 3 \\ -2 & 2 & -4 \end{vmatrix}$

$$= \begin{vmatrix} 2 & 3 \\ 2 & -4 \end{vmatrix}\mathbf{i} - \begin{vmatrix} 1 & 3 \\ -2 & -4 \end{vmatrix}\mathbf{j} + \begin{vmatrix} 1 & 2 \\ -2 & 2 \end{vmatrix}\mathbf{k}$$
$$= (-8 - 6)\mathbf{i} - (-4 + 6)\mathbf{j} + (2 + 4)\mathbf{k} = -14\mathbf{i} - 2\mathbf{j} + 6\mathbf{k}$$
is perpendicular to both **a** and **b**. All
perpendicular vectors will have the form
$c(-14\mathbf{i} - 2\mathbf{j} + 6\mathbf{k})$ where c is a real number.

5. $\mathbf{u} = \langle 3 - 1, -1 - 3, 2 - 5 \rangle = \langle 2, -4, -3 \rangle$ and
$\mathbf{v} = \langle 4 - 1, 0 - 3, 1 - 5 \rangle = \langle 3, -3, -4 \rangle$ are in the
plane.

$$\mathbf{u} \times \mathbf{v} = \begin{vmatrix} \mathbf{i} & \mathbf{j} & \mathbf{k} \\ 2 & -4 & -3 \\ 3 & -3 & -4 \end{vmatrix}$$

$$= \begin{vmatrix} -4 & -3 \\ -3 & -4 \end{vmatrix}\mathbf{i} - \begin{vmatrix} 2 & -3 \\ 3 & -4 \end{vmatrix}\mathbf{j} + \begin{vmatrix} 2 & -4 \\ 3 & -3 \end{vmatrix}\mathbf{k}$$
$$= (16 - 9)\mathbf{i} - (-8 + 9)\mathbf{j} + (-6 + 12)\mathbf{k}$$
$$= \langle 7, -1, 6 \rangle \text{ is perpendicular to the plane.}$$

$$\pm \frac{1}{\sqrt{49 + 1 + 36}}\langle 7, -1, 6 \rangle = \pm\left\langle \frac{7}{\sqrt{86}}, -\frac{1}{\sqrt{86}}, \frac{6}{\sqrt{86}} \right\rangle$$
are the vectors perpendicular to the plane.

7. $\mathbf{a} \times \mathbf{b} = \begin{vmatrix} \mathbf{i} & \mathbf{j} & \mathbf{k} \\ -1 & 1 & -3 \\ 4 & 2 & -4 \end{vmatrix}$

$$= \begin{vmatrix} 1 & -3 \\ 2 & -4 \end{vmatrix}\mathbf{i} - \begin{vmatrix} -1 & -3 \\ 4 & -4 \end{vmatrix}\mathbf{j} + \begin{vmatrix} -1 & 1 \\ 4 & 2 \end{vmatrix}\mathbf{k}$$

$= (-4 + 6)\mathbf{i} - (4 + 12)\mathbf{j} + (-2 - 4)\mathbf{k}$

$= 2\mathbf{i} - 16\mathbf{j} - 6\mathbf{k}$

Area of parallelogram $= |\mathbf{a} \times \mathbf{b}|$

$\qquad = \sqrt{4 + 256 + 36} = 2\sqrt{74}$

9. $\mathbf{a} = \langle 2-3,\ 4-2,\ 6-1 \rangle = \langle -1,\ 2,\ 5 \rangle$ and

$\mathbf{b} = \langle -1-3, 2-2, 5-1 \rangle = \langle -4, 0, 4 \rangle$ are adjacent

sides of the triangle. The area of the triangle is
half the area of the parallelogram with $\mathbf{a}$ and $\mathbf{b}$ as
adjacent sides.

$\mathbf{a} \times \mathbf{b} = \begin{vmatrix} \mathbf{i} & \mathbf{j} & \mathbf{k} \\ -1 & 2 & 5 \\ -4 & 0 & 4 \end{vmatrix}$

$= \begin{vmatrix} 2 & 5 \\ 0 & 4 \end{vmatrix}\mathbf{i} - \begin{vmatrix} -1 & 5 \\ -4 & 4 \end{vmatrix}\mathbf{j} + \begin{vmatrix} -1 & 2 \\ -4 & 0 \end{vmatrix}\mathbf{k}$

$= (8 - 0)\mathbf{i} - (-4 + 20)\mathbf{j} + (0 + 8)\mathbf{k} = \langle 8, \text{-}16, 8 \rangle$

Area of triangle $= \dfrac{1}{2}\sqrt{64 + 256 + 64} = \dfrac{1}{2}\left(8\sqrt{6}\right)$

$= 4\sqrt{6}$

11. $\mathbf{u} = \langle 0-1, 3-3, 0-2 \rangle = \langle -1, 0, -2 \rangle$ and

$\mathbf{v} = \langle 2-1, 4-3, 3-2 \rangle = \langle 1, 1, 1 \rangle$ are in the plane.

$\mathbf{u} \times \mathbf{v} = \begin{vmatrix} \mathbf{i} & \mathbf{j} & \mathbf{k} \\ -1 & 0 & -2 \\ 1 & 1 & 1 \end{vmatrix}$

$= \begin{vmatrix} 0 & -2 \\ 1 & 1 \end{vmatrix}\mathbf{i} - \begin{vmatrix} -1 & -2 \\ 1 & 1 \end{vmatrix}\mathbf{j} + \begin{vmatrix} -1 & 0 \\ 1 & 1 \end{vmatrix}\mathbf{k}$

$= (0 + 2)\mathbf{i} - (-1 + 2)\mathbf{j} + (-1 - 0)\mathbf{k} = \langle 2, -1, -1 \rangle$

The plane through $(1, 3, 2)$ with normal
$\langle 2, -1, -1 \rangle$ has equation

$2(x - 1) - 1(y - 3) - 1(z - 2) = 0$ or

$2x - y - z = -3$.

13. The plane's normals will be perpendicular to the
normals of the other two planes. Then a normal is
$\langle 1, -3, 2 \rangle \times \langle 2, -2, -1 \rangle = \langle 7, 5, 4 \rangle$. An equation
of the plane is $7(x + 1) + 5(y + 2) + 4(z - 3) = 0$
or $7x + 5y + 4z = -5$.

15. Each vector normal to the plane we seek is
parallel to the line of intersection of the given
planes. Also, the cross product of vectors normal
to the given planes will be parallel to both planes,
hence parallel to the line of intersection. Thus,
$\langle 4, -3, 2 \rangle \times \langle 3, 2, -1 \rangle = \langle -1, 10, 17 \rangle$ is normal to
the plane we seek. An equation of the plane is
$-1(x - 6) + 10(y - 2) + 17(z + 1) = 0$ or
$-x + 10y + 17z = -3$.

17. Volume $= |\langle 2, 3, 4 \rangle \cdot (\langle 0, 4, -1 \rangle \times \langle 5, 1, 3 \rangle)| = |\langle 2, 3, 4 \rangle \cdot \langle 13, -5, -20 \rangle| = |-69| = 69$

19. a. Volume $= |\mathbf{u} \cdot (\mathbf{v} \times \mathbf{w})| = |\langle 3, 2, 1 \rangle \cdot \langle -3, -1, 2 \rangle| = |-9| = 9$

   b. Area $= |\mathbf{u} \times \mathbf{v}| = |\langle 3, -5, 1 \rangle| = \sqrt{9 + 25 + 1} = \sqrt{35}$

   c. Let θ be the angle. Then θ is the complement of the smaller angle between $\mathbf{u}$ and $\mathbf{v} \times \mathbf{w}$.

   $\cos\left(\dfrac{\pi}{2} - \theta\right) = \sin\theta = \dfrac{|\mathbf{u} \cdot (\mathbf{v} \times \mathbf{w})|}{|\mathbf{u}||\mathbf{v} \times \mathbf{w}|} = \dfrac{9}{\sqrt{9+4+1}\sqrt{9+1+4}} = \dfrac{9}{14},\ \theta \approx 40.01°$

21. Choice (c) does not make sense because $(\mathbf{a} \cdot \mathbf{b})$ is a scalar and can't be crossed with a vector. Choice (d) does not
make sense because $(\mathbf{a} \times \mathbf{b})$ is a vector and can't be added to a constant.

23. Let $\mathbf{b}$ and $\mathbf{c}$ determine the (triangular) base of the tetrahedron. Then the area of the base is $\dfrac{1}{2}|\mathbf{b} \times \mathbf{c}|$ which is half of

the area of the parallelogram determined by $\mathbf{b}$ and $\mathbf{c}$. Thus,

$\dfrac{1}{3}(\text{area of base})(\text{height}) = \dfrac{1}{3}\left[\dfrac{1}{2}(\text{area of corresponding parallelogram})(\text{height})\right]$

$= \dfrac{1}{6}(\text{area of corresponding parallelpiped}) = \dfrac{1}{6}|\mathbf{a} \cdot (\mathbf{b} \times \mathbf{c})|$

25. Let $\mathbf{u} = \langle u_1, u_2, u_3 \rangle$ and $\mathbf{v} = \langle v_1, v_2, v_3 \rangle$ then

$$\mathbf{u} \times \mathbf{v} = \langle u_2 v_3 - u_3 v_2, \; u_3 v_1 - u_1 v_3, \; u_1 v_2 - u_2 v_1 \rangle$$

$$|\mathbf{u} \times \mathbf{v}|^2 = u_2^2 v_3^2 - 2 u_2 u_3 v_2 v_3 + u_3^2 v_2^2 + u_3^2 v_1^2 - 2 u_1 u_3 v_1 v_3 + u_1^2 v_3^2 + u_1^2 v_2^2 - 2 u_1 u_2 v_1 v_2 + u_2^2 v_1^2$$

$$= u_1^2 (v_1^2 + v_2^2 + v_3^2) - u_1^2 v_1^2 + u_2^2 (v_1^2 + v_2^2 + v_3^2) - u_2^2 v_2^2 + u_3^2 (v_1^2 + v_2^2 + v_3^2)$$
$$- u_3^2 v_3^2 - 2 u_2 u_3 v_2 v_3 - 2 u_1 u_3 v_1 v_3 - 2 u_1 u_2 v_1 v_2$$

$$= (u_1^2 + u_2^2 + u_3^2)(v_1^2 + v_2^2 + v_3^2) - (u_1^2 v_1^2 + u_2^2 v_2^2 + u_3^2 v_3^2 + 2 u_2 u_3 v_2 v_3 + 2 u_1 u_3 v_1 v_3 + 2 u_1 u_2 v_1 v_2)$$

$$= (u_1^2 + u_2^2 + u_3^2)(v_1^2 + v_2^2 + v_3^2) - (u_1 v_1 + u_2 v_2 + u_3 v_3)^2 \; = |\mathbf{u}|^2 |\mathbf{v}|^2 - (\mathbf{u} \cdot \mathbf{v})^2$$

27. $(\mathbf{v} + \mathbf{w}) \times \mathbf{u} = -[\mathbf{u} \times (\mathbf{v} + \mathbf{w})] = -[(\mathbf{u} \times \mathbf{v}) + (\mathbf{u} \times \mathbf{w})] = -(\mathbf{u} \times \mathbf{v}) - (\mathbf{u} \times \mathbf{w}) = (\mathbf{v} \times \mathbf{u}) + (\mathbf{w} \times \mathbf{u})$

29. $\overrightarrow{PQ} = \langle -a, b, 0 \rangle$, $\overrightarrow{PR} = \langle -a, 0, c \rangle$,

The area of the triangle is half the area of the parallelogram with $\overrightarrow{PQ}$ and $\overrightarrow{PR}$ as adjacent sides, so area

$$(\Delta PQR) = \frac{1}{2} |\langle -a, b, 0 \rangle \times \langle -a, 0, c \rangle| \; = \frac{1}{2} |\langle bc, ac, ab \rangle| = \frac{1}{2} \sqrt{b^2 c^2 + a^2 c^2 + a^2 b^2} \; .$$

31. From Problem 29, $D^2 = \frac{1}{4}(b^2 c^2 + a^2 c^2 + a^2 b^2) \; = \left(\frac{1}{2}bc\right)^2 + \left(\frac{1}{2}ac\right)^2 + \left(\frac{1}{2}ab\right)^2 = A^2 + B^2 + C^2$.

33. The area of the triangle is $A = \frac{1}{2}|\mathbf{a} \times \mathbf{b}|$. Thus,

$$A^2 = \frac{1}{4}|\mathbf{a} \times \mathbf{b}|^2 = \frac{1}{4}\left(|\mathbf{a}|^2 |\mathbf{b}|^2 - (\mathbf{a} \cdot \mathbf{b})^2\right) \; = \frac{1}{4}\left[|\mathbf{a}|^2 |\mathbf{b}|^2 - \frac{1}{4}\left(|\mathbf{a}|^2 + |\mathbf{b}|^2 - |\mathbf{a} - \mathbf{b}|^2\right)^2\right]$$

$$= \frac{1}{16}\left[4 a^2 b^2 - (a^2 + b^2 - c^2)^2\right] \; = \frac{1}{16}(2 a^2 b^2 - a^4 + 2 a^2 c^2 - b^4 + 2 b^2 c^2 - c^4).$$

Note that $s - a = \frac{1}{2}(b + c - a)$,

$s - b = \frac{1}{2}(a + c - b)$, and $s - c = \frac{1}{2}(a + b - c)$.

$$s(s-a)(s-b)(s-c) = \frac{1}{16}(a+b+c)(b+c-a)(a+c-b)(a+b-c) \; = \frac{1}{16}(2 a^2 b^2 - a^4 + 2 a^2 c^2 - b^4 + 2 b^2 c^2 - c^4)$$

which is the same as was obtained above.

14.4 Concepts Review

1. $1 + 4t; \; -3 - 2t; \; 2 - t$

3. $2t\mathbf{i} - 3\mathbf{j} + 3t^2\mathbf{k}$

Problem Set 14.4

1. A parallel vector is
$\mathbf{v} = \langle 4 - 1, \, 5 + 2, \, 6 - 3 \rangle = \langle 3, 7, 3 \rangle$.
$x = 1 + 3t, \; y = -2 + 7t, \; z = 3 + 3t$

3. A parallel vector is
$\mathbf{v} = \langle 6 - 4, \, 2 - 2, \, -1 - 3 \rangle = \langle 2, 0, -4 \rangle$ or $\langle 1, 0, -2 \rangle$.
$x = 4 + t, \; y = 2, \; z = 3 - 2t$

5. $x = 4 + 3t, \; y = 5 + 2t, \; z = 6 + t$
$$\frac{x-4}{3} = \frac{y-5}{2} = \frac{z-6}{1}$$

7. Another parallel vector is $\langle 1, 10, 100 \rangle$.
$x = 1 + t, \; y = 1 + 10t, \; z = 1 + 100t$
$$\frac{x-1}{1} = \frac{y-1}{10} = \frac{z-1}{100}$$

9. Set $z = 0$. Solving $4x + 3y = 1$ and
$10x + 6y = 10$ yields $x = 4, \; y = -5$. Thus
$P_1(4, -5, 0)$ is on the line. Set $y = 0$. Solving
$4x - 7z = 1$ and $10x - 5z = 10$ yields
$x = \frac{13}{10}, \; z = \frac{3}{5}$. Thus $P_2\left(\frac{13}{10}, 0, \frac{3}{5}\right)$ is on the line.

$\overrightarrow{P_1P_2} = \left\langle \frac{13}{10} - 4, 0 - (-5), \frac{3}{5} - 0 \right\rangle = \left\langle -\frac{27}{10}, 5, \frac{3}{5} \right\rangle$ is a

direction vector. Thus,

$\langle 27, -50, -6 \rangle = -10\overrightarrow{P_1P_2}$ is also a direction

vector. The symmetric equations are thus

$$\frac{x-4}{27} = \frac{y+5}{-50} = \frac{z}{-6}$$

11. $\mathbf{u} = \langle 1, 4, -2 \rangle$ and $\mathbf{v} = \langle 2, -1, -2 \rangle$ are both

perpendicular to the line, so $\mathbf{u} \times \mathbf{v} =$
$\langle -10, -2, -9 \rangle$, and hence $\langle 10, 2, 9 \rangle$ is parallel to
the line.
With $y = 0$, $x - 2z = 13$ and $2x - 2z = 5$ yield

$\left(-8, 0, -\frac{21}{2} \right)$. The symmetric equations are

$$\frac{x+8}{10} = \frac{y}{2} = \frac{z + \frac{21}{2}}{9}$$

13. $\langle 1, -5, 2 \rangle$ is a vector in the direction of the line.

$$\frac{x-4}{1} = \frac{y}{-5} = \frac{z-6}{2}$$

15. The point of intersection on the z-axis is $(0, 0, 4)$.

A vector in the direction of the line is
$\langle 5 - 0, -3 - 0, 4 - 4 \rangle = \langle 5, -3, 0 \rangle$. Parametric
equations are $x = 5t$, $y = -3t$, $z = 4$.

17. Using $t = 0$ and $t = 1$, two points on the first line

are $(-2, 1, 2)$ and $(0, 5, 1)$. Using $t = 0$, a point on
the second line is $(2, 3, 1)$. Thus, two nonparallel
vectors in the plane are
$\langle 0 + 2, 5 - 1, 1 - 2 \rangle = \langle 2, 4, 1 \rangle$ and
$\langle 2 + 2, 3 - 1, 1 - 2 \rangle = \langle 4, 2, -1 \rangle$
Hence, $\langle 2, 4, -1 \rangle \times \langle 4, 2, -1 \rangle = \langle -2, -2, -12 \rangle$ is a
normal to the plane, and so is $\langle 1, 1, 6 \rangle$. An
equation of the plane is
$1(x + 2) + 1(y - 1) + 6(z - 2) = 0$ or
$x + y + 6z = 11$.

19. Using $t = 0$, another point in the plane is

$(1, -1, 4)$ and $\langle 2, 3, 1 \rangle$ is parallel to the plane.
Another parallel vector is
$\langle 1 - 1, -1 + 1, 5 - 4 \rangle = \langle 0, 0, 1 \rangle$. Thus,
$\langle 2, 3, 1 \rangle \times \langle 0, 0, 1 \rangle = \langle 3, -2, 0 \rangle$ is a normal to

the plane. An equation of the plane is
$3(x - 1) - 2(y + 1) + 0(z - 5) = 0$ or $3x - 2y = 5$.

21. a. With $t = 0$ in the first line, $x = 2 - 0 = 2$,

$y = 3 + 4 \cdot 0 = 3$, $z = 2 \cdot 0 = 0$, so $(2, 3, 0)$ is
on the first line.

b. $\langle -1, 4, 2 \rangle$ is parallel to the first line, while

$\langle 1, 0, 2 \rangle$ is parallel to the second line, so
$\langle -1, 4, 2 \rangle \times \langle 1, 0, 2 \rangle = \langle 8, 4, -4 \rangle = 4\langle 2, 1, -1 \rangle$
is normal to both. Thus, π has equation
$2(x - 2) + 1(y - 3) - 1(z - 0) = 0$ or
$2x + y - z = 7$, and contains the first line.

c. With $t = 0$ in the second line,

$x = -1 + 0 = -1$, $y = 2$, $z = -1 + 2 \cdot 0 = -1$, so
$Q(-1, 2, -1)$ is on the second line.

d. From Example 6 of Section 14.2, the

distance from Q to π is
$$\frac{|2(-1) + (2) - (-1) - 7|}{\sqrt{4 + 1 + 1}} = \frac{6}{\sqrt{6}} = \sqrt{6} \approx 2.449.$$

23. $\mathbf{r}\left(\frac{\pi}{3} \right) = \mathbf{i} + 3\sqrt{3}\mathbf{j} + \frac{\pi}{3}\mathbf{k}$, so $\left(1, 3\sqrt{3}, \frac{\pi}{3} \right)$ is on the

tangent line.
$\mathbf{r}'(t) = -2\sin t \mathbf{i} + 6\cos t \mathbf{j} + \mathbf{k}$, so
$\mathbf{r}'\left(\frac{\pi}{3} \right) = -\sqrt{3}\mathbf{i} + 3\mathbf{j} + \mathbf{k}$ is parallel to the tangent

line at $t = \frac{\pi}{3}$. The symmetric equations of the

line are $\dfrac{x-1}{-\sqrt{3}} = \dfrac{y - 3\sqrt{3}}{3} = \dfrac{z - \frac{\pi}{3}}{1}$.

25. The curve is given by $\mathbf{r}(t) = 3t\mathbf{i} + 2t^2\mathbf{j} + t^5\mathbf{k}$.

$\mathbf{r}(-1) = -3\mathbf{i} + 2\mathbf{j} - \mathbf{k}$, so $(-3, 2, -1)$ is on the
plane.
$\mathbf{r}'(t) = 3\mathbf{i} + 4t\mathbf{j} + 5t^4\mathbf{k}$, so $\mathbf{r}'(-1) = 3\mathbf{i} - 4\mathbf{j} + 5\mathbf{k}$ is
in the direction of the curve at $t = -1$, hence
normal to the plane. An equation of the plane is
$3(x + 3) - 4(y - 2) + 5(z + 1) = 0$ or
$3x - 4y + 5z = -22$.

27. a. $[x(t)]^2 + [y(t)]^2 + [z(t)]^2 = (\sin t \cos t)^2 + (\sin^2 t)^2 + \cos^2 t = \sin^2 t \cos^2 t + \sin^4 t + \cos^2 t$

$= \sin^2 t(\cos^2 t + \sin^2 t) + \cos^2 t = \sin^2 t + \cos^2 t = 1$

Thus, the curve lies on the sphere $x^2 + y^2 + z^2 = 1$
whose center is at the origin.

b. $r\left(\dfrac{\pi}{6}\right) = \left(\dfrac{1}{2}\right)\left(\dfrac{\sqrt{3}}{2}\right)i + \dfrac{1}{4}j + \dfrac{\sqrt{3}}{2}k = \dfrac{\sqrt{3}}{4}i + \dfrac{1}{4}j + \dfrac{\sqrt{3}}{2}k$, so $\left(\dfrac{\sqrt{3}}{4}, \dfrac{1}{4}, \dfrac{\sqrt{3}}{2}\right)$ is on the tangent line.

$r'(t) = (\cos^2 t - \sin^2 t)i + 2\cos t \sin t\, j - \sin t\, k$ so $r'\left(\dfrac{\pi}{6}\right) = \dfrac{1}{2}i + \dfrac{\sqrt{3}}{2}j - \dfrac{1}{2}k$ is parallel to the line.

The line has equations $x = \dfrac{\sqrt{3}}{4} + t,\ y = \dfrac{1}{4} + \sqrt{3}t,\ z = \dfrac{\sqrt{3}}{2} - t$.

The line intersects the xy-plane when $z = 0$, so $t = \dfrac{\sqrt{3}}{2}$, hence $x = \dfrac{\sqrt{3}}{4} + \dfrac{\sqrt{3}}{2} = \dfrac{3\sqrt{3}}{4},\ y = \dfrac{1}{4} + \dfrac{3}{2} = \dfrac{7}{4}$.

The point is $\left(\dfrac{3\sqrt{3}}{4}, \dfrac{7}{4}, 0\right)$.

29. Let $\overrightarrow{PR}$ be the scalar projection of $\overrightarrow{PQ}$ on n. Then $\left|\overrightarrow{PQ}\right|^2 = \left|\overrightarrow{PR}\right|^2 + d^2$ so

$$d^2 = \left|\overrightarrow{PQ}\right|^2 - \left|\overrightarrow{PR}\right|^2 = \left|\overrightarrow{PQ}\right|^2 - \dfrac{\left|\overrightarrow{PQ}\cdot n\right|^2}{|n|^2}$$

$$= \dfrac{\left|\overrightarrow{PQ}\right|^2 |n|^2 - \left(\overrightarrow{PQ}\cdot n\right)^2}{|n|^2} = \dfrac{\left|\overrightarrow{PQ}\times n\right|^2}{|n|^2}$$

by Lagrange's Identity. Thus, $d = \dfrac{\left|\overrightarrow{PQ}\times n\right|}{|n|}$.

a. $P(3, -2, 1)$ is on the line, so $\overrightarrow{PQ} = \langle 1-3,\ 0+2,\ -4-1\rangle = \langle -2,\ 2,\ -5\rangle$ while $n = \langle 2, -2, 1\rangle$, so

$$d = \dfrac{\left|\langle -2,\ 2,\ -5\rangle \times \langle 2,\ -2,\ 1\rangle\right|}{\sqrt{4+4+1}} = \dfrac{\left|\langle -8,\ -8,\ 0\rangle\right|}{3}$$

$$= \dfrac{8\sqrt{2}}{3} \approx 3.771$$

b. $P(1, -1, 0)$ is on the line, so $\overrightarrow{PQ} = \langle 2-1,\ -1+1,\ 3-0\rangle = \langle 1,\ 0,\ 3\rangle$ while $n = \langle 2, 3, -6\rangle$.

$$d = \dfrac{\left|\langle 1,\ 0,\ 3\rangle \times \langle 2,\ 3,\ -6\rangle\right|}{\sqrt{4+9+36}} = \dfrac{\left|\langle -9,\ 12,\ 3\rangle\right|}{7}$$

$$= \dfrac{3\sqrt{26}}{7} \approx 2.185$$

14.5 Concepts Review

1. $r'(t);\ r''(t)$

3. parallel; concave

Problem Set 14.5

1. $v(t) = r'(t) = 4i + 10tj + 2k$
$a(t) = r''(t) = 10j$
$v(1) = 4i + 10j + 2k;\ a(1) = 10j$;
$s(1) = \sqrt{16+100+4} = 2\sqrt{30} \approx 10.954$

3. $\mathbf{v}(t) = -\dfrac{1}{t^2}\mathbf{i} - \dfrac{2t}{(t^2-1)^2}\mathbf{j} + 5t^4\mathbf{k}$

$\mathbf{a}(t) = \dfrac{2}{t^3}\mathbf{i} + \dfrac{2+6t^2}{(t^2-1)^3}\mathbf{j} + 20t^3\mathbf{k}$

$\mathbf{v}(2) = -\dfrac{1}{4}\mathbf{i} - \dfrac{4}{9}\mathbf{j} + 80\mathbf{k}$; $\mathbf{a}(2) = \dfrac{1}{4}\mathbf{i} + \dfrac{26}{27}\mathbf{j} + 160\mathbf{k}$;

$s(2) = \sqrt{\dfrac{1}{16} + \dfrac{16}{81} + 6400} = \dfrac{\sqrt{8,294,737}}{36}$

≈ 80.002

5. $\mathbf{v}(t) = t^2\mathbf{j} + \dfrac{2}{3}t^{-1/3}\mathbf{k}$; $\mathbf{a}(t) = 2t\mathbf{j} - \dfrac{2}{9}t^{-4/3}\mathbf{k}$

$\mathbf{v}(2) = 4\mathbf{j} + \dfrac{2^{2/3}}{3}\mathbf{k}$; $\mathbf{a}(2) = 4\mathbf{j} - \dfrac{2^{-1/3}}{9}\mathbf{k}$

$s(2) = \sqrt{16 + \dfrac{2^{4/3}}{9}} \approx 4.035$

$\mathbf{v}(2) = 4\mathbf{j} + \dfrac{2^{2/3}}{3}\mathbf{k}$; $\mathbf{a}(2) = 4\mathbf{j} - \dfrac{1}{9\sqrt[3]{2}}\mathbf{k}$;

$s(2) = \sqrt{16 + \dfrac{2^{4/3}}{9}} \approx 4.035$

7. $\mathbf{v}(t) = -\sin t\,\mathbf{i} + \cos t\,\mathbf{j} + \mathbf{k}$
$\mathbf{a}(t) = -\cos t\,\mathbf{i} - \sin t\,\mathbf{j}$
$\mathbf{v}(\pi) = -\mathbf{j} + \mathbf{k}$; $\mathbf{a}(\pi) = \mathbf{i}$;
$s(\pi) = \sqrt{1+1} = \sqrt{2} \approx 1.414$

9. $\mathbf{v}(t) = \sec^2 t\,\mathbf{i} + 3e^t\mathbf{j} - 4\sin 4t\,\mathbf{k}$

$\mathbf{a}(t) = 2\sec^2 t \tan t\,\mathbf{i} + 3e^t\mathbf{j} - 16\cos 4t\,\mathbf{k}$

$\mathbf{v}\left(\dfrac{\pi}{4}\right) = 2\mathbf{i} + 3e^{\pi/4}\mathbf{j}$; $\mathbf{a}\left(\dfrac{\pi}{4}\right) = 4\mathbf{i} + 3e^{\pi/4}\mathbf{j} + 16\mathbf{k}$;

$s\left(\dfrac{\pi}{4}\right) = \sqrt{4 + 9e^{\pi/2}} \approx 6.877$

11. $\mathbf{v}(t) = (\pi t\cos\pi t + \sin\pi t)\mathbf{i} + (\cos\pi t - \pi t\sin\pi t)\mathbf{j} - e^{-t}\mathbf{k}$

$\mathbf{a}(t) = (2\pi\cos\pi t - \pi^2 t\sin\pi t)\mathbf{i} + (-2\pi\sin\pi t - \pi^2 t\cos\pi t)\mathbf{j} + e^{-t}\mathbf{k}$

$\mathbf{v}(2) = 2\pi\mathbf{i} + \mathbf{j} - e^{-2}\mathbf{k}$; $\mathbf{a}(2) = 2\pi\mathbf{i} - 2\pi^2\mathbf{j} + e^{-2}\mathbf{k}$;

$s(2) = \sqrt{4\pi^2 + 1 + e^{-4}} \approx 6.364$

13. If $|\mathbf{v}| = C$, then $|\mathbf{v}|^2 = \mathbf{v}\cdot\mathbf{v} = C^2$. Differentiate implicitly to get $D_t(\mathbf{v}\cdot\mathbf{v}) = 2\mathbf{v}\cdot\mathbf{v}' = 0$. Thus, $\mathbf{v}\cdot\mathbf{v}' = \mathbf{v}\cdot\mathbf{a} = 0$, so $\mathbf{a}$ is perpendicular to $\mathbf{v}$.

15. $s = \displaystyle\int_0^2 \sqrt{1^2 + \cos^2 t + (-\sin t)^2}\,dt = \int_0^2 \sqrt{1 + \cos^2 t + \sin^2 t}\,dt = \int_0^2 \sqrt{1+1}\,dt = \sqrt{2}\int_0^2 dt = \sqrt{2}(2-0) = 2\sqrt{2}$

17. $s = \displaystyle\int_3^6 \sqrt{24t^2 + 4t^4 + 36}\,dt = \int_3^6 2\sqrt{t^4 + 6t^2 + 9}\,dt$

$\displaystyle\int_3^6 2(t^2+3)\,dt = 2\left[\dfrac{t^3}{3} + 3t\right]_3^6 = 2[72 + 18 - (9+9)] = 144$

19. $s = \displaystyle\int_0^1 \sqrt{9t^4 + 36t^4 + 324t^4}\,dt = \int_0^1 3\sqrt{41}t^2\,dt = \left[\sqrt{41}t^3\right]_0^1 = \sqrt{41} \approx 6.403$

21. $s = \displaystyle\int_0^{\pi/2} \sqrt{9\cos^4 t\sin^2 t + 9\sin^4 t\cos^2 t}\,dt = \int_0^{\pi/2} 3\cos t\sin t\sqrt{\cos^2 t + \sin^2 t}\,dt$

$= \displaystyle\int_0^{\pi/2} 3\cos t\sin t\,dt = \left[\dfrac{3}{2}\sin^2 t\right]_0^{\pi/2} = \dfrac{3}{2}$

23. $s = \displaystyle\int_0^\pi \sqrt{\cosh^2 t + \sinh^2 t + 1}\,dt = \int_0^\pi \sqrt{2\cosh^2 t}\,dt = \int_0^\pi \sqrt{2}\cosh t\,dt = \left[\sqrt{2}\sinh t\right]_0^\pi = \sqrt{2}\sinh\pi \approx 16.332$

25. $\mathbf{r}'(t) = 2t\mathbf{i} + 2\mathbf{j} + (2t-4)\mathbf{k}$
$\mathbf{r}''(t) = 2\mathbf{i} + 2\mathbf{k}$

$$\mathbf{T}(2) = \frac{\mathbf{r}'(2)}{|\mathbf{r}'(2)|} = \frac{4\mathbf{i} + 2\mathbf{j}}{\sqrt{16 + 4}} = \frac{2}{\sqrt{5}}\mathbf{i} + \frac{1}{\sqrt{5}}\mathbf{j}$$

$$a_T(2) = \frac{\mathbf{r}'(2) \cdot \mathbf{r}''(2)}{|\mathbf{r}'(2)|} = \frac{(4\mathbf{i} + 2\mathbf{j}) \cdot (2\mathbf{i} + 2\mathbf{k})}{2\sqrt{5}} = \frac{8}{2\sqrt{5}} = \frac{4}{\sqrt{5}}$$

$$a_N(2) = \frac{|\mathbf{r}'(2) \times \mathbf{r}''(2)|}{|\mathbf{r}'(2)|} = \frac{|(4\mathbf{i} + 2\mathbf{j}) \times (2\mathbf{i} + 2\mathbf{k})|}{2\sqrt{5}} = \frac{|4\mathbf{i} - 8\mathbf{j} - 4\mathbf{k}|}{2\sqrt{5}} = \frac{4\sqrt{6}}{2\sqrt{5}} = 2\sqrt{\frac{6}{5}}$$

$$\mathbf{N}(2) = \frac{\mathbf{r}''(2) - a_T(2)\mathbf{T}(2)}{a_N(2)} = \frac{(2\mathbf{i} + 2\mathbf{k}) - \frac{4}{\sqrt{5}}\left(\frac{2}{\sqrt{5}}\mathbf{i} + \frac{1}{\sqrt{5}}\mathbf{j}\right)}{2\sqrt{\frac{6}{5}}} = \frac{\sqrt{5}}{2\sqrt{6}}\left(\frac{2}{5}\mathbf{i} - \frac{4}{5}\mathbf{j} + 2\mathbf{k}\right) = \frac{1}{\sqrt{30}}\mathbf{i} - \frac{2}{\sqrt{30}}\mathbf{j} + \frac{5}{\sqrt{30}}\mathbf{k}$$

$$\kappa(2) = \frac{|\mathbf{r}'(2) \times \mathbf{r}''(2)|}{|\mathbf{r}'(2)|^3} = \frac{4\sqrt{6}}{\left(2\sqrt{5}\right)^3} = \frac{\sqrt{6}}{10\sqrt{5}}$$

$$\mathbf{B}(2) = \mathbf{T}(2) \times \mathbf{N}(2) = \left(\frac{2}{\sqrt{5}}\mathbf{i} + \frac{1}{\sqrt{5}}\mathbf{j}\right) \times \left(\frac{1}{\sqrt{30}}\mathbf{i} - \frac{2}{\sqrt{30}}\mathbf{j} + \sqrt{\frac{5}{6}}\mathbf{k}\right) = \frac{1}{\sqrt{6}}\mathbf{i} - \frac{2}{\sqrt{6}}\mathbf{j} - \frac{1}{\sqrt{6}}\mathbf{k}$$

27. $\mathbf{r}'(t) = t\mathbf{i} + \mathbf{j} + t^2\mathbf{k}$

$\mathbf{r}''(t) = \mathbf{i} + 2t\mathbf{k}$

$$\mathbf{T}(2) = \frac{\mathbf{r}'(2)}{|\mathbf{r}'(2)|} = \frac{2\mathbf{i} + \mathbf{j} + 4\mathbf{k}}{\sqrt{4 + 1 + 16}} = \frac{2}{\sqrt{21}}\mathbf{i} + \frac{1}{\sqrt{21}}\mathbf{j} + \frac{4}{\sqrt{21}}\mathbf{k}$$

$$a_T(2) = \frac{\mathbf{r}'(2) \cdot \mathbf{r}''(2)}{|\mathbf{r}'(2)|} = \frac{(2\mathbf{i} + \mathbf{j} + 4\mathbf{k}) \cdot (\mathbf{i} + 4\mathbf{k})}{\sqrt{21}} = \frac{18}{\sqrt{21}}$$

$$a_N(2) = \frac{|\mathbf{r}'(2) \times \mathbf{r}''(2)|}{|\mathbf{r}'(2)|} = \frac{|(2\mathbf{i} + \mathbf{j} + 4\mathbf{k}) \times (\mathbf{i} + 4\mathbf{k})|}{\sqrt{21}} = \frac{1}{\sqrt{21}}|4\mathbf{i} - 4\mathbf{j} - \mathbf{k}| = \frac{\sqrt{33}}{\sqrt{21}} = \sqrt{\frac{11}{7}}$$

$$\mathbf{N}(2) = \frac{\mathbf{r}''(2) - a_T(2)\mathbf{T}(2)}{a_N(2)} = \frac{(\mathbf{i} + 4\mathbf{k}) - \frac{18}{\sqrt{21}}\left(\frac{2}{\sqrt{21}}\mathbf{i} + \frac{1}{\sqrt{21}}\mathbf{j} + \frac{4}{\sqrt{21}}\mathbf{k}\right)}{\sqrt{\frac{11}{7}}} = \sqrt{\frac{7}{11}}\left(-\frac{15}{21}\mathbf{i} - \frac{18}{21}\mathbf{j} + \frac{12}{21}\mathbf{k}\right)$$

$$= \sqrt{\frac{7}{11}}\left(-\frac{5}{7}\mathbf{i} - \frac{6}{7}\mathbf{j} + \frac{4}{7}\mathbf{k}\right) = -\frac{5}{\sqrt{77}}\mathbf{i} - \frac{6}{\sqrt{77}}\mathbf{j} + \frac{4}{\sqrt{77}}\mathbf{k}$$

$$\kappa(2) = \frac{|\mathbf{r}'(2) \times \mathbf{r}''(2)|}{|\mathbf{r}'(2)|^3} = \frac{\sqrt{33}}{\left(\sqrt{21}\right)^3} = \sqrt{\frac{33}{9261}} = \sqrt{\frac{11}{3087}} = \frac{\sqrt{11}}{21\sqrt{7}}$$

$$\mathbf{B}(2) = \mathbf{T}(2) \times \mathbf{N}(2) = \left(\frac{2}{\sqrt{21}}\mathbf{i} + \frac{1}{\sqrt{21}}\mathbf{j} + \frac{4}{\sqrt{21}}\mathbf{k}\right) \times \left(-\frac{5}{\sqrt{77}}\mathbf{i} - \frac{6}{\sqrt{77}}\mathbf{j} + \frac{4}{\sqrt{77}}\mathbf{k}\right) = \frac{4}{\sqrt{33}}\mathbf{i} - \frac{4}{\sqrt{33}}\mathbf{j} - \frac{1}{\sqrt{33}}\mathbf{k}$$

29. $\mathbf{r}(t) = \langle 7\sin 3t,\ 7\cos 3t,\ 14t \rangle$

$\mathbf{r}'(t) = \langle 21\cos 3t,\ -21\sin 3t,\ 14 \rangle$

$\mathbf{r}''(t) = \langle -63\sin 3t,\ -63\cos 3t,\ 0 \rangle$

$$\mathbf{T}\left(\frac{\pi}{3}\right) = \frac{\mathbf{r}'\left(\frac{\pi}{3}\right)}{\left|\mathbf{r}'\left(\frac{\pi}{3}\right)\right|} = \frac{\langle -21,\ 0,\ 14 \rangle}{\sqrt{441 + 196}} = \frac{1}{7\sqrt{13}}\langle -21,\ 0,\ 14 \rangle = \left\langle -\frac{3}{\sqrt{13}},\ 0,\ \frac{2}{\sqrt{13}} \right\rangle$$

$$a_T\left(\frac{\pi}{3}\right) = \frac{\mathbf{r}'\left(\frac{\pi}{3}\right) \cdot \mathbf{r}''\left(\frac{\pi}{3}\right)}{\left|\mathbf{r}'\left(\frac{\pi}{3}\right)\right|} = \frac{\langle -21,\ 0,\ 14 \rangle \cdot \langle 0,\ -63,\ 0 \rangle}{7\sqrt{13}} = 0$$

$$a_N\left(\frac{\pi}{3}\right) = \frac{\left|\mathbf{r}'\left(\frac{\pi}{3}\right) \times \mathbf{r}''\left(\frac{\pi}{3}\right)\right|}{\left|\mathbf{r}'\left(\frac{\pi}{3}\right)\right|} = \frac{|\langle -21, 0, 14\rangle \times \langle 0, -63, 0\rangle|}{7\sqrt{13}} = \frac{|\langle 882, 0, -1323\rangle|}{7\sqrt{13}} = \frac{441\sqrt{13}}{7\sqrt{13}} = 63$$

$$\mathbf{N}\left(\frac{\pi}{3}\right) = \frac{\mathbf{r}''\left(\frac{\pi}{3}\right) - a_T\left(\frac{\pi}{3}\right)\mathbf{T}\left(\frac{\pi}{3}\right)}{a_N\left(\frac{\pi}{3}\right)} = \frac{1}{63}\langle 0, 63, 0\rangle = \langle 0, 1, 0\rangle$$

$$\kappa\left(\frac{\pi}{3}\right) = \frac{\left|\mathbf{r}\left(\frac{\pi}{3}\right) \times \mathbf{r}''\left(\frac{\pi}{3}\right)\right|}{\left|\mathbf{r}'\left(\frac{\pi}{3}\right)\right|^3} = \frac{441\sqrt{13}}{\left(7\sqrt{13}\right)^3} = \frac{9}{91}$$

$$\mathbf{B}\left(\frac{\pi}{3}\right) = \mathbf{T}\left(\frac{\pi}{3}\right) \times \mathbf{N}\left(\frac{\pi}{3}\right) = \left\langle -\frac{3}{\sqrt{13}}, 0, \frac{2}{\sqrt{13}}\right\rangle \times \langle 0, 1, 0\rangle = \left\langle -\frac{2}{\sqrt{13}}, 0, -\frac{3}{\sqrt{13}}\right\rangle$$

31. $\mathbf{r}'(t) = \sinh\frac{t}{3}\mathbf{i} + \mathbf{j}$

$\mathbf{r}''(t) = \frac{1}{3}\cosh\frac{t}{3}\mathbf{i}$

$$\mathbf{T}(1) = \frac{\mathbf{r}'(1)}{|\mathbf{r}'(1)|} = \frac{\sinh\frac{1}{3}\mathbf{i} + \mathbf{j}}{\sqrt{\sinh^2\frac{1}{3}+1}} = \frac{\sinh\frac{1}{3}\mathbf{i}+\mathbf{j}}{\cosh\frac{1}{3}} = \tanh\frac{1}{3}\mathbf{i} + \text{sech}\frac{1}{3}\mathbf{j}$$

$$a_T(1) = \frac{\mathbf{r}'(1)\cdot\mathbf{r}''(1)}{|\mathbf{r}'(1)|} = \frac{\left(\sinh\frac{1}{3}\mathbf{i}+\mathbf{j}\right)\cdot\left(\frac{1}{3}\cosh\frac{1}{3}\mathbf{i}\right)}{\cosh\frac{1}{3}} = \frac{1}{3}\sinh\frac{1}{3}$$

$$a_N(1) = \frac{|\mathbf{r}'(1)\times\mathbf{r}''(1)|}{|\mathbf{r}'(1)|} = \frac{\left|\left(\sinh\frac{1}{3}\mathbf{i}+\mathbf{j}\right)\times\left(\frac{1}{3}\cosh\frac{1}{3}\mathbf{i}\right)\right|}{\cosh\frac{1}{3}} = \frac{\left|-\frac{1}{3}\cos\frac{1}{3}\mathbf{k}\right|}{\cosh\frac{1}{3}} = \frac{1}{3}$$

$$\mathbf{N}(1) = \frac{\mathbf{r}''(1) - a_T(1)\mathbf{T}(1)}{a_N(1)} = 3\left[\frac{1}{3}\cosh\frac{1}{3}\mathbf{i} - \frac{1}{3}\sinh\frac{1}{3}\left(\tanh\frac{1}{3}\mathbf{i}+\text{sech}\frac{1}{3}\mathbf{j}\right)\right] = \left(\cosh\frac{1}{3} - \frac{\sinh^2\frac{1}{3}}{\cosh\frac{1}{3}}\right)\mathbf{i} - \tanh\frac{1}{3}\mathbf{j}$$

$$= \text{sech}\frac{1}{3}\mathbf{i} - \tanh\frac{1}{3}\mathbf{j}$$

$$\kappa(1) = \frac{|\mathbf{r}'(1)\times\mathbf{r}''(1)|}{|\mathbf{r}'(1)|^3} = \frac{\frac{1}{3}\cosh\frac{1}{3}}{\cosh^3\frac{1}{3}} = \frac{1}{3}\text{sech}^2\frac{1}{3}$$

$$\mathbf{B}(1) = \mathbf{T}(1)\times\mathbf{N}(1) = \left(\tanh\frac{1}{3}\mathbf{i}+\text{sech}\frac{1}{3}\mathbf{j}\right)\times\left(\text{sech}\frac{1}{3}\mathbf{i}-\tanh\frac{1}{3}\mathbf{j}\right) = \left(-\text{sech}^2\frac{1}{3}-\tanh^2\frac{1}{3}\right)\mathbf{k} = -\mathbf{k}$$

33. $\mathbf{r}'(t) = -2e^{-2t}\mathbf{i} + 2e^{2t}\mathbf{j} + 2\sqrt{2}\mathbf{k}$

$\mathbf{r}''(t) = 4e^{-2t}\mathbf{i} + 4e^{2t}\mathbf{j}$

$$\mathbf{T}(0) = \frac{\mathbf{r}'(0)}{|\mathbf{r}'(0)|} = \frac{-2\mathbf{i}+2\mathbf{j}+2\sqrt{2}\mathbf{k}}{\sqrt{4+4+8}}$$

$$= -\frac{1}{2}\mathbf{i} + \frac{1}{2}\mathbf{j} + \frac{\sqrt{2}}{2}\mathbf{k}$$

$$a_T(0) = \frac{\mathbf{r}'(0)\cdot\mathbf{r}''(0)}{|\mathbf{r}'(0)|}$$

$$= \frac{(-2\mathbf{i}+2\mathbf{j}+2\sqrt{2}\mathbf{k})\cdot(4\mathbf{i}+4\mathbf{j})}{4} = 0$$

$$a_N(0) = \frac{|\mathbf{r}'(0)\times\mathbf{r}''(0)|}{|\mathbf{r}'(0)|} = \frac{|-8\sqrt{2}\mathbf{i}+8\sqrt{2}\mathbf{j}-16\mathbf{k}|}{4}$$

$$= \frac{16\sqrt{2}}{4} = 4\sqrt{2}$$

$$\mathbf{N}(0) = \frac{\mathbf{r}''(0) - a_T(0)\mathbf{T}(0)}{a_N(0)} = \frac{4\mathbf{i}+4\mathbf{j}}{4\sqrt{2}}$$

$$= \frac{1}{\sqrt{2}}\mathbf{i} + \frac{1}{\sqrt{2}}\mathbf{j}$$

$$\kappa(0) = \frac{|\mathbf{r}'(0) \times \mathbf{r}''(0)|}{|\mathbf{r}'(0)|^3} = \frac{16\sqrt{2}}{64} = \frac{\sqrt{2}}{4}$$

$$\mathbf{B}(0) = \mathbf{T}(0) \times \mathbf{N}(0)$$

$$= \left(-\frac{1}{2}\mathbf{i} + \frac{1}{2}\mathbf{j} + \frac{\sqrt{2}}{2}\mathbf{k}\right) \times \left(\frac{1}{\sqrt{2}}\mathbf{i} + \frac{1}{\sqrt{2}}\mathbf{j}\right)$$

$$= -\frac{1}{2}\mathbf{i} + \frac{1}{2}\mathbf{j} - \frac{1}{\sqrt{2}}\mathbf{k}$$

35. $\mathbf{r}'(t) = \mathbf{i} + 3\mathbf{j} + 2t\mathbf{k}$

$\mathbf{r}''(t) = 2\mathbf{k}$

$$a_T(t) = \frac{(\mathbf{i} + 3\mathbf{j} + 2t\mathbf{k}) \cdot (2\mathbf{k})}{\sqrt{1 + 9 + 4t^2}} = \frac{4t}{\sqrt{10 + 4t^2}}$$

$$a_N(t) = \frac{|(\mathbf{i} + 3\mathbf{j} + 2t\mathbf{k}) \times (2\mathbf{k})|}{\sqrt{10 + 4t^2}} = \frac{|6\mathbf{i} - 2\mathbf{j}|}{\sqrt{10 + 4t^2}}$$

$$= \frac{\sqrt{36 + 4}}{\sqrt{10 + 4t^2}} = 2\sqrt{\frac{10}{10 + 4t^2}} = 2\sqrt{\frac{5}{5 + 2t^2}}$$

37. $\mathbf{r}(t) = \left\langle e^{-t}, 2t, e^t \right\rangle$

$\mathbf{r}'(t) = \left\langle -e^{-t}, 2, e^t \right\rangle$

$\mathbf{r}''(t) = \left\langle e^{-t}, 0, e^t \right\rangle$

$\mathbf{r}'(t) \cdot \mathbf{r}''(t) = -e^{-2t} + e^{2t}$

$|\mathbf{r}'(t)| = \sqrt{e^{-2t} + 4 + e^{2t}}$

$|\mathbf{r}'(t) \times \mathbf{r}''(t)| = \left|\left\langle 2e^t, 2, -2e^{-t} \right\rangle\right|$

$= \sqrt{4e^{2t} + 4 + 4e^{-2t}} = 2\sqrt{e^{2t} + 1 + e^{-2t}}$

$$a_T(t) = \frac{e^{2t} - e^{-2t}}{\sqrt{e^{2t} + 4 + e^{-2t}}}$$

$$a_N(t) = 2\sqrt{\frac{e^{2t} + 1 + e^{-2t}}{e^{2t} + 4 + e^{-2t}}}$$

39. $\mathbf{r}'(t) = (1 - t^2)\mathbf{i} - (1 + t^2)\mathbf{j} + \mathbf{k}$

$\mathbf{r}''(t) = -2t\mathbf{i} - 2t\mathbf{j}$

$\mathbf{r}'(t) \cdot \mathbf{r}''(t) = -2t(1 - t^2) + 2t(1 + t^2) = 4t^3$

$|\mathbf{r}'(t)| = \sqrt{(1 - t^2)^2 + (1 + t^2)^2 + 1} = \sqrt{2t^4 + 3}$

$|\mathbf{r}'(t) \times \mathbf{r}''(t)| = |2t\mathbf{i} - 2t\mathbf{j} - 4t\mathbf{k}|$

$= \sqrt{4t^2 + 4t^2 + 16t^2} = 2\sqrt{6}|t|$

$$a_T(t) = \frac{4t^3}{\sqrt{2t^4 + 3}}$$

$$a_N(t) = \frac{2\sqrt{6}|t|}{\sqrt{2t^4 + 3}}$$

41. $\mathbf{r}'(t) = \cot t\,\mathbf{i} - \tan t\,\mathbf{j} + \mathbf{k}$

$\mathbf{r}''(t) = -\csc^2 t\,\mathbf{i} - \sec^2 t\,\mathbf{j}$

$\mathbf{r}'(t) \cdot \mathbf{r}''(t) = -\frac{\cos t}{\sin^3 t} + \frac{\sin t}{\cos^3 t} = \frac{-\cos^4 t + \sin^4 t}{\sin^3 t \cos^3 t}$

$= \tan t \sec^2 t - \cot t \csc^2 t$

$|\mathbf{r}'(t)| = \sqrt{\cot^2 t + \tan^2 t + 1}$

$|\mathbf{r}'(t) \times \mathbf{r}''(t)| = |\sec^2 t\,\mathbf{i} - \csc^2 t\,\mathbf{j} - 2\csc t \sec t\,\mathbf{k}|$

$= \sqrt{\sec^4 t + \csc^4 t + 4\csc^2 t \sec^2 t}$

$$a_T(t) = \frac{\tan t \sec^2 t - \cot t \csc^2 t}{\sqrt{\cot^2 t + \tan^2 t + 1}}$$

$$a_N(t) = \frac{\sqrt{\sec^4 t + \csc^4 t + 4\csc^2 t \sec^2 t}}{\sqrt{\cot^2 t + \tan^2 t + 1}}$$

43. $\mathbf{r}(t) = \left\langle t, t^2, \frac{2}{3}t^3 \right\rangle$

$\mathbf{r}'(t) = \left\langle 1, 2t, 2t^2 \right\rangle$

$\mathbf{r}''(t) = \langle 0, 2, 4t \rangle$

$\mathbf{r}'(1) = \langle 1, 2, 2 \rangle$

$\mathbf{r}''(1) = \langle 0, 2, 4 \rangle$

$\mathbf{T}(1) = \frac{1}{\sqrt{1 + 4 + 4}}\langle 1, 2, 2 \rangle = \left\langle \frac{1}{3}, \frac{2}{3}, \frac{2}{3} \right\rangle$

$a_T(1) = \frac{1}{3}(0 + 4 + 8) = 4$

$a_N(1) = \frac{1}{3}|\langle 4, -4, 2 \rangle| = \frac{1}{3}\sqrt{16 + 16 + 4} = \frac{6}{3} = 2$

$\mathbf{N}(1) = \frac{1}{2}\left(\langle 0, 2, 4 \rangle - 4\left\langle \frac{1}{3}, \frac{2}{3}, \frac{2}{3} \right\rangle\right)$

$= \frac{1}{2}\left\langle -\frac{4}{3}, -\frac{2}{3}, \frac{4}{3} \right\rangle = \left\langle -\frac{2}{3}, -\frac{1}{3}, \frac{2}{3} \right\rangle$

$\mathbf{B}(1) = \left\langle \frac{1}{3}, \frac{2}{3}, \frac{2}{3} \right\rangle \times \left\langle -\frac{2}{3}, -\frac{1}{3}, \frac{2}{3} \right\rangle = \left\langle \frac{2}{3}, -\frac{2}{3}, \frac{1}{3} \right\rangle$

45. $\mathbf{r}'(t) = \cosh\dfrac{t}{c}\mathbf{i} + \mathbf{k}$

$\mathbf{r}''(t) = \dfrac{1}{c}\sinh\dfrac{t}{c}\mathbf{i}$

$\mathbf{r}'\left(\dfrac{\pi}{6}\right) = \cosh\dfrac{\pi}{6c}\mathbf{i} + \mathbf{k}$

$\mathbf{r}''\left(\dfrac{\pi}{6}\right) = \dfrac{1}{c}\sinh\dfrac{\pi}{6c}\mathbf{i}$

$\mathbf{T}\left(\dfrac{\pi}{6}\right) = \dfrac{1}{\sqrt{\cosh^2\frac{\pi}{6c}+1}}\left(\cosh\dfrac{\pi}{6c}\mathbf{i} + \mathbf{k}\right)$

$a_T\left(\dfrac{\pi}{6}\right) = \dfrac{\frac{1}{c}\cosh\frac{\pi}{6c}\sinh\frac{\pi}{6c}}{\sqrt{\cosh^2\frac{\pi}{6c}+1}}$

$a_N\left(\dfrac{\pi}{6}\right) = \dfrac{1}{\sqrt{\cosh^2\frac{\pi}{6c}+1}}\left|\dfrac{1}{c}\sinh\dfrac{\pi}{6c}\mathbf{j}\right| = \dfrac{\frac{1}{c}\sinh\frac{\pi}{6c}}{\sqrt{\cosh^2\frac{\pi}{6c}+1}}$

$\mathbf{N}\left(\dfrac{\pi}{6}\right) = \dfrac{\sqrt{\cosh^2\frac{\pi}{6c}+1}}{\frac{1}{c}\sinh\frac{\pi}{6c}}\left[\dfrac{1}{c}\sinh\dfrac{\pi}{6c}\mathbf{i} - \dfrac{\frac{1}{c}\cosh\frac{\pi}{6c}\sinh\frac{\pi}{6c}}{\cosh^2\frac{\pi}{6c}+1}\left(\cosh\dfrac{\pi}{6c}\mathbf{i}+\mathbf{k}\right)\right] = \dfrac{1}{\sqrt{\cosh^2\frac{\pi}{6c}+1}}\left(\mathbf{i} - \cosh\dfrac{\pi}{6c}\mathbf{k}\right)$

$\mathbf{B}\left(\dfrac{\pi}{6}\right) = \mathbf{T}\left(\dfrac{\pi}{6}\right) \times \mathbf{N}\left(\dfrac{\pi}{6}\right) = \mathbf{j}$

47. $\mathbf{r}'(t) = (\cos t - t\sin t)\mathbf{i} + (t\cos t + \sin t)\mathbf{j} + 2t\mathbf{k}$

$\mathbf{r}''(t) = (-t\cos t - 2\sin t)\mathbf{i} + (2\cos t - t\sin t)\mathbf{j} + 2\mathbf{k}$

$\mathbf{r}'(\pi) = -\mathbf{i} - \pi\mathbf{j} + 2\pi\mathbf{k}$

$\mathbf{r}''(\pi) = \pi\mathbf{i} - 2\mathbf{j} + 2\mathbf{k}$

$\mathbf{T}(\pi) = \dfrac{1}{\sqrt{1+5\pi^2}}(-\mathbf{i} - \pi\mathbf{j} + 2\pi\mathbf{k})$

$a_T(\pi) = \dfrac{1}{\sqrt{1+5\pi^2}}(2\pi - \pi + 4\pi) = \dfrac{5\pi}{\sqrt{1+5\pi^2}}$

$a_N(\pi) = \dfrac{1}{\sqrt{1+5\pi^2}}\left|2\pi\mathbf{i} + (2+2\pi^2)\mathbf{j} + (2+\pi^2)\mathbf{k}\right| = \sqrt{\dfrac{8+16\pi^2+5\pi^4}{1+5\pi^2}}$

$\mathbf{N}(\pi) = \sqrt{\dfrac{1+5\pi^2}{8+16\pi^2+5\pi^4}}\left[(\pi\mathbf{i} - 2\mathbf{j} + 2\mathbf{k}) - \dfrac{5\pi}{1+5\pi^2}(-\mathbf{i} - \pi\mathbf{j} + 2\pi\mathbf{k})\right]$

$= \dfrac{1}{\sqrt{(8+16\pi^2+5\pi^4)(1+5\pi^2)}}\left[(5\pi^3 + 6\pi)\mathbf{i} + (-2 - 5\pi^2)\mathbf{j} + 2\mathbf{k}\right]$

$\mathbf{B}(\pi) = \dfrac{1}{\sqrt{8+16\pi^2+5\pi^4}}\left(2\pi\mathbf{i} + (2+2\pi^2)\mathbf{j} + (2+\pi^2)\mathbf{k}\right)$

49. $\mathbf{r}'(t) = e^t\mathbf{i} + e^t(\cos t - \sin t)\mathbf{j} + e^t(\cos t + \sin t)\mathbf{k}$

$\mathbf{r}''(t) = e^t\mathbf{i} - 2e^t\sin t\mathbf{j} + 2e^t\cos t\mathbf{k}$

$\mathbf{r}'\left(\dfrac{\pi}{3}\right) = e^{\pi/3}\mathbf{i} + \dfrac{e^{\pi/3}}{2}\left(1 - \sqrt{3}\right)\mathbf{j} + \dfrac{e^{\pi/3}}{2}\left(1 + \sqrt{3}\right)\mathbf{k}$

$\mathbf{r}''\left(\dfrac{\pi}{3}\right) = e^{\pi/3}\mathbf{i} - \sqrt{3}e^{\pi/3}\mathbf{j} + e^{\pi/3}\mathbf{k}$

$$\mathbf{T}\left(\frac{\pi}{3}\right) = \frac{1}{\sqrt{3}e^{\pi/3}}\left[e^{\pi/3}\mathbf{i} + \frac{e^{\pi/3}}{2}\left(1-\sqrt{3}\right)\mathbf{j} + \frac{e^{\pi/3}}{2}\left(1+\sqrt{3}\right)\mathbf{k}\right] = \frac{1}{\sqrt{3}}\mathbf{i} + \frac{1}{2}\left(\frac{1}{\sqrt{3}}-1\right)\mathbf{j} + \frac{1}{2}\left(\frac{1}{\sqrt{3}}+1\right)\mathbf{k}$$

$$a_T\left(\frac{\pi}{3}\right) = \frac{1}{\sqrt{3}e^{\pi/3}}\left[e^{2\pi/3} - \frac{\sqrt{3}}{2}e^{2\pi/3}\left(1-\sqrt{3}\right) + \frac{e^{2\pi/3}}{2}\left(1+\sqrt{3}\right)\right] = \frac{1}{\sqrt{3}e^{\pi/3}}\left(3e^{2\pi/3}\right) = \sqrt{3}e^{\pi/3}$$

$$a_N\left(\frac{\pi}{3}\right) = \frac{1}{\sqrt{3}e^{\pi/3}}\left|2e^{2\pi/3}\mathbf{i} + \frac{e^{2\pi/3}}{2}\left(\sqrt{3}-1\right)\mathbf{j} + \frac{e^{2\pi/3}}{2}\left(-\sqrt{3}-1\right)\mathbf{k}\right| = \frac{1}{\sqrt{3}e^{\pi/3}}\left(\sqrt{6}e^{2\pi/3}\right) = \sqrt{2}e^{\pi/3}$$

$$\mathbf{N}\left(\frac{\pi}{3}\right) = \frac{1}{\sqrt{2}e^{\pi/3}}\left[e^{\pi/3}\left(\mathbf{i} - \sqrt{3}\mathbf{j} + \mathbf{k}\right) - \sqrt{3}e^{\pi/3}\left(\frac{1}{\sqrt{3}}\mathbf{i} + \frac{1}{2}\left(\frac{1}{\sqrt{3}}-1\right)\mathbf{j} + \frac{1}{2}\left(\frac{1}{\sqrt{3}}+1\right)\mathbf{k}\right)\right]$$

$$= \frac{1}{\sqrt{2}}\left(-\frac{1+\sqrt{3}}{2}\mathbf{j} + \frac{1-\sqrt{3}}{2}\mathbf{k}\right) = -\frac{1+\sqrt{3}}{2\sqrt{2}}\mathbf{j} + \frac{1-\sqrt{3}}{2\sqrt{2}}\mathbf{k}$$

$$\mathbf{B}\left(\frac{\pi}{3}\right) = \mathbf{T}\left(\frac{\pi}{3}\right) \times \mathbf{N}\left(\frac{\pi}{3}\right) = \sqrt{\frac{2}{3}}\mathbf{i} + \frac{1}{12}\left(3\sqrt{2}-\sqrt{6}\right)\mathbf{j} - \frac{1}{12}\left(3\sqrt{2}+\sqrt{6}\right)\mathbf{k}$$

51. $\mathbf{r}'(t) = \cosh t\,\mathbf{i} + \mathbf{j} + \sinh t\,\mathbf{k}$

$\mathbf{r}''(t) = \sinh t\,\mathbf{i} + \cosh t\,\mathbf{k}$

$$\mathbf{r}'\left(\frac{\pi}{3}\right) = \cosh\frac{\pi}{3}\mathbf{i} + \mathbf{j} + \sinh\frac{\pi}{3}\mathbf{k}$$

$$\mathbf{r}''\left(\frac{\pi}{3}\right) = \sinh\frac{\pi}{3}\mathbf{i} + \cosh\frac{\pi}{3}\mathbf{k}$$

$$\mathbf{T}\left(\frac{\pi}{3}\right) = \frac{1}{\sqrt{\cosh^2\frac{\pi}{3} + 1 + \sinh^2\frac{\pi}{3}}}\left(\cosh\frac{\pi}{3}\mathbf{i} + \mathbf{j} + \sinh\frac{\pi}{3}\mathbf{k}\right) = \frac{1}{\sqrt{2}\cosh\frac{\pi}{3}}\left(\cosh\frac{\pi}{3}\mathbf{i} + \mathbf{j} + \sinh\frac{\pi}{3}\mathbf{k}\right)$$

$$= \frac{1}{\sqrt{2}}\mathbf{i} + \frac{\operatorname{sech}\frac{\pi}{3}}{\sqrt{2}}\mathbf{j} + \frac{\tanh\frac{\pi}{3}}{\sqrt{2}}\mathbf{k}$$

$$a_T\left(\frac{\pi}{3}\right) = \frac{1}{\sqrt{2}\cosh\frac{\pi}{3}}\left(2\cosh\frac{\pi}{3}\sinh\frac{\pi}{3}\right) = \sqrt{2}\sinh\frac{\pi}{3}$$

$$a_N\left(\frac{\pi}{3}\right) = \frac{1}{\sqrt{2}\cosh\frac{\pi}{3}}\left|\cosh\frac{\pi}{3}\mathbf{i} - \mathbf{j} - \sinh\frac{\pi}{3}\mathbf{k}\right| = 1$$

$$\mathbf{N}\left(\frac{\pi}{3}\right) = \left(\sinh\frac{\pi}{3}\mathbf{i} + \cosh\frac{\pi}{3}\mathbf{k}\right) - \sqrt{2}\sinh\frac{\pi}{3}\left(\frac{1}{\sqrt{2}}\mathbf{i} + \frac{\operatorname{sech}\frac{\pi}{3}}{\sqrt{2}}\mathbf{j} + \frac{\tanh\frac{\pi}{3}}{\sqrt{2}}\mathbf{k}\right) = -\tanh\frac{\pi}{3}\mathbf{j} + \operatorname{sech}\frac{\pi}{3}\mathbf{k}$$

$$\mathbf{B}\left(\frac{\pi}{3}\right) = \mathbf{T}\left(\frac{\pi}{3}\right) \times \mathbf{N}\left(\frac{\pi}{3}\right) = \frac{1}{\sqrt{2}}\mathbf{i} - \frac{\operatorname{sech}\frac{\pi}{3}}{\sqrt{2}}\mathbf{j} - \frac{\tanh\frac{\pi}{3}}{\sqrt{2}}\mathbf{k}$$

53. a. The motion of the planet with respect to the sun can be given by $x = R_p\cos t$, $y = R_p\sin t$.

Assume that when $t = 0$, both the planet and the moon are on the x-axis.
Since the moon orbits the planet 10 times for every time the planet orbits the sun, the motion of the moon with respect to the planet can be given by $x = R_m\cos 10t$, $y = R_m\sin 10t$.
Combining these equations, the motion of the moon with respect to the sun is given by
$x = R_p\cos t + R_m\cos 10t$, $y = R_p\sin t + R_m\sin 10t$.

b. $x'(t) = -R_p\sin t - 10R_m\sin 10t$

$y'(t) = R_p\cos t + 10R_m\cos 10t$

The moon is motionless with respect to the sun when $x'(t)$ and $y'(t)$ are both 0.

Solve $x'(t) = 0$ for $\sin t$ and $y'(t) = 0$ for $\cos t$ to get $\sin t = -\dfrac{10R_m}{R_p}\sin 10t$, $\cos t = -\dfrac{10R_m}{R_p}\cos 10t$.

Since $\sin^2 t + \cos^2 t = 1$

$$1 = \frac{100R_m^2}{R_p^2}\sin^2 10t + \frac{100R_m^2}{R_p^2}\cos^2 10t$$

$$= \frac{100R_m^2}{R_p^2}.$$ Thus, $R_p^2 = 100R_m^2$ or $R_p = 10R_m$. Substitute this into $x'(t) = 0$ and $y'(t) = 0$ to get

$$-R_p(\sin t + \sin 10t) = 0 \text{ and } R_p(\cos t + \cos 10t) = 0.$$

If $0 \le t \le \dfrac{\pi}{2}$, then to have $\sin t + \sin 10t = 0$ and $\cos t + \cos 10t = 0$ it must be that

$$10t = \pi + t \text{ or } t = \frac{\pi}{9}.$$

Thus, when the radius of the planet's orbit around the sun is ten times the radius of the moon's orbit around the planet and $t = \dfrac{\pi}{9}$, the moon is motionless with respect to the sun.

55. a. Winding upward around the right circular cylinder $x = \sin t$, $y = \cos t$ as t increases.

b. Same as part a, but winding faster/slower by a factor of $3t^2$.

c. With standard orientation of the axes, the motion is winding to the right around the right circular cylinder $x = \sin t$, $z = \cos t$.

d. Spiraling upward, with increasing radius, along the spiral $x = t \sin t$, $y = t \cos t$.

e. Spiraling upward, with decreasing radius, along the spiral $x = \dfrac{1}{t^2}\sin t$, $y = \dfrac{1}{t^2}\cos t$.

f. Spiraling to the right, with increasing radius, along the spiral $x = t^2 \sin(\ln t)$,

$z = t^2 \cos(\ln t)$.

57. Let

$P_5(x) = a_0 + a_1 x + a_2 x^2 + a_3 x^3 + a_4 x^4 + a_5 x^5$.

$P_5(0) = 0 \Rightarrow a_0 = 0$

$P_5'(x) = a_1 + 2a_2 x + 3a_3 x^2 + 4a_4 x^3 + 5a_5 x^4$, so

$P_5'(0) = 0 \Rightarrow a_1 = 0$.

$P''(x) = 2a_2 + 6a_3 x + 12a_4 x^2 + 20a_5 x^3$, so

$P_5''(0) = 0 \Rightarrow a_2 = 0$.

Thus, $P_5(x) = a_3 x^3 + a_4 x^4 + a_5 x^5$,

$P_5'(x) = 3a_3 x^2 + 4a_4 x^3 + 5a_5 x^4$, and

$P_5''(x) = 6a_3 x + 12a_4 x^2 + 20a_5 x^3$.

$P_5(1) = 1$, $P_5'(1) = 0$, and $P_5''(1) = 0 \Rightarrow$

$a_3 + a_4 + a_5 = 1$

$3a_3 + 4a_4 + 5a_5 = 0$

$6a_3 + 12a_4 + 20a_5 = 0$

The simultaneous solution to these equations is

$a_3 = 10$, $a_4 = -15$, $a_5 = 6$, so

$P_5(x) = 10x^3 - 15x^4 + 6x^5$.

59. $\dfrac{d\mathbf{B}}{ds} = \dfrac{d}{ds}(\mathbf{T} \times \mathbf{N}) = \dfrac{d\mathbf{T}}{ds} \times \mathbf{N} + \mathbf{T} \times \dfrac{d\mathbf{N}}{ds}$

Since $\mathbf{N} = \dfrac{\frac{d\mathbf{T}}{ds}}{\left|\frac{d\mathbf{T}}{ds}\right|}$, $\dfrac{d\mathbf{T}}{ds} \times \mathbf{N} = 0$, so $\dfrac{d\mathbf{B}}{ds} = \mathbf{T} \times \dfrac{d\mathbf{N}}{ds}$.

Thus $\mathbf{T} \cdot \dfrac{d\mathbf{B}}{ds} = \mathbf{T} \cdot \left(\mathbf{T} \times \dfrac{d\mathbf{N}}{ds}\right) = (\mathbf{T} \times \mathbf{T}) \cdot \dfrac{d\mathbf{N}}{ds} = 0$, so

$\dfrac{d\mathbf{B}}{ds}$ is perpendicular to $\mathbf{T}$.

61. Let $ax + by + cz + d = 0$ be the equation of the plane containing the curve. Since $\mathbf{T}$ and $\mathbf{N}$ lie in the plane $\mathbf{B} = \pm \dfrac{a\mathbf{i} + b\mathbf{j} + c\mathbf{k}}{\sqrt{a^2 + b^2 + c^2}}$. Thus, $\mathbf{B}$ is a constant vector and $\dfrac{d\mathbf{B}}{ds} = \mathbf{0}$, so $\tau(s) = 0$, since $\mathbf{N}$ will not necessarily be 0 everywhere.

63. $\mathbf{r}(t) = 6\cos \pi t\mathbf{i} + 6\sin \pi t\mathbf{j}, + 2t\mathbf{k},\ t > 0$

Let $(6\cos \pi t)^2 + (6\sin \pi t)^2 + (2t)^2 = 100$.

Then $36(\cos^2 \pi t + \sin^2 \pi t) + 4t^2 = 100;\ 4t^2 = 64;\ t = 4$.

$\mathbf{r}(4) = 6\mathbf{i} + 8\mathbf{k}$, so the fly will hit the sphere at the point $(6, 0, 8)$.

$\mathbf{r}'(t) = -6\pi\sin \pi t\mathbf{i} + 6\pi\cos \pi t\mathbf{j} + 2\mathbf{k}$, so the fly will have traveled

$$\int_0^4 \sqrt{(-6\pi\sin \pi t)^2 + (6\pi\cos \pi t)^2 + (2)^2}\,dt = \int_0^4 \sqrt{36\pi^2 + 4}\,dt = \sqrt{36\pi^2 + 4}(4-0)$$

$$= 8\sqrt{9\pi^2 + 1} \approx 75.8214$$

65. a. $\mathbf{r}(t) = $ constant, so $\mathbf{r} = \mathbf{r}_0$.

b. $\mathbf{r}(t) = \mathbf{c}t + \mathbf{C}$
$\mathbf{r}(0) = \mathbf{r}_0 \Rightarrow \mathbf{r}(t) = \mathbf{c}t + \mathbf{r}_0$

c. $\mathbf{r}'(t) = \mathbf{c}t + \mathbf{C}_1$
$\mathbf{r}'(0) = \mathbf{v}_0 \Rightarrow \mathbf{r}'(t) = \mathbf{c}t + \mathbf{v}_0$

$\mathbf{r}(t) = \dfrac{1}{2}\mathbf{c}t^2 + \mathbf{v}_0 t + \mathbf{C}_2$

$\mathbf{r}(0) = \mathbf{r}_0 \Rightarrow \mathbf{r}(t) = \dfrac{1}{2}\mathbf{c}t^2 + \mathbf{v}_0 t + \mathbf{r}_0$

d. Say $\mathbf{r}(t) = x(t)\mathbf{i} + y(t)\mathbf{j} + z(t)\mathbf{k}$.

Then $\dfrac{d\mathbf{r}}{dt} = c\mathbf{r}$ is equivalent to $\dfrac{dx}{dt} = cx$,

$\dfrac{dy}{dt} = cy,\ \dfrac{dz}{dt} = cz$.

These equations have the solutions

$x(t) = x_0 e^{ct},\ y(t) = y_0 e^{ct},\ z(t) = z_0 e^{ct}$.

Hence, $\mathbf{r}(t) = \mathbf{r}_0 e^{ct}$.

67. It left the path at the point $\mathbf{r}(4\pi) = \langle 4\pi, 0, 4\pi\rangle$;

and $\mathbf{r}'(4\pi) = \langle 1,\ 4\pi,\ 1\rangle$.

Hence, parametric equations of the tangent line are:

$x = 4\pi + t,\ y = 4\pi t,\ z = 4\pi + t$.

Then the point of intersection of the tangent line and the plane $x + y = 30$

occurs where $(4\pi + t) + (4\pi t) = 30$;

$t = \dfrac{30 - 4\pi}{4\pi + 1}$.

The bee hits the plane at the point with coordinates:

$x = 4\pi + \dfrac{30 - 4\pi}{4\pi + 1} = \dfrac{2(8\pi^2 + 15)}{4\pi + 1} \approx 13.85$,

$y = 4\pi\left[\dfrac{30 - 4\pi}{4\pi + 1}\right] = \dfrac{8\pi(15 - 2\pi)}{4\pi + 1} \approx 16.15,\ z \approx 13.85$

(same as x).

69. $\dfrac{d}{dt}[\mathbf{r}(t) \times \mathbf{r}'(t)] = \mathbf{r}(t) \times \mathbf{r}''(t)$

$= \mathbf{r}(t) \times c\mathbf{r}(t) = c[\mathbf{r}(t) \times \mathbf{r}(t)] = 0$

Hence, $\mathbf{r}(t) \times \mathbf{r}'(t) = \mathbf{c}$, a constant vector, with $\mathbf{r}'(t)$ perpendicular to $\mathbf{c}$ for all t. Thus, the motion is in a plane with $\mathbf{c}$ as a normal.

71. $\mathbf{r} = r\cos \theta\mathbf{i} + r\sin \theta\mathbf{j}$ where $r = r(t)$ and $\theta = \theta(t)$ are functions of t.

$\mathbf{v}(t) = \mathbf{r}'(t) = (r'\cos \theta - r\sin \theta\theta')\mathbf{i} + (r'\sin \theta + r\cos \theta\theta')\mathbf{j}$

Thus $\mathbf{L}(t) = m\mathbf{r}(t) \times \mathbf{v}(t) = \begin{vmatrix} \mathbf{i} & \mathbf{j} & \mathbf{k} \\ mr\cos \theta & mr\sin \theta & 0 \\ r'\cos \theta - r\sin \theta\theta' & r'\sin \theta + r\cos \theta\theta' & 0 \end{vmatrix}$

$= 0\mathbf{i} - 0\mathbf{j} + [mr\cos \theta(r'\sin \theta + r\cos \theta\theta') - mr\sin \theta(r'\cos \theta - r\sin \theta\theta')]\mathbf{k}$

$= (mr^2\cos^2 \theta\theta' + mr^2\sin^2 \theta\theta')\mathbf{k} = mr^2\theta'\mathbf{k} = mr^2\dfrac{d\theta}{dt}\mathbf{k}$

73. $\mathbf{B} \times \mathbf{T} = (\mathbf{T} \times \mathbf{N}) \times \mathbf{T} = -[\mathbf{T} \times (\mathbf{T} \times \mathbf{N})] = -[(\mathbf{T} \cdot \mathbf{N})\mathbf{T} - (\mathbf{T} \cdot \mathbf{T})\mathbf{N}] = \mathbf{N}$

since $\mathbf{T}$ and $\mathbf{N}$ are perpendicular unit vectors, hence $\mathbf{T} \cdot \mathbf{N} = 0$ and $\mathbf{T} \cdot \mathbf{T} = |\mathbf{T}|^2 = 1^2 = 1$.

14.6 Concepts Review

1. traces; cross sections

3. ellipsoid

Problem Set 14.6

1. $\dfrac{x^2}{36} + \dfrac{y^2}{4} = 1$

Elliptic cylinder

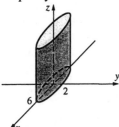

3. Plane

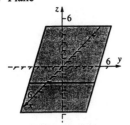

5. $(x-4)^2 + (y+2)^2 = 7$

Circular cylinder

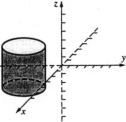

7. $\dfrac{x^2}{441} + \dfrac{y^2}{196} + \dfrac{z^2}{36} = 1$

Ellipsoid

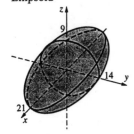

9. $z = \dfrac{x^2}{8} + \dfrac{y^2}{2}$

Elliptic paraboloid

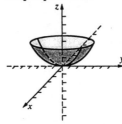

11. Cylinder

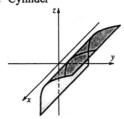

13. Hyperbolic paraboloid

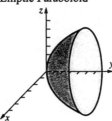

15. $y = \dfrac{x^2}{4} + \dfrac{z^2}{9}$

Elliptic Paraboloid

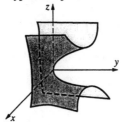

17. Plane

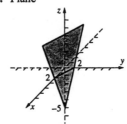

19. Hemisphere

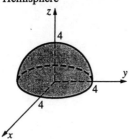

21. **a.** Replacing x by $-x$ results in an equivalent equation.

    b. Replacing x by $-x$ and y by $-y$ results in an equivalent equation.

    c. Replacing y by $-y$ and z by $-z$ results in an equivalent equation.

    d. Replacing x by $-x$, y by $-y$, and z by $-z$, results in an equivalent equation.

23. At $y = x$, the revolution generates a circle of radius $x = \sqrt{\dfrac{y}{2}} = \sqrt{\dfrac{k}{2}}$. Thus, the cross section in the plane $y = k$ is the circle $x^2 + z^2 = \dfrac{k}{2}$ or $2x^2 + 2z^2 = k$. The equation of the surface is $y = 2x^2 + 2z^2$.

25. At $y = k$, the revolution generates a circle of radius $x = \sqrt{3 - \dfrac{3}{4}y^2} = \sqrt{3 - \dfrac{3}{4}k^2}$. Thus, the cross section in the plane $y = k$ is the circle $x^2 + z^2 = 3 - \dfrac{3}{4}k^2$ or $12 - 4x^2 - 4z^2 = 3k^2$. The equation of the surface is $4x^2 + 3y^2 + 4z^2 = 12$.

27. When $z = 4$ the equation is $4 = \dfrac{x^2}{4} + \dfrac{y^2}{9}$ or $1 = \dfrac{x^2}{16} + \dfrac{y^2}{36}$, so $a^2 = 36, b^2 = 16$, and $c^2 = a^2 - b^2 = 20$, hence $c = \pm 2\sqrt{5}$. The major axis of the ellipse is on the y-axis so the foci are at $\left(0, \pm 2\sqrt{5}, 4\right)$.

29. When $z = h$, the equation is $\dfrac{x^2}{a^2} + \dfrac{y^2}{b^2} + \dfrac{h^2}{c^2} = 1$ or $\dfrac{x^2}{a^2} + \dfrac{y^2}{b^2} = \dfrac{c^2 - h^2}{c^2}$ which is equivalent to $\dfrac{x^2}{\frac{a^2(c^2 - h^2)}{c^2}} + \dfrac{y^2}{\frac{b^2(c^2 - h^2)}{c^2}} = 1$, which is $\dfrac{x^2}{A^2} + \dfrac{y^2}{B^2} = 1$ with $A = \dfrac{a}{c}\sqrt{c^2 - h^2}$ and $B = \dfrac{b}{c}\sqrt{c^2 - h^2}$. Thus, the area is

$$\pi\left(\dfrac{a}{c}\sqrt{c^2 - h^2}\right)\left(\dfrac{b}{c}\sqrt{c^2 - h^2}\right) = \dfrac{\pi ab(c^2 - h^2)}{c^2}$$

31. Equating the expressions for y, $4 - x^2 = x^2 + z^2$ or $1 = \dfrac{x^2}{2} + \dfrac{z^2}{4}$ which is the equation of an ellipse in the xz-plane with major diameter of $2\sqrt{4} = 4$ and minor diameter $2\sqrt{2}$.

33. $(t\cos t)^2 + (t\sin t)^2 - t^2 = t^2(\cos^2 t + \sin^2 t) - t^2$
$= t^2 - t^2 = 0$, hence every point on the spiral is on the cone.
For $\mathbf{r} = 3t\cos t\,\mathbf{i} + t\sin t\,\mathbf{j} + t\mathbf{k}$, every point satisfies $x^2 + 9y^2 - 9z^2 = 0$ so the spiral lies on the elliptical cone.

14.7 Concepts Review

1. circular cylinder; sphere

3. $\rho^2 = r^2 + z^2$

Problem Set 14.7

1. Cylindrical to Spherical:
$$\rho = \sqrt{r^2 + z^2}$$
$$\cos\phi = \dfrac{z}{\sqrt{r^2 + z^2}}$$
$$\theta = \theta$$

Spherical to Cylindrical:
$$r = \rho\sin\phi$$
$$z = \rho\cos\phi$$
$$\theta = \theta$$

3. **a.** $x = 6\cos\left(\dfrac{\pi}{6}\right) = 3\sqrt{3}$
$$y = 6\sin\left(\dfrac{\pi}{6}\right) = 3$$
$$z = -2$$

b. $x = 4\cos\left(\dfrac{4\pi}{3}\right) = -2$

$y = 4\sin\left(\dfrac{4\pi}{3}\right) = -2\sqrt{3}$

$z = -8$

5. a. $\rho = \sqrt{x^2 + y^2 + z^2} = \sqrt{4 + 12 + 16} = 4\sqrt{2}$

$\tan\theta = \dfrac{y}{x} = \dfrac{-2\sqrt{3}}{2} = -\sqrt{3}$ and (x, y) is in the

4th quadrant so $\theta = \dfrac{5\pi}{3}$.

$\cos\phi = \dfrac{z}{\rho} = \dfrac{4}{4\sqrt{2}} = \dfrac{\sqrt{2}}{2}$ so $\phi = \dfrac{\pi}{4}$.

Spherical: $\left(4\sqrt{2},\ \dfrac{5\pi}{3},\ \dfrac{\pi}{4}\right)$

b. $\rho = \sqrt{2 + 2 + 12} = 4$

$\tan\theta = \dfrac{\sqrt{2}}{-\sqrt{2}} = -1$ and (x, y) is in the 2nd

quadrant so $\theta = \dfrac{3\pi}{4}$.

$\cos\phi = \dfrac{2\sqrt{3}}{4} = \dfrac{\sqrt{3}}{2}$ so $\dfrac{\pi}{6}$.

Spherical: $\left(4,\ \dfrac{3\pi}{4},\ \dfrac{\pi}{6}\right)$

7. $r = 5$

Cylinder

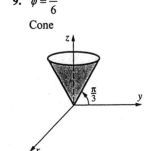

9. $\phi = \dfrac{\pi}{6}$

Cone

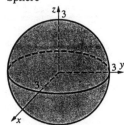

11. $r = 3\cos\theta$

Circular cylinder

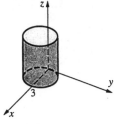

13. $\rho = 3\cos\phi$

$x^2 + y^2 + \left(z - \dfrac{3}{2}\right)^2 = \dfrac{9}{4}$

Sphere

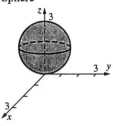

15. $r^2 + z^2 = 9$

$x^2 + y^2 + z^2 = 9$

Sphere

17. $x^2 + y^2 = 9;\ r^2 = 9;\ r = 3$

19. $r^2 + 4z^2 = 10$

21. $(x^2 + y^2 + z^2) - 3z^2 = 0;\ \rho^2 - 3\rho^2\cos^2\phi = 0;$

$\cos^2\phi = \dfrac{1}{3}$ (pole is not lost); $\cos^2\phi = \dfrac{1}{3}$ (or

$\sin^2\phi = \dfrac{2}{3}$ or $\tan^2\phi = 2$)

23. $(r^2 + z^2) + z^2 = 4;\ \rho^2 + \rho^2\cos^2\phi = 4;$

$\rho^2 = \dfrac{4}{1 + \cos^2\phi}$

25. $r\cos\theta + r\sin\theta = 4;\ r = \dfrac{4}{\sin\theta + \cos\theta}$

27. $(x^2+y^2+z^2)-z^2=9;$ $\rho^2-\rho^2\cos^2\phi=9;$
$\rho^2(1-\cos^2\phi)=9;$ $\rho^2\sin^2\phi=9$

29. $r^2\cos 2\theta=z;$ $r^2(\cos^2\theta-\sin^2\theta)=z;$
$(r\cos\theta)^2-(r\sin\theta)^2=z;$ $x^2-y^2=z$

31. $z=2x^2+2y^2=2(x^2+y^2)$ (Cartesian); $z=2r^2$
(cylindrical)

33. For St. Paul:
$\rho=3960,$
$\theta=360°-93.1°=266.9°\approx 4.6583$ rad

$\phi=90°-45°=45°=\dfrac{\pi}{4}$ rad

$x=3960\sin\dfrac{\pi}{4}\cos 4.6583\approx -151.4$

$y=3960\sin\dfrac{\pi}{4}\sin 4.6583\approx -2796.0$

$z=3960\cos\dfrac{\pi}{4}\approx 2800.1$

For Oslo:
$\rho=3960,$ $\theta=10.5°\approx 0.1833$ rad,
$\phi=90°-59.6°=30.4°\approx 0.5306$ rad
$x=3960\sin 0.5306\cos 0.1833\approx 1970.4$
$y=3960\sin 0.5306\sin 0.1833\approx 365.3$
$z=3960\cos 0.5306\approx 3415.5$
As in Example 7,

$$\cos\gamma\approx\frac{(-151.4)(1970.4)+(-2796.0)(365.3)+(2800.1)(3415.5)}{3960^2}\approx 0.5257$$

so $\gamma\approx 1.0173$ and the great-circle distance is
$d\approx 3960(1.0173)\approx 4029$ mi

35. From Problem 33, the coordinates of St. Paul are $P(-151.4,-2796.0,2800.1)$.
For Turin:

$\rho=3960,$ $\theta=7.4°\approx 0.1292$ rad, $\phi=\dfrac{\pi}{4}$ rad

$x=3960\sin\dfrac{\pi}{4}\cos 0.1292\approx 2776.8$

$y=3960\sin\dfrac{\pi}{4}\sin 0.1292\approx 360.8$

$z=3960\cos\dfrac{\pi}{4}\approx 2800.1$

$$\cos\gamma\approx\frac{(-151.4)(2776.8)+(-2796.0)(360.8)+(2800.1)(2800.1)}{3960^2}\approx 0.4088$$

so $\gamma\approx 1.1497$ and the great-circle distance is $d\approx 3960(1.1497)\approx 4553$ mi

37. Let St. Paul be at $P_1(-151.4,-2796.0,2800.1)$ and Turin be at $P_2(2776.8,360.8,2800.1)$ and O be the center of

the earth. Let β be the angle between the z-axis and the plane determined by O, P_1, and P_2. $\overrightarrow{OP_1}\times\overrightarrow{OP_2}$ is normal

to the plane. The angle between the z-axis and $\overrightarrow{OP_1}\times\overrightarrow{OP_2}$ is complementary to β. Hence

$$\beta = \frac{\pi}{2} - \cos^{-1}\left(\frac{\left(\overrightarrow{OP_1} \times \overrightarrow{OP_2}\right) \cdot \mathbf{k}}{\left|\overrightarrow{OP_1} \times \overrightarrow{OP_2}\right|\left|\mathbf{k}\right|}\right) \approx \frac{\pi}{2} - \cos^{-1}\left(\frac{7.709 \times 10^6}{1.431 \times 10^7}\right) \approx 0.5689 .$$

The distance between the North Pole and the St. Paul-Turin great-circle is $3960(0.5689) \approx 2253$ mi

39. Let P_1 be (a_1, θ_1, ϕ_1) and P_2 be (a_2, θ_2, ϕ_2). If γ is the angle between $\overrightarrow{OP_1}$ and $\overrightarrow{OP_2}$ then the great-circle distance between P_1 and P_2 is $a\gamma$. $|OP_1| = |OP_2| = a$ while the straight-line distance between P_1 and P_2 is (from Problem 38) $d^2 = (a-a)^2 + 2a^2[1 - \cos(\theta_1 - \theta_2)\sin\phi_1 \sin\phi_2 - \cos\phi_1 \cos\phi_2]$
$= 2a^2\{1 - [\cos(\theta_1 - \theta_2)\sin\phi_1 \sin\phi_2 + \cos\phi_1 \cos\phi_2]\}.$
Using the Law Of Cosines on the triangle OP_1P_2,
$d^2 = a^2 + a^2 - 2a^2 \cos\gamma = 2a^2(1 - \cos\gamma).$
Thus, γ is the central angle and $\cos\gamma = \cos(\theta_1 - \theta_2)\sin\phi_1 \sin\phi_2 + \cos\phi_1 \cos\phi_2$.

41. a. New York $(-74°, 40.4°)$; Greenwich $(0°, 51.3°)$
$\cos\gamma = \cos(-74° - 0°)\cos(40.4°)\cos(51.3°) + \sin(40.4°)\sin(51.3°) \approx 0.637$
Then $\gamma \approx 0.880$ rad, so $d \approx 3960(0.8801) \approx 3485$ mi.

b. St. Paul $(-93.1°, 45°)$; Turin $(7.4°, 45°)$
$\cos\gamma = \cos(-93.1° - 7.4°) \cos(45°) \cos(45°) + \sin(45°) \sin(45°) \approx 0.4089$
Then $\gamma \approx 1.495$ rad, so $d \approx 3960(1.1495) \approx 4552$ mi.

c. South Pole $(7.4°, -90°)$; Turin $(7.4°, 45°)$
Note that any value of α can be used for the poles.
$\cos\gamma = \cos 0° \cos(-90°) \cos(45°) + \sin(-90°)\sin(45°) = -\frac{1}{\sqrt{2}}$
thus $\gamma = 135° = \frac{3\pi}{4}$ rad, so $d = 3960\left(\frac{3\pi}{4}\right) \approx 9331$ mi.

d. New York $(-74°, 40.4°)$; Cape Town $(18.4°, -33.9°)$
$\cos\gamma = \cos(-74° - 18.4°) \cos(40.4°) \cos(-33.9°) + \sin(40.4°) \sin(-33.9°) \approx -0.3880$
Then $\gamma \approx 1.9693$ rad, so $d \approx 3960(1.9693) \approx 7798$ mi.

e. For these points $\alpha_1 = 100°$ and $\alpha_2 = -80°$ while $\beta_1 = \beta_2 = 0$, hence $\cos\gamma = \cos 180°$ and $\gamma = \pi$ rad, so $d = 3960\pi \approx 12,441$ mi.

14.8 Chapter Review

Concepts Test

1. True: The coordinates are defined in terms of distances from the coordinate planes in such a way that they are unique.

3. True: See Section 14.2.

5. False: The distance between $(0, 0, 3)$ and $(0, 0, -3)$ (a point from each plane) is 6, so the distance between the planes is less than or equal to 6 units.

7. True: Let $t = \frac{1}{2}$.

9. True: $\|u\|u\| = |u|\|u\| = |u|^2$

11. True: $|u \times v| = |-v \times u| = |-1|\|v \times u\| = |v \times u|$

13. False: Obviously not true if $u = v$. (More generally, it is only true when u and v are also perpendicular.)

15. True: $\dfrac{|\mathbf{u}\times\mathbf{v}|}{(\mathbf{u}\cdot\mathbf{v})}=\dfrac{|\mathbf{u}||\mathbf{v}|\sin\theta}{|\mathbf{u}||\mathbf{v}|\cos\theta}=\tan\theta$

17. True: $|(2\mathbf{i}\times2\mathbf{j})\cdot(\mathbf{j}\times\mathbf{i})|=4|(\mathbf{k})\cdot(-\mathbf{k})|$
$=4(\mathbf{k}\cdot\mathbf{k})=4$

19. True: Since $\langle b_1,\,b_2,\,b_3\rangle$ is normal to the plane.

21. True: $\kappa=\dfrac{|\mathbf{r}'\times\mathbf{r}''|}{|\mathbf{r}'|^3}=0\Rightarrow|\mathbf{r}'\times\mathbf{r}''|$
$=|\mathbf{r}'||\mathbf{r}''|\sin\theta=0$. Thus, either $\mathbf{r}'$ and $\mathbf{r}''$ are parallel or either $\mathbf{r}'$ or $\mathbf{r}''$ is $\mathbf{0}$, which implies that the path is a straight line.

23. False: κ depends only on the shape of the curve.

25. True: $\mathbf{T}$ depends only upon the shape of the curve, hence $\mathbf{N}$ and $\mathbf{B}$ also.

27. False: If $\mathbf{r}(t)=a\cos t\mathbf{i}+a\sin t\mathbf{j}+ct\mathbf{k}$ then $\mathbf{v}$ is perpendicular to $\mathbf{a}$, but the path of motion is a circular helix, not a circle.

29. True: The curves are identical, although the motion of an object moving along the curves would be different.

31. True: The parametrization affects only the rate at which the curve is traced out.

33. False: For $\mathbf{r}(t)=\cos t\mathbf{i}+\sin t\,\mathbf{j}$, $|\mathbf{r}(t)|=1$, but $\mathbf{r}'(t)\neq\mathbf{0}$.

35. False: It is the part of the z-axis with $z\geq0$.

37. False: The origin, $\rho=0$, has infinitely many spherical coordinates, since any value of θ and ϕ can be used.

Sample Test Problems

1. The center of the sphere is the midpoint $\left(\dfrac{-2+4}{2},\dfrac{3+1}{2},\dfrac{3+5}{2}\right)=(1,\,2,\,4)$ of the diameter.
The radius is $r=\sqrt{(1+2)^2+(2-3)^2+(4-3)^2}$
$=\sqrt{9+1+1}=\sqrt{11}$. The equation of the sphere is $(x-1)^2+(y-2)^2+(z-4)^2=11$

3.

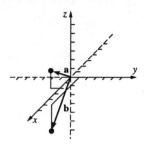

a. $|\mathbf{a}|=\sqrt{4+1+4}=3;\,|\mathbf{b}|=\sqrt{25+1+9}=\sqrt{35}$

b. $\dfrac{\mathbf{a}}{|\mathbf{a}|}=\dfrac{2}{3}\mathbf{i}-\dfrac{1}{3}\mathbf{j}+\dfrac{2}{3}\mathbf{k}$, direction cosines $\dfrac{2}{3},-\dfrac{1}{3}$, and $\dfrac{2}{3}$.
$\dfrac{\mathbf{b}}{|\mathbf{b}|}=\dfrac{5}{\sqrt{35}}\mathbf{i}+\dfrac{1}{\sqrt{35}}\mathbf{j}-\dfrac{3}{\sqrt{35}}\mathbf{k}$, direction cosines $\dfrac{5}{\sqrt{35}},\dfrac{1}{\sqrt{35}},-\dfrac{3}{\sqrt{35}}$

c. $\dfrac{2}{3}\mathbf{i}-\dfrac{1}{3}\mathbf{j}+\dfrac{2}{3}\mathbf{k}$

d. $\cos\theta=\dfrac{\mathbf{a}\cdot\mathbf{b}}{|\mathbf{a}||\mathbf{b}|}=\dfrac{10-1-6}{3\sqrt{35}}=\dfrac{3}{3\sqrt{35}}=\dfrac{1}{\sqrt{35}}$
$\theta=\cos^{-1}\dfrac{1}{\sqrt{35}}\approx1.4010\approx80.27°$

5. $c\langle3,3,-1\rangle\times\langle-1,-2,4\rangle=c\langle10,-11,-3\rangle$ for any c in R.

7. **a.** $y=7$, since y must be a constant.

b. $x=-5$, since it is parallel to the yz-plane.

c. $z=-2$, since it is parallel to the xy-plane.

d. $3x-4y+z=-45$, since it can be expressed as $3x-4y+z=D$ and D must satisfy $3(-5)-4(7)+(-2)=D$, so $D=-45$.

9. If the planes are perpendicular, their normals will also be perpendicular. Thus $0=\langle1,\,5,\,C\rangle\cdot\langle4,\,-1,\,1\rangle=4-5+C$, so $C=1$.

11. A vector in the direction of the line is $\langle8,1,-8\rangle$. Parametric equations are $x=-2+8t,\,y=1+t$, $z=5-8t$.

13. $(0,\,25,\,16)$ and $(-50,\,0,\,16)$ are on the line, so $\langle50,25,0\rangle=25\langle2,1,0\rangle$ is in the direction of the line. Parametric equations are $x=0+2t$,

$y = 25 + 1t, z = 16 + 0t$ or $x = 2t, y = 25 + t,$
$z = 16.$

15. $\langle 5, -4, -3 \rangle$ is a vector in the direction of the line,
and $\langle 2, -2, 1 \rangle$ is a position vector to the line.
Then a vector equation of the line is
$\mathbf{r}(t) = \langle 2, -2, 1 \rangle + t \langle 5, -4, -3 \rangle$.

17. $\mathbf{r}'(t) = \left\langle 1, t, t^2 \right\rangle, \mathbf{r}'(2) = \langle 1, 2, 4 \rangle$ and
$\mathbf{r}(2) = \left\langle 2, 2, \dfrac{8}{3} \right\rangle$. Symmetric equations for the

tangent line are $\dfrac{x-2}{1} = \dfrac{y-2}{2} = \dfrac{z - \frac{8}{3}}{4}$. Normal

plane is $1(x-2) + 2(y-2) + 4\left(z - \dfrac{8}{3}\right) = 0$ or

$3x + 6y + 12z = 50.$

19. $\mathbf{r}'(t) = e^t \langle \cos t + \sin t, -\sin t + \cos t, 1 \rangle$
$|\mathbf{r}'(t)| = \sqrt{3}e^t$
Length is
$\int_1^5 \sqrt{3}e^t \, dt = \left[\sqrt{3}e^t \right]_1^5 = \sqrt{3}(e^5 - e) \approx 252.3509.$

21. $\mathbf{v}(t) = \left\langle 1, 2t, 3t^2 \right\rangle; \mathbf{a}(t) = \langle 0, 2, 6t \rangle$
$\mathbf{v}(1) = \langle 1, 2, 3 \rangle; |\mathbf{v}(1)| = \sqrt{14}; \mathbf{a}(1) = \langle 0, 2, 6 \rangle$
$a_T = \dfrac{\mathbf{v} \cdot \mathbf{a}}{|\mathbf{v}|} = \dfrac{0 + 4 + 18}{\sqrt{14}} = \dfrac{22}{\sqrt{14}} \approx 5.880;$
$a_N = \dfrac{|\mathbf{v} \times \mathbf{a}|}{|\mathbf{v}|} = \dfrac{|\langle 6, -6, 2 \rangle|}{\sqrt{14}} = \dfrac{2\sqrt{19}}{\sqrt{14}} \approx 2.330$

23. Sphere

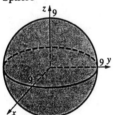

25. Circular paraboloid

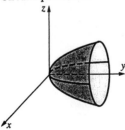

27. Plane

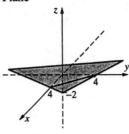

29. $\dfrac{x^2}{12} + \dfrac{y^2}{9} + \dfrac{z^2}{4} = 1$
Ellipsoid

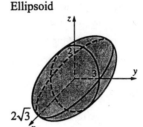

31. a. $r^2 = 9; r = 3$

b. $(x^2 + y^2) + 3y^2 = 16$
$r^2 + 3r^2 \sin^2 \theta = 16, r^2 = \dfrac{16}{1 + 3\sin^2 \theta}$

c. $r^2 = 9z$

d. $r^2 + 4z^2 = 10$

33. a. $\rho^2 = 4; \rho = 2$

b. $x^2 + y^2 + z^2 - 2z^2 = 0;$
$\rho^2 - 2\rho^2 \cos^2 \phi = 0; \rho^2(1 - 2\cos^2 \phi) = 0;$
$1 - 2\cos^2 \phi = 0; \cos^2 \phi = \dfrac{1}{2}; \phi = \dfrac{\pi}{4}$ or
$\phi = \dfrac{3\pi}{4}.$
Any of the following (as well as others)
would be acceptable:
$\left(\phi - \dfrac{\pi}{4}\right)\left(\phi - \dfrac{3\pi}{4}\right) = 0$
$\cos^2 \phi = \dfrac{1}{2}$
$\sec^2 \phi = 2$
$\tan^2 \phi = 1$

c. $2x^2 - (x^2 + y^2 + z^2) = 1$;

$2\rho^2 \sin^2 \phi \cos^2 \theta - \rho^2 = 1$;

$$\rho^2 = \frac{1}{2\sin^2 \phi \cos^2 \theta - 1}$$

d. $x^2 + y^2 = z$;

$\rho^2 \sin^2 \phi \cos^2 \theta + \rho^2 \sin^2 \phi \sin^2 \theta = \rho \cos \phi$

$\rho^2 \sin^2 \phi (\cos^2 \theta + \sin^2 \theta) = \rho \cos \phi$;

$\rho \sin^2 \phi = \cos \phi$; $\rho = \cot \phi \csc \phi$

(Note that when we divided through by ρ in part c and d we did not lose the pole since it is also a solution of the resulting equations.)

35. (2, 0, 0) is a point of the first plane. The distance between the planes is

$$\frac{\left|2(2) - 3(0) + \sqrt{3}(0) - 9\right|}{\sqrt{4+9+3}} = \frac{5}{\sqrt{16}} = 1.25$$

37. If speed $= \dfrac{ds}{dt} = c$, a constant, then

$$\mathbf{a} = \frac{d^2 s}{dt^2}\mathbf{T} + \left(\frac{ds}{dt}\right)^2 \kappa \mathbf{N} = c^2 \kappa \mathbf{N} \text{ since } \frac{d^2 s}{dt^2} = 0. \text{ } \mathbf{T}$$

is in the direction of $\mathbf{v}$, while $\mathbf{N}$ is perpendicular to $\mathbf{T}$ and hence to $\mathbf{v}$ also. Thus, $\mathbf{a} = c^2 \kappa \mathbf{N}$ is perpendicular to $\mathbf{v}$.

CHAPTER 15

The Derivative in *n*-Space

15.1 Concepts Review

1. real-valued function of two real variables

3. concentric circles

Problem Set 15.1

1. a. 5

   b. 0

   c. 6

   d. $a^6 + a^2$

   e. $2x^2, \; x \neq 0$

   f. Undefined

The natural domain is the set of all (x, y) such that y is nonnegative.

3. a. $\sin(2\pi) = 0$

   b. $4\sin\left(\dfrac{\pi}{6}\right) = 2$

   c. $16\sin\left(\dfrac{\pi}{2}\right) = 16$

   d. $\pi^2 \sin(\pi^2) \approx -4.2469$

   e. $1.44 \sin [(3.1)(4.2)] \approx 0.6311$

5. $F(t \cos t, \sec^2 t) = t^2 \cos^2 t \sec^2 t = t^2, \;\; \cos t \neq 0$

7. $z = 6$ is a plane.

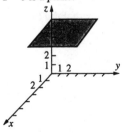

9. $x + 2y + z = 6$ is a plane.

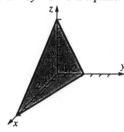

11. $x^2 + y^2 + z^2 = 16$, $z \geq 0$ is a hemisphere.

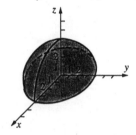

13. $z = 3 - x^2 - y^2$ is a paraboloid.

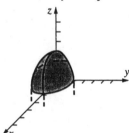

15. $z = \exp[-(x^2 + y^2)]$

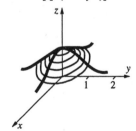

17. $x^2 + y^2 = 2z; \; x^2 + y^2 = 2k$

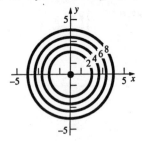

19. $x^2 = zy, \; y \neq 0; \; x^2 = ky, \; y \neq 0$

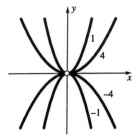

21. $z = \dfrac{x^2 + y}{x + y^2}, \; x \neq -y^2$

$k = 0: \; y = -x^2$
Parabola except $(0, 0)$ and $(-1, -1)$

$k = 1: \; x^2 + y = x + y^2$

$\left(x - \dfrac{1}{2}\right)^2 - \left(y - \dfrac{1}{2}\right)^2 = 0$

$y = x$ or $y = -x + 1$
Intersecting lines except $(0, 0)$ and $(-1, -1)$

$k = 2: \; x^2 + y = 2x + 2y^2$

$\dfrac{(x-1)^2}{\frac{7}{8}} - \dfrac{\left(y - \frac{1}{4}\right)^2}{\frac{7}{16}} = 1$

Hyperbola except $(0, 0)$ and $(-1, -1)$

$k = 4: \; x^2 + y = 4x + 4y^2$

$\dfrac{(x-2)^2}{\frac{63}{16}} - \dfrac{\left(y - \frac{1}{8}\right)^2}{\frac{63}{64}} = 1$

Hyperbola except $(0, 0)$ and $(-1, -1)$

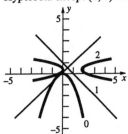

23. $x = 0$, if $T = 0$:

$y^2 = \left(\dfrac{1}{T} - 1\right) x^2$, if $y \neq 0$.

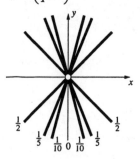

25. a. San Francisco and St. Louis had a temperature between 70 and 80 degrees Fahrenheit.

b. Drive northwest to get to cooler temperatures, and drive southeast to get warmer temperatures.

c. Since the level curve for 70 runs southwest to northeast, you could drive southwest or northeast and stay at about the same temperature.

27. $x^2 + y^2 + z^2 \geq 16$; the set of all points on and outside the sphere of radius 4 that is centered at the origin

29. $\dfrac{x^2}{9} + \dfrac{y^2}{16} + \dfrac{z^2}{1} \leq 1$; points inside and on the ellipsoid

31. Since the argument to the natural logarithm function must be positive, we must have $x^2 + y^2 + z^2 > 0$. This is true for all (x, y, z) except $(x, y, z) = (0, 0, 0)$. The domain consists all points in $\mathbb{R}^3$ except the origin.

33. $x^2 + y^2 + z^2 = k, \; k > 0$; set of all spheres centered at the origin

35. $\dfrac{x^2}{\frac{1}{16}} + \dfrac{y^2}{\frac{1}{4}} - \dfrac{z^2}{1} = k$; the elliptic cone

$\dfrac{x^2}{9} + \dfrac{y^2}{9} = \dfrac{z^2}{16}$ and all hyperboloids (one and two sheets) with z-axis for axis such that $a{:}b{:}c$ is

$\left(\dfrac{1}{4}\right) : \left(\dfrac{1}{4}\right) : \left(\dfrac{1}{3}\right)$ or $3{:}3{:}4$.

37. $4x^2 - 9y^2 = k$, k in R; $\dfrac{x^2}{\frac{k}{4}} - \dfrac{y^2}{\frac{k}{9}} = 1$, if $k \neq 0$;

planes $y = \pm\dfrac{2x}{3}$ (for $k = 0$) and all hyperbolic

cylinders parallel to the z-axis such that the ratio

$a{:}b$ is $\left(\dfrac{1}{2}\right){:}\left(\dfrac{1}{3}\right)$ or 3:2 (where a is associated

with the x-term)

39. a. All (w, x, y, z) except $(0,0,0,0)$, which
would cause division by 0.

   b. All $(x_1, x_2, \ldots, x_n)$ in n-space.

   c. All $(x_1, x_2, \ldots, x_n)$ that satisfy
   $x_1^2 + x_2^2 + \cdots + x_n^2 \leq 1$; other values of
   $(x_1, x_2, \ldots, x_n)$ would lead to the square
   root of a negative number.

41. a. AC is the least steep path and BC is the most
steep path between A and C since the level
curves are farthest apart along AC and
closest together along BC.

   b. $|AC| \approx \sqrt{(5750)^2 + (3000)^2} \approx 6490$ ft

   $|BC| \approx \sqrt{(580)^2 + (3000)^2} \approx 3060$ ft

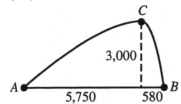

43.

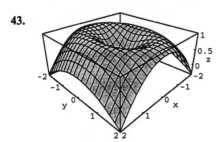

$\sin\sqrt{2x^2 + y^2}$

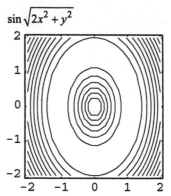

45.

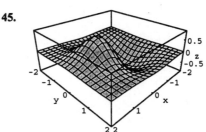

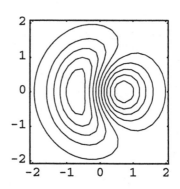

$\dfrac{\sin x \sin y}{(1 + x^2 + y^2)}$

15.2 Concepts Review

1. $\lim\limits_{h \to 0} \dfrac{[(f(x_0 + h, y_0) - f(x_0, y_0)]}{h}$; partial
derivative of f with respect to x

3. $\dfrac{\partial^2 f}{\partial y \partial x}$

Problem Set 15.2

1. $f_x(x, y) = 8(2x - y)^3$; $f_y(x, y) = -4(2x - y)^3$

3. $f_x(x, y) = \dfrac{(xy)(2x) - (x^2 - y^2)(y)}{(xy)^2} = \dfrac{x^2 + y^2}{x^2 y}$

$f_y(x, y) = \dfrac{(xy)(-2y) - (x^2 - y^2)(x)}{(xy)^2}$

$\qquad = -\dfrac{(x^2 + y^2)}{xy^2}$

5. $f_x(x, y) = e^y \cos x;\ f_y(x, y) = e^y \sin x$

7. $f_x(x, y) = x(x^2 - y^2)^{-1/2};$
$\quad f_y(x, y) = -y(x^2 - y^2)^{-1/2}$

9. $g_x(x, y) = -ye^{-xy};\ g_y(x, y) = -xe^{-xy}$

11. $f_x(x, y) = 4[1 + (4x - 7y)^2]^{-1};$
$\quad f_y(x, y) = -7[1 + (4x - 7y)^2]^{-1}$

13. $f_x(x, y) = -2xy \sin(x^2 + y^2);$
$\quad f_y(x, y) = -2y^2 \sin(x^2 + y^2) + \cos(x^2 + y^2)$

15. $F_x(x, y) = 2\cos x \cos y;\ F_y(x, y) = -2\sin x \sin y$

17. $f_x(x, y) = 4xy^3 - 3x^2 y^5;$
$\quad f_{xy}(x, y) = 12xy^2 - 15x^2 y^4$
$\quad f_y(x, y) = 6x^2 y^2 - 5x^3 y^4;$
$\quad f_{yx}(x, y) = 12xy^2 - 15x^2 y^4$

19. $f_x(x, y) = 6e^{2x} \cos y;\ f_{xy}(x, y) = -6e^{2x} \sin y$
$\quad f_y(x, y) = -3e^{2x} \sin y;\ f_{yx}(x, y) = -6e^{2x} \sin y$

21. $F_x(x, y) = \dfrac{(xy)(2) - (2x - y)(y)}{(xy)^2} = \dfrac{y^2}{x^2 y^2} = \dfrac{1}{x^2};$

$\quad F_x(3, -2) = \dfrac{1}{9}$

$\quad F_y(x, y) = \dfrac{(xy)(-1) - (2x - y)(x)}{(xy)^2} = \dfrac{-2x^2}{x^2 y^2} = -\dfrac{2}{x^2};$

$\quad F_y(3, -2) = -\dfrac{1}{2}$

23. $f_x(x, y) = -y^2(x^2 + y^4)^{-1};$
$\quad f_x\left(\sqrt{5}, -2\right) = -\dfrac{4}{21} \approx -0.1905$

$f_y(x, y) = 2xy(x^2 + y^4)^{-1};$

$\quad f_y\left(\sqrt{5}, -2\right) = -\dfrac{4\sqrt{5}}{21} \approx -0.4259$

25. Let $z = f(x, y) = \dfrac{x^2}{9} + \dfrac{y^2}{4}.$

$\quad f_y(x, y) = \dfrac{y}{2}$
$\quad$ The slope is $f_y(3, 2) = 1.$

27.
$\quad z = f(x, y) = \left(\dfrac{1}{2}\right)(9x^2 + 9y^2 - 36)^{1/2}.$

$\quad f_x(x, y) = \dfrac{9x}{2(9x^2 + 9y^2 - 36)^{1/2}}$
$\quad f_x(2, 1) = 3$

29. $V_r(r, h) = 2\pi rh;$
$\quad V_r(6, 10) = 120\pi \approx 376.99$ in.^{2}

31. $P(V, T) = \dfrac{kT}{V}$

$\quad P_T(V, T) = \dfrac{k}{V};$

$\quad P_T(100, 300) = \dfrac{k}{100}$ lb/in.^{2} per degree

33. $f_x(x, y) = 3x^2 y - y^3;\ f_{xx}(x, y) = 6xy;$
$\quad f_y(x, y) = x^3 - 3xy^2;\ f_{yy}(x, y) = -6xy$
$\quad$ Therefore, $f_{xx}(x, y) + f_{yy}(x, y) = 0.$

35. $F_y(x, y) = 15x^4 y^4 - 6x^2 y^2;$
$\quad F_{yy}(x, y) = 60x^4 y^3 - 12x^2 y;$
$\quad F_{yyy}(x, y) = 180x^4 y^2 - 12x^2$

37. a. $\dfrac{\partial^3 f}{\partial y^3}$

b. $\dfrac{\partial^3 y}{\partial y \partial x^2}$

c. $\dfrac{\partial^4 y}{\partial y^3 \partial x}$

39. a. $f_x(x, y, z) = 6xy - yz$

b. $f_y(x, y, z) = 3x^2 - xz + 2yz^2$;

$f_y(0, 1, 2) = 8$

c. Using the result in a, $f_{xy}(x, y, z) = 6x - z$.

41. $f_x(x, y, x) = -yze^{-xyz} - y(xy - z^2)^{-1}$

43. If $f(x, y) = x^4 + xy^3 + 12$, $f_y(x, y) = 3xy^2$;

$f_y(1, -2) = 12$. Therefore, along the tangent line
$\Delta y = 1 \Rightarrow \Delta z = 12$, so $\langle 0, 1, 12 \rangle$ is a tangent
vector (since $\Delta x = 0$). Then parametric equations

of the tangent line are $\begin{cases} x = 1 \\ y = -2 + t \\ z = 5 + 12t \end{cases}$. Then the

point of xy-plane at which the bee hits is
$(1, 0, 29)$ [since $y = 0 \Rightarrow t = 2 \Rightarrow x = 1, z = 29$].

45. Domain: (Case $x < y$)
The lengths of the sides are then x, $y - x$, and
$1 - y$. The sum of the lengths of any two sides
must be greater than the length of the remaining
side, leading to three inequalities:

$x + (y - x) > 1 - y \Rightarrow y > \dfrac{1}{2}$

$(y - x) + (1 - y) > x \Rightarrow x < \dfrac{1}{2}$

$x + (1 - y) > y - x \Rightarrow y < x + \dfrac{1}{2}$

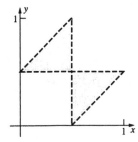

The case for $y < x$ yields similar inequalities
(x and y interchanged). The graph of D_A, the
domain of A is given above. In set notation it is

$D_A = \left\{ (x, y) : x < \dfrac{1}{2}, y > \dfrac{1}{2}, y < x + \dfrac{1}{2} \right\}$

$\cup \left\{ (x, y) : x > \dfrac{1}{2}, y < \dfrac{1}{2}, x < y + \dfrac{1}{2} \right\}$.

Range: The area is greater than zero but can be
arbitrarily close to zero since one side can be
arbitrarily small and the other two sides are
bounded above. It seems that the area would be
largest when the triangle is equilateral. An

equilateral triangle with sides equal to $\dfrac{1}{3}$ has

area $\dfrac{\sqrt{3}}{36}$. Hence the range of A is $\left(0, \dfrac{\sqrt{3}}{36} \right]$. (In

Sections 8 and 9 of this chapter methods will be
presented which will make it easy to prove that
the largest value of A will occur when the triangle
is equilateral.)

47. a. Moving parallel to the y-axis from the point
$(1, 1)$ to the nearest level curve and

approximating $\dfrac{\Delta z}{\Delta y}$, we obtain

$f_y(1, 1) = \dfrac{4 - 5}{1.25 - 1} = -4$.

b. Moving parallel to the x-axis from the point
$(-4, 2)$ to the nearest level curve and

approximating $\dfrac{\Delta z}{\Delta x}$, we obtain

$f_x(-4, 2) \approx \dfrac{1 - 0}{-2.5 - (-4)} = \dfrac{2}{3}$.

c. Moving parallel to the x-axis from the point
$(-5, -2)$ to the nearest level curve and

approximately $\dfrac{\Delta z}{\Delta x}$, we obtain

$f_x(-4, -5) \approx \dfrac{1 - 0}{-2.5 - (-5)} = \dfrac{2}{5}$.

d. Moving parallel to the y-axis from the point
$(0, -2)$ to the nearest level curve and

approximating $\dfrac{\Delta z}{\Delta y}$, we obtain

$f_y(0, 2) \approx \dfrac{0 - 1}{\frac{-19}{8} - (-2)} = \dfrac{8}{3}$.

49. a. $f_y(x, y, z)$

$= \lim_{\Delta y \to 0} \dfrac{f(x, y + \Delta y, z) - f(x, y, z)}{\Delta y}$

b. $f_z(x, y, z)$

$= \lim_{\Delta z \to 0} \dfrac{f(x, y, z + \Delta z) - f(x, y, z)}{\Delta z}$

c. $G_x(w, x, y, z)$

$= \lim_{\Delta x \to 0} \dfrac{G(w, x + \Delta x, y, z) - G(w, x, y, z)}{\Delta x}$

d. $\dfrac{\partial}{\partial z}\lambda(x,y,z,t)$

$$= \lim_{\Delta z \to 0} \dfrac{\lambda(x,y,z+\Delta z,t) - \lambda(x,y,z,t)}{\Delta z}$$

e. $\dfrac{\partial}{\partial b_2} S(b_0, b_1, b_2, \ldots, b_n) =$

$$= \lim_{\Delta b_2 \to 0} \left(\dfrac{\begin{array}{c} S(b_0, b_1, b_2 + \Delta b_2, \ldots, b_n) \\ - S(b_0, b_1, b_2, \ldots, b_n) \end{array}}{\Delta b_2} \right)$$

15.3 Concepts Review

1. $3;\ (x, y)$ approaches $(1, 2)$.

3. contained in S

Problem Set 15.3

1. -18

3. $\displaystyle\lim_{(x,\ y)\to(2,\ \pi)}\left[x\cos^2 xy - \sin\left(\dfrac{xy}{3}\right) \right]$

$$= 2\cos^2 2\pi - \sin\left(\dfrac{2\pi}{3}\right) = 2 - \dfrac{\sqrt{3}}{2} \approx 1.1340$$

5. $\dfrac{1}{3}$

7. The limit does not exist since the function is not defined anywhere along the line $y = x$. That is, there is no neighborhood of the origin in which the function is defined everywhere except possibly at the origin.

9. The entire plane since $x^2 + y^2 + 1$ is never zero.

11. Require $y - x^2 \neq 0$. S is the entire plane except the parabola $y = x^2$.

13. Require $x - y + 1 \geq 0;\ y \leq x + 1$. S is the region below and on the line $y = x + 1$.

15. Along x-axis $(y = 0)$: $\displaystyle\lim_{(x,\ y)\to(0,\ 0)}\dfrac{0}{x^2 + 0} = 0$.

Along $y = x$:
$$\lim_{(x,\ y)\to(0,\ 0)}\dfrac{x^2}{2x^2} = \lim_{(x,\ y)\to(0,\ 0)}\dfrac{1}{2} = \dfrac{1}{2}.$$
Hence, the limit does not exist because for some points near the origin $f(x, y)$ is getting closer to 0, but for others it is getting closer to $\dfrac{1}{2}$.

17. **a.** $\displaystyle\lim_{x\to 0}\dfrac{x^2(mx)}{x^4 + (mx)^2} = \lim_{x\to 0}\dfrac{mx^3}{x^4 + m^2 x^2}$

$$= \lim_{x\to 0}\dfrac{mx}{x^2 + m^2} = 0$$

b. $\displaystyle\lim_{x\to 0}\dfrac{x^2(x^2)}{x^4 + (x^2)^2} = \lim_{x\to 0}\dfrac{x^4}{2x^4} = \lim_{x\to 0}\dfrac{1}{2} = \dfrac{1}{2}$

c. $\displaystyle\lim_{(x,\ y)\to(0,\ 0)}\dfrac{x^2 y}{x^4 + y}$ does not exist.

19. The boundary consists of the points that form the outer edge of the rectangle. The set is closed.

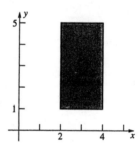

21. The boundary consists of the circle and the origin. The set is neither open (since, for example, $(1, 0)$ is not an interior point), nor closed (since $(0, 0)$ is not in the set).

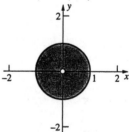

23. The boundary consists of the graph of
$$y = \sin\left(\dfrac{1}{x}\right) \text{ along with the part of the } y\text{-axis for}$$

which $y \leq 1$. The set is open.

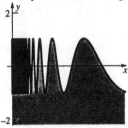

25. $\dfrac{x^2 - 4y^2}{x - 2y} = \dfrac{(x+2y)(x-2y)}{x-2y} = x + 2y$ (if $x \neq 2y$)

We want

$$g(x) = x + 2\left(\frac{x}{2}\right)\left(\text{if } x = 2y, \text{ or } y = \frac{x}{2}\right) = 2x$$

27. Note: $(x, y) \to (0, 0)$ is equivalent to $r \to 0$.

a. $f(x, y) = \dfrac{(r\cos\theta)(r\sin\theta)}{\sqrt{r^2}} = |r|\sin\theta\cos\theta$

$= \dfrac{|r|\sin 2\theta}{2} \to 0$ as $r \to 0$.

b. $f(x, y) = \dfrac{(r\cos\theta)(r\sin\theta)}{r^2} = \dfrac{\sin 2\theta}{2}$ which

does not approach 0 as $r \to 0$.

c. $f(x, y) = \dfrac{r^{7/3}\cos^{7/3}\theta}{r^2} = r\cos^{7/3}\theta \to 0$ as

$r \to 0$.

d. $f(x, y) = (r\cos\theta)(r\sin\theta)(\cos 2\theta)$ (See introduction to this problem for third factor.)

$= \dfrac{r^2\sin 2\theta\cos 2\theta}{2} = \dfrac{r^2\sin 4\theta}{4} \to 0$ as $r \to 0$.

e. $f(x, y) = \dfrac{r^4\cos^2\theta\sin^2\theta}{r^2\cos^2\theta + r^4\sin^4\theta}$

$= r^2\left(\dfrac{\cos^2\theta\sin^2\theta}{\cos^2\theta + r^2\sin^4\theta}\right)$

$= r^2\left(\dfrac{\sin^2\theta}{1 + r^2\sin^2\theta\tan^2\theta}\right)$ if $\theta \neq \pm\dfrac{\pi}{2}$. This

31. a. $f(x, y) = \begin{cases} (x^2 + y^2)^{1/2} + 1 & \text{if } y \neq 0 \\ |x - 1| & \text{if } y = 0 \end{cases}$. Check discontinuities where $y = 0$.

As $y = 0$, $(x^2 + y^2)^{1/2} + 1 = |x| + 1$, so f is continuous if $|x| + 1 = |x - 1|$. Squaring each side and simplifying

yields $|x| = -x$, so f is continuous for $x \leq 0$. That is, f is discontinuous along the positive x-axis.

converges to 0 as $r \to 0$ since the fraction is bounded (the numerator is in $[0, 1]$ and the denominator is greater than or equal to 1). If

$$\theta = \pm\frac{\pi}{2}, f(x, y) = 0.$$

f. This one is not easier in polar coordinates. Here is a Cartesian coordinates solution.

Along curve $x = y^2$:

$$\dfrac{xy^2}{x^2 + y^4} = \dfrac{(y^2)y^2}{(y^2)^2 + y^4} = \dfrac{1}{2}$$ which does not approach 0.

Conclusion: The functions of parts a, c, d, and e are continuous at the origin. Those of parts b and f are discontinuous at the origin.

29. a. $\{(x, y, z) : x^2 + y^2 = 1, z \text{ in } [1, 2]\}$ [For $x^2 + y^2 < 1$, the particle hits the hemisphere and then slides to the origin (or bounds toward the origin); for $x^2 + y^2 = 1$, it bounces up; for $x^2 + y^2 > 1$, it falls straight down.]

b. $\{(x, y, z) : x^2 + y^2 = 1, z = 1\}$ (As one moves at a level of $z = 1$ from the rim of the bowl toward any position away from the bowl there is a change from seeing all of the interior of the bowl to seeing none of it.)

c. $\{(x, y, z) : z = 1\}$ [$f(x, y, z)$ is undefined (infinite) at $(x, y, 1)$.]

d. ϕ (Small changes in points of the domain result in small changes in the shortest path from the points to the origin.)

b. Let $P = (u, v)$ and $Q = (x, y)$.

$$f(u,v,x,y) = \begin{cases} |OP| + |OQ| & \text{if } P \text{ and } Q \text{ are not on same ray from the origin and neither is the origin} \\ |PQ| & \text{otherwise} \end{cases}.$$

This means that in the first case one travels from P to the origin and then to Q; in the second case one travels directly from P to Q without passing through the origin, so f is discontinuous on the set

$\{(u, v, x, y) : \langle u, v \rangle = k \langle x, y \rangle \text{ for some } k > 0, \langle u, v \rangle \neq 0, \langle x, y \rangle \neq 0\}$.

33.

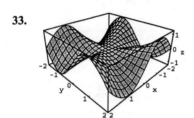

35.

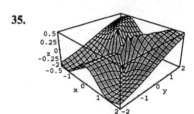

37. If we approach the point $(0,0,0)$ along a straight path from the point (x,x,x), we have

$$\lim_{(x,x,x) \to (0,0,0)} \frac{x(x)(x)}{x^3 + x^3 + x^3} = \lim_{(x,x,x) \to (0,0,0)} \frac{x^3}{3x^3} = \frac{1}{3}$$

Since the limit does not equal to $f(0,0,0)$, the function is not continuous at the point $(0,0,0)$.

15.4 Concepts Review

1. gradient

3. $\dfrac{\partial f}{\partial x}(\mathbf{p})\mathbf{i} + \dfrac{\partial f}{\partial y}(\mathbf{p})\mathbf{j}$; $y^2\mathbf{i} + 2xy\mathbf{j}$

Problem Set 15.4

1. $\left\langle 2xy + 3y, \ x^2 + 3x \right\rangle$

3. $\nabla f(x, y) = \left\langle (x)(e^{xy} y) + (e^{xy})(1), \ xe^{xy} x \right\rangle$
   $= e^{xy} \left\langle xy + 1, \ x^2 \right\rangle$

5. $x(x+y)^{-2} \left\langle y(x+2), \ x^2 \right\rangle$

7. $(x^2 + y^2 + z^2)^{-1/2} \left\langle x, \ y, \ z \right\rangle$

9. $\nabla f(x, y) = \left\langle (x^2 y)(e^{x-z}) + (e^{x-z})(2xy), \ x^2 e^{x-z}, \ x^2 y e^{x-z}(-1) \right\rangle = xe^{x-z} \left\langle y(x+2), \ x, \ -xy \right\rangle$

11. $\nabla f(x, y) = \left\langle 2xy - y^2, \ x^2 - 2xy \right\rangle$; $\nabla f(-2, 3) = \left\langle -21, \ 16 \right\rangle$
    $z = f(-2, 3) + \left\langle -21, 16 \right\rangle \cdot \left\langle x + 2, \ y - 3 \right\rangle = 30 + (-21x - 42 + 16y - 48)$
    $z = -21x + 16y - 60$

13. $\nabla f(x, y) = \left\langle -\pi \sin(\pi x) \sin(\pi y), \ \pi \cos(\pi x) \cos(\pi y) + 2\pi \cos(2\pi y) \right\rangle$
    $\nabla f\left(-1, \dfrac{1}{2}\right) = \left\langle 0, \ -2\pi \right\rangle$

$$z = f\left(-1, \frac{1}{2}\right) + \langle 0, -2\pi\rangle \cdot \left\langle x+1, y-\frac{1}{2}\right\rangle = -1 + (0 - 2\pi y + \pi);$$
$$z = -2\pi y + (\pi - 1)$$

15. $\nabla f(x, y, z) = \left\langle 6x + z^2, -4y, 2xz\right\rangle$, so $\nabla f(1, 2, -1) = \langle 7, -8, -2\rangle$

Tangent hyperplane:
$$w = f(1, 2, -1) + \nabla f(1, 2, -1) \cdot \langle x-1, y-2, z+1\rangle = -4 + \langle 7, -8, -2\rangle \cdot \langle x-1, y-2, z+1\rangle$$
$$= -4 + (7x - 7 - 8y + 16 - 2z - 2)$$
$$w = 7x - 8y - 2z + 3$$

17. $\nabla\left(\dfrac{f}{g}\right) = \dfrac{\langle gf_x - fg_x, gf_y - fg_y, gf_z - fg_z\rangle}{g^2} = \dfrac{g\langle f_x, f_y, f_z\rangle - f\langle g_x, g_y, g_z\rangle}{g^2} = \dfrac{g\nabla f - f\nabla g}{g^2}$

19. Let $F(x, y, z) = x^2 - 6x + 2y^2 - 10y + 2xy - z = 0$
$$\nabla F(x, y, z) = \langle 2x - 6 + 2y, 4y - 10 + 2x, -1\rangle$$
The tangent plane will be horizontal if $\nabla F(x, y, z) = \langle 0, 0, k\rangle$, where $k \neq 0$. Therefore, we have the following system of equations:
$$2x + 2y - 6 = 0$$
$$2x + 4y - 10 = 0$$
Solving this system yields $x = 1$ and $y = 2$. Thus, there is a horizontal tangent plane at $(x, y) = (1, 2)$.

21. a. The point $(2, 1, 9)$ projects to $(2, 1, 0)$ on the xy plane. The equation of a plane containing this point and parallel to the x-axis is given by $y = 1$. The tangent plane to the surface at the point $(2, 1, 9)$ is given by
$$z = f(2, 1) + \nabla f(2, 1) \cdot \langle x-2, y-1\rangle$$
$$= 9 + \langle 12, 10\rangle\langle x-2, y-1\rangle$$
$$= 12x + 10y - 25$$
The line of intersection of the two planes is the tangent line to the surface, passing through the point $(2, 1, 9)$, whose projection in the xy plane is parallel to the x-axis. This line of intersection is parallel to the cross product of the normal vectors for the planes. The normal vectors are $\langle 12, 10, -1\rangle$ and $\langle 0, 1, 0\rangle$ for the tangent plane and vertical plane respectively. The cross product is given by
$$\langle 12, 10, -1\rangle \times \langle 0, 1, 0\rangle = \langle 1, 0, 12\rangle$$
Thus, parametric equations for the desired tangent line are
$$x = 2 + t$$
$$y = 1$$
$$z = 9 + 12t$$

b. Using the equation for the tangent plane from the previous part, we now want the vertical plane to be parallel to the y-axis, but still pass through the projected point $(2, 1, 0)$. The vertical plane now has equation $x = 2$. The normal equations are given by $\langle 12, 10, -1\rangle$ and $\langle 1, 0, 0\rangle$ for the tangent and vertical planes respectively. Again we find the cross product of the normal vectors:
$$\langle 12, 10, -1\rangle \times \langle 1, 0, 0\rangle = \langle 0, 10, 10\rangle$$
Thus, parametric equations for the desired tangent line are
$$x = 2$$
$$y = 1 + 10t$$
$$z = 9 + 10t$$

c. Using the equation for the tangent plane from the first part, we now want the vertical plane to be parallel to the line $y = x$, but still pass through the projected point $(2, 1, 0)$. The vertical plane now has equation $y - x + 1 = 0$. The normal equations are given by $\langle 12, 10, -1\rangle$ and $\langle 1, -1, 0\rangle$ for the tangent and vertical planes respectively. Again we find the cross product of the normal vectors:
$$\langle 12, 10, -1\rangle \times \langle 1, -1, 0\rangle = \langle -1, -1, -22\rangle$$
Thus, parametric equations for the desired tangent line are
$$x = 2 - t$$
$$y = 1 - t$$
$$z = 9 - 22t$$

23. $\nabla f(x, y) = \left\langle -10\left(\dfrac{1}{2\sqrt{|xy|}}\dfrac{|xy|}{xy}y\right), -10\left(\dfrac{1}{2\sqrt{|xy|}}\dfrac{|xy|}{xy}x\right)\right\rangle = \dfrac{-5xy}{|xy|^{3/2}}\langle y, x\rangle$ $\left[\text{Note that } \dfrac{|a|}{a} = \dfrac{a}{|a|}.\right]$

$\nabla f(1, -1) = \langle -5, 5\rangle$

Tangent plane:

$z = f(1, -1) + \nabla f(1, -1) \cdot \langle x-1, y+1\rangle = -10 + \langle -5, 5\rangle \cdot \langle x-1, y+1\rangle = -10 + (-5x+5+5y+5)$

$z = -5x + 5y$

25. $\nabla f(\mathbf{p}) = \nabla g(\mathbf{p}) \Rightarrow \nabla[f(\mathbf{p}) - g(\mathbf{p})] = \mathbf{0}$

$\Rightarrow f(\mathbf{p}) - g(\mathbf{p})$ is a constant.

(ii) $\nabla[\alpha f] = \dfrac{\partial[\alpha f]}{\partial x}\mathbf{i} + \dfrac{\partial[\alpha f]}{\partial y}\mathbf{j} + \dfrac{\partial[\alpha f]}{\partial z}\mathbf{k}$

$\qquad = \alpha\dfrac{\partial[f]}{\partial x}\mathbf{i} + \alpha\dfrac{\partial[f]}{\partial y}\mathbf{j} + \alpha\dfrac{\partial[f]}{\partial z}\mathbf{k}$

$\qquad = \alpha\nabla f$

27.

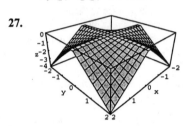

$-|xy|$

(iii)

$\nabla[fg] = \dfrac{\partial(fg)}{\partial x}\mathbf{i} + \dfrac{\partial(fg)}{\partial y}\mathbf{j} + \dfrac{\partial(fg)}{\partial z}\mathbf{k}$

$\qquad = \left(f\dfrac{\partial g}{\partial x} + g\dfrac{\partial f}{\partial x}\right)\mathbf{i} + \left(f\dfrac{\partial g}{\partial y} + g\dfrac{\partial f}{\partial y}\right)\mathbf{j}$

$\qquad + \left(f\dfrac{\partial g}{\partial z} + g\dfrac{\partial f}{\partial z}\right)\mathbf{k}$

$\qquad = f\left(\dfrac{\partial g}{\partial x}\mathbf{i} + \dfrac{\partial g}{\partial y}\mathbf{j} + \dfrac{\partial g}{\partial z}\mathbf{k}\right)$

$\qquad + g\left(\dfrac{\partial f}{\partial x}\mathbf{i} + \dfrac{\partial f}{\partial y}\mathbf{j} + \dfrac{\partial f}{\partial z}\mathbf{k}\right)$

$\qquad = f\nabla g + g\nabla f$

a. The gradient points in the direction of greatest increase of the function.

b. No. If it were, $|0+h| - |0| = 0 + |h|\delta(h)$ where $\delta(h) \to 0$ as $h \to 0$, which is possible.

29. a. (i)

$\nabla[f + g] = \dfrac{\partial(f+g)}{\partial x}\mathbf{i} + \dfrac{\partial(f+g)}{\partial y}\mathbf{j} + \dfrac{\partial(f+g)}{\partial z}\mathbf{k}$

$\qquad = \dfrac{\partial f}{\partial x}\mathbf{i} + \dfrac{\partial g}{\partial x}\mathbf{i} + \dfrac{\partial f}{\partial y}\mathbf{j} + \dfrac{\partial g}{\partial y}\mathbf{j} + \dfrac{\partial f}{\partial z}\mathbf{k} + \dfrac{\partial g}{\partial z}\mathbf{k}$

$\qquad = \dfrac{\partial f}{\partial x}\mathbf{i} + \dfrac{\partial f}{\partial y}\mathbf{j} + \dfrac{\partial f}{\partial z}\mathbf{k} + \dfrac{\partial g}{\partial x}\mathbf{i} + \dfrac{\partial g}{\partial y}\mathbf{j} + \dfrac{\partial g}{\partial z}\mathbf{k}$

$\qquad = \nabla f + \nabla g$

b. (i)

$$\nabla[f+g] = \frac{\partial(f+g)}{\partial x_1}\mathbf{i}_1 + \frac{\partial(f+g)}{\partial x_2}\mathbf{i}_2$$
$$+ \cdots + \frac{\partial(f+g)}{\partial x_n}\mathbf{i}_n$$

$$= \frac{\partial f}{\partial x_1}\mathbf{i}_1 + \frac{\partial g}{\partial x_1}\mathbf{i}_1 + \frac{\partial f}{\partial x_2}\mathbf{i}_2 + \frac{\partial g}{\partial x_2}\mathbf{i}_2$$
$$+ \cdots + \frac{\partial f}{\partial x_n}\mathbf{i}_n + \frac{\partial g}{\partial x_n}\mathbf{i}_n$$

$$= \frac{\partial f}{\partial x_1}\mathbf{i}_1 + \frac{\partial f}{\partial x_2}\mathbf{i}_2 + \cdots + \frac{\partial f}{\partial x_n}\mathbf{i}_n$$
$$+ \frac{\partial g}{\partial x_1}\mathbf{i}_1 + \frac{\partial g}{\partial x_2}\mathbf{i}_2 + \cdots + \frac{\partial g}{\partial x_n}\mathbf{i}_n$$

$$= \nabla f + \nabla g$$

(ii)

$$\nabla[\alpha f] = \frac{\partial[\alpha f]}{\partial x_1}\mathbf{i}_1 + \frac{\partial[\alpha f]}{\partial x_2}\mathbf{i}_2 + \cdots + \frac{\partial[\alpha f]}{\partial x_n}\mathbf{i}_n$$

$$= \alpha\frac{\partial[f]}{\partial x_1}\mathbf{i}_1 + \alpha\frac{\partial[f]}{\partial x_2}\mathbf{i}_2 + \cdots + \alpha\frac{\partial[f]}{\partial x_n}\mathbf{i}_n$$

$$= \alpha\nabla f$$

(iii)

$$\nabla[fg] = \frac{\partial(fg)}{\partial x_1}\mathbf{i}_1 + \frac{\partial(fg)}{\partial x_2}\mathbf{i}_2 + \cdots + \frac{\partial(fg)}{\partial x_n}\mathbf{i}_n$$

$$= \left(f\frac{\partial g}{\partial x_1} + g\frac{\partial f}{\partial x_1}\right)\mathbf{i}_1 + \left(f\frac{\partial g}{\partial x_2} + g\frac{\partial f}{\partial x_2}\right)\mathbf{i}$$

$$+ \cdots + \left(f\frac{\partial g}{\partial x_n} + g\frac{\partial f}{\partial x_n}\right)\mathbf{i}_n$$

$$= f\left(\frac{\partial g}{\partial x_1}\mathbf{i}_1 + \frac{\partial g}{\partial x_2}\mathbf{i}_2 + \cdots + \frac{\partial g}{\partial x_n}\mathbf{i}_n\right)$$

$$+ g\left(\frac{\partial f}{\partial x_1}\mathbf{i}_1 + \frac{\partial f}{\partial x_2}\mathbf{i}_2 + \cdots + \frac{\partial f}{\partial x_n}\mathbf{i}_n\right)$$

$$= f\nabla g + g\nabla f$$

15.5 Concepts Review

1. $\dfrac{[f(\mathbf{p}+h\mathbf{u}) - f(\mathbf{p})]}{h}$

3. greatest increase

Problem Set 15.5

1. $D_{\mathbf{u}}f(x, y) = \left\langle 2xy, x^2 \right\rangle \cdot \left\langle \dfrac{3}{5}, -\dfrac{4}{5} \right\rangle$; $D_{\mathbf{u}}f(1, 2) = \dfrac{8}{5}$

3. $D_{\mathbf{u}}f(x, y) = f(x, y) \cdot \mathbf{u} \left(\text{where } u = \dfrac{\mathbf{a}}{|\mathbf{a}|} \right)$

$$= \left\langle 4x+y, x-2y \right\rangle \cdot \frac{\langle 1, -1 \rangle}{\sqrt{2}};$$

$$D_{\mathbf{u}}f(3, -2) = \langle 10, 7 \rangle \cdot \frac{\langle 1, -1 \rangle}{\sqrt{2}} = \frac{3}{\sqrt{2}} \approx 2.1213$$

5. $D_{\mathbf{u}}f(x, y) = e^x \left\langle \sin y, \cos y \right\rangle \cdot \left[\left(\dfrac{1}{2} \right)\langle 1, \sqrt{3} \rangle \right]$;

$$D_{\mathbf{u}}f\left(0, \frac{\pi}{4}\right) = \frac{(\sqrt{2}+\sqrt{6})}{4} \approx 0.9659$$

7. $D_{\mathbf{u}}f(x, y, z) =$

$$= \left\langle 3x^2 y, x^3 - 2yz^2, -2y^2 z \right\rangle \cdot \left[\left(\frac{1}{3} \right)\langle 1, -2, 2 \rangle \right];$$

$$D_{\mathbf{u}}f(-2, 1, 3) = \frac{52}{3}$$

9. f increases most rapidly in the direction of the gradient. $\nabla f(x, y) = \left\langle 3x^2, -5y^4 \right\rangle$;

$$\nabla f(2, -1) = \langle 12, -5 \rangle$$

$\dfrac{\langle 12, -5 \rangle}{13}$ is the unit vector in that direction. The rate of change of $f(x, y)$ in that direction at that point is the magnitude of the gradient.
$|\langle 12, -5 \rangle| = 13$

11. $\nabla f(x, y, z) = \left\langle 2xyz, x^2 z, x^2 y \right\rangle$;

$$f(1, -1, 2) = \langle -4, 2, -1 \rangle$$

A unit vector in that direction is

$\left(\dfrac{1}{\sqrt{21}} \right)\langle -4, 2, -1 \rangle$. The rate of change in that

direction is $\sqrt{21} \approx 4.5826$.

13. $-\nabla f(x, y) = 2\langle x, y \rangle$; $-\nabla f(-1, 2) = 2\langle -1, 2 \rangle$ is the direction of most rapid decrease. A unit vector in that direction is $\mathbf{u} = \left(\dfrac{1}{\sqrt{5}}\right)\langle -1, 2 \rangle$.

15. The level curves are $\dfrac{y}{x^2} = k$. For $\mathbf{p} = (1, \ 2)$,

$k = 2$, so the level curve through $(1, 2)$ is $\dfrac{y}{x^2} = 2$

or $y = 2x^2$ $(x \neq 0)$.

$\nabla f(x, y) = \langle -2yx^{-3}, x^{-2} \rangle$

$\nabla f(1, 2) = \langle -4, 1 \rangle$, which is perpendicular to the parabola at $(1, 2)$.

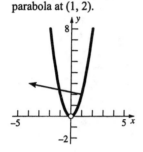

17. $u = \left\langle \dfrac{2}{3}, -\dfrac{2}{3}, \dfrac{1}{3} \right\rangle$

$D_{\mathbf{u}} f(x, y, z) = \langle y, x, 2z \rangle \cdot \left\langle \dfrac{2}{3}, -\dfrac{2}{3}, \dfrac{1}{3} \right\rangle$

$D_{\mathbf{u}} f(1, 1, 1) = \dfrac{2}{3}$

19. a. Hottest if denominator is smallest; i.e., at the origin.

b. $\nabla T(x, y, z) = \dfrac{-200\langle 2x, 2y, 2z \rangle}{(5 + x^2 + y^2 + z^2)^2}$;

$\nabla T(1, -1, 1) = \left(-\dfrac{25}{4}\right)\langle 1, -1, 1 \rangle$

$\langle -1, 1, -1 \rangle$ is one vector in the direction of greatest increase.

c. Yes

21.

$\nabla f(x, y, z) = \left\langle x\left(x^2 + y^2 + z^2\right)^{-\frac{1}{2}} \cos \sqrt{x^2 + y^2 + z^2}, \right.$

$y\left(x^2 + y^2 + z^2\right)^{-\frac{1}{2}} \cos \sqrt{x^2 + y^2 + z^2},$

$\left. z\left(x^2 + y^2 + z^2\right)^{-\frac{1}{2}} \cos \sqrt{x^2 + y^2 + z^2} \right\rangle$

$= \left(\left(x^2 + y^2 + z^2\right)^{-\frac{1}{2}} \cos \sqrt{x^2 + y^2 + z^2}\right)\langle x, y, z \rangle$

which either points towards or away from the origin.

23. He should move in the direction of

$-\nabla f(\mathbf{p}) = -\langle f_x(\mathbf{p}), f_y(\mathbf{p}) \rangle = -\left\langle -\dfrac{1}{2}, -\dfrac{1}{4} \right\rangle$

$= \left(\dfrac{1}{4}\right)\langle 2, 1 \rangle$. Or use $\langle 2, 1 \rangle$. The angle α formed

with the East is $\tan^{-1}\left(\dfrac{1}{2}\right) \approx 26.57°$ (N63.43°E).

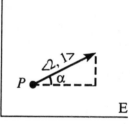

25. The climber is moving in the direction of

$\mathbf{u} = \left(\dfrac{1}{\sqrt{2}}\right)\langle -1, 1 \rangle$. Let

$f(x, y) = 3000e^{-(x^2 + 2y^2)/100}$.

$\nabla f(x, y) = 3000e^{-(x^2 + 2y^2)/100}\left\langle -\dfrac{x}{50}, -\dfrac{y}{25} \right\rangle$;

$f(10, 10) = -600e^{-3}\langle 1, 2 \rangle$

She will move at a slope of

$D_u(10, 10) = -600e^{-3}\langle 1, 2 \rangle \cdot \left(\dfrac{1}{\sqrt{2}}\right)\langle -1, 1 \rangle$

$= \left(-300\sqrt{2}\right)e^{-3} \approx -21.1229$.

She will descend. Slope is about -21.

27. $\nabla T(x, y) = \langle -4x, -2y \rangle$

$\dfrac{dx}{dt} = -4x, \dfrac{dy}{dt} = -2y$

$\dfrac{\dfrac{dx}{dt}}{-4x} = \dfrac{\dfrac{dy}{dt}}{-2y}$ has solution $|x| = 2y^2$. Since the
particle starts at $(-2, 1)$, this simplifies to
$x = -2y^2$.

29. a. $\nabla T(x, y, z) = \left\langle -\dfrac{10(2x)}{(x^2 + y^2 + z^2)^2}, -\dfrac{10(2y)}{(x^2 + y^2 + z^2)^2}, -\dfrac{10(2z)}{(x^2 + y^2 + z^2)^2} \right\rangle$

$= -\dfrac{20}{(x^2 + y^2 + z^2)^2} \langle x, y, z \rangle$

$\mathbf{r}(t) = \langle t \cos \pi t, t \sin \pi t, t \rangle$, so $\mathbf{r}(1) = \langle -1, 0, 1 \rangle$. Therefore, when $t = 1$, the bee is at $(-1, 0, 1)$, and
$\nabla T(-1, 0, 1) = -5 \langle -1, 0, 1 \rangle$.

$\mathbf{r}'(t) = \langle \cos \pi t - \pi t \sin \pi t, \sin \pi t + \pi t \cos \pi t, 1 \rangle$, so $\mathbf{r}'(1) = \langle -1, -\pi, 1 \rangle$.

$U = \dfrac{\mathbf{r}'(1)}{|\mathbf{r}'(1)|} = \dfrac{\langle -1, -\pi, 1 \rangle}{2 + \pi^2}$ is the unit tangent vector at $(-1, 0, 1)$.

$D_{\mathbf{u}}T(-1, 0, 1) = \mathbf{u} \cdot \nabla T(-1, 0, 1)$

$= \dfrac{\langle -1, -\pi, 1 \rangle \cdot \langle 5, 0, -5 \rangle}{\sqrt{2 + \pi^2}} = -\dfrac{10}{\sqrt{2 + \pi^2}} \approx -2.9026$

Therefore, the temperature is decreasing at about 2.9°C per meter traveled when the bee is at $(-1, 0, 1)$; i.e.,
when $t = 1$ s.

b. Method 1: (First express T in terms of t.)

$T = \dfrac{10}{x^2 + y^2 + z^2} = \dfrac{10}{(t \cos \pi t)^2 + (t \sin \pi t)^2 + (t)^2} = \dfrac{10}{2t^2} = \dfrac{5}{t^2}$

$T(t) = 5t^{-2}; T'(t) = -10t^{-3}; t'(1) = -10$

Method 2: (Use Chain Rule.)

$D_t T(t) = \dfrac{dT}{ds}\dfrac{ds}{dt} = (D_{\mathbf{u}}T)(|\mathbf{r}'(t)|)$, so $D_t T(t) = [D_{\mathbf{u}}T(-1, 0, 1)](|\mathbf{r}'(1)|) = -\dfrac{10}{\sqrt{2 + \pi^2}}\left(\sqrt{2 + \pi^2}\right) = -10$

Therefore, the temperature is decreasing at about 10°C per second when the bee is at $(-1, 0, 1)$; i.e., when
$t = 1$ s.

31.

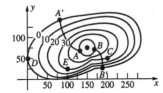

a. $A'(100, 120)$

b. $B'(190, 25)$

c. $f_x(C) \approx \dfrac{20 - 30}{230 - 200} = -\dfrac{1}{3}; f_y(D) = 0; D_{\mathbf{u}}f(E) \approx \dfrac{40 - 30}{25} = \dfrac{2}{5}$

33. Leave: (−0.1, −5)

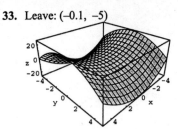

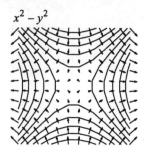

$x^2 - y^2$

35. Leave: (3, 5)

15.6 Concepts Review

1. $\dfrac{\partial z}{\partial x}\dfrac{dx}{dt} + \dfrac{\partial z}{\partial y}\dfrac{dy}{dt}$

3. $\dfrac{\partial z}{\partial x}\dfrac{\partial x}{\partial t} + \dfrac{\partial z}{\partial y}\dfrac{\partial y}{\partial t}$

Problem Set 15.6

1. $\dfrac{dw}{dt} = (2xy^3)(3t^2) + (3x^3y^2)(2t)$

$= (2t^9)(3t^2) + (3t^{10})(2t) = 12t^{11}$

3. $\dfrac{dw}{dt} = (e^x \sin y + e^y \cos x)(3) + (e^x \cos y + e^y \sin x)(2) = 3e^{3t} \sin 2t + 3e^{2t} \cos 3t + 2e^{3t} \cos 2t + 2e^{2t} \sin 3t$

5. $\dfrac{dw}{dt} = [yz^2(\cos(xyz^2))](3t^2) + [xz^2 \cos(xyz^2)](2t) + [2xyz \cos(xyz^2)](1)$

$= (3yz^2t^2 + 2xz^2t + 2xyz)\cos(xyz^2) = (3t^6 + 2t^6 + 2t^6)\cos(t^7) = 7t^6 \cos(t^7)$

7. $\dfrac{\partial w}{\partial t} = (2xy)(s) + (x^2)(-1) = 2st(s-t)s - s^2t^2 = s^2t(2s - 3t)$

9. $\dfrac{\partial w}{\partial t} = e^{x^2+y^2}(2x)(s\cos t) + e^{x^2+y^2}(2y)(\sin s) = 2e^{x^2+y^2}(xs\cos t + y\sin s)$

$= 2(s^2 \sin t \cos t + t \sin^2 s)\exp(s^2 \sin^2 t + t^2 \sin^2 s)$

11. $\dfrac{\partial w}{\partial t} = \dfrac{x(-s\sin st)}{(x^2 + y^2 + z^2)^{1/2}} + \dfrac{y(s\cos st)}{(x^2 + y^2 + z^2)^{1/2}} + \dfrac{z(s^2)}{(x^2 + y^2 + z^2)^{1/2}} = s^4t(1 + s^4t^2)^{-1/2}$

13. $\dfrac{\partial z}{\partial t} = (2xy)(2) + (x^2)(-2st) = 4(2t + s)(1 - st^2) - 2st(2t + s)^2; \left.\left(\dfrac{\partial z}{\partial t}\right)\right|_{(1,\,-2)} = 72$

15. $\dfrac{dw}{dx} = (2u - \tan v)(1) + (-u\sec^2 v)(\pi) = 2x - \tan \pi x - \pi x \sec^2 \pi x$

$\left.\dfrac{dw}{dx}\right|_{x=\frac{1}{4}} = \left(\dfrac{1}{2}\right) - 1 - \left(\dfrac{\pi}{2}\right) = -\dfrac{1+\pi}{2} \approx -2.0708$

17. $V(r, h) = \pi r^2 h, \dfrac{dr}{dt} = 0.5$ in./yr,

$\dfrac{dh}{dt} = 8$ in./yr

$\dfrac{dV}{dt} = (2\pi rh)\left(\dfrac{dr}{dt}\right) + (\pi r^2)\left(\dfrac{dh}{dt}\right)$;

$\left(\dfrac{dV}{dt}\right)\Big|_{(20,\,400)} = 11200\pi$ in.^{3}/yr

$= \dfrac{11200\pi \text{ in.}^3}{1 \text{ yr}} \times \dfrac{1 \text{ board ft}}{144 \text{ in.}^3} \approx 244.35$ board

ft/yr

19. The stream carries the boat along at 2 ft/s with respect to the boy.

$\dfrac{dx}{dt} = 2, \dfrac{dy}{dt} = 4, s^2 = x^2 + y^2$

$2s\left(\dfrac{ds}{dt}\right) = 2x\left(\dfrac{dx}{dt}\right) + 2y\left(\dfrac{dy}{dt}\right)$

$\dfrac{ds}{dt} = \dfrac{(2x + 4y)}{s}$

When $t = 3$, $x = 6$, $y = 12$, $s = 6\sqrt{5}$. Thus,

$\left(\dfrac{ds}{dt}\right)\Big|_{t=3} = \sqrt{20} \approx 4.47$ ft/s

21. Let $F(x, y) = x^3 + 2x^2 y - y^3 = 0$.

Then $\dfrac{dy}{dx} = \dfrac{-\frac{\partial F}{\partial x}}{\frac{\partial F}{\partial y}} = -\dfrac{(3x^2 + 4xy)}{2x^2 - 3y^2} = \dfrac{3x^2 + 4xy}{3y^2 - 2x^2}$.

23. Let $F(x, y) = x \sin y + y \cos x = 0$.
$\dfrac{dy}{dx} = -\dfrac{(\sin y - y \sin x)}{x \cos y + \cos x} = \dfrac{y \sin x - \sin y}{x \cos y + \cos x}$

25. Let $F(x, y, z) = 3x^2 z + y^3 - xyz^3 = 0$.

$\dfrac{\partial z}{\partial x} = -\dfrac{(6xz - yz^3)}{3x^2 - 3xyz^2} = \dfrac{yz^3 - 6xz}{3x^2 - 3xyz^2}$

27. $\dfrac{\partial T}{\partial s} = \dfrac{\partial T}{\partial x}\dfrac{\partial x}{\partial s} + \dfrac{\partial T}{\partial y}\dfrac{\partial y}{\partial s} + \dfrac{\partial T}{\partial z}\dfrac{\partial z}{\partial s} + \dfrac{\partial T}{\partial w}\dfrac{\partial w}{\partial s}$

29. $y = \left(\dfrac{1}{2}\right)[f(u) + f(v)]$,

where $u = x - ct$, $v = x + ct$.

$y_x = \left(\dfrac{1}{2}\right)[f'(u)(1) + f'(v)(1)] = \left(\dfrac{1}{2}\right)[f'(u) + f'(v)]$

$y_{xx} = \left(\dfrac{1}{2}\right)[f''(u)(1) + f''(v)(1)]$

$= \left(\dfrac{1}{2}\right)[f''(u) + f''(v)]$

$y_t = \left(\dfrac{1}{2}\right)[f'(u)(-c) + f'(v)(c)]$

$= \left(-\dfrac{c}{2}\right)[f'(u) - f'(u)]$

$y_{tt} = \left(-\dfrac{c}{2}\right)[f''(u)(-c) - f''(v)(c)]$

$= \left(\dfrac{c^2}{2}\right)[f''(u) + f''(v)] = c^2 y_{xx}$

31. Let $w = \int_x^y f(u)\,du = -\int_y^x f(u)\,du$, where $x = g(t)$,

$y = h(t)$.
Then

$\dfrac{dw}{dt} = \dfrac{\partial w}{\partial x}\dfrac{dx}{dt} + \dfrac{\partial w}{\partial y}\dfrac{dy}{dt} = -f(x)g'(t) + f(y)h'(t)$

$= f(h(t))h'(t) - f(g(t))g'(t)$.

Thus, for the particular function given

$F'(t) = \sqrt{9 + (t^2)^4}\,(2t)$

$\qquad - \sqrt{9 + \left(\sin\sqrt{2}\pi t\right)^4}\left(\sqrt{2}\pi\cos\sqrt{2}\pi t\right)$;

$F'\left(\sqrt{2}\right) = (5)\left(2\sqrt{2}\right) - (3)\left(\sqrt{2}\pi\right)$

$= 10\sqrt{2} - 3\sqrt{2}\pi \approx 0.8135$.

33. $c^2 = a^2 + b^2 - 2ab\cos 40°$ (Law of Cosines) where a, b, and c are functions of t.

$2cc' = 2aa' + 2bb' - 2(a'b + ab')\cos 40°$ so $c' = \dfrac{aa' + bb' - (a'b + ab')\cos 40°}{c}$.

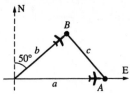

When $a = 200$ and $b = 150$, $c^2 = (200)^2 + (150)^2 - 2(200)(150)\cos 40° = 62{,}500 - 60{,}000\cos 40°$.

It is given that $a' = 450$ and $b' = 400$, so at that instant,

$$c' = \frac{(200)(450) + (150)(400) - [(450)(150) + (200)(400)]\cos 40°}{\sqrt{62{,}500 - 60{,}000\cos 40°}} \approx 288.$$

Thus, the distance between the airplanes is increasing at about 288 mph.

15.7 Concepts Review

1. perpendicular

3. $x - +4(y-1) + 6(z-1) = 0$

Problem Set 15.7

1. $\nabla F(x, y, z) = 2\langle x, y, z\rangle$;

$\nabla F\left(2, 3, \sqrt{3}\right) = 2\left\langle 2, 3, \sqrt{3}\right\rangle$

Tangent Plane:

$2(x-2) + 3(y-3) + \sqrt{3}\left(z - \sqrt{3}\right) = 0$, or

$2x + 3y + \sqrt{3}z = 16$

3. Let $F(x, y, z) = x^2 - y^2 + z^2 + 1 = 0$.

$\nabla F(x, y, z) = \langle 2x, -2y, 2z\rangle = 2\langle x, -y, z\rangle$

$\nabla F\left(1, 3, \sqrt{7}\right) = 2\left\langle 1, -3, \sqrt{7}\right\rangle$, so $\left\langle 1, -3, \sqrt{7}\right\rangle$ is

normal to the surface at the point. Then the tangent plane is

$1(x-1) - 3(y-3) + \sqrt{7}\left(z - \sqrt{7}\right) = 0$, or more

simply, $x - 3y + \sqrt{7}z = -1$.

5. $\nabla f(x, y) = \left(\dfrac{1}{2}\right)\langle x, y\rangle; \nabla f(2, 2) = \langle 1, 1\rangle$

Tangent plane: $z - 2 = 1(x-2) + 1(y-2)$, or

$x + y - z = 2$.

7. $\nabla f(x, y) = \left\langle -4e^{3y}\sin 2x, 6e^{3y}\cos 2x\right\rangle$;

$\nabla f\left(\dfrac{\pi}{3}, 0\right) = \left\langle -2\sqrt{3}, -3\right\rangle$

Tangent plane: $z + 1 = -2\sqrt{3}\left(x - \dfrac{\pi}{3}\right) - 3(y - 0)$,

or $2\sqrt{3}x + 3y + z = \dfrac{\left(2\sqrt{3}\pi - 3\right)}{3}$.

9. Let $z = f(x, y) = 2x^2 y^3$;

$dz = 4xy^3\,dx + 6x^2 y^2\,dy$. For the points given,

$dx = -0.01,\ dy = 0.02$,

$dz = 4(-0.01) + 6(0.02) = 0.08$.

$\Delta z = f(0.99, 1.02) - f(1, 1)$

$= 2(0.99)^2(1.02)^3 - 2(1)^2(1)^3 \approx 0.08017992$

11. $dz = 2x^{-1}dx + y^{-1}dy = (-1)(0.02) + \left(\dfrac{1}{4}\right)(-0.04)$

$= -0.03$

$\Delta z = f(-1.98, 3.96) - f(-2.4)$

$= \ln[(-1.98)^2(3.96)] - \ln 16 \approx -0.030151$

13. Let

$F(x, y, z) = x^2 - 2xy - y^2 - 8x + 4y - z = 0$;

$\nabla F(x, y, z) = \langle 2x - 2y - 8, -2x - 2y + 4, -1\rangle$

Tangent plane is horizontal if $\nabla F = \langle 0, 0, k\rangle$ for

any $k \neq 0$.

$2x - 2y - 8 = 0$ and $-2x - 2y + 4 = 0$ if $x = 3$ and

$y = -1$. Then $z = -14$. There is a horizontal

tangent plane at $(3, -1, -14)$.

15. For $F(x, y, z) = x^2 + 4y + z^2 = 0$,

$\nabla F(x, y, z) = \langle 2x, 4, 2z\rangle = 2\langle x, 2, z\rangle$.

$F(0, -1, 2) = 0$, and

$\nabla F(0, -1, 2) = 2\langle 0, 2, 2\rangle = 4\langle 0, 1, 1\rangle$.

For $G(x, y, z) = x^2 + y^2 + z^2 - 6z + 7 = 0$,

$\nabla G(x, y, z) = \langle 2x, 2y, 2z - 6\rangle = 2\langle x, y, z - 3\rangle$.

$G(0, -1, 2) = 0$, and

$\nabla G(0, -1, 2) = 2\langle 0, -1, -1\rangle = -2\langle 0, 1, 1\rangle$.

$\langle 0, 1, 1\rangle$ is normal to both surfaces at

$(0, -1, 2)$ so the surfaces have the same tangent

plane; hence, they are tangent to each other at

$(0, -1, 2)$.

17. Let $F(x, y, z) = x^2 + 2y^2 + 3z^2 - 12 = 0$;

$\nabla F(x, y, z) = 2\langle x, 2y, 3z \rangle$ is normal to the plane.

A vector in the direction of the line,

$\langle 2, 8, -6 \rangle = 2\langle 1, 4, -3 \rangle$, is normal to the plane.

$\langle x, 2y, 3z \rangle = k\langle 1, 4, -3 \rangle$ and (x, y, z) is on the surface for points $(1, 2, -1)$ [when $k = 1$] and $(-1, -2, 1)$ [when $k = -1$].

19. $\nabla f(x, y, z) = 2\langle 9x, 4y, 4z \rangle$;

$\nabla f(1, 2, 2) = 2\langle 9, 8, 8 \rangle$

$\nabla g(x, y, z) = 2\langle 2x, -y, 3z \rangle$;

$\nabla f(1, 2, 2) = 4\langle 1, -1, 3 \rangle$

$\langle 9, 8, 8 \rangle \times \langle 1, -1, 3 \rangle = \langle 32, -19, -17 \rangle$

Line: $x = 1 + 32t, y = 2 - 19t, z = 2 - 17t$

21. $dS = S_A dA + S_W dW$

$= -\dfrac{W}{(A-W)^2} dA + \dfrac{A}{(A-W)^2} dW = \dfrac{-WdA + AdW}{(A-W)^2}$

At $W = 20, A = 36$:

$dS = \dfrac{-20dA + 36dW}{256} = -\dfrac{5dA + 9dW}{64}$.

Thus, $|dS| \le \dfrac{5|dA| + 9|dW|}{64} \le \dfrac{5(0.02) + 9(0.02)}{64}$

$= 0.004375$

23. $V = \pi r^2 h, dV = 2\pi rh\, dr + \pi r^2 dh$

$|dV| \le 2\pi rh|dr| + \pi r^2 |dh| \le 2\pi rh(0.02r) + \pi r^2(0.03h)$

$= 0.04\pi r^2 h + 0.03\pi r^2 h = 0.07V$

Maximum error in V is 7%.

25. Solving for R, $R = \dfrac{R_1 R_2}{R_1 + R_2}$, so

$\dfrac{\partial R}{\partial R_1} = \dfrac{R_2^2}{(R_1 + R_2)^2}$ and $\dfrac{\partial R}{\partial R_2} = \dfrac{R_1^2}{(R_1 + R_2)^2}$.

29. $f(x, y) = (x^2 + y^2)^{1/2}; f(3, 4) = 5$

$f_x(x, y) = x(x^2 + y^2)^{-1/2}; f_x(3, 4) = \dfrac{3}{5} = 0.6$

$f_y = (x, y) = y(x^2 + y^2)^{-1/2}; f_x(3, 4) = \dfrac{4}{5} = 0.8$

$f_{xx}(x, y) = y^2(x^2 + y^2)^{-3/2}; \ f_x(3, 4) = \dfrac{16}{125} = 0.128$

$f_{xy}(x, y) = -xy(x^2 + y^2)^{-3/2}; \ f_{xy}(3, 4) = -\dfrac{12}{125} = -0.096$

Therefore, $dR = \dfrac{R_2^2 dR_1 + R_1^2 dR_2}{(R_1 + R_2)^2}$;

$|dR| \le \dfrac{R_2^2 |dR_1| + R_1^2 |dR_2|}{(R_1 + R_2)^2}$. Then at $R_1 = 25$,

$R_2 = 100, dR_1 = dR_2 = 0.5, \ R = \dfrac{(25)(100)}{25 + 100} = 20$

and $|dR| \le \dfrac{(100)^2(0.5) + (25)^2(0.5)}{(125)^2} = 0.34$.

27. Let $F(x, y, z) = xyz = k$; let (a, b, c) be any point on the surface of F.

$\nabla F(x, y, z) = \langle yz, xz, xy \rangle = \left\langle \dfrac{k}{x}, \dfrac{k}{y}, \dfrac{k}{z} \right\rangle$

$= k\left\langle \dfrac{1}{x}, \dfrac{1}{y}, \dfrac{1}{z} \right\rangle$

$\nabla F(a, b, c) = k\left\langle \dfrac{1}{a}, \dfrac{1}{b}, \dfrac{1}{c} \right\rangle$

An equation of the tangent plane at the point is

$\left(\dfrac{1}{a}\right)(x-a) + \left(\dfrac{1}{b}\right)(x-b) + \left(\dfrac{1}{c}\right)(x-c) = 0$, or

$\dfrac{x}{a} + \dfrac{y}{b} + \dfrac{z}{c} = 3$.

Points of intersection of the tangent plane on the coordinate axes are $(3a, 0, 0)$, $(0, 3b, 0)$, and $(0, 0, 3c)$.

The volume of the tetrahedron is

$\left(\dfrac{1}{3}\right)$(area of base)(altitude)$= \dfrac{1}{3}\left(\dfrac{1}{2}|3a||3b|\right)(|3c|)$

$= \dfrac{9|abc|}{2} = \dfrac{9|k|}{2}$ (a constant).

$$f_{yy} = x^2(x^2+y^2)^{-3/2}; \quad f_{xx}(3, 4) = \frac{9}{125} = 0.072$$

Therefore, the second order Taylor approximation is

$$f(x, y) = 5 + 0.6(x-3) + 0.8(y-4) + 0.5[0.128(x-3)^2 + 2(-0.096)(x-3)(y-4) + 0.072(y-4)^2]$$

a. First order Taylor approximation: $f(x, y) = 5 + 0.6(x-3) + 0.8(y-4)$.
    Thus, $f(3.1, 3.9) \approx 5 + 0.6(0.1) + 0.8(-0.1) = 4.98$.

b. $f(3.1, 3.9) \approx 5 + 0.6(-0.1) + 0.8(0.1) + 0.5[0.128(0.1)^2 + 2(-0.096)(0.1)(-0.1) + 0.072(-0.1)^2] = 4.98196$

c. $f(3.1, 3.9) \approx 4.9819675$

15.8 Concepts Review

1. closed bounded

3. $\nabla f(x_0, y_0) = \mathbf{0}$

Problem Set 15.8

1. $\nabla f(x, y) = \langle 2x-4, 8y \rangle = \langle 0, 0 \rangle$ at $(2, 0)$, a
    stationary point.
    $D = f_{xx}f_{yy} - f_{xy}^2 = (2)(8) - (0)^2 = 16 > 0$ and
    $f_{xx} = 2 > 0$. Local minimum at $(2, 0)$.

3. $\nabla f(x, y) = \langle 8x^3 - 2x, 6y \rangle = \langle 2x(4x^2-1), 6y \rangle$
    $= (0, 0)$, at $(0,0), (0.5, 0), (-0.5, 0)$ all stationary
    points.
    $f_{xx} = 24x^2 - 2;$
    $D = f_{xx}f_{yy} - f_{xy}^2 = (24x^2-2)(6) - (0)^2$
    $= 12(12x^2 - 1).$
    At $(0,0)$: $D = -12$, so $(0, 0)$ is a saddle point.
    At $(0.5, 0)$ and $(-0.5, 0)$: $D = 24$ and $f_{xx} = 6$,
    so local minima occur at these points.

5. $\nabla f(x, y) = \langle y, x \rangle = \langle 0, 0 \rangle$ at $(0, 0)$, a stationary
    point.
    $D = f_{xx}f_{yy} - f_{xy}^2 = (0)(0) - (1)^2 = -1$, so $(0, 0)$ is
    a saddle point.

7. $\nabla f(x, y) = \left\langle \dfrac{x^2y-2}{x^2}, \dfrac{xy^2-4}{y^2} \right\rangle = \langle 0, 0 \rangle$ at $(1, 2)$.
    $D = f_{xx}f_{yy} - f_{xy}^2 = (4x-3)(8y-3) - (1)^2$
    $= 32x^{-3}y^{-3} - 1, f_{xx} = 4x^{-3}$
    At $(1, 2)$: $D > 0$, and $f_{xx} > 0$, so a local
    minimum at $(1, 2)$.

9. Let $\nabla f(x, y)$
    $= \langle -\sin x - \sin(x+y), -\sin y - \sin(x+y) \rangle$
    $= \langle 0, 0 \rangle$
    Then $\begin{pmatrix} -\sin x - \sin(x+y) = 0 \\ \sin y + \sin(x+y) = 0 \end{pmatrix}$. Therefore,

    $\sin x = \sin y$, so $x = y = \dfrac{\pi}{4}$. However, these
    values satisfy neither equation. Therefore, the
    gradient is defined but never zero in its domain,
    and the boundary of the domain is outside the
    domain, so there are no critical points.

11. We do not need to use calculus for this one. $3x$ is
    minimum at 0 and $4y$ is minimum at -1. $(0, -1)$ is
    in S, so $3x + 4y$ is minimum at $(0, -1)$; the
    minimum value is -4. Similarly, $3x$ and $4y$ are
    each maximum at 1. $(1, 1)$ is in S, so $3x + 4y$ is
    maximum at $(1, 1)$; the maximum value is 7. (Use
    calculus techniques and compare.)

13. $\nabla f(x, y) = \langle 2x, -2y \rangle = \langle 0, 0 \rangle$ at $(0, 0)$.
    $D = f_{xx}f_{yy} - f_{xy}^2 = (2)(-2) - (0)2 < 0$, so $(0, 0)$ is
    a saddle point. A parametric representation of the
    boundary of S is $x = \cos t, y = \sin t, t$ in
    $[0, 2\pi]$.

    $f(x, y) = f(x(t), y(t)) = \cos^2 t - \sin^2 t + 1$
    $= \cos 2t - 1$
    $\cos 2t - 1$ is maximum if $\cos 2t = 1$, which occurs
    for $t = 0, \pi, 2\pi$. The points of the curve are
    $(\pm 1, 0). f(\pm 1, 0) = 2$
    $f(x, y) = \cos 2t - 1$ is minimum if $\cos 2t = -1$,
    which occurs for $t = \dfrac{\pi}{2}, \dfrac{3\pi}{2}$. The points of the
    curve are $(0, \pm 1). f(0, \pm 1) = 0$. Global minimum
    of 0 at $(0, \pm 1)$; global maximum of 2 at $(\pm 1, 0)$.

15. Let x, y, z denote the numbers, so $x + y + z = N$.
    Maximize
    $P = xyz = xy(N - x - y) = Nxy - x^2y - xy^2$.

386   Section 15.8   Student Solutions Manual

Let $\nabla P(x, y) = \langle Ny - 2xy - y^2, Nx - x^2 - 2xy \rangle$

$= \langle 0, 0 \rangle$.

Then $\begin{pmatrix} Ny - 2xy - y^2 = 0 \\ Nx - x^2 - 2xy = 0 \end{pmatrix}$.

$N(x, -y) = x^2 - y^2 = (x + y)(x - y)$. $x = y$ or $N = x + y$.

Therefore, $x = y$ (since $N = x + y$ would mean that $P = 0$, certainly not a maximum value).

Then, substituting into $Nx - x^2 - 2xy = 0$, we

obtain $Nx - x^2 - 2x^2 = 0$, from which we obtain

$x(N - 3x) = 0$, so $x = \dfrac{N}{3}$ (since $x = 0 \Rightarrow P = 0$).

$P_{xx} = -2y$;

$D = P_{xx}P_{yy} - P_{xy}^2$

$= (-2y)(-2x) - (N - 2x - 2y)^2 = 4xy - (N - 2x - 2y)^2$

At $x = y = \dfrac{N}{3} : D = \dfrac{N^2}{3} > 0, P_{xx} = -\dfrac{2N}{3} < 0$ (so

local maximum)

If $x = y = \dfrac{N}{3}$, then $z = \dfrac{N}{3}$.

Conclusion: Each number is $\dfrac{N}{3}$. (If the intent is

to find three distinct numbers, then there is no
maximum value of P that satisfies that
condition.)

17. Let S denote the surface area of the box with
dimensions x, y, z.

$S = 2xy + 2xz + 2yz$ and $V_0 = xyz$, so

$S = 2(xy + V_0 y^{-1} + V_0 x^{-1})$.

Minimize $f(x, y) = xy + V_0 y^{-1} + V_0 x^{-1}$ subject
to $x > 0, y > 0$.

$\nabla f(x, y) = \langle y - V_0 x^{-2}, x - V_0 y^{-2} \rangle = \langle 0, 0 \rangle$ at

$(V_0^{1/3}, V_0^{1/3})$.

$D = f_{xx}f_{yy} - f_{xy}^2 = 4V_0^2 x^{-3} y^{-3} - 1$,

$f_{xx} = 2V_0 x^{-3}$.

At $\left(V_0^{1/3}, V_0^{1/3} \right): D = 3 > 0, f_{xx} = 2 > 0$, so

local minimum.

Conclusion: The box is a cube with edge $V_0^{1/3}$.

19. Let S denote the area of the sides and bottom of
the tank with base l by w and depth h.

$S = lw + 2lh + 2wh$ and $lwh = 256$.

$S(l, w) = lw + 2l\left(\dfrac{256}{lw} \right) + 2w\left(\dfrac{256}{lw} \right), \ w > 0, l > 0$.

$S(l \ w) = \left\langle w - 5121^{-2}, l - 512w^{-2} \right\rangle = \langle 0, 0 \rangle$ at

$(8, 8)$. $h = 4$ there. At $(8, 8)$ $D > 0$ and $S_{11} > 0$,
so local minimum. Dimensions are 8' $\times$ 8' $\times$ 4'.

21. Let $\langle x, y, z \rangle$ denote the vector; let S be the sum of its components.

$x^2 + y^2 + z^2 = 81$, so $z = (81 - x^2 - y^2)^{1/2}$.

Maximize $S(x, y) = x + y + (81 - x^2 - y^2)^{1/2}, \ 0 \le x^2 + y^2 \le 9$.

Let $\nabla S(x, y) = \left\langle 1 - x(81 - x^2 - y^2)^{-1/2}, 1 - y(81 - x^2 - y^2)^{-1/2} \right\rangle = \langle 0, 0 \rangle$.

Therefore, $x = (81 - x^2 - y^2)^{1/2}$ and $y = (81 - x^2 - y^2)^{1/2}$. We then obtain $x = y = 3\sqrt{3}$ as the only stationary
point. For these values of x and y, $z = 3\sqrt{3}$ and $S = 9\sqrt{3} \approx 15.59$.
The boundary needs to be checked. It is fairly easy to check each edge of the boundary separately. The largest

value of S at a boundary point occurs at three places and turns out to be $\dfrac{18}{\sqrt{2}} \approx 12.73$.

Conclusion: the vector is $3\sqrt{3} \langle 1, 1, 1 \rangle$.

23. $A = \left(\dfrac{1}{2} \right)[y + (y + 2x\sin\alpha)](x\cos\alpha)$ and

$2x + y = 12$. Maximize $A(x, \alpha) = 12x\cos\alpha - 2x^2 \cos\alpha + \left(\dfrac{1}{2} \right)x^2 \sin 2\alpha$, x in $(0, 6]$, a in $\left(0, \dfrac{\pi}{2} \right)$.

$A(x, \alpha) = \left\langle 12\cos\alpha - 4x\cos\alpha + 2x\sin\alpha\cos\alpha, -12x\sin\alpha + 2x^2 \sin\alpha + x^2 \cos 2\alpha \right\rangle = \langle 0, 0 \rangle$ at $\left(4, \dfrac{\pi}{6} \right)$.

At $\left(4, \dfrac{\pi}{6}\right)$, $D > 0$ and $A_{xx} < 0$, so local maximum, and $A = 12\sqrt{3} \approx 20.78$. At the boundary point of $x = 6$, we get

$\alpha = \dfrac{\pi}{4}$, $A = 18$. Thus, the maximum occurs for width of turned-up sides = 4", and base angle $= \dfrac{\pi}{2} + \dfrac{\pi}{6} = \dfrac{2\pi}{3}$.

25. Let M be the maximum value of $f(x, y)$ on the polygonal region, P. Then $ax + by + (c - M) = 0$ is a line that either contains a vertex of P or divides P into two subregions. In the latter case $ax + by + (c - M)$ is positive in one of the regions and negative in the other. $ax + by + (c - M) > 0$ contradicts that M is the maximum value of $ax + by + c$ on P. (Similar argument for minimum.)

a.

x	y	$2x + 3y + 4$
-1	2	8
0	1	7
1	0	6
-3	0	-2
0	-4	-8

Maximum at $(-1, 2)$

b.

x	y	$-3x + 2y + 1$
-3	0	10
0	5	11
2	3	1
4	0	-11
1	-4	-10

Minimum at $(4, 0)$

27. $z(x, y) = y^2 - x^2$

$z(x, y) = \langle -2x, 2y \rangle = \langle 0, 0 \rangle$ at $(0, 0)$.

There are no stationary points and no singular points, so consider boundary points.

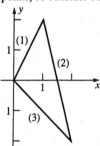

On side 1:

$y = 2x$, so $z = 4x^2 - x^2 = 3x^2$

$z'(x) = 6x = 0$ if $x = 0$.

Therefore, $(0, 0)$ is a candidate.

On side 2:

$y = -4x + 6$, so

$z = (-4x + 6)^2 - x^2 = 15x^2 - 48x + 36$.

$z'(x) = 30x - 48 = 0$ if $x = 1.6$.

Therefore, $(1.6, -0.4)$ is a candidate.

On side 3:

$y = -x$, so $z = (-x)^2 - x^2 = 0$.

Also, all vertices are candidates.

x	y	z
0	0	0
1.6	-0.4	-2.4
2	-2	0
1	2	3

Minimum value of -2.4; maximum value of 3

29.

x_i	y_i	x_i^2	$x_i y_i$
3	2	9	6
4	3	16	12
5	4	25	20
6	4	36	24
7	5	49	35
$\Sigma_{i=1}^5$ 25	18	135	97

$m(135) + b(25) = (97)$ and $m(25) + (5)b = (18)$.
Solve simultaneously and obtain $m = 0.7$, $b = 0.1$.
The least-squares line is $y = 0.7x + 0.1$.

31. $T(x, y) = 2x^2 + y^2 - y$

$\nabla T = \langle 4x, 2y - 1 \rangle = \mathbf{0}$

If $x = 0$ and $y = \dfrac{1}{2}$, so $\left(0, \dfrac{1}{2}\right)$ is the only interior critical point.

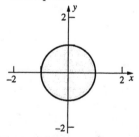

On the boundary $x^2 = 1 - y^2$, so $\mathbf{T}$ is a function of y there.

$T(y) = 2(1 - y^2) + y^2 - y = 2 - y - y^2$,

$y = [-1, 1]$

$T'(y) = -1 - 2y = 0$ if $y = -\dfrac{1}{2}$, so on the boundary, critical points occur where y is

$-1, -\dfrac{1}{2}, 1.$

Thus, points to consider are $\left(0, \dfrac{1}{2}\right)$, $(0, -1)$,

$\left(\dfrac{\sqrt{3}}{2}, -\dfrac{1}{2}\right), \left(-\dfrac{\sqrt{3}}{2}, -\dfrac{1}{2}\right)$ and $(0, 1)$. Substituting these into $T(x, y)$ yields that the coldest spot is

$\left(0, \dfrac{1}{2}\right)$ where the temperature is $-\dfrac{1}{4}$, and there is a tie for the hottest spot at $\left(\pm\dfrac{\sqrt{3}}{2}, -\dfrac{1}{2}\right)$ where the temperature is $\dfrac{9}{4}$.

33. Without loss of generality we will assume that $\alpha \le \beta \le \gamma$. We will consider it intuitively clear that for a triangle of maximum area the center of the circle will be inside or on the boundary of the triangle; i.e., α, β, and γ are in the interval $[0, \pi]$. Along with $\alpha + \beta + \gamma = 2\pi$, this implies that $\alpha + \beta \ge \pi$.

The area of an isosceles triangle with congruent sides of length r and included angle θ is $\dfrac{1}{2}r^2 \sin\theta$.

$$\text{Area}(\triangle ABC) = \dfrac{1}{2}r^2 \sin\alpha + \dfrac{1}{2}r^2 \sin\beta + \dfrac{1}{2}r^2 \sin\gamma$$

$$= \dfrac{1}{2}r^2(\sin\alpha + \sin\beta + \sin[2\pi - (\alpha + \beta)])$$

$$= \dfrac{1}{2}r^2[\sin\alpha + \sin\beta - \sin(\alpha + \beta)]$$

Area$(\triangle ABC)$ will be maximum if (*) $A(\alpha, \beta) = \sin\alpha + \sin\beta - \sin(\alpha + \beta)$ is maximum.

Restrictions are $0 \le \alpha \le \beta \le \pi$, and $\alpha + \beta \ge \pi$.

Three critical points are the vertices of the triangular domain of A : $\left(\dfrac{\pi}{2}, \dfrac{\pi}{2}\right)$, $(0, \pi)$, and (π, π). We will now search for others.

$\Delta A(\alpha, \beta) = \langle \cos\alpha - \cos(\alpha + \beta), \cos\beta - \cos(\alpha + \beta) \rangle = 0$ if $\cos\alpha = \cos(\alpha + \beta) = \cos\beta$.

Therefore, $\cos\alpha = \cos\beta$, so $\alpha = \beta$ [due to the restrictions stated]. Then $\cos\alpha = \cos(\alpha + \alpha) = \cos 2\alpha = 2\cos^2\alpha - 1$, so $\cos\alpha = 2\cos^2\alpha - 1$.

Solve for α: $2\cos^2\alpha - \cos\alpha - 1 = 0$; $(2\cos\alpha + 1)(\cos\alpha - 1) = 0$;

$\cos\alpha = -\dfrac{1}{2}$ or $\cos\alpha = 1$; $\alpha = \dfrac{2\pi}{3}$ or $\alpha = 0$.

(We are still in the case where $\alpha = \beta$.) $\left(\dfrac{2\pi}{3}, \dfrac{2\pi}{3}\right)$ is a new critical point, but $(0, 0)$ is out of the domain of A. There are no critical points in the interior of the domain of A.

On the $\beta = \pi$ edge of the domain of A;

$A(\alpha) = \sin\alpha - \sin(\alpha - \pi) = 2\sin\alpha$ so $A'(\alpha) = 2\cos\alpha$.

$A'(\alpha) = 0$ if $\alpha = \dfrac{\pi}{2}$. $\left(\dfrac{\pi}{2}, \pi\right)$ is a new critical point.

On the $\beta = \pi - \alpha$ edge of the domain of A:

$A(\alpha) = \sin\alpha + \sin(\pi - \alpha) - \sin(2\alpha - \pi) = 2\sin\alpha + \sin 2\alpha$, so

$A'(\alpha) = 2\cos\alpha + 2\cos 2\alpha = 2[\cos\alpha + (2\cos^2\alpha - 1)] = 2(2\cos\alpha - 1)(\cos\alpha + 1)$.

$A'(\alpha) = 0$ if $\cos\alpha = \dfrac{1}{2}$ or $\cos\alpha = -1$, so $\alpha = \dfrac{\pi}{3}$ or $\alpha = \pi$.

$\left(\dfrac{\pi}{3}, \dfrac{2\pi}{3}\right)$ and $(\pi, 0)$ are outside the domain of A.

(The critical points are indicated on the graph of the domain of A.)

α	β	A	
$\dfrac{\pi}{2}$	$\dfrac{\pi}{2}$	2	
0	π	0	
π	π	0	
$\dfrac{2\pi}{3}$	$\dfrac{2\pi}{3}$	$\dfrac{3\sqrt{3}}{2}$	Maximum value of A. The triangle is equilateral.
$\dfrac{\pi}{2}$	π	2	

35. Local max: $f(1.75, 0) = 1.15$
Global max: $f(-3.8, 0) = 2.30$

37. Global min: $f(0, 1) = f(0, -1) = -0.12$

39. Global max: $f(1.13, 0.79) = f(1.13, -0.79) = 0.53$
Global min: $f(-1.13, 0.79) = f(-1.13, -0.79)$
$= -0.53$

41. Global max: $f(3,3) = f(-3,3) \approx 74.9225$
Global min: $f(1.5708, 0) = f(-1.5708, 0) = -8$

43. Global max: $f(0.67, 0) = 5.06$
Global min: $f(-0.75, 0) = -3.54$

45. Global max: $f(2.1, 2.1) = 3.5$
Global min: $f(4.2, 4.2) = -3.5$

15.9 Concepts Review

1. free; constrained

3. $g(x, y) = 0$

Problem Set 15.9

1. $\langle 2x, 2y \rangle = \lambda \langle y, x \rangle$

$2x = \lambda y$, $2y = \lambda x$, $xy = 3$
Critical points are
$\left(\pm\sqrt{3}, \pm\sqrt{3}\right)$, $f\left(\pm\sqrt{3}, \pm\sqrt{3}\right) = 6$.

It is not clear whether 6 is the minimum or maximum, so take any other point on $xy = 3$, for example $(1, 3)$. $f(1, 3) = 10$, so 6 is the minimum value.

3. Let $\nabla f(x, y) = \lambda\nabla g(x, y)$, where

$g(x, y) = x^2 + y^2 - 1 = 0$.

$\langle 8x - 4y, -4x + 2y \rangle = \lambda\langle 2x, 2y \rangle$

1. $4x - 2y = \lambda x$
2. $-2x + y = \lambda y$
3. $x^2 + y^2 = 1$
4. $0 = \lambda x + 2\lambda y$ (From equations 1 and 2)
5. $\lambda = 0$ or $x + 2y = 0$ (4)

$\lambda = 0$:   6. $y = 2x$ (1)

          7. $x = \pm\dfrac{1}{\sqrt{5}}$ (6, 3)

          8. $y = \pm\dfrac{2}{\sqrt{5}}$ (7, 6)

$x + 2y = 0$:   9. $x = -2y$

10. $y = \pm \dfrac{1}{\sqrt{5}}$ (9, 3)

11. $x = \dfrac{2}{\sqrt{5}}$ (10, 9)

Critical points: $\left(\dfrac{1}{\sqrt{5}}, \dfrac{2}{\sqrt{5}} \right)$, $\left(-\dfrac{1}{\sqrt{5}}, -\dfrac{2}{\sqrt{5}} \right)$,

$\left(\dfrac{2}{\sqrt{5}}, -\dfrac{1}{\sqrt{5}} \right)$, $\left(-\dfrac{2}{\sqrt{5}}, \dfrac{1}{\sqrt{5}} \right)$

$f(x, y)$ is 0 at the first two critical points and 5 at the last two. Therefore, the maximum value of $f(x, y)$ is 5.

5. $\langle 2x, 2y, 2z \rangle = \lambda \langle 1, 3, -2 \rangle$

$2x = \lambda,\ 2y = 3\lambda,\ 2z = -2\lambda,\ x + 3y - 2z = 12$

Critical point is $\left(\dfrac{6}{7}, \dfrac{18}{7}, -\dfrac{12}{7} \right)$.

$f\left(\dfrac{6}{7}, \dfrac{18}{7}, -\dfrac{12}{7} \right) = \dfrac{72}{7}$ is the minimum (e.g., $f(12, 0, 0) = 144$).

7. Let l and w denote the dimensions of the base, h denote the depth. Maximize $V(l, w, h) = lwh$ subject to $g(l, w, h) = lw + 2lh + 2wh = 48$.

$\langle wh, lh, lw \rangle = \lambda \langle w + 2h, l + 2h, 2l + 2w \rangle$

$wh = \lambda(w + 2h),\ lh = \lambda(l + 2h),\ lw = \lambda(2l + 2w),$
$lw + 2lh + 2wh = 48$
Critical point is (4, 4, 2).
$V(4, 4, 2) = 32$ is the maximum. ($V(11, 2, 1) = 22$, for example.)

9. Let l and w denote the dimensions of the base, h the depth. Maximize $V(l, w, h) = lwh$ subject to $0.601w + 0.20(lw + 2lh + 2wh) = 12$, which simplifies to $21w + lh + wh = 30$, or $g(l, w, h) = 2lw + lh + wh - 30$.
Let $\nabla V(l, w, h) = \lambda \nabla g(l, w, h)$;

$\langle wh, lh, lw \rangle = \lambda \langle 2w + h, 2l + h, l + w \rangle$.

1. $wh = \lambda(2w + h)$
2. $lh = \lambda(2l + h)$
3. $lw = \lambda(l + w)$
4. $2lw + lh + wh = 30$
5. $(w - l)h = 2\lambda(w - l)$ (1, 2)
6. $w = l$ or $h = 2\lambda$
$w = l$:
7. $l = 2\lambda = w$ (3) Note: $w \neq 0$, for then $V = 0$.
8. $h = 4\lambda$ (7, 2)
9. $\lambda = \dfrac{\sqrt{5}}{2}$ (7, 8, 4)
10. $l = w = \sqrt{5},\ h = 2\sqrt{5}$ (9, 7, 8)

$h = 2\lambda$:
11. $\lambda = 0$ (2)
12. $l = w = h = 0$ (11, 1 – 3)
(Not possible since this does not satisfy 4.)
$\left(\sqrt{5}, \sqrt{5}, 2\sqrt{5} \right)$ is a critical point and

$V\left(\sqrt{5}, \sqrt{5}, 2\sqrt{5} \right) = 10\sqrt{5} \approx 22.36$ ft^3 is the maximum volume (rather than the minimum volume since, for example, $g(1, 1, 14) = 30$ and $V(1, 1, 14) = 14$ which is less than 22.36).

11. Maximize $f(x, y, z) = xyz$, subject to
$g(x, y, z) = b^2c^2x^2 + a^2c^2y^2 + a^2b^2z^2 - a^2b^2c^2$
     $= 0$
$\langle yz, xz, xy \rangle = \lambda \langle 2b^2c^2x, 2a^2c^2y, 2a^2b^2z \rangle$

$yz = 2b^2c^2x,\ xz = 2a^2c^2y,\ xy = 2a^2b^2z,$
$b^2c^2x^2 + a^2c^2y^2 + a^2b^2z^2 = a^2b^2c^2$

Critical point is $\left(\dfrac{a}{\sqrt{3}}, \dfrac{b}{\sqrt{3}}, \dfrac{c}{\sqrt{3}} \right)$.

$V\left(\dfrac{a}{\sqrt{3}}, \dfrac{b}{\sqrt{3}}, \dfrac{c}{\sqrt{3}} \right) = \dfrac{8abc}{3\sqrt{3}}$, which is the maximum.

13. A different hint, which will be used here, is to let $Ax + By + Cz = 1$ be the plane. (See write-up for Problem 34, Section 15.8.)
Maximize $f(A, B, C) = ABC$ subject to

$g(A, B, C) = aA + bB + cC - 1 = 0$.
Let $\langle BC, AC, BA \rangle = \lambda \langle a, b, c \rangle$.

Then $BC = \lambda a,\ AC = \lambda b,\ BA = \lambda c,$
$aA + bB + cC = 1$.
Therefore, $\lambda aA = \lambda bB = \lambda cC$ (since each equals ABC), so $aA = bB = cC$ (since $\lambda = 0$ implies $A = B = C = 0$ which doesn't satisfy the constraint equation).

Then $3aA = 1$, so $A = \dfrac{1}{3a}$; similarly $B = \dfrac{1}{3b}$ and

$C = \dfrac{1}{3c}$. The rest follows as in the solution for Problem 34, Section 15.8.

15. Let $\alpha + \beta + \gamma = 1$, $\alpha > 0$, $\beta > 0$, and $\gamma > 0$.

Maximize $P(x, y, z) = kx^\alpha y^\beta z^\gamma$, subject to $g(x, y, z) = ax + by + cz - d = 0$.

Let $\nabla P(x, y, z) = \lambda \nabla g(x, y, z)$. Then $\left\langle k\alpha x^{\alpha-1} y^\beta z^\gamma,\ k\beta x^\alpha y^{\beta-1} z^\gamma,\ k\gamma x^\alpha y^\beta z^{\gamma-1} \right\rangle = \lambda \langle a, b, c \rangle$.

Therefore, $\dfrac{\lambda a x}{\alpha} = \dfrac{\lambda b y}{\beta} = \dfrac{\lambda c z}{\gamma}$ (since each equals $kx^\alpha y^\beta z^\gamma$).

$\lambda \neq 0$ since $\lambda = 0$ would imply $x = y = z = 0$ which would imply $P = 0$.

Therefore, $\dfrac{ax}{\alpha} = \dfrac{by}{\beta} = \dfrac{cz}{\gamma}$ (*).

The constraints $ax + by + cz = d$ in the form $\alpha\left(\dfrac{ax}{\alpha}\right) + \beta\left(\dfrac{by}{\beta}\right) + \gamma\left(\dfrac{cz}{\gamma}\right) = d$ becomes

$\alpha\left(\dfrac{ax}{\alpha}\right) + \beta\left(\dfrac{ax}{\alpha}\right) + \gamma\left(\dfrac{ax}{\alpha}\right) = d$, using (*).

Then $(\alpha + \beta + \gamma)\left(\dfrac{ax}{\alpha}\right) = d$, or $\dfrac{ax}{\alpha} = d$ (since $\alpha + \beta + \gamma = 1$).

$x = \dfrac{\alpha d}{a}$ (**); $y = \dfrac{\beta d}{b}$ and $z = \dfrac{\gamma d}{c}$ then following using (*) and (**).

Since there is only one interior critical point, and since P is 0 on the boundary, P is maximum when

$x = \dfrac{\alpha d}{a}$, $y = \dfrac{\beta d}{b}$, $z = \dfrac{\gamma d}{c}$.

17. $\langle -1, 2, 2 \rangle = \lambda \langle 2x, 2y, 0 \rangle + \mu \langle 0, 1, 2 \rangle$

$-1 = 2\lambda x$, $2 = 2\lambda y + \mu$, $2 = 2\mu$, $x^2 + y^2 = 2$,
$y + 2z = 1$
Critical points are $(-1, 1, 0)$ and $(1, -1, 1)$.
$f(-1, 1, 0) = 3$, the maximum value;
$f(1, -1, 1) = -1$, the minimum value.

19. Let $\langle a_1, a_2, \ldots a_n \rangle = \lambda \langle 2x_1, 2x_2, \ldots, 2x_n \rangle$.

Therefore, $a_i = 2\lambda x_i$, for each $i = 1, 2, \ldots, n$
(since $\lambda = 0$ implies $a_i = 0$, contrary to the hypothesis).

$\dfrac{x_i}{a_i} = \dfrac{x_j}{a_j}$ for all i, j $\left(\text{since each equals } \dfrac{1}{2\lambda}\right)$.

The constraint equation can be expressed

$a_1^2\left(\dfrac{x_1}{a_1}\right)^2 + a_2^2\left(\dfrac{x_2}{a_2}\right)^2 + \ldots + a_n^2\left(\dfrac{x_n}{a_n}\right)^2 = 1$.

Therefore, $\left(a_1^2 + a_2^2 + \ldots + a_n^2\right)\left(\dfrac{x_1}{a_1}\right)^2 = 1$.

$x_1^2 = \dfrac{a_1^2}{a_1^2 + \ldots + a_n^2}$; similar for each other x_i^2.

The function to be maximized in a hyperplane with positive coefficients and constant (so intercepts on all axes are positive), and the constraint is a hypersphere of radius 1, so the maximum will occur where each x_i is positive. There is only one such critical point, the one obtained from the above by taking the principal square root to solve for x_i.

Then the maximum value of w is

$a_1\left(\dfrac{a_1}{\sqrt{A}}\right) + a_2\left(\dfrac{a_2}{\sqrt{A}}\right) + \ldots + a_n\left(\dfrac{a_n}{\sqrt{A}}\right) = \dfrac{A}{\sqrt{A}} = \sqrt{A}$

where $A = a_1^2 + a_2^2 + \ldots + a_n^2$.

21. Min: $f(4, 0) = -4$

23. Min: $f(0, 3) = f(0, -3) = -0.99$

15.10 Chapter Review

Concepts Test

1. True: Except for the trivial case of $z = 0$, which gives a point.

3. True: Since $g'(0) = f_x(0, 0)$

5. True: Use "Continuity of a Product" Theorem.

7. False: See Problem 25, Section 15.4.

9. True: Since $\langle 0, 0, -1 \rangle$ is normal to the tangent plane

11. True: It will point in the direction of greatest increase of heat, and at the origin, $\nabla T(0, 0) = \langle 1, 0 \rangle$ is that direction.

13. True: Along the x-axis, $f(x, 0) \to \pm\infty$ as $x \to \pm\infty$.

15. True: $-D_{\mathbf{u}} f(x, y) = -[\nabla f(x, y) \cdot \mathbf{u}]$
$= \nabla f(x, y) \cdot (-\mathbf{u}) = D_{-\mathbf{u}} f(x, y)$

17. True: By the Min-Max Existence Theorem

19. False: $f\left(\dfrac{\pi}{2}, 1\right) = \sin\left(\dfrac{\pi}{2}\right) = 1$, the maximum value of f, and $(\pi/2, 1)$ is in the set.

Sample Test Problems

1. a. $x^2 + 4y^2 - 100 \geq 0$

$\dfrac{x^2}{100} + \dfrac{y^2}{25} \geq 1$

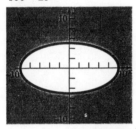

b. $2x - y - 1 \geq 0$

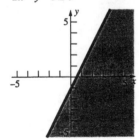

3. $f_x(x, y) = 12x^3 y^2 + 14xy^7$
$f_{xx}(x, y) = 36x^2 y^2 + 14y^7$
$f_{xy}(x, y) = 24x^3 y + 98xy^6$

5. $f_x(x, y) = e^{-y} \sec^2 x$
$f_{xx}(x, y) = 2e^{-y} \sec^2 x \tan x$
$f_{xy}(x, y) = -e^{-y} \sec^2 x$

7. $F_y(x, y) = 30x^3 y^5 - 7xy^6$
$F_{yy}(x, y) = 150x^3 y^4 - 42xy^5$
$F_{yyx}(x, y) = 450x^2 y^4 - 42y^5$

9. $z_y(x, y) = \dfrac{y}{2}$; $z_y(2, 2) = \dfrac{2}{2} = 1$

11. No. On the path $y = x$, $\displaystyle\lim_{x \to 0} \dfrac{x - x}{x + x} = 0$. On the path $y = 0$, $\displaystyle\lim_{x \to 0} \dfrac{x - 0}{x + 0} = 1$.

13. a. $\nabla f(x, y, z) = \left\langle 2xyz^3, x^2 z^3, 3x^2 yz^2 \right\rangle$
$f(1, 2, -1) = \langle -4, -1, 6 \rangle$

b. $\nabla f(x, y, z)$
$= \left\langle y^2 z \cos xz, 2y \sin xz, xy^2 \cos xz \right\rangle$
$\nabla f(1, 2, -1) = -4 \langle \cos(1), \sin(1), -\cos(1) \rangle$

$\approx \langle -2.1612, -3.3659, 2.1612 \rangle$

15. $z = f(x, y) = x^2 + y^2$
$\left\langle 1, -\sqrt{3}, 0 \right\rangle$ is horizontal and is normal to the vertical plane that is given. By inspection, $\left\langle \sqrt{3}, 1, 0 \right\rangle$ is also a horizontal vector and is perpendicular to $\left\langle 1, -\sqrt{3}, 0 \right\rangle$ and therefore is parallel to the vertical plane. Then $\mathbf{u} = \left\langle \dfrac{\sqrt{3}}{2}, \dfrac{1}{2} \right\rangle$
is the corresponding 2-dimensional unit vector.
$D_{\mathbf{u}} f(x, y) = \nabla f(x, y) \cdot \mathbf{u}$
$= \langle 2x, 2y \rangle \cdot \left\langle \dfrac{\sqrt{3}}{2}, \dfrac{1}{2} \right\rangle = \sqrt{3}x + y$

$D_{\mathbf{u}} f(1, 2) = \sqrt{3} + 2 \approx 3.7321$ is the slope of the tangent to the curve.

17. a. $f(4, 1) = 9$, so $\dfrac{x^2}{2} + y^2 = 9$, or $\dfrac{x^2}{18} + \dfrac{y^2}{9} = 1$.

b. $\nabla f(x, y) = \langle x, 2y \rangle$, so $f(4, 1) = \langle 4, 2 \rangle$.

c.

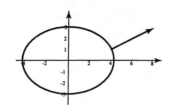

19. $f_x = f_u u_x + f_v u_y = \left(\frac{1}{v}\right)(2x) + \left(-\frac{u}{v^2}\right)(yz)$

$\qquad = x^{-2} y^{-1} z^{-1}(x^2 + 3y - 4z)$

$\quad f_y = f_u u_y + f_v v_y = \left(\frac{1}{v}\right)(-3) + \left(-\frac{u}{v^2}\right)(xz)$

$\qquad = -x^{-1} y^{-2} z^{-1}(x^2 + 4z)$

$f_z = f_u u_z + f_v v_z = \left(\frac{1}{v}\right)(4) + \left(-\frac{u}{v^2}\right)(xy)$

$\quad = x^{-1} y^{-1} z^{-2}(3y - x^2)$

21. $F_t = F_x x_t + F_y y_t + F_z z_t$

$= \left(\frac{10xy}{z^3}\right)\left(\frac{3t^{1/2}}{2}\right) + \left(\frac{5x^2}{z^3}\right)\left(\frac{1}{t}\right) + \left(-\frac{15x^2 y}{z^4}\right)(3e^{3t})$

$= \dfrac{15xy\sqrt{t}}{z^3} + \dfrac{5x^2}{z^3 t} - \dfrac{45x^2 y e^{3t}}{z^4}$

23. Let $F(x, y, z) = 9x^2 + 4y^2 + 9z^2 - 34 = 0$

$\nabla F(x, y, z) = \langle 18x, 8y, 18z \rangle$, so $\nabla f(1, 2, -1) = 2\langle 9, 8, -9 \rangle$.

Tangent plane is $9(x - 1) + 8(y - 2) - 9(z + 1) = 0$, or $9x + 8y - 9z = 34$.

25. $df = y^2(1 + z^2)^{-1} dx + 2xy(1 + z^2)^{-1} dy - 2xy^2 z(1 + z^2)^{-2} dz$

If $x = 1, y = 2, z = 2, dx = 0.01, dy = -0.02, dz = 0.03$, then $df = -0.0272$.

Therefore, $f(1.01, 1.98, 2.03) \approx f(1, 2, 2) + df = 0.8 - 0.0272 = 0.7728$

27. Let (x, y, z) denote the coordinates of the 1st octant vertex of the box. Maximize

$f(x, y, z) = xyz$ subject to

$\quad g(x, y, z) = 36x^2 + 4y^2 + 9z^2 - 36 = 0$

(where $x, y, z > 0$ and the box's volume is

$V(x, y, z) = f(x, y, z)$.

Let $\nabla f(x, y, z) = \lambda \nabla g(x, y, z)$.

$\langle yz, xz, xy \rangle 8 = \lambda \langle 72x, 8y, 18z \rangle$

1. $8yz = 72\lambda x$

2. $8xz = 8\lambda y$

3. $8xy = 18\lambda z$

4. $36x^2 + 4y^2 + 9z^2 = 36$

5. $\dfrac{yz}{xz} = \dfrac{72\lambda x}{8\lambda y}$, so $y^2 = 9x^2$. $\qquad$ (1, 2)

6. $\dfrac{yz}{xz} = \dfrac{72\lambda x}{18\lambda y}$, so $z^2 = 4x^2$. $\qquad$ (1, 3)

7. $36x^2 + 36x^2 + 36x^2 = 36$, so $x = \dfrac{1}{\sqrt{3}}$.

$\qquad\qquad\qquad\qquad\qquad$ (5, 6, 4)

8. $y = \dfrac{3}{\sqrt{3}}, z = \dfrac{2}{\sqrt{3}}$ $\qquad$ (7, 5, 6)

$V\left(\dfrac{1}{\sqrt{3}}, \dfrac{3}{\sqrt{3}}, \dfrac{2}{\sqrt{3}}\right) = 8\left(\dfrac{1}{\sqrt{3}}\right)\left(\dfrac{3}{\sqrt{3}}\right)\left(\dfrac{2}{\sqrt{3}}\right)$

$= \dfrac{16}{\sqrt{3}} \approx 9.2376$

The nature of the problem indicates that the critical point yields a maximum value rather than a minimum value.

(For a generalization of this problem, see Problem 11 or Section 15.9.)

29. Maximize $V(r, h) = \pi r^2 h$, subject to

$\quad S(r, h) = 2\pi r^2 + 2\pi r h - 24\pi = 0$.

$\langle 2\pi r h, \pi r^2 \rangle = \lambda \langle 4\pi r + 2\pi h, 2\pi r \rangle$

$rh = \lambda(2r + h), r = 2\lambda, r^2 + rh = 12$

Critical point is (2, 4). The nature of the problem indicates that the critical point yields a maximum value rather than a minimum value. Conclusion: The dimensions are radius of 2 and height of 4.

16

The Integral in
n-Space

16.1 Concepts Review

1. $\sum_{k=1}^{n} f(\bar{x}_k, \bar{y}_k)\Delta A_k$

3. continuous

Problem Set 16.1

1. $\iint_{R_1} 2\,dA + \iint_{R_2} 3\,dA = 2A(R_1) + 3A(R_2)$
$= 2(4) + 3(2) = 14$

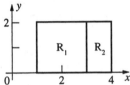

3. $\iint_R f(x, y)\,dA = \iint_{R_1} 2\,dA + \iint_{R_2} 1\,dA + \iint_{R_3} 3\,dA$
$= 2A(R_1) + 1A(R_2) + 3A(R_3)$
$= 2(2) + 1(2) + 3(2) = 12$

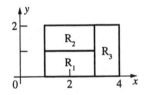

5. $3\iint_R f(x,y)\,dA - \iint_R g(x,y)\,dA = 3(3) - (5) = 4$

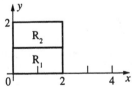

7. $\iint_R g(x,y)\,dA - \iint_{R_1} g(x,y)\,dA = (5) - (2) = 3$

9.

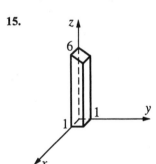

$[f(1, 1) + f(3, 1) + f(5, 1) + f(1, 3) + f(3, 3)$
$+ f(5, 3)](4) = [(10) + (8) + (6) + (8) + (6)$
$+ (4)](4) = 168$

11. $4(3 + 11 + 27 + 19 + 27 + 43) = 520$

13. $4\left(\sqrt{2} + \sqrt{4} + \sqrt{6} + \sqrt{4} + \sqrt{6} + \sqrt{8}\right) \approx 52.5665$

15.

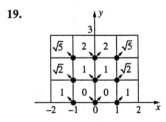

$z = 6 - y$ is a plane parallel to the x-axis. Let T be the area of the front trapezoidal face; let D be the distance between the front and back faces.

$\iint_R (6 - y)\,dA = $ volume of solid $= (T)(D)$

$= \left[\left(\frac{1}{2}\right)(6 + 5)\right](1) = 5.5$

17. $\iint_R 0\,dA = 0A(R) = 0$
The conclusion follows.

19.

For c, take the sample point in each square to be the point of the square that is closest to the origin. Then $c = 2\sqrt{5} + 2\sqrt{2} + 2(2) + 4(1) \approx 15.3006$

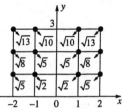

For C, take the sample point in each square to be the point of the square that is farthest from the origin. Then,

$C = 2\sqrt{13} + 2\sqrt{10} + 2\sqrt{8} + 4\sqrt{5} + 2\sqrt{2} \approx 30.9652.$

21. The values of $[\![x]\!][\![y]\!]$ and $[\![x]\!] + [\![y]\!]$ are indicated in the various square subregions of R. In each case the value of the integral on R is the sum of the values in the squares since the area of each square is 1.

a. The integral equals –6.

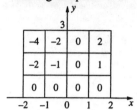

b. The integral equals 6.

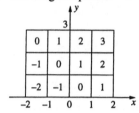

23. Total rainfall in Colorado in 1999; average rainfall in Colorado in 1999

25. To begin, we divide the region R (we will use the outline of the contour plot) into 16 equal squares. Then we can approximate the volume by

$$V = \iint\limits_{R} f(x, y)\,dA \approx \sum_{k=1}^{16} f(\bar{x}_k, \bar{y}_k)\Delta A_k\,.$$

Each square will have $\Delta A = (1\cdot 1) = 1$ and we will use the height at the center of each square as $f(\bar{x}_k, \bar{y}_k)$. Therefore, we get

$$V \approx \sum_{k=1}^{16} f(\bar{x}_k, \bar{y}_k) = 20 + 21 + 24 + 29 + 22 + 23 + 26 + 32 + 26 + 27 + 30 + 35 + 32 + 33 + 36 + 42$$

$$= 458 \text{ cubic units}$$

16.2 Concepts Review

1. iterated

3. signed; plus; minus

Problem Set 16.2

1. $\int_0^2 \left[\left(\frac{1}{2}\right)x^2 y^2\right]_{y=1}^{3}\,dx = \int_0^2 4x^2\,dx = \frac{32}{3}$

3. $\int_1^2 \left[\frac{x^2 y}{2} + xy^2\right]_{x=0}^{3}\,dx = \int_1^2 \left(\frac{9y}{2} + 3y^2\right)dy$

$= \left[\frac{9y^2}{4} + y^3\right]_1^2 = 17 - \frac{13}{4} = \frac{55}{4} = 13.75$

5. $\int_0^\pi \left[\left(\frac{1}{2}\right)x^2 \sin y\right]_{x=0}^{1}\,dy = \int_0^\pi \left(\frac{1}{2}\right)\sin y\,dy = 1$

7. $\int_0^{\pi/2} [-\cos xy]_{y=0}^{1}\,dx = \int_0^{\pi/2} (1 - \cos x)\,dx$

$= \frac{\pi}{2} - 1 \approx 0.5708$

9. $\int_0^3 \left[\frac{2(x^2 + y)^{3/2}}{3}\right]_{x=0}^{1}\,dy$

$= \int_0^3 \frac{2[(1 + y)^{3/2} - y^{3/2}]}{3}\,dy$

$= \left[\frac{4[(1 + y)^{5/2} - y^{5/2}]}{15}\right]_0^3 = \frac{4(32 - 9\sqrt{3}) - 4}{15}$

$= \frac{4(31 - 9\sqrt{3})}{15} \approx 4.1097$

11. $\int_0^{\ln 3}\left[\left(\frac{1}{2}\right)\exp(xy^2)\right]_{y=0}^1 dx = \int_0^{\ln 3}\left(\frac{1}{2}\right)(e^x-1)dx$

$= 1-\left(\frac{1}{2}\right)\ln 3 \approx 0.4507$

13. $\int_0^1 0\,dx = 0$ (since xy^3 defines an odd

function in y).

15. $\int_0^{\pi/2}\int_0^{\pi/2}\sin(x+y)dx\,dy$

$= \int_0^{\pi/2}[-\cos(x+y)]_{x=0}^{\pi/2}\,dy$

$= \int_0^{\pi/2}\left[-\cos\left(\frac{\pi}{2}+y\right)+\cos y\right]dy$

$= \int_0^{\pi/2}(\sin y+\cos y)dy = [-\cos y+\sin y]_0^{\pi/2}$

$= (0+1)-(-1+0) = 2$

17. $z = \dfrac{x}{2}$ is a plane.

$x-2z = 0$

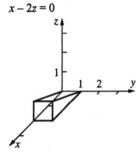

19. $z = x^2+y^2$ is a paraboloid opening upward with

z-axis for axis.

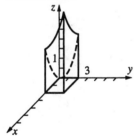

21. $\int_1^3\int_0^1(x+y+1)dx\,dy = \int_1^3\left[\left(\frac{1}{2}\right)x^2+yx+x\right]_{x=0}^1 dy$

$= \int_1^3\left(y+\frac{3}{2}\right)dy = 7$

23. $x^2+y^2+2 > 1$

$\int_{-1}^1\int_0^1[(x^2+y^2+2)-1]dy\,dx$

$= \int_{-1}^1\left[x^2 y+\left(\frac{1}{3}\right)y^3+y\right]_{y=0}^1 dx$

$= \int_{-1}^1\left(x^2+\frac{4}{3}\right)dx = \frac{10}{3}$

25. $\int_a^b\int_c^d g(x)h(y)dy\,dx = \int_a^b g(x)\int_c^d h(y)dy\,dx$

$= \int_c^d h(y)dy\int_a^b g(x)dx$

(First step used linearity of integration with respect to y; second step used linearity of integration with respect to x; now commute.)

27. $\int_0^1\int_0^1 xye^{x^2}e^{y^2}dy\,dx = \left(\int_0^1 xe^{x^2}dx\right)\left(\int_0^1 ye^{y^2}dy\right)$

$= \left(\int_0^1 xe^{x^2}dx\right)^2$ (Changed the dummy variable y

to the dummy variable x.)

$= \left(\left[\frac{e^{x^2}}{2}\right]_0^1\right)^2 = \left(\frac{e-1}{2}\right)^2 \approx 0.7381$

29. $\int_{-2}^2 x^2\,dx\int_{-1}^1|y^3|dy = 2\int_0^2 x^2\,dx\,2\int_0^1 y^3\,dy$

$= 2\left(\frac{8}{3}\right)2\left(\frac{1}{4}\right) = \frac{8}{3}$

31. $\int_{-2}^2[x^2]dx\int_{-1}^1|y^3|dy = 2\int_0^2[x^2]dx\,2\int_0^1 y^3\,dy$

$= 2\left[\int_0^1 0\,dx+\int_1^{\sqrt2}1\,dx+\int_{\sqrt2}^{\sqrt3}2\,dx+\int_{\sqrt3}^2 3\,dx\right]\left[2\left(\frac{1}{4}\right)\right]$

$= 2\left[0+\left(\sqrt2-1\right)+2\left(\sqrt3-\sqrt2\right)+3\left(2-\sqrt3\right)\right]\left[\frac{1}{2}\right]$

$= 5-\sqrt3-\sqrt2 \approx 1.8537$

33. $0 \le \int_a^b \int_a^b [f(x)g(y) - f(y)g(x)]^2 \, dx \, dy = \int_a^b \int_a^b [f^2(x)g^2(y) - 2f(x)g(x)f(y)g(y) + f^2(y)g^2(x)] \, dx \, dy$

$= \int_a^b f^2(x)dx \int_a^b g^2(y)dy - 2 \int_a^b f(x)g(x)dx \int_a^b f(y)g(y)dy + \int_a^b f^2(y)dy \int_a^b g^2(x)dx$

$= 2\int_a^b f^2(x)dx \int_a^b g^2(x)dx - 2\left[\int_a^b f(x)g(x)dx\right]^2$

Therefore, $\left[\int_a^b f(x)g(x)dx\right]^2 \le \int_a^b f^2(x)dx \int_a^b g^2(x)dx.$

16.3 Concepts Review

1. A rectangle containing S; 0

3. $\int_a^b \int_{\phi_1 x}^{\phi_2(x)} f(x, y) dy \, dx$

Problem Set 16.3

1. $\int_0^1 [x^2 y]_{y=0}^{3x} \, dx = \int_0^1 3x^3 dx = \dfrac{3}{4}$

3. $\int_{-1}^3 \left[\dfrac{x^3}{3} + y^2 x\right]_{x=0}^{3y} dy = \int_{-1}^3 (9y^3 + 3y^3) dy$

$= [3y^4]_{-1}^3 = 243 - 3 = 240$

5. $\int_1^3 \left[\left(\dfrac{1}{2}\right)x^2 \exp(y^3)\right]_{x=-y}^{2y} dy = \int_1^3 \left(\dfrac{3}{2}\right) y^2 \exp(y^3) dy$

$= \left(\dfrac{1}{2}\right)(e^{27} - e) \approx 2.660 \times 10^{11}$

7. $\int_{1/2}^1 [y\cos(\pi x^2)]_{y=0}^{2x} \, dx = \int_{1/2}^1 2x\cos(\pi x^2) dx$

$= -\dfrac{\sqrt{2}}{2\pi} \approx -0.2251$

9. $\int_0^{\pi/9} [\tan\theta]_{\theta=\pi/4}^{3r} \, dr = \int_0^{\pi/9} (\tan 3r - 1) dr$

$= \left[-\dfrac{\ln|\cos 3r|}{3} - r\right]_0^{\pi/9}$

$= \left(\dfrac{-\ln\left(\frac{1}{2}\right)}{3} - \dfrac{\pi}{9}\right) - \left(-\dfrac{\ln(1)}{3} - 0\right)$

$= \dfrac{3\ln 2 - \pi}{9} \approx -0.1180$

11. $\int_0^2 \left[xy + \left(\dfrac{1}{2}\right)y^2\right]_{y=0}^{\sqrt{4-x^2}} dx$

$= \int_0^2 \left[x(4-x^2)^{1/2} + 2 - \left(\dfrac{1}{2}\right)x^2\right] dx = \dfrac{16}{3}$

13. $\int_{-1}^1 \int_{x^2}^1 xy \, dy \, dx = 0$

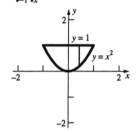

15.

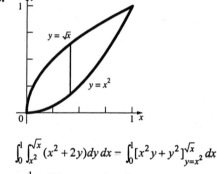

$\int_0^1 \int_{x^2}^{\sqrt{x}} (x^2 + 2y) dy \, dx = \int_0^1 [x^2 y + y^2]_{y=x^2}^{\sqrt{x}} dx$

$= \int_0^1 [(x^{5/2} + x) - (x^4 + x^4)] dx$

$= \left[\dfrac{2x^{7/2}}{7} + \dfrac{x^2}{2} - \dfrac{2x^5}{5}\right]_0^1 = \dfrac{2}{7} + \dfrac{1}{2} - \dfrac{2}{5}$

$= \dfrac{27}{70} \approx 0.3857$

17. $\int_0^2 \int_x^2 2(1+x^2)^{-1} dy \, dx = 4\tan^{-1} 2 - \ln 5 \approx 2.8192$

19.

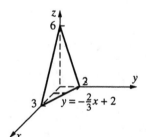

$$\int_0^3 \int_0^{(-2/3)x+2} (6-2x-3y)\,dy\,dx = 6$$

21.

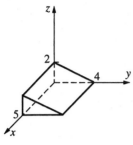

$$\int_0^5 \int_0^4 \frac{4-y}{2}\,dy\,dx = \left(\int_0^5 1\,dx\right)\left(\int_0^4 \frac{4-y}{2}\,dy\right)$$

$$= 5\left[2y - \frac{y^2}{4}\right]_0^4 = 5(8-4) = 20$$

23.

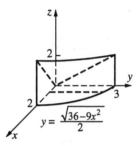

$$y = \frac{\sqrt{36-9x^2}}{2}$$

$$\int_0^2 \int_0^{(1/2)\sqrt{36-9x^2}} \left(\frac{1}{6}\right)(9x+4y)\,dy\,dx = 10$$

25.

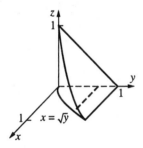

$$x = \sqrt{y}$$

$$\int_0^1 \int_0^{\sqrt{y}} (1-y)\,dx\,dy = \frac{4}{15}$$

27.

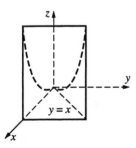

$$y = x$$

$$\int_0^1 \int_0^x \tan x^2\,dy\,dx = \int_0^1 [y \tan x^2]_{y=0}^x\,dx$$

$$= \int_0^1 x \tan x^2\,dx = \left[-\frac{\ln|\cos x^2|}{2}\right]_0^1 = \left(-\frac{1}{2}\right)\ln(\cos 1)$$

$$\approx 0.3078$$

29.

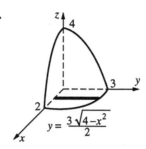

$$y = \frac{3\sqrt{4-x^2}}{2}$$

$$\int_0^2 \int_0^{(3/2)\sqrt{4-x^2}} \left[4-x^2 - \left(\frac{4}{9}\right)y^2\right]\,dy\,dx = 3\pi$$

$$\approx 9.4248$$

Making use of symmetry, the volume is

$$2\iint_{R_1} (16-x^2)^{1/2}\,dA = 2\int_0^4 \int_0^x (16-x^2)^{1/2}\,dy\,dx$$

$$= 2\int_0^4 [(16-x^2)^{1/2}\,y]_{y=0}^x\,dx$$

$$= 2\int_0^4 (16-x^2)^{1/2}\,x\,dx = \left[\frac{-2(16-x^2)^{3/2}}{3}\right]_0^4$$

$$= 0 + \frac{2(64)}{3} = \frac{128}{3} \approx 42.6667$$

31. $\displaystyle\int_0^1 \int_y^1 f(x,\,y)\,dx\,dy$

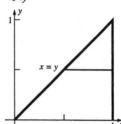

$x = y$

33. $\displaystyle\int_0^1 \int_{y^4}^{\sqrt{y}} f(x,\,y)\,dx\,dy$

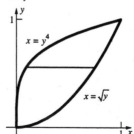

$x = y^4$

$x = \sqrt{y}$

35. $\int_{-1}^{0}\int_{-x}^{1} f(x, y)\,dy\,dx + \int_{0}^{1}\int_{x}^{1} f(x, y)\,dy\,dx$

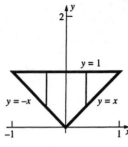

37. $\int_{0}^{2}\int_{0}^{2-y} xy^2\,dx\,dy + \int_{2}^{4}\int_{0}^{y-2} xy^2\,dx\,dy = \dfrac{256}{15}$

≈ 17.0667

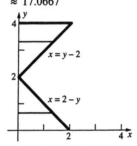

39.

$\int_{0}^{2}\int_{0}^{y^2} \sin(y^3)\,dx\,dy = \int_{0}^{2}[x\sin(y^3)]_{x=0}^{y^2}\,dy$

$= \int_{0}^{2} y^2 \sin(y^3)\,dy = \left[-\dfrac{\cos(y^3)}{3}\right]_{0}^{2}$

$= \dfrac{1-\cos 8}{3} \approx 0.3818$

41. The integral over S of $x^4 y$ is 0 (see Problem 40). Therefore,

$\iint_{S}(x^2 + x^4 y)\,dA = \iint_{S} x^2\,dA$

$= 4\left(\iint_{S_1} x^2\,dA + \iint_{S_2} x^2\,dA\right)$

$= 4\left(\int_{0}^{1}\int_{\sqrt{1-x^2}}^{\sqrt{4-x^2}} x^2\,dy\,dx + \int_{1}^{2}\int_{0}^{\sqrt{4-x^2}} x^2\,dy\,dx\right)$

$= 4\left(\int_{0}^{2} x^2\sqrt{4-x^2}\,dx - \int_{0}^{1} x^2\sqrt{1-x^2}\,dx\right)$

$= 4\left(16\int_{0}^{\pi/2} \sin^2\theta\cos^2\theta\,d\theta - \int_{0}^{\pi/2} \sin^2\phi\cos^2\phi\,d\phi\right)$

(using $x = 2\sin\theta$ in 1st integral; $x = \sin\phi$ in 2nd)

$= 60\int_{0}^{\pi/2} \sin^2\theta\cos^2\theta\,d\theta = \dfrac{15\pi}{4}$

(See work in Problem 42.)

≈ 11.7810

43. We first slice the river into eleven 100' sections parallel to the bridge. We will assume that the cross-section of the river is roughly the shape of an isosceles triangle and that the cross-sectional area is uniform across a slice. We can then approximate the volume of the water by

$V \approx \sum_{k=1}^{11} A_k(y_k)\Delta y = \sum_{k=1}^{11}\dfrac{1}{2}(w_k)(d_k)100$

$= 50\sum_{k=1}^{11}(w_k)(d_k)$

where w_k is the width across the river at the left side of the kth slice, and d_k is the center depth of the river at the left side of the kth slice. This gives

$V \approx 50[300\cdot40 + 300\cdot39 + 300\cdot35 + 300\cdot31$

$+ 290\cdot28 + 275\cdot26 + 250\cdot25 + 225\cdot24$

$+ 205\cdot23 + 200\cdot21 + 175\cdot19]$

$= 4{,}133{,}000 \text{ ft}^3$

16.4 Concepts Review

1. $a \le r \le b; \alpha \le \theta \le \beta$

3. $\int_{0}^{\pi}\int_{0}^{2} r^3\,dr\,d\theta$

Problem Set 16.4

1. $\int_{0}^{\pi/2}\left[\left(\dfrac{1}{3}\right)r^3\sin\theta\right]_{r=0}^{\cos\theta}\,d\theta$

$= \int_{0}^{\pi/2}\left(\dfrac{1}{3}\right)\cos^3\theta\sin\theta\,d\theta$

$= \dfrac{1}{12} \approx 0.0833$

3. $\displaystyle\int_0^\pi \left[\frac{r^3}{3}\right]_{r=0}^{\sin\theta} d\theta = \int_0^\pi \frac{\sin^3\theta}{3}\,d\theta$

$\displaystyle = \int_0^\pi \frac{(1-\cos^2\theta)\sin\theta}{3}\,d\theta$

$\displaystyle = \left[\frac{-\cos\theta}{3} + \frac{\cos^3\theta}{9}\right]_0^\pi$

$\displaystyle = \left(\frac{1}{3} - \frac{1}{9}\right) - \left(-\frac{1}{3} + \frac{1}{9}\right) = \frac{4}{9}$

5.

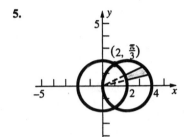

$\displaystyle 2\int_0^{\pi/3}\int_2^{4\cos\theta} r\,dr\,d\theta = 2\left[\frac{2\pi}{3} + \sqrt{3}\right] \approx 7.6529$

7.

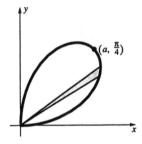

$\displaystyle \int_0^{\pi/2}\int_0^{a\sin 2\theta} r\,dr\,d\theta = \frac{a^2\pi}{8}$

9.

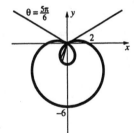

$\displaystyle 2\int_{5\pi/6}^{3\pi/2}\int_0^{2-4\sin\theta} r\,dr\,d\theta = 2\int_{5\pi/6}^{3\pi/2}\left[\frac{r^2}{2}\right]_0^{2-4\sin\theta} d\theta$

$\displaystyle = 2\int_{5\pi/6}^{3\pi/2}(6-8\sin\theta - 4\cos 2\theta)\,d\theta$

$= 2[6\theta + 8\cos\theta - 2\sin 2\theta]_{5\pi/6}^{3\pi/2}$

$= 2\left(4\pi + 3\sqrt{3}\right) \approx 35.525$

11.

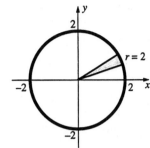

$\displaystyle 2\int_0^\pi\int_0^2 e^{r^2} r\,dr\,d\theta = \pi(e^4 - 1) \approx 168.3836$

13. $\displaystyle \int_0^{\pi/4}\int_0^2 (4+r^2)^{-1} r\,dr\,d\theta = \left(\frac{\pi}{8}\right)\ln 2 \approx 0.2722$

15.

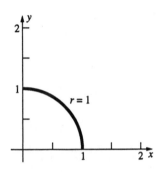

$\displaystyle \int_0^{\pi/2}\int_0^1 (4-r^2)^{-1/2} r\,dr\,d\theta$

$\displaystyle = \int_0^{\pi/2}[-(4-r^2)^{1/2}]_0^1\,d\theta$

$\displaystyle = \int_0^{\pi/2}\left(-\sqrt{3}+2\right)d\theta = \left(-\sqrt{3}+2\right)\left(\frac{\pi}{2}\right) \approx 0.4209$

17. $\displaystyle \int_{\pi/4}^{\pi/2}\int_0^{\csc\theta} r^2\cos^2\theta\, r\,dr\,d\theta = \frac{1}{12} \approx 0.0833$

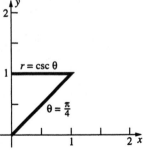

19.

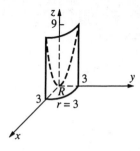

$$\iint_R (x^2 + y^2)\,dA = \int_0^{\pi/2} \int_0^3 r^2 r\,dr\,d\theta$$

$$= \frac{81\pi}{8} \approx 31.8086$$

21.

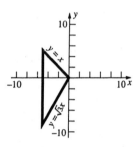

$$\int_{-5}^0 \int_{\sqrt{3}x}^x (y^2)\,dy\,dx = \int_{-5}^0 \left[\frac{y^3}{3}\right]_{\sqrt{3}x}^{-x} dx$$

$$= \int_{-5}^0 \frac{-1-3\sqrt{3}}{3} x^3\,dx = \left[\frac{\left(-1-3\sqrt{3}\right)x^4}{12}\right]_{-5}^0$$

$$= \frac{\left(1+3\sqrt{3}\right)625}{12} \approx 322.7163$$

23. This can be done by the methods of this section, but an easier way to do it is to realize that the intersection is the union of two congruent segments (of one base) of the spheres, so (see

Problem 20, Section 6.2, with $d = h$ and $a = r$) the volume is $2\left[\left(\frac{1}{3}\right)\pi d^2 (3a - d)\right] = 2\pi d^2 \frac{(3a-d)}{3}$.

25.

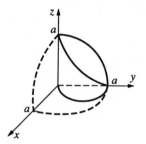

$$\text{Volume} = 4\int_0^{\pi/2} \int_0^{a\sin\theta} \sqrt{a^2 - r^2}\,r\,dr\,d\theta$$

$$= \int_0^{\pi/2} \left[\left(-\frac{1}{3}\right)(a^3 \cos^3\theta - a^3)\right]d\theta$$

$$= \left(-\frac{4}{3}\right)a^3 \left[\frac{2}{3} - \frac{\pi}{2}\right] = \left(\frac{2}{9}\right)a^3(3\pi - 4)$$

27. Choose a coordinate system so the center of the sphere is the origin and the axis of the part removed is the z-axis.
Volume (Ring) = Volume (Sphere of radius a) − Volume (Part removed)

$$= \frac{4}{3}\pi a^3 - 2\int_0^{2\pi} \int_0^{\sqrt{a^2-b^2}} \sqrt{a^2 - r^2}\,r\,dr\,d\theta = \frac{4}{3}\pi a^3 - 2(2\pi)\int_0^{\sqrt{a^2-b^2}} (a^2 - r^2)^{1/2} r\,dr$$

$$= \frac{4}{3}\pi a^3 - 4\pi \left[\frac{1}{3}(a^2 - r^2)^{3/2}\right]_0^{\sqrt{a^2-b^2}} = \frac{4}{3}\pi a^3 + 4\pi\frac{1}{3}(b^3 - a^3) = \frac{4}{3}\pi b^3$$

29. $\int_0^{\pi/2} \left[\lim_{b\to\infty} \int_0^b (1+r^2)^{-2} r\,dr\right]d\theta = \int_0^{\pi/2}\left(\lim_{b\to\infty}\left[\left(-\frac{1}{2}\right)(1+b^2)^{-1} - \left(-\frac{1}{2}\right)\right]\right)d\theta = \int_0^{\pi/2}\left(\frac{1}{2}\right)d\theta = \frac{\pi}{4} \approx 0.7854$

31. From symmetry we have

$$\int_{-\infty}^{\infty}\frac{1}{\sigma\sqrt{2\pi}}e^{-(x-\mu)^2/2\sigma^2}\,dx$$

$$=2\int_{0}^{\infty}\frac{1}{\sigma\sqrt{2\pi}}e^{-(x-\mu)^2/2\sigma^2}\,dx.$$

Using the substitution $u=\dfrac{x-\mu}{\sigma\sqrt{2}}$ we get

$du=\dfrac{dx}{\sigma\sqrt{2}}$. Our integral then becomes

$$\int_{-\infty}^{\infty}\frac{1}{\sigma\sqrt{2\pi}}e^{-(x-\mu)^2/2\sigma^2}\,dx$$

$$=\frac{2}{\sqrt{\pi}}\int_{0}^{\infty}e^{-u^2}\,du$$

Using the result from Example 4, we see that

$\displaystyle\int_{0}^{\infty}e^{-u^2}\,du=\frac{\sqrt{\pi}}{2}$. Thus we have

$$\int_{-\infty}^{\infty}\frac{1}{\sigma\sqrt{2\pi}}e^{-(x-\mu)^2/2\sigma^2}\,dx$$

$$=\frac{2}{\sqrt{\pi}}\cdot\frac{\sqrt{\pi}}{2}=1.$$

16.5 Concepts Review

1. $\displaystyle\iint_S x^2 y^4\,dA$

3. $\displaystyle\iint_S x^4 y^4\,dA$

Problem Set 16.5

1.

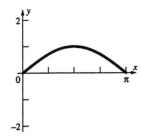

$m=\displaystyle\int_0^3\int_0^4 (y+1)\,dx\,dy=30$

$M_y=\displaystyle\int_0^3\int_0^4 x(y+1)\,dx\,dy=60$

$M_x=\displaystyle\int_0^3\int_0^4 y(y+1)\,dx\,dy=54$

$(\bar{x},\bar{y})=(2,1.8)$

3.

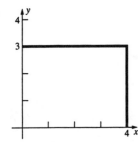

$m=\displaystyle\int_0^\pi\int_0^{\sin x} y\,dy\,dx=\int_0^\pi\left[\frac{y^2}{2}\right]_0^{\sin x}dx$

$=\displaystyle\int_0^\pi\frac{\sin^2 x}{2}\,dx=\int_0^\pi\frac{1-\cos 2x}{4}\,dx$

$=\left[\dfrac{x}{4}-\dfrac{\sin 2x}{8}\right]_0^\pi=\dfrac{\pi}{4}$

$M_x=\displaystyle\int_0^\pi\int_0^{\sin x} yy\,dy\,dx=\int_0^\pi\left[\frac{y^3}{3}\right]_0^{\sin x}dx$

$=\displaystyle\int_0^\pi\frac{\sin^3 x}{3}\,dx=\frac{1}{3}\int_0^\pi(1-\cos^2 x)\sin x\,dx$

$=\dfrac{1}{3}\left[-\cos x+\dfrac{\cos^3 x}{3}\right]_0^\pi=\dfrac{4}{9}$

$\bar{y}=\dfrac{M_x}{m}=\dfrac{\frac{4}{9}}{\frac{\pi}{4}}=\dfrac{16}{9\pi}\approx 0.5659;$

$\bar{x}=\dfrac{\pi}{2}$ (by symmetry)

Thus, $M_y=\bar{x}\cdot m=\dfrac{\pi}{4}\cdot\dfrac{\pi}{2}=\dfrac{\pi^2}{8}$

5.

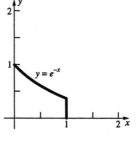

$m=\displaystyle\int_0^1\int_0^{e^{-x}} y^2\,dy\,dx=\left(\frac{1}{9}\right)(1-e^{-3})$

$$M_x = \int_0^1 \int_0^{e^{-x}} y^3 \, dy \, dx = \left(\frac{1}{16}\right)(1 - e^{-4})$$

$$M_y = \int_0^1 \int_0^{e^{-x}} xy^2 \, dy \, dx = \left(\frac{1}{27}\right)(1 - 4e^{-3}) \approx 0.1056$$

$$(\overline{x}, \overline{y}) =$$

$$\left(\left(\frac{1}{3}\right)(e^3 - 4)(e^3 - 1)^{-1}, \left(\frac{9}{16}\right)e^{-1}(e^4 - 1)(e^3 - 1)^{-1}\right)$$

$$\approx (0.2809, 0.5811)$$

7.

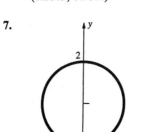

$$m = \int_0^\pi \int_0^{2\sin\theta} r \, r \, dr \, d\theta = \frac{32}{9}$$

$$M_x = \int_0^\pi \int_0^{2\sin\theta} (r\sin\theta) r \, r \, dr \, d\theta = \frac{64}{15}$$

$$M_y = 0 \text{ (symmetry)}$$

$$(\overline{x}, \overline{y}) = (0, 1.2)$$

9.

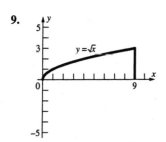

$$I_x = \int_0^3 \int_{y^2}^9 y^2 (x + y) \, dx \, dy$$

$$= \int_0^3 \left(\frac{81y^2}{2} + 9y^3 - \frac{y^6}{2} - y^5\right) dy = \frac{7533}{28} \approx 269$$

$$I_y = \int_0^9 \int_0^{\sqrt{x}} x^2 (x + y) \, dy \, dx = \int_0^9 \left(x^{7/2} + \frac{x^3}{2}\right) dx$$

$$= \frac{41553}{8} \approx 5194$$

$$I_z = I_x + I_y = \frac{305937}{56} \approx 5463$$

11.

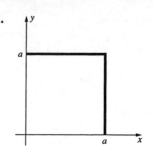

$$I_x = \int_0^a \int_0^a (x + y)y^2 \, dx \, dy = \left(\frac{5}{12}\right)a^5$$

$$I_y = \left(\frac{5}{12}\right)a^5$$

$$I_z = \left(\frac{5}{6}\right)a^5$$

13. $m = \int_0^a \int_0^a (x + y) \, dx \, dy = a^3$

$$\overline{r} = \left(\frac{I_x}{m}\right)^{1/2} = \left(\frac{5}{12}\right)^{1/2} a \approx 0.6455a$$

15.

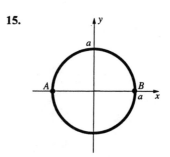

$$m = \delta\pi a^2$$

The moment of inertia about diameter AB is

$$I = I_x = \int_0^{2\pi} \int_0^a \delta r^2 \sin^2\theta \, r \, dr \, d\theta$$

$$= \int_0^{2\pi} \frac{\delta a^4 \sin^2\theta}{4} \, d\theta = \frac{\delta a^4}{8} \int_0^{2\pi} (1 - \cos 2\theta) \, d\theta$$

$$= \frac{\delta a^4}{8} \left[\theta - \frac{\sin 2\theta}{2}\right]_0^{2\pi} = \frac{\delta a^4 \pi}{4}$$

$$\overline{r} = \left(\frac{I}{m}\right)^{1/2} = \left(\frac{\frac{\delta a^4 \pi}{4}}{\delta\pi a^2}\right)^{1/2} = \frac{a}{2}$$

17.

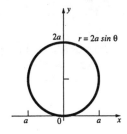

$$I_x = \iint_S \delta y^2 \, dA$$

$$= 2\delta \int_0^{\pi/2} \int_0^{2a\sin\theta} (r\sin\theta)^2 r \, dr \, d\theta$$

$$= 2\delta \int_0^{\pi/2} 4a^4 \sin^6\theta \, d\theta$$

$$= 8a^4\delta \frac{(1)(3)(5)}{(2)(4)(6)} \frac{\pi}{2} = \frac{5a^4\delta\pi}{4}$$

19.

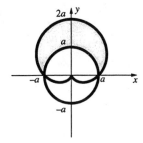

$x = 0$ (by symmetry)

$$M_x = \iint_S ky \, dA = 2k \int_{-\pi/2}^{\pi/2} \int_0^{a(1+\sin\theta)} (r\sin\theta) r \, dr \, d\theta = 2k \int_0^{\pi/2} \left(\frac{a^3}{3}\right)(3\sin^2\theta + 3\sin^3\theta + \sin^4\theta)d\theta$$

$$= \left(\frac{2}{3}\right)ka^3\left[\frac{(15\pi+32)}{16}\right] = \left(\frac{1}{24}\right)ka^3(15\pi+32)$$

$$m = \iint_S k \, dA = 2k \int_{-\pi/2}^{\pi/2} \int_0^{a(1+\sin\theta)} r \, dr \, d\theta = 2k \int_0^{\pi/2} \left(\frac{1}{2}\right)a^2(2\sin\theta+\sin^2\theta)d\theta = ka^2\left[\frac{(8+\pi)}{4}\right] = \left(\frac{1}{4}\right)ka^2(\pi+8)$$

Therefore, $\bar{y} = \dfrac{M_x}{m} = \dfrac{\left(\frac{1}{24}\right)ka^3(15\pi+32)}{\left(\frac{1}{4}\right)ka^2(\pi+8)} = \dfrac{a(15\pi+32)}{6(\pi+8)} \approx 1.1836a$

21. a.

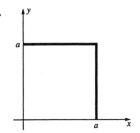

$$m = \iint_S (x+y)dA = \int_0^a \int_0^a (x+y)dx \, dy$$

$$= \int_0^a \left(\left[\frac{x^2}{2} + xy\right]_{x=0}^a\right)dy = \int_0^a \left(\frac{a^2}{2} + ay\right)dy$$

$$= \left[\frac{a^2y}{2} + \frac{ay^2}{2}\right]_0^a = a^3$$

b. $M_y = \iint_S x(x+y)dA = \int_0^a \int_0^a (x^2+xy)dy\,dx$

$= \int_0^a \left(\left[x^2 y + \frac{xy^2}{2} \right]_{y=0}^a \right) dx = \int_0^a \left(ax^2 + \frac{a^2 x}{2} \right) dx$

$= \left[\frac{ax^3}{3} + \frac{a^2 x^2}{4} \right]_0^a = \frac{7a^4}{12}$

Therefore, $\bar{x} = \dfrac{M_y}{m} = \dfrac{7a}{12}$.

c.

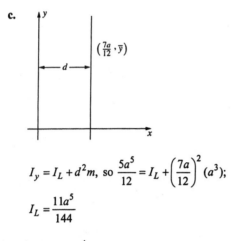

$\left(\frac{7a}{12}, \bar{y} \right)$

$I_y = I_L + d^2 m$, so $\dfrac{5a^5}{12} = I_L + \left(\dfrac{7a}{12} \right)^2 (a^3)$;

$I_L = \dfrac{11a^5}{144}$

23. $I_x = 2[I_{15}] = \dfrac{ka^4 \pi}{2}$

$I_y = 2[I_{15} + md^2]$

$= 2[0.25a^4 \pi + (k\pi a^2)(2a)^2] = 8.5ka^4 \pi$

$I_z = I_x + I_y = 9ka^4 \pi$

25. $M_y = \iint_{S_1 \cup S_2} x\delta(x,y)dA$

$= \iint_{S_1} x\delta(x,y)dA + \iint_{S_2} x\delta(x,y)dA$

$= \dfrac{m_1 \iint_{S_1} x\delta(x,y)dA}{m_1} + \dfrac{m_2 \iint_{S_2} x\delta(x,y)dA}{m_2}$

$= m_1 \bar{x}_1 + m_2 \bar{x}_2$

16.6 Concepts Review

1. $|\mathbf{u} \times \mathbf{v}|$

3. $\displaystyle\int_{-a}^a \int_{-\sqrt{a^2-x^2}}^{\sqrt{a^2-x^2}} \left(\frac{a}{\sqrt{a^2-x^2-y^2}} \right) dy\,dx$

$= \displaystyle\int_0^{2\pi} \int_0^a \left(\frac{ar}{\sqrt{a^2-r^2}} \right) dr\,d\theta; \ 2\pi a^2$

Thus, $\bar{x} = \dfrac{M_y}{m} = \dfrac{m_1 x_1 + m_2 x_2}{m_1 + M_2}$ which is equal to what we are to obtain and which is what we would obtain using the center of mass formula for two point masses. (Similar result can be obtained for $\bar{y}$.)

27.

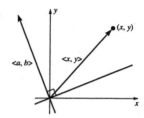

$\langle a,b \rangle$ is perpendicular to the line $ax + by = 0$. Therefore, the (signed) distance of (x, y) to L is the scalar projection of $\langle x,y \rangle$ onto $\langle a,b \rangle$, which

is $d(x, y) = \dfrac{\langle x, y \rangle \cdot \langle a, b \rangle}{|\langle a, b \rangle|} = \dfrac{ax+by}{|\langle a, b \rangle|}$.

$M_L = \iint_S d(x, y)\delta(x, y)dA$

$= \iint_S \dfrac{ax+by}{|\langle a, b \rangle|} \delta(x, y)dA$

$= \dfrac{a}{|\langle a, b \rangle|} \iint_S x\delta(x, y)dA + \dfrac{b}{|\langle a, b \rangle|} \iint_S y\delta(x, y)dA$

$= \dfrac{a}{|\langle a, b \rangle|}(0) + \dfrac{b}{|\langle a, b \rangle|}(0) = 0$

[since $(\bar{x}, \bar{y}) = (0, 0)$]

Problem Set 16.6

1.

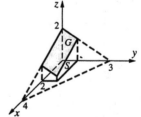

$$z = 2 - \frac{1}{2}x - \frac{2}{3}y$$

$$f_x(x, y) = -\frac{1}{2}; f_y(x, y) = -\frac{2}{3}$$

$$A(G) = \int_0^2 \int_0^1 \sqrt{\frac{1}{4} + \frac{4}{9} + 1}\, dy\, dx$$

$$= \frac{\sqrt{61}}{3} \approx 2.6034$$

3.

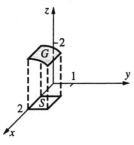

$$z = f(x, y) = (4 - y^2)^{1/2}; f_x(x, y) = 0;$$

$$f_y(x, y) = -y(4 - y^2)^{-1/2}$$

$$A(G) = \int_0^1 \int_1^2 \sqrt{y^2(4 - y^2)^{-1} + 1}\, dx\, dy$$

$$= \int_0^1 \int_1^2 \frac{2}{\sqrt{4 - y^2}}\, dx\, dy$$

$$= \int_0^1 \frac{2}{\sqrt{4 - y^2}}\, dy = \left[2\sin^{-1}\left(\frac{y}{2}\right)\right]_0^1 = 2\left(\frac{\pi}{6}\right) - 2(0)$$

$$= \frac{\pi}{3} \approx 1.0472$$

$$A(G) = \int_0^2 \int_0^{\sqrt{4-y^2}} 2(4 - y^2)^{-1/2}\, dx\, dy = 4$$

(See problem 3 for the integrand.)

5.

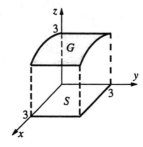

Let $z = f(x, y) = (9 - x^2)^{1/2}$.

$$f_x(x, y) = -x(9 - x^2)^{-1/2}, f_y(x, y) = 0$$

$$A(G) = \int_0^2 \int_0^3 [x^2(9 - x^2)^{-1} + 1]\, dy\, dx$$

$$= \int_0^2 \int_0^3 3(9 - x^2)^{-1/2}\, dx\, dy = 9\sin^{-1}\left(\frac{2}{3}\right)$$

$$\approx 6.5675$$

7.

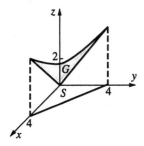

$$z = f(x, y) = (x^2 + y^2)^{1/2}$$

$$f_x(x, y) = x(x^2 + y^2)^{-1/2}, f_y(x, y) = y(x^2 + y^2)^{-1/2}$$

$$A(G) = \int_0^4 \int_0^{4-x} [x^2(x^2 + y^2)^{-1} + y^2(x^2 + y^2)^{-1} + 1]^{1/2}\, dy\, dx = \int_0^4 \int_0^{4-x} \sqrt{2}\, dy\, dx = 8\sqrt{2}$$

9.

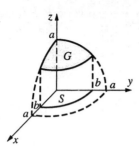

$$f_x(x,y) = \frac{x^2}{a^2 - x^2 - y^2}; f_y(x,y) = \frac{y^2}{a^2 - x^2 - y^2}$$

(See Example 3.)

$$A(G) = 8 \iint_S \frac{a}{\sqrt{a^2 - x^2 - y^2}} \, dA$$

$$= 8 \int_0^{\pi/2} \int_0^b \frac{a}{\sqrt{a^2 - r^2}} \, r \, dr \, d\theta$$

$$= 8a\left(\frac{\pi}{2}\right) \int_0^b (a^2 - r^2)^{-1/2} r \, dr$$

$$= -4a\pi \left[(a^2 - r^2)^{1/2}\right]_0^b = 4\pi a\left(a - \sqrt{a^2 - b^2}\right)$$

11.

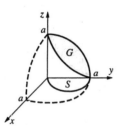

$$f_x(x,y) = \frac{x^2}{a^2 - x^2 - y^2}; f_y(x,y) = \frac{y^2}{a^2 - x^2 - y^2}$$

(See Example 3.)

$$A(G) = 4 \int_0^{\pi/2} \int_0^{a\sin\theta} \frac{a}{\sqrt{a^2 - r^2}} \, r \, dr \, d\theta$$

$$= 4a^2 \int_0^{\pi/2} (1 - \cos\theta) d\theta = 2a^2(\pi - 2)$$

13.

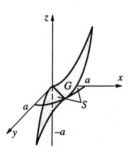

Let $F(x,y,z) = x^2 - y^2 - az$.

$$f_x(x,y) = \frac{2x}{a}; f_y(x,y) = \frac{-2y}{a}$$

$$A(G) = \int_0^{2\pi} \int_0^a \frac{\sqrt{4r^2 + a^2}}{a} r \, dr \, d\theta$$

$$= \frac{2\pi}{a} \int_0^a (4r^2 + a^2)^{1/2} r \, dr = \frac{\pi a^2\left(5\sqrt{5} - 1\right)}{6}$$

15. $\bar{x} = \bar{y} = 0$ (by symmetry)

Let $h = \dfrac{h_1 + h_2}{2}$. Planes $z = h_1$ and $z = h$ cut out the same surface area as planes $z = h$ and $z = h_2$. Therefore, $\bar{z} = h$, the arithmetic average of h_1 and h_2.

17. a. $A = \pi b^2$

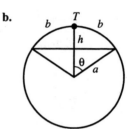

b.

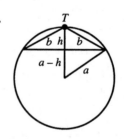

$$B = 2\pi a^2(1 - \cos\phi) \quad \text{(Problem 16)}$$

$$= 2\pi a^2 \left[1 - \cos\left(\frac{b}{a}\right)\right]$$

$$= 2\pi a^2 \left[\frac{b^2}{2!a^2} - \frac{b^4}{4!a^4} + \frac{b^6}{6!a^4} + \cdots\right]$$

$$= \pi b^2 \left[1 - \frac{b^2}{12a^2} + \frac{b^4}{360a^4} - \cdots\right] \le \pi b^2$$

c.

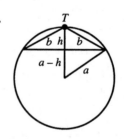

$a^2 - (a-h)^2 = b^2 - h^2$, so $h = \dfrac{b^2}{2a}$.

Thus, $C = 2\pi ah$

$$= 2\pi a\left(\frac{b^2}{2a}\right) = \pi b^2.$$

d.

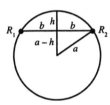

$D = 2\pi ah$

$$= 2\pi a\left(a - \sqrt{a^2 - b^2}\right) = \frac{2\pi a[a^2 - (a^2 - b^2)]}{a + \sqrt{a^2 - b^2}}$$

$$= \frac{2\pi ab^2}{a + \sqrt{a^2 - b^2}} > \pi b^2$$

Therefore, $B < A = C < D$.

19. In the following, each double integral is over S_{xy}

$A(S_{xy})f(\overline{x}, \overline{y}) = A(S_{xy})(a\overline{x} + b\overline{y} + c)$

$$= \iint dA\left[a\frac{\iint x\,dA}{\iint dA} + b\frac{\iint y\,dA}{\iint dA} + c\right]$$

$$= a\iint x\,dA + b\iint y\,dA + c\iint dA$$

$$= \iint(ax + by + c)dA$$

$$= \text{Volume of solid cylinder under } S_{xy}$$

21. Let G denote the surface of that part of the plane $z = Ax + By + C$ over the region S. First, suppose that S is the rectangle $a \le x \le b, c \le y \le d$. Then the vectors $\mathbf{u}$ and $\mathbf{v}$ that form the edge of the parallelogram G are $\mathbf{u} = (b-a)\mathbf{i} + 0\mathbf{j} + A(b-a)\mathbf{k}$ and $\mathbf{v} = 0\mathbf{i} + (d-c)\mathbf{j} + B(d-c)\mathbf{k}$. The surface area of G is thus

$|\mathbf{u} \times \mathbf{v}| =$

$\left|-A(b-a)(d-c)\mathbf{i} - B(b-a)(d-c)\mathbf{j} + (b-a)(d-c)\right|$

$= (b-a)(d-c)\sqrt{A^2 + B^2 + 1}$

A normal vector to the plane is $\mathbf{n} = -A\mathbf{i} - B\mathbf{j} + \mathbf{k}$.

Thus,

$$\cos\gamma = \frac{\mathbf{n}\cdot\mathbf{k}}{|\mathbf{n}||\mathbf{k}|} = \frac{\langle-A,-B,1\rangle\cdot\langle0,0,1\rangle}{\sqrt{A^2 + B^2 + 1}\cdot 1} = \frac{1}{\sqrt{A^2 + B^2 + 1}}$$

$$\sec\gamma = \frac{1}{\cos\gamma} = \sqrt{A^2 + B^2 + 1} = |\mathbf{u} \times \mathbf{v}| = A(G).$$

If S is not a rectangle, then make a partition of S with rectangles $R_1, R_2, \ldots, R_n$. The Riemann sum will be

$$\sum_{m=1}^{n} A(G_m)\Delta x_m \Delta y_m = \sum_{m=1}^{n} \sec\gamma\, \Delta x_m \Delta y_m .$$

$$= \sec\gamma\sum_{m=1}^{n} A(R_m). \text{ As we take the limit as}$$

$|P| \to 0$ the sum converges to the area of S.

Thus the surface area will

be $A(G) = \lim_{|P|\to 0} \sec\gamma\sum_{m=1}^{n} A(R_m) = \sec\gamma A(S)$.

G is thus

$|\mathbf{u} \times \mathbf{v}| =$

$\left|-A(b-a)(d-c)\mathbf{i} - B(b-a)(d-c)\mathbf{j} + (b-a)(d-c)\right|$

$= (b-a)(d-c)\sqrt{A^2 + B^2 + 1}$

A normal vector to the plane is $\mathbf{n} = -A\mathbf{i} - B\mathbf{j} + \mathbf{k}$.

Thus,

$$\cos\gamma = \frac{\mathbf{n}\cdot\mathbf{k}}{|\mathbf{n}||\mathbf{k}|} = \frac{\langle-A,-B,1\rangle\cdot\langle0,0,1\rangle}{\sqrt{A^2 + B^2 + 1}\cdot 1} = \frac{1}{\sqrt{A^2 + B^2 + 1}}$$

$$\sec\gamma = \frac{1}{\cos\gamma} = \sqrt{A^2 + B^2 + 1} = |\mathbf{u} \times \mathbf{v}| = A(G).$$

If S is not a rectangle, then make a partition of S with rectangles $R_1, R_2, \ldots, R_n$. The Riemann sum will be

$$\sum_{m=1}^{n} A(G_m)\Delta x_m \Delta y_m = \sum_{m=1}^{n} \sec\gamma\, \Delta x_m \Delta y_m .$$

$$= \sec\gamma\sum_{m=1}^{n} A(R_m). \text{ As we take the limit as}$$

$|P| \to 0$ the sum converges to the area of S.

Thus the surface area will be

$$A(G) = \lim_{|P|\to 0} \sec\gamma\sum_{m=1}^{n} A(R_m) = \sec\gamma A(S).$$

16.7 Concepts Review

1. volume

3. $y; \sqrt{y}$

Problem Set 16.7

1. $\int_{-3}^{7} \int_{0}^{2x} (x - 1 - y)\,dy\,dx = \int_{-3}^{7} -2x\,dx = -40$

3. $\int_{1}^{4} \int_{z-1}^{2z} \int_{0}^{y+2z} dx\,dy\,dz = \int_{1}^{4} \int_{z-1}^{2z} (y + 2z)\,dy\,dz$

$= \int_{1}^{4} \left[\frac{y^2}{2} + 2yz \right]_{y=z-1}^{2z} dz$

$= \int_{1}^{4} \left(\frac{7z^2}{2} + 3z - \frac{1}{2} \right) dz$

$= \left[\frac{7z^3}{6} + \frac{3z^2}{2} - \frac{z}{2} \right]_{1}^{4} = \frac{189}{2} = 94.5$

5. $\int_{0}^{2} \int_{1}^{z} x^2\,dx\,dz = \int_{0}^{2} \left(\frac{1}{3} \right)(z^3 - 1)\,dz = \frac{2}{3}$

7. $\int_{-2}^{4} \int_{x-1}^{x+1} 3y^2\,dy\,dx = \int_{-2}^{4} (6x^2 + 2)\,dx = 156$

9. $\int_{0}^{1} \int_{0}^{3} \int_{0}^{(12-3x-2y)/6} f(x, y, z)\,dz\,dy\,dx$

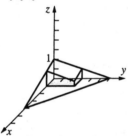

11. $\int_{0}^{2} \int_{0}^{4} \int_{0}^{y/2} f(x, y, z)\,dx\,dy\,dz$

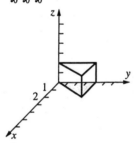

13. $\int_{0}^{2.4} \int_{x/3}^{(4-x)/2} \int_{0}^{4-x-2z} f(x, y, z)\,dy\,dz\,dx$

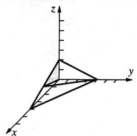

15.

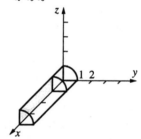

Using the cross product of vectors along edges, it is easy to show that $\langle 2, 6, 9 \rangle$ is normal to the upward face. Then obtain that its equation is $2x + 6y + 9z = 18$.

$\int_{0}^{3} \int_{2x/3}^{(9-x)/3} \int_{0}^{(18-2x-6y)/9} f(x, y, z)\,dz\,dy\,dx$

17. $\int_{1}^{4} \int_{0}^{1} \int_{0}^{\sqrt{1-y^2}} f(x, y, z)\,dz\,dy\,dx$

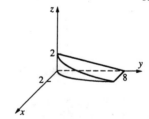

19. $\int_{0}^{2} \int_{2x^2}^{8} \int_{0}^{2-y/4} 1\,dz\,dy\,dx = \frac{128}{15}$

21. $V = 4 \int_0^1 \int_0^{\sqrt{y}} \int_0^{\sqrt{y}} 1\, dz\, dx\, dy = 4 \int_0^1 \int_0^{\sqrt{y}} \sqrt{y}\, dx\, dy$

$= 4 \int_0^1 \sqrt{y}\sqrt{y}\, dy = [2y^2]_0^1 = 2$

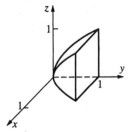

23. Let $\delta(x, y, z) = x + y + z$. (See note with next problem.)

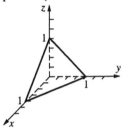

$m = \int_0^1 \int_0^{1-x} \int_0^{1-x-y} (x+y+z)\, dz\, dy\, dx = \dfrac{1}{8}$

$M_{yz} = \int_0^1 \int_0^{1-x} \int_0^{1-x-y} x(x+y+z)\, dz\, dy\, dx = \dfrac{1}{30}$

$\bar{x} = \dfrac{4}{15}$

Then $\bar{y} = \bar{z} = \dfrac{4}{15}$ (symmetry).

25. Let $\delta(x, y, z) = 1$. (See note with previous problem.)

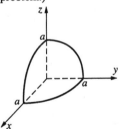

31.

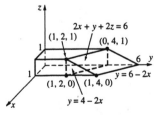

a. $\int_0^1 \int_0^{4-2x} \int_0^1 dz\, dy\, dx + \int_0^1 \int_{4-2x}^{6-2x} \int_0^{3-x-y/2} dz\, dy\, dx = 3 + 1 = 4$

$m = \left(\dfrac{1}{8}\right)(\text{volume of sphere}) = \left(\dfrac{\pi}{6}\right)a^3$

$M_{xy} = \int_0^a \int_0^{\sqrt{a^2-x^2}} \int_0^{\sqrt{a^2-x^2-y^2}} z\, dz\, dy\, dx$

$= \int_0^{\pi/2} \int_0^a \int_0^{\sqrt{a^2-r^2}} zr\, dz\, dr\, d\theta = \left(\dfrac{\pi}{16}\right)a^4$

$\bar{x} = \left(\dfrac{3}{8}\right)a$

$\bar{y} = \bar{z} = \left(\dfrac{3}{8}\right)a$ (by symmetry)

27.

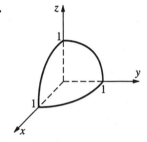

The limits of integration are those for the first octant part of a sphere of radius 1.

$\int_0^1 \int_0^{\sqrt{1-x^2}} \int_0^{\sqrt{1-x^2-y^2}} f(x, y, z)\, dz\, dy\, dx$

29. $\int_0^2 \int_0^{2-z} \int_0^{9-x^2} f(x, y, z)\, dy\, dx\, dz$

Figure is same as for Problem 30 except that the solid doesn't need to be divided into two parts.

b. $\int_0^1 \int_0^1 \int_0^{6-2x-2z} 1 \, dy \, dx \, dz = 4$

c. $A(S_{xz}) f(\bar{x}, \bar{z})$ (S_{xz} is the unit square in the corner of xz-plane; and $(\bar{x}, \bar{z}) = \left(\dfrac{1}{2}, \dfrac{1}{2}\right)$ is the centroid of S_{xz}.)

$$= (1)\left[6 - 2\left(\frac{1}{2}\right) - 2\left(\frac{1}{2}\right)\right] = 4$$

33. $\int_0^1 \int_0^1 \int_0^{6-2x-2z} (30-z) \, dy \, dx \, dz = \int_0^1 \int_0^1 (30-z)(6-2x-2z) \, dx \, dz = \int_0^1 ([(30-z)(6x - x^2 - 2xz)]_{x=0}^1) \, dz$

$$= \int_0^1 (30-z)(5-2z) \, dz = \int_0^1 (150 - 65z + 2z^2) \, dz = \left[150z - \frac{65z^2}{2} + \frac{2z^3}{3}\right]_0^1 = \frac{709}{6}$$

The volume of the solid is 4 (from Problem 31).

Hence, the average temperature of the solid is $\dfrac{\frac{709}{6}}{4} = \dfrac{709}{24} \approx 29.54°$.

$$\int_0^1 \int_0^1 \int_0^{6-2x-2z} z \, dy \, dx \, dz = \int_0^1 \int_0^1 z(6-2x-2z) \, dx \, dz = \int_0^1 ([z(6x - x^2 - 2xz)]_{x=0}^1) \, dz$$

$$= \int_0^1 (5z - 2z^2) \, dz = \left[\frac{5z^2}{2} - \frac{2z^3}{3}\right]_0^1 = \frac{11}{6}$$

Hence, $\bar{z} = \dfrac{\frac{11}{6}}{4} = \dfrac{11}{24} \approx 0.4583$.

35. The result obtained from Mathematica is:

$$\int_0^c \int_0^{b\sqrt{1-z^2/c^2}} \int_0^{a\sqrt{1-y^2/b^2-z^2/c^2}} 8(xy + xz + yz) \, dx \, dy \, dz = \frac{8}{15} a^2 b^2 c + \frac{8}{15} a^2 bc^2 + \frac{8}{15} ab^2 c^2$$

$$= \frac{8}{15} acb(ca + cb + ab)$$

16.8 Concepts Review

1. $r \, dz \, dr \, d\theta; \ \rho^2 \sin\phi \, d\rho \, d\theta \, d\phi$

3. $\int_0^\pi \int_0^{2\pi} \int_0^1 \rho^4 \cos^2\phi \sin\phi \, d\rho \, d\theta \, d\phi$

Problem Set 16.8

1. $\int_0^{2\pi} \int_0^2 \int_{r^2}^4 r \, dz \, dr \, d\theta = 8\pi \approx 25.1327$

3. $\int_0^{2\pi} \int_0^2 \int_{r^2/4}^{\sqrt{5-r^2}} r \, dz \, dr \, d\theta$

$$= \int_0^{2\pi} \int_0^2 \left[4r(5-r^2)^{1/2} - \frac{r^3}{4}\right] dr \, d\theta$$

$$= \int_0^{2\pi} \frac{5^{3/2} - 4}{3} \, d\theta = \frac{2\pi(5^{3/2} - 4)}{3} \approx 15.0385$$

5. Let $\delta(x, y, z) = 1$.
(See write-up of Problem 24, Section 16.7.)

$$m = \int_0^{2\pi} \int_0^2 \int_{r^2}^{12-2r^2} r \, dz \, dr \, d\theta = 24\pi$$

$$M_{xy} = \int_0^{2\pi} \int_0^2 \int_{r^2}^{12-2r^2} zr \, dz \, dr \, d\theta = 128\pi$$

$$\bar{z} = \frac{16}{3}$$

$\bar{x} = \bar{y} = 0$ (by symmetry)

7. Let $\delta(x, y, z) = k\rho$

$$m = \int_0^\pi \int_0^{2\pi} \int_a^b k\rho\rho^2 \sin\phi \, d\rho \, d\theta \, d\phi = k\pi(b^4 - a^4)$$

9. Let $\delta(x, y, z) = \rho$.
(Letting $k = 1$ -- see comment at the beginning of the write-up of Problem 24 of the previous section.)

$$m = \int_0^{\pi/2} \int_0^{2\pi} \int_0^a \rho^3 \sin\phi \, d\rho \, d\theta \, d\phi$$

$$= \int_0^{\pi/2} \int_0^{2\pi} \frac{a^4 \sin\phi}{4} \, d\theta \, d\phi$$

$$= \int_0^{\pi/2} \frac{\pi a^4 \sin\phi}{2}\,d\phi = \frac{\pi a^4}{2}$$

$$M_{xy} = \int_0^{\pi/2} \int_0^{2\pi} \int_0^a \rho^4 \sin\phi\cos\phi\,d\rho\,d\theta\,d\phi$$

$(z = \rho\cos\phi)$

$$= \int_0^{\pi/2} \int_0^{2\pi} \frac{a^5 \sin 2\phi}{10}\,d\theta\,d\phi$$

$$= \int_0^{\pi/2} \frac{\pi a^5 \sin 2\phi}{5}\,d\phi = \left(\frac{\pi}{5}\right) a^5$$

$$\overline{z} = \frac{\frac{\pi a^5}{5}}{\frac{\pi a^4}{2}} = \frac{2}{5}a; \overline{x} = \overline{y} = 0 \ \text{(by symmetry)}$$

11. $I_z = \iint_S (x^2 + y^2)k(x^2 + y^2)^{1/2}\,dV$

$$= \int_0^{\pi/2} \int_0^{2\pi} \int_0^a k\rho^5 \sin^4\phi\,d\rho\,d\theta\,d\phi = \left(\frac{k}{16}\right)\pi^2 a^6$$

13. $\int_0^{\pi} \int_0^{\pi/6} \int_0^1 \rho^2 \sin\phi\,d\rho\,d\theta\,d\phi = \frac{\pi}{9} \approx 0.3491$

15. Volume $= \int_0^{\pi} \int_0^{\sin\theta} \int_{r^2}^{r\sin\theta} r\,dz\,dr\,d\theta$

$$= \int_0^{\pi} \int_0^{\sin\theta} r(r\sin\theta - r^2)\,dr\,d\theta = \int_0^{\pi} \frac{\sin^4\theta}{12}\,d\theta$$

$$= \frac{1}{48} \int_0^{\pi} \left[1 - 2\cos 2\theta + \frac{1+\cos 4\theta}{2}\right] d\theta$$

$$= \frac{\pi}{32} \approx 0.0982$$

17. a. Position the ball with its center at the origin. The distance of (x, y, z) from the origin is $(x^2 + y^2 + z^2)^{1/2} = \rho$.

$$\iiint_S (x^2 + y^2 + z^2)^{1/2}\,dV = 8\int_0^{\pi/2} \int_0^{\pi/2} \int_0^a \rho(\sin\theta\rho^2)\,d\rho\,d\theta\,d\phi = \pi a^4$$

Then the average distance from the center is $\dfrac{\pi a^4}{\left[\left(\frac{4}{3}\right)\pi a^3\right]} = \dfrac{3a}{4}$.

b. Position the ball with its center at the origin and consider the diameter along the z-axis. The distance of (x, y, z) from the z-axis is $(x^2 + y^2)^{1/2} = \rho\sin\phi$.

$$\iiint_S (x^2 + y^2)^{1/2}\,dV = 8\int_0^{\pi/2} \int_0^{\pi/2} \int_0^a (\rho\sin\phi)(\rho^2\sin\theta)\,d\rho\,d\theta\,d\phi = \frac{a^4\pi^2}{4}$$

Then the average distance from a diameter is $\dfrac{\left[\frac{a^4\pi^2}{4}\right]}{\left[\left(\frac{4}{3}\right)\pi a^3\right]} = \dfrac{3\pi a}{16}$.

c. Position the sphere above and tangent to the xy-plane at the origin and consider the point on the boundary to be the origin. The equation of the sphere is $\rho = 2a\cos\phi$, and the distance of (x, y, z) from the origin is ρ.

$$\iiint_S (x^2 + y^2 + z^2)^{1/2}\,dV = \int_0^{\pi/2} \int_0^{2\pi} \int_0^{2a\cos\phi} \rho(\rho^2\sin\theta)\,d\rho\,d\theta\,d\phi = \frac{8\pi a^4}{5}$$

Then the average distance from the origin is $\dfrac{\left[\frac{8\pi a^4}{5}\right]}{\left[\left(\frac{4}{3}\right)\pi a^3\right]} = \dfrac{6a}{5}$.

19. a. $M_{yz} = \iiint_S kx\,dV = 4k\int_0^{\pi/2} \int_0^{\alpha} \int_0^a (\rho\sin\phi\cos\theta)(\rho^2\sin\phi)\,d\rho\,d\theta\,d\phi = ka^4\pi\dfrac{(\sin\alpha)}{4}$

$m = \iiint_S k\,dV = 4k\int_0^{\pi/2} \int_0^{\alpha} \int_0^a \rho^2\sin\phi\,d\rho\,d\theta\,d\phi = \dfrac{4a^3 k\alpha}{3}$

Therefore, $\bar{x} = \dfrac{\left[\dfrac{ka^4\pi(\sin\alpha)}{4}\right]}{\left[\dfrac{4a^3ka}{3}\right]} = \dfrac{3a\pi(\sin\alpha)}{16\alpha}$.

b. $\dfrac{3\pi a}{16}$ (See Problem 17b.)

21. Let m_1 and m_2 be the masses of the left and right balls, respectively. Then $m_1 = \dfrac{4}{3}\pi a^3 k$ and $m_2 = \dfrac{4}{3}\pi a^3 (ck)$, so $m_2 = cm_1$.

$$\bar{y} = \dfrac{m_1(-a-b) + m_2(a+b)}{m_1 + m_2}$$

$$= \dfrac{m_1(-a-b) + cm_1(a+b)}{m_1 + cm_1} = \dfrac{-a-b+c(a+b)}{1+c}$$

$$= \dfrac{(a+b)(-1+c)}{1+c} = \dfrac{c-1}{c+1}(a+b)$$

(Analogue) $\bar{y} = \dfrac{m_1\bar{y}_1 + m_2\bar{y}_2}{m_1 + m_2} = \bar{y}_1\dfrac{m_1}{m_1+m_2} + \bar{y}_2\dfrac{m_2}{m_1+m_2}$

23. $x = \rho\sin\phi\cos\theta$, $y = \rho\sin\phi\sin\theta$, $z = \rho\cos\phi$

$$J(\rho,\phi,\theta) = \begin{vmatrix} x_\rho & x_\phi & x_\theta \\ y_\rho & y_\phi & y_\theta \\ z_\rho & z_\phi & z_\theta \end{vmatrix} = \begin{vmatrix} \sin\phi\cos\theta & \rho\cos\phi\cos\theta & -\rho\sin\phi\sin\theta \\ \sin\phi\sin\theta & \rho\cos\phi\sin\theta & \rho\sin\phi\cos\theta \\ \cos\phi & -\rho\sin\phi & 0 \end{vmatrix}$$

$$= (\cos\phi)(\rho^2\cos\phi\sin\phi)(\cos^2\theta + \sin^2\theta) + (\rho\sin\phi)(\rho\sin^2\phi)(\cos^2\theta + \sin^2\theta)$$

$$= \rho^2\sin\phi(\cos^2\phi + \sin^2\phi) = \rho^2\sin\phi$$

(Expansion was along the third row of the determinant.)

16.9 Chapter Review

Concepts Test

1. True: Use result of Problem 25, Section 16.2, and then change dummy variable y to dummy variable x.

3. True: Inside integral is 0 since $\sin(x^3y^2)$ is an odd function in x.

5. True: It is less than or equal to $\int_1^2 \int_0^2 1\,dx\,dy$ which equals 2.

7. False: Let $f(x,y) = x$, $g(x,y) = x^2$, $R = \{(x,y): x \text{ in } [0,2], y \text{ in } [0,1]\}$. The inequality holds for the integrals but $f(0.5, 0) > g(0.5, 0)$.

9. True: See the write-up of Problem 24, Section 16.7.

11. True: The integral is the volume between concentric spheres of radii 4 and 1. That volume is 84π.

13. False: There are 6.

15. True: $|\nabla f|$ is the magnitude of the greatest increase in f.

$$|D_{\mathbf{u}}f| = |\nabla f \cdot \mathbf{u}| = |\langle f_x, f_y\rangle \cdot \mathbf{u}|$$

$$= \sqrt{f_x^2 + f_y^2}\,(1)\cos\theta \le \sqrt{4+4} = \sqrt{8}$$

Therefore,
Area$(G) \le$ Area(R) max$\{\sec\gamma\}$
$\le (1)\sec\left(\tan^{-1}\sqrt{8}\right) = 3$.

Sample Test Problems

1. $\int_0^1 \left(\frac{1}{2}\right)(x^2 - x^3)dx = \frac{1}{24} \approx 0.0417$

3. $\int_0^{\pi/2}\left[\frac{r^2\cos\theta}{2}\right]_{r=0}^{2\sin\theta} d\theta = \int_0^{\pi/2} 2\sin^2\theta\cos\theta\,d\theta$

$= \left[\frac{2\sin^3\theta}{3}\right]_0^{\pi/2} = \frac{2}{3}$

5. $\int_0^1 \int_0^y f(x,\,y)dx\,dy$

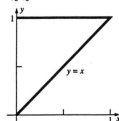

$y = x$

7. $\int_0^{1/2}\int_0^{1-2y}\int_0^{1-2y-z} f(x,\,y,\,z)dx\,dz\,dy$

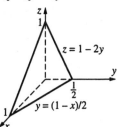

$z = 1 - 2y$

$y = (1-x)/2$

9. a. $8\int_0^a \int_0^{\sqrt{a^2-x^2}} \int_0^{\sqrt{a^2-x^2-y^2}} dz\,dy\,dx$

b. $8\int_0^{\pi/2}\int_0^a \int_0^{\sqrt{a^2-r^2}} r\,dz\,dr\,d\theta$

c. $8\int_0^{\pi/2}\int_0^{\pi/2}\int_0^a \rho^2\sin\phi\,d\rho\,d\theta\,d\phi$

11. $8\int_0^1 \int_0^x \int_0^{1-y^2} z^2\,dz\,dy\,dx = \frac{31}{35} \approx 0.8857$

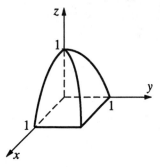

13. $m = \int_0^2 \int_1^3 xy^2\,dx\,dy = \frac{32}{3}$

$M_x = \int_0^2 \int_1^3 xy^3\,dx\,dy = 16$

$M_y = \int_0^2 \int_1^3 x^2 y^2\,dx\,dy = \frac{208}{9}$

$(\overline{x},\,\overline{y}) = \left(\frac{13}{6},\frac{3}{2}\right)$

15. $z = f(x,\,y) = (9-y^2)^{1/2};\ f_x(x,\,y) = 0;$

$f_y(x,\,y) = -y(9-y^2)^{-1/2}$

$\text{Area} = \int_0^3 \int_{y/3}^y \sqrt{y^2(9-y^2)^{-1}+1}\,dx\,dy$

$= \int_0^2 \int_{y/3}^y 3(9-y^2)^{-1/2}\,dx\,dy$

$= \int_0^3 (9-y^2)^{-1/2}(2y)dy = [-2(9-y^2)^{1/2}]_0^3 = 6$

17. $\delta(x,\,y,\,z) = k\rho$

$m = \int_0^\pi \int_0^{2\pi}\int_1^3 k\rho\,\rho^2\sin\phi\,d\rho\,d\theta\,d\phi = 80\pi k$

19. $m = \int_0^a \int_0^{(b/a)(a-x)}\int_0^{(c/ab)(ab-bx-ay)} kx\,dz\,dy\,dx$

$= \left(\frac{k}{24}\right)a^2 bc$

17

Vector Calculus

17.1 Concepts Review

1. vector-valued function of three real variables or a vector field

3. gravitational fields; electric fields

Problem Set 17.1

1.

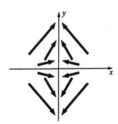

3.

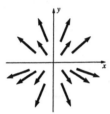

5.

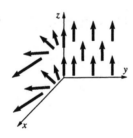

7. $\langle 2x - 3y, -3x, 2 \rangle$

9. $f(x, y, z) = \ln|x| + \ln|y| + \ln|z|$;
$\nabla f(x, y, z) = \langle x^{-1}, y^{-1}, z^{-1} \rangle$

11. $e^y \langle \cos z, x\cos z, -x\sin z \rangle$

13. $\operatorname{div} \mathbf{F} = 2x - 2x + 2yz = 2yz$
$\operatorname{curl} \mathbf{F} = \langle z^2, 0, -2y \rangle$

15. $\operatorname{div} \mathbf{F} = \nabla \cdot \mathbf{F} = 0 + 0 + 0 = 0$
$\operatorname{curl} \mathbf{F} = \nabla \times \mathbf{F} = \langle x - x, \, y - y, \, z - z \rangle = 0$

17. $\operatorname{div} \mathbf{F} = e^x \cos y + e^x \cos y + 1 = 2e^x \cos y + 1$
$\operatorname{curl} \mathbf{F} = \langle 0, 0, 2e^x \sin y \rangle$

19. **a.** meaningless

   b. vector field

   c. vector field

   d. scalar field

   e. vector field

   f. vector field

   g. vector field

   h. meaningless

   i. meaningless

   j. scalar field

   k. meaningless

21. Let $f(x, y, z) = -c|\mathbf{r}|^{-3}$, so
$$\operatorname{grad}(f) = 3c|\mathbf{r}|^{-4} \frac{\mathbf{r}}{|\mathbf{r}|} = 3c|\mathbf{r}|^{-5} \mathbf{r}.$$
Then $\operatorname{curl} \mathbf{F} = \operatorname{curl}\left[\left(-c|\mathbf{r}|^{-3}\right)\mathbf{r}\right]$
$$= \left(-c|\mathbf{r}|^{-3}\right)(\operatorname{curl} \mathbf{r}) + \left(3c|\mathbf{r}|^{-5} r\right) \times \mathbf{r} \text{ (by 20d)}$$
$$= \left(-c|\mathbf{r}|^{-3}\right)(\mathbf{0}) + \left(3c|\mathbf{r}|^{-5}\right)(\mathbf{r} \times \mathbf{r}) = 0 + 0 = 0$$
$\operatorname{div} \mathbf{F} = \operatorname{div}\left[\left(-c|\mathbf{r}|^{-3}\right)\mathbf{r}\right]$
$$= \left(-c|\mathbf{r}|^{-3}\right)(\operatorname{div} \mathbf{r}) + \left(3c|\mathbf{r}|^{-5} \mathbf{r}\right) \cdot \mathbf{r} \text{ (by 20c)}$$

$$= \left(-c|\mathbf{r}|^{-3}\right)(1+1+1) + \left(3c|\mathbf{r}|^{-5}\right)|\mathbf{r}|^2$$

$$= \left(-3c|\mathbf{r}|^{-3}\right) + 3c|\mathbf{r}|^{-3} = 0$$

23. $\operatorname{grad} f = \left\langle f'(r)xr^{-1/2}, f'(r)yr^{-1/2}, f'(r)zr^{-1/2}\right\rangle$

(if $r \neq 0$)

$= f'(r)r^{-1/2}\langle x, y, z\rangle = f'(r)r^{-1/2}\mathbf{r}$

$\operatorname{curl} \mathbf{F} = [f(r)][\operatorname{curl} \mathbf{r}] + [f'(r)r^{-1/2}\mathbf{r}]\times\mathbf{r}$

$= [f(r)][\operatorname{curl} \mathbf{r}] + [f'(r)r^{-1/2}\mathbf{r}]\times\mathbf{r}$

$= 0 + 0 = 0$

25. a. Let $P = (x_0, y_0)$.

div $\mathbf{F} =$ div $\mathbf{H} = 0$ since there is no tendency toward P except along the line $x = x_0$, and along that line the tendencies toward and away from P are balanced; div $\mathbf{G} < 0$ since there is no tendency toward P except along the line $x = x_0$, and along that line there is more tendency toward than away from P; div $\mathbf{L} > 0$ since the tendency away from P is greater than the tendency toward P.

b. No rotation for $\mathbf{F}$, $\mathbf{G}$, $\mathbf{L}$; clockwise rotation for $\mathbf{H}$ since the magnitudes of the forces to the right of P are less than those to the left.

c. div $\mathbf{F} = 0$; curl $\mathbf{F} = \mathbf{0}$

div $\mathbf{G} = -2ye^{-y^2} < 0$ since $y > 0$ at P; curl $\mathbf{G} = \mathbf{0}$

div $\mathbf{L} = (x^2 + y^2)^{-1/2}$; curl $\mathbf{L} = \mathbf{0}$

div $\mathbf{H} = 0$; curl $\mathbf{H} = \left\langle 0, 0, -2xe^{-x^2}\right\rangle$ which

points downward at P, so the rotation is clockwise in a right-hand system.

27. $\nabla f(x, y, z) = \dfrac{1}{2}m\omega^2\langle 2x, 2y, 2z\rangle = m\omega^2\langle x, y, z\rangle$

$= \mathbf{F}(x, y, z)$

Hence, each is harmonic.

29. a. $\mathbf{F}\times\mathbf{G} = (f_y g_z - f_z g_y)\mathbf{i} - (f_x g_z - f_z g_x)\mathbf{j} + (f_x g_z - f_z g_x)\mathbf{k}$

Therefore,

$$div(\mathbf{F}\times\mathbf{G}) = \frac{\partial}{\partial x}(f_y g_z - f_z g_y) - \frac{\partial}{\partial y}(f_x g_z - f_z g_x) + \frac{\partial}{\partial z}(f_x g_y - f_y g_x).$$

Using the product rule for partials and some algebra gives

$$div(\mathbf{F}\times\mathbf{G}) = g_x\left[\frac{\partial f_z}{\partial y} - \frac{\partial f_y}{\partial z}\right] + g_y\left[\frac{\partial f_x}{\partial z} - \frac{\partial f_z}{\partial x}\right] + g_z\left[\frac{\partial f_y}{\partial x} - \frac{\partial f_x}{\partial y}\right]$$

$$+ f_x\left[\frac{\partial g_z}{\partial y} - \frac{\partial g_y}{\partial z}\right] + f_y\left[\frac{\partial g_x}{\partial z} - \frac{\partial g_z}{\partial x}\right] + f_z\left[\frac{\partial g_y}{\partial x} - \frac{\partial g_x}{\partial y}\right]$$

$$= \mathbf{G}\cdot curl(\mathbf{F}) - \mathbf{F}\cdot curl(\mathbf{G})$$

b, $\nabla f\times\nabla g = \left(\dfrac{\partial f}{\partial y}\dfrac{\partial g}{\partial z} - \dfrac{\partial f}{\partial z}\dfrac{\partial g}{\partial y}\right)\mathbf{i} - \left(\dfrac{\partial f}{\partial x}\dfrac{\partial g}{\partial z} - \dfrac{\partial f}{\partial z}\dfrac{\partial g}{\partial x}\right)\mathbf{j} + \left(\dfrac{\partial f}{\partial x}\dfrac{\partial g}{\partial y} - \dfrac{\partial f}{\partial y}\dfrac{\partial g}{\partial x}\right)\mathbf{k}$

Therefore,

$$div(\nabla f\times\nabla g) = \frac{\partial}{\partial x}\left(\frac{\partial f}{\partial y}\frac{\partial g}{\partial z} - \frac{\partial f}{\partial z}\frac{\partial g}{\partial y}\right) - \frac{\partial}{\partial y}\left(\frac{\partial f}{\partial x}\frac{\partial g}{\partial z} - \frac{\partial f}{\partial z}\frac{\partial g}{\partial x}\right) + \frac{\partial}{\partial z}\left(\frac{\partial f}{\partial x}\frac{\partial g}{\partial y} - \frac{\partial f}{\partial y}\frac{\partial g}{\partial x}\right).$$

Using the product rule for partials and some algebra will yield the result

$div(\nabla f\times\nabla g) = 0$

17.2 Concepts Review

1. Increasing values of t

3. $f(x(t), y(t))\sqrt{[x'(t)]^2 + [y'(t)]^2}$

Problem Set 17.2

1. $\int_0^1 (27t^3 + t^3)(9 + 9t^4)^{1/2}\,dt = 14\left(2\sqrt{2} - 1\right)$

≈ 25.5980

3. Let $x = t, y = 2t$, t in $[0, \pi]$.

Then

$$\int_C (\sin x + \cos y)ds = \int_0^\pi (\sin t + \cos 2t)\sqrt{1+4}\,dt$$

$$= 2\sqrt{5} \approx 4.4721$$

5. $\int_0^1 (2t + 9t^3)(1 + 4t^2 + 9t^4)^{1/2}\,dt = \left(\frac{1}{6}\right)(14^{3/2} - 1)$

≈ 8.5639

7. $\int_0^2 [(t^2 - 1)(2) + (4t^2)(2t)]dt = \dfrac{100}{3}$

9. $\int_C y^3 dx + x^3 dy = \int_{C_1} y^3 dx + x^3 dy$

$$+ \int_{C_2} y^3 dx + x^3 dy$$

$$= \int_1^{-2} (-4)^3 dy + \int_{-4}^2 (-2)^3 dx = 192 + (-48) = 144$$

11. $y = -x + 2$

$$\int_1^3 ([x + 2(-x+2)](1) + [x - 2(-x+2)](-1))dx = 0$$

13. $\langle x, y, z \rangle = \langle 1, 2, 1 \rangle + t\langle 1, -1, 0 \rangle$

$$\int_0^1 [(4-t)(1) + (1+t)(-1) - (2 - 3t + t^2)(-1)]dt = \frac{17}{6}$$

≈ 2.8333

15. On C_1 : $y = z = dy = dz = 0$

On C_2 : $x = 2, z = dx = dz = 0$

On C_3 : $x = 2, y = 3, dx = dy = 0$

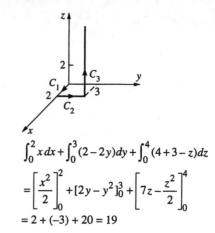

$$\int_0^2 x\,dx + \int_0^3 (2 - 2y)dy + \int_0^4 (4 + 3 - z)dz$$

$$= \left[\frac{x^2}{2}\right]_0^2 + [2y - y^2]_0^3 + \left[7z - \frac{z^2}{2}\right]_0^4$$

$$= 2 + (-3) + 20 = 19$$

17. $m = \int_C k|x|ds = \int_{-2}^2 k|x|(1 + 4x^2)^{1/2}\,dx$

$$= \left(\frac{k}{6}\right)(17^{3/2} - 1) \approx 11.6821k$$

19 $\int_C (x^3 - y^3)dx + xy^2 dy$

$$= \int_{-1}^0 [(t^6 - t^9)(2t) + (t^2)(t^6)(3t^2)]dt$$

$$= -\frac{7}{44} \approx -0.1591$$

21. $W = \int_C \mathbf{F} \cdot d\mathbf{r} = \int_C (x + y)dx + (x - y)dy = \int_0^{\pi/2} [(a\cos t + b\sin t)(-a\sin t) + (a\cos t - b\sin t)(b\cos t)]dt$

$$= \int_0^{\pi/2} [-(a^2 + b^2)\sin t \cos t + ab(\cos^2 t - \sin^2 t)]dt = \int_0^{\pi/2} \frac{-(a^2 + b^2)\sin 2t}{2} + ab\cos 2t\,dt$$

$$= \left[\frac{(a^2 + b^2)\cos 2t}{4} + \frac{ab\sin 2t}{2}\right]_0^{\pi/2} = \frac{a^2 + b^2}{-2}$$

23. $\int_0^\pi \left[\left(\frac{\pi}{2}\right)\sin\left(\frac{\pi t}{2}\right)\cos\left(\frac{\pi t}{2}\right) + \pi t\cos\left(\frac{\pi t}{2}\right) + \sin\left(\frac{\pi t}{2}\right) - t\right]dt = 2 - \frac{2}{\pi} \approx 1.3634$

25. $\int_C \left(1 + \frac{y}{3}\right)ds = \int_0^2 (1 + 10\sin^3 t)[(-90\cos^2 t \sin t)^2 + (90\sin^2 t \cos t)^2]^{1/2}\,dt = 225$

Christy needs $\dfrac{450}{200} = 2.25$ gal of paint.

27. $C: x + y = a$

Let $x = t, y = a - t$, t in $[0, a]$.

Cylinder: $x + y = a$; $(x + y)^2 = a^2$;

$x^2 + 2xy + y^2 = a^2$

Sphere: $x^2 + y^2 + z^2 = a^2$

The curve of intersection satisfies:

$z^2 = 2xy; z = \sqrt{2xy}$.

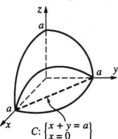

$C: \begin{cases} x + y = a \\ x = 0 \end{cases}$

$\text{Area} = 8\int_C \sqrt{2xy}\,ds = 8\int_0^a \sqrt{2t(a-t)}\sqrt{(1)^2 + (-1)^2}\,dt$

$= 16\int_0^a \sqrt{at - t^2}\,dt$

$= 16\left[\dfrac{t - \frac{a}{2}}{2}\sqrt{at - t^2} + \dfrac{\left(\frac{a}{2}\right)^2}{2}\sin^{-1}\left(\dfrac{t - \frac{a}{2}}{\frac{a}{2}}\right)\right]_0^a$

$= 16\left[\left[0 + \left(\dfrac{a^2}{8}\right)\left(\dfrac{\pi}{2}\right)\right] - \left[0 + \left(\dfrac{a^2}{8}\right)\left(\dfrac{-\pi}{2}\right)\right]\right] = 2a^2\pi$

Trivial way: Each side of the cylinder is part of a plane that intersects the sphere in a circle. The radius of each circle is the value of z in

$z = \sqrt{2xy}$ when $x = y = \dfrac{a}{2}$. That is, the radius is

$\sqrt{2\left(\dfrac{a}{2}\right)\left(\dfrac{a}{2}\right)} = \dfrac{a\sqrt{2}}{2}$. Therefore, the total area of

the part cut out is $r\left[\pi\left(\dfrac{a\sqrt{2}}{2}\right)^2\right] = 2a^2\pi.$

29. Note that $r = a \cos \theta$ along C.

Then $(a^2 - x^2 - y^2)^{1/2} = (a^2 - r^2)^{1/2} = a\cos\theta$.

Let $\begin{cases} x = r\cos\theta = (a\sin\theta)\cos\theta \\ y = r\sin\theta = (a\sin\theta)\sin\theta \end{cases}$, θ in $\left[0, \dfrac{\pi}{2}\right]$.

Therefore, $x'(\theta) = a\cos 2\theta$; $y(\theta) = a\sin 2\theta$.

Then $\text{Area} = 4\int_C (a^2 - x^2 - y^2)^{1/2}\,ds$

$= 4\int_0^{\pi/2} (a\cos\theta)[(a\sin 2\theta)^2 + (a\cos 2\theta)^2]^{1/2}\,d\theta$

$= 4a^2$.

31. a. $\int_C x^2 y\,ds = \int_0^{\pi/2}(3\sin t)^2(3\cos t)[(3\cos t)^2 + (-3\sin t)^2]^{1/2}\,dt = 81\int_0^{\pi/2}\sin^2 t\cos t\,dt = 81\left[\left(\dfrac{1}{3}\right)\sin^3 t\right]_0^{\pi/2} = 27$

b. $\int_{C_4} xy^2\,dx + xy^2\,dy = \int_0^3(3-t)(5-t)^2(-1)dt + \int_0^3(3-t)(5-t)^2(-1)dt = 2\int_0^3(t^3 - 13t^2 + 55t - 75)dt = -148.5$

17.3 Concepts Review

1. $f(b) - f(a)$

3. $0; 0$

Problem Set 17.3

1. $M_y = -7 = N_x$, so $\mathbf{F}$ is conservative.

$f(x, y) = 5x^2 - 7xy + y^2 + C$

3. $M_y = 90x^4 y - 36y^5 \neq N_x$ since

$N_x = 90x^4 y - 12y^5$, so $\mathbf{F}$ is not conservative.

5. $M_y = \left(-\dfrac{12}{5}\right)x^2 y^{-3} = N_x$, so $\mathbf{F}$ is conservative.

$f(x, y) = \left(\dfrac{2}{5}\right)x^3 y^{-2} + C$

7. $M_y = 2e^y - e^x = N_x$ so $\mathbf{F}$ is conservative.

$f(x, y) = 2xe^y - ye^x + C$

9. $M_y = 0 = N_x, M_z = 0 = P_x$, and $N_z = 0 = P_y$, so $\mathbf{F}$ is conservative. f satisfies

$f_x(x, y, z) = 3x^2$, $f_y(x, y, z) = 6y^2$, and

$f_z(x, y, z) = 9z^2$.

Therefore, f satisfies

1. $f(x, y, z) = x^3 + C_1(y, z)$,

2. $f(x, y, z) = 2y^3 + C_2(x, z)$, and

3. $f(x, y, z) = 3z^3 + C_3(x, y)$.

A function with an arbitrary constant that satisfies 1, 2, and 3 is

$f(x, y, z) = x^3 + 2y^3 + 3z^3 + C.$

11. $M_y = 2y + 2x = N_x$, so integral is path independent. $f(x, y) = xy^2 + x^2 y$

$\int_{(-1,2)}^{(3,1)} (y^2 + 2xy)dx + (x^2 + 2xy)dy = [xy^2 + x2y]_{(-1,2)}^{(3,1)}$

$= 14$ (Or use paths.)

13. $M_y = 18xy^2 = N_x, M_z = 4x = P_x, \ N_z = 0 = P_y$.
By paths $(0, 0, 0)$ to $(1, 0, 0)$; $(1, 0, 0)$ to $(1, 1, 0)$; $(1, 1, 0)$ to $(1, 1, 1)$

$\int_0^1 0\,dx + \int_0^1 9y^2 dy + \int_0^1 (4z+1)dz = 6$

(Or use $f(x, y) = 3x^2 y^3 + 2xz^2 + z$.)

15. $M_y = 1 = N_x, M_z = 1 = P_x, \ N_z = 1 = P_y$ (so path independent). From inspection observe that $f(x, y, z) = xy + xz + yz$ satisfies $f = \langle y+z, x+z, x+y \rangle$, so the integral equals

$[xy + xz + yz]_{(0,0,0)}^{(-1,0,\pi)} = -\pi$. (Or use line segments $(0,1,\ 0)$ to $(1,1,0)$, then $(1,1,0)$ to $(1,1,1)$.)

17. $f_x = M, f_y = N, f_z = P$
$f_{xy} = M_y$, and $f_{yx} = N_x$, so $M_y = N_x$.
$f_{xz} = M_z$ and $f_{zx} = P_x$, so $M_z = P_x$.
$f_{yz} = N_z$ and $f_{zy} = P_y$, so $N_z = P_y$.

19. $\mathbf{F}(x, y, z) = k|\mathbf{r}|\dfrac{\mathbf{r}}{|\mathbf{r}|} = k\mathbf{r} = k\langle x, y, z \rangle$

$f(x, y, z) = \left(\dfrac{k}{2}\right)(x^2 + y^2 + z^2)$ works.

17.4 Concepts Review

1. $\dfrac{\partial N}{\partial x} - \dfrac{\partial M}{\partial y}$

3. source; sink

Problem Set 17.4

1.

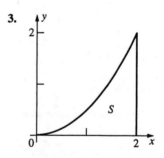

21. $\int_C \mathbf{F} \cdot d\mathbf{r} = \int_a^b (m\mathbf{r}'' \cdot \mathbf{r}')dt$

$= m\int_a^b (x'x' + y'y' + z'z')dt$

$= m\left[\dfrac{(x')^2}{2} + \dfrac{(y')^2}{2} + \dfrac{(z')^2}{2}\right]_a^b$

$= \dfrac{m}{2}\left[|\mathbf{r}'(t)|^2\right]_a^b = \dfrac{m}{2}\left[|\mathbf{r}'(b)|^2 - |\mathbf{r}'(a)|^2\right]$

23. $f(x, y, z) = -gmz$ satisfies
$\nabla f(x, y, z) = \langle 0, 0, -gm \rangle = \mathbf{F}$. Then, assuming the path is piecewise smooth,

Work $= \int_C \mathbf{F} \cdot d\mathbf{r} = [-gmz]_{(x_1, y_1, z_1)}^{(x_2, y_2, z_2)}$

$= -gm(z_2 - z_1) = gm(z_1 - z_2)$.

25. a. $M = \dfrac{y}{(x^2 + y^2)}; M_y = \dfrac{(x^2 - y^2)}{(x^2 + y^2)^2}$

$N = -\dfrac{x}{(x^2 + y^2)}; N_x = \dfrac{(x^2 - y^2)}{(x^2 + y^2)^2}$

b. $M = \dfrac{y}{(x^2 + y^2)} = \dfrac{(\sin t)}{(\cos^2 t + \sin^2 t)} = \sin t$

$N = -\dfrac{x}{(x^2 + y^2)} = \dfrac{(-\cos t)}{(\cos^2 t + \sin^2 t)} = -\cos t$

$\int_C \mathbf{F} \cdot d\mathbf{r} = \int_C M\,dx + N\,dy$

$= \int_0^{2\pi} [(\sin t)(-\sin t) + (-\cos t)(\cos t)]dt$

$= -\int_0^{2\pi} 1\,dt = -2\pi \neq 0$

$\oint_C 2xy\,dx + y^2 dy = \iint_S (0 - 2x)dA$

$= \int_0^2 \int_{y^2}^{2y} -2x\,dx\,dy = -\dfrac{64}{15} \approx -4.2667$

3.

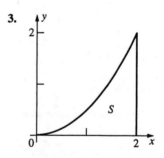

$$\oint_C (2x+y^2)dx+(x^2+2y)dy = \iint_S (2x-2y)dA$$

$$= \int_0^2 \int_0^{x^3/4}(2x-2y)dy\,dx = \int_0^2\left[\frac{x^4}{2}-\frac{x^6}{16}\right]dx$$

$$= \frac{16}{5}-\frac{8}{7}=\frac{72}{35}\approx 2.0571$$

5.

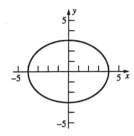

$$\oint_C(x^2+4xy)dx+(2x^2+3y)dy = \iint_S(4x-4x)dA$$
$$= 0$$

7.

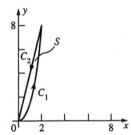

$$A(S)=\left(\frac{1}{2}\right)\oint_C x\,dy - y\,dx$$

$$=\left(\frac{1}{2}\right)\int_0^2[4x^2-2x^2]dx+\left(\frac{1}{2}\right)\int_2^0[4x-4x]dx=\frac{8}{3}$$

9.

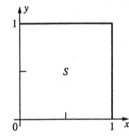

a. $\displaystyle\iint_S \text{div }\mathbf{F}\,dA = \iint_S(M_x+N_y)dA$
$$= \iint_S(0+0)dA = 0$$

b. $\displaystyle\iint_S(\text{curl }\mathbf{F})\cdot\mathbf{k}\,dA = \iint_S(N_x-M_y)dA$
$$= \iint_S(2x-2y)dA = \int_0^1\int_0^1(2x-2y)dx\,dy$$
$$= \int_0^1(1-2y)dy = 0$$

11. a. $\displaystyle\iint_S(0+0)dA = 0$

b. $\displaystyle\iint_S(3x^2-3y^2)dA = 0,$ since for the
integrand, $f(y,x)=-f(x,y)$.

13. $\displaystyle\iint_S(\text{curl }\mathbf{F})\cdot\mathbf{k}dA = \int_{C_1}\mathbf{F}\cdot\mathbf{T}ds - \int_{C_2}\mathbf{F}\cdot\mathbf{T}ds$
$$= 30-(-20)=50$$

15.

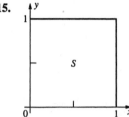

$$W=\oint_C\mathbf{F}\cdot\mathbf{T}ds = \iint_S(N_x-M_y)dA$$
$$= \iint_S(-2y-2y)dA = \int_0^1\int_0^1 -4y\,dx\,dy$$
$$= \int_0^1 -4y\,dy = -2$$

17. $\mathbf{F}$ is a constant, so $N_x = M_y = 0.$
$$\oint_C\mathbf{F}\cdot\mathbf{T}ds = \iint_S(N_x-M_y)dA = 0$$

19. a. Each equals $(x^2-y^2)(x^2+y^2)^{-2}.$

b. $\displaystyle\oint_C y(x^2+y^2)^{-1}dx - x(x^2+y^2)^{-1}dy = \int_0^{2\pi}(-\sin^2 t - \cos^2 t)dt$
$$= \int_0^{2\pi}-1\,dt$$

c. M and N are discontinuous at $(0,0).$

21. Use Green's Theorem with $M(x, y) = -y$ and $N(x, y) = 0$.

$$\oint_C (-y)dx = \iint_S [0 - (-1)]dA = A(S)$$

Now use Green's Theorem with $M(x, y) = 0$ and $N(x, y) = x$.

$$\oint_C x\,dy = \iint_S (1 - 0)dA = A(S)$$

23. $A(S) = \left(\dfrac{1}{2}\right)\oint_C x\,dy - y\,dx = \left(\dfrac{1}{2}\right)\int_0^{2\pi}[(a\cos^3 t)(3a\sin^2 t)(\cos t) - (a\sin^3 t)(3a\cos^2 t)(-\sin t)]dt = \left(\dfrac{3}{8}\right)a^2\pi$

25. a. $\mathbf{F}\cdot\mathbf{n} = \dfrac{x^2 + y^2}{(x^2 + y^2)^{3/2}} = \dfrac{1}{(x^2 + y^2)^{1/2}} = \dfrac{1}{a}$

Therefore, $\displaystyle\int_C \mathbf{F}\cdot\mathbf{n}\,ds = \frac{1}{a}\int_C 1\,ds = \frac{1}{a}(2\pi a) = 2\pi.$

b. $\text{div } \mathbf{F} = \dfrac{(x^2 + y^2)(1) - (x)(2x)}{(x^2 + y^2)^2} + \dfrac{(x^2 + y^2)(1) - (y)(2y)}{(x^2 + y^2)^2} = 0$

c. $M = \dfrac{x}{(x^2 + y^2)}$ is not defined at $(0, 0)$ which is inside C.

d. If origin is outside C, then $\displaystyle\oint_C \mathbf{F}\cdot\mathbf{n}\,ds = \iint_S \text{div } \mathbf{F}\,dA = \iint_S 0\,dA = 0.$

If origin is inside C, let C' be a circle (centered at the origin) inside C and oriented clockwise. Let S be the region between C and C'.

Then $0 = \displaystyle\iint_S \text{div } \mathbf{F}\,dA$ (by "origin outide C" case)

$= \displaystyle\int_C \mathbf{F}\cdot\mathbf{n}\,ds - \int_{C'} \mathbf{F}\cdot\mathbf{n}\,ds$ (by Green's Theorem)

$= \displaystyle\int_C \mathbf{F}\cdot\mathbf{n}\,ds - 2\pi$ (by part a), so $\displaystyle\int_C \mathbf{F}\cdot\mathbf{n}\,ds = 2\pi.$

27. a. div $\mathbf{F} = 4$

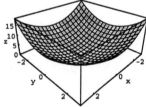

$x^2 + y^2$

b. $4(36) = 144$

29. a. div $\mathbf{F} = -2\sin x \sin y$
div $\mathbf{F} < 0$ in quadrants I and III

div $\mathbf{F} > 0$ in quadrants II and IV

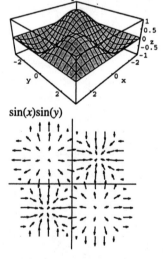

$\sin(x)\sin(y)$

b. Flux across boundary of S is 0.
Flux across boundary T is $-2(1 - \cos 3)^2$.

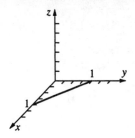

17.5 Concepts Review

1. surface integral

3. $\sqrt{f_x^2 + f_y^2 + 1}$

Problem Set 17.5

1. $\iint_R [x^2 + y^2 + (x + y + 1)](1 + 1 + 1)^{1/2} dA$

$= \int_0^1 \int_0^1 \sqrt{3}(x^2 + y^2 + x + y + 1)dx\,dy = \dfrac{8\sqrt{3}}{3}$

3 $\iint_R (x + y)\sqrt{[-x(4 - x^2)^{-1/2}]^2 + 0 + 1}\,dA$

$= \int_0^{\sqrt{3}} \int_0^1 \dfrac{2(x + y)}{(4 - x^2)^{1/2}} dy\,dx$

$= \int_0^{\sqrt{3}} \dfrac{2x + 1}{(4 - x^2)^{1/2}} dx$

$= \left[-2(4 - x^2)^{1/2} + \sin^{-1}\left(\dfrac{x}{2}\right) \right]_0^{\sqrt{3}}$

$= \dfrac{\pi + 6}{3} = 2 + \dfrac{\pi}{3} \approx 3.0472$

5. $\int_0^\pi \int_0^{\sin\theta} (4r^2 + 1)r\,dr\,d\theta = \left(\dfrac{5}{8}\right)\pi \approx 1.9635$

7. $\iint_R (x + y)(0 + 0 + 1)^{1/2} dA$

Bottom $(z = 0)$: $\int_0^1 \int_0^1 (x + y)dx\,dy = 1$

Top $(z = 1)$: Same integral

Left side $(y = 0)$: $\int_0^1 \int_0^1 (x + 0)dx\,dz = \dfrac{1}{2}$

Right side $(y = 1)$: $\int_0^1 \int_0^1 (x + 1)dx\,dz = \dfrac{3}{2}$

Back $(x = 0)$: $\int_0^1 \int_0^1 (0 + y)dy\,dz = \dfrac{1}{2}$

Front $(x = 1)$: $\int_0^1 \int_0^1 (1 + y)dy\,dz = \dfrac{3}{2}$

Therefore, the integral equals

$1 + 1 + \dfrac{1}{2} + \dfrac{3}{2} + \dfrac{1}{2} + \dfrac{3}{2} = 6.$

$\iint_G \mathbf{F} \cdot \mathbf{n}\,ds = \iint_R (-Mf_x - Nf_y + P)dA$

$= \int_0^1 \int_0^{1-y} (8y + 4x + 0)dx\,dy$

$= \int_0^1 [8(1 - y)y + 2(1 - y)^2]dy$

$= \int_0^1 (-6y^2 + 4y + 2)dy = 2$

11. $\int_0^5 \int_{-1}^1 [-xy(1 - y^2)^{-1/2} + 2]dy\,dx = 20$

(In the inside integral, note that the first term is odd in y.)

13. $m = \iint_G kx^2 ds = \iint_R kx^2 \sqrt{3}\,dA$

$= \sqrt{3}k \int_0^a \int_0^{a-x} x^2 dy\,dx = \left(\dfrac{\sqrt{3}k}{12}\right)a^4$

15.

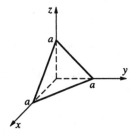

Let $\delta = 1$.

$m = \iint_S 1\,ds = \iint_R (1 + 1 + 1)^{1/2} dA$

$= \sqrt{3}\int_0^a \int_0^{a-y} dx\,dy = \sqrt{3}\int_0^a (a - y)dy = \dfrac{a^2\sqrt{3}}{2}$

$M_{xy} = \iint_S z\,ds = \iint_R (a - x - y)\sqrt{3}\,dA$

$= \sqrt{3}\int_0^a \int_0^{a-y} (a - x - y)dx\,dy$

$= \sqrt{3}\int_0^a \left[a(a - y) - \dfrac{(a - y)^2}{2} - y(a - y) \right]dy$

$= \sqrt{3}\int_0^a \left(\dfrac{a^2}{2} - ay + \dfrac{y^2}{2} \right)dy = \dfrac{a^3\sqrt{3}}{6}$

$\bar{z} = \dfrac{M_{xy}}{m} = \dfrac{a}{3}$; then $\bar{x} = \bar{y} = \dfrac{a}{3}$ (by symmetry).

17. a. 0 (By symmetry, since
$g(x, y, -z) = -g(x, y, z)$.)

b. 0 (By symmetry, since
$g(x, y, -z) = -g(x, y, z)$.)

c. $\iint_G (x^2 + y^2 + z^2)dS = \iint_G a^2 dS$
$= a^2 \text{Area}(G) = a^2(4\pi a^2) = 4\pi a^4$

d. Note:
$\iint_G (x^2 + y^2 + z^2)dS = \iint_G x^2 dS + \iint_G y^2 dS$
$= \iint_G z^2 dS = 3\iint_G x^2 dS$
(due to symmetry of the sphere with respect to the origin.)
Therefore,
$\iint_G x^2 dS = \left(\frac{1}{3}\right)\iint_G (x^2 + y^2 + z^2)dS$
$= \left(\frac{1}{3}\right)4\pi a^4 = \frac{4\pi a^4}{3}$.

e. $\iint_G (x^2 + y^2)dS = \left(\frac{2}{3}\right)4\pi a^4 = \frac{8\pi a^4}{3}$

19. a. Place center of sphere at the origin.
$F = \iint_G k(a - z)dS = ka\iint_G 1 dS - k\iint_G z dS$
$= ka(4\pi a^2) - 0 = 4\pi a^3 k$

b. Place hemisphere above xy-plane with center at origin and circular base in xy-plane.
F = Force on hemisphere + Force on circular

base
$= \iint_G k(a - z)dS + ka(\pi a^2)$
$= ka\iint_G 1 dS - k\iint_G z dS + \pi a^3 k$
$= ka(2\pi a^2) - k\iint_R z\sqrt{\dfrac{a^2}{a^2 - x^2 - y^2}}\, dA + \pi a^3 k$
$= 3\pi a^3 k - k\iint_R z\dfrac{a}{z}\, dA$
$= 3\pi a^3 k - ka(\pi a^2) = 2\pi a^3 k$

c. Place the cylinder above xy-plane with circular base in xy-plane with the center at the origin.
F = Force on top + Force on cylindrical side + Force on base
$= 0 + \iint_G k(h - z)dS + kh(\pi a^2)$
$= kh\iint_G 1 dS - k\iint_G z dS + \pi a^2 hk$
$= kh(2\pi ah) - 4k\iint_R z\sqrt{\dfrac{a^2}{a^2 - y^2}} + 0 + 1\, dA + \pi a^2$
(where R is a region in the yz-plane:
$0 \le y = a, 0 \le z \le h$)
$= 2\pi ah^2 k + \pi a^2 hk - 4k\int_0^a\int_0^h \dfrac{az}{\sqrt{a^2 - y^2}}\, dz\, dy$
$= 2\pi ah^2 k + \pi a^2 hk - \pi kah^2$
$= \pi ah^2 k + \pi a^2 hk = \pi ahk(h + a)$

17.6 Concepts Review

1. boundary; ∂S

3. div **F**

Problem Set 17.6

1. $\iiint_S (0 + 0 + 0)dV = 0$

3. $\iint_{\partial S} \mathbf{F} \cdot \mathbf{n}\, dS = \iiint_S (M_x + N_y + P_z)dV$
$= \int_0^c\int_0^b\int_0^a (2xyz + 2xyz + 2xyz)dx\, dy\, dz$
$= \int_0^c\int_0^b 3a^2 yz\, dy\, dz = \int_0^c \dfrac{3a^2 b^2 z}{2}\, dz$
$= \dfrac{3a^2 b^2 c^2}{4}$

5. $2\iiint_S (x+y+z)dV = 2\int_0^{2\pi}\int_0^2\int_0^{4-r^2}(r\cos\theta+r\sin\theta+z)r\,dz\,dr\,d\theta = \dfrac{64\pi}{3} \approx 67.02$

7. $\iiint_S (1+1+0)dV = 2(\text{volume of cylinder}) = 2\pi(1)^2(2) = 4\pi \approx 12.5664$

9. $\iiint_S (M_x + N_y + P_z)dV = \iiint_S (2+3+4)dV$

$= 9(\text{Volume of spherical shell})$

$= 9\left(\dfrac{4\pi}{3}\right)(5^3 - 3^3) = 1176\pi \approx 3694.51$

$\text{Volume} = \left(\dfrac{1}{3}\right)\iint_S \mathbf{F}\cdot\mathbf{n}\,dS \quad (\text{where } \mathbf{F} = \langle x, y, z\rangle).$

$= \left(\dfrac{1}{3}\right)\iint_R \mathbf{F}\cdot\mathbf{n}\,dS \quad (\text{by Note 3})$

$= \left(\dfrac{1}{3}\right)\iint_R \dfrac{(ax+by+cz)}{\sqrt{a^2+b^2+c^2}}\,dS$

$= \dfrac{dD}{3\sqrt{a^2+b^2+c^2}}$

11. $\left(\dfrac{1}{3}\right)\iiint_S (1+1+1)dV = V(S)$

13. Note:

1. $\iint_R (ax+by+cz)dS = \iint_R d\,dS = dD$ (R is the slanted face.)

2. $\mathbf{n} = \dfrac{\langle a, b, c\rangle}{(a^2+b^2+c^2)^{1/2}}$ (for slanted face)

3. $\mathbf{F}\cdot\mathbf{n} = 0$ on each coordinate-plane face.

15. **a.** div $\mathbf{F} = 2+3+2z = 5+2z$

$\iint_{\partial S}\mathbf{F}\cdot\mathbf{n}\,dS = \iiint_S (5+2z)dV = \iiint_S 5\,dV + 2\iiint_S z\,dV = 5\,(\text{Volume of } S) + 2M_{xy}$

$= 5\left(\dfrac{4\pi}{3}\right) + 2z\,(\text{Volume of } S) = \dfrac{20\pi}{3} + 2(0)(\text{Volume}) = \dfrac{20\pi}{3}$

b. $\mathbf{F}\cdot\mathbf{n} = (x^2+y^2+z^2)^{3/2}\langle x, y, z\rangle \cdot \dfrac{\langle x, y, z\rangle}{\sqrt{x^2+y^2+z^2}} = (x^2+y^2+z^2)^2 = 1$ on ∂S.

$\iint_{\partial S}\mathbf{F}\cdot\mathbf{n}\,dS = \iiint_S 1\,dV = 4\pi(1)^2 = 4\pi$

c. div $\mathbf{F} = 2x+2y+2z$

$\iint_{\partial S}\mathbf{F}\cdot\mathbf{n}\,dS = \iiint_S 2(x+y+z)dV$

$= 2\iiint_S x\,dV \quad (\text{Since } \bar{x} = \bar{z} = 0 \text{ as in a.})$

$= 2M_{yz} = 2(\bar{x})(\text{Volume of } S) = 2(2)\left(\dfrac{4\pi}{3}\right) = \dfrac{16\pi}{3}$

d. $\mathbf{F}\cdot\mathbf{n} = 0$ on each face except the face R in the plane $x = 1$.

$\iint_{\partial S}\mathbf{F}\cdot\mathbf{n}\,dS = \iint_R \mathbf{F}\cdot\mathbf{n}\,dS = \iint_R \langle 1, 0, 0\rangle \cdot \langle 1, 0, 0\rangle dS = \iint_R 1\,dS = (1)^2 = 1$

e. div $\mathbf{F} = 1+1+1 = 3$

$\iint_{\partial S}\mathbf{F}\cdot\mathbf{n}\,dS = \iiint_S 3\,dV = 3(\text{Volume of } S) = 3\left(\dfrac{1}{3}\left[\dfrac{1}{2}(4)(3)\right](6)\right) = 36$

f.

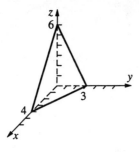

$\text{div } \mathbf{F} = 3x^2 + 3y^2 + 3z^2 = 3(x^2 + y^2 + z^2) = 3 \text{ on } \partial S.$

$\iint_{\partial S} \mathbf{F} \cdot \mathbf{n} \, dS = 3\iiint_S (x^2 + y^2 + z^2) dV = 3\left(\dfrac{3}{2}\dfrac{8\pi}{15}\right) = \dfrac{12\pi}{5}$

(That answer can be obtained by making use of symmetry and a change to spherical coordinates. Or you could go to the solution for Problem 24, Section 16.8, and realize that the value of the integral in this problem is $\dfrac{3}{2}$ (there are three terms instead of two) times the answer obtained there for I_z, letting $kabc = 1$.)

g. $\mathbf{F} \cdot \mathbf{n} = [\ln(x^2 + y^2)]\langle x, y, 0\rangle \cdot \langle 0, 0, 1\rangle = 0$ on top and bottom.

$\mathbf{F} \cdot \mathbf{n} = (\ln 4)\langle x, y, 0\rangle \cdot \dfrac{\langle x, y, 0\rangle}{\sqrt{x^2 + y^2}} = (\ln 4)\sqrt{x^2 + y^2} = (\ln 4)\sqrt{4} = 2\ln 4 = 4\ln 2$ on the side.

$\iint_{\partial S} \mathbf{F} \cdot \mathbf{n} \, dS = \iint_R 4\ln 2 \, dS = (4\ln 2)[2\pi(2)(2)] = 32\pi \ln 2$

17. $\iint_{\partial S} D_{\mathbf{n}} f \, dS = \iint_{\partial S} \nabla f \cdot \mathbf{n} \, dS = \iiint_S \text{div}(\nabla f) dV = \iiint_S \nabla^2 f \, dV$ (See next problem.)

19. $\iint_{\partial S} f D_{\mathbf{n}} g \, dS = \iint_{\partial S} f(\nabla g \cdot \mathbf{n}) dS = \iint_{\partial S} (f\nabla g) \cdot \mathbf{n} \, dS = \iiint_S \text{div}(f\nabla g) dV$ (Gauss)

$= \iiint_S [f(\text{div } \nabla g) + (\nabla f) \cdot (\nabla g)] dV = \iiint_S (f\nabla^2 g + \nabla f \cdot \nabla g) dV$ (See Problem 20c, Section 17.1.)

17.7 Concepts Review

1. $(\text{curl } \mathbf{F}) \cdot \mathbf{n}$

3. $\iint\limits_S (\text{curl } \mathbf{F}) \cdot \mathbf{n} \, dS$

Problem Set 17.7

1. $\oint_{\partial S} \mathbf{F} \cdot \mathbf{T} \, dS = \iint_R (N_x - M_y) dA = \iint_R 0 \, dA = 0$

3.

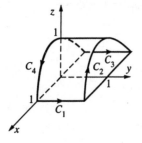

$$\iint_S (\text{curl } \mathbf{F}) \cdot \mathbf{n}\, dS = \oint_{\partial S} \mathbf{F} \cdot \mathbf{T}\, ds = \oint_{\partial S} (y+z)dx + (x^2+z^2)dy + y\, dz$$

$$= \int_0^1 1\, dt + \int_0^\pi [(1+\sin t)(-\sin t) + \cos t]dt + \int_0^1 -1\, dt + \int_0^\pi \sin^2 t\, dt \quad (*)$$

$$= \int_0^\pi (-\sin t + \cos t)dt = -2$$

The result at (*) was obtained by integrating along S by doing so along C_1, C_2, C_3, C_4 in that order.

Along C_1: $x = 1$, $y = t$, $z = 0$, $dx = dz = 0$, $dy = dt$, t in $[0, 1]$

Along C_2: $x = \cos t$, $y = 1$, $z = \sin t$, $dx = -\sin dt$, $dy = 0$, $dz = \cos t\, dt$, t in $[0, \pi]$

Along C_3: $x = -1$, $y = 1 - t$, $z = 0$, $dx = dz = 0$, $dy = dt$, t in $[0, 1]$

Along C_4: $x = -\cos t$, $y = 0$, $z = \sin t$, $dx = \sin t\, dt$, $dy = 0$, $dz = \cos t\, dt$, t in $[0, \pi]$

5. ∂S is the circle $x^2 + y^2 = 12$, $z = 2$.

Parameterization of circle:

$x = \sqrt{12}\cos t$, $y = \sqrt{12}\sin t$, $z = 2$, t in $[0, 2\pi]$

$$\oint_{\partial S} \mathbf{F} \cdot \mathbf{T}\, ds = \oint_{\partial S} yz\, dx + 3xz\, dy + z^2 dz$$

$$= \int_0^{2\pi} (24\sin^2 t - 72\cos^2 t)dt = -48\pi \approx -150.80$$

7. $(\text{curl }\mathbf{F}) \cdot \mathbf{n} = \langle 3, 2, 1\rangle \cdot \left\langle \left(\dfrac{1}{\sqrt{2}}\right)1, 0, -1\right\rangle = \sqrt{2}$

$$\oint_{\partial S} \mathbf{F} \cdot \mathbf{T}\, ds = \iint_S \sqrt{2}\, dS = \sqrt{2}A(S)$$

$$= \sqrt{2}[\sec(45°)](\text{Area of a circle}) = 8\pi \approx 25.1327$$

9. $(\text{curl }\mathbf{F}) = \langle -1+1, 0-1, 1-1\rangle = \langle 0, -1, 0\rangle$

The unit normal vector that is needed to apply Stokes' Theorem points downward. It is

$$n = \frac{\langle -1, -2, -1\rangle}{\sqrt{6}}.$$

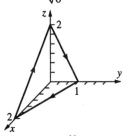

$$\oint_C \mathbf{F} \cdot \mathbf{T}\, ds = \iint_S (\text{curl }\mathbf{F}) \cdot \mathbf{n}\, dS$$

$$= \iint_S \left(\frac{2}{\sqrt{6}}\right) dS = \iint_R \left(\frac{2}{\sqrt{6}}\right)(1+4+1)^{1/2}\, dA$$

$$= \iint_R 2\, dA = 2(\text{Area of triangle in } xy-\text{plane})$$

$$= 2(1) = 2$$

11. $(\text{curl }\mathbf{F}) \cdot \mathbf{n} = \langle 0, 0, 1\rangle \cdot \langle x, y, z\rangle = z$

$$\iint_S z\, dS = \iint_R 1\, dA = \text{Area of } R$$

$$\pi\left(\frac{1}{2}\right)^2 = \frac{\pi}{4} \approx 0.7854$$

13. Let $H(x, y, z) = z - g(x, y) = 0$.

Then $\mathbf{n} = \dfrac{\nabla H}{|\nabla H|} = \dfrac{\langle -g_x, g_y, 1\rangle}{\sqrt{1+g_x^2+g_y^2}}$ points upward.

Thus,

$$\iint_S (\text{curl }\mathbf{F}) \cdot \mathbf{n}\, dS = \iint_{S_{xy}} (\text{curl }\mathbf{F}) \cdot \mathbf{n}\sec\gamma\, dA$$

$$= \iint_{S_{xy}} (\text{curl }\mathbf{F}) \frac{\langle -g_x, -g_y, 1\rangle}{\sqrt{g_x^2+g_y^2+1}}\sqrt{g_x^2+g_y^2+1}\, dA$$

(Theorem A, Section 17.5)

$$= \iint_{S_{xy}} (\text{curl }\mathbf{F}) \cdot \langle -g_x, -g_y, 1\rangle\, dA$$

15. $\text{curl }F = \langle 0-x, 0-0, z-0\rangle = \langle -x, 0, z\rangle$

$$\oint_C \mathbf{F} \cdot \mathbf{T}\, ds = \iint_S (\text{curl }\mathbf{F}) \cdot \mathbf{n}\, dS$$

$$= \iint_{S_{xy}} (\text{curl }\mathbf{F}) \cdot \langle -g_x, -g_y, 1\rangle\, dA \text{ where}$$

$z = g(x, y) = xy^2$. (Problem 13)

$$= \iint_{S_{xy}} \langle -x, 0, z\rangle \cdot \langle -y^2, -2xy, 1\rangle\, dA$$

$$= \int_0^1 \int_0^1 (xy^2 + 0 + xy^2)dx\, dy$$

$$= \int_0^1 \left([x^2 y^2]_{x=0}^1\right) dy = \int_0^1 y^2 dy = \frac{1}{3}$$

17. $$\oint_{\partial S} \mathbf{F} \cdot \mathbf{T}\, ds = \iint_S (\text{curl }\mathbf{F}) \cdot \mathbf{n}\, dS$$

$$= \iint_{S_{xy}} (\text{curl }\mathbf{F}) \cdot \langle -g_x, -g_y, 1\rangle\, dA$$

$$= \iint_{S_{xy}} \langle 2, 2, 0\rangle \cdot \left[\frac{\langle x, y, (a^2-x^2-y^2)^{-1/2}\rangle}{(a^2-x^2-y^2)^{-1/2}}\right] dA$$

$$= 2\iint_{S_{xy}} (x+y)(a^2 - x^2 - y^2)^{-1/2}\,dA$$

$$= 2\iint_{S_{xy}} y(a^2 - x^2 - y^2)^{-1/2}\,dA$$

$$= 4\int_0^{\pi/2}\int_0^{a\sin\theta} (r\sin\theta)(a^2 - r^2)^{-1/2}\,dr\,d\theta$$

$$= \frac{4a^2}{3}\ \text{joules}$$

19. **a.** Let C be any piecewise smooth simple closed oriented curve C that separates the

17.8 Chapter Review

Concepts Test

1. **True:** See Example 4, Section 17.1

3. **False:** grad(curl $\mathbf{F}$) is not defined since curl $\mathbf{F}$ is not a scalar field.

5. **True:** See the three equivalent conditions in Section 17.3.

7. **False:** $N_z = 0 \neq z^2 = P_y$

9. **True:** It is the case in which the surface is in a plane.

11. **True:** See discussion on text page 777.

Sample Test Problems

1.

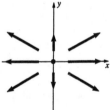

9. **a.**

"nice" surface into two "nice" surfaces, S_1 and S_2.

$$\iint_{\partial S} (\text{curl } \mathbf{F})\cdot\mathbf{n}\,dS$$

$$= \iint_{S_1} (\text{curl } \mathbf{F})\cdot\mathbf{n}\,dS + \iint_{S_2} (\text{curl } \mathbf{F})\cdot\mathbf{n}\,dS$$

$$= \oint_C \mathbf{F}\cdot\mathbf{T}\,ds + \oint_{-C} \mathbf{F}\cdot\mathbf{T}\,ds = 0\ \ (-C \text{ is } C \text{ with}$$

opposite orientation.)

b. div(curl $\mathbf{F}$) = 0 (See Problem 20, Section 17.1.) Result follows.

2. div $\mathbf{F} = 2yz - 6y + 2y^2$

curl $\mathbf{F} = \langle 4yz, 2xy, -2xz \rangle$

grad(div $\mathbf{F}$) $= \langle 0, 2z - 6 + 4y, 2y \rangle$

div(curl $\mathbf{F}$) = 0

(See 20a, Section 17.1.)

3. curl$(f\nabla f) = (f)(\text{curl }\nabla f) + (\nabla f \times \nabla f)$

$= (f)(\mathbf{0}) + \mathbf{0} = \mathbf{0}$

5. **a.** Parameterization is $x = \sin t$, $y = -\cos t$, t in

$$\left[0, \frac{\pi}{2}\right].$$

$$\int_0^{\pi/2} (1 - \cos^2 t)(\sin^2 t + \cos^2 t)^{1/2}\,dt = \frac{\pi}{4}$$

$$\approx 0.7854$$

b $\int_0^{\pi/2} [t\cos t - \sin^2 t\cos t + \sin t\cos t]\,dt$

$$= \frac{(3\pi - 5)}{6} \approx 0.7375$$

7. $[xy^2]_{(1,\,1)}^{(3,\,4)} = 47$

$$\int_0^1 0\,dx + \int_0^1 (1+y^2)\,dy + \int_1^0 x\,dx + \int_1^0 y^2\,dy = 0 + \frac{4}{3} - \frac{1}{2} - \frac{1}{3} = \frac{1}{2}$$

b.

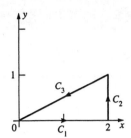

A vector equation of C_3 is $\langle x, y \rangle = \langle 2,1 \rangle + t\langle -2,-1 \rangle$ for t in $[0, 1]$, so let $x = 2 - 2t$, $y = 1 - t$ for t in $[0, 1]$ be parametric equations of C_3.

$$\int_0^2 0\,dx + \int_0^1 (4 + y^2)\,dy + \int_0^1 [2(1-t)^2(-2) + 5(1-t)^2(-1)]\,dt = 0 + \frac{13}{3} - 3 = \frac{4}{3}$$

c.

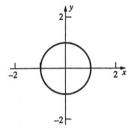

$x = \cos t;\ y = \sin t;\ t$ in $[2, \pi]$

$$\int_0^{2\pi} [(\cos t)(\sin t)(-\sin t) + (\cos^2 t + \sin^2 t)(\cos t)]\,dt = \int_0^{2\pi} (1 - \sin^2 t)\cos t\,dt = \left[\sin t - \frac{\sin^3 t}{3}\right]_0^{2\pi} = 0$$

11. Let $f(x, y) = (1 - x^2 - y^2)$ and $g(x, y) = -(1 - x^2 - y^2)$, the upper and lower hemispheres.

Then Flux $= \iint_G \mathbf{F} \cdot \mathbf{n}\,dS = \iint_R [-Mf_x - Nf_y + P]\,dA + \iint_R [-Mg_x - Ng_y + P]\,dA$

$= \iint_R 2P\,dA$ (since $f_x = -g_x$ and $f_y = -g_y$) $= \iint_R 6\,dA = 6$ (Area of R, the circle $x^2 + y^2 = 1$, $z = 0$)

$= 6\pi \approx 18.8496$

13. ∂S is the circle $x^2 + y^2 = 1$, $z = 1$. A parameterization of the circle is $x = \cos t$, $y = \sin t$, $z = 1$, t in $[0, 2\pi]$.

$$\oint_{\partial S} \mathbf{F} \cdot \mathbf{T}\,ds = \oint_{\partial S} \left[x^3 y\,dx + e^y\,dy + z\tan\left(\frac{xyz}{4}\right)dz \right] = \oint_{\partial S} (x^3 y + e^y\,dy)$$

$$= \int_0^{2\pi} [(\cos t)^3(\sin t)(-\sin t) + (e^{\sin t})(\cos t)]\,dt = 0$$

15. curl $\mathbf{F} = \langle 3-0, 0-0, -1-1 \rangle = \langle 3, 0, -2 \rangle$; $\mathbf{n} = \dfrac{\langle a, b, 1 \rangle}{\sqrt{a^2 + b^2 + 1}}$

$$\oint_C \mathbf{F} \cdot \mathbf{T}\,ds = \iint_S (\text{curl }\mathbf{F}) \cdot \mathbf{n}\,dS = \iint_S \frac{3a - 2}{\sqrt{a^2 + b^2 + 1}}\,dS = \frac{3a - 2}{\sqrt{a^2 + b^2 + 1}}[A(S)]$$

$$= \frac{3a - 2}{\sqrt{a^2 + b^2 + 1}}(9\pi) \quad (S \text{ is a circle of radius } 3.)$$

$$= \frac{9\pi(3a - 2)}{\sqrt{a^2 + b^2 + 1}}$$

18.1 Concepts Review

1. $r^2 + a_1 r + a_2 = 0$; complex conjugate roots

3. $(C_1 + C_2 x)e^x$

Problem Set 18.1

1. Roots are 2 and 3. General solution is
$y = C_1 e^{2x} + C_2 e^{3x}$.

3. Auxiliary equation: $r^2 + 6r - 7 = 0$,
$(r + 7)(r - 1) = 0$ has roots –7, 1.
General solution: $y = C_1 e^{-7x} + C_2 e^x$
$y' = -7C_1 e^{-7x} + C_2 e^x$
If $x = 0, y = 0$, $y' = 4$, then $0 = C_1 + C_2$ and

$4 = -7C_1 + C_2$, so $C_1 = -\dfrac{1}{2}$ and $C_2 = \dfrac{1}{2}$.

Therefore, $y = \dfrac{e^x - e^{-7x}}{2}$.

5. Repeated root 2. General solution is
$y = (C_1 + C_2 x)e^{2x}$.

7. Roots are $2 \pm \sqrt{3}$. General solution is
$y = e^{2x}(C_1 e^{\sqrt{3}x} + C_2 e^{-\sqrt{3}x})$.

9. Auxiliary equation: $r^2 + 4 = 0$ has roots $\pm 2i$.
General solution: $y = C_1 \cos 2x + C_2 \sin 2x$

If $x = 0$ and $y = 2$, then $2 = C_1$; if $x = \dfrac{\pi}{4}$ and

$y = 3$, then $3 = C_2$.
Therefore, $y = 2\cos 2x + 3\sin 2x$.

11. Roots are $-1 \pm i$. General solution is
$y = e^{-x}(C_1 \cos x + C_2 \sin x)$.

13. Roots are 0, 0, –4, 1.
General solution is
$y = C_1 + C_2 x + C_3 e^{-4x} + C_4 e^x$.

15. Auxiliary equation: $r^4 + 3r^2 - 4 = 0$,
$(r + 1)(r - 1)(r^2 + 4) = 0$ has roots –1, 1, $\pm 2i$.
General solution:
$y = C_1 e^{-x} + C_2 e^x + C_3 \cos 2x + C_4 \sin 2x$

17. Roots are –2, 2. General solution is
$y = C_1 e^{-2x} + C_2 e^{2x}$.
$y = C_1(\cosh 2x - \sinh 2x) + C_2(\sinh 2x + \cosh 2x)$
$= (-C_1 + C_2)\sinh 2x + (C_1 + C_2)\cosh 2x$
$= D_1 \sinh 2x + D_2 \cosh 2x$

19. Repeated roots $\left(-\dfrac{1}{2}\right) \pm \left(\dfrac{\sqrt{3}}{2}\right)i$.

General solution is $y = e^{-x/2}\left[(C_1 + C_2 x)\cos\left(\dfrac{\sqrt{3}}{2}\right)x + (C_3 + C_4 x)\sin\left(\dfrac{\sqrt{3}}{2}\right)x\right]$.

21. $(*) x^2 y'' + 5xy' + 4y = 0$
Let $x = e^z$. Then $z = \ln x$;
$y' = \dfrac{dy}{dx} = \dfrac{dy}{dz}\dfrac{dz}{dx} = \dfrac{dy}{dz}\dfrac{1}{x}$;

$y'' = \dfrac{dy'}{dx} = \dfrac{d}{dx}\left(\dfrac{dy}{dz}\dfrac{1}{x}\right) = \dfrac{dy}{dz}\dfrac{-1}{x^2} + \dfrac{1}{x}\dfrac{d^2 y}{dz^2}\dfrac{dz}{dx}$

$= \dfrac{dy}{dz}\dfrac{-1}{x^2} + \dfrac{1}{x}\dfrac{d^2 y}{dz^2}\dfrac{1}{x}$

$$\left(-\frac{dy}{dz}+\frac{d^2y}{dz^2}\right)+\left(5\frac{dy}{dz}\right)+4y=0$$

(Substituting y' and y'' into (*))

$$\frac{d^2y}{dz^2}+4\frac{dy}{dz}+4y=0$$

Auxiliary equation: $r^2+4r+4=0$, $(r+2)^2=0$ has roots $-2, -2$.

General solution: $y=(C_1+C_2z)e^{-2z}$,

$y=(C_1+C_2\ln x)e^{-2\ln x}$

$y=(C_1+C_2\ln x)x^{-2}$

23. We need to show that $y''+a_1y'+a_2y=0$ if r_1 and r_2 are distinct real roots of the auxiliary equation.

We have,

$y'=C_1r_1e^{r_1x}+C_2r_2e^{r_2x}$

$y''=C_1r_1^2e^{r_1x}+C_2r_2^2e^{r_2x}$

When put into the differential equation, we obtain

$y''+a_1y'+a_2y=C_1r_1^2e^{r_1x}+C_2r_2^2e^{r_2x}$
$$+a_1\left(C_1r_1e^{r_1x}+C_2r_2e^{r_2x}\right)+a_2\left(C_1e^{r_1x}+C_2e^{r_2x}\right)$$

The solutions to the auxiliary equation are given by

$$r_1=\frac{1}{2}\left(-a_1-\sqrt{a_1^2-4a_2}\right)\text{ and}$$

$$r_2=\frac{1}{2}\left(-a_1+\sqrt{a_1^2-4a_2}\right).$$

Putting these values into (*) and simplifying yields the desired result: $y''+a_1y'+a_2y=0$.

25. a. $e^{bi}=1+(bi)+\dfrac{(bi)^2}{2!}+\dfrac{(bi)^3}{3!}+\dfrac{(bi)^4}{4!}+\dfrac{(bi)^5}{5!}+\ldots=\left(1-\dfrac{b^2}{2!}+\dfrac{b^4}{4!}-\dfrac{b^6}{6!}+\right)\ldots+i\left(b-\dfrac{b^3}{3!}+\dfrac{b^5}{5!}-\dfrac{b^7}{7!}+\ldots\right)$

$=\cos(b)+i\sin(b)$

b. $e^{a+bi}=e^ae^{bi}=e^a[\cos(b)+i\sin(b)]$

c. $D_x\left[e^{(\alpha+\beta i)x}\right]=D_x[e^{\alpha x}(\cos\beta x+i\sin\beta x)]=\alpha e^{\alpha x}(\cos\beta x+i\sin\beta x)+e^{\alpha x}(-i\beta\sin\beta x+i\beta\cos\beta x)$

$=e^{\alpha x}[(\alpha+\beta i)\cos\beta x+(\alpha i-\beta)\sin\beta x]$

$(\alpha+\beta)e^{(\alpha+\beta i)x}=(\alpha+\beta i)[e^{\alpha x}(\cos\beta x+i\sin\beta x)]=e^{\alpha x}[(\alpha+\beta i)\cos\beta x+(\alpha i-\beta)\sin\beta x]$

Therefore, $D_x[e^{(\alpha+\beta i)x}]=(\alpha+i\beta)e^{(\alpha+\beta i)x}$

27. $y=0.5e^{5.16228x}+0.5e^{-1.162278x}$

29. $y=1.29099e^{-0.25x}\sin(0.968246x)$

18.2 Concepts Review

1. particular solution to the nonhomogeneous equation; homogeneous equation

3. $y=Ax^2+Bx+C$

Problem Set 18.2

1. $y_h=C_1e^{-3x}+C_2e^{3x}$

$y_p=\left(-\dfrac{1}{9}\right)x+0$

$y=\left(-\dfrac{1}{9}\right)x+C_1e^{-3x}+C_2e^{3x}$

3. Auxiliary equation: $r^2-2r+1=0$ has roots $1, 1$.

$y_h=(C_1+C_2x)e^x$

Let $y_p = Ax^2 + Bx + C$; $y_p' = 2Ax + B$;
$y_p'' = 2A$.

Then $(2A) - 2(2Ax + B) + (Ax^2 + Bx + C) = x^2 + x$.

$Ax^2 + (-4A + B)x + (2A - 2B + C) = x^2 + x$

Thus, $A = 1$, $-4A + B = 1$, $2A - 2B + C = 0$, so
$A = 1$, $B = 5$, $C = 8$.

General solution: $y = x^2 + 5x + 8 + (C_1 + C_2 x)e^x$

5. $y_h = C_1 e^{2x} + C_2 e^{3x} \cdot y_p = \left(\dfrac{1}{2}\right)e^x \cdot y$

$= \left(\dfrac{1}{2}\right)e^x + C_1 e^{2x} + C_2 e^{3x}$

7. $y_h = C_1 e^{-3x} + C_2 e^{-x}$

$y_p = \left(-\dfrac{1}{2}\right)xe^{-3x}$

$y = \left(-\dfrac{1}{2}\right)xe^{-3x} + C_1 e^{-3x} + C_2 e^{-x}$

9. Auxiliary equation: $r^2 - r - 2 = 0$,
$(r + 1)(r - 2) = 0$ has roots -1, 2.
$y_h = C_1 e^{-x} + C_2 e^{2x}$
Let $y_p = B\cos x + C\sin x$; $y_p' = -B\sin x + C\cos x$; $y_p'' = -B\cos x - C\sin x$.
Then $(-B\cos x - C\sin x) - (-B\sin x + C\cos x)$
$\qquad -2(B\cos x + C\sin x) = 2\sin x$.

$(-3B - C)\cos x + (B - 3C)\sin x = 2\sin x$, so $-3B - C = 0$ so $-3B - C = 0$ and $B - 3C = 2$; $B = \dfrac{1}{5}$; $C = \dfrac{-3}{5}$.

General solution: $\left(\dfrac{1}{5}\right)\cos x - \left(\dfrac{3}{5}\right)\sin x + C_1 e^{2x} + C_2 e^{-x}$

11. $y_h = C_1 \cos 2x + C_2 \sin 2x$

$y_p = (0)x\cos 2x + \left(\dfrac{1}{2}\right)x\sin 2x$

$y = \left(\dfrac{1}{2}\right)x\sin 2x + C_1 \cos 2x + C_2 \sin 2x$

13. $y_h = C_1 \cos 3x + C_2 \sin 3x$

$y_p = (0)\cos x + \left(\dfrac{1}{8}\right)\sin x + \left(\dfrac{1}{13}\right)e^{2x}$

$y = \left(\dfrac{1}{8}\right)\sin x + \left(\dfrac{1}{13}\right)e^{2x} + C_1 \cos 3x + C_2 \sin 3x$

15. Auxiliary equation: $r^2 - 5r + 6 = 0$ has roots 2 and 3, so $y_h = C_1 e^{2x} + C_2 e^{3x}$.

Let $y_p = Be^x$; $y_p' = Be^x$; $y_p'' = Be^x$.

Then $(Be^x) - 5(Be^x) + 6(Be^x) = 2e^x$; $2Be^x = 2e^x$; $B = 1$.

General solution: $y = e^x + C_1 e^{2x} + C_2 e^{3x}$

$y' = e^x + 2C_1 e^{2x} + 3C_2 e^{3x}$

If $x = 0, y = 1$, $y' = 0$, then $1 = 1 + C_1 + C_2$ and $0 = 1 + 2C_1 + 3C_2$; $C_1 = 1, C_2 = -1$.

Therefore, $y = e^x + e^{2x} - e^{3x}$.

17. $y_h = C_1 e^x + C_2 e^{2x}$

$y_p = \left(\dfrac{1}{4}\right)(10x + 19)$

$y = \left(\dfrac{1}{4}\right)(10x + 19) + C_1 e^x + C_2 e^{2x}$

19. $y_h = C_1 \cos x + C_2 \sin x$

$y_p = -\cos \ln|\sin x| - \cos x - x \sin x$

$y = -\cos x \ln|\sin x| - x \sin x + C_3 \cos x + C_2 \sin x$ (combined cos x terms)

21. Auxiliary equation: $r^2 - 3r + 2 = 0$ has roots 1, 2, so $y_h = C_1 e^x + C_2 e^{2x}$.

Let $y_p = v_1 e^x + v_2 e^{2x}$ subject to $v_1' e^x + v_2' e^{2x} = 0$, and $v_1'(e^x) + v_2'(2e^{2x}) = e^x (ex + 1)^{-1}$.

Then $v_1' = \dfrac{-e^x}{e^x (e^x + 1)}$ so $v_1 = \int \dfrac{-e^x}{e^x (e^x + 1)} dx = \int \dfrac{-1}{u(u+1)} du$

$= \int \left(\dfrac{-1}{u} + \dfrac{1}{u+1}\right) du = -\ln u + \ln(u+1) = \ln\left(\dfrac{u+1}{u}\right) = \ln\dfrac{e^x + 1}{e^x} = \ln(1 + e^{-x})$

$v_2' = \dfrac{e^x}{e^{2x}(e^x + 1)}$ so $v_2 = -e^{-x} + \ln(1 + e^{-x})$

(similar to finding v_1)

General solution: $y = e^x \ln(1 + e^{-x}) - e^x + e^{2x} \ln(1 + e^{-x}) + C_1 e^x + C_2 e^{2x}$

$$y = (e^x + e^{2x}) \ln(1 + e^{-x}) + D_1 e^x + D_2 e^{2x}$$

23. $L(y_p) = (v_1 u_1 + v_2 u_2)'' + b(v_1 u_1 + v_2 u_2)' + c(v_1 u_1 + v_2 u_2)$

$= (v_1' u_1 + v_1 u_1' + v_2' u_2 + v_2 u_2')' + b(v_1' u_1 + v_1 u_1' + v_2' u_2 + v_2 u_2') + c(v_1 u_1 + v_2 u_2)$

$= (v_1'' u_1 + v_1' u_1' + v_1' u_1' + v_1 u_1'' + v_2'' u_2 + v_2' u_2' + v_2' u_2' + v_2 u_2'') + b(v_1' u_1 + v_1 u_1' + v_2' u_2 + v_2 u_2') + c(v_1 u_1 + v_2 u_2)$

$= v_1(u_1'' + bu_1' + cu_1) + v_2(u_2'' + bu_2' + cu_2) + b(v_1' u_1 + v_2' u_2) + (v_1'' u_1 + v_1' u_1' + v_2'' u_2 + v_2' u_2') + (v_1' u_1' + v_2' u_2')$

$= v_1(u_1'' + bu_1' + cu_1) + v_2(u_2'' + bu_2' + cu_2) + b(v_1' u_1 + v_2' u_2) + (v_1' u_1 + v_2' u_2)' + (v_1' u_1' + v_2' u_2')$

$= v_1(0) + v_2(0) + b(0) + (0) + k(x) = k(x)$

18.3 Concepts Review

1. $3; \pi$ **3.** 0

Problem Set 18.3

1. $k = 250, m = 10, B^2 = k/m = 250/10 = 25, B = 5$

(the problem gives the mass as $m = 10\,\text{kg}$)

Thus, $y'' = -25y$. The general solution is $y = C_1 \cos 5t + C_2 \sin 5t$. Apply the initial condition to get $y = 0.1\cos 5t$.

The period is $\dfrac{2\pi}{5}$ seconds.

3. $y = 0.1\cos 5t = 0$ whenever $5t = \dfrac{\pi}{2} + \pi k$ or $t = \dfrac{\pi}{10} + \dfrac{\pi}{5}k$.

$$\left| y'\left(\dfrac{\pi}{10} + \dfrac{\pi}{5}k\right) \right| = 0.5 \left| \sin 5\left(\dfrac{\pi}{10} + \dfrac{\pi}{5}k\right) \right|$$

$$= 0.5 \left| \sin\left(\dfrac{\pi}{2} + \pi k\right) \right|$$

$$= 0.5 \text{ meters per second}$$

5. $k = 20$ lb/ft; $w = 10$ lb; $y_0 = 1$ ft; $q = \dfrac{1}{10}$ s-lb/ft, $B = 8$, $E = 0.32$

$E^2 - 4B^2 < 0$, so there is damped motion. Roots of auxiliary equation are approximately $-0.16 \pm 8i$.

General solution is $y \approx e^{-0.16t}(C_1 \cos 8t + C_2 \sin 8t)$. $y \approx e^{-0.16t}(\cos 8t + 0.02\sin 8t)$ satisfies the initial conditions.

7. Original amplitude is 1 ft. Considering the contribution of the sine term to be negligible due to the 0.02 coefficient, the amplitude is approximately $e^{-0.16t}$.

$e^{-0.16t} \approx 0.1$ if $t \approx 14.39$, so amplitude will be about one-tenth of original in about 14.4 s.

9. $LQ'' + RQ' + \dfrac{Q}{C} = E(t)$; $10^6 Q' + 10^6 Q = 1$; $Q' + Q = 10^{-6}$

Integrating factor: e^t

$D[Qe^t] = 10^{-6} e^t$; $Qe^t = 10^{-6} e^t + C$;

$Q = 10^{-6} + Ce^{-t}$

If $t = 0$, $Q = 0$, then $C = -10^{-6}$.

Therefore, $Q(t) = 10^{-6} - 10^{-6}e^{-t} = 10^{-6}(1 - e^{-t})$.

11. $\dfrac{Q}{[2(10^{-6})]} = 120\sin 377t$

a. $Q(t) = 0.00024 \sin 377t$

b. $I(t) = Q'(t) = 0.09048 \cos 377t$

13. $3.5Q'' + 1000Q + \dfrac{Q}{[2(10^{-6})]} = 120\sin 377t$

(Values are approximated to 6 significant figures for the remainder of the problem.)

$Q'' + 285.714Q' + 142857Q = 34.2857\sin 377t$

Roots of the auxiliary equation are
$-142.857 \pm 349.927i$.

$Q_h = e^{-142.857t}(C_1 \cos 349.927t + C_2 \sin 349.927t)$

$Q_p = -3.18288(10^{-4})\cos 377t + 2.15119(10^{-6})\sin 377t$

Then, $Q = -3.18288(10^{-4})\cos 377t + 2.15119(10^{-6})\sin 377t + Q_h$.

$I = Q' = 0.119995\sin 377t + 0.000810998\cos 377t + Q_h'$

$0.000888 \cos 377t$ is small and $Q_h' \to 0$ as $t \to \infty$, so the steady-state current is $I \approx 0.12\sin 377t$.

15. $A\sin(\beta t + \gamma) = A(\sin \beta t \cos \gamma + \cos \beta t \sin \gamma)$

$= (A\cos\gamma)\sin\beta t + (A\sin\gamma)\cos\beta t$

$= C_1 \sin\beta t + C_2 \cos\beta t$, where $C_1 = A\cos\gamma$ and

$C_2 = A\sin\gamma$.

[Note that

$C_1^2 + C_2^2 = A^2 \cos^2\gamma + A^2 \sin^2\gamma = A^2$.)

17. The magnitudes of the tangential components of the forces acting on the pendulum bob must be equal.

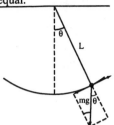

18.4 Chapter Review

Concepts Test

1. False: y^2 is not linear in y.

3. True: $y' = \sec^2 x + \sec x \tan x$

$$2y' - y^2 = (2\sec^2 x + 2\sec x \tan x)$$
$$-(\tan^2 x + 2\sec x \tan x + \sec^2 x)$$
$$= \sec^2 x - \tan^2 x = 1$$

5. True: D^2 adheres to the conditions for linear operators.

$$D^2(kf) = kD^2(f)$$
$$D^2(f+g) = D^2 f + D^2 g$$

7. True: -1 is a repeated root, with multiplicity 3, of the auxiliary equation.

9. False: That is the form of y_h. y_p should have the form $Bx \cos 3x + Cx \sin 3x$.

Sample Test Problems

1. $u' + 3u = e^x$. Integrating factor is e^{3x}.

$$D[ue^{3x}] = e^{4x}$$
$$y = \left(\frac{1}{4}\right)e^x + C_1 e^{-3x}$$
$$y' = \left(\frac{1}{4}\right)e^x + C_1 e^{-3x}$$
$$y = \left(\frac{1}{4}\right)e^x + C_3 e^{-3x} + C_2$$

3. (Second order homogeneous)

The auxiliary equation, $r^2 - 3r + 2 = 0$, has roots 1, 2. The general solution is $y = C_1 e^x + C_2 e^{2x}$.

$$y' = C_1 e^x + 2C_2 e^{2x}$$

Therefore, $-m\dfrac{d^2 s}{dt^2} = mg \sin \theta$.

$s = L\theta$, so $\dfrac{d^2 s}{dt^2} = L\dfrac{d^2\theta}{dt^2}$.

Therefore, $-mL\dfrac{d^2\theta}{dt^2} = mg \sin \theta$.

Hence, $\dfrac{d^2\theta}{dt^2} = -\dfrac{g}{L}\sin\theta$.

If $x = 0, y = 0, \; y' = 3$, then $0 = C_1 + C_2$ and $3 = C_1 + 2C_2$, so $C_1 = -3, C_2 = 3$.

Therefore, $y = -3e^x + 3e^{2x}$.

5. $y_h = C_1 e^{-x} + C_2 e^x$ (Problem 2)

$$y_p = -1 + C_1 e^{-x} + C_2 e^x$$

7. $y_h = (C_1 + C_2 x)e^{-2x}$ (Problem 12)

$$y_p = \left(\frac{1}{2}\right)x^2 e^{-2x}$$
$$y = \left[\left(\frac{1}{2}\right)x^2 + C_1 + C_2 x\right]e^{-2x}$$

9. (Second-order homogeneous)

The auxiliary equation, $r^2 + 6r + 25 = 0$, has roots $-3 \pm 4i$. General solution:

$$y = e^{-3x}(C_1 \cos 4x + C_2 \sin 4x)$$

11. Roots are $-4, 0, 2$. $y = C_1 e^{-4x} + C_2 + C_3 e^{2x}$

13. Repeated roots $\pm\sqrt{2}$

$$y = (C_1 + C_2 x)e^{-\sqrt{2}x} + (C_3 + C_4 x)e^{\sqrt{2}x}$$

15. (Simple harmonic motion)

$k = 5; w = 10; y_0 = -1; B = \sqrt{\dfrac{(5)(32)}{10}} = 4$

Then the equation of motion is $y = -\cos 4t$.

The amplitude is $|-1| = 1$; the period is $\dfrac{2\pi}{4} = \dfrac{\pi}{2}$.

17. $Q'' + 2Q' + 2Q = 1$

Roots are $-1 \pm i$.

$Q_h = e^{-t}(C_1 \cos t + C_2 \sin t)$ and $Q_p = \dfrac{1}{2}$;

$$Q = e^{-t}(C_1 \cos t + C_2 \sin t) + \frac{1}{2}$$
$$I(t) = Q'(t) = -e^{-t}[(C_1 - C_2)\cos t + (C_1 + C_2)\sin t]$$
$$I(t) = e^{-t}\sin t \text{ satisfies the initial conditions.}$$

Maple and Mathematica Code for some Technology Projects

Technology Project 12.1 Rotations in the Plane

Maple:

```
with(plots):
f := (x,y) ->x^2 + 24*x*y + 8*y^2 - 136;
implicitplot(f(x,y) = 0,x=-16..16,y=-16..16);
```

Mathematica:

```
<<Graphics`ImplicitPlot`
f[x_,y_] = x^2 + 24 x y + 8 y^2 -136;
ImplicitPlot[ f[x,y]==0 , {x,-16,16} , {y,-16,16} ]
```

Technology Project 12.1 Roses

Maple:

```
rect := (theta,r) ->[ r*cos(theta) , r*sin(theta) ] :
n := 5:
d := 49:
degr := 2*Pi/360:
rose := k ->evalf( rect( k*degr , sin(n*k*degr) ) ):
path := [ seq ( rose(d*k) , k=1..360 ) ]:
PLOT( CURVES(path) );
```

Mathematica:

```
rect[theta_,r_] := N[{r Cos[theta],r Sin[theta]}]
Attributes[rose]={Listable}
n = 5; d = 13;
rose[k_] = rect[k Degree,Sin[n k Degree] ];
path = Line[ rose[Range[0,360 d,d]] ];
Show[Graphics[{Thickness[.0005],path}],AspectRatio->1]
```

Technology Project 13.2 Measuring Home Run Distance

Maple:

```
degr := 2*Pi/360:
g := 3:
d := 430:
h := 80:
```

```
phi := 30*degr:
theta := arctan( tan(phi) + 2*(h-g)/d );
v0 := evalf( sqrt( 32*d / ((cos(theta))^2*(tan(phi)+tan(theta))) ) );
y := x ->-16*x^2/(v0^2 * (cos(theta))^2) + (tan(theta))*x + g;
plot(y(x), x=0..d, scaling=constrained);
HRdist := evalf(
( - tan(theta) - sqrt( (tan(theta))^2 + 64*g/(v0^2 * (cos(theta))^2) ) ) /
(-32/(v0^2 * (cos(theta))^2)) );
```

Technology Project 15.1   Newton's Method for Two Equations in Two Unknowns

Maple:

```
f := (x,y) ->x^2 + y^2 - 48:
g := (x,y) ->x + y - 6:
dfx := unapply(diff(f(x,y),x),(x,y)):
dfy := unapply(diff(f(x,y),y),(x,y)):
dgx := unapply(diff(g(x,y),x),(x,y)):
dgy := unapply(diff(g(x,y),y),(x,y)):
x1 := 2:  y1 := 0.5:
for n from 1 to 5 do
 x2 := x1 - ( f(x1,y1)*dgy(x1,y1) - g(x1,y1)*dfy(x1,y1) ) /
 ( dfx(x1,y1)*dgy(x1,y1) - dgx(x1,y1)*dfy(x1,y1) ):1
 y2 := y1 - ( g(x1,y1)*dfx(x1,y1) - f(x1,y1)*dgx(x1,y1) ) /
 ( dfx(x1,y1)*dgy(x1,y1) - dgx(x1,y1)*dfy(x1,y1) ):
 x1 := x2:
 y1 := y2:
od;
```

Mathematica:

```
f[x_,y_] = x^2 +y^2 - 48;
g[x_,y_] = x + y - 6;
dfx[x_,y_] = D[f[x,y],x];
dfy[x_,y_] = D[f[x,y],y];
dgx[x_,y_] = D[g[x,y],x];
dgy[x_,y_] = D[g[x,y],y];
x1 = 2; y1 = 0.5;
Do[ x2 = x1 - ( f[x1,y1] dgy[x1,y1] - g[x1,y1] dfy[x1,y1] ) /
  ( dfx[x1,y1] dgy[x1,y1] - dgx[x1,y1] dfy[x1,y1] );
  y2 = y1 - ( g[x1,y1] dfx[x1,y1] - f[x1,y1] dgx[x1,y1] ) /
  ( dfx[x1,y1] dgy[x1,y1] - dgx[x1,y1] dfy[x1,y1] );
  x1 = x2;
  y1 = y2;
```

```
   Print[x2,'' '',y2] ,
 {n,1,7} ]
```

Technology Project 15.2   Visualizing the Directional Derivative

Maple
```
with(plots):
f := (x,y) ->3*x + 4*y + 4:
theta := -70:  phi := 53:
x0 := 1:  y0 := -1:
e := 0.2:
a := t ->(f(x0+e*cos(t),y0+e*sin(t))-f(x0,y0)) / e:
h := t ->f(x0+e*cos(t),y0+e*sin(t)):
p1 := plot3d( f(x,y) , x=-2..2 , y=-2..2 , axes=frame , color=grey ,
orientation=[theta,phi] ):
p2 := spacecurve( {[x0+e*cos(t),y0+e*sin(t) , h(t)]} , t=0..2*Pi ,
axes=FRAME , thickness=2 , color=black , orientation=[theta,phi]):
p3 := contourplot( f(x,y) , x=-2..2 , y=-2..2 , grid=[21,21] ,
color=black , scaling=constrained , contours=20 ):
p4 := plot( [ (x0+e*cos(t),y0+e*sin(t)) , t=0..2*Pi] ,
scaling=constrained , color=black , thickness=2 ):
display([p1,p2]);
display([p3,p4]);
plot( h(t) , t=0..2*Pi );
plot( a(t) , t=0..2*Pi );
evalf( int(a(t),t=0..2*Pi) );
```

Mathematica
```
f[x_,y_] = 3 x + 4 y + 4;
x0 = 1; y0 = -1;
e = 0.2;
a[t_] = ( f[x0+e*Cos[t],y0+e*Sin[t]] - f[x0,y0] ) / e;
h[t_] = f[x0+e*Cos[t],y0+e*Sin[t]];
p1 = Plot3D[ f[x,y] , {x,-2,2} , {y,-2,2} , AxesLabel->{''x'',''y'',''z''} ,
 Boxed->True , ViewPoint ->{5, -15, 8} , DisplayFunction->Identity ];
p2 = ParametricPlot3D[ {x0+e*Cos[t] , y0+e*Sin[t] , h[t]} , {t,0,2Pi} ,
 DisplayFunction->Identity ];
p3 = ContourPlot[ f[x,y] , {x,-2,2} , {y,-2,2} , PlotPoints->21 ,
 ContourShading->False , DisplayFunction->Identity ];
p4 = ParametricPlot[ {x0+e*Cos[t] , y0+e*Sin[t]} , {t,0,2Pi} ,
 DisplayFunction->Identity ];
Show[ {p1,p2} , DisplayFunction->$DisplayFunction ];
Show[ {p3,p4} , DisplayFunction->$DisplayFunction];
```

```
Plot[ h[t] , {t,0,2Pi} ];
Plot[ a[t] , {t,0,2Pi} ];
NIntegrate[ a[t] , {t,0,2Pi} ]
```

Technology Project 16.2   Monte Carlo Integration

Mathematica:
```
f[x_] := x (1-x) (Sin[100 x (1-x)])^2
xmin = 0; xmax = 1;
ymin = 0; ymax = 0.25;
numtrials = 1000;
p1 := Plot[f[x],{x,0,1},PlotStyle->Thickness[0.012],
 DisplayFunction->Identity]
x = Table[ Random[Real,{xmin,xmax}],{numtrials} ];
y = Table[ Random[Real,{ymin,ymax}],{numtrials} ];
data = Transpose[{x,y}];
p2 := ListPlot[data,PlotStyle->PointSize[0.012],
 DisplayFunction->Identity];
Show[p1,p2,DisplayFunction->$DisplayFunction]
numsucc = 0;
Do[If[ y[[k]] <f[x[[k]]],numsucc=numsucc+1] ,
 {k,1,numtrials}]
areaestimate = (xmax-xmin)(ymax-ymin)numsucc/numtrials;
Print[ areaestimate ]
```

Maple
```
with(plots):
with(stats):
with(linalg):
f := x ->x*(1-x)*sin(100*x*(1-x))^2:
xmin := 0:   xmax := 1:
ymin := 0:   ymax := 0.25:
numtrials := 1000:
p1 := plot( f(x) , x=xmin..xmax , y=ymin..ymax ):
randomize(Seed);
xx := [ stats[random, uniform](numtrials) ]:
yy := [ stats[random, uniform](numtrials) ]:
xx := [seq(xmin,i=1..numtrials)] + (xmax-xmin)*xx:
yy := [seq(ymin,i=1..numtrials)] + (ymax-ymin)*yy:
p2 := pointplot( transpose( [xx,yy] ) ):
display( [ p1,p2 ] );
numsucc := 0:
for i from 1 to numtrials do
```

```
if yy[i] <f(xx[i]) then numsucc := numsucc+1 fi:
od;
areaestimate := (xmax-xmin)*(ymax-ymin)*numsucc/numtrials;
```

Technology Project 17.2   Parametrized Surfaces

Maple

```
plot3d( [sin(s)*cos(t) , sin(s)*sin(t) , cos(s)] , s=0..Pi , t=0..2*Pi );
```

Mathematica

```
ParametricPlot3D[ { Sin[s]Cos[t] , Sin[s]Sin[t] , Cos[s] } ,
{s,0,Pi} , {t,0,2 Pi} , PlotPoints->31 ]
```